AF581082

R. von Steiger • G. Gloeckler • G.M. Mason
Editors

# The Composition of Matter

Symposium honouring Johannes Geiss on the occasion of his 80th birthday

Introduction by R. von Steiger, G. Gloeckler and G.M. Mason

Previously published in *Space Science Reviews* Volume 130, Issues 1–4, 2007

Rudolf von Steiger
International Space Science Institute (ISSI),
Bern, Switzerland

George Gloeckler
University of Michigan
Ann Arbor MI, USA

Glenn M. Mason
APL, Johns Hopkins University
Laurel, MD, USA

*Cover illustration:* Courtesy of NASA. http://www.apolloarchive.com/apollo_gallery.html

Library of Congress Control Number: 2007937516

ISBN-978-0-387-74183-3 e-ISBN-978-0-387-74184-0

Printed on acid-free paper.



9 8 7 6 5 4 3 2 1

springer.com

# Contents

Space Sci Rev (2007) 130: 1–2
DOI 10.1007/s11214-007-9241-z

# Foreword

Published online: 18 July 2007

It was at the Fall AGU Meeting in 2005—at which Johannes Geiss received the Bowie Medal—that representatives of both the ACE mission and of ISSI first came together to explore ideas for a Symposium on the Composition of Matter. An important aspect of the symposium was to honour Johannes' lifetime achievements on the occasion of his 80th birthday by bringing together all communities working on composition, one of the principal topics of his work. This included not only the ACE science team but also Ulysses, SoHO, Genesis, Stardust, and many other missions. The symposium was to be organised by ISSI, which is another brainchild of Johannes.

The purpose of the symposium was to explore new insights into the composition of solar-system and galactic matter, and fractionation processes affecting samples of this matter. These new findings have been brought about by recent space missions, ground-based studies, and theoretical advances. The symposium was convened by G.M. Mason (chair), P. Bochsler, J.J. Connell, G. Gloeckler, M.H. Israel, R.A. Mewaldt, R. von Steiger, M.E. Wiedenbeck, and T.H. Zurbuchen. The convenors compiled a program of overview, invited, contributed, and poster presentations that was organised into five sessions: Linking primordial to solar composition, Planetary samples, Solar sources and fractionation processes, Interstellar gas, and Cosmic rays.

The symposium was held in Grindelwald in the Swiss Alps on September 6–10, 2006, and attended by some 70 participants. One afternoon was devoted to a special session highlighting Johannes' lifetime contributions to studies of the composition of matter, with speakers H. Reeves, F. Begemann, and G. Gloeckler, who were among his closest collaborators during different phases of his career. The session was followed by an excursion and dinner, where a movie by T.H. Zurbuchen and B. Grimm about the scientific life of Johannes was screened for the first time.

The structure of the volume at hand largely follows the structure of the symposium. The editors are happy that it includes almost all of the overview and invited papers and many of the contributed ones; we thank the authors for their timely work. All papers in the volume have been thoroughly reviewed, and the excellent referee reports have contributed significantly to the quality of the papers. We are grateful to the referees for their generally underacknowledged, yet very important work. The volume is concluded with a special paper about Johannes' scientific life as featured in the movie mentioned above; the movie itself is linked as an electronic supplement to that paper and can be viewed via the Springer website.

It is our pleasure to thank all those who have made this Symposium possible and successful. First of all we thank E.C. Stone, Principal Investigator of the ACE Team, and R.M. Bonnet, Executive Director of ISSI, for their generous sponsorship. We also thank the local organisation team led by B. Gerber and S. Wenger for their professional work that guaranteed a smooth development of the entire program. Further we thank all symposium participants for giving inspiring presentations and contributing to the lively discussions. But above all we thank Johannes for providing us with the occasion to gather in his name and for shaping and supporting the careers of so many of us.

May 2007
R. von Steiger, G. Gloeckler, G.M. Mason

Space Sci Rev (2007) 130: 3–4
DOI 10.1007/s11214-007-9240-0

# Acknowledgement

Published online: 21 July 2007


Early in 2006, Edward C. Stone, Principal Investigator of NASA's Advanced Composition Explorer (ACE); Len A. Fisk, Chair of the Space Studies Board of the US National Research Council; and the Directorate of the International Space Science Institute (ISSI), Roger M. Bonnet, Andre Balogh and Rudolf von Steiger, agreed to hold a joint ACE/ISSI Symposium on the "Composition of Matter", at the occasion of my 80th birthday. I feel deeply honoured by this exceptional distinction and was thrilled to participate, because for more than 50 years my research has been centred around measurements of the composition of matter of various origins, reaching—in geocentric coordinates—from deep-sea sediments to the limits of solar influence, where comets come from and where the solar wind meets the interstellar medium. The symposium at Grindelwald, beautifully located at the foot of the Bernese Alps, was organised by Ruedi von Steiger, Silvia Wenger and Barbara Gerber, who created the special atmosphere for which ISSI meetings have become known.

The presentations at the symposium and the discussions achieved what the Programme Committee, chaired by Glenn Mason, had intended: to explore new insights into composition and evolution of solar-system and galactic matter that have been brought about by ACE and other recent spacecraft measurements, and by ground-based observations. The Symposium showed that composition measurements, if they are to be fully exploited, require interdisciplinary interpretation. The authors and the editors in particular are to be congratulated for clearly bringing this out in this volume.

My special thanks go to Hubert Reeves, Friedrich Begemann and George Gloeckler who talked about my contributions to the study of the composition of matter. Some of the most successful of these studies were done jointly with one or the other of these three friends, and to this day I cherish the spirit by which we searched for new insights and sometimes even found them. The dinner at Giessbach, above the lake of Brienz, was preceded by a very nice film, created and professionally produced by Thomas Zurbuchen and Brian Grimm, about the many stages in my professional life. A splendid birthday present! Before, during and after dinner, Roger Bonnet, Peter Creola, Len Fisk, Chris Gloeckler, Ed Stone, Heinz Völk and Ewald Weibel gave a set of elegant, thoughtful, witty, and humorous speeches. These speeches and the film drew a picture of me and my efforts and achievements in science, science policy and politics that were certainly too complimentary. But I thoroughly liked and enjoyed them, and so did Carmen and the colleagues and friends that were assembled. Many thanks to all who contributed to this memorable symposium!

June 2007
Johannes Geiss

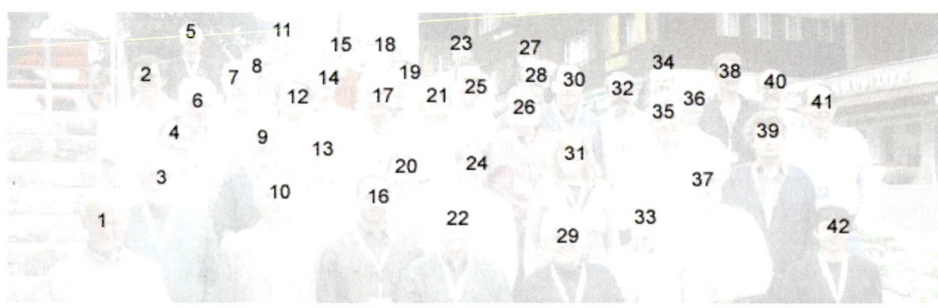


1. A. Balogh
2. V. Izmodenov
3. L. Fisk
4. U. Ott
5. R. von Steiger
6. H. Völk
7. G. Gloeckler
8. E. Möbius
9. R. Marsden
10. B. Klecker
11. N. Grevesse
12. S. Krimigis
13. R. Mewaldt
14. H. Krüger
15. R. Lallement
16. M. Desai
17. W. Binns
18. G. Mason
19. M. Lee
20. A. Kilchenmann
21. P. Frisch
22. A. Cummings
23. M. Wiedenbeck
24. S. Delanoye
25. T. Zurbuchen
26. G. Flynn
27. R. Karrer
28. R. Leske
29. J. Geiss
30. M. Israel
31. C. Cohen
32. R. Murphy
33. E. Stone
34. F. Allegrini
35. G. Zank
36. A. Westphal
37. J. Waddington
38. A. Grimberg
39. R. Kallenbach
40. T. von Rosenvinge
41. P. Bochsler
42. C. Giammanco

Space Sci Rev (2007) 130: 5–26
DOI 10.1007/s11214-007-9235-x

# Linking Primordial to Solar and Galactic Composition

**Johannes Geiss · George Gloeckler**

Received: 30 March 2007 / Accepted: 11 June 2007 / Published online: 25 July 2007


**Abstract** Evolution and composition of baryonic matter is influenced by the evolution of other forms of matter and energy in the universe. At the time of primordial nucleosynthesis the universal expansion and thus the decrease of the density and temperature of baryonic matter were controlled by leptons and photons. Non-baryonic dark matter initiated the formation of clusters and galaxies, and to this day, dark matter largely determines the dynamics and geometries of these baryonic structures and indirectly influences their chemical evolution. Chemical analyses and isotopic abundance measurements in the solar system established the composition in the protosolar cloud (PSC). The abundances of nuclear species in the PSC led to the discovery of the magic numbers and the nuclear shell model, and they allowed the identification of nucleosynthetic sites and processes. To this day, we know the abundances of the $\sim$300 stable and long-lived nuclides infinitely better in the PSC than in any other sample of matter in the universe. Thus, we know the exact composition of a Galactic sample of intermediate age, allowing us to check on theories of Galactic evolution before and after the formation of the solar system. This paper specifically discusses the nucleosynthesis in the early universe and the Galactic evolution during the last 5 Gyr.

**Keywords** Cosmology: Big Bang · Galaxy: Galactic evolution · Interstellar Medium: composition

J. Geiss (✉)
International Space Science Institute, Hallerstrasse 6, 3012 Bern, Switzerland
e-mail: geiss@issibern.ch

G. Gloeckler
Department of Oceanic, Atmospheric and Space Sciences, University of Michigan, Ann Arbor, MI 48109-2143, USA
e-mail: gglo@umich.edu

G. Gloeckler
Department of Physics, University of Maryland, College Park, ML 20742-0001, USA

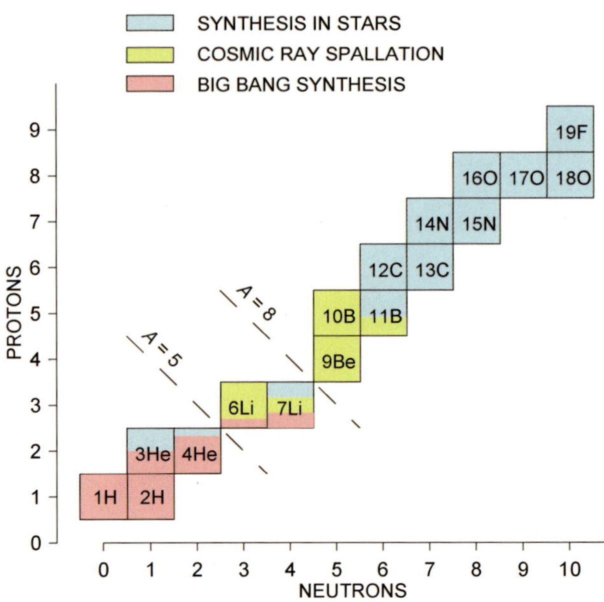


**Fig. 1** Origin of nuclei: products from the three principal sites of nucleosynthesis. For species with mixed origin, such as $^7$Li, the relative proportions change with time and location (after Geiss and von Steiger 1997)

## 1 Introduction

The chemical elements and their isotopes are synthesized at three principal sites: (a) the Big Bang, which yielded the major part of the light nuclei H, D, $^3$He, $^4$He and some $^7$Li; (b) the stars, which synthesize C and all the heavier elements; and finally (c) high-energy cosmic rays, which yield very rare nuclei, in particular $^9$Be, $^6$Li and $^{10}$B (Fig. 1).

Our concepts of nucleosynthesis are largely based on abundance measurements in the solar system. The protosolar cloud (PSC) represents a sample of Galactic matter frozen-in (in terms of nuclear evolution) 4.6 billion years ago. The abundances of elements and isotopes are uniquely well determined in this Galactic sample of intermediate age, and therefore PSC abundances are indispensable for understanding nucleosynthesis from the Big Bang to the present. In this paper, we discuss in some detail the primordial nucleosynthesis and the chemical evolution in the Galaxy following the formation of the Sun.

Non-baryonic forms of matter and energy not only control the expansion of the universe, but they also play a major role in the evolution of baryonic matter. Helium and other light nuclides are produced in the early universe in a contest between reaction rates and the expansion rate that is determined by the photon and lepton populations. The major structures in the universe, the galaxies and clusters of galaxies could not have been formed and preserved without the help of non-baryonic dark matter, and in the present epoch, dark energy is beginning to weaken these structures. We include in this paper, therefore, a brief account of the evolution of all known forms of matter and energy in the universe.

## 2 The Expanding Universe

Early in the 20th century, Albert Einstein laid the foundation for a self-consistent physical cosmology. In 1905, he introduced the equivalence of mass and energy as a general principle. Thus mass density, $\rho_i$, and energy density, $\varepsilon_i$, are related by $\varepsilon_i = \rho_i c^2$ for all forms of matter and energy. With the creation in 1915 of the theory of general relativity, space and time became an object of scientific study, along with matter. As John Wheeler formulated: "Matter tells space-time how to curve, and space-time tells matter how to move."

We live in a universe that on the largest scale is homogeneous and isotropic and that has been expanding from a hot early stage, the "Big Bang." This is very well based on the cosmic microwave background (CMB) and other observations.

In a homogeneous and isotropic universe, the space-time metric has only one parameter, the curvature parameter $k$—the universe is closed for positive $k$ and open for negative $k$. The 3D space is Euclidean or "flat" for the limiting case of $k = 0$.

The expansion equation for a homogeneous universe,

$$H^2 = \left(\frac{\dot{R}}{R}\right)^2 = \frac{8\pi G_N \varepsilon}{3c^2} - \frac{k}{R^2}, \tag{1}$$

was derived from Einstein's equations by Friedmann in 1922 (e.g., Eidelman et al. 2004). $R$ is the cosmological scale factor, $H(t)$ is the Hubble function, its present value $H_0$ is called the Hubble constant. Its best value is 73 $\mathrm{km\,s^{-1}\,Mpc^{-1}}$ (e.g., Bennett et al. 2003). $G_N$ is Newton's gravitational constant, $c$ is the speed of light, and $\varepsilon = \sum \varepsilon_i$ is the sum of the energy densities of all species in the expanding medium.

The expansion proceeds adiabatically throughout the time interval dealt with in this paper (0.03 s to ~14 Gyr), and the first law of thermodynamics reads

$$d(\varepsilon R^3) + p\,d(R^3) = 0 \tag{2}$$

with the total pressure $p = \sum p_i$.

There are no observations that would indicate non-adiabatic expansion after 0.03 s. $\Delta S = 0$ even holds to a good approximation during the epoch (around 1 s) of $e^+e^-$ annihilation (see Weinberg 1972).

The acceleration of expansion $R$ is obtained from (1) and (2)

$$\frac{\ddot{R}}{R} = -\frac{4\pi G_N}{3c^2}(\varepsilon + 3p). \tag{3}$$

Equation (3) is valid for all $k$ and independent of the equation of state.

With $k = 0$ in (1), one defines the critical density $\rho_c$ and the critical density parameter $\Omega$,

$$\rho_c = \frac{3H^2}{8\pi G_N}, \qquad \Omega = \Sigma\Omega_i = \rho/\rho_c = \Sigma\rho_i/\rho_c. \tag{4}$$

For relativistic particles and for cold matter, we have the most simple equations of state

$$p_i/\varepsilon_i = w_i = \text{constant}. \tag{5}$$

Insertion into (1) and integration gives

$$\varepsilon_i \propto R^{-3(1+w_i)}, \qquad w_i = \text{constant} \tag{6}$$

and, for constant total $w$ and a flat universe ($k = 0$), integration of (1) gives

$$R \propto t^{2/3(1+w)}, \qquad w = \text{constant}. \tag{7}$$

## 3 Curved Space

Carl Friedrich Gauss was the first to test the flatness of physical 3D space. In 1823 he measured the angles in a triangle of 2,800 $km^2$ between three landmarks in the region of Göttingen (Fig. 2), using his least-squares method for "data reduction." A pioneering experiment but, as we now know, the deviation from 180 degrees for the sum of the three angles in his triangle, as caused by Earth's gravity, is only of the order of $10^{-8}$ arc seconds, far too small to be measurable. On a cosmic scale, Gauss' experiment or another test of Euclid's 5th axiom would be even more difficult to perform. Nevertheless, during the early decades of the 19th century, Gauss, J. Bolyai, N.I. Lobachevsky, and B. Riemann developed the non-Euclidean geometries that are adequate for describing curved space in the cosmos on any scale. High matter densities cause strong warping of space. Gauss' experiment, performed above the surface of a neutron star of 1.5 solar masses and a radius of $\sim$12 km would have given 183 degrees for a triangle of ten square kilometers (see Geiss and Gloeckler 2005).

The total energy density in the universe—that is, the contributions from dark energy, dark matter and baryonic matter taken together—is presently estimated to lie within 10% of the critical density (Eq. 4). Thus, on large scales, space is Euclidean or flat, at least approximately. This does not mean, of course, that the pre-Einstein physics would provide an adequate tool for cosmology. Whereas, Newton's physics *presupposes* space to be Euclidean, Einstein's general relativity, on the basis of observation, *concludes* that space is flat on the largest scale, at least approximately, but in the vicinity of mass concentrations, space is significantly curved and non-Euclidean.

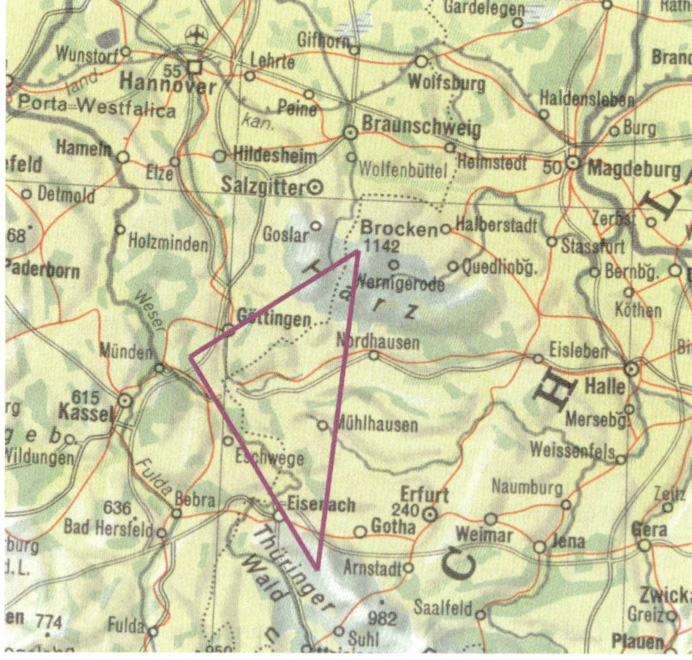


**Fig. 2** The triangle used by Carl Friedrich Gauss in 1823 for his test for a possible curvature of 3D space

**Table 1** Single component Friedmann universe

| Population | Particles (examples) | $p/\varepsilon$ | Expansion law | Energy density | Acceleration coefficient | Type of universe |
|---|---|---|---|---|---|---|
| Relativistic $m_0c^2 \ll kT^*$ | Photons neutrinos | 1/3 | $R \propto t^{1/2}$ | $\varepsilon \propto 1/R^4$ | $q = -1$ | Radiation dominated (early universe) |
| Cold matter | Baryons WIMPs | 0 | $R \propto t^{2/3}$ | $\varepsilon \propto 1/R^3$ | $q = -1/2$ | Einstein–deSitter ($\sim 10^6$ to $\sim 3 \cdot 10^9$ years) |
| Dark energy | a | −1/3 | $R \propto t$ | $\varepsilon \propto 1/R^2$ | $q = 0$ | "Empty universe" Lemaitre (∼6–7 Gyr) |
| Dark energy | b | −2/3 | $R \propto t^2$ | $\varepsilon \propto 1/R$ | $q = +1/2$ | Accelerating universe (expands forever) |
| Dark energy | c | −1 | $R \propto e^{H_0 t}$ | $\varepsilon \propto \text{const}$ | $q = +1$ | Steady state universe Bondi, Gold and Hoyle |

*In early universe

## 4 The Major Forms of Energy and Matter in the Universe

Table 1 lists the major forms of matter and energy that have been populating the universe after the dissolution of the quark-gluon plasma, and Fig. 3 presents the changing composition from a cosmic time of 0.1 s to the present time of ∼14 Gyr. The energy density $\varepsilon(R)$, as given in (6), is not valid at the time of $e^+$ $e^-$ annihilation (∼1 s in cosmic time). This is the only universal phase transition we know of during the time span covered by Fig. 3. Phase transitions involving dark matter may *not* have been negligible, but so far have remained undetected. Implicit in (3) is the acceleration coefficient $f = \ddot{R}R/\dot{R}^2$ $(k = 0)$.

To this day, the curvature term in Friedmann's equation (1) is small compared to the energy-density term and, therefore, matter and energy content control the geometry and expansion of the universe. In the early universe, the density was dominated by relativistic particles. Their influence on the expansion has become negligible in the present epoch, and baryons, non-baryonic dark matter and dark energy dominate the large-scale dynamics and geometry of the universe today (Fig. 3). Baryons are the well-known constituents of ordinary matter. For the existence of the other two components, we have less, but increasingly compelling evidence.

Although the influence of baryons on the overall dynamics and geometry of the present universe is relatively minor, their physical properties are unique. Among the major forms of matter and energy that populate the present universe, only Baryonic Matter participates in all the physical forces known to us; that is, the strong, electromagnetic, weak, and gravitational forces. These four physical modes of interaction enable baryons to self-organize, form a multitude of microscopic and macroscopic structures and, indeed, to create all the variety and beauty that we observe in the world.

## 5 Dark Matter

In 1937 Fritz Zwicky discovered that the visible mass of the galaxies in large clusters was not sufficient to keep them gravitationally bound, and he concluded that these clusters were

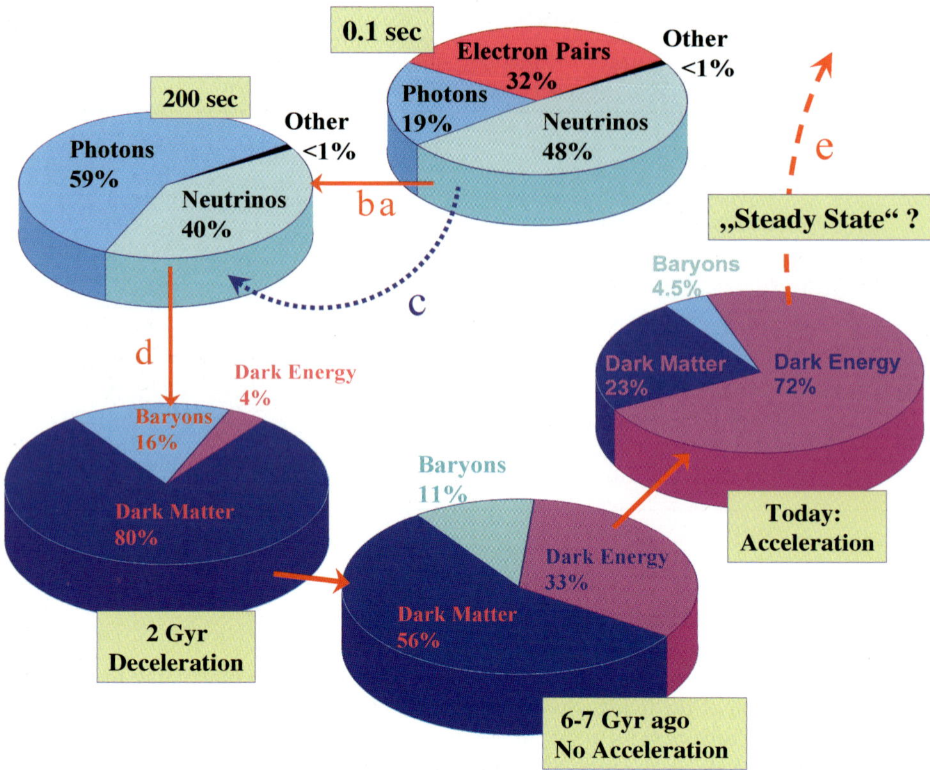


**Fig. 3** The dominant matter and energy components in the Universe. The equation of state determines the decrease in energy density (6 and 7): $\varepsilon \propto R^{-4}$ for relativistic particles, $\varepsilon \propto R^{-3}$ for cold matter, constant density for dark energy (if $p/\varepsilon = -1$); see Table 1. Some important events are indicated: (a) decoupling of neutrinos; (b) $e^+ e^-$ annihilation; (c) primordial nucleosynthesis; (d) baryon–photon decoupling; (e) if $p/\varepsilon = -1$ for dark energy persists into the distant future, the universe becomes a steady state universe of the type proposed in 1948 by Hermann Bondi, Thomas Gold and Fred Hoyle

held together by a surplus of dark matter that astronomers could not readily account for. During the last decades of the 20th century, it became increasingly clear that the universe harbors more gravitational attraction than could possibly be generated by the $\sim 0.2$ atoms/m$^3$ of matter that was derived from D and $^3$He abundances (see the following). The rotation curves of galaxies including our own, as well as $\gamma$-ray and X-ray observations (see Figs. 4 and 5) demonstrate that non-baryonic matter contributes most of the gravitational forces on the scale of galaxies and clusters of galaxies (e.g., Böhringer 2002).

When small fluctuations in the cosmic microwave background (CMB) were discovered and measured with the COBE satellite, from the ground and with the Wilkinson Microwave Anisotropy Probe (WMAP), it was demonstrated that in addition to baryonic matter, a non-baryonic form of matter must already have existed in the early universe (e.g., Smoot et al. 1992; Rebolo 2002; Bennett et al. 2003).

Dark matter, not being affected by electromagnetic interactions, decoupled from the photon gas very early and initiated the growth of cosmic structure long before baryons could have done this. When at a cosmic time of $\sim 10^5$ years baryons decoupled from photons, they were rapidly drawn into already existing blobs of dark matter and began to form the structures we observe. In places of strong enough concentration, baryonic matter, contract-

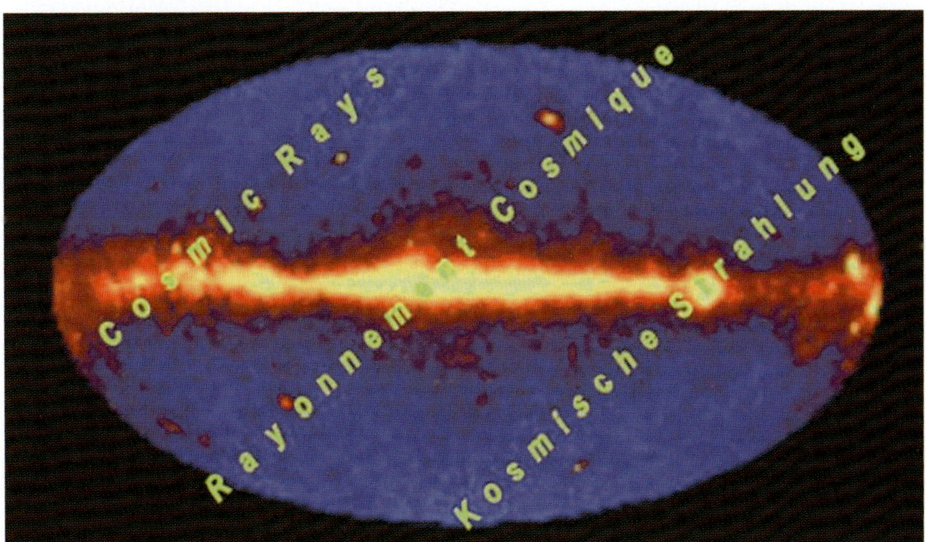


**Fig. 4** Cosmic rays produce gamma-rays whenever they hit baryonic matter. This map, from NASA's Compton Gamma-Ray Observatory, shows that gamma-ray sources are essentially confined to the Galactic disk. Bright spots outside the disk are active Galactic nuclei and a quasar, located far away from our Galaxy. Indirect evidence shows that the cosmic rays are not confined just to the Galactic disk but fill a large halo. The absence, shown in this map, of gamma-ray sources outside the disk implies that there is very little gas or dust of baryonic matter in the halo

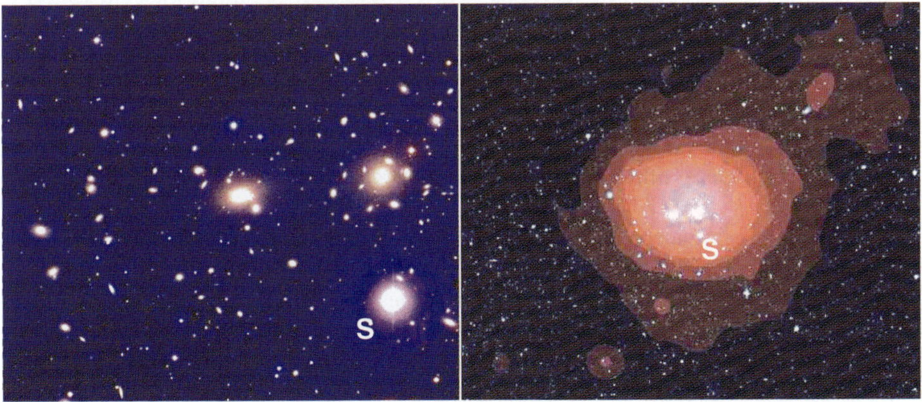


**Fig. 5** *Left*: The central region of the coma cluster. This cluster of galaxies is 300 million years away from us, and consists of 2,000 galaxies of various sizes. The two brightest of them at the center are much more massive than the Milky Way (S is a nearby star). *Right*: X-ray image of the coma cluster from the Rosat All-Sky Survey (Böhringer 2002). The optical image (from the Palomar Sky Survey) is superimposed. The density of the hot intergalactic medium is not sufficient to bind these galaxies. Typical distribution of matter in this and other large clusters is 5% in galaxies, ~20% in the luminous gas, and ~75% non-baryonic dark matter (Böhringer 2002; Evrard 1997)

ing under its own weight, formed stars that then produced carbon and heavier elements, essential ingredients of complex molecules and crystals. These highly organized systems of baryonic matter are the crucial building blocks of comets, solid planets and life.

The particles of dark matter have not yet been identified. Weakly interacting massive particles (WIMPs), but also virtually non-interacting light particles (axions) are being considered. Experiments to detect WIMPs produced by accelerators or natural WIMPs are under way (Pretzl 2002, 2007). Such measurements could give information on the mass and interaction properties of dark matter particles, and on their temperature in the solar neighborhood. These properties of dark matter allow predictions of the evolution of medium-scale structures, such as the number of dwarf galaxies in relation to fully grown galaxies, the amount of baryonic matter falling quasi-continuously into our own Galaxy, or the concentration of matter towards the center of galaxies and clusters.

## 6 Dark Energy

Observations of Type Ia super novae explosions revealed that, surprisingly, the expansion of the universe has been speeding up during the past several billion years (Perlmutter et al. 1999; Riess et al. 1998). Dark energy, with an equation of state that combines positive energy with negative pressure was postulated to account for this kinematic discovery. Observations so far are compatible with $p/\varepsilon = -1$, but lower values and also the case of variable $p/\varepsilon$ are being discussed under the name of quintessence (see Kirshner 2003; Wetterich 2002). In all these cases, (1), (2) and (3) are applicable. In Table 1 we have included three examples for the physical properties of dark energy. A pressure/energy density ratio of $p/\varepsilon = -1$ produces a constant energy density. This case corresponds to Einstein's cosmological constant, and the expansion is as in the Steady State Theory advanced in the 1940s by Herman Bondi, Thomas Gold and Fred Hoyle.

Since the densities of dark and baryonic matter decrease more rapidly than the density of dark energy, deceleration of the expanding universe turned into acceleration several billion years ago, and so it will continue to expand into the distant future, as long as we have $w = p/\varepsilon < -1/3$. That is our present understanding.

## 7 Primordial Nucleosynthesis (0.1 s to 3 min in cosmic time)

The quark-gluon plasma epoch ended when the temperature approached $\sim 10^{12}$ K at a cosmic time of $\sim 10^{-4}$ s. In a major phase transition, quarks and anti-quarks formed mesons, baryons and anti-baryons. Within microseconds mesons decayed while baryons and anti-baryons annihilated. Because of some symmetry breaking in the early universe, a very tiny fraction of the baryons was spared, but this was enough for populating the world with all the stars we see. The lifetime of the proton far exceeds the age of the universe, and thus the number of baryons has not decreased by spontaneous decay since these early times.

Baryon anti-baryon annihilation is a strong interaction process. Therefore, in a homogeneous universe, only a totally insignificant amount of primordial anti-baryons should have survived. However, interactions of cosmic rays with matter produce proton-antiproton pairs. The fraction of antiprotons found in cosmic rays is compatible with such a secondary origin.

During the epoch of primordial nucleosynthesis, lasting from $\sim 100$ ms to $\sim 3$ min in cosmic time, the physics are well known, so that we can make quantitative predictions for microscopic and macroscopic processes. Wagoner et al. (1967) formulated the theory of Standard Big Bang Nucleosynthesis (SBBN), assuming a homogeneous and isotropic universe during the epoch of nucleosynthesis, and neglecting degeneracy of leptons. The number of neutrino flavors was determined to be three ($N_\nu = 3$), and neutrino oscillation experiments confirm that the rest masses of all these neutrinos are negligible during the nucleosynthesis epoch. Thus, the baryonic density remains *the only* free parameter in the framework of the SBBN theory.

The sequence of events during the epoch of primordial nucleosynthesis is as follows: At a cosmic age of ~100 ms the temperature had decreased to $10^{11}$ K. Mesons and heavier leptons had virtually all decayed, while protons and neutrons, the lightest variety of baryons, remained. As a result, energy density and expansion rate were determined by relativistic particles—that is, photons, neutrinos and electrons (Fig. 3)—with protons and neutrons being minor constituents. Since neutrons are heavier than protons, the neutron/proton ratio decreased with decreasing temperature until, at a cosmic time of ~1 s and a temperature of ~$10^{10}$ K, the weak interaction became ineffective, and the neutron/proton ratio was frozen-in at a value of 0.20. Afterwards, beta decay of the neutrons slowly decreased this ratio somewhat further until all remaining neutrons were bound in stable nuclei.

Nucleosynthesis—that is, the fusion of protons and neutrons into deuterium and heavier nuclei—effectively began when the temperature decreased to $10^9$ K at a cosmic time of ~150 s, and it was completed 200 s later. Effective production did not go beyond the isotopes of the lightest three elements (Fig. 1). Of these, only deuterium (D or $^2$H) the heavy isotope of hydrogen was created exclusively (>99%) during the first few minutes in the life of the universe.

## 8 The Origin of Complex Nuclei

Primordial nucleosynthesis practically does not go beyond the isotopes of the lightest three elements (only tiny fractions of Be and B are produced in SBBN), because all nuclei of atomic mass $A = 5$ and $A = 8$ are extremely short lived.

These gaps in the sequence of stable nuclei (see Fig. 1) are overcome by the 3-alpha nuclear reaction, producing $^{12}$C, the major isotope of carbon. Being a reaction involving three partners, this process needs a high density to become effective, and this condition is fulfilled only when stars have evolved into red giants, with high enough central densities and temperatures of ~100 million degrees. Once $^{12}$C is present in a star, the synthesis continues to heavier elements in a multitude of nuclear reactions as the star contracts further and increases its core temperature. The integral effect of all the nucleosynthetic processes leads to a continuous increase of metallicity in galaxies, that is the increase of the heavier nuclides relative to hydrogen.

The very rare nuclides $^6$Li, $^9$Be and $^{10}$B are so instable that, similar to deuterium, they are not synthesized, but destroyed by stars. To solve the puzzle of their existence, Fowler et al. (1962) proposed a local origin for these rare nuclides. They argued that in the early solar system a large number of meter-sized objects were irradiated by energetic particles of local origin, producing the Li, Be and B isotopes and deuterium. A local production should lead to local composition variations, and several authors checked experimentally for such variations. Balsiger et al. (1968) determined precisely the $^6$Li/$^7$Li ratio in different meteorite classes and components of meteorites of potentially different origin. Finding $^6$Li/$^7$Li to be constant with a standard deviation of less than 2%, they argued strongly against a local origin. The definite solution of the Li Be B puzzle came from Hubert Reeves and his co-workers (e.g., Meneguzzi et al. 1971) who showed that collisions of the Galactic cosmic rays with interstellar matter would produce the observed quantities of $^6$Li, $^9$Be and $^{10}$B.

Recently, a new $^6$Li puzzle has become apparent (see Prantzos 2007; Reeves 2007). In the early universe, a small amount of $^6$Li was produced that is difficult to accommodate in the production scheme shown in Fig. 1.

## 9 The Universal Density of Baryonic Matter

Nuclides that are produced fully or partly in the Big Bang are indicated in Fig. 1. More than 99.9% of the total mass resides in $^{1}$H and $^{4}$He. D and $^{3}$He are rare and their yields depend inversely on baryonic density. This is analogous to chemical reactions, where the relative yields of intermediate products decrease with increasing supply of reacting partners.

Primordial abundances of D and $^{3}$He are derived from the isotopic composition of H and He in the Protosolar Cloud (PSC), the Local Interstellar Cloud (LIC) and very distant, low-metallicity clouds (see Figs. 6 and 7). The current best values are plotted in Fig. 8 as a function of the nucleosynthetic age. The decrease of the D abundance in the Galaxy is compensated by an increase in the $^{3}$He abundance so that (D+$^{3}$He)/H has approximately remained constant over Galactic history. This demonstrates that the principal effect of stellar processing is the conversion of deuterium into $^{3}$He with the sum, D+$^{3}$He remaining nearly

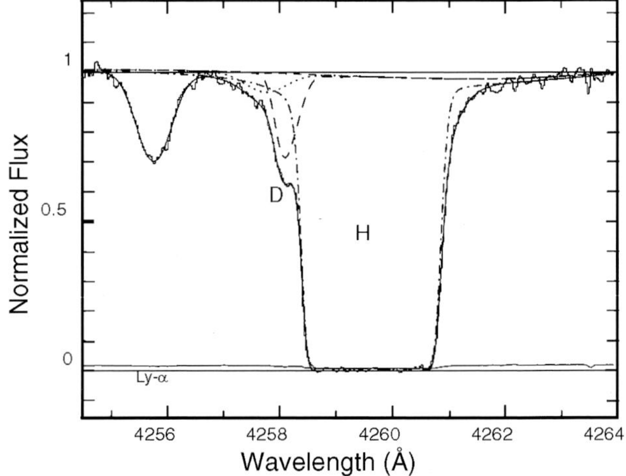


**Fig. 6** Lyman-alpha absorption by a very distant cloud observed with the Keck 10-meter telescope on Mauna Kea, Hawaii (Burles and Tytler 1998). The Lyman-$\alpha$ lines of H and D are shifted from far ultraviolet to visible wavelengths and can be observed from the ground. From several such distant clouds a primordial deuterium/hydrogen ratio of $3 \times 10^{-5}$ was derived (O'Meira et al. 2001, 2006)

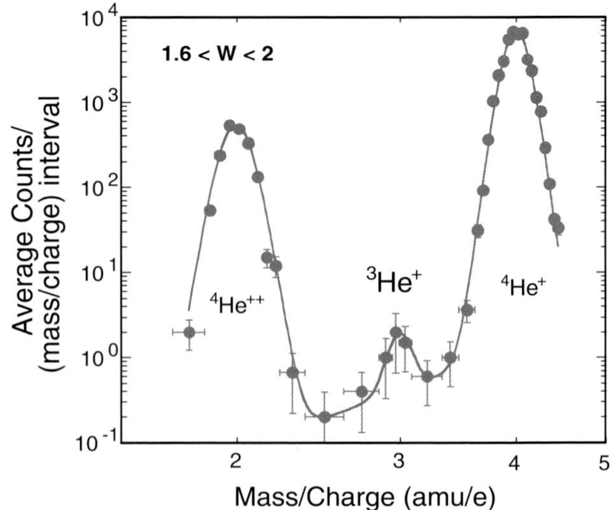


**Fig. 7** Neutral helium of the Local Interstellar Cloud (LIC) penetrates deep into the heliosphere where it can be directly investigated by spacecraft. The mass spectrum shown here was obtained with the Solar Wind Ion Composition Spectrometer on Ulysses (Gloeckler and Geiss 1996, 1998). The LIC is the only present-day Galactic sample for which both the $^{3}$He and deuterium (Linsky and Wood 2000) abundances have been determined

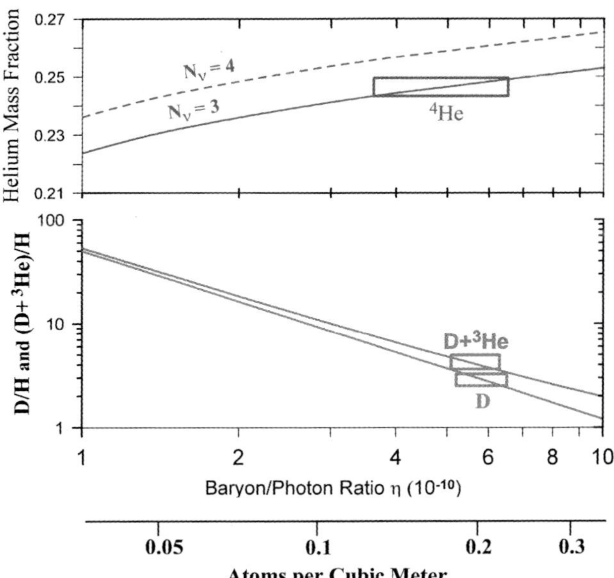


**Fig. 8** Predicted (*solid lines*) and observed (*boxes*) primordial helium mass fraction (*top*) and the D/H and (D+$^3$He)/H ratios (*bottom*), as a function of the baryon/photon ratio and of the baryonic density. The *solid line* labeled $N_\nu = 3$ corresponds to the SBBN prediction (3 neutrino flavors). The *dashed line* labeled $N_\nu = 4$ is calculated for the hypothetical case of 4 neutrino flavors (see text)

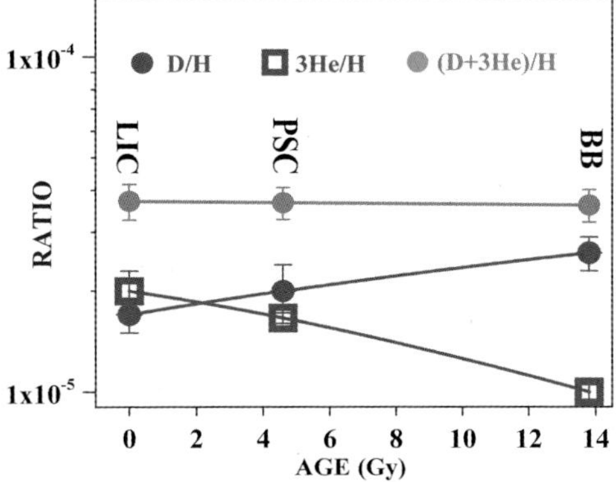


**Fig. 9** The abundance ratios relative to hydrogen of (D/H), ($^3$He/H) and (D+$^3$He)/H in the Local Interstellar Cloud (LIC), the Protosolar Cloud (PSC), and very distant clouds that approximately represent matter released from the Big Bang (BB) (after Geiss and Gloeckler 2005). Data are from Linsky and Wood (2000); Linsky et al. (2006); Gloeckler and Geiss (1996, 1998); Busemann et al. (2006) for the LIC; Mahaffy et al. (1998) and Gloeckler and Geiss (2000) for the PSC; O'Meira et al. (2001) and Bania et al. (2002) for the BB. Deuterium is exclusively produced in the Big Bang (see Fig. 1) and converted thereafter into $^3$He in stars. The net effect on these two species by other nuclear processes is found to be relatively small, so that throughout Galactic history the (D+$^3$He)/H ratio remained relatively unchanged

constant (Fig. 9). This is supported by theoretical work (Charbonnel 1995, 1998; Tosi 1998) showing that the combined effects of $^3$He production from incomplete hydrogen burning and $^3$He destruction has a limited effect on the chemical evolution in the Galaxy.

The PSC (D +$^3$He)/H ratio can be directly determined from $^3$He in the solar wind, because D is converted into $^3$He that cannot have been destroyed in the matter of the outer convective zone of the Sun (Geiss and Reeves 1972; Reeves et al. 1973; see Fig. 10).

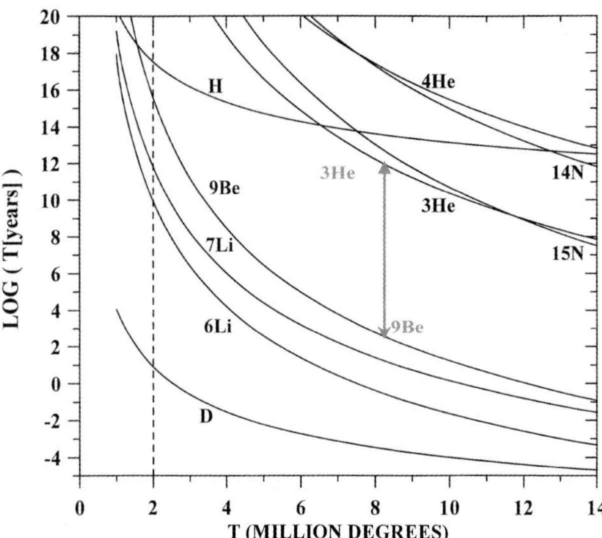


**Fig. 10** Lifetimes—corresponding to the fastest reaction of nuclei—as a function of temperature, with density and composition normalized to the present condition at the bottom of the Outer Convective Zone of the Sun (Geiss and Gloeckler 2003). Since $^3$He is much more stable than $^9$Be, the presence of $^9$Be in the photosphere assures that $^3$He can be used for deriving (D+$^3$He)/H in the original Sun (Geiss and Reeves 1972)

The best current estimates of primordial D/H and (D+$^3$He)/H are compared in Fig. 8 with the theoretically predicted dependence on the universal baryon/photon ratio $\eta \equiv \rho_B/\rho_\gamma$. It is evident that both ratios give a consistent value of $\eta = (5.8 \pm 0.6) \times 10^{-10}$, and a present-day universal density of baryonic matter of $\rho_B = (4.1 \pm 0.4) \times 10^{-31}$ g/cm$^3$ or about 0.2 atoms per cubic meter. This density determination from deuterium and $^3$He is very specific for baryonic matter. If, for example, all the dark matter were baryonic, the abundance of deuterium would be 200 times lower, and the D absorption line in Fig. 6 or the $^3$He peak in Fig. 7 would not be noticeable.

The baryon/photon ratio is one of the fundamental numbers of cosmology. So far, it is known only empirically. Any theory of the earliest phases of the Big Bang will have to predict a value that is compatible with the number derived from deuterium and $^3$He.

Since (D+$^3$He)/H is approximately independent of Galactic evolution, the primordial baryonic density can be derived from this sum with little, if any, extrapolation. Thus, at the time of primordial nucleosynthesis, the baryonic densities in the far-away regions of these clouds and in our part of the universe were the same, providing evidence for a homogeneous universe at the time of primordial nucleosynthesis.

## 10 Primordial $^4$He Abundance: Test of Conditions during the SBBN Epoch

The primordial abundance of $^4$He is obtained by extrapolating the helium abundance to "zero metallicity." We adopted a mass fraction of 24.5%, obtained by Thuan and Izotov (1998).

The Big Bang production of $^4$He depends only weakly on the baryonic density. Thus, by using the baryon/photon ratio as determined earlier, primordial $^4$He can be used for testing the validity of the SBBN theory or, to express it more generally, the validity of the laws of physics under the extreme conditions prevailing in the early universe. We mention two examples:

1. The measured primordial abundance of $^4$He is not compatible with the curve labeled $N_\nu = 4$ in Fig. 8, showing that, at a cosmic time of 1 s, the universe was populated

by no other relativistic particles than photons, electrons and three neutrino flavors. This exclusion holds for light particles of any kind, provided they are covered by Einstein's equivalence principle and had a similar temperature as the known particles.

2. The agreement between the predicted and observed primordial helium abundances shows that the relative strengths of the strong, weak and gravitational forces were the same at a cosmic time of 1 s as those measured in laboratories on Earth. This is a remarkable invariance considering that at a cosmic time of one second, the total density was $10^{35}$ times higher than it is in the present universe.

## 11 Galactic Evolution During the Last 5 Gyr

There is no sample in the present-day Galaxy for which the composition is nearly as well known as in the PSC. The Local Interstellar Cloud (LIC) comes closest, because elements and isotopes are determined by several methods: optical and UV spectroscopy inside (Lallement 2001) and outside (Linsky and Wood 2000; Linsky et al. 2006) the heliosphere, pickup ion mass spectrometry (Gloeckler and Geiss 1996, 2000; see Fig. 7) and Anomalous Cosmic Ray measurements (Cummings and Stone 1996; Cummings et al. 1999; Leske et al. 1996, 2000). So far, we have measurements for the LIC with reasonable precision for hydrogen, the important pair D and $^{3}$He, and the "metals" $^{14}$N and $^{16}$O. For $^{17}$O, $^{18}$O, $^{12}$C, $^{13}$C and $^{15}$N we use the abundances in the Local Interstellar Medium (LISM). The $^{17}$O/$^{18}$O ratio, important for our following discussion, does not appreciably vary from 4 kpc to 12 kpc Galactocentric distance (see discussion below), and thus the LISM value should be a good proxy for $^{17}$O/$^{18}$O in the LIC. A similar constancy in the ISM is not observed for the C isotopes or $^{15}$N. We have taken this into account by assigning large enough uncertainties to the LISM abundances of these nuclides.

We study the chemical evolution of the Galaxy during the last $\sim$5 Gyr by comparing two samples: the Protosolar Cloud (PSC) that existed $\sim$5 Gyr ago, and the LIC, a Galactic cloud that happens to surround the solar system at the present time. The comparison reveals that the composition of the interstellar medium in the solar neighborhood could not have evolved from matter with a PSC composition in a closed system environment. There are four observations, in particular, that defy a closed system interpretation.

1. The D/H ratio in the solar ring has not decreased as much as would be expected in a closed system galaxy. The explanation, now generally accepted, is that infall into the Milky Way of moderately processed material has limited the decrease of the D/H ratio in the ISM (Tosi 1988).
2. Within the limits of uncertainty, the metallicity is the same in the PSC and the present-day ISM in the solar neighborhood. The apparent lack of growth in metallicity could be explained by increased infall into the Galaxy of moderately processed matter or by Galactic winds provided that they lead to loss of heavy elements (Völk 1991).
3. The relative abundances of the three oxygen isotopes in the PSC and ISM are very different, and it is difficult to find a chemical evolution model that explains this "$^{18}$O-puzzle" (Prantzos et al. 1996). Particularly puzzling is the large difference between the $^{18}$O/$^{17}$O ratio in the PSC of 5.34 and that in the interstellar medium of $3.6 \pm 0.3$.
4. $^{14}$N is—at least in part—a secondary nucleus, and it is produced in significant amounts in low mass stars. Therefore, in a closed system, the $^{14}$N/$^{16}$O ratio should increase with time, contrary to measurements that indicate a lower N/O ratio in the LIC than in the PSC (Gloeckler and Geiss 2001; Geiss et al. 2002; Gloeckler 2005).

One could try to account for these observations by simply assuming a special history of the Sun or an anomalous chemical evolution of the PSC. The problem is that concrete proposals for such a special history tend to create new inconsistencies.

For instance, points 1 and 2 could be explained by assuming that, relative to its present position, the Sun's birthplace was significantly closer to the Galactic center. This assumption, however, could not explain the observed constancy of $^{18}O/^{17}O = 3.6 \pm 0.3$ over galactocentric distances from 4 kpc to 12 kpc (Kahane 1995; Prantzos et al. 1996).

If the Sun was born in an OB association it would have incorporated an extra amount of freshly processed material ejected by massive stars. It has been considered (e.g., Henkel and Mauersberger 1993) that this could have led to a general increase of metallicity in the PSC and also to some exceptional isotopic abundances such as a high $^{18}O/^{17}O$ ratio. However, since OB associations typically last for 30 Myr (e.g., Preibisch et al. 2006), this hypothesis is difficult to reconcile with the low abundance of now extinct radioactive nuclides in the PSC that may be determined from the abundances of their decay products measured in meteorites. If, in the PSC, a significant fraction of *stable* nuclides from massive stars was synthesized in the last OB association, a very high concentration of heavy *extinct radioactive* nuclides with $T_{1/2}$ ~10 Myr to ~100 Myr should have been present in the PSC and would have been incorporated into meteorites. However, nuclides such as $^{244}$Pu ($T_{1/2} = 82$ Myr), $^{182}$Hf ($T_{1/2} = 9$ Myr), $^{146}$Sm ($T_{1/2} = 10^3$ Myr), $^{129}$I ($T_{1/2} = 17$ Myr) or $^{107}$Pd ($T_{1/2} = 6.5$ Myr) are found to have had only approximately "uniform Galactic production" abundances (e.g., Shukoliukov and Begemann 1996; Lee and Halliday 1996; Podosek and Nichols 1997), implying that no significant excess of $^{18}$O and other *stable* products from massive stars could have been synthesized during the $10^7$–$10^8$ yr that preceded the formation of the solar system (Geiss et al. 2007).

Here we do not assume an atypical history of the Sun for explaining differences between the composition of the PSC and the ISM. Instead we seek to find the cause for these differences in processes that affected a large portion of the Galaxy. We assume that during the last 5 Gyr a significant fraction of the infall into the Milky Way came from dwarf galaxies, considering that infall could very well be changing in intensity and composition on a time scale of $10^9$–$10^{10}$ yr. Using the mixing model of Geiss et al. (2002, 2007), we investigate changes in the composition of the LISM that would have been caused by infall of matter that carries the nucleosynthetic signature of dwarf galaxies.

The Large Magellanic Cloud (LMC) is the dwarf galaxy for which we have the best composition data. In Fig. 11, abundance ratios observed in the LISM, LMC and PSC are compared. Relative to the PSC, the LMC abundances are high for $^{13}$C, $^{15}$N and $^{17}$O nuclides that are produced by high-temperature hydrogen burning that occurs during the RGB (Red Giant Branch star) and AGB (Asymptotic Giant Branch star) phases of Intermediate Mass Stars (IMS), and also during nova explosions (Boothroyd and Sackmann 1999; Marigo 2001; Ventura et al. 2002; Romano and Matteucci 2003). On the other hand, $^{14}$N and $^{18}$O are rare in the LMC. Massive stars produce $^{18}$O (e.g., Prantzos et al. 1996) and $^{14}$N is made by low-mass stars as well as massive stars (Chiappini et al. 2005). We propose that the abundance pattern in the LMC, particularly the high abundance of $^{13}$C, $^{15}$N and $^{17}$O is due to two factors:

(1) Products of massive stars like $^{18}$O and $^{14}$N are preferentially lost from dwarf galaxies through super novae and star bursts (d'Ercole and Brighenti 1999; Veilleux et al. 2005).
(2) Products of low-mass stars, including $^{14}$N or $^3$He are underrepresented because the LMC contains a large fraction of young stars (Dopita 1991) that have not yet left the main sequence.

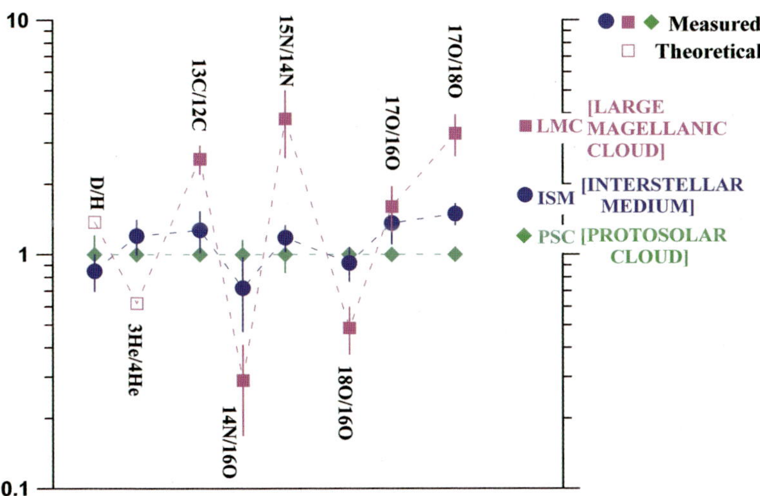


**Fig. 11** Abundance ratios in the Local Galactic Interstellar Medium (LISM) and in the Large Magellanic Cloud (LMC), normalized to the protosolar ratios (for LMC abundances, see Chin et al. 1999, and references quoted therein)

Combined, these two factors lead to an excessive abundance of $^{13}$C, $^{15}$N and $^{17}$O and a low abundance of $^{18}$O and $^{14}$N relative to $^{16}$O.

As a rule, dwarf galaxies should lose matter from starbursts or isolated super novae more effectively than the Galaxy does. Also, if tidal forces trigger star formation, dwarfs approaching the Galaxy would contain an excess of young stars. Thus, an overabundance of $^{13}$C, $^{15}$N and $^{17}$O and an under abundance of products from low-mass stars, as is found in the LMC, may be quite common in dwarf galaxies that are close to the Galaxy.

## 12 The Mixing Model

In this section we compare the PSC, a Galactic sample that existed 4.6 Gyr ago, with the LIC, a present-day Galactic sample. We choose the LIC because D and $^3$He as well as several elemental abundances were determined in this cloud. LIC abundances are determined by several methods: optical and UV spectroscopy inside (Lallement 2001) and outside (Linsky and Wood 2000; Linsky et al. 2006) the heliosphere, pickup ion mass spectrometry (Gloeckler and Geiss 1996, 2000; see Fig. 7) and anomalous cosmic ray measurements (Cummings and Stone 1996; Cummings et al. 1999; Leske et al. 1996, 2000; Leske 2000; see Fig. 12). When LIC abundance ratios are not-yet-available, we use abundances measured in the local interstellar medium (LISM).

For studying the effect on Galactic evolution during the last 5 Gyr of infall of matter from dwarf galaxies, Geiss et al. (2002, 2007) developed a two-component mixing model with "$PSC_0$" and "excess-infall" as the two components. $PSC_0$ represents a hypothetical cloud that had a PSC composition 4.6 Gyr ago and has since continued to evolve under nucleosynthetic, infall, and Galactic-wind conditions that are consistent with the evolution of the matter in the PSC prior to 4.6 Gyr. For the second component, the "excess-infall," they assumed a composition of the type found in dwarf galaxies.

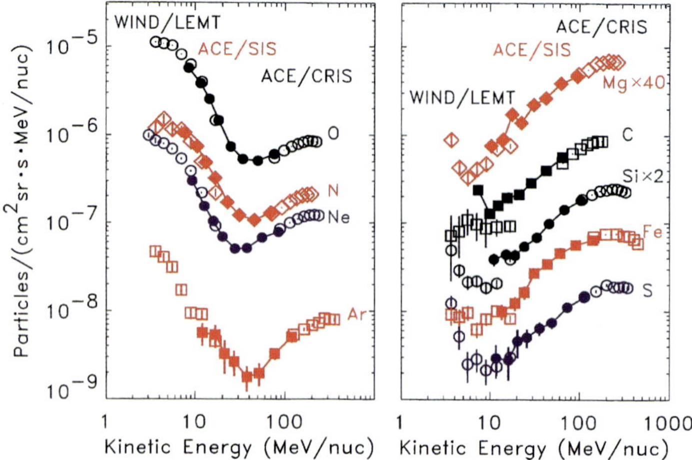


**Fig. 12** Anomalous Cosmic Rays (ACR), after Leske et al. (2000). At solar minimum, the ions below ~30 MeV/nucleon (*left panel*) are mainly locally accelerated interstellar atoms (Fisk et al. 1974) that had entered the heliosphere. Composition measurements of ACRs as well as of interstellar pickup ions measured in the inner heliosphere and accelerated pickup ions observed in the heliosheath are important for determining elemental and isotopic abundances in the LIC

If $(\mathrm{A/H})_{\mathrm{PSC}_0}$, $(\mathrm{A/H})_{\text{excess-infall}}$ and $(\mathrm{A/H})_{\mathrm{LIC}}$ are the abundances, relative to hydrogen of a nuclide A in the $\mathrm{PSC}_0$, in the "excess-infall" and in the LIC, respectively, we have

$$(\mathrm{A/H})_{\mathrm{LIC}} = (1 - X)(\mathrm{A/H})_{\mathrm{PSC}_0} + X(\mathrm{A/H})_{\text{excess-infall}}, \tag{8}$$

where $X$ is the mixing ratio.

Völk (1991) pointed out that loss by Galactic winds, even from sizable galaxies, could significantly influence the chemical evolution, if products of massive stars are lost preferentially. Geiss et al. (2007) generalized the mixing model equations to include the possibility of significant loss by winds from our Galaxy. In this paper, however, we do not consider loss, but only infall of external matter into our Galaxy. Solving (8) for $X$ and expressing abundance ratios relative to PSC values by braces, { }, the mixing ratio $X$ obtained for the O/H ratio measured in the LIC is given by Geiss et al. (2007):

$$X(\mathrm{O/H}) = \frac{\{\mathrm{O/H}\}_{\mathrm{PSC}_0} - \{\mathrm{O/H}\}_{\mathrm{LIC}}}{\{\mathrm{O/H}\}_{\mathrm{PSC}_0} - \{\mathrm{O/H}\}_{\text{excess-infall}}}. \tag{9}$$

The $X$-values for other nuclides relative to H are obtained in an analogous way.

Mixing ratios can also be calculated for the ratio of two nuclides, especially for an abundance ratio of isotopes of a given element. The corresponding equation is obtained by writing (8) for the two species and dividing the two resulting equations. For the case of $^{17}\mathrm{O}/^{18}\mathrm{O}$ one obtains (Geiss et al. 2007)

$$\{^{17}\mathrm{O}/^{18}\mathrm{O}\}_{\mathrm{LISM}} = \frac{(1 - X)\{^{17}\mathrm{O/H}\}_{\mathrm{PSC}_0} + X\{^{17}\mathrm{O/H}\}_{\text{excess-infall}}}{(1 - X)\{^{18}\mathrm{O/H}\}_{\mathrm{PSC}_0} + X\{^{18}\mathrm{O/H}\}_{\text{excess-infall}}}. \tag{10}$$

The mixing ratio $X$ can be determined from $^{18}\mathrm{O}/^{17}\mathrm{O}$ measured in the LIC or LISM and the given model parameters contained in the right-hand side of (10). This equation is

**Table 2** $PSC_0$, excess-infall and LIC or LISM abundances, and the mixing ratio $X$ obtained from model 1 and model 2

| Model[a] | Ratio | $PSC_0$ | Excess-infall | LIC or LISM | Mixing ratio $X$ |
|---|---|---|---|---|---|
| 1 | $\{D/H\}^b$ | $0.6^c$ | $1.375^c$ | $0.85 \pm 0.15^d$ | $0.32 \pm 0.25$ |
| 1 | $\{^3He/H\}$ | $1.5^c$ | $0.62^c$ | $1.20 \pm 0.20^d$ | $0.34 \pm 0.23$ |
| 1 | $\{^{12}C/H\}$ | 1.4 | 0.25 | $1.0 \pm 0.2^e$ | $0.35 \pm 0.17$ |
| 1 | $\{^{13}C/H\}$ | 1.4 | 0.64 | $1.27 \pm 0.30^e$ | $0.17 \pm 0.33$ |
| 1 | $\{^{13}C/^{12}C\}$ | 1.0 | 2.56 | $1.27 \pm 0.20^e$ | $0.54^{+0.17}_{-0.27}$ |
| 1 | $\{^{14}N/H\}$ | 1.4 | 0.1 | $0.76 \pm 0.25^d$ | $0.49 \pm 0.27$ |
| 1 | $\{^{15}N/H\}$ | 1.4 | 0.38 | $0.90 \pm 0.25^e$ | $0.49 \pm 0.27$ |
| 1 | $\{^{16}O/H\}$ | 1.4 | 0.35 | $1.06 \pm 0.30^d$ | $0.32 \pm 0.20$ |
| 1 | $\{^{18}O/H\}$ | 1.4 | 0.17 | $0.97 \pm 0.35^e$ | $0.35 \pm 0.28$ |
| 1 | $\{^{17}O/H\}$ | 1.4 | 0.56 | $1.44 \pm 0.52^e$ | $-0.05 \pm 0.57$ |
| 1 | $\{^{17}O/^{18}O\}$ | 1.0 | 3.29 | $1.484 \pm 0.114^e$ | $0.69 \pm 0.07$ |
| 2 | $\{^{13}C/H\}$ | 1.4 | 0.67 | $1.27 \pm 0.30^e$ | $0.18 \pm 0.33$ |
| 2 | $\{^{13}C/^{12}C\}$ | 1.0 | 2.69 | $1.27 \pm 0.20^e$ | $0.515^{+0.17}_{-0.27}$ |
| 2 | $\{^{17}O/H\}$ | 1.4 | 1.16 | $1.44 \pm 0.52^e$ | $-0.17 \pm 2.33$ |
| 2 | $\{^{17}O/^{18}O\}$ | 1.0 | 6.82 | $1.484 \pm 0.114^e$ | $0.43 \pm 0.07$ |

[a]Model 1: excess-infall = LMC, $(^{18}O/^{17}O)_{\text{excess-infall}} = 1.65$; model 2: excess-infall = LMC + 0.5% IMS, $(^{18}O/^{17}O)_{\text{excess-infall}} = 0.80$

[b]Abundances enclosed by { } are relative to PSC values

[c]By interpolation, see text

[d]LIC abundance

[e]LISM abundance

independent of the abundance of hydrogen. In cases where the ratio of the two species is well determined and better known than their LISM abundances relative to hydrogen, (10) yields mixing ratios with smaller errors than (9).

In this paper we add the isotopes of C and N to the species included earlier (Geiss et al. 2007). Table 2 shows the LIC (or LISM) and LMC abundances, relative to the respective PSC abundance used in this paper. We have adopted the N/H, O/H and oxygen isotope abundances used by Gloeckler (2005) and Geiss et al. (2007), who give the original references. The $(^{15}N/^{14}N)_{PSC}$ ratio was adopted from Owen et al. (2001). The $\{D/H\}_{LMC}$ and $\{^3He/H\}_{LMC}$ ratios were obtained by interpolation between the primordial and the PSC abundances (see Fig. 11), assuming that D/H and $^3$He/H change in proportion to metallicity (Geiss et al. 2002). For the metal licity in the LMC, we used 0.25, an average of the elemental abundances of the CNO group relative to protosolar abundance (Pagel 2003). The C and N isotope abundances were taken from Prantzos et al. (1996), and Chin et al. (1999) for the LISM, and from Chin (1999), Chin et al. (1999), Heikkilä and Johansson (1999), and Pagel (2003) for the LMC.

The $\{D/H\}_{PSC_0}$ and $\{^3He/H\}_{PSC_0}$ abundances were derived from Galactic evolution models (see Geiss et al. 2002). We adopted $\{A/H\}_{PSC_0} = 1.4$ for all CNO nuclides (column 3 in Table 2), thus assuming equal evolution from PSC to $PSC_0$.

We discuss here the two models introduced by Geiss et al. (2007). In model 1 the excess-infall composition is identical to the LMC composition. In model 2 we combine 99.5% of matter with LMC composition with 0.5% of ejecta from intermediate mass stars, using the

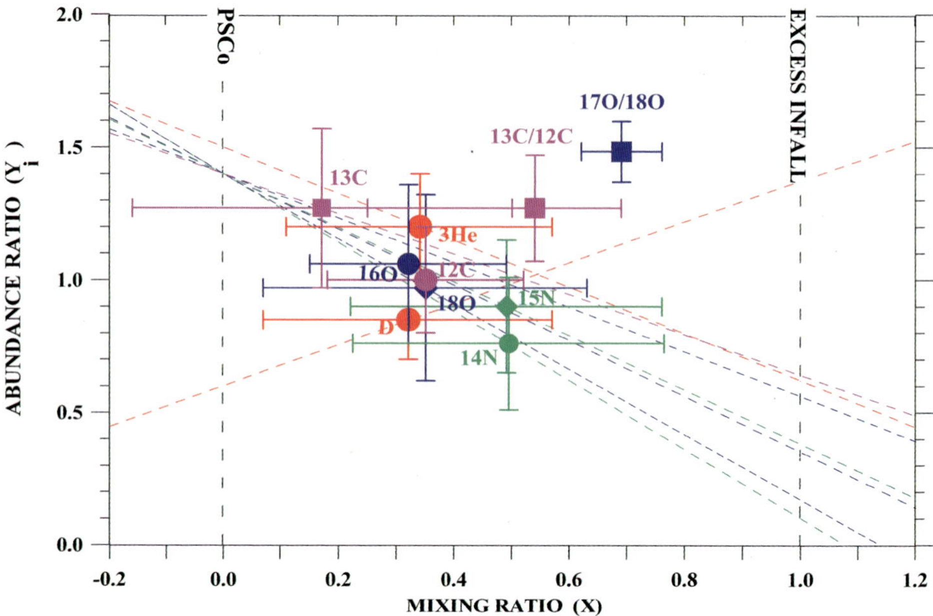


**Fig. 13** Results obtained with mixing model 1. Measured abundances, relative to H and normalized to the protosolar ratios, are plotted versus the respective mixing ratios $X$ computed using (9). All abundance ratios are normalized to PSC values. The figure shows the mixing lines connecting the values of a given abundance ratio in the two defining reservoirs, $PSC_0$ (at $X = 0$) and "excess-infall" (at $X = 1$); see (8–10) and Table 2. Except for $^{17}O/^{18}O$, the error limits of all data points cover the "range of compatibility" $0.25 < X < 0.50$

oxygen isotope yields calculated by Ventura et al. (2002) for stars with initial mass between 3.5 $M_{Sun}$ and 5.5 $M_{Sun}$ and $Z = 0.01$. For the isotopes of C and N we adopted the yields given by Marigo (2001).

Mixing ratios $X$ calculated from (9) are given in the last column of Table 2. The measured LIC or LISM data points $\{A/H\}_{LIC}$ are plotted in Figs. 13 and 14 on the respective mixing line at the calculated $X$-value. These mixing lines are defined by the composition of the two components, $PSC_0$ and excess-infall. In the case of $\{^{17}O/H\}$ we obtain negative mixing ratios $X$ with large errors that cover the whole "range of compatibility" defined in the following (see Table 2). We did not include the $\{^{17}O/H\}$ data points in Figs. 13 and 14, because the mixing ratios for $\{^{17}O/^{18}O\}$ have much smaller errors and therefore, are much more significant as a test of the assumptions going into mixing models 1 and 2.

Mixing ratios obtained from (10) are superior to those obtained from (9) whenever isotopic abundances are given with high enough precision. Since this is the case for $^{18}O/^{17}O$ and, to a lesser extent, for $^{13}C/^{12}C$, we include in Figs. 13 and 14 the $X$-values for these two isotopic ratios.

Model 1 gives mixing ratios in the "compatibility range" of $0.25 < X < 0.50$. In this range, the error bars overlap for all the ratios in Table 2, except $\{^{17}O/^{18}O\}$ that gives a well-defined mixing ratio of $X = 0.69 \pm 0.07$.

This discrepancy is eliminated in model 2. The admixture of ~0.5% of IMS ejecta to the LMC-like matter decreases the $(^{18}O/^{17}O)_{\text{excess-infall}}$ ratio from 1.65 to 0.80, corresponding to $X = 0.43 \pm 0.07$, well inside the "compatibility range" of model 1. Mixing ratios inside the "compatibility range"; that is, $X$-values between 0.5 and 0.25 are obtained by admixing 0.3% to 1.3% of ISM ejecta. These limits do not take uncertainties in the $^{17}O$ yields from

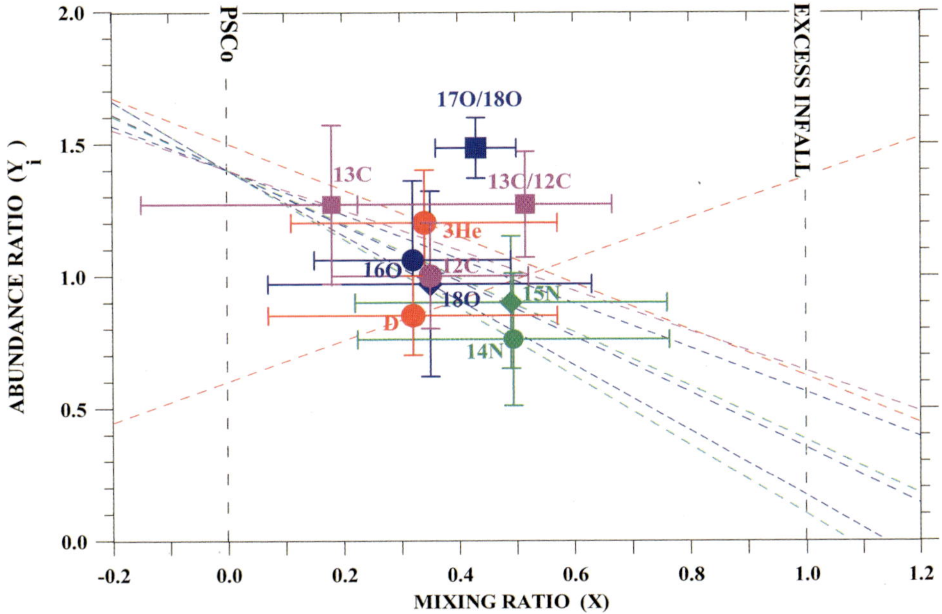


**Fig. 14** Results obtained with mixing model 2. For this model all abundance ratios are compatible with the "range of compatibility" given above, implying that Galactic matter in the LIC and LISM can be represented by a mixture of PSC matter (extrapolated to the present) and an excess-infall with the composition assumed for model 2

intermediate mass stars into account. Intermediate mass stars are very specific $^{17}$O producers, the effects of the addition of 0.5% IMS ejecta for other nuclides is minor. Thus $^{13}$C is increased by only about 5%, resulting in a minor shift of $^{13}$C/H and $^{13}$C/$^{12}$C towards the center of "the range of compatibility" (see Table 2, Figs. 13 and 14).

Ratios of $^{18}$O/$^{17}$O as low as 0.80 are not unrealistic. Much lower values are predicted for IMS ejecta (Marigo 2001; Ventura et al. 2002). Moreover, ratios $^{18}\mathrm{O}/^{17}\mathrm{O} \leq 1$ are typically observed in the spectra of red giant stars, and they occur abundantly in the pre-solar grains of meteorites (see, e.g., Ott 2007; Geiss et al. 2007 and references therein).

The theoretical yields (Ventura et al. 2002; Marigo 2001) for IMS ejecta give virtually identical excess-infall abundances and mixing ratios for models 1 and 2 for all the species discussed here, except for values involving $^{17}$O and $^{13}$C. Of course, theoretical yields of the rare isotopes of CNO are rather uncertain (e.g., Romano and Matteucci 2003), but a particularly high $^{17}$O abundance in IMS ejecta seems to be pretty well established.

The relatively high abundances of $^{15}$N and $^{13}$C in the LMC could in part be due to ejecta from novae (Romano and Matteucci 2003). We have not considered adding such ejecta, because for model 2 all species give $X$-values in the "range of compatibility."

By analyzing spectra obtained with the FUSE (Far Ultraviolet Spectroscopic Explorer) satellite and earlier data, Linsky et al. (2006) found a surprisingly wide range of D/H ratios for the interstellar gas in the Galactic disk beyond the local bubble (see also Romano et al. 2006). The authors explained this variability by depletion of deuterium onto dust grains, and they concluded that the most representative ratio in the interstellar medium (gas plus dust) within 1 kpc of the Sun is $\geq 23.1 \pm 2.4$ ppm ($1\sigma$). In spite of these important new data and interesting interpretations, we decided not (yet) to change $\{\mathrm{D/H}\}_{\mathrm{LIC}} = 0.85 \pm 0.15$ (D/H = 17 ppm), the value used by Geiss et al. (2007). At least some of the observed D/H variability could very well be due to variations in space, time and composition of infall (Steigman et al. 2007). Such local differences in D/H would not affect the conclusions of our mixing model, because there we compare two specific Galactic samples that differ by ~5 Gyr in age. In any case, if D/H in the LIC would really be as high as in the PSC, we would have $\{D/H\} = 1.0 \pm 0.15$ and a mixing ratio $X = 0.51 \pm 0.19$ which is still in reasonable agreement with the results of model 2 (see Fig. 14).

The results obtained with the mixing model as presented in Table 2 and Fig. 13 point to dwarf galaxies as an important source of infall into the Galaxy during the last 5 Gyr. Improving crucial composition data, such as N/O, or the isotopic composition of oxygen and carbon in the PSC, LISM, LMC and other dwarf galaxies would further consolidate this result. Further improvements of LIC composition are expected from pickup ions, accelerated pickup ions and anomalous cosmic ray measurements.

As important, however, would be to find complementary dynamical evidence. The approach and eventual accommodation of dwarf galaxy material into the Galaxy is a complex process that is difficult to model. The discussion of this process by Geiss et al. (2007) suggests that a high abundance of IMS products carried by the dwarf galaxy would not be lost during the accommodation process.

If small galaxies were originally more abundant than presently recognized (e.g., Ostriker and Steinhardt 2003), the probability increases that dwarfs have in the past fallen into the Galaxy. Evidence for ongoing infall of matter from small galaxies into the Milky Way is provided by the Magellanic Stream and the Sagittarius dwarf galaxy. Recently, several small dwarf spheroidals were discovered close to the Galaxy by the Sloan Digital Sky Survey (Belokurov et al. 2007). The Andromeda galaxy has two centers, probably resulting from an impact of a small galaxy (Lauer et al. 1993; Gerssen et al. 1995). Perhaps the warp in the Galactic rotation curve at ~14 kpc (e.g., Honma and Sofue 1997) is due to dark matter that is a remnant from the infall of a small galaxy in the past (de Boer et al. 2005). Improved knowledge about the distribution of dark matter around our Galaxy and beyond would allow one to better extrapolate into the past, because it is the dark matter that primarily controls the dynamics in the local group (including the Galaxy, the Andromeda galaxy, and many smaller galaxies).

## 13 Concluding Remarks

We understand the evolution of baryonic matter much better for the three-minute epoch of primordial nucleosynthesis than for the last 5 Gyr in the life of the Galaxy. The reason is, of course, that in the early epoch we perfectly know the physics of the dominant forms of matter in the universe (electron pairs, photons, and neutrinos; see Fig. 3 and Table 2), and we have reliable evidence for their homogeneous distribution and expansion. During this early epoch, baryonic matter is a minor component participating in, but not influencing, the homogeneous expansion, and as a consequence theoretical predictions agree with the well-determined primordial abundances.

The situation is different for the highly heterogeneous, evolved universe. Infall into the Galaxy has caused its chemical evolution to change only slightly during the last 5 Gyr, so that differences in composition are difficult to establish. On the other hand, infall into the Galaxy depends on gravitational fields and dynamics in the local group, both being largely determined by the distribution of Dark Matter of unknown identity. Thus, to better understand the chemical evolution of the Galaxy during the last 5 Gyr, we need improved abundances of crucial elements and isotopes in the PSC and LISM, but we should also know the identity, physical properties and distribution of dark matter around the Galaxy and in the region of the local group.

**Acknowledgements** We thank Donatella Romano, Monica Tosi, Klaus Pretzl, Heinrich Leutwyler and Heinz Völk for discussions and suggestions, the anonymous referee for constructive criticism and Chris Gloeckler for help with the manuscript. This work was supported, in part, by NASA contract NAGR-10975, and by JPL contract 1237843.

## References

H. Balsiger, J. Geiss, G. Grögler, A. Wyttenbach, Earth Planet. Sci. Lett. **5**, 17–22 (1968)
T.M. Bania, R.T. Rood, D.S. Balser, Nature **415**, 54 (2002)
V. Belokurov et al., Astrophys. J. **654**, 897 (2007)
C.L. Bennett et al., Astrophys. J. Suppl. Ser. **148**, 1–27 (2003)
H. Böhringer, Space Sci. Rev. **100**, 49 (2002)
A.I. Boothroyd, I.-J. Sackmann, Astrophys. J. **510**, 232 (1999)
S. Burles, B. Tytler, Space Sci. Rev. **84**, 65 (1998)
H. Busemann, F. Buehler, A. Grimberg, V.S. Heber, Y.N. Agafonov, H. Baur, P. Bochsler, N.A. Eismont, R. Wieler, G.N. Zastenker, Astrophys. J. **639**, 246–258 (2006)
C. Charbonnel, Astrophys. J. **453**, L41 (1995)
C. Charbonnel, Space Sci. Rev. **84**, 199–206 (1998)
C. Chiappini, F. Matteucci, S.K. Ballero, Astron. Astrophys. **437**, 429C (2005)
Y. Chin, in *New Views of the Magellanic Clouds*. IAU Symposium, vol. 190 (Kluwer Academic, Dordrecht, 1999), pp. 279–281, and references given therein
Y.-N. Chin, C. Henkel, N. Langer, R. Mauersberger, Astropys. J. **512**, L143 (1999)
A.C. Cummings, E.C. Stone, Space Sci. Rev. **78**, 43–52 (1996)
A.C. Cummings, E.C. Stone, C.D. Steenberg, Proc. 26th Intl. Cosmic Ray Conf. **7**, 531 (1999)
W. de Boer, C. Sander, V. Zhukov, A.V. Gladyshev, D.I. Kazakov, Astron. Astrophys. **444**, 51–67 (2005)
A. d'Ercole, F. Brighenti, Mon. Not. R. Astron. Soc. **309**, 941 (1999)
M.A. Dopita, in *The Magellanic Clouds*, ed. by R. Haynes, D. Milne. IAU Symposium, vol. 148 (Kluwer Academic, Dordrecht, 1991), p. 299
S. Eidelman et al., Phys. Lett. B **592**, 1 (2004)
A.E. Evrard, Mon. Not. R. Astron. Soc. **292**, 289 (1997)
L.A. Fisk, B. Kozlovsky, R. Ramaty, Astrophys. J. Lett. **190**, L35 (1974)
J.L. Fowler, W.A. Greenstein, F. Hoyle, Geophys. J. **6**, 148 (1962)
J. Geiss, G. Gloeckler, Space Sci. Rev. **106**, 13–18 (2003), with references to the origin of the data
J. Geiss, G. Gloeckler, in *The Solar System and Beyond: Ten Years of ISSI*, ed. by J. Geiss, B. Hultqvist. ISSI Scientific Report SR-003 (ESA-ESTEC Publ. Division., Noordwijk, 2005), pp. 53–68
J. Geiss, H. Reeves, Astron. Astrophys. **18**, 126 (1972)
J. Geiss, R. von Steiger, in *Fundamental Physics in Space*, Proc. Alpbach Summer School 1997, ESA SP-420 (1997)
J. Geiss, G. Gloeckler, C. Charbonnel, Astrophys. J. **578**, 562 (2002)
J. Geiss, G. Gloeckler, L.A. Fisk, in *The Physics of the Heliospheric Boundaries*, ed. by V. Izmodenov, R. Kallenbach. ISSI Scientific Report No. 5 (ESA-ESTEC, Paris, 2007), pp. 137–181
J. Gerssen, K. Kuijken, M.R. Merrifield, Mon. Not. R. Astron. Soc. **277**, L21–L24 (1995)
G. Gloeckler, presented at the LoLaGE ISSI Team Meeting (2005)
G. Gloeckler, J. Geiss, Nature **386**, 210 (1996)
G. Gloeckler, J. Geiss, Space Sci. Rev. **84**, 275–284 (1998)
G. Gloeckler, J. Geiss, in *The Light Elements and Their Evolution, Deuterium and Helium-3 in the Protosolar Cloud*, ed. by L. da Silva, M. Spite, J.R. de Medeiros. IUA Symposium, vol. 198 (Kluwer Academic, Dordrecht, 2000), pp. 224–233
G. Gloeckler, J. Geiss, in *Solar and Galactic Composition*, ed. by R.F. Wimmer-Schweingruber. AIP Conf. Proc., vol. 598 (2001), pp. 281–289
A. Heikkilä, L.E.B. Johansson, in *New Views of the Magellanic Clouds*, ed. by Y.-H. Chu et al. IAU Symposium, vol. 190 (Kluwer Academic, Dordrecht, 1999), p. 275
C. Henkel, R. Mauersberger, Astron. Astrophys. **274**, 730–742 (1993)
M. Honma, Y. Sofue, Publ. Astron. Soc. Japan **49**, 453–460 (1997)
C. Kahane, in *Nuclei in the Cosmos III*, ed. by N. Busso et al. AIP, vol. 327 (1995), p. 19

R.P. Kirshner, Science **300**, 1914–1918 (2003)
R. Lallement, in *The Century of Space Science*, ed. by J. Bleeker, J. Geiss, M.C.E. Huber (Kluwer Academic, Dordrecht, 2001), p. 1191
T.R. Lauer et al., Astron. J. **106**, 1436 (1993)
D.-C. Lee, A.N. Halliday, Science **274**, 1876 (1996)
R.A. Leske, in *AIP Conference Proceedings, 26th International Cosmic Ray Conference*, ed. by B.L. Dingus, D.B. Kieda, M.H. Salamon. AIP, vol. 516 (2000), pp. 274–282
R.A. Leske, R.A. Mewaldt, A.C. Cummings, E.C. Stone, T.T. van Rosenvinge, Space Sci. Ser. of ISSI **1** and Space Sci. Rev. **78**, 49 (1996)
R.A. Leske, R.A. Mewaldt, E.R. Christian, C.M.S. Cohen, A.C. Cummings, P.L. Slocum, E.C. Stone, T.T. van Rosenvinge, M.E. Wiedenbeck, in *Acceleration and Transport of Energetic Particles Observed in the Heliosphere*, ed. by R.A. Mewaldt et al. AIP Conf. Proc., vol. 528 (2000), p. 293
J.L. Linsky, B.E. Wood, in *The Light Elements and Their Evolution*, ed. by L. da Silva, M. Spite, J.R. de Medeiros. IUA Symposium, vol. 1998 (Kluwer Academic, Dordrecht, 2000), pp. 141–150
J.L. Linsky et al., Astrophys. J. **647**, 1106 (2006)
P.R. Mahaffy, T.M. Donahue, S.K. Atreya, T.C. Owen, H.B. Niemann, Space Science Series of ISSI **4** and Space Sci. Rev. **84**, 251–263 (1998)
P. Marigo, Astron. Astrophys. **370**, 194–217 (2001)
M. Meneguzzi, J. Audouze, H. Reeves, Astron. Astrophys. **15**, 337 (1971)
J.M. O'Meira, D. Tytler, D. Kirkman et al., Astrophys. J. **552**, 718–730 (2001)
J.M. O'Meira, S. Burles, J.X. Prochaska, G.E. Prochter, R.A. Bernstein, K.M. Burgess, Astrophys. J. **649**, L61 (2006)
J.P. Ostriker, P. Steinhardt, Science **300**, 1909–1913 (2003)
U. Ott, Space Sci. Rev. **130** (2007), this volume, doi:10.1007/s11214-007-9159-5
T. Owen, P.R. Mahaffy, H.B. Niemann, S. Atreya, M. Wong, Astrophys. J. **553**, L77 (2001)
B.E.J. Pagel, in *CNO in the Universe*, ed. by C. Charbonnel, D. Schaerer, G. Meynet. ASP Conf. Series, vol. 304 (2003), p. 187
S. Perlmutter et al., Astrophys. J. **517**, 565 (1999)
F.A. Podosek, R.H. Nichols, in *Astropysicl Implications of the Laboratory Study of Presolar Materials*, ed. by T.J. Bernatovicz, E.K. Zinner. AIP Conf. Proc., vol. 402 (1997), p. 617
N. Prantzos, Space Sci. Rev. **130** (2007), this volume
N. Prantzos, O. Aubert, J. Audouze, Astron. Astrophys. **309**, 760 (1996)
T. Preibisch, C. Briceno, H. Zinnecker, F. Walter, E. Mamajek, R. Mathieu, B. Sherry, Power-Point presentation (2006), www.as.utexas.edu/astronomy/people/scalo/IMFat50.Scalo5.6.ppt
K. Pretzl, Space Sci. Rev. **100**, 209–220 (2002)
K. Pretzl, Space Sci. Rev. **130** (2007), this volume, doi:10.1007/s11214-007-9151-0
R. Rebolo, Space Sci. Rev. **100**, 15–28 (2002)
H. Reeves, Space Sci. Rev. **130** (2007), this volume, doi:10.1007/s11214-007-9199-x
H. Reeves, J. Audouze, W.A. Fowler, D.N. Schramm, Astrophys. J. **179**, 909 (1973)
A.G. Riess et al., Astron. J. **116**, 109 (1998)
D. Romano, F. Matteucci, Mon. Not. R. Astron. Soc. **342**, 185 (2003)
D. Romano, M. Tosi, C. Chiappini, F. Matteucci, Mon. Not. R. Astron. Soc. **369**, 295 (2006)
A. Shukoliukov, F. Begemann, Geochim. Cosmochim. Acta **60**, 2453 (1996)
G.F. Smoot et al., Astrophys. J. **396**, L1 (1992)
G. Steigman, D. Romano, M. Tosi, Mon. Not. R. Astron. Soc. **378**, 576S (2007)
T.X. Thuan, Y.I. Izotov, Space Sci. Rev. **84**, 83–94 (1998)
M. Tosi, Astron. Astrophys. **197**, 47–51 (1988)
M. Tosi, Space Sci. Rev. **84**, 207–218 (1998)
M. Tosi, in *CNO in the Universe*, ed. by C. Charbonnel, D. Schaerer, G. Meynet. ASP Conf. Ser., vol. 304 (2003), pp. 1–6
S. Veilleux, G. Cecil, J. Bland-Hawthorn, Ann. Rev. Astron. Astrophys. **43**, 769–826 (2005)
P. Ventura, F. D'Antona, I. Mazzitelli, Astron. Astrophys. **393**, 215 (2002)
H. Völk, in *The Interstellar Disk-Halo Connection in Galaxies*, ed. by H. Bloemen (Kluwer Academic, Dordrecht, 1991), pp. 345–353
W.A. Wagoner R.V. Fowler, F. Hoyle, Astrophys. J. **447**, 680–685 (1967)
S. Weinberg, *Gravitation and Cosmology* (Wiley, New York, 1972)
C. Wetterich, Space Sci. Rev. **100**, 195–208 (2002)

Space Sci Rev (2007) 130: 27–42
DOI 10.1007/s11214-007-9183-5

# Origin and Evolution of the Light Nuclides

**N. Prantzos**

Received: 23 January 2007 / Accepted: 30 March 2007 /
Published online: 15 May 2007


**Abstract** After a short historical (and highly subjective) introduction to the field, I discuss our current understanding of the origin and evolution of the light nuclides D, $^3$He, $^4$He, $^6$Li, $^7$Li, $^9$Be, $^{10}$B and $^{11}$B. Despite considerable observational and theoretical progress, important uncertainties still persist for each and every one of those nuclides. The present-day abundance of D in the local interstellar medium is currently uncertain, making it difficult to infer the recent chemical evolution of the solar neighborhood. To account for the observed quasi-constancy of $^3$He abundance from the Big Bang to our days, the stellar production of that nuclide *must* be negligible; however, the scarce observations of its abundance in planetary nebulae seem to contradict this idea. The observed Be and B evolution as primaries suggests that the source composition of cosmic rays has remained ~constant since the early days of the Galaxy, a suggestion with far reaching implications for the origin of cosmic rays; however, the main idea proposed to account for that constancy, namely that superbubbles are at the source of cosmic rays, encounters some serious difficulties. The best explanation for the mismatch between primordial Li and the observed "Spite-plateau" in halo stars appears to be depletion of Li in stellar envelopes, by some yet poorly understood mechanism. But this explanation impacts on the level of the recently discovered early "$^6$Li plateau", which (if confirmed), seriously challenges current ideas of cosmic ray nucleosynthesis.

**Keywords** Light elements · Chemical evolution · Early Galaxy · Metal-poor stars · Cosmic rays

## 1 Introduction

In their monumental study on "Synthesis of the Elements in Stars", Burbidge et al. (1957; B$^2$FH) recognized the difficulty of finding a nuclear process able to synthesize the light nuclides D, $^6$Li and $^7$Li, $^9$Be, $^{10}$B and $^{11}$B. Indeed, these nuclides are so fragile (as revealed by

N. Prantzos (✉)
Institut d'Astrophysique de Paris, 98bis Bd Arago, 75014 Paris, France
e-mail: prantzos@iap.fr

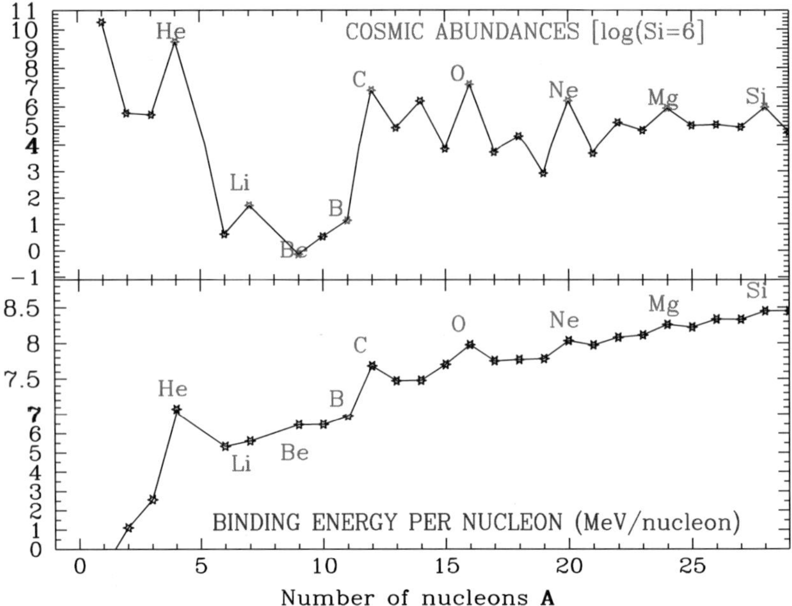


**Fig. 1** *Top:* Cosmic abundances of the light nuclides, from H to Si, in the log (Si) $= 6$ scale. *Symbols* indicate major isotopes of a given element; alpha-nuclides (C-12, O-16, etc.) dominate always their neighbors, up to Ca-40. *Bottom:* Binding energies of the light nuclides (note the change in the *vertical scale* at 7 MeV/nucleon). D, $^3$He, and LiBeB isotopes are more fragile than neighboring nuclei; that fragility is clearly reflected in the *cosmic abundance curve*

their binding energies in Fig. 1) that they are consumed in stellar interiors, once hydrogen-rich material is brought to temperatures higher than 0.6 MK for D, 2 MK for $^6$Li, 2.5 MK for $^7$Li, 3.5 MK for $^9$Be and 5 MK for the boron isotopes[1].

B$^2$FH argued that the "x-process" (as they called the unknown nucleosynthetic mechanism) should occur in low-density, low-temperature environments. They discussed stellar atmospheres (of active, magnetized stars) and gaseous nebulae (traversed by energetic particles) as possible sites, and they concluded that, most probably, D originates from a different process than the Li, Be and B (hereafter LiBeB) isotopes.

The synthesis of the He isotopes ($^3$He and $^4$He) drew very little attention in B$^2$FH, where it was flatly attributed to stellar H-burning with no further comments. This (most surprising) neglect of B$^2$FH was corrected in Hoyle and Tayler (1964), who demonstrated that H-burning stars of the Milky Way (MW), releasing an energy of $\varepsilon(\mathrm{H} \rightarrow {}^4\mathrm{He}) = 6 \times 10^{18}$ erg g$^{-1}$, having a total mass $M_{\mathrm{MW}} = 10^{11}$ M$_\odot$ and shining collectively with a luminosity $L_{\mathrm{MW}} = 6 \times 10^{43}$ erg s$^{-1}$ for $T = 10^{10}$ yr, could produce a mass fraction of $^4$He of only a few per cent; this is about 10 times less than the observed abundance of $^4$He (mass fraction $X(^4\mathrm{He}) \sim 0.25$), which requires then another nucleosynthesis site, like the hot early universe (or, in Hoyle's views, high temperature explosions of extremely massive pre-galactic stars).

[1] In such temperatures, and for densities comparable to those encountered in the bottom of the outer convective zones of low mass stars, like the Sun, the lifetimes of light nuclides against proton captures are smaller than a few Gyr.

After the discovery of the cosmic microwave background (CMB) by Penzias and Wilson (1965), which strongly supported the Big Bang model for the origin of the Universe, calculations of primordial nucleosynthesis by Peebles (1966) and Wagoner et al. (1967) showed that D, $^3$He and $^4$He could be produced in large amounts (i.e. comparable to their present-day measured values) in the hot early Universe. Moreover, the latter work showed that significant amounts of $^7$Li could also be generated in that rapidly cooling environment.

In the early 1970ies, the pre-solar abundances of D and $^3$He were accurately established (Black 1971; Geiss and Reeves 1972) and it was convincingly argued that D could be produced in no realistic astrophysical site other than the hot early Universe (Reeves et al. 1973; Epstein 1976). Since D can only be destroyed by astration after the Big Bang, it can then be used as a "baryometer" (Reeves et al. 1973), revealing that the cosmic baryonic density is smaller than the critical value (e.g. Gott et al. 1974), i.e. that baryons cannot "close" the Universe. The precise value of the baryonic density was pinpointed only 30 years later, from converging measurements of the CMB anisotropies by the WMAP satellite and observations of (presumably primordial) D abundances in remote gas clouds (Fig. 2 and texts by Tytler, Geiss, Reeves in this volume). However, the exact amount of D astration during galactic evolution remains unknown at present, due to uncertainties on its present-day local ISM value (see Sect. 2.1 and text by Linsky in this volume).

The case of $^3$He turned out to be much more complex than the one of D, since $^3$He can be produced not only in the Big Bang, but also in stars (from burning of primordial D: D + p $\rightarrow$ D *and* from the p–p chains), whereas it may also be destroyed in stellar zones hotter than $10^7$ K. In the first comprehensive Galactic chemical evolution model ever made (Truran and Cameron 1971), it was shown that, *if* standard stellar nucleosynthesis prescriptions are adopted for $^3$He (e.g. from models by Iben 1967), then that nuclide is largely overproduced during galactic evolution. Thirty six years later, the issue is not satisfactorily settled, despite theoretical and observational developments (see Sect. 2.2 and text by Bania in this volume).

The suggestion of B$^2$FH that substantial production of light nuclides can occur on the surfaces of active stars, further elaborated in Fowler et al. (1962), was refuted by Ryter et al. (1970) on the grounds of energetics arguments: the total amount of available (gravitational) energy, mostly in the active T Tauri phase of stellar youth, is insufficient for that. Noting that the LiBeB/CNO ratio in galactic cosmic rays (GCR) is $\sim 10^4$ times higher than in the ISM, Reeves et al. (1970) proposed that LiBeB isotopes are produced by spallation reactions on CNO nuclei, occurring during the propagation of GCR in the interstelar medium (ISM) of the Galaxy. The process was definitely modeled by Meneguzzi et al. (1971, MAR), who found that the pre-solar abundances of $^6$Li, $^9$Be and $^{10}$B, $\sim$20% of $^7$Li and $\sim$60% of $^{11}$B can be produced that way, after 10 Gyr of galactic evolution. The majority of $^7$Li should originate in another (presumably stellar) site, unidentified as yet. AGB stars, where the Cameron and Fowler (1971) process may operate, appear as an attractive possibility (supported by observations of Li-rich evolved stars), but explosive H-burning in novae remains an interesting alternative.

The major development of the 80ies was the discovery (Spite and Spite 1982) that the Li abundance in metal-poor halo stars remains constant, at about 0.05 of its pre-solar value (the "Spite plateau"). This behaviour, shared by no other metal, suggests that early Li is primordial and gives further support to the theory of the Big Bang. However, the cosmic baryonic density inferred from WMAP measurements of CMB corresponds to a Li abundance 2–3 times higher than the Spite plateau (Fig. 2) and makes the statement "the Li plateau is primordial" sound rather strange. After a flurry of possible explanations, it appears now that the fault lies within the stars themselves (see Sect. 4.1 and text by Gratton, this volume), able to transform the primordial plateau into another, lower lying, one.

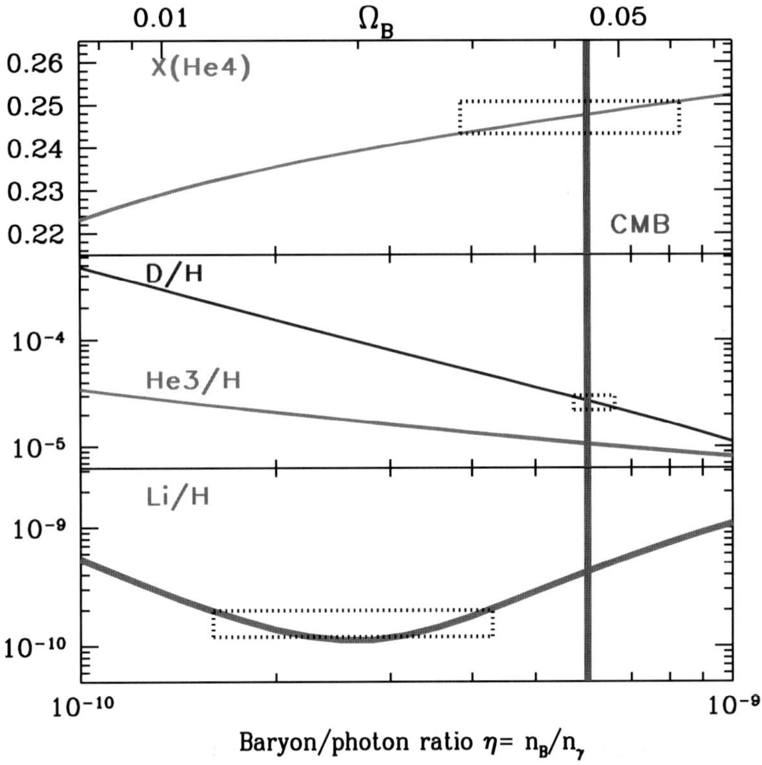


**Fig. 2** Results of standard Big Bang nucleosynthesis (SBBN) calculations vs. cosmic baryon/photon ratio $\eta$ (*bottom horizontal axis*) or cosmic baryon density (in units of the critical density, *top horizontal axis*). A Hubble constant of $H_0 = 70$ km/s/Mpc is assumed. The width of the curves reflects 1-$\sigma$ statistical uncertainties in the 12 main nuclear reaction rates of SBBN, still important in the case of $^7$Li. Abundances are by number, except for $^4$He, for which the mass fraction is given. The *vertical line across all panels* corresponds to the baryonic density determined by analysis of the Cosmic Microwave Background anisotropies, detected by WMAP. *Boxes* indicate observed, or observationally inferred, primordial abundances of the light nuclides (see text). The agreement with WMAP is extremely good in the case of D and poor in the case of $^7$Li (a factor of 2–3), while determination of primordial $^4$He suffers from considerable systematic uncertainties

Two major developments were made in the field in the 90ies. First, it was realized that the two LiBeB isotopes underproduced in the standard GCR scenario of MAR ($^{11}$B and—to a smaller extent—$^7$Li), can also be produced by neutrino-induced nucleosynthesis during core-collapse supernova explosions (Woosley et al. 1990); however, various uncertainties (neutrino spectra, structure of progenitor star, explosion mechanism, etc.) make yield predictions for those nuclides unreliable by rather large factors. Secondly, observations of Be and B in metal-poor stars (Ryan et al. 1992; Duncan et al. 1992) showed that those elements behave as *primaries*, i.e. the Be/Fe and B/Fe ratios remain constant with metallicity. A *secondary* behaviour was a priori expected (since the yields of those nuclides, being proportional to splatted CNO, should increase with time/metallicity), and this expectation is not modified even by invoking extreme features for the GCR propagation in the early Galaxy (Prantzos et al. 1993a). The B data may be fixed by its neutrino production in supernovae, but for Be one has to assume that the CNO content of cosmic rays does not increase with time/metallicity (Duncan et al. 1992; Prantzos et al. 1993b). That bold conjecture is, in fact, the only possibility, as shown by Ramaty et al. (1997) on the grounds of energetics: if the CNO content of GCR were lower in the past, much more energy in GCR would be required to compensate for that and to keep the Be/Fe ratio constant; but the required energy is much larger than available from supernovae explosions, which are the main energy source of GCR (Sect. 3.1 and Fig. 7a). The implications of that discovery for the (still debated) origin of GCR are not clear yet (see Sect. 3.2 and text by Binns, this volume).

A new twist to the LiBeB saga came in the 2000s, with the discovery of $^6$Li in metal-poor halo stars (Asplund et al. 2006), after several unconvincing attempts in the 90ies. Be was expected to behave as secondary and found to behave as primary. $^6$Li was expected to behave as primary (since it is mostly produced by metallicity independent $\alpha+\alpha$ reactions in the early Galaxy, as argued by Steigman and Walker 1992) and found to display a "plateau", at a level $\sim$20 times lower than the Spite plateau of Li. The $^6$Li plateau lies above the well-constrained contribution of standard GCR, calling for other explanations for its origin (Sect. 4.2 and Fig. 9).

In the following sections I discuss each one of the light nuclides (except for $^4$He), insisting on recent developments and current issues.[2]

## 2 Deuterium and Helium-3

The lives of D and $^3$He are intimately, but not totally, coupled: they are both produced in the Big Bang and D is rapidly turned into $^3$He inside stars. In the 80ies, a lot of effort was devoted to find how much of this $^3$He survived and was rejected in the ISM (e.g. Dearborn et al. 1986), in order to use the sum of D + $^3$He to constrain the baryonic density from SBBN (e.g. Yang et al. 1984; Walker et al. 1991). However, such attempts were futile, due to the (well known at the time) fact that stars can produce their own $^3$He (i.e. independently of any initial D), but also they can destroy D *and* $^3$He (i.e. without producing any $^3$He). In other terms, it is not a priori known whether the sum of D + $^3$He (used in such studies) has to stay constant, to decrease or to increase during galactic evolution. The evolution of the two nuclides should then be considered independently.

**Table 1** Abundances of primordial nuclides (*from references in parenthesis*)

| Nuclide | SBBN + WMAP<br>−13.7 Gyr | Observed earliest<br>−(10–13) Gyr | Pre-solar<br>−4.6 Gyr | Local ISM<br>Today |
|---|---|---|---|---|
| D/H ($10^{-5}$) | $2.56\pm0.18^{(1)}$ | $2.6\pm0.4^{(1)}$ | $2.\pm0.35^{(2)}$ | $2.3\pm0.24^{(3)}$<br>$0.98\pm0.19^{(4)}$ |
| $^3$He/H ($10^{-5}$) | $1.04\pm0.04^{(1)}$ | | $1.6\pm0.06^{(2)}$ | $2.4\pm0.7^{(5)}$<br>$1.7\pm0.7^{(6)}$ |
| $^4$He ($Y_P$) | $0.2482\pm0.0007^{(1)}$ | $0.2472\pm0.0035^{(1)}$ | $0.274^{(7)}$ | |
| $^7$Li/H ($10^{-10}$) | $4.44\pm0.57^{(1)}$ | $1.1–2.^{(8)}$ | $22.8^{(7)}$ | |
| $^6$Li/H ($10^{-10}$) | $0.0001^{(9)}$ | $0.08^{(10)}$ | $1.73^{(7)}$ | |

$^{(1)}$Steigman (2006) and references therein (note the discussion on $Y_{P,OBS}$); $^{(2)}$Geiss and Gloeckler (2002); $^{(3)}$Linsky et al. (2006); $^{(4)}$Hebrard et al. (2005); $^{(5)}$Gloeckler and Geiss (1996); $^{(6)}$Salerno et al. (2003); $^{(7)}$Lodders (2003); $^{(8)}$Gratton (this volume); $^{(9)}$Serpico et al. (2004); $^{(10)}$Asplund et al. (2006)

[2]For comprehensive reviews see: Reeves (1994), Prantzos et al. (1998) and Steigman (2006).

## 2.1 Deuterium

Modelling the Galactic chemical evolution (GCE) of deuterium is a most straightforward enterprise, since this fragile isotope is 100% destroyed in stars of all masses and has no known source of substantial production other than BBN. If the boundary conditions of its evolution (namely the primordial abundance resulting from BBN and the present day one) were precisely known, the degree of astration, which depends on the adopted stellar initial mass function (IMF) and star formation rate, should be severely constrained.

The difficulty to determine the primordial D abundance in the 90ies pushed researchers to turn the problem upside down and try to determine that abundance through reasonable models of local GCE (assuming that the present day abundance is precisely known). Those efforts concluded that reasonable GCE models, reproducing the major observational constraints in the solar neighbourhood, result only in moderate D depletion, by less than a factor of two (Prantzos 1996; Dearborn et al. 1996; Prantzos and Silk 1998; Tosi et al. 1998; Chiappini et al. 2002).

The primordial abundance of D is now well determined (Table 1), since observations of D in high redshift gas clouds agree with abundances derived from observations of the Cosmic Microwave Background combined to SBBN calculations; it points to a small D depletion up to solar system formation 4.5 Gyr ago (Fig. 3a). However, the present day abundance of D in the local ISM is now under debate. Indeed, UV measurements of the FUSE satellite along various lines of sight suggest substantial differences (a factor of two to three) in D abundance between the Local Bubble and beyond it (see Table 1 and Fig. 3a). Until the origin of that discrepancy is found (see Hebrard et al. 2005 and Linsky, this volume), the local GCE of D in the past few Gyr will remain poorly understood: naively, one may expect that a high value would imply strong late infall of primordial composition, while a low value would imply strong late astration (e.g. Geiss et al. 2002; Romano et al. 2006). In any case, corresponding models should also satisfy all other local observables, like the overall metallicity evolution and the G-dwarf metallicity distribution, which is not an easy task. It should be noted that the FUSE data may also be interpreted as suggesting an inhomogeneous composition for

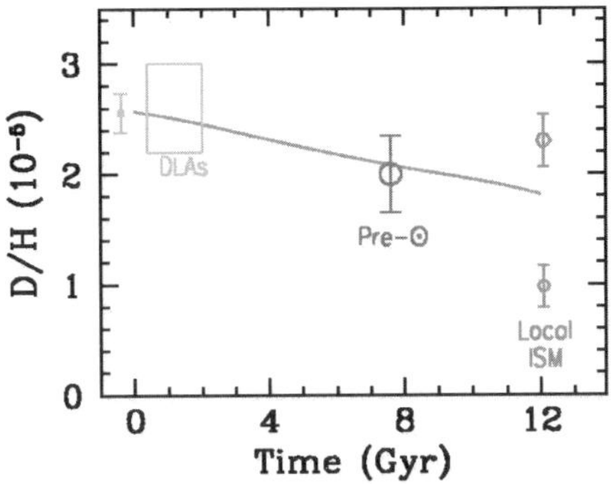


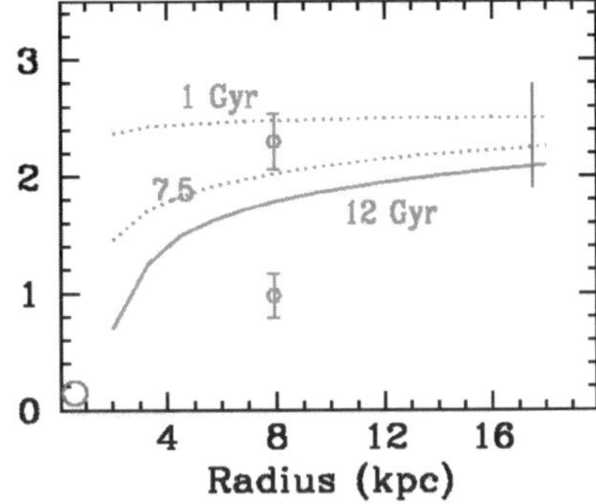


**Fig. 3** *Left:* Evolution of deuterium in the solar neighborhood, as a function of time. The adopted model satisfies all major local observational constraints, but is not unique (i.e. other satisfactory solutions may be found, where D is slightly more destroyed, e.g. with a different IMF). Data are from Table 1. *Right:* Evolution of the deuterium abundance profile in the Milky Way disk; *curves* correspond to 1 Gyr, 7.5 Gyr (Sun's birth) and 12 Gyr (today), *from top to bottom*; the latter is to be compared to data for the present-day ISM. Data are from Table 1 for local values (at a radius of 8 kpc), from Rogers et al. (2005) at 16 kpc and from Lubowich et al. (2000) in the inner Galaxy

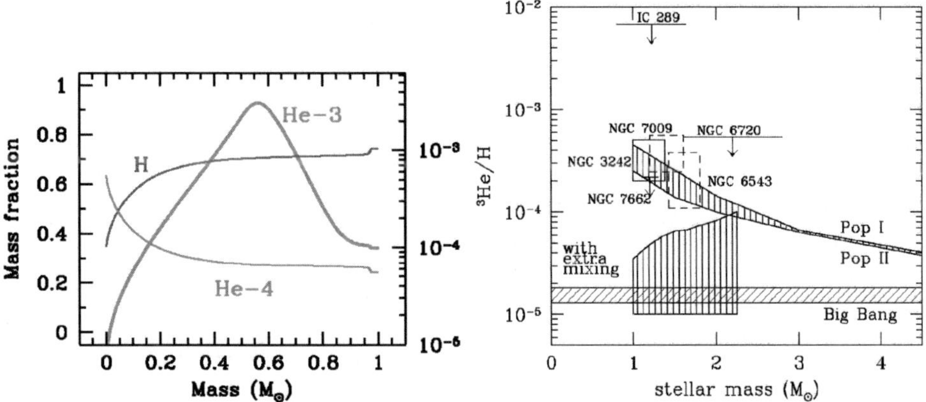


**Fig. 4** *Left:* Abundance profiles of H, $^4$He (*left axis*) and $^3$He (*right axis*) in the present-day Sun, as a function of the mass coordinate; when low mass stars become red giants, the convective envelope reaches regions of enhanced $^3$He abundance and brings it to the surface. *Right:* Abundance of $^3$He in planetary nebulae (from Galli 2005). *Upper shaded area* indicates predictions of standard models, in agreement with observations (*within rectangles*); such high abundances lead to overproduction of $^3$He during galactic evolution (*upper curves* in Fig. 5). *Lower shaded area* indicates required level of production in order to avoid overproduction of $^3$He during galactic evolution (*lower curves* in Fig. 5); such a reduced yield may result from extra-mixing in red giants, also required on other observational grounds (Charbonnel 1995). It should affect 90–95% of all stars below 2 $M_\odot$, while current observations of $^3$He in such stars would concern then the remaining 5–10%

the local ISM, on scales of ~500 pc (for D, but also O and N, see e.g. Knauth et al. 2006), which does not seem to be the case for other elements (Cartledge et al. 2006).

The evolution of D in the Galactic disk was considered originally with analytical models by Ostriker and Tinsley (1975), who found that D should be largely depleted in the inner disk. Using numerical models (satisfying all the major observational constraints for the disk) Prantzos (1996) confirmed that finding (Fig. 3b) and showed that the D/O profile of the disk offers a most sensitive test of its past history; unfortunately, such a profile has not been established in realiable way yet.

## 2.2 $^3$He

Since the pioneering work of Iben (1967) stars are known to produce substantial amounts of $^3$He, through the action of p–p chains on the main sequence (see Fig. 4a). The net $^3$He yield varies steeply with mass (roughly $\propto M^{-2}$), since the p–p chains are less effective in more massive stars. In standard stellar models, 1–2 $M_\odot$ stars are the most prolific producers.

Combining those yields with simple GCE models, Truran and Cameron (1971) and Rood et al. (1976) found that local abundances of $^3$He are largely overproduced. Indeed, the current ISM abundance of $^3$He/H ~ (1–2 $\times$ $10^{-5}$ is not very different from the pre-solar value (see Table 1 and Bania, this volume). In other terms, observations show that $^3$He abundance remained ~constant through the ages, while standard stellar models combined to GCE models (e.g. Prantzos 1996; Dearborn et al. 1996; Galli et al. 1997; Romano et al. 2003) point to a large increase (Fig. 5, upper curves).

A possible solution to the problem was suggested by Hogan (1995) and Charbonnel (1995). It postulates destruction of $^3$He in the red giant phase of Low mass stars through some "extra-mixing" mechanism, which brings $^3$He in H-burning zones. The "bonus" is a

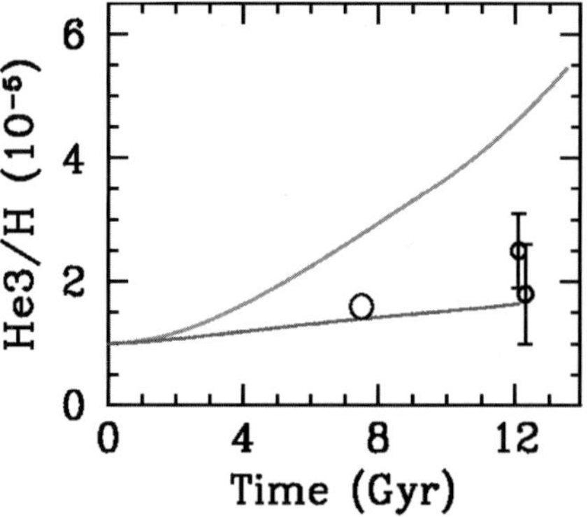


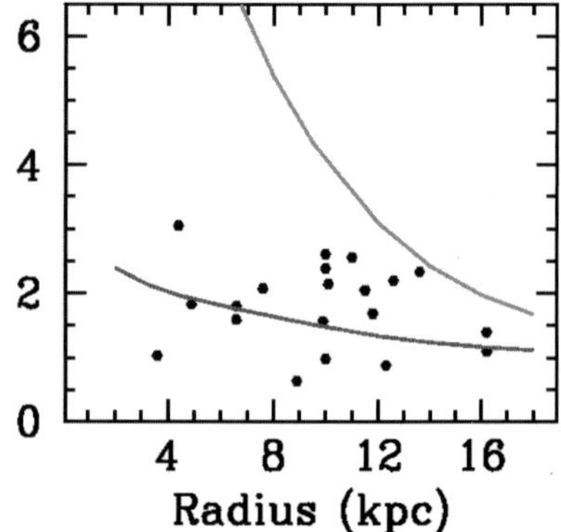


**Fig. 5** Evolution of the abundance of $^3$He in the solar neighborhood as a function of time (*left*) and present day profile of $^3$He/H in the Milky Way disk (*right*). In both cases, the *upper curves* are calculated with standard $^3$He yields from low mass stars (and clearly overproduce $^3$He) while the *lower ones* by assuming that 95% of the $^3$He of low mass stars is destroyed by some non-standard mechanism; this latter assumption allows to satisfy observational constraints, but is not supported by the rare observations of $^3$He in planetary nebulae of presumably known mass (see Fig. 4b). In the *left panel* pre-solar $^3$He (*large circle*) and present day values in the local ISM (*small circles*) are from Table 1. In the *right panel*, ISM values are from Bania et al. (2002)

concomitant modification of the $^{12}$C/$^{13}$C isotopic ratio in red giants, in excellent agreement with observations (Charbonnel and Do Nascimento 1998).

Thus, low and intermediate mass stars should destroy in the red giant phase whatever $^3$He they produce on the main sequence. A possible drawback to the idea is that observations in (at least one) planetary nebulae of known mass are in full agreement with standard model predictions, i.e. with no extra-mixing (see Fig. 4b and Galli 2005). GCE requires that in $>$90% of the stars, $^3$He produced on the main sequence must be destroyed in the red giant phase, in order to avoid overproduction (Fig. 5, lower curves). It may well be that current detections of $^3$He in planetary nebulae (see Bania, this volume) concern only the remaining $<$10% of the stars, but it is still early to draw definitive conclusions.

## 3 Beryllium and Boron

The recent evolution of Be and B in the solar neighborhood was considered already in Reeves and Meyer (1978). However, only in the 90ies spectroscopic observations of Be and B in metal poor halo stars of the MW became feasible, revealing an unexpected primary behaviour for those elements (Fig. 6a).[3]

### 3.1 Primary Be and B vs. Energetics of GCR

As pointed out in MAR, there are two components in the production of BeB from GCR (Fig. 6b): component A (fast protons and alphas impinging on ISM CNO) and component B

[3] The case of boron is more complicated. As found in MAR, standard GCR (i.e. standard equilibrium GCR spectra folded with well-known spallation cross-sections of CNO nuclei) produce a $^{11}$B/$^{10}$B ratio of 2.4, instead of the pre-solar value of 4. Thus, 40% of solar $^{11}$B should originate from another source, and this might well be $\nu$-nucleosynthesis in core collapse supernovae (Woosley et al. 1990; Olive et al. 1994); this produces primary B and could explain the observed linearity of B vs. Fe (but not the one of Be).

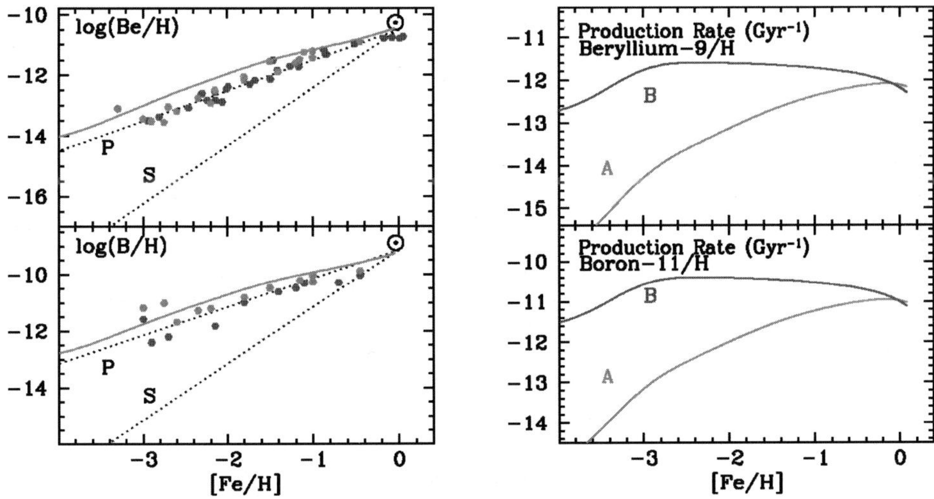


**Fig. 6** *Left:* Evolution of Be/H and B/H (*solid curves*) assuming that the GCR composition is independent of time (or ISM metallicity); Be and B are then produced as primaries, in agreement with observations. *Dotted curves* indicate primary (P) and secondary (S) behaviour with respect to Fe (while Be and B are produced from CNO, behaving not exactly as Fe). Note that ∼40% of solar $^{11}$B has to be produced by a source other than standard GCR, like e.g. $\nu$-nucleosynthesis in supernovae, which is a primary process (this is not included in the figure). *Right:* Production rates of Be and B as a function of metallicity, for GCR components A (fast protons and alphas impinging on ISM CNO) and B (fast CNOs impinging on ISM H and He). Component A (producing secondary BeB) slightly dominates today, but component B (*assumed to be metallicity independent*) largely dominated during the halo phase, i.e. at [Fe/H] $< -1$, producing primary BeB. Standard GCR injection spectra, energetics and confinement are assumed throughout galactic evolution in this figure

(fast CNOs impinging on ISM H and He). Component A has, in principle, a production rate proportional to the steadily increasing CNO abundances of the ISM and produces secondary BeB. To boost the BeB production of component A at low metallicities, one may assume either that GCR (part of which is "leaking" out of the Galactic disk today) were much better confined in the early Galaxy (Prantzos et al. 1993a) or that their total energy content was much larger than today; in both cases, the number of induced reactions with ISM CNO (per unit H atom) is increased with respect to its current value. The former option was shown to be inefficient (although improving the situation, it cannot produce a linear BeB vs. Fe relation, as found in Prantzos et al. 1993a) where the latter was shown to be unrealistic by Ramaty et al. (1997): already a large fraction (10–20%) of the kinetic energy of SN goes into acceleration of GCR, and it is simply impossible to increase that fraction by a factor of, say, 100–1000 (as to compensate for the 100–1000 times lower CNO content of the ISM in the early Galaxy). This "energetics barrier" to the early production of Be is illustrated in Fig. 7a.

This leaves component B (presently sub-dominant) as the only option, at least for Be, which has no other known source. As suggested in Duncan et al. (1992) one has to *assume that the CNO content of GCR remained* ∼*constant since the earliest days of the Galaxy.* This bold assumption is tightly related to the, yet unsettled, question of the GCR source composition (see Meyer et al. 1997 and references therein).

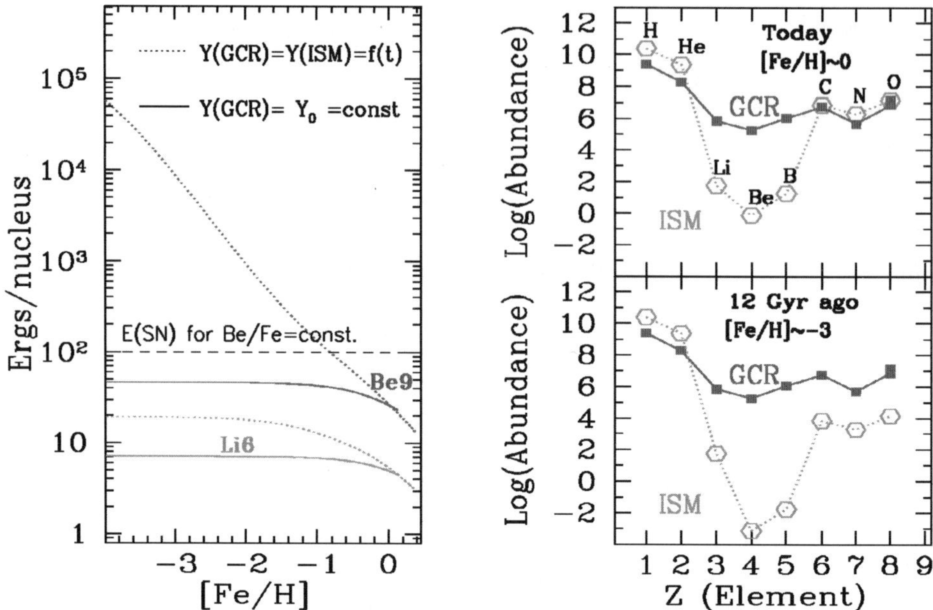


**Fig. 7** *Left:* Energy requirements (energy in accelerated particles per light nucleus produced) for the production of $^9$Be (*upper curves*) and $^6$Li (*lower curves*) through the interaction of GCR with the ISM, as a function of the ISM metallicity. $^6$Li is mostly produced by $\alpha + \alpha$ reactions and the energetics is quasi-independent of metallicity. This also happens in the case of $^9$Be, *if* it is assumed that the GCR composition is independent of time (*solid curves*); if the GCR composition scales with the one of the ISM (*dotted curves*), then extremely large energies are required for the production of $^9$Be below [Fe/H] $\sim -1$. Such energies are larger than available from a typical SN, as indicated by the *horizontal dashed line*; it is assumed here that SN turn 20% of their kinetic energy $E_{\rm KIN} = 1.5 \times 10^{51}$ erg into GCR, produce 0.1 $M_\odot$ of Fe and induce the production of $10^{-7}$ $M_\odot$ of Be, as to keep the Be/Fe ratio always constant (in agreement with observations). A time (or metallicity) dependent GCR composition should then be excluded. *Right:* Composition of GCR (*solid curves, filled symbols*) and ISM (*dotted curves, open symbols*), today (*upper panel*, observed) and in the early Galaxy (*lower panel, inferred*); GCR *had to be much more metallic than the ambient ISM* in those early days, to account for the observed primary behaviour of Be

## 3.2 On the GCR Source Composition

A ~constant abundance of C and O in GCR can naturally be understood if SN accelerate their own ejecta. However, the absence of unstable $^{59}$Ni (decaying through e$^-$-capture within $10^5$ yr) from observed GCR suggests that acceleration occurs $>10^5$ yr after the explosion (Wiedenbeck et al. 1999), when SN ejecta are presumably diluted in the ISM. Obviously then, SN do not accelerate their own ejecta. However, they can certainly accelerate the ejecta of their neighbours. Higdon et al. (1998) suggested that this happens in *superbubbles* (SB), enriched by the ejecta of many SN (Binns et al. 2005) as to have a large and ~constant metallicity. Since then, this became *by default*, the "standard" scenario for the production of primary Be and B by GCR, invoked in almost every work on that topic.

However, the SB scenario suffers from several problems. First, core collapse SN are observationally associated to HII regions (van Dyk et al. 1996) and it is well known that the metallicity of HII regions reflects the one of the *ambient ISM* (i.e. it can be very low, as in IZw18) rather than the one of SN. Secondly, the scenario requires that SB in the early Galaxy retain most of their metals, to the point of being much more metallic than the ambient

ISM (say, by factor of $\sim$1000, see Fig. 7b). But this goes against current wisdom: indeed, observations and hydrodynamical simulations suggest that small galactic units (such as those that merged to form the Galactic halo in the framework of the hierarchical merging scenario) are metal poor, possibly because their weak gravity cannot retain the hot, metal-rich ejecta of supernovae. It is hard to understand then why SB in the galactic sub-units forming the early Galaxy would be so much more metal-rich than their environment (and, on top of that, with *always the same* quasi-solar metallicity).

Finally, Higdon et al. (1998) evaluated the time interval $\Delta t$ between SN explosions in a SB to a comfortable $\Delta t \sim 3 \times 10^5$ yr, leaving enough time to $^{59}$Ni to decay before the next SN explosion and subsequent acceleration. However, SB are constantly powered not only by SN but also by the strong *winds of massive stars* (with integrated energy and acceleration efficiency similar to the SN one, e.g. Parizot et al. 2004), which should continuously accelerate $^{59}$Ni, as soon as it is ejected from SN explosions. $^{59}$Ni should then be observed in GCR, which is not the case (Prantzos 2005). Thus, SB suffer exactly from the same problem that plagued SN as accelerators of their own, metal rich, ejecta. Note that a loophole in the latter argument is suggested by Binns (this volume): only the most massive stars ($>$30 $M_\odot$) display strong winds; such stars live for less than 6 Myr, i.e. less than the first 1/4 of the SB lifetime, implying that the largest fraction of GCR (*if* accelerated in SB) should be free of $^{59}$Ni, in agreement with observations. It remains to be seen whether the argument holds quantitatively, but even in that case the first two objections against the SB idea still hold.

The problem of the source and acceleration site of GCR, so crucial for the observed linearity of Be and B vs. Fe (but also for our understanding of GCR in general) has not found a satisfactory explanation yet (at least to the opinion of the author of this paper).

## 4 The Li Isotopes

The isotope $^7$Li holds a unique position among the $\sim$315 naturally occurring nuclides, since it is produced by more than two nucleosynthesis sites: the hot early Universe, galactic cosmic rays (not only by p + CNO but also by $\alpha + \alpha$ reactions), AGB stars, novae, and $\nu$-nucleosynthesis in core collapse SN. The contribution of the first two processes is relatively well known, while the remaining ones are hard to quantify at present (see e.g. Romano et al. 2003; Travaglio et al. 2001, for such attempts).

### 4.1 From Primordial $^7$Li to the "Spite Plateau"

The Li abundance of the "Spite plateau" (Li/H $\sim$ (1–2) $\times 10^{-10}$ $\sim$const. for halo stars, down to the lowest metallicities) is a factor of 2–3 lower than the WMAP + SBBN value (Table 1 and Fig. 8a). Barring systematic errors (see Gratton, this volume), the conclusion is that primordial Li has been depleted, either (a) *before* getting into the stars it is observed today, or (b) *during* the lifetime of those stars. Two "depletion agents" have been proposed in the former case: decaying supersymmetric particles (Jedamzik 2004) and astration in a first generation of exclusively massive (mass range $m_* = 10$–$40\ M_\odot$) Pop. III stars (Piau et al. 2006). The latter idea, however, suffers from a serious flaw, since in that case the metallicity of the ISM (out of which the next stellar generation would form with depleted Li) would rise to levels much higher than those observed in EMP stars. This can be seen as follows (Prantzos 2006b):

Assuming that the current halo stellar mass ($M_{\rm H} = 2 \times 10^9\ M_\odot$) was initially in the form of gas, a fraction $f$ of which was astrated through massive stars, the resulting Li mass

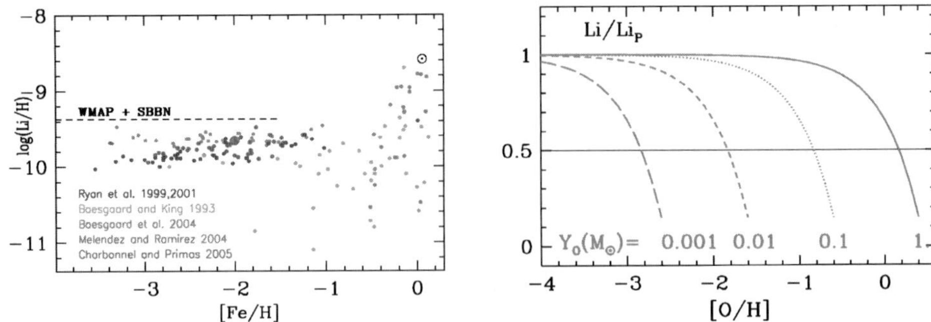


**Fig. 8** *Left:* Observations of Li in halo and disk stars of the Milky Way. The primordial Li value, obtained from the baryonic density of WMAP and calculations of SBBN, is indicated by a *horizontal dashed line*. The observed Li plateau at low metallicities depends sensitively on assumed stellar temperature, and differs from the WMAP value by factors 2–3. *Right:* Illustration of the problem encountered by the idea that the discrepancy is due to astration of Li by a Pop. III composed of exclusively massive stars, in the 10–40 $M_\odot$ range (Piau et al. 2006): such stars necessarily eject metals, either through their winds (e.g. nitrogen, in case of rotating stars) or through the final supernova explosion (e.g. oxygen). Mixing the ejecta with various proportions of primordial material would result in Li depletion (by a factor of ∼2 in case of a 50–50 mixture), but the resulting metallicity would be much higher than the one of halo stars, unless abnormally low oxygen yields were assumed (i.e. the *curve* in Fig. 8b parametrized with a yield $Y_O = 0.001\ M_\odot$, see text)

fraction is $X_{Li} = (1 - f)X_{Li,P}$ where $X_{Li,P}$ is the primordial Li abundance (assuming a return fraction $R \sim 1$ for the astrating and metal producing stars). Similarly, the resulting oxygen abundance would be $X_O = m_O/M_H$, where the mass of oxygen $m_O = N_{SN}\ Y_O$ is produced by a number of supernovae $N_{SN} = f M_H/m_*$, each one with a typical oxygen yield of $Y_O$ (in $M_\odot$, to be discussed below). Then:

$$\frac{X_{Li}}{X_{Li,P}} = 1 - 0.28\frac{(X_O/0.007)(m_*/40\ M_\odot)}{Y_O}.$$

That relation appears in Fig. 8 (right panel) as a function of $\log(X_O/X_{O,\odot})$, with adopted solar abundance $X_{O,\odot} = 0.007$. The four curves correspond to different assumptions about the typical oxygen yield of a massive star of $Z = 0$, ranging from 0.001 to 1 $M_\odot$; only the first of those yields leads to large Li astration at low metallicities, but (as discussed in Prantzos 2006b), stellar models produce generically more than 1 $M_\odot$ of oxygen per massive star. Another way to eject astrated material by $Z \sim 0$ massive stars is through stellar winds, which require rapidly rotating stars (radiative pressure being inefficient at low metallicities); but rotating massive stars produce large amounts of nitrogen (which may in fact help explaining the observed primary-like N in EMP stars, e.g. Meynet et al. 2006), thus the problem of metal overproduction is not avoided in that case either. Astration in massive Pop. III stars cannot solve the Li discrepancy between the Spite plateau and WMAP + SBBN:[4] even a small Li depletion should be accompanied by excessive metal enhancement.

Several mechanisms were proposed over the years to account for case (b) above, i.e. depletion *during* the stellar evolution within the stellar envelope: rotational mixing, gravity waves, microscopic diffusion etc. (e.g. Charbonnel and Primas 2005 and references therein).

[4] At least, not the 10–40 $M_\odot$ stars suggested in Piau et al. 2006 (provided that current nucleosynthesis models for such stars are correct); 100 $M_\odot$ stars collapsing to black holes would be better candidates (provided they eject a substantial fraction of their astrated envelope only, but not of their metal-rich core).

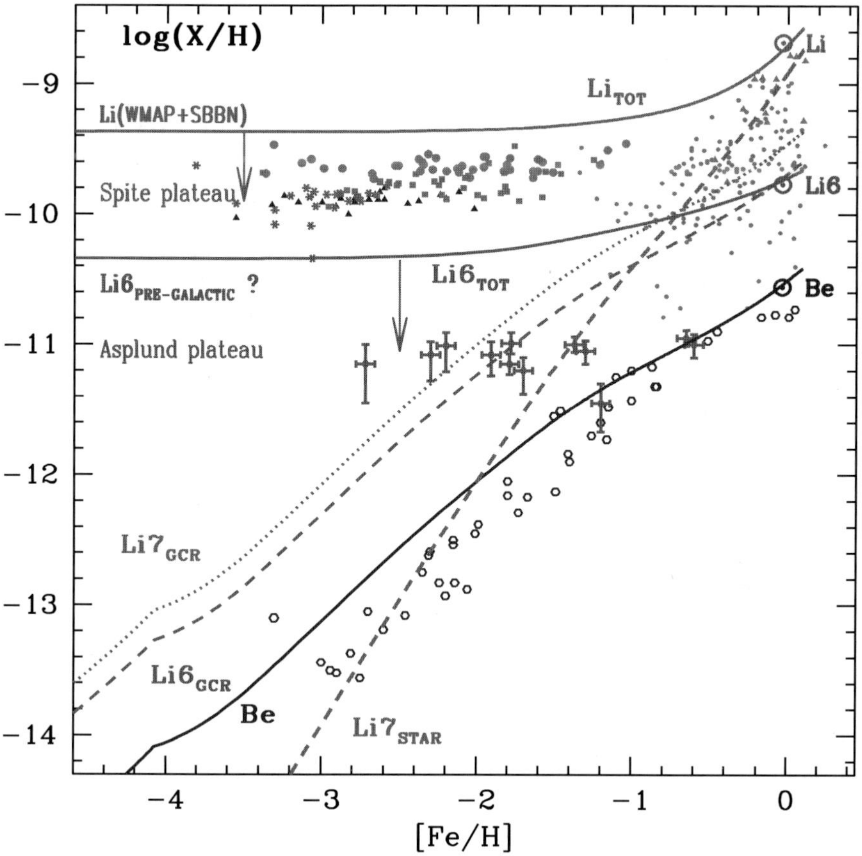

**Fig. 9** Evolution of $^7$Li, $^6$Li and Be: observations vs. (a possible) theoretical picture. Observations are for total Li (*upper set of filled symbols*, the "Spite plateau"), $^6$Li (*points with error bars in the middle of the diagram*, the "Asplund plateau"—to be confirmed) and Be (*lower set of open symbols*). The Spite plateau lies below the primordial value inferred from WMAP (*early part of upper solid curve*), most probably due to depletion of Li inside the stellar envelopes by ∼0.3 dex (*vertical arrow*). The more fragile $^6$Li should then suffer from even larger depletion (*vertical arrow*), its true value lying perhaps ∼0.7 dex above the Asplund plateau (*solid curve labeled* $\mathrm{Li6_{TOT}}$). The contribution of GCR to $^6$Li (*dashed curve labeled* $\mathrm{Li6_{GCR}}$) is well constrained once the evolution of Be (*solid curve labeled* Be) is reproduced; it can explain pre-solar $^6$Li (as already found in MAR), but it is clearly insufficient to explain the "Asplund plateau" and even less so the "undepleted" $^6$Li plateau. The GCR component of $^7$Li (*dotted curve labeled* $\mathrm{Li7_{GCR}}$) is also well defined and contributes marginally to total Li. Finally, the stellar (AGB or nova) $^7$Li component (*dotted curve labeled* $\mathrm{Li7_{STAR}}$) is required to explain ∼60% of pre-solar Li; note that only the late part of that component (at $[\mathrm{Fe/H}] > -0.7$) is constrained, by the upper envelope of the Li data (Note: $\mathrm{Li_{TOT}} = \mathrm{Li7_{TOT}} + \mathrm{Li6_{TOT}}$ and $\mathrm{Li7_{TOT}} = \mathrm{Li7_{BBN}} + \mathrm{Li7_{GCR}} + \mathrm{Li7_{STAR}}$)

The main difficulty is to obtain a uniform Li depletion of ∼0.3 dex over the whole metallicity range of the plateau, with negligible dispersion. Richard et al. (2005) proposed a model with a few ingredients (microscopic diffusion coupled to levitation due to radiation pressure, and moderated by turbulent diffusion at the base of the convective envelope) which reproduces satisfactorily that feature. Such models are supported by recent spectroscopic observations of stars in the metal-poor globular cluster NGC6397, revealing trends of atmospheric abundance with evolutionary stage for various elements (Korn et al. 2006).

### 4.2 Early $^6$Li: Primordial, Pre-galactic or Simply Stellar?

The surprising detection of a $^6$Li "plateau" in metal-poor stars of the Galaxy (Asplund et al. 2006) challenges our understanding of the origin of light nuclides. The value of the "Asplund plateau", $^6\mathrm{Li/H} \sim 10^{-11}$, is $\sim$10 times higher than the one contributed from standard GCR (accounting for the observed evolution of Be) at $[\mathrm{Fe/H}] = -3$ (see Fig. 9), and $\sim$1000 times larger than resulting from SBBN in the early Universe. The discrepancy is even more serious if Li isotopes are depleted in stellar atmospheres (see Sect. 4.1), since $^6$Li is more fragile than $^7$Li; its true abundance could be then (considerably) higher than $10^{-11}$.

Several ideas have been put forward to explain the production of a high early abundance of $^6$Li:

(a) primordial, in non-standard Big Bang nucleosynthesis involving the decay/annihilation of massive particle (Jedamzik 2004). The bonus of that idea is that such particles could also destroy part of the primordial $^7$Li (releasing the tension between WMAP + SBBN and observations) but this is less appealing now, since observations favour stellar depletion (Sect. 4.1).
(b) pre-galactic, by fusion of $\alpha$-particles; these could be accelerated by the energy released (i) during structure formation (Suzuki and Inoue 2002) or (ii) from accretion onto supermassive black holes or (iii) from an early generation of Pop. III massive stars (Reeves 2005 and this volume). A critique of those ideas, based on a careful evaluation of the energetics of $^6$Li production from energetic particles (see Fig. 7a) is made in Prantzos (2006a); the first two appear much less promising than the last one.
(c) stellar, by in situ reactions of energetic particles (mostly $^3\mathrm{He} + {}^4\mathrm{He}$ and assuming an enhanced $^3$He abundance) in the atmosphere of the stars during the 10 Gyr of their evolution (Tatischeff and Thibaud 2006). The stars are required to be very active in accelerating particles ($\sim 3 \times 10^4$ times the activity of the present-day Sun in their early main sequence), that activity being attributed to their rapid rotation. The values of the "Asplund plateau" can then be reproduced in some extreme cases.

The early $^6$Li plateau is the latest (but probably not the last) twist in the saga of the light nuclides. More data and a thorough understanding of the stellar properties are required before concluding whether the answer to the puzzle lies among (a), (b) or (c) above, or it is something completely different.

## 5 Summary

The *x-process* turned out to be the most complex of all the nucleosynthetic processes envisioned in B$^2$FH. Despite 50 years of progress in theory and observation, it is still unknown where most of $^3$He and $^7$Li and a large fraction of $^{11}$B come from. The origin of early $^6$Li remains equally mysterious, while the degree of astration of D in the solar neighborhood is poorly known. This is certainly good news: we shall have exciting things to discuss for Johannes' 90ieth anniversary!

**Acknowledgements** I am grateful to the organizers for their kind invitation and for giving me the opportunity to participate in such an interesting meeting, celebrating the contributions of Johannes Geiss to our understanding of the origin of the light nuclides. *Bon anniversaire* Johannes!

## References

M. Asplund, D. Lambert, P. Nissen, F. Primas, V. Smith, Astrophys. J. **644**, 229–259 (2006)
T. Bania, R. Rood, D. Balser, Nature **415**, 54–57 (2002)
W.R. Binns, M.E. Wiedenbeck, M. Arnould et al., Astrophys. J. **634**, 351–364 (2005)
D.C. Black, Nature **234**, 148 (1971)
M. Burbidge, G. Burbidge, W. Fowler, F. Hoyle, Rev. Mod. Phys. **29**, 547–650 (1957)
A.W.G. Cameron, W. Fowler, Astrophys. J. **164**, 111–114 (1971)
S. Cartledge, J. Lauroesch, D. Meyer, U. Sofia, Astrophys. J. **641**, 327–346 (2006)
C. Charbonnel, Astrophys. J. **453**, L41–L44 (1995)
C. Charbonnel, J. Do Nascimento, Astron. Astrophys. **336**, 915–919 (1998)
C. Charbonnel, F. Primas, Astron. Astrophys. **442**, 961–992 (2005)
C. Chiappini, A. Renda, F. Matteucci, Astron. Astrophys. **395**, 789–801 (2002)
D. Dearborn, D. Schramm, G. Steigman, Astrophys. J. **302**, 35–38 (1986)
D. Dearborn, G. Steigman, M. Tosi, Astrophys. J. **465**, 887–898 (1996)
D. Duncan, D. Lambert, M. Lemke, Astrophys. J. **401**, 584–595 (1992)
R. Epstein, Nature **263**, 198–202 (1976)
W. Fowler, J. Greenstein, F. Hoyle, PASP **73**, 326–326 (1962)
D. Galli, in *Chemical Abundances and Mixing*, ed. by S. Randich, L. Pasquini (Springer, 2005), pp. 343–348
D. Galli, L. Stanghellini, M. Tosi, F. Palla, Astrophys. J. **477**, 218–218 (1997)
J. Geiss, H. Reeves, Astron. Astrophys. **18**, 126 (1972)
J. Geiss, G. Gloeckler, Space Sci. Rev. **106**, 3–18 (2002)
J. Geiss, G. Gloeckler, C. Charbonnel, Astrophys. J. **578**, 862–867 (2002)
G. Gloeckler, J. Geiss, Nature **381**, 210–212 (1996)
J.R. Gott, III, D., Schramm, B. Tinsley, J. Gunn, Astrophys. J. **194**, 543–553 (1974)
G. Hebrard, T. Tripp, P. Chayer et al., Astrophys. J. **635**, 1136–1150 (2005)
J. Higdon, R. Lingenfelter, R. Ramaty, Astrophys. J. **509**, L33–L36 (1998)
G. Hogan, Astrophys. J. **441**, L17–L20 (1995)
F. Hoyle, R. Tayler, Nature **203**, 1108–1110 (1964)
I. Iben Jr., Astrophys. J. **147**, 624–649 (1967)
K. Jedamzik, Phys. Rev. D **70**, 063524 (2004)
D. Knauth, D. Meyer, J. Lauroesch, Astrophys. J. **647**, L115–L118 (2006)
A. Korn, F. Grundahl, O. Richard et al., Nature **442**, 657–659 (2006)
J. Linsky, B. Draine, H. Moos et al., Astrophys. J. **647**, 1106–1124 (2006)
K. Lodders, Astrophys. J. **591**, 1220–1247 (2003)
D. Lubowich, J. Pasachoff, T. Balonek et al., Nature **405**, 1025–1027 (2000)
M. Meneguzzi, J. Audouze, H. Reeves, Astron. Astrophys. **15**, 337–356 (1971)
J.P. Meyer, D. Elisson, L. Drury, Astrophys. J. **487**, 182–196 (1997)
G. Meynet, S. Ekström, A. Maeder, Astron. Astrophys. **447**, 623–639 (2006)
K. Olive, N. Prantzos, E. Vangioni-Flam, Astrophys. J. **424**, 666–670 (1994)
J. Ostriker, B. Tinsley, Astrophys. J. **201**, L51–54 (1975)
E. Parizot, A. Marcowith, E. van der Swaluw et al., Astron. Astrophys. **424**, 747–760 (2004)
P.J.E. Peebles, Astrophys. J. **146**, 542–552 (1966)
A. Penzias, R. Wilson, Astrophys. J. **142**, 419–421 (1965)
L. Piau, T. Beers, D. Balsara et al., Astrophys. J. **653**, 300–315 (2006)
N. Prantzos, Astron. Astrophys. **310**, 106–114 (1996)
N. Prantzos, in *Chemical Abundances and Mixing*, ed. by S. Randich, L. Pasquini (Springer, 2005), pp. 351–357
N. Prantzos, Astron. Astrophys. **448**, 665–675 (2006a)
N. Prantzos, (2006b). astro-ph//0612633
N. Prantzos, J. Silk, Astrophys. J. **507**, 229–240 (1998)
N. Prantzos, M. Casse, E. Vangioni-Flam, Astrophys. J. **403**, 630–643 (1993a)
N. Prantzos, M. Casse, E. Vangioni-Flam, in *Origin and Evolution of the Elements*, ed. by N. Prantzos et al. (Cambridge, 1993b), pp. 156–167
N. Prantzos, M. Tosi, R. von Steiger, in *Primordial Nuclides and Their Galactic Evolution*. ISSI Sp. Sc. Series (Kluwer, 1998)
R. Ramaty, B. Kozlovsky, R. Lingenfelter, H. Reeves, Astrophys. J. **488**, 730–748 (1997)
H. Reeves, Rev. Mod. Phys. **66** , 193–216 (1994)
H. Reeves, EAS Publ. Ser. **17**, 15–19 (2005)
H. Reeves, J.P. Meyer, Astrophys. J. **226**, 613–631 (1978)
H. Reeves, W. Fowler, F. Hoyle, Nature **226**, 727–729 (1970)

H. Reeves, J. Audouze, W. Foowler, D. Schramm, Astrophys. J. **179**, 909–930 (1973)
O. Richard, G. Michaud, J. Richer, Astrophys. J. **619**, 538–548 (2005)
A. Rogers, K. Duvedoir, J. Carter et al., Astrophys. J. **630**, L41–L44 (2005)
D. Romano, M. Tosi, F. Matteucci, C. Chiappini, MNRAS **346**, 295–303 (2003)
D. Romano, M. Tosi, C. Chiappini, F. Matteucci, MNRAS **369**, 295–304 (2006)
R. Rood, G. Steigman, B. Tinsley, Astrophys. J. **207**, L57–L60 (1976)
S. Ryan, J. Norris, M. Bessell, C. Deliyannis, Astrophys. J. **388**, 184–189 (1992)
C. Ryter, H. Reeves, E. Gradsztajn, J. Audouze, Astron. Astrophys. **8**, 389–397 (1970)
E. Salerno, F. Bhler, P. Bochsler et al., Astrophys. J. **585**, 840–849 (2003)
M. Spite, F. Spite, Astron. Astrophys. **115**, 357–366 (1982)
P.D. Serpico, S. Esposito, F. Iocco et al., J. Cosmol. Astropart. Phys. **12**, 010 (2004)
G. Steigman, Int. J. Mod. Phys. E **15**, 1–35 (2006)
G. Steigman, T.P. Walker, Astrophys. J. **385**, L13–L16 (1992)
T.K. Suzuki, S. Inoue, Astrophys. J. **573**, 168–173 (2002)
V. Tatischeff, J.-P. Thibaud (2006). astro-ph/0610756
M. Tosi, G. Steigman, F. Matteucci, C. Chiappini, Astrophys. J. **498**, 226–235 (1998)
C. Travaglio, S. Randich, D. Galli et al., Astrophys. J. **559**, 909–924 (2001)
J. Truran, A.G.W. Cameron, Astrophys. Space Sci. **14**, 179–222 (1971)
S. van Dyk, M. Hamuy, A. Fillipenko, Astrophys. J. **111**, 2017–2027 (1996)
R. Wagoner, W. Fowler, F. Hoyle, Astrophys. J. **148**, 3–49 (1967)
T.P. Walker, G. Steigman, H.-S. Kang et al., Astrophys. J. **376**, 51–69 (1991)
M. Wiedenbeck, W.R. Binns, E.R. Christian et al., Astrophys. J. **523**, L61–L64 (1999)
S. Woosley, D. Hartmann, R. Hoffman, W. Haxton, Astrophys. J. **356**, 272–301 (1990)
J. Yang, M. Turner, D. Schramm et al., Astrophys. J. **281**, 493–511 (1984)

Space Sci Rev (2007) 130: 43–52
DOI 10.1007/s11214-007-9243-x

# Abundances of Light Elements

**R.G. Gratton**

Received: 3 January 2007 / Accepted: 5 June 2007 / Published online: 2 September 2007


**Abstract** This paper briefly reviews a few relevant features about the abundances of light elements (D, $^4$He, $^6$Li, $^7$Li, $^9$Be) in the Milky Way. It places special emphasis on metal-poor stars. Observational concerns are discussed. The use of $^7$Li and $^6$Li as cosmological probes and of $^9$Be as a chronometer for the early evolution of our Galaxy are discussed.

**Keywords** The Galaxy: evolution of · Abundances: light elements

## 1 Introduction

The light elements $^6$Li, $^7$Li, $^9$Be play a very special and important role in a large variety of astrophysical issues. This is due to a number of important properties: their production site is atypical with respect to most of the other heavier elements and they can be easily destroyed in stars, so that their surface abundances provide interesting insights into details of stellar evolution. In this paper, I will briefly review a few issues concerning their role in Big Bang Nucleosynthesis, in galactic evolution, and in the understanding of the brief but significant chemical evolution of globular clusters.

## 2 Standard Big Bang Nucleosynthesis

An excellent introductory review on standard Big Bang Nucleosynthesis (BBN) can be found in Steigman (2002); updates with WMAP results can be found in Spergel et al. (2003) and Barger et al. (2003). Here we recall only a few basic notions. BBN started when the Universe was a few minutes old and the temperature dropped below 80 keV. It stopped when the Universe was 20 minutes old and the temperature dropped below 30 keV. The main results of BBN are that neutrons and protons combine to form D, $^3$H, $^3$He and $^4$He. Since there is a gap at mass = 5 (no stable nucleus), only charged particles could be used to form more massive

R.G. Gratton (✉)
INAF—Osservatorio Astronomico di Padova, Vicolo dell'Osservatorio 5, 35122 Padova, Italy
e-mail: raffaele.gratton@oapd.inaf.it

nuclei: hence these are rare. Li is produced by two mechanisms: by $^3H(\alpha, \gamma)^7Li$ reactions at low baryon densities, and by $^3He(\alpha, \gamma)^7Be$ then decaying to $^7Li$ at high baryon densities. In standard BBN, $^6Li$ is expected to be about two orders of magnitudes less abundant than $^7Li$. Since there is a new gap at mass = 8, virtually no heavier nuclei are produced.

Standard BBN is an elegant theory, that has simple and robust features. Virtually all neutrons are incorporated into $^4He$ (the most stable nucleus). The $^4He$ production depends only on the neutron production, due to the competition between the weak interaction (depending on effective number of neutrinos $g_{\rm eff}$) and the early Universe expansion rate (at about 1 s). Production of D, $^3He$, $^7Li$ are rate limited, depending on the competition between nuclear reaction and Universe expansion (at about 1,000 s). Hence they are potential baryometers because their abundances are sensitive to the density of nucleons, which is usually expressed as

$$\eta_{10} = 10^{10}(n_N/n_\gamma) = 274\Omega_B h^2. \tag{1}$$

Note that $\eta$ is constant after $e^\pm$ annihilation.

BBN has several observable features. These include the abundances of D, $^3He$, $^4He$, $^7Li$: as a rule we should expect plateaus for the abundances of these elements due to BBN.

Also results from Cosmic Microwave Background (CMB) (at recombination, that is after few hundred thousand years) can be used to (strongly) constrain $\eta$ and (weakly) constrain $g_{\rm eff}$. To understand this point, we should consider that the early Universe was dominated by relativistic particles (radiation). When the Universe expands and cools (after $10^5$ yr), non-relativistic particles (matter) dominate. Pre-existing perturbations will grow due to gravity. Oscillations (sound waves) develop. At recombination, CMB photons are free to travel, and preserve the record of oscillations as temperature fluctuations. The CMB power spectrum is due to competition between gravitational potential and pressure gradients. The redshift of matter-radiation equality affects the time duration over which this competition occurs. An increase in the relativistic content (decrease of $\eta$, increase of $g_{\rm eff}$) causes the Universe to be younger at recombination, with a corresponding smaller sound horizon $s*$. The location of the $n$th peak of the CMB spectrum scales as $n\pi D*/s*$ ($D*$ is the co-moving angular diameter). Hence the peaks shift to smaller angular separation (larger $l$). Note that there is a degeneracy with $H_0$. The location and height of the peaks also depend on the history after recombination ($H_0$ and $\Omega_\Lambda$). The amplification of the power in the lowest $l$ is a probe of the dark energy. One possible overall solution can be obtained using maximum likelihood methods. Assuming a flat Universe, $H_0 = 72 \pm 8$ km/s/Mpc (from the HST Key Program, Freedman et al. 2001) and an age of the Universe >11 Gyr (from observation of globular clusters), Barger et al. (2003) found $\eta_{10} = 6.30^{+0.96}_{-0.72}$ (2-$\sigma$ range), $\Omega_B = 0.0230^{+0.035}_{-0.026}$ (2-$\sigma$ range), and $g_{\rm eff} = 2.8^{+5.5}_{-1.9}$ (2-$\sigma$ range). Similar results, with smaller error bars, have been very recently obtained by Spergel et al. (2006) from the three-year WMAP results: $\Omega_B h^2 = 0.0223^{+0.0007}_{-0.0009}$, and $\eta_{10} = 6.0965 \pm 0.2055$ (1-$\sigma$ range).

Although the predictions by the CMB are very precise, it is still very important to verify how elemental abundances compare with these standard BBN predictions. The most successful results have been obtained for Deuterium. Deuterium can only be destroyed in stars. To find primordial D, it is best to look where most matter was still not elaborated in stars, hence in low metallicity environments. The Deuterium spectrum is very similar to that of H (shifted by only 81 km/s). Since H lines are very broad in the stellar spectra, the best observations are in QSO Absorbing Line Systems. In that case, the main problem is related to the forest of H lines, which may make detection of the much weaker D lines uncertain. Table 1 contains a summary of determinations of D abundances from absorption lines in QSO spectra. While there is a reasonable agreement between the various determinations over a

**Table 1** Determination of D abundances from absorption lines in QSO spectra

| Object | $z_{abs}$ | D/H ($\times 10^5$) | Reference |
|---|---|---|---|
| PKS 1937-1009 | 3.572 | 3.3±0.3 | Burles and Tytler (1998) |
| 1009+256 | 2.504 | 4.0±0.7 | Burles and Tytler (1998) |
| HS0105+1619 | 2.536 | 2.54±0.23 | O'Meara et al. (2001) |
| Q2206-199 | 2.076 | 1.65±0.35 | Pettini and Bowen (2001) |
| Q0347-3819 | 3.025 | 2.24±0.67 | D'Odorico et al. (2001) |
| | | 3.75±0.25 | Levshakov et al. (2002) |
| Q1243+3047 | 2.526 | 2.4±0.3 | Kirkman et al. (2003) |
| PKS 1937-1009 | 3.256 | 1.6±0.3 | Crighton et al. (2004) |

value of a few times $10^{-5}$, it should be noticed that the scatter of individual determinations is much larger than the internal errors. It is then possible that these errors are underestimated.

From a subset of these data (six absorption lines in QSO spectra), Kirkman et al. (2003) obtained an average abundance of $n(\mathrm{D}) = 2.78^{+0.44}_{-0.38} \times 10^{-5}$. However, this is based on a logarithmic average; furthermore, dispersion is much larger than internal errors ($\chi^2 = 15.3$ with four degrees of freedom). From a reevaluation of the data, Barger et al. concluded $n(\mathrm{D}) = 2.6 \pm 0.4 \times 10^{-5}$. For standard BBN (with $g_{\mathrm{eff}} = 3$) this corresponds to $\eta_{10} = 6.1^{+0.7}_{-0.5}$, that is $\Omega_B = 0.022 \pm 0.002$, which is in excellent agreement with the CMB results.

Determinations of the abundance of $^3$He were reviewed by Tosi (2000). $^3$He is observable only in the local ISM; it may be both produced and destroyed in stars and its evolution is difficult to predict at present. Hence it cannot be used to constrain BBN.

$^4$He is also produced by stars. In order to derive constraints on BBN, it has then to be observed in metal-poor environments. Practically, a correlation with metal abundance is observed. Since abundance of He is best determined from HII regions, the most suitable objects are extragalactic (metal-poor) HII regions. $^4$He abundance $Y$ is only weakly sensitive on the baryon density. Very high precision abundance determinations are required to obtain significant limits. In order to get these very high-precision values, a number of subtle effects should be considered, including corrections for underlying absorption lines; ionization, temperature and density structure of the HII regions (that are not spatially resolved at the distance of the nearby metal-poor extragalactic regions); collisional excitation; optical thickness of the He I triplet lines; atomic parameters; and uncertainties in the He–metal abundance ratios. The differences between authors does not then come as a surprise; a list of recent values include: $Y = 0.2452 \pm 0.0015 \pm 0.0070$ (Izotov and Thuan 2004); $Y = 0.234 \pm 0.003$ (Olive and Skillman 2001); $Y = 0.2374 \pm 0.0035 \pm 0.0010$ (Peimbert et al. 2003). A completely independent determination of $Y$ comes from counts of the number of stars in different branches of the colour-magnitude diagrams of globular clusters, the so-called R-method. From a recent application of this method, Cassisi et al. (2003) obtained $Y = 0.244 \pm 0.006$. While this value agrees well with the expectations from BBN, it should be noticed that this method depends on the accuracy of stellar models, which may cause a not-well-determined error.

When comparing determinations of the baryon density $\eta$ from different abundance indicators, it should be recalled that the higher $\eta$, the faster primordial D is destroyed; hence, D is anticorrelated with $\eta$. On the other hand, the faster the Universe expands (larger $g_{\mathrm{eff}}$), the less time there is available for D destruction; hence, D is (weakly) positively correlated with $g_{\mathrm{eff}}$. For what concerns $^4$He, we recall that it incorporates all available neutrons; hence, $^4$He is quite insensitive to $\eta$. The neutron-to-proton ratio is sensitive to the Universe expansion rate near $e^{\pm}$ annihilation, so the faster (the largest $g_{\mathrm{eff}}$) the Universe expands, the more neutrons are available. For this reason the abundance of $^4$He is strongly correlated with $g_{\rm eff}$. Strict constraints on the two main parameters of BBN ($\eta$ and $g_{\rm eff}$) can then be obtained by combining determinations of D and $^4$He abundances. If we consider current determinations of the abundances of these elements with the predictions from the CMB, the He abundance results are lower than expected. In principle, this discrepancy could be solved by non-standard BBN (i.e. obtained by relaxing some of the approximations used in standard models). However, the discrepancy between He and D abundances is only at about $2\sigma$ level, and it cannot be considered a very robust result in view of the large uncertainties still present in He abundance determinations. Taken at their face value, current results favour a value of $g_{\rm eff} < 3$, and even one extra fully thermalized neutrino is strongly disfavored. Other possibilities are still open, e.g. non-minimally coupled fields or higher dimensional phenomena (see references in Barger et al. 2003). However, these modifications of the standard scenario are not well justified.

## 3 Cosmological $^7$Li

$^7$Li is the only nuclide heavier than $^4$He expected to be produced in significant (i.e. detectable) amounts by Big Bang Nucleosynthesis (BBN; see e.g. Steigman 2002; Spergel et al. 2003; Barger et al. 2003). Li is both produced and destroyed in stars, therefore primordial Li must be observed in metal-poor stars. Since the pioneering observations by Spite and Spite (1982), who found a roughly constant abundance of $^7$Li in metal-poor stars near the turn-off (the so-called Spite's plateau), a large number of authors tried to relate these abundances with BBN (for a few quite recent such attempts see e.g. Ryan et al. 1999; Asplund et al. 2003; Bonifacio et al. 2002; Bonifacio 2002; Charbonnel and Primas 2004). The average Li abundance obtained by these studies is in the range $\log n(\mathrm{Li}) = 2.0$–$2.3$, well below the value predicted by standard BBN using the value of $\eta$ from CMB determinations (that is $\log n(\mathrm{Li}) \simeq 2.6$). This discrepancy is much larger than possible measurement errors, and requires some explanation.

Almost all authors agree on the presence of a significant trend of Li abundances with overall metal abundance. For instance, Ryan et al. (1999) obtained $\log n(\mathrm{Li}) = (2.447 \pm 0.066) + (0.118 \pm 0.023)[\mathrm{Fe/H}]$. A slope quite similar to that found by Ryan was obtained by Asplund et al. (2003): $\log n(\mathrm{Li}) = (2.409 \pm 0.020) + (0.103 \pm 0.010)[\mathrm{Fe/H}]$. More recently, the whole issue was reexamined by Charbonnel and Primas (2004), with similar findings. These results suggest an evolution of the Li abundances; since it is the lowest value obtained at very low metal abundances that should be compared with the BBN predictions, consideration of such an evolution of the Li abundances exacerbate the Li discrepancy.

There are various concerns in the determination of the primordial Li abundance. We will briefly discuss some of them. First of all, various mechanisms may lead to depletion of Li in the stellar atmospheres. However, the observed scatter is small (Ryan et al. 1999; Bonifacio et al. 2002) and the few stars with large Li depletion are likely the result of the evolution in binary systems (Ryan et al. 2002). This leads to strong constraints on mechanisms of Li depletion ($<$0.1–0.2 dex) due to rotational mixing, which are expected to be variable from star-to-star (Pinsonneault et al. 2002). Better results are obtained with mechanisms such as gravity waves (Talon and Charbonnel 2004), that are expected to act similarly in all stars. Even more important might be the impact of microscopic diffusion, which should be coupled with levitation due radiation pressure; however, that is negligible for Li (but not for other elements like e.g. Fe). The impact of microscopic diffusion is constrained by comparisons of the photospheric abundances for stars at the turn off and those on the early subgiant branch of metal-poor globular clusters (Gratton et al. 2001; Korn et al. 2006). In fact, for stars at turn-off the impact of sedimentation due to diffusion is expected to be the largest because of the small size of the outward convective region. For stars at the base of the subgiant branch, the inward penetration of the outer convective envelope should cancel the effects of sedimentation. Results from these analyses (in particular the most recent paper by Korn et al., which is based on higher quality spectra) suggest the need for turbulent diffusion at the base of the outer convective envelope (as modeled e.g. by Richard et al. 2002) to explain the observations for both Fe and Li. When turbulence is included in the models, a moderate depletion of Li ($\sim$0.2 dex) is predicted for stars on the Spite plateau: this goes quite a long way in the right direction toward solving the discrepancy between BBN predicted from CMB data, and the observed level of the Spite's plateau (see also Charbonnel and Primas 2004).

An additional issue concerns the temperature scale adopted in the photospheric abundance analysis. Li abundance is sensitive to the adopted temperatures ($\Delta \log n(\mathrm{Li}) = 0.00065 \Delta T_{\mathrm{eff}}$). Different temperature scales have been adopted, causing some differences in Li abundances: uncertainties in the best temperature scale available (Alonso et al. 1996) should be about 50 K, causing an uncertainty of 0.03 dex in the Li abundances. However, recently a much higher temperature scale resulting in higher Li abundances (about $\log n(\mathrm{Li}) \sim 2.4$) was proposed by Ramirez and Melendez (2005). While not enough by itself to solve the Li discrepancy, adoption of such a high temperature scale would help a lot. However, the issue is still open: see e.g. the careful discussion in Charbonnel and Primas (2004), who in particular called attention to the somewhat neglected point of the impact of accurate estimates for the interstellar reddening. Charbonnel and Primas favoured a temperature scale lower than that suggested by Ramirez and Melendez, and closer to the classical one suggested by Alonso et al.

A third important issue is the model atmospheres used in the abundance derivations and the assumptions that lines form in LTE. These issues were considered in detail by Asplund et al. (2003). Li abundances are sensitive to departures from LTE, mainly overionization, and to the structure of the atmosphere because the lines form in cool outer layers. However, once these are estimated with state-of-the-art models (3D models + non-LTE), the abundance corrections with respect to standard LTE + 1D models are small ($<$0.1 dex). The corrections are positive (i.e. larger abundances).

The conclusion is that the photospheric Li abundance measured in most metal-poor stars is $\log n(\mathrm{Li}) = 2.1$–$2.2$. Assuming a further correction of 0.1 dex for the 3D and non-LTE effects, the value is $\log n(\mathrm{Li}) = 2.2$–$2.3$. This is about 0.4 dex below the value expected from the value expected from BBN based on CMB. Agreement can be obtained assuming a factor of 2.5 depletion. The depletion mechanism should be fine tuned, in order to provide both a small scatter and the right temperature dependence. Combination of microscopic diffusion and turbulence at the base of the convective envelope might provide the required depletion. Agreement between theoretical predictions and observations for several elements might possibly be obtained by slightly adjusting the maximum penetration of turbulence. However, for NGC 6397 (the cluster used for fine tuning the parameters) this agreement is critically dependent on the temperature difference between stars at TO and at the base of the RGB. Uncertainties regarding this difference are still not negligible.

## 4 The Mysterious $^6$Li

$^6$Li is very fragile and easily destroyed in stars. Its presence at the surface of metal-poor, turn-off stars would then be a critical test of the depletion mechanism(s) proposed for $^7$Li.

However, $^6$Li is not produced in significant amounts by the BBN; it is currently thought to be produced mainly by spallation mechanisms, the same ones that should be responsible for Be production. Since this mechanism requires cosmic rays, that are thought to be produced mainly by SN explosions, elements produced by spallation mechanism should cumulate with star formation, and their abundance should then be roughly proportional to metal abundance. $^6$Li is then expected to be detectable only in stars that are not very metal-poor. On the other hand, it can be easily destroyed during the pre-main sequence phase in small mass stars: hence, we expect to observe it only in a few metal-poor stars just near the turn-off (that is the most massive among the surviving metal-poor stars). In these stars we expect to observe an abundance which is roughly proportional to overall metal abundance.

In the last years there have been several claims of detection of $^6$Li in metal-poor stars (Smith et al. 1993, 1998; Hobbs and Thornburn 1994, 1997; Cayrel et al. 1999); however, this is a very difficult observation, since it is based on atomic lines (mainly the resonance doublet at 6707.8 Å). Isotopic splitting (due to the different reduced molecular weight) is small, about 0.02 Å, corresponding to a radial velocity shift of less than 1 km/s, which is much less than the line broadening due to temperature and velocity fields in the stellar atmospheres. Hence, (i) very high-quality observations (high resolution, $>100{,}000$, and S/N, $>300$) and (ii) accurate modeling of the line profile (due to fine and hyperfine structure, instrumental effects and velocity fields) are required. The impact of potential blends should also be carefully considered.

Asplund et al. (2006) published a very careful analysis of $^6$Li abundance in about 20 metal-poor stars, based on extremely high-quality spectra taken with UVES at VLT. They accurately considered all the most relevant problems (potential blends, instrumental profile, stellar velocity fields using 3D model atmospheres, non-LTE effects). The Li abundances obtained agree well with those determined from previous, much more extended studies, although those studies were based on lower quality spectra. They were then able to derive Li isotopic ratios with typical errors of about 2–3%. They detected $^6$Li at more than 2$\sigma$ level in about half of their stars, all of them with effective temperatures $T_{\rm eff} > 5{,}900$ K; some of these stars confirmed earlier detection, but their observations are by far more accurate and reliable.

The most surprising result obtained by Asplund et al. is that their data suggest the presence of a $^6$Li plateau, similar to the Spite's plateau for $^7$Li. In general, their $^6$Li abundances are much larger than those expected if production is due only to spallation on cosmic rays. This discrepancy is exacerbated if depletion mechanisms (that should be very important for $^6$Li) are considered.

The result obtained by Asplund et al. is really surprising: in fact, since $^6$Li is destroyed at much lower temperatures than $^7$Li, $^6$Li should be much more depleted than $^7$Li. Are the detections of $^6$Li compatible with depletion of primordial Li? What is the mechanism producing all this $^6$Li? The same presence of a plateau suggests a primordial origin for $^6$Li, but BBN does not produce enough $^6$Li. Hence, where it comes from? Is it produced by annihilation of neutralinos (Jedamzik 2004) or decay of gravitinos and axions (Ellis et al. 2005), as considered by Asplund et al.? This last explanation would have the further advantage of simultaneously reducing the $^7$Li discrepancy, since in this case $^6$Li would be produced at the expense of $^7$Li nuclei. Alternative explanations include production by cosmological cosmic rays (Rollinde et al. 2005; Fields and Prodanovic 2005), Pop. III massive stars (Reeves 2005; Rollinde et al. 2006), etc.

Before concluding this section, we should warn the reader that even if the $^6$Li analysis by Asplund et al. is perhaps the most careful paper on stellar abundance analysis published thus far, its detection is only at a few sigma, and only in a fraction of the stars: systematic errors at this level cannot be excluded.

## 5 Be as a Cosmochronometer

Be is produced by the interaction of Galactic Cosmic Rays (GCR) with the interstellar medium, through the spallation of heavy element nuclei, most noticeably carbon, nitrogen and oxygen. Due to its origin, and in particular to the fact that GCRs are generated and transported globally on a Galactic scale, Be is expected to be characterized by a smaller dispersion than the products of core collapse supernovae, whose abundances in the early Galaxy are affected by the dispersed character of star formation and inefficient mixing of gas. This led to the suggestion that Be represents a "cosmic clock" (Suzuki and Yoshii 2001; Beers et al. 2000). It should be noticed here that recent results (Carretta et al. 2002; Cayrel et al. 2004; Spite et al. 2005) obtain very homogeneous ratios between alpha-elements and Fe in metal-poor stars. Although this might be indication of an efficient mixing of the early interstellar medium, it might also indicate a rather homogeneous production ratio in the particular SNe that were efficient at old epochs, since the alpha–Fe ratio is observed to vary by large amounts at higher metallicities (see e.g. Nissen and Schuster 1997, and many other recent references). An alternative view from Be is then useful.

To test this hypothesis Pasquini et al. (2004) carried out the first measurements of Be abundances in a globular cluster; namely they observed Be in two turn-off stars of the metal-poor ([Fe/H] $= -2$) cluster NGC 6397, for which an independent age estimate was available (Gratton et al. 2003a). Be was detected in both stars at a level consistent with that of stars in the field with the same [Fe/H] abundance. By comparing their Be values with models of galactic evolution of Be as a function of time, they concluded that the cluster formed about 0.2–0.3 Gyr after the onset of star formaton in the halo, in very good agreement with the cluster age derived from main sequence fitting.

This approach can be extended to test if samples of halo and thick disk stars are coeval, using the Be abundance as an equivalent time scale. Such a test was conducted by Pasquini et al. (2005b) on two groups of stars identified to belong to these two populations by Gratton et al. (2003b). These authors considered a kinematical class composed of a population with galactic rotation velocity larger than 40 km/s and apogalactic distance of less than 15 kpc, that was called dissipative collapse component because it broadly corresponds to the classical Eggen et al. (1962) dissipative collapse population. It includes stars from the thick disk and the classical halo. The second kinematical class was composed of non-rotating or counter-rotating stars, and contains mainly stars of the classical halo. It was called the accretion component, because it can be roughly identified with the accreted population first proposed by Searle and Zinn (1978) to explain the formation of the halo.

These two components differ not only in their kinematical properties, but also in their chemical composition: the dissipation component has a very well-defined trend of [$\alpha$/Fe] ratios with metallicity and kinematics (galactic rotation velocity), with very small scatter around the mean relation. The accretion component have on average a smaller excess of $\alpha$-elements, and a much larger scatter around the average value (see Gratton et al. 2003b).

Pasquini et al. (2005b) showed that the dissipative component also has a very well-defined correlation between [$\alpha$/Fe] and Be abundances, with a very small scatter: this agrees with the consideration that both of them describe time. At a given Be abundance (i.e. time) the accretion population has a larger scatter in [$\alpha$/Fe], and seems to follow a distinct relation, consistent with a lower star formation rate.

## 6 Li and Be and Self-Pollution in Globular Clusters

As discussed by Pasquini et al. (2004), Be and Li measurements in globular clusters provide important constraints on the hotly debated issue of cluster formation. Detailed studies of chemical abundances in globular cluster stars have revealed that anomalies are present in all the clusters studied, showing that they are not homogeneous populations as far as the chemical composition of the stellar atmospheres is concerned (see Gratton et al. 2004 and references therein). More specifically, whereas globular clusters appear to be extremely homogeneous in Fe and Fe-peak elements (with star-to-star variations $<10\%$; see Carretta et al. 2006, 2007), star-to-star variations are seen for those elements involved in H-burning at high temperature. This behaviour is not present among field stars, which do not exhibit significant star-to-star variations in abundance ratios like [O/Na] or [Mg/Al] at a given metallicity. This intriguing abundance pattern, and in particular the fact that abundance anomalies are seen not only among evolved globular cluster stars but also in unevolved stars at the turn-off, suggests that part of the gas forming the stars that we observe today has been processed and ejected by a previous generation of stars. Which kind of stars were the polluters is actually debated: they may be either massive AGB stars ($5 < M < 7\ M_{\odot}$) where the products of hot-bottom burning are brought to the surface by the Cameron & Fowler (1971) mechanism (see e.g. Ventura et al. 2001); or fast-rotating massive stars, where heavy mass-loss peels the stars, exposing the regions of H-burning (WR stars; see e.g. Charbonnel and Prantzos 2006). Both these explanations have pros and cons, and there is not at present any clear-cut observations that allow us to select the best model.

In this respect, it is useful to note that $^7$Li and $^9$Be are both destroyed at relatively low temperatures ($\sim$2.5 and $\sim$3.5 million K, respectively) in stellar interiors. These temperatures are significantly below those at which typical reactions responsible for the chemical anomalies in globular clusters occur. Thus, in the hypothesis that a fraction of globular cluster stars is born from the ejecta of a previous generation of stars, these stars should in principle show low Be and Li abundances. Note, however, that Li and Be form through different mechanisms. Whereas $^9$Be can be produced only by GCR spallation on the ISM, $^7$Li is produced by BBN, and the Galaxy is later enriched by several mechanisms, such as GCRs spallation, AGB stars, and possibly Novae (e.g. Travaglio et al. 2001).

In particular, observations of Be in O-poor, Li-poor stars of globular clusters are potentially very helpful, because they might tell us directly how much time has elapsed between the formation of the first and second generation of stars in globular clusters.[1] This would in turn provide hints on the type of polluter: massive stars (short delay, of the order of a few million years, and then very low Be abundances) or massive AGB stars (long delay, tens or hundred million years, and then higher, detectable Be abundance). The reason this test can work is that in the case of massive AGB stars, a rather long time ($>10^8$ yr) is required for accumulation of enough polluted gas, because stars with rather different lifetimes are to be considered.

Observational results can be summarized as follows: the Li abundance is constant in NGC6397 (Bonifacio et al. 2002), while Li is found to vary from star-to-star in NGC6752 and 47 Tuc. In both these clusters, Li abundances are correlated with O and anti-correlated with Na abundances (Pasquini et al. 2005a; Bonifacio et al. 2007).

As for Be, the mere existence of Be in the two turn-off stars of NGC6397 suggests that the gas which formed the stars we now observe must have been sitting for at least a few hundred million years in the ISM exposed to GCR spallation before the stars formed. NGC

[1] Some original Be could be observed also in O-poor stars, if there has been mixing between O-depleted material and some original undepleted one. While this requires a reservoir of pristine gas, its availability depends on the adopted scenario for the second-generation star. Also, in this case, Be should be observed to be strongly depleted in those stars which are very poor in O—the Be depletion should in this case follow the O-depletion.

6397 could on the other hand be a somewhat "special" case, because it is one of the globular clusters where chemical anomalies are present at the lowest level and, as mentioned above, the Li abundance is strikingly constant.

Pasquini et al. (2007) then looked for Be in two turn-off stars in NGC6752, a cluster which exhibits an extended O–Na anticorrelation. The stars were selected having different composition: one representative of the O-rich, Li-rich component, and the other of the Li-poor, O-poor one. Be observation is very difficult, being at the extreme UV edge of the optical window usable from ground. Observations are quite inefficient at this wavelength, even with the best optimized spectrographs like UVES at VLT: only rather low S/N spectra could then be obtained for faint stars at the turn-off of globular clusters. As expected, Be lines were indeed detected on the spectrum of the O-rich, Li-rich star; unfortunately, only an upper limit could be obtained for the O-poor, Li-poor one. This upper limit is enough to conclude that this second star has not more Be than the other (and hence cannot be much younger), but it is not enough to severely constrain the type of polluters that were active in this star.

## References


A. Alonso, S. Arribas, C. Martinez-Roger, Astron. Astrophys. **313**, 873 (1996)

M. Asplund, M. Carlsson, A.V. Botnen, Astron. Astrophys. **399**, L31 (2003)

M. Asplund, D.L. Lambert, P.E. Nissen, F. Primas, V.V. Smith, Astrophys. J. **644**, 229 (2006)

V. Barger, J.P. Kneller, H.-S. Lee, D. Marfatia, G. Steigman, Phys. Lett. **566**, 8 (2003)

T.C. Beers, T.K. Suzuki, Y. Yoshii, in *The Light Elements and Their Evolution*, ed. by L. Da Silva, M. Spite, J.R. de Medeiros. IAU Symp., vol. 198 (ASP, San Francisco, 2000), p. 425

P. Bonifacio, Astron. Astrophys. **395**, 515 (2002)

P. Bonifacio et al., Astron. Astrophys. **390**, 91 (2002)

P. Bonifacio et al., Astron. Astrophys. **470**, 153 (2007)

S. Burles, D. Tytler, Astrophys. J. **507**, 732 (1998)

A.G.W. Cameron, W.A. Fowler, Astrophys. J. **164**, 111 (1971)

E. Carretta, R. Gratton, J.G. Cohen, T.C. Beers, N. Christlieb, Astron. J. **124**, 481 (2002)

E. Carretta, A. Bragaglia, R.G. Gratton, F. Leone, A. Recio-Blanco, S. Lucatello, Astron. Astrophys. **450**, 523 (2006)

E. Carretta et al., Astron. Astrophys. (2007, in press)

S. Cassisi et al., Astrophys. J. **588**, 862 (2003)

R. Cayrel, M. Spite, F. Spite, E. Vangioni-Flam, M. Cassé, J. Audouze, Astron. Astrophys. **343**, 923 (1999)

R. Cayrel, E. Depagne, M. Spite, V. Hill, F. Spite et al., Astron. Astrophys. **416**, 1117 (2004)

C. Charbonnel, F. Primas, Astron. Astrophys. **442**, 961 (2004)

C. Charbonnel, N. Prantzos (2006), astro-ph/06-06220

N.H.M. Crighton, J.K. Webb, A. Ortiz-Gil, A. Fernández-Soto, Mon. Not. Roy. Astron. Soc. **355**, 1042 (2004)

S. D'Odorico, M. Dessauges-Zavadsky, P. Molaro, Astron. Astrophys. **368**, L21 (2001)

O.C. Eggen, D. Lynden-Bell, A.R. Sandage, Astrophys. J. **136**, 748 (1962)

J. Ellis, K.A. Olive, E. Vangioni, Phys. Lett. B **619**, 30 (2005)

B.D. Fields, T. Prodanovic, Astrophys. J. **623**, 877 (2005)

W. Freedman et al., Astrophys. J. **553**, 47 (2001)

R.G. Gratton et al., Astron. Astrophys. **369**, 87 (2001)

R.G. Gratton, A. Bragaglia, E. Carretta et al., Astron. Astrophys. **408**, 529 (2003a)

R.G. Gratton, E. Carretta, S. Desidera et al., Astron. Astrophys. **406**, 131 (2003b)

R.G. Gratton, C. Sneden, E. Carretta, Annu. Rev. Astron. Astrophys. **42**, 385 (2004)

L.M. Hobbs, J.A. Thornburn, Astrophys. J. **428**, L25 (1994)

L.M. Hobbs, J.A. Thornburn, Astrophys. J. **491**, 772 (1997)

Y.I. Izotov, T.X. Thuan, Astrophys. J. **602**, 200 (2004)

K. Jedamzik, Phys. Rev. D **70**, 063524 (2004)

D. Kirkman, D. Tytler, N. Suzuki, J.M. O'Meara, D. Lubin, Astrophys. J. Suppl. Ser. **149**, 1 (2003)

A.J. Korn et al., Nature **442**, 657 (2006)

S.A. Levshakov, M. Dessauges-Zavadsky, S. D'Odorico, P. Molaro, Astrophys. J. **565**, 696 (2002)
P.E. Nissen, W.J. Schuster, Astron. Astrophys. **326**, 751 (1997)
K.A. Olive, E. Skillman, New Astron. **6**, 119 (2001)
J.M. O'Meara, D. Tytler, D. Kirkman, N. Suzuki, J.X. Prochaska, D. Lubin, A.M. Wolfe, Astrophys. J. **552**, 718 (2001)
L. Pasquini, P. Bonifacio et al., Astron. Astrophys. **426**, 651 (2004)
L. Pasquini, P. Bonifacio et al., Astron. Astrophys. **441**, 549 (2005a)
L. Pasquini, D. Galli et al., Astron. Astrophys. **436**, L57 (2005b)
L. Pasquini, P. Bonifacio et al., Astron. Astrophys. (2007, in press)
M. Peimbert, A. Peimbert, V. Luridiana, M.T. Ruiz, in *Star Formation Through Time*, ed. by E. Perez, R.M. Gonzalez Delgado, G. Tenorio-Tagle. ASP Conf. Ser., vol. 297 (ASP, San Farncisco, 2003), p. 81
M. Pettini, D.V. Bowen, Astrophys. J. **560**, 41 (2001)
M.H. Pinsonneault, G. Steigman, T.P. Walker, V.K. Narayanan, Astrophys. J. **574**, 398 (2002)
I. Ramirez, J. Melendez, Astrophys. J. **626**, 446 (2005)
H. Reeves, EAS Publ. Ser. **17**, 15 (2005)
O. Richard, G. Michaud, J. Richer et al., Astrophys. J. **568**, 979 (2002)
E. Rollinde, E. Vangioni, K.A. Olive, Astrophys. J. **626**, 666 (2005)
E. Rollinde, E. Vangioni, K.A. Olive, Astrophys. J. **651**, 658 (2006)
S.G. Ryan, J.E. Norris, T.C. Beers, Astrophys. J. **523**, 654 (1999)
S.G. Ryan, S.G. Gregory, U. Kolb, T.C. Beers, T. Kajino, Astrophys. J. **571**, 501 (2002)
L. Searle, R. Zinn, Astrophys. J. **225**, 357 (1978)
V.V. Smith, D.L. Lambert, P.E. Nissen, Astrophys. J. **408**, 262 (1993)
V.V. Smith, D.L. Lambert, P.E. Nissen, Astrophys. J. **506**, 405 (1998)
D.N. Spergel et al., Astrophys. J. Suppl. Ser. **148**, 175 (2003)
D.N. Spergel et al. (2006), astro-ph/06-03449
M. Spite, F. Spite, Nature **297**, 483 (1982)
M. Spite, R. Cayrel, B. Plez, V. Hill, F. Spite et al., Astron. Astrophys. **430**, 655 (2005)
G. Steigman (2002), astro-ph/0208186
T.K. Suzuki, Y. Yoshii, Astrophys. J. **549**, 303 (2001)
S. Talon, C. Charbonnel, Astron. Astrophys. **418**, 1051 (2004)
M. Tosi, in *The Light Elements and their Evolution*, ed. by L. Da Silva, M. Spite, J.R. de Medeiros. IAU Symp., vol. 198 (ASP, San Francisco, 2000), p. 525
C. Travaglio, S. Randich, D. Galli et al., Astron. Astrophys. **559**, 909 (2001)
P. Ventura, F. D'Antona, I. Mazzitelli, R. Gratton, Astrophys. J. **550**, 65 (2001)

Space Sci Rev (2007) 130: 53–62
DOI 10.1007/s11214-007-9144-z

# The Milky Way 3-Helium Abundance

**T.M. Bania · R.T. Rood · D.S. Balser**

Received: 18 December 2006 / Accepted: 4 January 2007 /
Published online: 23 March 2007


**Abstract** We are making precise determinations of the abundance of the light isotope of helium, $^3$He. The $^3$He abundance in Milky Way sources impacts stellar evolution, chemical evolution, and cosmology. The abundance of $^3$He is derived from measurements of the hyperfine transition of $^3$He$^+$ which has a rest wavelength of 3.46 cm (8.665 GHz). As with all the light elements, the present interstellar $^3$He abundance results from a combination of Big Bang Nucleosynthesis (BBNS) and stellar nucleosynthesis. We are measuring the $^3$He abundance in Milky Way H II regions and planetary nebulae (PNe). The source sample is currently comprised of 60 H II regions and 12 PNe. H II regions are examples of zero-age objects that are young relative to the age of the Galaxy. Therefore their abundances chronicle the results of billions of years of Galactic chemical evolution. PNe probe material that has been ejected from low-mass ($M \leq 2M_\odot$) to intermediate-mass ($M \sim 2$–$5M_\odot$) stars to be further processed by future stellar generations. Because the Milky Way ISM is optically thin at centimeter wavelengths, our source sample probes a larger volume of the Galactic disk than does any other light element tracer of Galactic chemical evolution. The sources in our sample possess a wide range of physical properties (including object type, size, temperature, excitation, etc.). The $^3$He abundances we derive have led to what has been called "The $^3$He Problem".

**Keywords** Cosmology: cosmological parameters · The Galaxy: abundances, evolution · ISM: abundances, evolution, H II regions · Stars: AGB, post-AGB

T.M. Bania (✉)
Department of Astronomy, Institute for Astrophysical Research, Boston University, 725 Commonweath Ave., Boston, MA 02215, USA
e-mail: bania@bu.edu

R.T. Rood
Department of Astronomy, University of Virginia, Box 3818 University Station, Charlottesville, VA 22903, USA

D.S. Balser
National Radio Astronomy Observatory, P.O. Box 2, Green Bank, WV 24944, USA

## 1 The $^3$He Problem

We continue to derive the $^3$He abundance in Galactic H II regions and planetary nebulae using the $^3He^+$ hyperfine transition. The study of the origin and evolution of the elements is one of the cornerstones of modern astrophysics. For any given isotope it is crucial to determine its abundance and how that abundance varies temporally and spatially. Knowing the cosmic abundance of the $^3$He isotope has a broad interdisciplinary impact: $^3$He can be used to test the theory of stellar nucleosynthesis; $^3$He gives important information needed to evaluate models of Galactic chemical evolution; $^3$He can help constrain Big Bang Nucleosynthesis. Big Bang theory predicts that during the first ~1,000 s significant amounts of the light elements ($^2$H, $^3$He, $^4$He, and $^7$Li) were produced. The near concordance now reached among all four Big Bang isotope abundances and the WMAP satellite results is a triumph of modern astrophysics.

Theory predicts not only that common solar-type stars produce $^3$He but also that the mass lost from winds generated at advanced stages of their evolution and the final planetary nebulae should be substantially enriched in $^3$He. Planetary nebula $^3$He abundances are therefore important tests of stellar evolution theory since these low-mass, evolved objects are expected to be significant sources of $^3$He. We have confirmed the stellar production of $^3$He in the planetary nebulae NGC 3242 and J 320; their $^3$He/H abundances are consistent with the predictions of standard stellar models. Measurement of the present $^3$He abundance is an important diagnostic of chemical evolution in the Galaxy. H II regions sample the result of the chemical evolution of the Milky Way since its formation. The $^3$He/H abundance ratio is expected to grow with time and to be higher in those parts of the Galaxy where there has been substantial stellar processing. That our observations are inconsistent with these expectations leads to "The $^3$He Problem".

## 2 $^3$He Abundances in H II Regions

We have identified a special class of "simple" H II regions for which accurate $^3$He/H abundances can be determined. Surprisingly, we find these sources to be rather plentiful. Furthermore, we can detect $^3He^+$ in such H II regions up to 11 kpc *beyond* the Galactic Center. In fact, we can determine $^3$He abundances over a larger fraction of the Galactic disk than any other isotopic probe of stellar and Galactic chemical evolution.

This source sample provides a strong constraint on models for Galactic chemical evolution. Standard evolution models predict that: (1) the protosolar $^3$He/H value should be less than that found in the present ISM; (2) the $^3$He/H abundance should grow with source metallicity; and (3) there should be a $^3$He/H abundance gradient in the Galactic disk with the highest abundances occurring in the highly-processed inner Galaxy. None of these predictions is confirmed by observations (Balser et al. 1998; Bania et al. 2002 [BRB]).

Specifically, measurements of $^3$He/H in protosolar material (Geiss 1993), the local solar neighborhood (Gloecker and Geiss 1996), and Galactic H II regions (Rood et al. 1995) all indicate a value of $^3\mathrm{He/H} \sim 2 \times 10^{-5}$ by number. Thus the H II regions show no evidence for stellar $^3$He enrichment during the last 4.5 Gyr (Fig. 1). There is no significant $^3$He abundance gradient across the Milky Way (Fig. 2). And, finally, there is no trend of $^3$He abundance with source metallicity (Fig. 3).

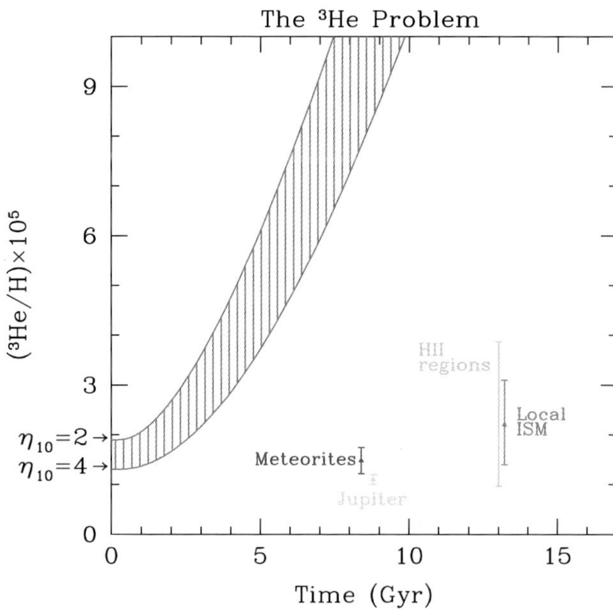


**Fig. 1** $^3$He abundances compared with Galactic chemical evolution models. The $^3He/H$ abundances derived for meteorites, Jupiter, the Local ISM, and H II regions are inconsistent with chemical evolution models that use standard stellar yields. This inconsistency is insensitive to the BBNS $^3$He production. The BBNS models are parameterized by the primordial baryon-to-photon ratio, $\eta$, expressed in units of $10^{-10}$. (Figure updated from Galli et al. 1995)

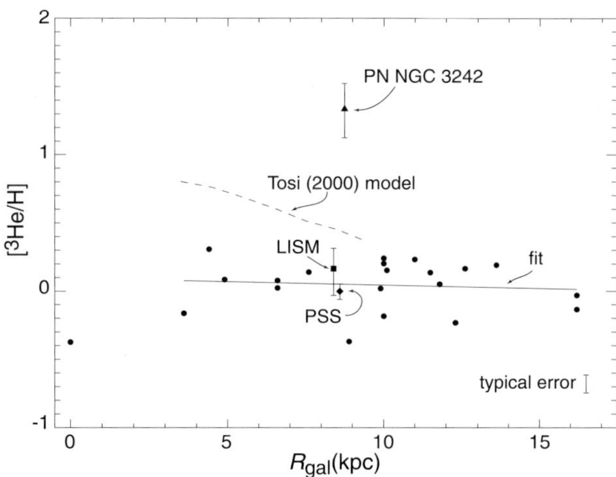


**Fig. 2** $^3He/H$ abundances as a function of Galactic radius (BRB). The [$^3He/H$] abundances by number for the BRB H II region sample are given with respect to the solar ratio. Shown also are the abundances for the planetary nebula NGC 3242 (*triangle*), the local interstellar medium (*LISM—square*), and protosolar material (*PSS—diamond*). There is no gradient in the $^3He/H$ abundance with Galactic position. To be compatible with this result Galactic chemical evolution models (Tosi 2000) require that ~90% of solar analog stars are non-producers of $^3$He

## 3 $^3$He Abundances in Planetary Nebulae

It was crucial to see if stars actually do produce $^3$He so we made a preliminary survey for $^3He^+$ in PNe (Balser et al. 1997). Working at the limit of the MPIfR 100 m telescope we made a detection of $^3He^+$ in NGC 3242, but did not have a definitive detection in any other single PN. There are, however, possible MPIfR 100 m detections in 2 additional PNe, and hints in 2 others. It was important to verify our detection with another telescope. Despite the fact that PNe with their small angular sizes were not inviting targets for the NRAO 140 Foot, we nonetheless were able to verify our NGC 3242 result. In one of the longest integrations ever made for a radio frequency spectrum (270 hours), we confirmed our MPIfR

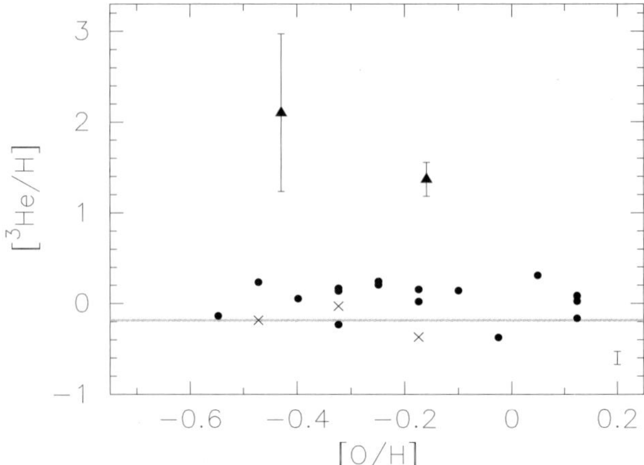


**Fig. 3** The $^3$He PLATEAU: [$^3$He/H] abundances as a function of source metallicity for the "simple" H II region sample. The gray line is the WMAP result. The ~0.15 dex typical error is shown in the right hand corner. The triangles denote abundances for the PNe J 320 (*left*) and NGC 3242. There is no trend in the $^3$He/H abundance with source metallicity

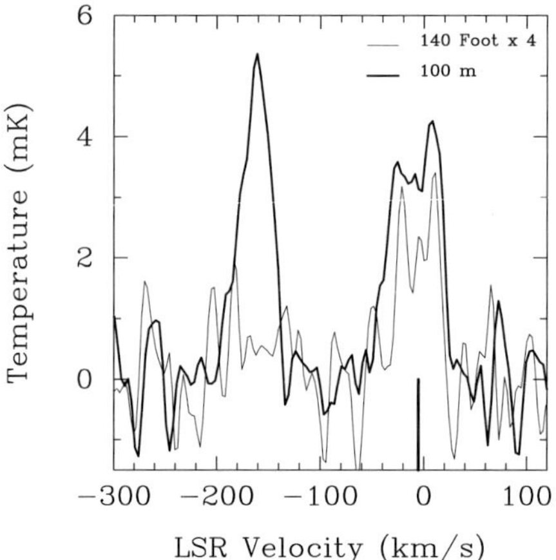


**Fig. 4** Brightness temperature spectra of $^3He^+$ emission from the planetary nebula NGC 3242 (BRB99). Shown are the MPIfR 100 m (*bold*) and NRAO 140 Foot (*thin*) spectra. The $^3He^+$ emission is flagged at $-5.3$ km s$^{-1}$; the prominent feature in the MPIfR spectrum is the H171$\eta$ recombination line which comes from clumped, dense gas that is invisible when diluted by the large 140 Foot beam

result (Fig. 4; Balser et al. 1999 [BRB99]). The $^3He^+$ line shape also confirmed our earlier suggestion that much of the $^3He^+$ emission comes from a large diffuse halo.

We have recently detected $^3He^+$ in the planetary nebula J 320 with the VLA (Fig. 5; Balser et al. 2005); the abundance we derive is in accord with standard stellar evolution. Thus it is clear that *some* (i.e., >2) PNe produce significant amounts of $^3$He that survives to the PN stage and enriches the ISM. Indeed, the *quantitative* agreement between theory and observation is quite reasonable (see Fig. 2 of Galli et al. 1997). Yet Fig. 1 shows the magnitude of the $^3$He Problem: the abundances predicted by standard chemical evolution models lie far above the observations. Our PN sample was purposefully biased to maximize the likelihood of finding $^3$He. The circumstantial evidence strongly suggests that our high $^3$He abundance PNe are atypical.

*Any* chemical evolution model that adopts standard stellar $^3$He nucleosynthesis overproduces $^3$He. All Galactic evolution models that match the other observational constraints (e.g., star formation rate, gas and total mass density, mass in-fall rate, IMF, etc.) predict $^3$He abundances that are inconsistent with those observed both locally and globally in the Milky Way. This can be seen in Fig. 2 where the dashed line shows the results of a chemical evolu-

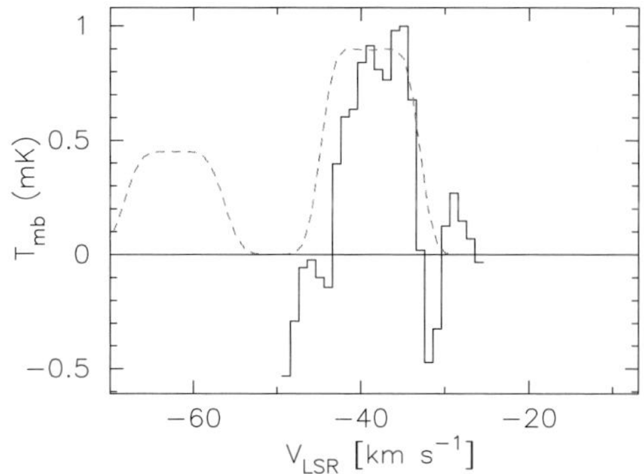


**Fig. 5** VLA $^{3}He^{+}$ detection for PN J 320 (Balser et al. 2005). Superimposed (*dashed line*) is a model for J 320 in which the nebula is taken to be a single component, spherical shell expanding at a velocity of 40 km s$^{-1}$. The bandwidth of the current VLA correlator is insufficient to include the H171$\eta$ transition (*left feature of the model spectrum*). The increased bandwidth of the new EVLA correlator will help make measurements such as this almost routine

tion model with standard yields (Tosi 2000). Such models reproduce observed values only if they adopt alternative nucleosyntheses with a strongly reduced $^{3}$He contribution from low- and intermediate-mass stars.

The standard view of $^{3}$He chemical evolution is relatively simple. It is indeed dominated by the net production of this element by low-mass stars thanks both to the destruction of D on the pre-main sequence and to the pp-chains that build up a $^{3}$He peak inside the main sequence star. This fresh $^{3}$He is then engulfed by the deepening convective envelope of the star during the first dredge-up on the lower RGB. Once in the convective layers of the red giant, $^{3}$He is preserved against nuclear destruction, and it is ejected into the ISM in the late stages of stellar evolution both through the stellar winds and the PN ejection. A substantial increase of the $^{3}$He abundance in the Galaxy is thus expected as soon as low-mass stars start to die and to pollute the ISM.

Rood et al. (1984) [RBW] suggested that the $^{3}$He problem could be related to striking chemical anomalies in red giant stars. Much accumulated observational evidence shows that low-mass RGB stars undergo an extra-mixing event. This extra-mixing adds to the standard first dredge-up to modify the surface abundances. In particular, Charbonnel (1994) and Charbonnel et al. (1998) used observations of the $^{12}C/^{13}C$ ratios to determine that this process occurs just after the so-called "bump" on the RGB. At this evolutionary point, the hydrogen burning shell crosses the discontinuity in molecular weight built by the convective envelope during the first dredge-up. Before the discontinuity, the molecular weight gradient probably acts as a barrier to mixing in the radiative zone. Beyond this point, however, no gradient of molecular weight exists above the hydrogen burning shell so the extra-mixing, whatever its nature, is free to act.

Several attempts have been made to simulate the extra-mixing in RGB stars. Charbonnel (1995) showed that rotation-induced mixing cannot only account for the observed behavior of the carbon isotopic ratio but also explain other abundance anomalies in low mass giants. When the extra-mixing begins to act, $^{3}$He is simultaneously and rapidly transported down to the regions where it burns by the $^{3}He(\alpha, \gamma)^{7}Be$ reaction. This leads to a decrease of the surface value of $^{3}He/H$ compared to the standard case. Under peculiar mixing conditions, a thermal instability can occur, which transports the resulting $^{7}$Be outwards and leads to an increase of the surface $^{7}$Li abundance during a very short period. A few giant stars with very high $^{7}$Li abundance have been discovered at the RGB bump. These so-called Li-rich stars are actually caught in the act of burning their $^{3}$He (Charbonnel and Balachandran 2000; Palacios et al. 2001).

Stars with different rotation and mass loss histories are expected to suffer different mixing efficiency and to display different chemical anomalies. *Significantly, all the relevant data indicate that the extra-mixing occurs in* $\sim$*90 to 95% of the low-mass stars* (Charbonnel and Do Nascimento 1998; Charbonnel 1998). This is the best current solution to the "$^3$He Problem". NGC 3242 and J 320 should belong to the $\sim$5 to 10% of stars which do not suffer from extra-mixing on the red giant branch. Such "standard" PNe should also show "normal" carbon isotopic ratios. This crucial test has already been verified for two PNe of the BRB99 sample. NGC 6720 has a $^{12}$C/$^{13}$C ratio of 23 which is in perfect agreement with the "standard" predictions (Bachiller et al. 1997). Furthermore, NGC 3242 itself has an HST-based limit on its $^{13}$C/$^{12}$C ratio which suggests that it is the outcome of the evolution of a standard, $^3$He-producing, low-mass star (Palla et al. 2002).

Because models with rotation-induced mixing are not yet available for stars with various initial masses and metallicities, uncertainties still remain on the actual $^3$He yields. In Charbonnel's preliminary Pop II models with rotation, $^3$He decreases by a large factor in the ejected envelope material. *Nonetheless these stars remain net producers of* $^3$He. Although these extra-mixing low mass stars are far less efficient in making $^3$He, in regard to chemical evolution such stars are *non-producers* rather than *destroyers* of $^3$He.

## 4 Current Status of the $^3$He Experiment

The $^3$He project requires extremely high sensitivity observations. The 3.5 cm (X-band in radioastronomy jargon) performance of all the telescopes that we use has recently been upgraded. The new Green Bank Telescope (GBT) is, moreover, now operating routinely at X-band. These spectrometers now have unprecedented sensitivity. The $^3$He project played an important role in commissioning the GBT and Arecibo: our $^3$He$^+$ spectra are very, very sensitive and thus are vital system performance tests. This new instrumental capability at X-band is taking the $^3$He observations to the next level of sensitivity.

The first GBT $^3$He$^+$ science observations were made in June 2004 and May 2005. Figure 6 shows the calibrated, but otherwise raw, spectrum for the S 209 H II region. Figure 7 shows this same spectrum processed in order to get the best possible line parameters for the $^3$He$^+$ transition. *We find that the GBT's spectral baselines are by far the best that we have ever seen with any single dish telescope.*

Our initial GBT observations have focussed on the S 209 H II region (Figs. 6 and 7) and a sample of 4 PNe. We now have tentative detections for two PNe but we need a few more observing epochs to assess possible systematic effects. Figure 8 shows the composite spectrum of the 4 PNe. This 125.7 receiver hour integration clearly shows that these PNe contain $^3$He$^+$. This GBT result confirms our conclusions based on a similar MPIfR 100 m composite spectrum (Balser et al. 1997). *But this GBT spectrum is an integration that is* 1/4 *that of the MPIfR spectrum*! The GBT spectral baselines are so much better that we have reproduced *in one GBT week* what it took us four times more integration (and 8 years!) to do in Effelsberg. The clear aperture design of the GBT has produced a spectrometer system that will enable us to measure $^3$He spectra with unprecedented sensitivity and speed.

Although 3.5 cm wavelength is at the high frequency limit of the Arecibo telescope, its vast collecting area can in principle produce unsurpassed X-band sensitivity for the $^3$He experiment. We assessed the X-band performance of Arecibo in May 2003 by making a survey of carbon recombination lines in a sample of 17 ultra-compact H II regions (Roshi et al. 2005). The Arecibo X-band spectral baselines are surprisingly good. NAIC subsequently made significant improvements in all aspects of the X-band spectrometer performance. Our

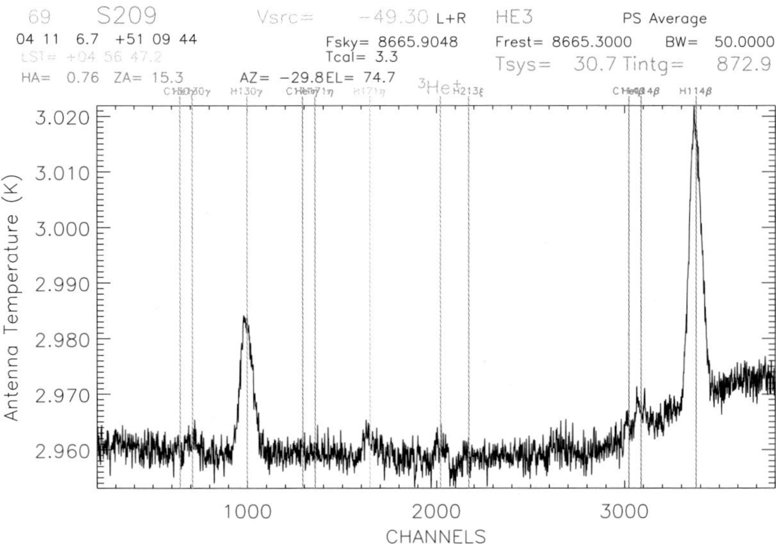


**Fig. 6** Raw GBT spectrum of the S209 H II region after a 14.5 receiver hour integration (average of two polarizations at 7.25 hr each). The $^3He^+$ and various recombination line transitions are flagged. The quality of the GBT X-band baselines in this calibrated, but otherwise unprocessed, spectrum is excellent

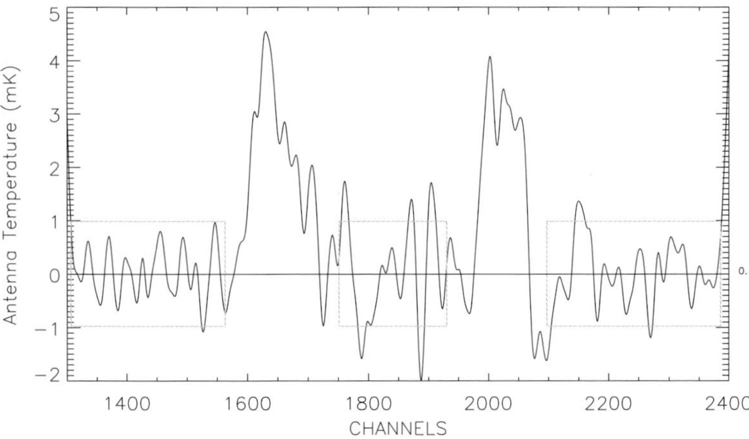


**Fig. 7** Zoom in of the $^3He^+$ spectral region for the Fig. 6 data after a polynomial baseline model was subtracted and the spectrum smoothed to 5 km s$^{-1}$ resolution. Emission lines from the H171$\eta$ recombination (*left*) and $^3He^+$ spin-flip (*right*) transitions are obvious. This GBT confirmation of our 140 ft $^3He^+$ measurement was made with only 10% the 140 ft integration time! The GBT has exceptional baselines

July 2005 $^3He^+$ observations of a PNe sample are shown in Fig. 9. They confirm the quality of the Arecibo X-band baselines and show that the long term program of tweaking the primary mirror surface figure has produced better, and more uniform, telescope gain.

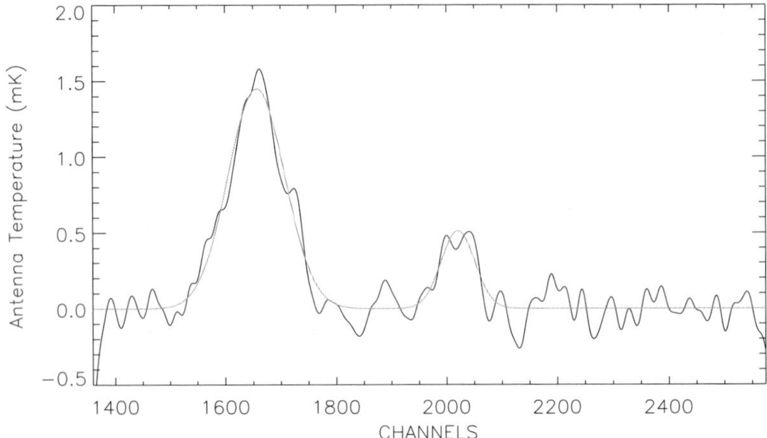


**Fig. 8** GBT $^3He^+$ spectrum for a composite average of a sample of 4 Galactic PNe. Shown is the average of spectra taken toward NGC 3242 + NGC 6543 + NGC 6826 + NGC 7009 aligned to a common LSR velocity. A polynomial baseline model has been removed and the spectrum smoothed to 5 km s$^{-1}$ resolution. Both the H171$\eta$ recombination (*left*) and $^3He^+$ spin-flip (*right*) transitions are clearly seen. Gaussian fits to these lines are superimposed

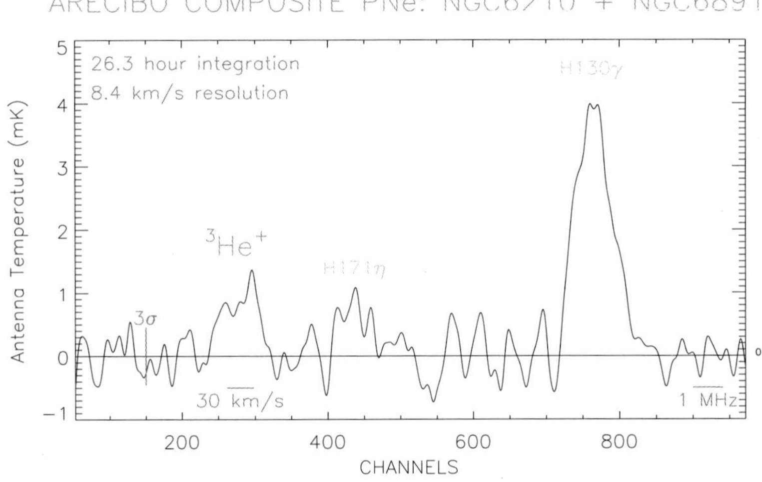


**Fig. 9** Arecibo $^3He^+$ spectrum for a composite average of a sample of 2 Galactic PNe. Shown is the average of spectra taken toward NGC 6210 + NGC 6891 aligned to a common LSR velocity. A polynomial baseline model has been removed and the spectrum smoothed to 8.4 km s$^{-1}$ resolution. The H130$\gamma$ recombination and $^3He^+$ spin-flip transitions are clearly seen

## 5 The Future of the $^3$He Experiment

We are the only group in the world making a systematic study of $^3$He throughout the Galaxy. We will observe a small number, ~10, of new $^3$He sources with the GBT, VLA, Arecibo, and ATNF Parkes 64 m telescopes. Our $^3$He results raise specific questions that can be answered by the enhanced spectroscopic sensitivity now available with these instruments.

We believe that just a few more, strategically chosen $^3$He sources can have a great impact in resolving our understanding of the cosmic $^3$He abundance. These new sources are: (1) H II regions strategically placed throughout the Galactic disk, enabling us to measure the disk $^3$He and metallicity abundance gradients, thus providing important constraints on Galactic chemical evolution; (2) Planetary nebulae that can directly test predictions made by new stellar extra-mixing models; and (3) Extragalactic H II regions in the LMC which will extend the metallicity range of sources with $^3$He abundances.

## 6 Summary

In sum, observations of $^3$He (ours, PSM and LISM), observations of the $^{12}$C/$^{13}$C abundance ratio in red giant stars, and rotating stellar models all indicate that: (1) the stellar contribution to the $^3$He abundance evolution is positive (i.e., $^3$He increases), and (2) $^3$He ISM enrichment is small compared to the size of the present observational errors. To be consistent with chemical evolution models, the existence of the $^3$He Plateau demands that the majority of PNe are not net producers of $^3$He. J 320 now joins NGC 3242 as only the second known example of those rare planetary nebulae that enrich the Galaxy in $^3$He. A larger sample of PN $^3$He abundances will provide a critical test of nonstandard stellar evolution theories.

Most theories put forward to resolve the $^3$He problem posit that some non-convective mixing process either prevents the buildup of $^3$He on the main sequence or destroys $^3$He along the upper red giant branch (RBW; Hogan 1995). We have already discussed Charbonnel's work which shows how this might arise from the rotationally driven mixing of Zahn (1992). Although this is our preferred explanation, other mechanisms, some entirely ad hoc, have been proposed (e.g., Denissenkov and Weiss 1996; Weiss et al. 1996; Wasserburg et al. 1995; Sackmann and Boothroyd 1999; Boothroyd and Sackmann 1999). These various scenarios argue $^3$He destruction is correlated with the long-standing $^{12}$C/$^{13}$C and other CNO isotopic ratio problems in RGB and AGB stars (as we suggested long ago in RBW Sect. 5c).

Eggleton et al. (2006) recently announced in *Science Express* the numerical discovery of what they claim is the resolution of the $^3$He problem. By modeling a red giant with a fully 3D code (Dearborn et al. 2006) they find that mixing arises in the supposedly stable zone between the hydrogen burning shell and the base of the convective envelope. This extra mixing is due to Rayleigh–Taylor instability within a zone just above the hydrogen-burning shell where a nuclear reaction lowers the mean molecular weight slightly. This mechanism should operate in *all* stars; there should be *no* $^3$He produced by *any* planetary nebulae. Our $^3$He detections in the PNe NGC 3242 and J 320, made with three different telescope spectrometer systems, are in violent disagreement with this prediction. A key part of the $^3$He problem is to find a physical mechanism wherein most, but not all, PNe fail to enrich the ISM in $^3$He. Rotational mixing is such a mechanism; the Eggleton et al. Rayleigh–Taylor instability apparently is not.

**Acknowledgements** We thank the international light element abundances community for their collegiality and support over the years. Our $^3$He research has been sporadically supported by the U.S. National Science Foundation. The most recent grants were AST 00-98047 to TMB and AST 00-0098449 to RTR.

## References

R. Bachiller, T. Forveille, P.J. Huggins, P. Cox, Astron. Astrophys. **324** 1123–1134 (1997)
D.S. Balser, T.M. Bania, R.T. Rood, T.L. Wilson, Astrophys. J. **483**, 320–334 (1997)

D.S. Balser, T.M. Bania, R.T. Rood, T.W. Wilson, Astrophys. J. **510**, 759–783 (1998)
D.S. Balser, R.T. Rood, T.M. Bania, Astrophys. J. **522**, L73–L76 (1999)
D.S. Balser, W.M. Goss, T.M. Bania, R.T. Rood, Astrophys. J. **640**, 360–368 (2005)
T.M. Bania, R.T. Rood, D.S. Balser, Nature **415**, 54–57 (2002)
A.I. Boothroyd, I.-J. Sackmann, Astrophys. J. **510**, 232–250 (1999)
C. Charbonnel, Astron. Astrophys. **282**, 811–820 (1994)
C. Charbonnel, Astrophys. J. **453**, L41–L44 (1995)
C. Charbonnel, Space Sci. Rev. **84**, 199–206 (1998)
C. Charbonnel, J.D. Do Nascimento, Astron. Astrophys. **336**, 915–919 (1998)
C. Charbonnel, J.A. Brown, G. Wallerstein, Astron. Astrophys. **332**, 204–214 (1998)
C. Charbonnel, S. Balachandran, Astron. Astrophys. **359**, 563–572 (2000)
D.S.P. Dearborn, J.C. Lattanzio, P.P. Eggleton, Astrophys. J. **639**, 405–415 (2006)
P.A. Denissenkov, A. Weiss, Astron. Astrophys. **308**, 773–784 (1996)
P.P. Eggleton, D.S.P. Dearborn, J.C. Lattanzio, Sci. Express (2006). 10.1126/science.1133065
D. Galli, F. Palla, F. Ferrini, U. Penco, Astrophys. J. **443**, 536–550 (1995)
D. Galli, L. Stanghellini, M. Tosi, F. Palla, Astrophys. J. **456**, 478–498 (1997)
J. Geiss, in *Origin and Evolution of Elements*, ed. by N. Prantzos, E. Vangioni-Flam, M. Casse (Cambridge Univ. Press, Cambridge, 1993), pp. 89–106
G. Gloecker, J. Geiss, Nature **381**, 210–212 (1996)
C.J. Hogan, Astrophys. J. **441**, L17–L20 (1995)
A. Palacios, C. Charbonnel, M. Forestini, Astron. Astrophys. **375**, L9–L13 (2001)
F. Palla, D. Galli, A. Marconi, L. Stanghellini, M. Tosi, Astrophys. J. **568**, L57–L60 (2002)
R.T. Rood, T.M. Bania, T.L. Wilson, Astrophys. J. **280**, 629–647 (1984)
R.T. Rood, T.M. Bania, T.L. Wilson, D.S. Balser, in *ESO/EIPC Workshop on the Light Elements*, ed. by P. Crane (Springer, Heidelberg, 1995), pp. 201–214
D.A. Roshi, D.S. Balser, T.M. Bania, W.M. Goss, C.G. De Pree, Astrophys. J. **625**, 181–193 (2005)
I.-J. Sackmann, A.I. Boothroyd, Astrophys. J. **510**, 217–231 (1999)
G.J. Wasserburg, A.I. Boothroyd, I.-J. Sackmann, Astrophys. J. **447**, L37–L40 (1995)
A. Weiss, J. Wagenhuber, P.A. Denissenkov, Astron. Astrophys. **313**, 581–590 (1996)
M. Tosi, in *The Light Elements and Their Evolution, Proceedings of IAU Symposium 198*, ed. by L. da Silva, M. Spite, J.R. de Medeiros (ASP, San Francisco, 2000), pp. 525–532
J.P. Zahn, Astron. Astrophys. **265**, 115–132 (1992)

Space Sci Rev (2007) 130: 63–72
DOI 10.1007/s11214-007-9151-0

# Dark Matter Searches

**K. Pretzl**

Received: 26 December 2006 / Accepted: 1 February 2007 /
Published online: 23 March 2007


**Abstract** According to our present knowledge the matter/energy budget of the universe consists of 74% dark energy, 22% dark matter and 4% ordinary (or so-called baryonic) matter. While the dark energy cannot be detected directly, searches for dark matter are performed with earth-bound and space-borne detection devices, assuming that the dark matter consists of weakly interacting massive particles, the so-called WIMPs. An overview of the present experimental situation is given.

**Keywords** Dark matter

## 1 Introduction

Since the discovery of dark matter in the Coma cluster by the Swiss astronomer Fritz Zwicky (1933) 73 years have passed and we still do not know what the real nature of it is. Dark matter shows its presence by gravitational interaction with ordinary matter. It holds numerous galaxies together in large clusters and it keeps stars rotating with practically constant velocities around the centers of spiral galaxies. With gravitational lensing, a technique which was originally proposed by Fritz Zwicky (1937) to determine the mass of galaxies and galaxy clusters, we are able to obtain today a very good picture of how the dark matter is distributed in the universe. We believe that the dark matter played an essential role during the formation of the galaxies and galaxy clusters. The dark matter provided the gravitational wells into which ordinary matter was drawn during the early evolution of the universe.

Recently much information about the matter/energy content of the universe was gained from several independent observations like: The Cosmic Microwave Background radiation (CMB), the Large Scale Structure surveys (LSS), the cluster searches and the Super Novae type 1a surveys (SN) (Spergel 2006). Combining the results of these observations the matter/energy budget of the universe turns out to be: $74 \pm 2\%$ dark energy, $26 \pm 2\%$ matter

K. Pretzl (✉)
Laboratory for High Energy Physics, University of Bern, Bern, Switzerland
e-mail: pretzl@lhep.unibe.ch

(including baryonic and dark matter) and $4.4 \pm 0.3\%$ baryonic matter. It is quite remarkable that similar baryonic matter densities could be derived from the light element abundances and the nuclear synthesis, to which Johannes Geiss and collaborators contributed very significantly. It was one of the main topics at this symposium (see contributions of J. Geiss, N. Prantzos, Th. Bania and H. Reeves).

The fact that 96% of the universe is of unknown nature provides enough motivation and challenge for scientists of different fields like astronomy, astrophysics, cosmology and particle physics to search for dark matter and dark energy in earth-bound and space-borne experiments. This article will concentrate on dark matter searches and will not report on developments to disentangle the mystery of the dark energy, which seems to be related to Einstein's cosmological constant. However, there are not many hints about the true nature of the dark matter other than it interacts gravitationally with ordinary matter and it must have been produced in an early phase of the universe. It may, however, also interact via very weak and so far unknown forces with matter. Still, particle physicists proposed some possible candidates:

1. Massive neutrinos. They would be natural candidates since they are among the most abundant particles in the universe. Neutrinos with a mass of a few eV $c^{-2}$ would significantly contribute to the missing mass. Since neutrinos are relativistic when they decouple from matter, they qualify as hot dark matter (HDM) candidates.

2. Weakly interacting massive particles, the so-called WIMPs. The most favored candidate is the neutralino, which is a particle predicted by Super Symmetry (SUSY). SUSY is the best studied extension of the Standard Model (SM) of particle physics. If the neutralino existed two problems could find a solution at the same time, namely the dark matter and the physics beyond the SM. Neutralinos would be candidates for cold dark matter (CDM) since they would be heavy and nonrelativistic when they decouple from matter in the early universe.

3. Axions. These particles, named after a laundry detergent, are associated with a pseudoscalar field, which was introduced by Peccei and Quinn (1977) to solve the charge conjugation and parity (CP) violation problem in Quantum Chromodynamics (QCD), the theory of strong interactions. Axions would be produced abundantly during the QCD phase transition in the early universe when hadrons were formed from quarks and gluons. Since axions are nonrelativistic at freeze out they would qualify as CDM candidates.

The direct detection of dark matter, if it exists in form of particles, is encouraged by the large expected particle flux which can be deduced under the following assumptions. In an isothermal dark matter halo model the velocity of particles in our galaxy is given by a Maxwell Boltzmann distribution with an average value of $\langle v \rangle = 230$ km s$^{-1}$ and an upper cutoff value of 575 km s$^{-1}$ corresponding to the escape velocity. The dark matter halo density in our solar neighborhood is estimated to be $\rho = 0.3$ GeV c$^{-2}$ cm$^{-3}$. From that one expects a flux of $\Phi = \rho\langle v \rangle / m_\chi \sim 7 \times 10^6 / m_\chi$ cm$^{-2}$ s$^{-1}$ with $m_\chi$ the mass of the dark matter particle in GeV c$^{-2}$. However, since neither the mass nor the interaction cross section of these particles are known one is forced to explore a very large parameter space, which requires very sensitive and efficient detection systems. In the following article some of the more recent dark matter searches are described.

## 2 Neutrinos, Do They Qualify for Dark Matter Candidates?

In most Big Bang models it is assumed that the relic abundance of neutrinos is comparable (9/11) to that of photons. Neutrinos come in 3 flavors: the electron neutrino $\nu_e$, the muon neutrino $\nu_\mu$ and the tau neutrino $\nu_\tau$. They contribute with $\Omega_\nu \cdot h^2 = \sum m_\nu / 94\ \mathrm{eV\,c^{-2}}$ to the matter density of the universe, where $\Omega_\nu$ is the neutrino density normalised to the critical density of the universe, h the Hubble parameter and $\sum m_\nu$ is the sum over the three neutrino masses in $\mathrm{eV\,c^{-2}}$. This illustrates that neutrinos with a mass of a few $\mathrm{eV\,c^{-2}}$ could make up most of the missing dark mass. We know from neutrino oscillation experiments that neutrinos must have a rest mass, but the absolute mass values are still missing. Direct neutrino mass measurements have so far resulted only in upper limits: $\leq 2.2\ \mathrm{eV\,c^{-2}}$, $\leq 160\ \mathrm{keV\,c^{-2}}$, $\leq 18\ \mathrm{MeV\,c^{-2}}$ for the electron, the muon and the tau neutrino respectively. From neutrino oscillation experiments with atmospheric, solar, reactor and accelerator neutrinos (Super-Kamiokande, SNO, KamLAND, K2K and MINOS) (Maltoni et al. 2004) only mass differences can be extracted. The combined results of these experiments yield mass differences for $\nu_e \rightarrow \nu_\mu$ oscillations of $\Delta m^2 = (8.1 \pm 0.3) \times 10^{-5}\ \mathrm{eV^2\,c^{-4}}$ and assuming $\nu_\mu \rightarrow \nu_\tau$ oscillations $\Delta m^2 = (2.3 \pm 0.6) \times 10^{-3}\ \mathrm{eV^2\,c^{-4}}$ leaving open the question of the mass hierarchy. Assuming a mass hierarchy for neutrinos similar to that of charged leptons which is inspired by Grand Unified Theories (GUT) the heaviest neutrino would have a mass of the order of $\leq 0.1\mathrm{eV\,c^{-2}}$. In this case neutrinos would very insignificantly contribute to the missing mass in the universe.

An upper limit for the sum of the neutrino masses can be obtained, when combining the results of CMB, LSS and SN experiments. Neutrinos being relativistic at freeze out are free streaming particles, which cluster preferentially at very large scales. Therefore massive neutrinos would enhance large-scale and suppress small-scale structure formations. From HDM and CDM model calculations fitting the LSS power spectrum one obtains a value for the ratio neutrino density to matter density $\Omega_\nu / \Omega_m$. From this value and $\Omega_m$ obtained from CMB and SN an upper limit for the sum of the neutrino masses $\sum m_\nu \leq 0.68\ \mathrm{eV\,c^{-2}}$ can be derived (Spergel 2006). It is interesting to note that from cosmology we have the best upper limit for the neutrino masses obtained so far. It strongly indicates, however, that neutrinos do not qualify for the dark matter. Model calculations using CDM and a cosmological constant $\Lambda$ for the dark energy ($\Lambda$CDM) seem to be in overall good agreement with the data.

## 3 Weakly Interacting Massive Particles, So-Called WIMPs

SUSY predicts the existence of new particles, of which the lightest and most stable can be a candidate for CDM. However, SUSY was not invented in the first place to solve the dark matter problem. Its main aim is to unify all forces in the universe in a GUT theory. SUSY predicts bosonic partners to the well known leptons and quarks, which are fermions, and fermionic partners to the known gauge bosons, which are transmitting the forces. If SUSY exists in nature, new particles and new forces would have to be discovered. The SUSY particle which is mostly favored as a candidate for the dark matter is the neutralino, which is neutrally charged as its name already indicates. In the Minimal Super Symmetric Model (MSSM) the neutralino is a linear superposition of four neutral spin $1/2$ gauginos:

$$\chi = n_{11}\tilde{W}^3 + n_{12}\tilde{B} + n_{13}\tilde{H}_1 + n_{14}\tilde{H}_2$$

with $\tilde{W}^3$ and $\tilde{B}$, the Wino and Bino, and $\tilde{H}_1$ and $\tilde{H}_2$, the two Higgsinos. Because of its complex structure the parameter space is large and makes mass and interaction cross section predictions difficult. However, the parameter space was constrained recently by the results from the electron positron collider (LEP) at CERN and from the measurements of the anomalous magnetic moment of the muon at Brookhaven. Within the Constrained Minimal

SUSY Model (CMSSM) the best estimate of the neutralino mass lies between 50 GeV $c^{-2}$ and 600 GeV $c^{-2}$ (Ellis et al. 2000). This mass range is well accessible to the Large Hadron Collider (LHC) at CERN, which is expected to be operational in 2007. The search for SUSY particles will be amongst the most prominent subjects for the LHC. In case neutralinos are first found at the LHC, they still need to be confirmed as the dark matter in the universe by their direct detection in WIMP experiments.

## 4 Direct Detection of Dark Matter

The direct detection of WIMPs is based on the measurement of nuclear recoils in elastic WIMP scattering processes. In the case of neutralinos, spin-independent coherent scatterings as well as spin-dependent scatterings are possible. The expressions for the corresponding cross sections can be found in Jungman (1996) and Pretzl (2002). In order to obtain good detection efficiencies, devices with high sensitivity to low nuclear recoil energies (eV) are needed. WIMP detectors can be categorized in conventional and cryogenic devices. Most of the conventional WIMP detectors use NaI, Ge crystals, liquid Xenon (LXe) or liquid Argon (LAr). Conventional devices have the advantage that large detector masses ($\sim$ton) can be employed, which makes them sensitive to annual modulations of the WIMP signal owing to the movement of the earth with respect to the dark halo rest frame. Annual modulation, if observed, would provide strong evidence for a WIMP signal, assuming it is not faked by spurious modulated background signals. However, due to quenching of the ionization signals, conventional detectors have lower nuclear recoil detection efficiencies than cryogenic devices.

Cryogenic detectors are able to measure small recoil energies with high efficiency because they measure the total deposited energy in form of ionization and heat (phonons). A small energy loss $\Delta$E can lead to an appreciable temperature increase $\Delta T = \Delta E/C$, provided the detector is operated at low enough temperatures (typically mK), where the heat capacity $C$ is small. A description of the various detection principles can be found in Pretzl (2000). Cryogenic detectors are made of many different materials, like Ge, Si, $TeO_2$, sapphire ($Al_2O_3$), LiF, $CaWO_4$ and BGO, including superconductors, like Sn, Zn, Al, etc. This turns out to be an advantage for the WIMP search, since the resulting recoil spectra are characteristically different for detectors with different materials, a feature which helps to effectively discriminate a WIMP signal against background. In comparison to conventional detectors, however, cryogenic detectors are rather limited in target mass ($\sim$kg). The most frequently used cryogenic devices are bolometers, which consist of an absorber and a sensitive thermometer attached to it. Another technique uses superheated superconducting granules (SSG). An SSG WIMP detector (ORPHEUS) consists of billions of spherical Sn granules with diameters of about 30 $\mu$m. The detector is operated in a magnetic field. WIMPs interacting in a granule can cause a phase transition from the superconducting to the normal state. This phase transition of individual granules can be detected by pickup loops which measure the flux change due to the disappearance of the Meissner–Ochsenfeld effect. The energy threshold of the detector is adjustable by setting the external magnetic field just below the phase transition boundary.

WIMP detectors are located in underground laboratories so as to be protected from cosmic ray background. In addition to a passive shielding against radioactivity from surrounding walls they need to be built from radio poor materials. This shielding alone provides only limited effectiveness and is expensive. Nevertheless, cryogenic bolometers, as used

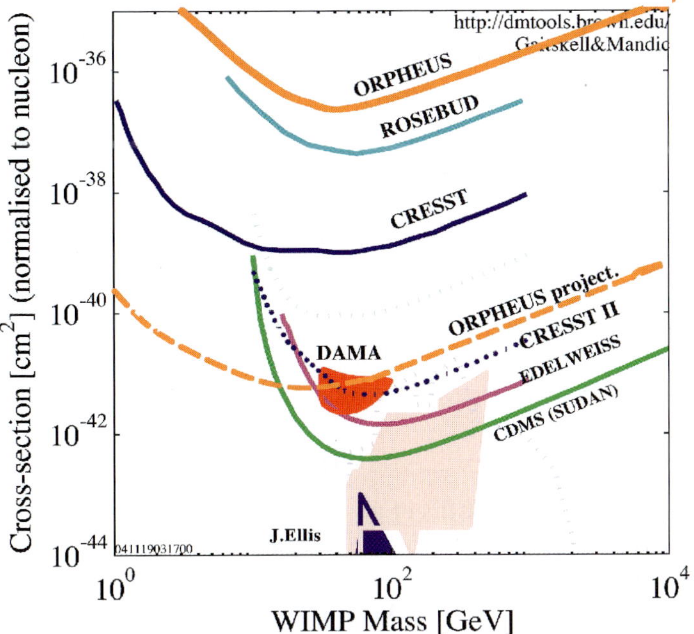


**Fig. 1** Exclusion plot for spin-independent WIMP interactions

by CDMS, EDELWEISS and CRESSTII, are able to distinguish nuclear recoils from minimum ionizing particles (Compton electrons) by measuring heat and ionization or photons, in the case of scintillating absorbers, for each event simultaneously, but separately. For the same deposited recoil energy, the ionization (or photon) signal from nuclear recoils is highly quenched compared to signals from electrons, a feature which allows to separate genuine nuclear recoils from electron background. This method turns out to be a very effective active background rejection. An active background rejection was also practiced with the ORPHEUS SSG detector, since minimum ionizing particles cause many granules to flip, while WIMPs cause only one granule to flip (flip meaning a transition from the superconducting to the normal state). More conventional detectors (NaI, LXe, LAr) rely on a signal shape analysis to reduce the background, which turns out to be less powerful.

The experimental results are usually presented as exclusion plots, which show the WIMP-nucleon cross section versus the WIMP mass. The exclusion plots from different experiments are shown in Fig. 1 for spin independent WIMP interactions (90% C.L.) and in Fig. 2 for spin dependent interactions.

So far the lowest cross section limits on spin independent interactions were obtained with cryogenic detectors like: CDMS (absorber: 1 kg Ge, 0.2 kg Si) at Soudan (2090 m.w.e.) (Akerib 2005), EDELWEISS (1 kg Ge) at Frejus (4800 m.w.e.) (Sanglard 2005) and CRESSTII (0.6 kg $CaWO_4$) at Gran Sasso (3800 m.w.e.) (Angloher 2005). The results of these experiments are not in agreement with the DAMA (100 kg NaI) experiment, which claimed to see an annual modulation signal consistent with a WIMP with a mass $m_x = (52^{+10}_{-8})$ GeV $c^{-2}$ (Bernabei 2000). Also shown in are the results of the first phase of the ORPHEUS (SSG, 0.2 kg Sn) experiment at the Bern underground laboratory (70 m.w.e.) (Borer 2004). With the same absorber mass, but some improvements on the detector and the shielding, limits as shown by ORPHEUS projected in Fig.1 can be reached in future. The main advantage

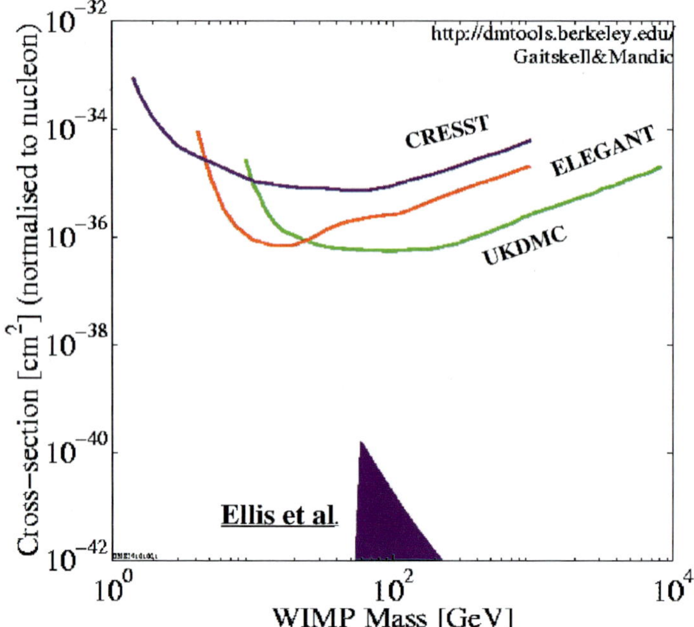


**Fig. 2** Exclusion plot for spin-dependent WIMP interactions

of SSG compared to bolometers is that they are potentially more sensitive to low WIMP masses due to the intrinsically lower recoil energy thresholds they can apply.

Exclusion plots for spin-dependent WIMP interactions from the experiments CRESST (0.262 kg $Al_2O_3$) (Angloher 2002), ELEGANT (660 kg NaI) at Kamioka (2700 m.w.e.) (Yoshida 2000) and UKDMC (6 kg NaI) at Boulby Mine (2530 m.w.e.) (Bravin 1999) are shown in Fig. 2. CDMS (Akerib 2006) has recently published upper limits for spin-dependent WIMP interaction cross sections normalized to protons and neutrons separately, which turn out to be slightly lower than the one shown in Fig. 2. The CMSSM model predictions of Ellis et al. (2000), also shown in Fig. 1 and Fig. 2, are still several orders of magnitude below the presently reached experimental sensitivities. However, a new generation of cryogenic WIMP detectors planning to employ absorber masses of 10–100 kg and conventional detectors, like GENIUS (Ge), with a mass of 1 ton are in preparation in order to gain several orders of magnitude in WIMP detection sensitivity in the near future.

## 5 Indirect Detection of Dark Matter

There is another strategy to look for WIMPs in the galactic halo. This strategy makes use of the fact that neutralinos annihilate mainly into quark anti-quark pairs in the final state, which then fragment into hadrons. The fragmentation products contain stable particles like protons, anti-protons, electrons, positrons, neutrinos, anti-neutrinos and photons. The challenge is to distinguish those particles coming from neutralino annihilations from background particles, which are generated by cosmic ray interactions in the interstellar gas or in the atmosphere. Neutral particles, like photons and neutrinos, have the advantage of carrying directional information in contrast to charged particles which are deflected by the magnetic fields of the galaxy. Since the neutralino velocities in the halo are of the order of $10^{-3}$ of the velocity of light, the annihilation can be considered at rest, which means that the energy equivalent to twice the neutralino mass will be distributed amongst the quark anti-quark pair in the final state.

In order to observe an appreciable flux of neutrinos and antineutrinos from neutralino annihilations one has to look at sources where neutralinos are accumulated in rather large concentrations. Such sources can be for example the sun, the earth and the center of our galaxy. They act as gravitational wells into which neutralinos are drawn and where they have accumulated since the time the source was created. The trapping rate of neutralinos within a source, i.e. the sun, depends on their interaction probability with the solar matter and their annihilation probability. Several neutrino telescopes like Kamiokande and Super-Kamiokande in Japan, AMANDA at the south pole and ANTARES in the Mediterranean sea looked for neutrinos coming from neutralino annihilations, but the limits they obtained so far are not very strong (Hooper and Silk 2004). A more sensitive experiment ICE CUBE is planned for the future.

Several balloon-borne experiments have been performed to measure anti-protons and positrons from neutralino annihilations. Some experiments have seen an excess of positrons and anti-protons over the expected rate from known sources. However, other experiments did not confirm these findings. In addition, the yield of low energy antiprotons (below 100 MeV) in cosmic ray interactions in the interstellar gas and in the atmosphere is not well known. This leaves the situation inconclusive for the moment. A discussion and a review of this matter can be found in Bergstrom (2000). However an effort is being made to collect more data on anti-protons and positrons in the near future by two space-borne experiments PAMELA, which has been launched in June this year, and AMS (Anti-Matter Spectrometer) which is still in preparation.

An excess of gamma rays above background (at energies above 1 GeV) has been observed by the EGRET telescope, which is one of the instruments on the Compton Gamma Ray Observatory. This excess, which was observed in every direction of the galaxy, was studied and interpreted by de Boer (2005) as gamma rays coming possibly from WIMP annihilations in the galactic halo. Assuming a WIMP annihilation at rest the data can be fitted with a WIMP mass of 60 GeV $c^{-2}$. The authors also claim to be able to reconstruct from these data the peculiar shape of the rotational velocity curve of our galaxy.

The observation of a bright 511 keV gamma ray line by INTEGRAL has been interpreted by Boehm (2004) as a result of low mass (1–100 MeV $c^{-2}$) WIMP annihilations into electron positron pairs in the galactic bulge. However, the WIMP interpretations of the EGRET and INTEGRAL data will remain subject to a great deal of debates until they can be substantiated, which seems to be inherently difficult with indirect detection methods.

## 6 Axion

The introduction of an axion was not motivated by cosmological considerations but by the fact that the general expression for the QCD Lagrangian contains a term that violates CP. Since the neutron electric dipole moment has so far escaped detection, the coefficient of this term must be tiny. This is referred to as the "strong CP puzzle". Peccei and Quinn (1977) showed that the puzzle can be solved by introducing a neutral pseudoscalar field which neutralizes the CP violation. This field is associated with a particle, the so-called axion. Axions would be abundantly produced in the early universe by a nonthermal mechanism and would qualify for CDM candidates. They are weakly interacting particles which couple to two photons with a coupling strength proportional to their mass $m_a$. Although the natural two-photon decay rate would be much too slow ($10^{40}$ years) to be detectable, by supplying one of the photons from a coherent source, such as a magnetic field, a significant conversion rate can be achieved. Since the virtual photon from the magnetic field has negligible energy the outgoing real photon carries an energy equal to the sum of the axion rest mass and its kinetic energy. The axion photon conversion of nonrelativistic axions with a mass around $m_a \approx 10^{-5}$ eV c$^{-2}$ produces photons in the GHz range. Sikivie (1983) proposed to detect galactic axions by their resonant conversion into photons using a microwave cavity in a strong magnetic field. The power induced by axions in the cavity is very tiny $P \sim V \cdot B^2 \cdot Q \cdot m_a \cdot \rho_a \sim 10^{-21}$ W, with $V$ the volume of the cavity, $B$ the magnetic field and $\rho_a$ the axion density. However, since the axion mass is unknown one has to scan over a large range of cavity frequencies with a long integration time at each frequency, which sets practical limits. Fortunately there are constraints which narrow the expected mass range of the axions.

The main constraints on the axion mass come from the stellar evolution and the duration of the supernova SN1987 neutrino signal. Axions can be produced in large quantities in the interior of stars via the Primakoff effect: $\gamma + Ze \rightarrow a + Ze$. In this process a photon interacts with the electric field of a nucleus to produce an axion. Since axions interact very weakly with matter they carry away large amounts of energy. This exotic loss of energy in stars would lead to observable modifications of the standard stellar evolution and is used to constrain the axion mass. Also the duration of the supernova SN1987 neutrino signal provides a limit on the axion mass. After its collapse the supernova core is so hot and dense that neutrinos will be trapped and escape only by diffusion. So it will take several seconds to cool an object of several solar masses. However, the emission of axions would quickly take energy away from the core due to their very weak interactions with matter and would influence the time delay of the observed neutrino signals. Taking all this into consideration we would expect axions to exist in the mass range from $10^{-6} \leq m_a \leq 10^{-3}$ eV c$^{-2}$ .

Pioneering experiments were performed in the 1980s by the University of Rochester, Brookhaven National Laboratory (BNL) and Fermilab (RBF) collaboration and the University of Florida (UF). As shown in Fig. 3, taken from van Bibber and Rosenberg (2006), they reached sensitivities which are still about two orders of magnitude below the KSVZ (Kim, Shifman, Vainshtein, Zakharov) and DFSZ (Dine, Fischler, Srednicki, Zhitnitsky) model predictions. The main difference between the two models is that in the latter axions couple to charged leptons in addition to nucleons and photons. The new generation galactic halo axion experiments employ larger cavity and magnetic field volumes as well as more sophisticated photon detection systems. The Axion Dark Matter Experiment (ADMX) at the Lawrence Livermore National Laboratory (LLNL) is using ultra low noise microwave technology as well as ultra low noise amplifiers. The microwave cavity is inside a 8 Tesla and 4 m long super-conducting magnet. The experiment of the Kyoto University is pursuing a different technique, namely detecting the photon as a particle rather than a wave. For this they developed a single quantum microwave detection system, which is based on a Rydberg excited atomic beam traversing the cavity. This technique allows a practically noise-free detection of photons (Tada 2006). As shown in Fig. 3 the ADMX experiment has already reached the sensitivity of the model predictions, but within a limited axion mass range. Since the resonance frequencies in the cavities increase with increasing axion mass this method becomes very difficult when searching for axions with masses above $10^{-4}$ eV c$^{-2}$. For higher mass axion searches the resonant cavity approach is replaced by experiments using X-ray detection systems. These offer an advantage over the time consuming frequency scanning, since they cover a large mass range at once, but they are much more limited by background.

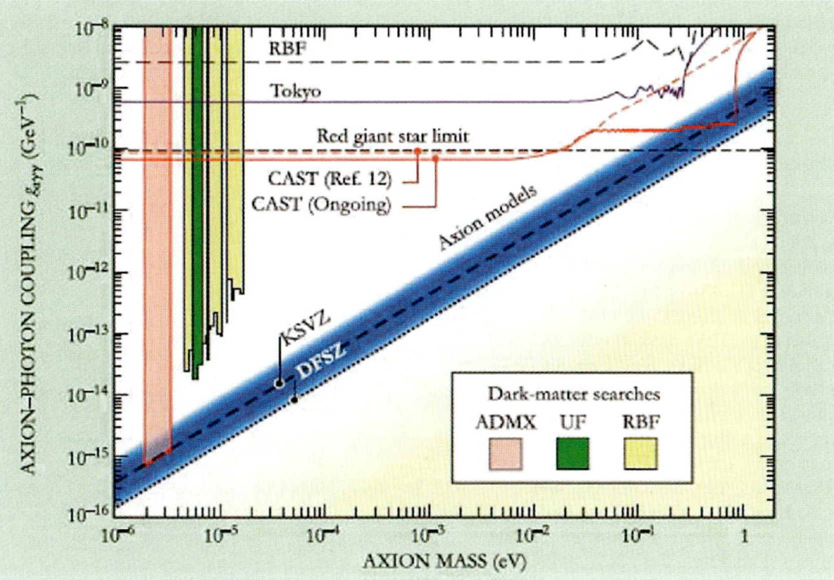


**Fig. 3** Exclusion plot of axion–photon coupling versus axion mass

Since axions can be produced in stars, the CAST (CERN Axion Solar Telescope) collaboration has decided to look for axions coming from the sun. Their telescope consists of a 9 m long, 9 Tesla superconducting dipole magnet (a prototype LHC magnet at CERN) and an X-ray detection system. The mean energy of solar axions is expected to be 4.2 keV. The results of phase one of their experiment are shown in Fig. 3 (Zioutas 2005). In the ongoing phase two they have filled the magnet bore with gas ($^3$He and $^4$He) with variable pressure in order to increase the mass range of the axions and to reach the sensitivity of the model predictions in the range of 0.1 to 1 eV $c^{-2}$.

Although no axions have so far been detected by these experiments an intriguing signal has been observed by the PVLAS (Polarization of the Vacuum with Laser) collaboration in Legnaro, Italy. They measured the rotation of the polarization of a laser beam when passing through a magnetic field. Their laser beam passed 44 000 times a 1 m long 5 Tesla magnet. They observed a rotation of $3.9 \times 10^{-12}$ rad per pass (Zavattini 2006). Provided all possible spurious effects can be excluded, the observed signal could find an explanation in the production of axions. However, the inferred mass and coupling of the axion from this observation contradicts the limits reached already by the other above described axion experiments. Further investigations are planned with experiments which are known as "shining the light through a wall" (van Bibber 1987).

## 7 Conclusions

There is compelling evidence that our universe consists of 96% dark matter/energy of unknown nature. However, with the turning-on of the Large Hadron Collider at CERN in 2007 and the new generation of direct and indirect detection experiments there is hope to bring soon light into the darkness. I would like to conclude with a quotation from Shakespeare: "There is no darkness, only ignorance".

**Acknowledgements** I would like to thank the organizers for inviting me to this symposium in honor of Johannes Geiss. I am grateful to Johannes Geiss for many discussions about dark matter and to Rainer Kotthaus for informing me about the latest developments of axion searches and of the CAST experiment.

## References

D.S. Akerib et al., Phys. Rev. D **72**, 052009 (2005)
D.S. Akerib et al., Phys. Rev. D **73**, 011102 (2006)
G. Angloher et al., Astropart. Phys. **18**, 43 (2002)
G. Angloher et al., Astropart. Phys. **23**, 325 (2005)
L. Bergstrom et al., Rep. Prog. Phys. **63**, 793 (2000)
R. Bernabei et al., Phys. Lett. B **480**, 23 (2000)
K. van Bibber et al., Phys. Rev. Lett. **59**, 759 (1987)
K. van Bibber, L.J. Rosenberg, Physics Today, Aug. 2006, pp. 30–35
C. Boehm et al., Phys. Rev. Lett. **92**, 101301 (2004)
K. Borer et al., Astropart. Phys. **22**, 199 (2004)
M. Bravin et al., Astropart. Phys. **12**, 107 (1999)
W. de Boer et al., Astron. Astrophys. **444**, 51 (2005). See also CERN Courrier Dec. 2005, pp. 17–19
J. Ellis, A. Ferstl, K.A. Olive, Phys. Lett. B **481**, 304 (2000)
D. Hooper, J. Silk, New J. Phys. **6**, 23 (2004)
G. Jungman et al., Phys. Rep. **267**, 195 (1996)
M. Maltoni, T. Schwetz, M. Tortola, J.W.F. Valle, New J. Phys. **6**, 122 (2004). See also MINOS hep-ex/0607088
R.D. Peccei, H.R. Quinn, Phys. Rev. Lett. **38**, 1440 (1977)
K. Pretzl, Nucl. Instr. Methods A **454**, 114 (2000)
K. Pretzl, Space Sci. Rev. **100**, 209 (2002)
V. Sanglard et al., Phys. Rev. D **71**, 122002 (2005)
P. Sikivie et al., Phys. Rev. Lett. **51**, 1415 (1983). See also Physics Today, Dec. 1996, pp. 22–27
D.N. Spergel et al., 2006. astro-ph/0603449
M. Tada et al., Phys. Lett. A **349**, 488 (2006)
S. Yoshida et al., Nucl. Phys. B (Proc. Suppl.) **87**, 58 (2000)
E. Zavattini et al., Phys. Rev. Lett. **96**, 110406 (2006)
K. Zioutas et al., Phys. Lett. **94**, 121301 (2005)
F. Zwicky, Helvetica Physica Acta **6**, 110 (1933)
F. Zwicky, Phys. Rev. **51**, 67 (1937)

Space Sci Rev (2007) 130: 73–78
DOI 10.1007/s11214-007-9210-6

# Comets and Chemical Composition

**S.N. Delanoye · J. De Keyser**

Received: 13 December 2006 / Accepted: 15 May 2007 / Published online: 15 August 2007


**Abstract** It is commonly believed that comets are made of primordial material. As a consequence, they can reveal more information about the origin of our solar system. To interpret the coma composition measurements of comet Churyumov–Gerasimenko that will be collected by the Rosetta mission, models of the coma chemistry have to be constructed. However, programming the chemistry of a cometary coma is extremely complex due to the large number of species and reactions involved. Moreover, such a program needs to be very flexible as one may want to extend, change, or update the set of species, reactions, and reaction rates. Therefore, we developed software to manage a database of species and reactions and to generate code automatically to compute source/loss balances. This database includes the data from the UMIST database and the ion–molecule reactions collected by V.G. Anicich. To use all these databases together, a lot of practical problems need to be solved, but the result is an enormous source of information about chemical reactions that can be used in chemical models, not only for comets but also for other applications.

**Keywords** Comets: chemical modeling · Comets: database

## 1 Introduction

People have always been fascinated by comets. Their bright appearance was believed to predict disasters and prosperity. Since science began to reveal some of their mysteries, the interest in comets continues to increase. Consisting of primordial material, including volatile components, comets can shed light on the history of our solar system. It is therefore not surprising that several missions to comets have been and will be carried out. One of these missions, Rosetta, intends to learn more about comets by studying comet 67P/Churyumov–Gerasimenko. Rosetta will reach the comet as it approaches the Sun and will orbit around the nucleus on the way to perihelion. This should provide the scientific world with an enormous amount of information.

S.N. Delanoye (✉) · J. De Keyser
Belgian Institute for Space Aeronomy, Ringlaan 3, 1180 Brussels, Belgium
e-mail: sofie.delanoye@aeronomie.be

A comet is a very complex system. The nucleus is made of dust and different ices. Water ice is most abundant, but also other volatile components are present such as CO, $CO_2$, $NH_3$, etc., and finally the nucleus also contains organic compounds. As the nucleus is heated by the Sun, the ices evaporate and the volatiles leave the nucleus with typical speeds between 0.5 and 1.0 $\mathrm{km\,s^{-1}}$. Some compounds, such as CO and $H_2CO$, are not only produced by the nucleus, but they also have an extended source in the coma. The composition of the evaporating gas depends on the distance from the Sun. Close to the nucleus, the gas is not in thermodynamic equilibrium. This boundary layer, called the Knudsen layer, is a few collisional mean-free paths thick. Outside the Knudsen layer, the outflow is essentially radial. Once the volatiles have left the nucleus, they undergo all kinds of reactions. Photodissociation and photoionization by solar ultraviolet radiation create new neutrals and ions and are the most important reactions together with proton exchange reactions with $H_3O^+$. But also charge-exchange reactions, ion–neutral reactions, and neutral–neutral reactions are possible. Dissociative recombination with thermal electrons is the major destruction path for ions (Häberli et al. 1995). Inside the contact surface, the border of the diamagnetic cavity close to the nucleus in which the solar wind cannot penetrate, the ions move with the same velocity as the neutral gas to which they are coupled due to ion–molecule collisions. Outside the contact surface, the cometary ions and the ions produced from cometary neutrals by photoionization or other reactions are picked up by the interplanetary magnetic field and are carried along with the solar wind. As a result, the dynamics of the coma become more complex. Also, the complexity of the chemistry in the coma increases. Solar wind ions participate in the different reactions, such as charge exchange, and electron impact ionization by suprathermal solar wind electrons forms an additional ionization source. The result of these interactions is an enormous mixture of neutral and ionized species for which the composition depends on the temperature of the different species, the flux of the photons, the electron profile, etc. Detailed models of the chemistry in a cometary coma are therefore a prerequisite for interpreting coma composition observations and to obtain information on the variability of the cometary composition as a function of the distance to the Sun and to the nucleus. These models can also be used to derive the composition of volatiles on the surface of the nucleus from the coma composition.

However, programming the chemistry of a cometary coma is extremely complex due to the large number of species and reactions involved. An example is shown in the following. Moreover, such a program needs to be very flexible as one may want to extend, change, or update the set of species, reactions, and reaction rates. Therefore, we have developed software to manage a database of species and reactions and to generate code automatically to compute source/loss balances. After solving some practical problems, the data from the UMIST database (Le Teuff et al. 2000) and the ion–molecule reactions collected by V.G. Anicich (Anicich 2003) have been included in the database. The result is an enormous source of information about chemical reactions that can be used in chemical models. Moreover, the application of the database is not limited to comets.

## 2 Reaction Databases and Chemical Modeling

### 2.1 Database

The software consists of two databases: a particle database and a reaction database. The particle database contains the list of the species that play a role in the chemical reactions: neutral atoms and molecules, ions, radicals, photons, and electrons. For each species the mass, the charge, and a graphical representation is defined.

As described earlier, different types of reactions can take place, such as photoreactions, ion–neutral reactions, electron impact reactions, etc. All the possible reaction types are included in the reaction database. Each reaction type is characterized by a generic reaction equation. For instance, for electron impact dissociation: $X + e^- \rightarrow U + V + e^-$.

The reaction types are used to construct actual reactions by filling in the species in the generic reaction equations. For each reaction, information about the reaction rate is collected. Also, additional information such as a reference to the literature can be provided.

From this list of reactions, the reactions that should be used during modeling can be chosen. The selected reactions are then used to automatically generate code to compute the source/loss balances for each species involved. This code can be produced in C, FORTRAN, or Matlab so that it can be used by any software for simulating the environment of a comet.

To make the available set of reactions as complete as possible, some databases available in the literature have been added. The properties of these databases and the difficulties encountered while integrating these databases are discussed in the following.

### 2.2 Databases Available in the Literature

The UMIST database for astrochemistry (Le Teuff et al. 2000) contains more than 4,000 gas-phase reactions important in comets and other astrochemical environments. The reactions involve 396 species, containing 12 elements. For each reaction the reaction rate and the temperature range in which the reaction rate is valid are given. For some reactions also the temperature dependence of the reaction rate is available. Different reaction types, such as photoreactions, electron impact reactions, etc. are included.

V.G. Anicich collected more than 28,000 bimolecular gas-phase ion–molecule reactions (Anicich 2003). Not all these reactions are different. Some reactions are included several times, studied at different temperatures, or taken from different literature sources. These reactions contain a few thousand species, including a lot of organic species and even amino acids. All elements of the periodic system are involved and for some species isotopic information is also available. For each reaction, the reaction rate and the temperature or temperature range for which the reaction rate is valid is specified.

### 2.3 Challenges

The UMIST database is available on the web (http://www.udfa.net) as an easily readable text file. The reactions are structured in a systematic way. Therefore, the UMIST reactions could be easily included in the general database.

The Anicich database is available only as a Word document. The reactions are not arranged in a table, but written as text in which each line describes a reaction, with the different fields—such as the reaction rate, temperature, etc.—separated by tabs. Moreover, special characters—such as Greek letters—were used and the representations of the species contain subscripts and superscripts. Software has been written to read this document. The program has been made intelligent enough to be able to detect and solve most inconsistencies that are present in the document, so that only a minimal amount of human intervention is needed in the process.

To use different databases together, a unique representation for each species is necessary. This is not trivial, especially in the case of isotopes. The most obvious way to represent isotopes is to add the mass number in front of the element symbol, but this gives rise to a rather cumbersome notation. A major difficulty is the lack of isotope-specific information concerning reaction rates.

Due to the combination of the different databases, some reactions appear several times, taken from different sources or valid in different temperature ranges. All the information about the reaction-rate measurements for such a reaction have been integrated to obtain a smooth piecewise linear spline fit for the reaction rate as a function of temperature. We also compute an estimate of the temperature-dependent uncertainty on the reaction rate.

## 2.4 Example

A simple example is a comet nucleus that contains only water. After evaporation of $H_2O$, the first reactions that take place are photodissociation and photoionization. In this way, $H_2O^+$, H, OH, $H_2$, and O are produced. These new species can then react themselves and produce other species. The result is a network of 11 species: H, $H^+$, $H_2$, $H_2^+$, O, $O^+$, OH, $OH^+$, $H_2O$, $H_2O^+$, and $H_3O^+$. Also, photons, electrons, and suprathermal electrons need to be added to the network. The combination of all these species leads to more than 70 different reactions of 15 different reaction types. This water database can then be used to generate code with source/loss balances for each species in the selected reactions. This code can be coupled to simulation software. As an illustration of the database's use, we developed a simple model for the chemistry inside the contact surface to calculate the abundance of the different species in a spherically symmetric coma as a function of the distance to the nucleus. The conditions in comet Halley at the moment of the Giotto encounter are used as input for the model (Häberli et al. 1995). We assume a radial outflow with a constant expansion velocity of 0.9 $\mathrm{km\,s^{-1}}$. A production rate of $5.5 \times 10^{29}$ particles $\mathrm{s^{-1}}$ for $H_2O$ is included. All species have the same temperature, 200 K. The rates of the photoreactions in the database are given for quiet solar conditions and a distance of 1 AU to the Sun. For the model calculations, the photorates are scaled to the distance of the Giotto encounter, 0.9 AU. Using linear interpolation, a solar activity index of 0.18 is obtained (0 corresponds to minimum and 1 to maximum solar activity). The photorates are also scaled to this solar activity. As we study the chemistry inside the contact surface, no suprathermal electrons are available. We therefore exclude all electron impact reactions from the list of operational reactions used for generating the code. Some resulting density profiles are shown in Fig. 1. The horizontal axis gives the radial distance from the center of the comet; the graph starts at the boundary of the Knudsen layer at $r = 75$ km. The $H_2O$ concentration decreases away from the nucleus because of the spherical expansion and because of photodissociation and photoionization reactions. Such photodissociation, for instance, produces equal amounts of H and OH, whose concentration is seen to rise quickly, reaching a maximum around $r =$ 150 km, while it decreases further out. Ions produced by the chemical network occur in much lower concentrations in the inner coma. The results are in good agreement with the Giotto data and the model results obtained by Häberli et al. (1995). The simulation program also allows us to focus on the chemistry at a certain distance. Figure 2 shows an example for a distance of 2,500 km from the nucleus. Looking at the balance for each species, it is clear that $H_2O$ is mainly consumed by photodissociation, producing H and OH, and to a lesser extent $H_2$ and O. Figure 3 displays the production and destruction reactions for $H_2O^+$ at a distance of 2,500 km, illustrating that photoionization of water is the major source, while proton transfer and electron recombination are the major destruction mechanisms.

When a more realistic comet composition is assumed, the problem becomes increasingly difficult. The model developed by Häberli et al. (1995) illustrates this. They took into account 7 source species, namely $H_2O$, CO, $H_2CO$, $CO_2$, $NH_3$, $CH_3OH$, and $CH_2$. This results in a much larger chemical network containing about 40 species. Including only the most important chemical processes leads to a set of more than 150 reactions. It is clear

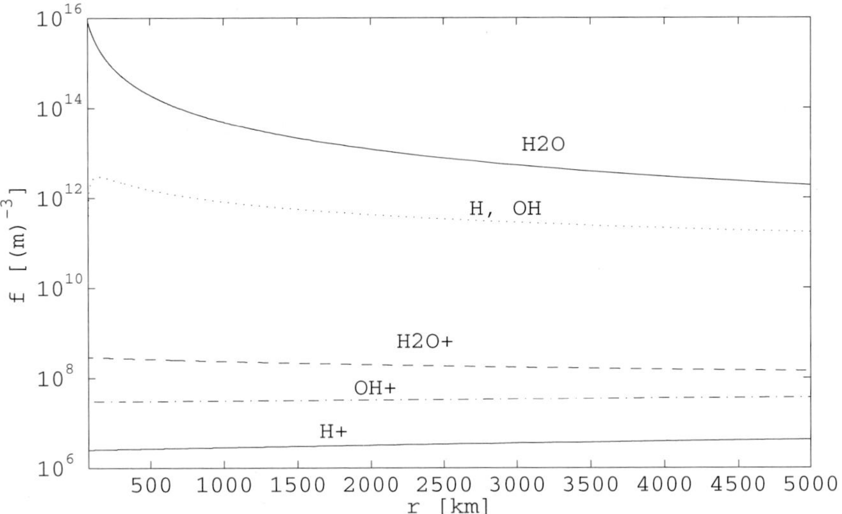


**Fig. 1** Density profiles for $H_2O$, $H_2O^+$, OH, $OH^+$, H, and $H^+$ as a function of distance from the nucleus

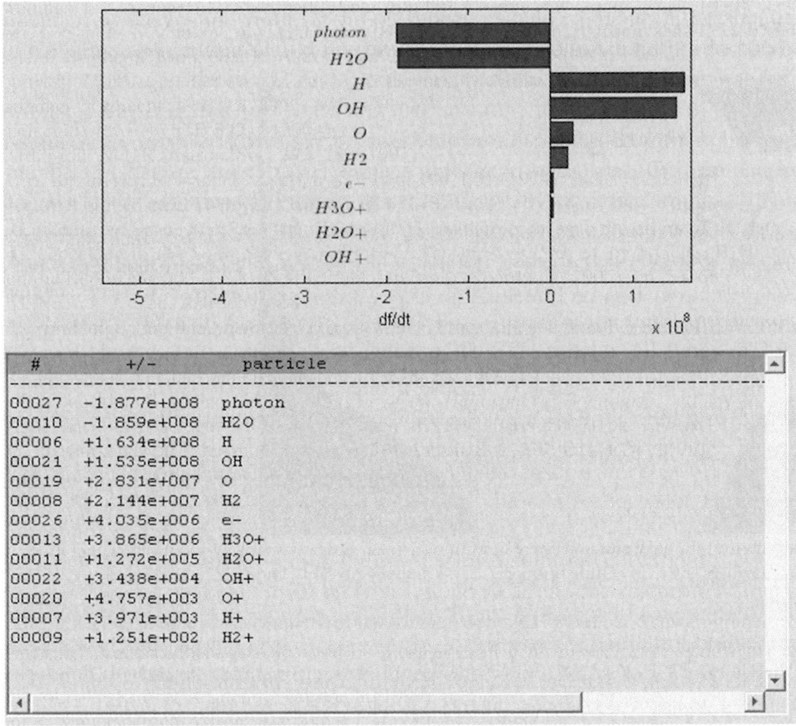


| # | +/- | particle |
|---|---|---|
| 00027 | -1.877e+008 | photon |
| 00010 | -1.859e+008 | H2O |
| 00006 | +1.634e+008 | H |
| 00021 | +1.535e+008 | OH |
| 00019 | +2.831e+007 | O |
| 00008 | +2.144e+007 | H2 |
| 00026 | +4.035e+006 | e- |
| 00013 | +3.865e+006 | H3O+ |
| 00011 | +1.272e+005 | H2O+ |
| 00022 | +3.438e+004 | OH+ |
| 00020 | +4.757e+003 | O+ |
| 00007 | +3.971e+003 | H+ |
| 00009 | +1.251e+002 | H2+ |

**Fig. 2** The major species and their changes at a distance of 2,500 km from the nucleus

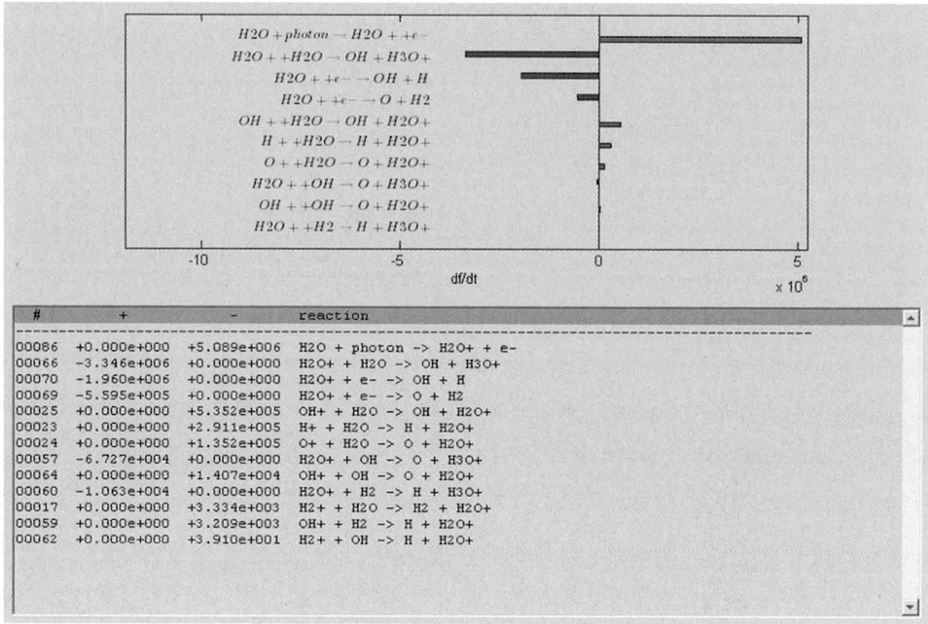


| # | + | - | reaction |
|---|---|---|---|
| 00086 | +0.000e+000 | +5.089e+006 | H2O + photon -> H2O+ + e- |
| 00066 | -3.346e+006 | +0.000e+000 | H2O+ + H2O -> OH + H3O+ |
| 00070 | -1.960e+006 | +0.000e+000 | H2O+ + e- -> OH + H |
| 00069 | -5.595e+005 | +0.000e+000 | H2O+ + e- -> O + H2 |
| 00025 | +0.000e+000 | +5.352e+005 | OH+ + H2O -> OH + H2O+ |
| 00023 | +0.000e+000 | +2.911e+005 | H+ + H2O -> H + H2O+ |
| 00024 | +0.000e+000 | +1.352e+005 | O+ + H2O -> O + H2O+ |
| 00057 | -6.727e+004 | +0.000e+000 | H2O+ + OH -> O + H3O+ |
| 00064 | +0.000e+000 | +1.407e+004 | OH+ + OH -> O + H2O+ |
| 00060 | -1.063e+004 | +0.000e+000 | H2O+ + H2 -> H + H3O+ |
| 00017 | +0.000e+000 | +3.334e+003 | H2+ + H2O -> H2 + H2O+ |
| 00059 | +0.000e+000 | +3.209e+003 | OH+ + H2 -> H + H2O+ |
| 00062 | +0.000e+000 | +3.910e+001 | H2+ + OH -> H + H2O+ |

**Fig. 3** The major production and destruction reactions for $H_2O^+$ at a distance of 2,500 km from the nucleus

that extending the number of source species makes modeling the chemistry very complex. Therefore, a systematic approach is necessary to keep the problem manageable.

## 3 Summary

The construction of a database containing reactions and reaction rate measurements is an important tool for gathering knowledge about the chemical processes in a cometary coma. Combining several existing reaction databases poses some practical problems, but results in an enormous collection of information and is therefore useful to study a wide range of compositional problems. Once such a database is available, our software is able to automatically generate the programming code for computing the changes in species abundances due to the chemistry. This is invaluable because it provides an easy path for extending, modifying, or updating the chemical network, while guaranteeing correctness and efficiency of the generated code, even for more complicated chemical systems. The database interface is flexible so that it can be easily extended and adapted for other chemical environments, such as planetary atmospheres.

## References

V.G. Anicich, JPL Publication 03-19, November (2003)
R.M. Häberli, K. Altwegg, H. Balsiger, J. Geiss, Astron. Astrophys. **297**, 881 (1995)
Y.H. Le Teuff, T.J. Millar, A.J. Markwick, Astron. Astrophys. Suppl. Ser. **146**, 157 (2000)

Space Sci Rev (2007) 130: 79–86
DOI 10.1007/s11214-007-9215-1

# Elemental Abundances of the Bulk Solar Wind: Analyses from Genesis and ACE

**D.B. Reisenfeld · D.S. Burnett · R.H. Becker · A.G. Grimberg · V.S. Heber · C.M. Hohenberg · A.J.G. Jurewicz · A. Meshik · R.O. Pepin · J.M. Raines · D.J. Schlutter · R. Wieler · R.C. Wiens · T.H. Zurbuchen**

Received: 8 February 2007 / Accepted: 15 May 2007 / Published online: 6 July 2007


**Abstract** Analysis of the Genesis samples is underway. Preliminary elemental abundances based on Genesis sample analyses are in good agreement with in situ-measured elemental abundances made by ACE/SWICS during the Genesis collection period. Comparison of these abundances with those of earlier solar cycles indicates that the solar wind composition is relatively stable between cycles for a given type of flow. ACE/SWICS measurements for the Genesis collection period also show a continuum in compositional variation as a function of velocity for the quasi-stationary flow that defies the simple binning of samples into their sources of coronal hole (CH) and interstream (IS).

**Keywords** Sun: etc.

D.B. Reisenfeld (✉)
Dept. of Physics and Astronomy, U. Montana, 32 Campus Dr., Missoula, MT 59812, USA
e-mail: dan.reisenfeld@umontana.edu

D.S. Burnett
California Institute of Technology, Pasadena, CA, USA

R.H. Becker · R.O. Pepin · D.J. Schlutter
Dept. of Physics, U. Minnesota, Minneapolis, MN, USA

A.G. Grimberg · V.S. Heber · R. Wieler
Isotope Geology and Mineral Resources, ETH Zurich, Zurich, Switzerland

C.M. Hohenberg
Dept. of Physics, Washington U., St. Louis, MO, USA

A.J.G. Jurewicz · A. Meshik
Center for Meteorite Studies, Arizona State U., Tempe, AZ, USA

J.M. Raines · T.H. Zurbuchen
Atmospheric, Oceanic, and Space Sciences, U. Michigan, Ann Arbor, MI, USA

R.C. Wiens
Space Science and Applications, Los Alamos National Laboratory, Los Alamos, NM, USA

## 1 Introduction

The NASA Genesis mission continues the exploration of solar wind composition begun by the Solar Wind Composition (SWC) experiments carried to the moon by the Apollo astronauts (Geiss et al. 2004 and references therein). Since then, a number of in situ spectroscopy experiments have flown, most notably the Solar Wind Ion Composition Spectrometer (SWICS) experiments on the Ulysses and Advanced Composition Explorer (ACE) missions, in which solar wind particles are analyzed on board and data are telemetered to Earth (Gloeckler et al. 1992, 1998). Genesis represents a return to the origins of the field, as Genesis is the first sample return mission since Apollo.

Genesis returned to Earth on September 8, 2004. Despite the hard landing that resulted from a failure of the avionics to deploy the parachute, many samples were returned in a condition that permit analyses. Analyses of these samples should give a far better understanding of the solar elemental and isotopic composition (Burnett et al. 2003). Further, the photospheric composition is thought to be representative of the solar nebula; thus, the Genesis mission will provide a new baseline for the primordial solar nebula composition with which to compare present-day compositions of planets, meteorites, and asteroids. Here we present preliminary analysis results, focusing on elemental abundances.

Although the Genesis samples were exposed to the solar wind over a period of 27 months, they represent only a composition snapshot of the solar wind, which is variable in both space and time. Therefore, to determine how representative the Genesis samples are of the average solar wind, we also present an inter-comparison between Genesis abundances and measurements from other solar wind missions.

## 2 Solar Wind Conditions During the Genesis Mission

The composition variability of the solar wind is not random, but is organized by indicators such as solar wind speed and structure. The solar wind consists of quasi-stationary flow punctuated by transient events, known as coronal mass ejections (CMEs). The quasi-stationary flow is a mixtures of two end-members: fast wind (>550 km/s), which originates in large coronal holes (CH), and slow wind (<450 km/s), also called interstream (IS) wind, which likely originates on the rapidly diverging field lines at the boundaries of coronal holes and on newly opened flux lines above or adjacent to streamers (Wang et al. 2000; Zurbuchen et al. 2002). CMEs are highly variable in composition, originating above magnetically active regions on the Sun (e.g., Neugebauer 1991).

To address how these three regimes are compositionally related to each other and to the photosphere, Genesis collected samples in each regime on separate collector arrays. To determine the regime, plasma and electron spectrometers continuously monitored the solar wind (Barraclough et al. 2003). Derived solar wind parameters were inputs into an onboard algorithm that autonomously determined the solar wind regime and directed the appropriate collector array to be deployed (Neugebauer et al. 2003). The algorithm takes into account the proton speed, proton temperature, alpha particle abundance, and the angular distribution of suprathermal electrons. Proton speed alone was used to determine the presence of CH or IS flow. A combination of indicators was used to determine the presence of CME material: a high alpha-to-proton density ratio, a low proton temperature, and/or the presence of counter-streaming electrons. We point out that solar wind speed regimes serve only as proxies for solar wind composition regimes, which are ultimately what we desire to isolate on

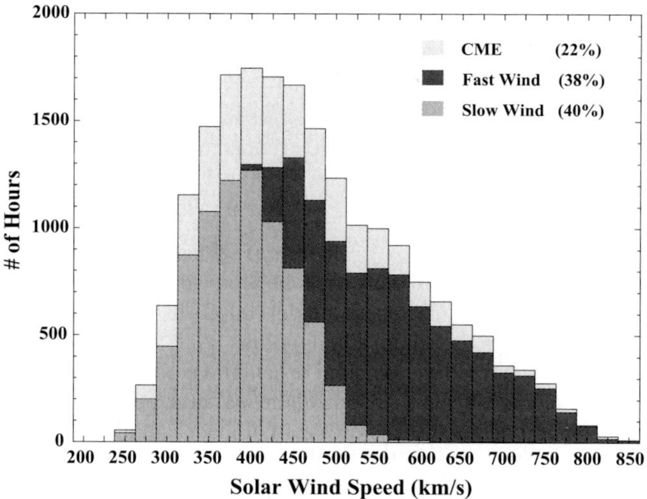


**Fig. 1** Distribution of solar wind velocities during Genesis sample collection period, designated by solar wind regime as determined by on-board algorithm

the different arrays. In the next section we discuss the success of the algorithm by comparison to heavy-ion composition data collected by ACE/SWICS during the Genesis collection interval.

At any given time, only one of the three regime-specific collector arrays was exposed to the solar wind. The overall exposure times for each of these arrays were 333.67 days (IS), 313.01 days (CH), and 193.25 days (CME), with nearly all of the CH exposure occurring in the second half of the period. Two bulk solar-wind arrays were continuously exposed for 852.83 days. The total number of collector deployment intervals for each regime was 146 (IS), 94 (CH), and 106 (CME). Figure 1 shows the measurement frequency for each of the three regimes binned by the measured proton speed.

Figure 1 also shows that there is a broad region of overlap between the IS and CH regimes. This reflects hysteresis logic used in the algorithm to determine the speed set point between IS and CH flow. Hysteresis was included in an attempt to improve the compositional purity of the regime samples by partially compensating for degradation of the correlation between solar wind speed and composition caused by hydrodynamic evolution from the photosphere to Genesis. In particular, on the leading edge of a stream interface observed at 1 AU, initially slow wind has been accelerated, but it will have an abundance consistent with a slower wind. Thus, when a slow-to-fast transition occurs, the speed set point for transitioning from IS to CH regimes is set relatively high: 525 km/s. Analogously, on the trailing side of a fast stream, originally fast wind has been decelerated due to the rarefaction caused by fast wind outrunning slow wind; thus, the speed set point for transitioning from CH to IS regimes is set relatively low: 425 km/s. The hysteresis algorithm is described in more detail in Neugebauer et al. (2003).

### 2.1 Solar Wind Composition Properties

It is well established that the elemental composition of the solar wind differs from that of the photosphere, and that it varies between solar-wind regimes. Elemental fractionation of the solar wind depends on the time it takes for an element to become ionized in the chromosphere, which in turn depends strongly on the element's first ionization potential (FIP). Low-FIP elements ($E_{\mathrm{FIP}} < 10$ eV) are enhanced over their photospheric abundance relative to high-FIP elements (Geiss 1982; von Steiger 2000).

Figure 2 shows selected elemental abundance ratios measured by ACE/SWICS separated according to the three solar wind regimes as determined by the Genesis onboard algorithm, and further sorted by solar wind helium speed measured by ACE. Shown are the ratios of iron and magnesium (two low-FIP elements) to oxygen (a high-FIP element), and the ratios of carbon and helium (two high-FIP elements), to oxygen (Von Steiger and Geiss 1993). For all elements a speed-dependent variation is observed. For the low-FIP elements the ratio with respect to oxygen decreases with increasing solar wind speed, as evident in the plots for the Fe/O and Mg/O ratios. This is consistent with earlier observations that the abundance of low-FIP elements decreases between IS and CH flow relative high-FIP elements (Zurbuchen et al. 2002; Geiss et al. 1995), but here we show that the variation is a continuous function of solar wind speed.

The plots of low-FIP elements also demonstrates the success of the hysteresis algorithm. A systematic separation in abundance is observed where the IS and CH regimes overlap in speed, demonstrating the use of recent flow history to help at least partially recover the abundance signature of the outflow source. A more complete analysis of the hysteresis algorithm's success is in progress.

For the high-FIP elements nitrogen ($E_{\mathrm{FIP}} = 14.53$ eV) and carbon ($E_{\mathrm{FIP}} = 11.26$ eV), their abundances relative to oxygen are essentially flat as a function of solar wind speed (with the exception of a slight rise for C/O between 250 and 350 $\mathrm{km\,s^{-1}}$). This flatness is likely due to the fact that carbon, nitrogen, and oxygen have nearly the same FIP, and presumably experience a similar acceleration history.

Figure 2 also demonstrates that CME material consistently exhibits Fe/O and Mg/O ratios higher than either the IS or CH wind for a given speed, a point first made by Reisenfeld et al. (2003) and further elaborated by Richardson and Cane (2004). For high-FIP elements, the difference between CME and quasi-stationary flow is not as obvious, although the CME data for C/O show a consistent skew to lower abundances. The overlap of abundance ratios for CMEs and quasi-stationary flow is in part due to the conservative design of the regime-selection algorithm. To minimize the risk that the CH and IS arrays were contaminated by CME material, the algorithm maintained the CME regime state for a minimum of 18 hours, plus an additional 6 hours after CME signatures are last detected. In this way, because CME signatures are often intermittent, enough persistence was designed into the algorithm to keep it from leaving the CME decision state prematurely. As a consequence, the CME array was unavoidably contaminated with IS and CH flow. Nevertheless, the clear distinction between the composition of the CME regime and the IS and CH regimes show that the algorithm successfully isolates CME material.

For all elements and regimes, we see a smooth variation with speed. This demonstrates that, although it is common to refer to different FIP fractionation values for IS and CH flows (as we have done here), there is not a sharp abundance boundary between high and low speed. Rather, fractionation varies smoothly with solar wind speed (Reisenfeld et al. 2003; Richardson and Cane 2004).

### 2.2 Composition Variation with Time

It is well known that the relative occurrence of IS, CH, and CME flow changes over the course of the solar cycle, posing the question of whether the composition of a given solar-wind regime also change over the course of the solar cycle. In other words, is the Genesis

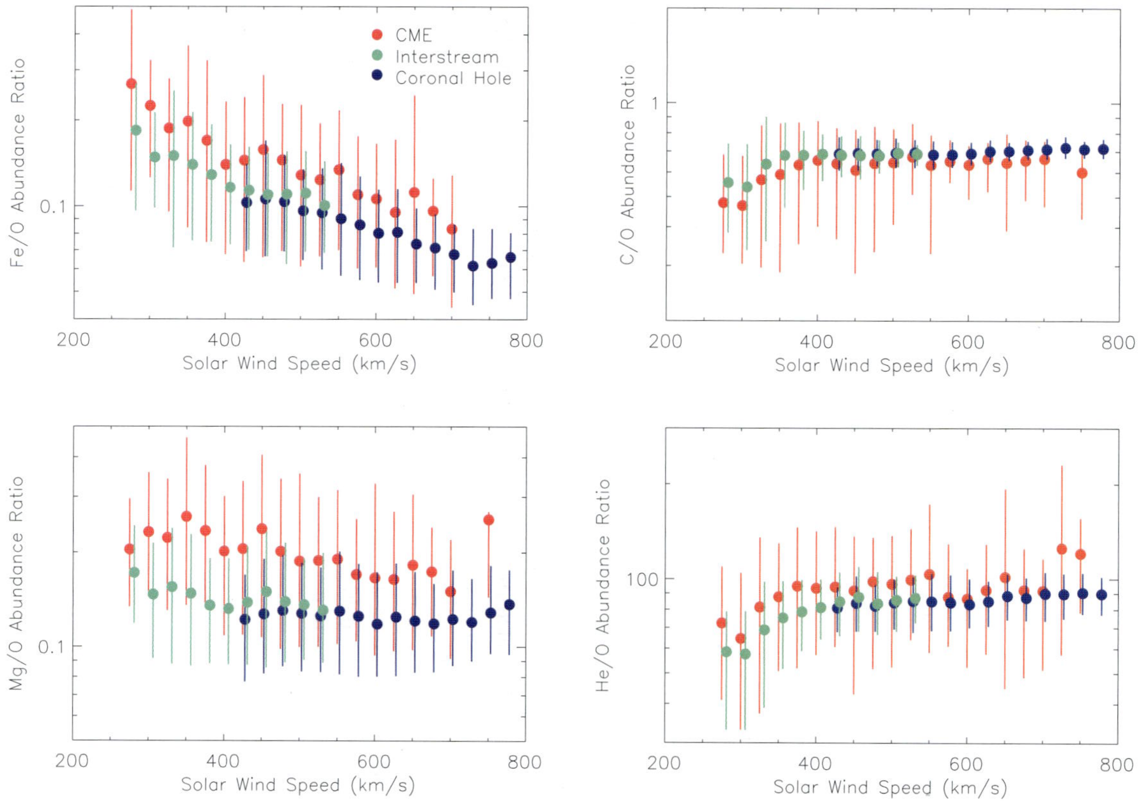

**Fig. 2** ACE/SWICS Elemental abundance ratios for (**a**) Fe/O, (**b**) C/O, (**c**) Mg/O, and (**d**) He/O as a function of the solar wind alpha speed (determined by ACE). The ratios are further color-coded by solar wind regime (CME, Interstream, or Coronal Hole) as determined by the Genesis onboard algorithm for the same period. The *points* indicate the mean of abundance values within bins 25 km s$^{-1}$ wide. The *vertical lines* span the 10- to 90-percentile ranges, indicating the observed spread. The uncertainty in the mean is typically a few percent or less

sample representative of solar wind in general, or just of the time for which it was collected? Elemental abundance ratios reported by the Ulysses/SWICS experiment indicate that the solar wind composition does not vary significantly between solar maximum and minimum. Von Steiger (2000) compared the FIP fractionation in the slow wind during the two equatorial phases of the Ulysses first orbit. The first (1991–1992) occurred near solar maximum, and the second (1997–1998) near solar minimum. The overall FIP enhancement can be quantified for the two periods by calculation of the FIP enhancement factor (von Steiger 2000), which is the ratio of low- to high-FIP element abundances:

$$f = \frac{\left(\frac{[\mathrm{Mg}]+[\mathrm{Si}]+[\mathrm{S}]+[\mathrm{Fe}]}{[\mathrm{C}]+[\mathrm{N}]+[\mathrm{O}]+[\mathrm{Ne}]}\right)_{\mathrm{SW}}}{\left(\frac{[\mathrm{Mg}]+[\mathrm{Si}]+[\mathrm{S}]+[\mathrm{Fe}]}{[\mathrm{C}]+[\mathrm{N}]+[\mathrm{O}]+[\mathrm{Ne}]}\right)_{\mathrm{Ph}}}, \tag{1}$$

where square brackets around element symbols denote elemental abundances. For the solar maximum period, $f = 2.0 \pm 0.4$, and for the solar minimum period, $f = 1.8 \pm 0.4$; there is no statistically significant difference. Note these are ratios relative to newly computed photospheric abundances (Asplund et al. 2005), thus the ratios reported here differ slightly from von Steiger (2000).

To investigate variations in composition *between* solar cycles, we have compared ACE/SWICS and Ulysses/SWICS composition data, which were collected during sequential solar cycles. As described earlier, the IS and CH periods used for ACE data are those determined by the Genesis onboard regime selection algorithm. The Ulysses/SWICS ratios are from von Steiger (2000).

The FIP enhancement factors for the ACE and Ulysses CH periods are $f = 1.5 \pm 0.3$ and $f = 1.4 \pm 0.2$, respectively; thus, the overall FIP enhancements are the same within uncertainties. For the IS periods, the ACE and Ulysses the factors are $f = 1.8 \pm 0.4$ and $f = 1.8 \pm 0.3$, respectively. Again, the FIP enhancements are the same. We mention that it was necessary to modify the definition of $f$ not to include the neon abundance, because we currently do not have an accurate neon measurement from ACE.

We conclude that, at this time, there is no clear evidence for systematic variation of composition between phases of the solar cycle or between different solar cycles, although this is an area of ongoing investigation. Between the ACE and Ulysses missions, we now have continuous composition data from 1991 through the present date, spanning over one and a half solar cycles, so it will be possible to explore this issue in detail.

## 3 Genesis Sample Composition Results

### 3.1 Elemental Abundances

The Genesis mission returned the longest-exposure solar wind sample to date, undertaken in particular for obtaining high isotopic precision. Elemental ratios have been measured in the solar wind by in-situ detectors for several decades. It is important to compare the results of the two analysis methods, as shown in Table 1, using the preliminary analyses from Genesis. For the noble gases, we also compare to the Apollo Solar Wind Experiment results. The initial fluence measurements all agree within measurement uncertainty. The Genesis fluence values listed in Table 1 have been corrected for backscatter (where necessary), but not for other sources of systematic error, such as the possibility of loss by diffusion.

The Genesis samples are being analyzed by a number of groups, using a variety of techniques. Magnesium and iron fluence measurements have been made using secondary ion

**Table 1** Comparison of solar-wind abundances derived from Genesis samples to those from the ACE/SWICS spectrometer and Solar Wind Composition Experiment (Apollo foils)

| | Genesis sample fluence ($cm^2$) | ACE-derived fluence ($cm^2$) | $X_{Gen}/X_{ACE}$ | Apollo SWC equivalent fluence ($cm^2$) | $X_{Gen}/X_{Ap}$ |
|---|---|---|---|---|---|
| Proton | $1.90 \times 10^{16}$[a] | – | – | – | – |
| $^4$He | $9.10 \times 10^{14}$ | $9.94 \times 10^{14}$ | $0.92 \pm 0.24$ | $8.79 \times 10^{14}$ | $1.04 \pm 0.11$ |
| $^{20}$Ne | $1.40 \times 10^{12}$ | – | – | $1.54 \times 10^{12}$ | $0.91 \pm 0.09$ |
| Mg | $1.54 \times 10^{12}$ | $1.84 \times 10^{12}$ | $0.84 \pm 0.24$ | – | – |
| $^{36}$Ar | $2.80 \times 10^{10}$ | – | – | $3.15 \times 10^{10}$ | $0.89 \pm 0.11$ |
| Fe | $1.30 \times 10^{12}$ | $1.45 \times 10^{12}$ | $0.90 \pm 0.26$ | – | – |

[a]From Genesis ion monitor

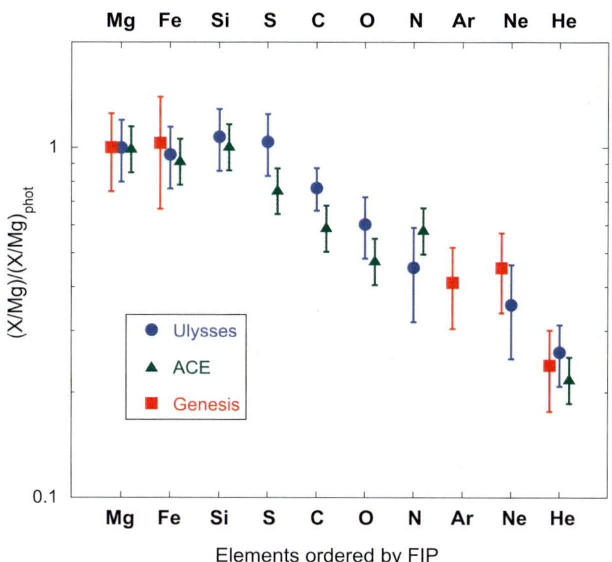


**Fig. 3** Comparison of abundance ratio measurements from Genesis, ACE, Ulysses, ordered by FIP. Abundances are normalized to Mg. No distinction is made for solar wind type. ACE values are based on data accumulated over the Genesis sample collection period; Ulysses values are averages of measurements reported by von Steiger (2000). *Error bars* show systematic error for spacecraft data, and statistical error for Genesis sample analyses (systematic errors have not yet been fully determined)

mass spectroscopy (SIMS) (Burnett et al. 2007). The noble gas fluence measurements have been made using the methods of pyrolysis, amalgamation, laser ablation, and close system sample etching (CSSE) (Mabry et al. 2007; Grimberg et al. 2007; Hohenberg et al. 2007; Pellin et al. 2007). The Genesis fluences reported in Table 1 are averages where multiple measurements have been made, and the measurement uncertainty reflects the scatter of these measurements. Detailed descriptions of the analysis techniques and their uncertainties are in preparation for publication by the individual groups. We expect more results to be forthcoming as additional elements are analyzed from the Genesis samples.

## 3.2 Genesis Samples vs. FIP

The preliminary analyses can also extend our understanding of FIP enrichment by comparing Genesis, ACE, and Ulysses measurements to photospheric abundances. Although

Genesis will eventually provide regime-specific abundances, at present, the majority of analyzed samples have been from the bulk collectors; thus we examine abundance ratios for the average solar wind, irrespective of regime.

Figure 3 presents elemental composition ordered according to FIP for all three missions. We have normalized to magnesium rather than oxygen (as is traditionally done), because we have yet to determine the oxygen fluence in the Genesis samples. The ACE abundances are averages for the times coincident with the Genesis bulk collection period. The Ulysses values are simple averages of the fast and slow abundances reported by von Steiger (2000). Within measurement uncertainty, all measurements are in very good agreement, indicating that the observed FIP enrichment is independent of (a) measurement technique and (b) solar cycle. We are led to conclude that we have a good understanding of the degree and the stability of the elemental fractionation of the solar wind, which is critical if we are to tie the Genesis samples to primordial abundances.

**Acknowledgements** The authors thank the NASA Discovery Program for their support of Genesis.

## References

M. Asplund, N. Grevesse, A.J. Sauval, in *Cosmic Abundances as Records of Stellar Evolution and Nucleosynthesis*, ed. by F.N. Bash, T.G. Barnes, ASP Conference Series, vol. 336 (2005), p. 25

B.L. Barraclough et al., Space Sci. Rev. **105**, 627 (2003)

D.S. Burnett et al., Space Sci. Rev. **105**, 509 (2003)

D.S. Burnett et al., *Lunar and Planetary Science Conference XXXVIII* (2007), p. 1843

J. Geiss, Space Sci. Rev. **33**, 201 (1982)

J. Geiss et al., Science **268**, 1033 (1995)

J. Geiss et al., Space Sci. Rev. **110**, 307 (2004)

G. Gloeckler et al., Space Sci. Rev. **86**, 497 (1998)

G. Gloeckler et al., Astron. Astrophys. Suppl. Ser. **92**, 267 (1992)

A. Grimberg et al., Space Sci. Rev. (2007), this issue. doi: 10.1007/s11214-007-9150-1

C.M. Hohenberg et al., Science (2007, submitted)

J.C. Mabry et al., *Lunar and Planetary Science Conference XXXVIII* (2007), p. 2412

M. Neugebauer, Science **252**, 404 (1991)

M. Neugebauer et al., Space Sci. Rev. **105**, 661 (2003)

M.J. Pellin et al., *Lunar and Planetary Science Conference XXXVIII* (2007), p. 2181

D.B. Reisenfeld et al., in *Solar Wind Ten*, ed. by M. Velli, R. Bruno, F. Malara. AIP Conf. Proc. vol. 679 (2003), p. 632

I.G. Richardson, H.V. Cane, J. Geophys. Res. **109**, A09104 (2004) doi:10:1029/2004JA010598

R. von Steiger, J. Geiss, Adv. Space Res. **13**, 63 (1993)

R. von Steiger, J. Geophys. Res. **105**, 27217 (2000)

Y.-M. Wang et al., J. Geophys. Res. **105**, 25133 (2000)

T.H. Zurbuchen, L.A. Fisk, G. Gloeckler, R. von Steiger, Geophys. Res. Lett. **29**, 1352 (2002)

Space Sci Rev (2007) 130: 87–95
DOI 10.1007/s11214-007-9159-5

# Presolar Grains in Meteorites and Their Compositions

**Ulrich Ott**

Received: 1 December 2006 / Accepted: 8 February 2007 /
Published online: 27 April 2007


**Abstract** Small amounts of pre-solar "stardust" grains have survived in the matrices of primitive meteorites and interplanetary dust particles. These grains—formed directly in the outflows of or from the ejecta of stars—include thermally and chemically refractory carbon materials such as diamond, graphite and silicon carbide; as well as refractory oxides and nitrides. Pre-solar silicates, which have only recently been identified, are the most abundant type except for possibly diamond. The detailed study with modern analytical tools, of isotopic signatures in particular, provides highly accurate and detailed information with regard to stellar nucleosynthesis and grain formation in stellar atmospheres. Important stellar sources are Red Giant (RG) and Asymptotic Giant Branch (AGB) stars, with supernova contributions apparently small. The survival of those grains puts constraints on conditions they were exposed to in the interstellar medium and in the early solar system.

**Keywords** Interstellar grains · Nucleosynthesis · Grain formation · Solar System · Meteorites

## 1 Introduction

The importance of dust in astrophysics considerably exceeds what might be expected from the small fraction of matter that it constitutes. A reason is, that, while only a percent or so of interstellar matter is contained in dust, it is host for almost all of the "metals". Hence, when looking for "The Composition of Matter" in the ISM one has to turn to the study of dust for most of the elements. An important, although special and biased type of interstellar dust is that found surviving in primitive meteorites and interplanetary dust particles. This material has become available about two decades ago and can be studied in detail with all the methods available in modern analytical laboratories.

It was the search for host phases of isotopically unusual noble gases that led to the first discovery in 1987 of surviving pre-solar minerals (diamond and silicon carbide) in primitive

U. Ott (✉)
Max-Planck-Institut für Chemie, Joh.-J.-Becher-Weg 27, 55128, Mainz, Germany
e-mail: ott@mpch-mainz.mpg.de

meteorites. These were followed by others (graphite, refractory oxides and nitrides, and finally silicates) in the years since. Pre-solar grains occur in even higher abundance than in meteorites in interplanetary dust particles (IDPs).

In the "classical" approach pioneered in the search for the noble gas carriers diamond, SiC and graphite, pre-solar grains are isolated by dissolving most of the meteorites (consisting mostly of silicate minerals) using strong acids, followed by further chemical and physical separation methods. For overviews at the early stage when only these types were known see reviews by Anders and Zinner (1993) and Ott (1993). More up-to-date reviews (although only barely including the only recently found pre-solar silicates) are by Zinner (1998), Hoppe and Zinner (2000) and Nittler (2003). Refractory oxides and nitrides were found in conjunction with the noble gas carrying minerals because of their similar chemical inertness. Identification of pre-solar silicates among the "sea of normal silicates", however, became possible only with the advent of a new generation of analytical instrumentation, the NanoSIMS (e.g., Hoppe et al. 2004) that allowed the imaging search for isotopically anomalous phases in situ, i.e. without the need for chemical/physical extraction.

Central to the identification of pre-solar minerals is the determination of isotopic compositions, which as a rule strongly deviate from solar. Isotopic composition is, as a matter of fact, the main criterion by which they can be identified; hence all our known "pre-solar grains" are "circum-stellar condensates" that carry the isotopic signatures of nucleosynthesis processes going on in specific parent stars. This criterion alone is not fool-proof, though, since isotope anomalies are also found in objects that formed within the Solar System (e.g. Ca–Al-rich inclusions, Loss et al. 1994), reflecting imperfect isotopic homogenization. Additional anomalies may have been created by energetic particle reactions in the Early Solar System (e.g., Gounelle et al. 2001; Chaussidon et al. 2006) and possibly by trapped cosmic rays (Desch et al. 2004). For the mineral phases described below, however, the combination of isotopic patterns in main and often trace elements they carry, sometimes combined with their mineralogical makeup, leaves little if any doubt that they are surviving pre-solar "stardust" (see reviews cited above).

In the following I will present an overview of the currently known (and generally accepted as such) types of stardust in primitive meteorites and address some topics on which the results of their study have borne.

## 2 Overview

An overview of the currently known inventory of circumstellar grains in meteorites is presented in Table 1, including characteristic isotopic signatures and inferred stellar sources. Note that the various mineral types are susceptible to and have been affected by thermal metamorphism or aqueous alteration in the solar nebula and/or meteorite parent bodies to different degrees. Hence I have chosen not to list a set of concentrations in a given meteorite type such as, e.g., the chemically most primitive CI meteorites. Instead the concentrations listed there represent the highest that have been reported in the literature. Abundances of silicates are definitely, those of the other pre-solar grains most likely, higher in IDPs. Below I am going to comment on the salient isotopic features observed in the various types of grains. This is followed by a discussion some of the implications of these results.

## 3 Isotopic Structures

### 3.1 Silicon Carbide

All SiC grains in primitive meteorites are of pre-solar origin, and they are the best characterized. This has been helped by their relatively high abundance, the fact that rather clean separates can be prepared and by their comparably high contents of minor and trace elements. The wealth of data has allowed identification of several subtypes based on isotopic composition: mainstream, A, B, X, Y and Z grains (e.g., Hoppe and Ott 1997; Hoppe and Zinner 2000).

Mainstream grains show enhancements of $^{13}C$ and $^{14}N$, which is in fact characteristic for most subtypes (Table 1; Fig. 1). They also show evidence for the former presence of $^{26}Al$ as indicated by overabundances of its daughter $^{26}Mg$, with inferred $^{26}Al/^{27}Al$ ratios up to $10^{-2}$ at the time of grain formation. Silicon isotopes typically show enrichments of the heavy isotopes $^{29}Si$ and $^{30}Si$ up to $\sim+200$‰ compared to normal silicon, i.e. $\delta^{29,30}Si/^{28}Si$ up to $\sim+200$‰ (Fig. 2). In a plot $\delta^{29}Si/^{28}Si$ vs. $\delta^{30}Si/^{28}Si$ data points follow an approximately linear relationship with slope $\sim$1.34. The trend is not due to nucleosynthesis in the parent star, but rather the result of galactic chemical evolution and the fact that the grains derive from a large number of individual stars (e.g., Timmes and Clayton 1996).

**Table 1** Overview of current knowledge on circum-stellar condensate grains in meteorites

| Mineral | Size [$\mu$m] abund. [ppm][1] | Isotopic signatures | Stellar sources | Contrib.[2] |
|---|---|---|---|---|
| diamond | $\sim$ 0.0026<br>$\sim$ 1500 | Kr-H, Xe-HL, Te-H | supernovae | ? |
| silicon carbide | $\sim$ 0.1–10<br>$\sim$ 30 | enhanced $^{13}C$, $^{14}N$, $^{22}Ne$, s-process elem. | AGB stars | >90% |
| | | low $^{12}C/^{13}C$, often enh. $^{15}N$ | J-type C-stars (?) | <5% |
| | | enhanced $^{12}C$, $^{15}N$, $^{28}Si$; extinct $^{26}Al$, $^{44}Ti$ | supernovae | 1% |
| | | low $^{12}C/^{13}C$, low $^{14}N/^{15}N$ | novae | 0.1% |
| graphite | $\sim$ 0.1–10<br>$\sim$ 10 | enh. $^{12}C$, $^{15}N$, $^{28}Si$; extinct $^{26}Al$, $^{41}Ca$, $^{44}Ti$ | SN (WR?) | <80 % |
| | | s-process elements | AGB stars | >10% |
| | | low $^{12}C/^{13}C$ | J-type C-stars (?) | <10% |
| | | low $^{12}C/^{13}C$; Ne-E(L) | novae | 2% |
| corundum/ spinel/ hibonite | $\sim$ 0.1–5<br>$\sim$ 50 | enhanced $^{17}O$, moderately depl. $^{18}O$ | RGB / AGB | >70% |
| | | enhanced $^{17}O$, strongly depl. $^{18}O$ | AGB stars | 20% |
| | | enhanced $^{16}O$ | supernovae | 1% |
| silicates | $\sim$ 0.1–1<br>$\sim$ 140 | similar to oxides above | | |
| silicon nitride | $\sim$ 1<br>$\sim$ 0.002 | enhanced $^{12}C$, $^{15}N$, $^{28}Si$; extinct $^{26}Al$ | supernovae | 100% |

[1] For abundances the highest abundances reported in the literature are listed. These apply for different meteorites for different grain types. Note uncertainty about actual fraction of diamonds that are pre-solar and for fraction of graphite attributed to SN and AGB stars (see text)

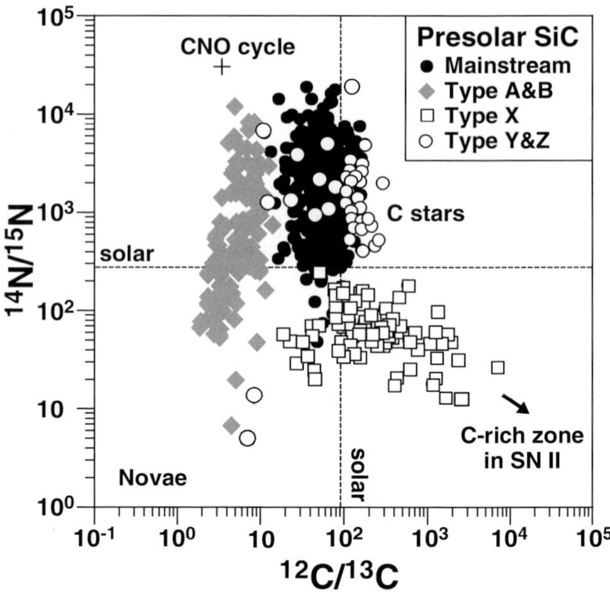


**Fig. 1** Carbon and nitrogen isotopic compositions measured in single pre-solar SiC grains. Solar isotopic compositions are indicated by *lines*. Most analyses plot in the mainstream field (cf. text and Table 1), displaced from normal in a direction indicating the influence of hydrogen burning via the CNO cycle. Also indicated are trends to be expected from novae and supernovae contributions. Graph courtesy of P. Hoppe (2001)

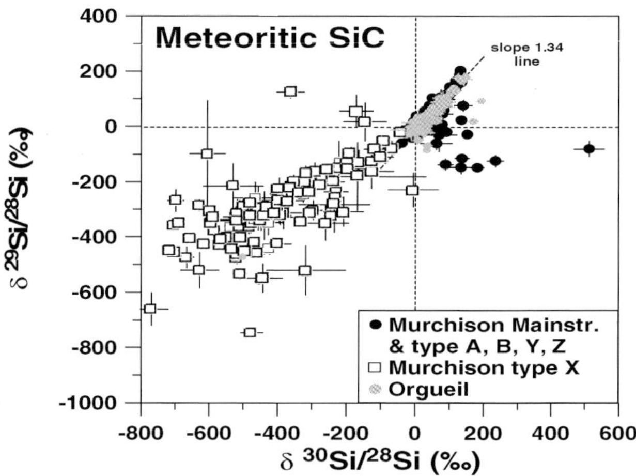


**Fig. 2** Si isotopic ratios measured in single presolar SiC grains. Ratios are shown as $\delta$-values, i.e. deviations from normal values in per mill. In this representation most grains plot within 200‰ of normal, close to a line with slope 1.34. X grains of likely supernova origin plot in the lower left showing large relative enhancements of $^{28}$Si. Graph courtesy of P. Hoppe

Among the noble gases mainstream SiC grains contain neon that is almost pure $^{22}$Ne [Ne-E(H)] and heavy elements showing the characteristic isotopic signature of the s-process. This quite clearly points to condensation out of the winds of low-mass (1–3 $M_{\mathrm{Sun}}$) carbon stars in the thermally pulsing asymptotic giant branch (TP-AGB) phase. Additional arguments include: a) carbon stars provide the necessary chemical environment ($C/O > 1$); b) AGB stars are the main contributors of carbonaceous dust to the interstellar medium (Whittet 1992; Henning and Salama 1998); c) AGB stars show in their atmospheres the 11.2 $\mu$m emission feature of SiC grains (e.g., Speck et al. 1999); d) the distribution of $^{12}C/^{13}C$ ratios in singly analyzed SiC grains is similar to that observed for carbon star atmospheres.

Calcium isotopes were mostly measured on aggregates of grains and the signatures are likely to be dominated by those of the mainstream grains. Enhancements at $^{42}$Ca and $^{43}$Ca are as expected from s-process contributions; the larger enhancements at $^{44}$Ca are likely to be caused by the presence of X grains from supernovae (see below) which carry radiogenic $^{44}$Ca from the decay of $^{44}$Ti ($T_{1/2} = 60$ a). In titanium typically lighter and heavier isotopes are enriched relative to $^{48}$Ti. As for the Si isotopes, the variations in Ti isotopes appear to be mostly due to galactic chemical evolution.

A and B grains are distinguished by their very low $^{12}C/^{13}C$ ratios <10 but otherwise share most of the properties of the mainstream grains. Likely sources are specific types of carbon stars, maybe J stars for grains that are s-process rich, and "born-again" AGB stars those that are s-process poor (Amari et al. 2001a).

Also Y and Z grains share many properties of the mainstream grains. Si isotopes mark the characteristic difference (Hoppe and Ott 1997; Fig. 2): both fall to the $^{30}$Si-rich side of the mainstream grains. Y grains have $\delta^{30}Si/^{28}Si > 0$ and $^{12}C/^{13}C > 100$. They probably come from AGB stars of low mass and lower than solar metallicity that experienced strong He shell dredge up (Hoppe and Zinner 2000; Amari et al. 2001b). Z grains, on the other hand, have always lower than solar $^{29}Si/^{28}Si$ and $^{12}C/^{13}C$. The most likely source are low-mass, low-metallicity AGB stars that experienced strong cool bottom processing during the red giant phase (Hoppe et al. 1997; Hoppe and Zinner 2000).

A percent or so of the presolar SiC grains in meteorites have a clearly distinct origin from the rest and are tied to supernovae: the X grains. They are characterized by high $^{12}$C, $^{28}$Si (Figs. 1, 2), typically low $^{14}$N and very high former abundances of $^{26}$Al ($T_{1/2} = 0.7$ Ma) as well as $^{44}$Ti ($T_{1/2} = 60$ a) seen as overabundances in the daughter nuclides $^{26}$Mg and $^{44}$Ca (Hoppe and Ott 1997; Hoppe and Zinner 2000). Ratios $^{26}Al/^{27}Al$ and $^{44}Ti/^{48}Ti$ at the time of grain formation approach unity. Interesting but not yet fully understood signatures are found in heavy trace elements such as molybdenum and barium (see below).

### 3.2 Graphite and Silicon Nitride

The case of graphite is more complex than that of silicon carbide. Ratios $^{12}C/^{13}C$ span the same wide range of $\sim 2$ to $\sim 7000$ as shown for SiC in Fig. 1, however the range in $^{14}N/^{15}N$ is from $\sim 30$ to $\sim 700$ only, being higher than solar in most grains. With these isotopic features, plus observed excesses of $^{18}$O, the often high inferred $^{26}Al/^{27}Al$ ratios (up to 0.15) at the time of grain formation, the fact that most grains show deficits in $^{29}$Si and $^{30}$Si (similar to type X SiC grains), as well as the large excesses in $^{44}$Ca from $^{44}$Ti decay (again, as SiC X grains) and $^{41}$K (from $^{41}$Ca, $T_{1/2} = 0.1$ Ma), type II supernovae have traditionally been to assumed to be the source of most presolar graphite grains (e.g., Zinner 1998; Hoppe and Zinner 2000). However, this percentage may have been overestimated. New single grain analyses indicate that refractory carbide grains rich in s-process elements are commonly found within the graphite spherules and so for many an AGB star seems more likely (Croat et al. 2005). The rare $Si_3N_4$ grains show isotopic signatures similar to SiC-X grains and supernova graphite grains and derive probably from supernovae as well.

### 3.3 Oxides and Silicates

Besides diamonds (see below) silicates—not unexpectedly—are the most abundant of the pre-solar grains that have been found. Isotopic information is much more limited than for SiC and graphite, mostly because of the lower contents of diagnostic trace elements. Ratios $^{16}O/^{17}O$ and $^{16}O/^{18}O$ range from $\sim 70$ to $\sim 30{,}000$ (0.025 to 11 $\times$ solar) and from 150 to 50,000 (0.3 to 100 $\times$ solar), respectively, and based on oxygen isotopes four populations have been recognized (Nittler et al. 1997; Fig. 3). Most group 1 and group 2 grains—constituting the majority (>70%; Table 1)—have lower than solar $^{16}O/^{17}O$ and higher than

**Fig. 3** Oxygen isotopic ratios measured in single pre-solar corundum grains. Based on oxygen isotopes most can be assigned to one of four groups (Nittler et al. 1997). Also shown are measured values for the atmospheres of red giant stars taken from the literature. The *line labeled "GCE"* represents the expected trend due to galactic chemical evolution (Clayton 1988; Hoppe and Zinner 2000). Figure courtesy Hoppe (2001)

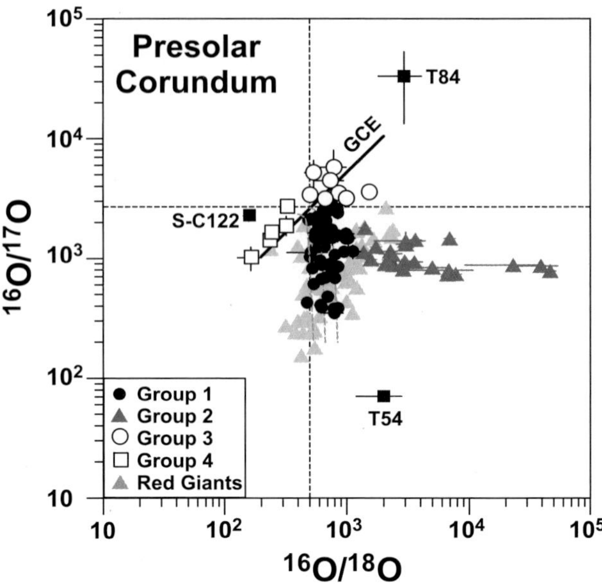


solar $^{16}O/^{18}O$ ratios. The composition of the former is similar to ratios observed in the atmospheres of Red Giant and AGB stars (1 to 9 $M_{Sun}$), which makes these the most likely stellar sources. Al and Mg have also been analyzed in many grains and inferred $^{26}Al/^{27}Al$ ratios at the time of grain formation show distinct distributions in the different populations, with the highest values (up to ~0.02) observed in type 1 and 2 grains. It is commonly assumed that grains without evidence for the former presence of $^{26}Al$ originate from RGB stars, those with $^{26}Al$ from AGB stars (Zinner 1998). The contribution from supernovae—as in the case of SiC—is low, on the order of a percent.

### 3.4 Nanodiamonds

In several ways these are the most enigmatic. Although discovered first, their pre-solar credentials are based solely on trace elements Te and noble gases, primarily the Xe-HL component (Fig. 4) that they carry. They are too small for individual analysis—each consisting of some 1000 carbon atoms only on average—and the carbon isotopic composition of "bulk samples" (i.e. many diamond grains) is within the range of Solar System materials. The case for nitrogen, the most abundant trace element (close to 1%; Russell et al. 1996), is similar. The ratio $^{14}N/^{15}N$ is ~400, i.e. $\delta^{15}N$ is $\sim -330$‰ relative to the standard, air, but close to what has been observed for the atmosphere of Jupiter by Owen et al. (2001).

Another complication arises from the noble gases themselves that constitute the evidence for the diamonds' presolar nature in the first place. Diamonds not just contain isotopically anomalous traces like Xe-HL, but also others which are essentially normal in their isotopic composition. The situation is best studied for xenon (Huss and Lewis 1994): diamonds contain the anomalous xenon (Xe-HL), an approximately normal xenon component (Xe-P3) and a third one that differs from both (Xe-P6). Ion implantation is the favored process by which the noble gases (Koscheev et al. 2001) and maybe other elements with supernova isotopic signature (Richter et al. 1997) were introduced. Generally, abundances of noble gases appear to correlate with grain size (Verchovsky et al. 1998), which has been used to estimate

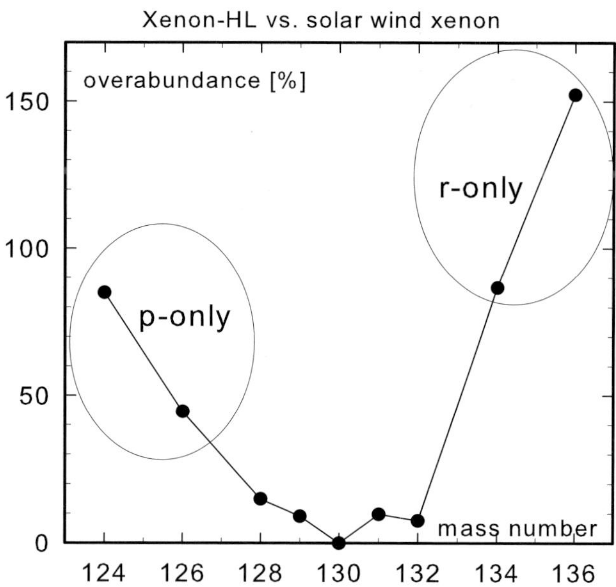


**Fig. 4** Xenon-HL of putative supernova origin shows overabundances in light (Xe-L) and heavy (Xe-H) isotopes of xenon. Shown are deviations of isotope ratios $^{i}Xe/^{130}Xe$ from solar xenon (in percent)

the energy of the implanted species (Verchovsky et al. 2003). There is also evidence that a minor fraction of the diamonds contains s-process xenon and thus could have an AGB star origin (Verchovsky et al. 2006).

## 4 Implications

### 4.1 Isotopic Structures and Nucleosynthesis

As isotopic structures are the key for establishing the grains as pre-solar, isotope studies are at the core of investigations that have been performed. Results from isotopic studies in turn are also those that bear strongest on astrophysics. For one, they allow to pinpoint the grains' stellar sources. In addition, given the precision of the laboratory isotopic analyses, which far exceeds what can be hoped for in remote analyses, they allow conclusions with regard to details of nucleosynthesis and mixing in the parent stars as well as of Galactic Chemical Evolution. They have borne strong on, e.g. the need for an extra mixing process (cool bottom processing) in Red Giants and provide detailed constraints on the operation of the s-process in AGB stars (e.g., Busso et al. 1999). A non-standard neutron capture process ("neutron burst") may be implied by the SiC-X grains from supernovae (Meyer et al. 2000) and possibly the trace Xe in the diamonds (e.g., Ott 2002).

### 4.2 Grain Formation

Chemical composition, sizes, and microstructures of grains constrain conditions during condensation in stellar winds and supernova ejecta. Condensation of SiC apparently occurred under close to equilibrium conditions (e.g., Lodders and Fegley 1998). Additional constraints are imposed by trace element contents both on average (Yin et al. 2006) as well as in individual grains (Amari et al. 1995). An important relevant observation is the occurrence of subgrains of primarily TiC within graphite (Croat et al. 2005).

### 4.3 The Lifecycle of Pre-Solar (and Maybe Interstellar in General) Grains

Interstellar grains are expected to be processed and eventually destroyed by sputtering or astration (e.g., Draine 2003), with an as yet unidentified formation process needed to account for the balance between formation and destruction. Pre-solar grains preserved in meteorites carry, in principle, a record on conditions they have been exposed to, which, however, is difficult to read. Determining an absolute age using long-lived radioisotopes is virtually ruled out by the fact that these systems use decay of rare constituents (e.g., K, Sr, Re, U) decaying into other rare elements with uncertain non-radiogenic compositions. However, appearance and microstructures of pristine (i.e. not chemically processed) SiC show little evidence for being processed, indicating either that they were surprisingly young when entering the forming Solar System or that they were protected (Bernatowicz et al. 2003); a similar situation is indicated by the lack of detectable spallation Xe produced by exposure to cosmic rays during residence in the ISM (Ott et al. 2005). The distribution, finally, among various types of meteorites, provides a measure of processing in the early Solar System.

## References

S. Amari, P. Hoppe, E. Zinner, R.S. Lewis, Meteoritics **30**, 679–693 (1995)
S. Amari, L.R. Nittler, E. Zinner, K. Lodders, R.S. Lewis, Astrophys. J. **559**, 463–483 (2001a)
S. Amari, L.R. Nittler, E. Zinner, R. Gallino, M. Lugaro, R.S. Lewis, Astrophys J. **546**, 248–266 (2001b)
E. Anders, E. Zinner, Meteoritics **28**, 490–514 (1993)
T.J. Bernatowicz, S. Messenger, O. Pravdivtseva, P. Swan, R.M. Walker, Geochimica Cosmochimica Acta **67**, 4679–4691 (2003)
M. Busso, R. Gallino, G.J. Wasserburg, Annu. Rev. Astron. Astrophys. **37**, 239–309 (1999)
M. Chaussidon, F. Robert, K.D. McKeegan, Geochimica Cosmochimica Acta **70**, 224–245 (2006)
D.D. Clayton, Astrophys. J. **334**, 191–195 (1988)
T.K. Croat, F.J. Stadermann, T.J. Bernatowicz, Astrophys. J. **631**, 976–987 (2005)
S.J. Desch, H.C. Connolly Jr., G. Srinivasan, Astrophys. J. **602**, 528–542 (2004)
B.T. Draine, Annu. Rev. Astron. Astrophys. **41**, 241–289 (2003)
T. Henning, F. Salama, Science **282**, 2204–2210 (1998)
M. Gounelle, F.H. Shuh, H. Shang, A.E. Glassgold, K.E. Rehm, T. Lee, Astrophys. J. **548**, 1051–1070 (2001)
P. Hoppe, Nucl. Phys. **A688**, 94c–101c (2001)
P. Hoppe, U. Ott, in: *Astrophysical Implications of the Laboratory Study of Presolar Materials*, ed. by T.J. Bernatowicz, E. Zinner (American Institute of Physics, Woodbury 1997) pp. 27–58
P. Hoppe, E. Zinner, Geophys. Res. **105**, 10371–10385 (2000)
P. Hoppe, P. Annen, R. Strebel, R. Eberhardt, P. Gallino, M. Lugaro, S. Amari, R.S. Lewis, Astrophys. J. **487**, L101–L104 (1997)
P. Hoppe, U. Ott, G.W. Lugmair, New Ast. Rev. **48**, 171–176 (2004)
G.R. Huss, R.S. Lewis, Meteoritics **29**, 791–810 (1994)
A.P. Koscheev, M.D. Gromov, R.K. Mohapatra, U. Ott, Nature **412**, 615–617 (2001)
K. Lodders, B. Fegley Jr., Meteorit. Planet. Sci. **33**, 871–880 (1998)
R.D. Loss, G.W. Lugmair, A.M. Davis, G.J. MacPherson, Astrophys. J. **436**, L193–L196 (1994)
B.S. Meyer, D.D. Clayton, The L.-S., Astrophys. J. **540**, L49–L52 (2000)
L.R. Nittler, Earth Planet. Sci. Lett. **209**, 259–273 (2003)
L.R. Nittler, C.M.O'D. Alexander, X. Gao, R.M. Walker, E. Zinner, Astrophys. J. **483**, 475–495 (1997)
U. Ott, Nature **364**, 25–33 (1993)
U. Ott, New Ast. Rev. **46**, 513–518 (2002)
U. Ott, M. Altmaier, U. Herpers, J. Kuhnhenn, S. Merchel, R. Michel, R.K. Mohapatra, Meteorit. Planet. Sci. **40**, 1635–1652 (2005)
T. Owen, P.R. Mahaffy, H.B. Niemann, S. Atreya, M. Wong, Astrophys. J. **553**, L77–L79 (2001)
S. Richter, U. Ott, F. Begemann, Lunar Planet. Sci. **XXVIII**, 1163–1164 (1997)
S.S. Russell, J.W. Arden, C.T. Pillinger, Meteorit. Planet. Sci. **31**, 343–355 (1996)
A.K. Speck, A.M. Hofmeister, M.J. Barlow, Astrophys. J. **513**, L87–L90 (1999)
F.X. Timmes, D.D. Clayton, Astrophys. J. **472**, 723–741 (1996)

A.B. Verchovsky, A.V. Fisenko, L.F. Semjonova, I.P. Wright, M.R. Lee, C.T. Pillinger, Science **281**, 1165–1168 (1998)
A.B. Verchovsky, I.P. Wright, C.T. Pillinger, Publ. Astron. Soc. Aust. **20**, 329–336 (2003)
A.B. Verchovsky, A.V. Fisenko, L.F. Semjonova, J. Bridges, M.R. Lee, I.P. Wright, Astrophys. J. **651**, 481–490 (2006)
D.C.B. Whittet, *Dust in the Galactic Environment* (Inst. Phys., New York, 1992), 295 pp.
Q.-Z. Yin, C.-T.A. Lee, U. Ott, Astrophys. J. **647**, 676–684 (2006)
E. Zinner, Annu. Rev. Earth Planet. Sci. **26**, 147–188 (1998)

Space Sci Rev (2007) 130: 97–104
DOI 10.1007/s11214-007-9233-z

# Planetary Atmospheres

**Tobias C. Owen**

Received: 26 March 2007 / Accepted: 7 June 2007 / Published online: 4 September 2007


**Abstract** The predominance of nitrogen in highly volatile forms and of carbon in solids set the abundance ratios of these elements in the inner planets, meteorites and comets. The absence of carbon compounds in an atmosphere then signals large deposits of carbon-bearing compounds in surface and/or subsurface deposits. In contrast, the icy planetesimals that contributed heavy elements to Jupiter must have had identical enrichments (relative to hydrogen) of both C and N, as well as other heavy elements that have been measured, compared to solar values. Capture of N and Ar suggests that the icy planetesimals that carried these elements must have formed at low temperatures, $<40$ K. New measurements of isotopes of nitrogen support this picture, but we must have more measurements in more atmospheres to be certain of this scenario.

**Keywords** Abundances · Atmospheres · Planets · Icy planetesimals · Carbon · Nitrogen

## 1 Introduction

Recent reviews have given comprehensive tables of abundances and isotopic ratios in the atmospheres of the planets and Titan (Mahaffy et al. 2000; Owen and Encrenaz 2003; Atreya et al. 2003; Wong et al. 2004). Rather than reproduce those compilations here, I will try to trace the sources and distributions of the major elements H, C, N, and O and their isotopes in planetary atmospheres. I will include some discussion of the noble gases and their isotopes where appropriate. Various mixtures of these elements and their compounds dominate the atmospheres of all the planets and the atmospheres of Titan, Pluto and Triton as well. In no case do we find an atmosphere with more neon or argon than any of the chemically active elements, despite the relatively high cosmic abundance of neon and the steady production of argon's radiogenic isotope $^{40}$Ar.

T.C. Owen (✉)
University of Hawaii, Institute for Astronomy, 2680 Woodlawn Drive, Honolulu, HI 96822, USA
e-mail: owen@ifa.hawaii.edu

## 2 The Interstellar Medium and the Solar Nebula

The discussion begins in the interstellar medium (ISM) where a fragment of an interstellar cloud of gas and dust contracted to form the solar nebula 4.6 BY ago. The abundances of the elements in this cloud fragment, perhaps "seasoned" by nearby supernova explosions, must have been identical to the abundances of the elements that formed the Sun.

Although the elemental abundances in the ISM cloud fragment were essentially solar, the compounds they formed at the low temperatures in the cloud, subject to ion–molecule reactions and gas–grain surface interactions, produced a fundamental difference between the reservoirs of nitrogen and carbon. Whereas carbon was primarily (>50%) in the form of solids, such as amorphous carbon, graphite, and organic compounds, nitrogen was primarily (>90%) in the form a highly volatile gas ($N_2$) or simply atomic nitrogen (Van Dishoeck et al. 1993). This difference dictates a major difference in the ease with which these two elements could subsequently be incorporated in planetesimals in the solar nebula. Carbon was easily trapped, whereas to capture nitrogen as $N_2$ efficiently, temperatures below 25 K would be required (Bar-Nun et al. 1988; Owen and Bar-Nun 1995, 2000). One therefore expects remnant planetesimals such as comets and meteorites to exhibit a ratio of C/N that is much greater than the solar value, and this is indeed the case (Fig. 1). The conversion of the atomic nitrogen component to $N_2$ and N-containing organics during the formation of the solar nebula remains a subject of active research (Charnley, private communication). The atomic N is probably responsible for the high $^{15}N/^{14}N$ found in some of the organics in some IDPs (Messenger 2000) and in cometary –CN (Arpigny et al. 2003).

## 3 Solid Body Atmospheres

Consequently, we anticipate that atmospheres formed primarily by outgassing from solid bodies that have accreted from these planetesimals or suffered a bombardment by them will exhibit the same deficiency in nitrogen compared to solar abundances found in the planetesimals as indeed they do. Among the inner planets, we see this most clearly on Venus,

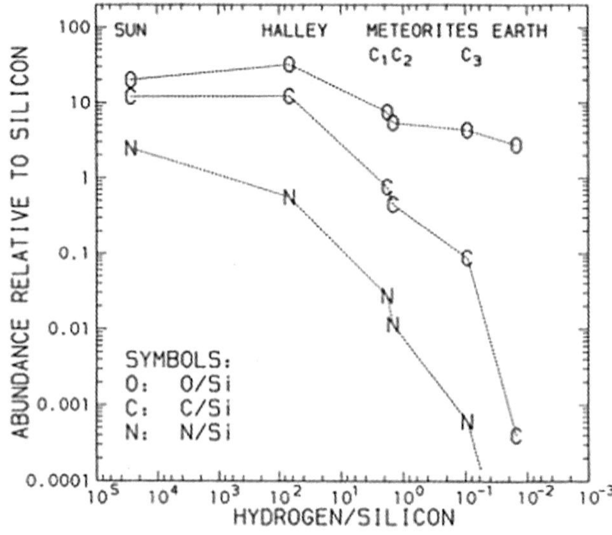


**Fig. 1** A reproduction of a figure from Geiss (1988) showing a comparison of major element abundances in Halley's comet, meteorites, and the Earth with solar values. The depletion of nitrogen in these solid bodies is evident

where most of the inventory of C and N is probably present in the atmosphere. Here we find C/N $\sim$ 55 (Donahue and Pollack 1983), $\sim$15$\times$ the solar value of Grevesse et al. (2005). On Mars we cannot assess inventories because of escape and deposition processes whose results are not well constrained. However, it appears that the $CO_2$ presently held in the polar caps is balanced by the amount of nitrogen that has escaped from the planet (Owen 1992). If so, the atmospheric ratio of C/N would be $\sim$18$\times$ solar. On Earth we find C/N $\sim$ 20 $\pm$ 10$\times$ solar if we account for the carbon bound up in carbonate rocks (Owen and Bar-Nun 1995). As is now well known, replacing this carbon as $CO_2$ in the terrestrial atmosphere would reproduce the abundances of $CO_2$ and $N_2$ now found on Venus.

These straightforward considerations lead to the following prediction: on any solid body with an atmosphere in which nitrogen is dominant, there must be a missing reservoir of carbon such that the total reservoir of C and N leads to C/N $\sim$ 15–20$\times$ solar.

There are two immediate applications of this prediction. On Mars, the presence of a predominantly (95%) $CO_2$ atmosphere suggests that there cannot be extensive deposits of carbonate rocks. If there were, the present atmosphere would be predominantly nitrogen (Owen and Bar-Nun 1995, 2000). An exception to this conclusion would occur if there were geological deposits of large amounts of nitrogen. In the absence of liquid water and a ready supply of oxygen, this seems highly unlikely. No evidence suggesting extensive deposits of either carbon dioxide or nitrogen minerals has been found in the Martian meteorites or by remote sensing of the Martian surface. The missing carbon—as $CO_2$—was evidently blown off the planet by impact erosion that would have removed the nitrogen and everything else (Melosh and Vickery 1989), a conclusion that is consistent with noble gas systematics (Owen and Bar-Nun 1995, 2000). On Titan, where we do find a predominantly nitrogen atmosphere, the missing carbon appears to be buried beneath the surface.

### 3.1 The Intriguing Case of Titan

We reach this conclusion about Titan's missing carbon by examining the rate at which methane, the dominant atmospheric carrier of this element, is currently being destroyed in the atmosphere by photochemical reactions (Strobel 1974; Yung et al. 1984). Assuming the rate has remained constant for the last 4.5 billion years, the equivalent of 1 to 2 km of hydrocarbons must have been precipitated onto Titan's surface. This corresponds to the destruction of $\sim$200 km atm of $CH_4$. But we now find 100 km atm of $N_2$ in Titan's atmosphere and from the observed fractionation of the nitrogen isotopes, we expect that approximately 5$\times$ this amount of nitrogen has escaped from the satellite (Lunine et al. 1999; Niemann et al. 2005). Therefore, to achieve the expected ratio of C/N in the satellite's volatile inventory, we expect an initial reservoir of $\sim 15\times5\times100$ km am$\times2 = 1.5\times10^4$ km atm of methane or equivalent carbon—almost all of which must be sequestered inside the satellite. In fact, the actual numbers must be less than this by a factor $\sim$2 given the small amount of escape-driven enrichment of $^{13}C/^{12}C$ (1.1 $\times$ Earth, compared with 1.5$\times$ Earth for $^{15}N/^{14}N$; (Niemann et al. 2005)).

Either the sub-crustal carbon has not yet been converted to a volatile compound—in this case methane (Owen et al. 2005; Niemann et al. 2005)—or an original supply of methane is trapped as clathrate hydrate at the surface of a putative sub-crustal ocean (Hersant et al. 2004, 2007; Tobie et al. 2006).

Both ideas have problems. Making the methane in situ requires resupply of catalysts and raw materials, while a clathrate reservoir requires that methane (and krypton, with similar volatility) brought to the satellite as clathrate escapes to the atmosphere during accretion, then reforms a thick clathrate layer at the surface of the putative sub-crustal ocean. At least a pathway for outgassing of methane from the interior has been established by the detection of $^{40}Ar$ in the atmosphere (Waite 2005; Niemann et al. 2005).

The origin of Titan's nitrogen seems more clear. The absence of detectable amounts of Kr and Xe and the orders of magnitude depletion of $^{36}Ar$ relative to $N_2$ in the atmosphere argue that nitrogen could not have *arrived* as $N_2$, but rather was delivered as condensable nitrogen compounds, primarily $NH_3$ (Owen et al. 2005, 2006; Niemann et al. 2005). The difficulty in trapping $N_2$ is shared by argon, and krypton is not far behind. Hence the very low abundance of $^{36}Ar$ in Titan's atmosphere signals that a very small amount of $N_2$ was captured in the icy planetesimals that accreted to form the satellite. One can make this argument quantitative in approximate terms by referring to the laboratory studies of Bar-Nun et al. (1988). On Titan today, we find $^{36}Ar/N_2 = 2.3 \times 10^{-8}$. Taking account the escape of $\sim 5\times$ the present atmospheric nitrogen, we have an original value of $^{36}Ar/N = 2.3 \times 10^{-9}$. In the sun, $^{36}Ar/N = 2 \times 10^{-2}$ (Grevesse et al. 2005). It is therefore necessary to reduce the amount of argon captured in the icy planetesimals by a factor $\sim 10^{-7}$ from its original value in the nebula. This can occur if the ice formed at a temperature of $\sim$100 K, virtually identical to the present-day temperature of Titan's surface. One can imagine that this temperature should be reached in the immediate vicinity of Saturn as the giant planet accreted, but careful modeling must check this. There is thus the possibility of a self-consistent picture for the origin of nitrogen: The ice that accreted to form Titan was accreting in the sub-nebula surrounding the planet, which was sufficiently warm that the forming ice could only trap a tiny fraction of the available $N_2$ and $^{36}Ar$. However, $NH_3$ could still be easily trapped along with other condensable nitrogen (and carbon) compounds and after degassing and photo-dissociation, these became the source of the atmospheric $N_2$ we see today (Atreya et al. 1978). This scenario also explains the low abundance of $^{36}Ar$, the non-detection of Xe and Kr and predicts that very little methane would be trapped by the satellite-forming planetesimals. The alternative idea that volatiles arrived as clathrate hydrates postulates a depletion of ice in the solar nebula at Saturn's orbital distance that would prevent the trapping of a solar amount of clathrated $N_2$. Then the absence of detectable Xe and Kr on Titan, originally predicted to be enhanced by this model (Hersant et al. 2004), is now seen as the result of the sequestration of dense clathrate hydrates of these two noble gases at the bottom of the putative ocean (Hersant et al. 2007). However, this scenario is hard to reconcile with the likely outgassing of planetesimal clathrates during accretion and the later degassing of ocean-bottom clathrates of Kr and Xe during the massive overturning of Titan's core postulated by Tobie et al. (2006).

## 4 Giant Planets

If accretion of solid material in the solar nebula forms a body with a mass of $\sim 10\ M_E$, gravitational collapse of the surrounding solar nebula gas can occur, producing a giant planet (Mizuno 1980; Pollack et al. 1996). The composition of the resulting atmosphere will therefore have two well-mixed components: a contribution from the outgassing and vaporization of the original solid material and impacting planetesimals dissolving in the gaseous envelope of the growing planet, plus the composition of the surrounding solar nebular gas, which will be predominantly $H_2$ and He. For the reasons described earlier, the accreted planetesmials were initially anticipated to be depleted in nitrogen, and thus the resulting giant planet atmospheres were also predicted to be deficient in this element (Pollack and Bodenheimer 1989; Owen and Bar-Nun 1995). Thus it was surprising to find from the Galileo Probe Mass Spectrometer measurements that not only nitrogen but argon as well are both present in

Jupiter's atmosphere with solar abundances relative to carbon (Niemann et al. 1998). Furthermore, all of the heavy elements except neon that could be measured were found to be enhanced compared with solar abundances relative to hydrogen. Using the solar abundances of Anders and Grevesse (1989), this appears to be a nearly uniform enrichment of $3 \pm 1\times$ solar (Owen et al. 1999). The revised abundances published by Grevesse et al. (2005, this conference) lead to a much less uniform enrichment of $4 \pm 2$. Nevertheless, the enrichment of both $^{36}$Ar and N remains a puzzle whose solution appears to require the early existence in the solar nebula of icy planetesimals or grains formed at temperatures below 25 K if trapped by amorphous ice (Owen et al. 1999) or cooled to $T < 40$ K if clathrates were formed from crystalline ice that condensed at higher temperatures (Gautier et al. 2001; Hersant et al. 2004, 2007).

Exactly how this low-temperature material formed and delivered the volatiles to the planets remains obscure. Clathration seems unlikely because it requires at least $3\times$ the solar value of O/C in order to provide enough water to form the crystalline cages that would trap the volatiles. In fact, a much greater excess of water would be required if the efficiency of clathrate formation is less than 100%, as seems likely (Miller 2003). Cuzzi and Zahnle (2004) suggested that grains formed at temperatures below 25 K beyond the orbit of Neptune in the early solar nebula could migrate inwards and accrete to form the icy planetesimals. Guillot and Hueso (2007) proposed that such grains would evaporate, but the released volatiles would be confined to the mid-plane of the nebula and provide the necessary concentration of heavy elements as hydrogen and helium evaporated from the outer edges of the nebula at Jupiter's distance from the Sun. We have suggested instead that low-temperature icy planetesimals formed at the mid-plane of the nebula in its very earliest phases, and accreted to form cores and enrich the envelopes of the giant planets (Owen and Encrenaz 2003, 2006). Obviously we need more data on more planets to determine which of these possibilities (or another yet to be devised) is correct.

To distinguish the low-temperature icy planetesimals from ordinary comets, we suggested the acronym SCIPs, for Solar Composition Icy Planetesimals (Owen and Encrenaz 2003, 2006). These objects are thus identical to the hypothetical comets of Type III postulated by Owen and Bar-Nun (1995) and may be represented today by members of the Kuiper Belt (KBO). The latter can then be thought of as the relics of a widespread population of low-temperature planetesimals that contributed excess heavy elements to the giant planets. If this proves to be the case, SCIPs would have been the most abundant solid material in the early solar nebula.

The scatter in the observed enrichment of $4 \pm 2$ may suggest that "solar composition" may be inappropriate, as the enrichments of the different elements are not as uniform with the new solar abundances as they first appeared using Anders and Grevesse (1989). However, the main causes of the spread in values are the low values of Kr and Xe, neither of which has been determined directly from solar observations. Thus the corrections that have been applied to the other abundances in response to more sophisticated modeling of the solar atmosphere and its spectrum cannot be (and have not been!) applied to the abundances of Kr and Xe (Grevesse et al. 2005). The "new" values are essentially identical to those given by Anders and Grevesse (1989). Furthermore, we note that the abundance of $^{36}$Ar reported at this conference by Gloeckler is $3\times$ lower than the value favored by Grevesse et al. (2005).

Clearly more work on these abundances is required before drawing any cosmological conclusions from the lack of uniformity in the elemental enrichments found on Jupiter. Perhaps the Jovian values are the "true" solar values, so Kr and Xe need to be adjusted to fit, and the correct enrichment is $5 \pm 1$.

But such conjectures are premature. Meanwhile, we can justify the requirement for low-temperature formation of these SCIPs simply on the basis of the nitrogen without invoking

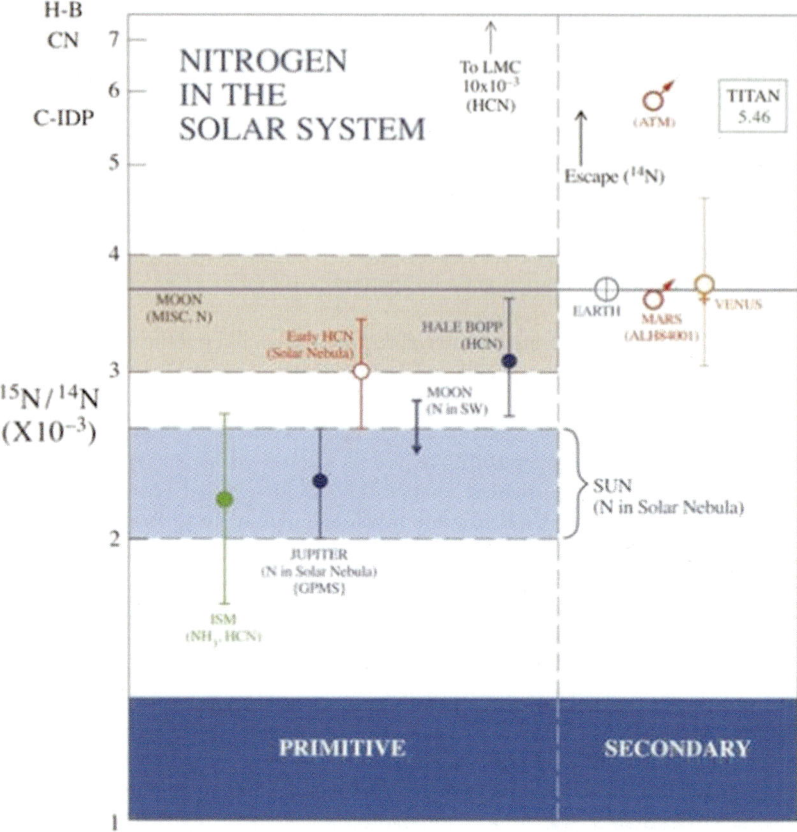


**Fig. 2** The figure illustrates the presence of the three distinct reservoirs of nitrogen in the solar system revealed by their isotopic ratios: –CN in comets (HB), and IDPs, more abundant nitrogen compounds in inner solar system solid bodies and HCN in an Oort cloud comet, and nitrogen on Jupiter, which is presumably the solar value (Owen et al. 2001), confirmed by an analysis of a high temperature inclusion in a meteorite (Meibom et al. 2007). Ratios produced by selective escape from Mars and Titan and interstellar values are also shown (see Owen et al. (2001) for discussion)

any of the noble gases. The fact that it was $N_2$ that was collected by the SCIPs, rather than condensed compounds such as $NH_3$ that dominated the delivery of N to Titan, the inner planets and meteorites is demonstrated by the nitrogen isotopes (Owen et al. 2001, Fig. 2). $^{15}N/^{14}N = 2.3 \pm 0.3 \times 10^{-3}$ on Jupiter, distinctly lower than the value on Venus, Mars, and the meteorites, which all exhibit ratios similar to the terrestrial value of $3.7 \times 10^{-3}$. This difference is close to the prediction from ion–molecule reactions in the ISM for isotope exchange between $N_2$ and $NH_3$ or HCN (Terzieva and Herbst 2000). (The ISM value for $^{15}N/^{14}N$ in $N_2$ has not been measured.) The Jovian isotope ratio is presumably identical to the solar value as in each case the ratio refers to the largest original reservoir of nitrogen in the solar nebula. Using Terzieva and Herbst (2000), we can calculate what the value of the isotope ratio in HCN in the solar nebula (= local ISM) would have been at the birth of the solar system, assuming the Jovian value of $2.3 \times 10^{-3}$ was indeed the value in the $N_2$. The result matches the value of $^{15}N/^{14}N$ found in HCN in comet Hale Bopp, lending support to this analysis (Fig. 2). Independent confirmation that the Jovian value represents the nitrogen in the sun was established by Meibom et al. (2007) from a nitrogen compound in a high-temperature inclusion in a meteorite. A third form of nitrogen that is present in the solar system is manifested by the high values of $^{15}N/^{14}N$ found in IDPs and in cometary –CN as mentioned above (Fig. 2).

But were SCIPs so abundant in the early Solar System? Does this analysis extend beyond Jupiter? Unfortunately, the phenomenally successful Cassini-Huygens mission does not include a Saturn probe, although it had been hoped originally that it would. The low temperature of Saturn ensures that the amount of ammonia in the sensible atmosphere is too low to permit a determination of the isotope ratio from the N in ammonia, as was done for Jupiter. (In the upper regions of these giant, hydrogen-rich atmospheres, $N_2$ is transformed to $NH_3$, just the reverse of the situation on Titan, where hydrogen easily escapes into space.) We therefore have only the abundance of methane to work with, and here indeed the SCIP model predicted the correct value of the enriched mixing ratio and hence C/H that was subsequently determined by the IR spectrometer (CIRS) on Cassini Huygens (Owen and Encrenaz 2003, 2006; Flasar et al. 2005).

For Uranus and Neptune, the situation is even worse because the remote sensing determinations of methane/hydrogen, that provide C/H in these atmospheres are very imprecise (Owen and Encrenaz 2006). The model is consistent with these approximate observations, but is not confirmed by them because of their low precision. Perhaps more significantly, the model predictions agree with the most recent determinations of D/H in the atmospheres of these two giants. In fact, it provides better agreement with the available data than models invoking a $100\times$ solar enrichment of water (Owen and Encrenaz 2006).

## 5 The Future

The next major advance in this field will come from a mission called Juno, being prepared for a trip to Jupiter in 2011. Juno will use microwave antennas to sense radiation from the planet's deep atmosphere down to 100 bars. This will enable the determination of the global water abundance on Jupiter, something the Galileo Probe was unable to accomplish because of its entry into a local "hot spot" depleted in condensable species (Niemann et al. 1998). In addition to the importance of determining the abundance of oxygen, the missing major element needed for models of the planet's interior, the global mixing ratio of $H_2O$ will also permit a distinction between clathrate hydrates and amorphous ice in planetesimals or grains as the major carrier of volatiles to Jupiter. As explained earlier, delivery by clathrates would lead to a large excess of water whereas the other carriers would provide a solar abundance of $H_2O$.

To move beyond this simple test, we need a *program* of atmospheric probes equipped with mass spectrometers for Saturn, Uranus and Neptune (Owen 2004; Bolton et al. 2003). We will then have abundances and isotope ratios we can compare with those of Jupiter that will allow us to distinguish among competing models for giant planet formation and their implications for time-dependent models for the solar nebula. Such a program would represent a major advance in our knowledge that would apply not only to our own planets but also to the hundreds of others now being discovered around other stars.

**Acknowledgements** I thank the organizers of this symposium for inviting me, an anonymous referee for an outstanding job of editing a very rough manuscript, Rudolf von Steiger for his seemingly endless patience and Johannes Geiss for his many fundamental contributions to the field of isotopic and elemental abundances as well as countless stimulating and generous discussions of their implications.

## References


E. Anders, N. Grevesse, Geochim. Cosmochim. Acta. **53**, 197–214 (1989)
C. Arpigny et al., Science **301**, 1522–1524 (2003)
S.K. Atreya, T.M. Donahue, W.R. Kuhn, Science **201**, 611–613 (1978)
S.K. Atreya, P.R. Mahaffy, H.B. Niemann, M.H. Wong, T.C. Owen, Planet. Space Sci. **51**, 105 (2003)
A. Bar-Nun, A.I. Kleinfeld, E. Kochavi, Phys. Rev. B **38**, 7749–7754 (1988)
S.J. Bolton et al., Bull. Am. Astron. Soc. DPS meeting #35, #41.08 (2003)
J.N. Cuzzi, K.J. Zahnle, Astrophys. J. **614**, 490 (2004)
E.F. Van Dishoeck et al., in *Protostars and Planets III*, ed. by E.H. Levy, J.I. Lunine (U. of Arizona Press, Tucson, 1993), pp. 163–241
T.M. Donahue, J.B. Pollack, in *Venus*, ed. by D.M. Hunten, L. Colin, T.M. Donahue, V.I. Moroz (U. of Ariz. Press, 1983), pp. 1003–1013
F.M. Flasar et al., Science **307**, 1247–1251 (2005)
D. Gautier, F. Hersant, O. Mousis, J.I. Lunine, Astrophys. J. **550**, L227 (2001) Erratum, Astrophys. J. **559**, L183
J. Geiss, Rev. Mod. Astron. **1**, 1–27 (1988)
N. Grevesse, M. Asplund, A.J. Sauval, in *Element Stratification in Stars, 40 years of Atomic Diffusion* ed. by G. Alecian, O. Richard, S. Vauclair, EAS Publication Series (EDP Sciences, 2005)
T. Guillot, R. Hueso, Mon. Not. Roy. Astron. Soc. (2007, in press)
F. Hersant, D. Gautier, J.I. Lunine, Planet. Space Sci. **52**, 623–641 (2004)
F. Hersant, D. Gautier, J.I. Lunine, Icarus (2007, in press)
J.I. Lunine, Y.L. Yung, R.D. Lorenz, Planet. Space Sci. **47**, 1291–1303 (1999)
P.R. Mahaffy et al., J. Geophys. Res. **105**(E6), 15061–15071 (2000)
A. Meibom et al., Astrophys. J. **656/2**, L33–L3 (2007)
H.J. Melosh, A.M. Vickery, Nature **338**, 487–489 (1989)
S. Messenger, Nature **404**, 968–971 (2000)
S.L. Miller, Personal communication (2003)
H. Mizuno, Prog. Theor. Phys. **64**, 544–557 (1980)
H.B. Niemann et al., J. Geophys. Res. **103**, 22831–22845 (1998)
H.B. Niemann et al., Nature **438**, 779–784 (2005)
T. Owen, in *Mars*, ed. by H.H. Kieffer, B.M. Jakosky, C.W. Snyder, M.S. Matthews (1992), pp. 818–834
T. Owen, *Atmospheric Probes: Needs and Prospects*. ESA SP-**544**, 2004, pp. 7–11
T. Owen, A. Bar-Nun, Icarus **116**, 215–226 (1995)
T. Owen, A. Bar-Nun, in *Origin of the Earth and Moon*, ed. by R. Canup, K. Righter (U. of Arizona Press, Tucson, 2000), pp. 459–474
T. Owen, T. Encrenaz, Space Sci. Rev. **106**(1), 121–138 (2003)
T. Owen, T. Encrenaz, Planet. Space Sci. **54**, 1188–1196 (2006)
T. Owen, S.K. Atreya, H.B. Niemann, Phys. Usp. **48**(6), 635–638 (2005)
T. Owen et al., Nature **402**, 269–270 (1999)
T. Owen et al., Astrophys. J. **553**, L77–L79 (2001)
T. Owen et al., Faraday Discuss. 387–391 (2006)
J.B. Pollack, P. Bodenheimer, in *Origin and Evolution of Planetary and Satellite Atmospheres*, ed. by J.B. Pollack, S.K. Atreya (U. of Arizona Press, Tucson, 1989), pp. 564–602
J.B. Pollack et al., Icarus **124**, 62–83 (1996)
D. Strobel, Icarus **21**, 466–470 (1974)
R. Terzieva, E. Herbst, Mon. Not. Roy. Astron. Soc. **317**, 563–568 (2000)
G. Tobie et al., Nature **440**, 61–64 (2006)
J.H. Waite Jr., et al., Science **308**, 982–986 (2005)
M.H. Wong et al., Icarus **171**, 153–170 (2004)
Y. Yung, M. Allen, J.P. Pinto, Astrophys. J. Suppl. Ser. **55**, 465–506 (1984)

Space Sci Rev (2007) 130: 105–114
DOI 10.1007/s11214-007-9173-7

# The Solar Chemical Composition

**N. Grevesse · M. Asplund · A.J. Sauval**

Received: 3 January 2007 / Accepted: 14 March 2007 /
Published online: 8 May 2007


**Abstract** We present our current knowledge of the solar chemical composition based on the recent significant downward revision of the solar photospheric abundances of the most abundant metals. These new solar abundances result from the use of a 3D hydrodynamic model of the solar atmosphere instead of the classical 1D hydrostatic models, accounting for departures from LTE, and improved atomic and molecular data. With these abundances, the new solar metallicity, $Z$, decreases to $Z = 0.012$, almost a factor of two lower than earlier widely used values. We compare our values with data from other sources and analyse a number of impacts of these new photospheric abundances. While resolving a number of longstanding problems, the new 3D-based solar photospheric composition also poses serious challenges for the standard solar model as judged by helioseismology.

**Keywords** Sun: abundances, photosphere, corona

## 1 Introduction

New generation of three-dimensional (3D) hydrodynamic models of the solar lower atmosphere have been applied, for the first time, to the analysis of the solar photospheric

N. Grevesse (✉)
Centre Spatial de Liège, Université de Liège, avenue Pré Aily, 4031 Angleur-Liège, Belgium
e-mail: nicolas.grevesse@ulg.ac.be

N. Grevesse
Institut d'Astrophysique et de Géophysique, Université de Liège, Allée du 6 Août, 17, B5C, 4000 Liège, Belgium

M. Asplund
Research School of Astronomy and Astrophysics, Australian National University, Cotter Road, Weston 2611, Australia
e-mail: martin@mso.anu.edu.au

A.J. Sauval
Observatoire Royal de Belgique, avenue Circulaire, 3, 1180 Bruxelles, Belgium
e-mail: jacques.sauval@oma.be

spectrum, rather than the classical 1D photospheric models used during more than four decades. This new approach, combined with considerations of non-LTE effects in the line formation, leads to significant downward revisions of the abundances. This is a totally new situation. Indeed, the reasons for abundance changes among older abundance tables, for example Grevesse and Sauval (1998) versus Grevesse and Noels (1993) versus Anders and Grevesse (1989), were essentially due to the use of more accurate atomic data, especially transition probabilities rather than improved models of the solar atmosphere.

The main results of the new analyses, concerning the most abundant elements, have been described in detail in a series of papers entitled "Line formation in solar granulation" (Asplund et al. 2000a, 2000b; Asplund 2000; Asplund et al. 2004; Asplund 2004; Asplund et al. 2005b; Scott et al. 2006) and in two recent reviews (Asplund et al. 2005a; Grevesse et al. 2005). We shall therefore describe the main advantages of the use of the new 3D model only briefly, and then discuss the new photospheric abundance results of C, N, O, Na to Ca and Fe as well as Ne and Ar, compare these results with data from other sources and comment on the various consequences of these new solar element abundances.

## 2 Model Atmospheres: 3D Versus 1D

The visible surface layer of the Sun is just on top of the convection zone. Therefore the solar granulation strongly influences the photospheric spectrum. We not only see the solar granulation but spectral lines do show, through their shapes (widths, shifts and asymmetries), that matter motions are present in the photosphere as well.

The 3D model atmosphere of the solar granulation results from the solution of the hydrodynamic equations of mass, momentum and energy conservation coupled to the equation of radiative transfer (see e.g. Asplund et al. 2000a and references therein). These models do not invoke any free parameters adjusted to agree with observational constraints as was earlier the case with the micro- and macro-turbulence parameters required with the 1D models.

The simulations using the 3D model successfully reproduce key observational facts, such as the granulation topology and statistics, the helioseismological constraints, the brightness contrast and, last but not least, the shapes, shifts and asymmetries of the photospheric spectral lines. Actually, for the first time, we are able to fit nearly perfectly a predicted line profile with the observed one.

## 3 Photospheric Abundances

Table 1 presents a compilation of the most reliable solar and meteoritic abundances; they are given in the logarithmic scale relative to hydrogen adopted by astronomers, $A_{\rm el} = \log N_{\rm el}/N_{\rm H} + 12.0$, where $N_{\rm el}$ is the abundance of a given element by number. Meteoritic values are taken from the compilation of Lodders (2003) but they are placed on a slightly different absolute abundance scale. Since the reference element is silicon in the meteoritic scale and since our recommended Si value is 0.03 dex lower than that advocated by Lodders (2003), we correspondingly adjusted all meteoritic abundances by that amount (−0.03 dex).

The present-day photospheric abundance of helium adopted is obtained from inversion of helioseismic data by Basu and Antia (2004): $Y$, the abundance by mass of He, is $Y = 0.2485$. Although this value is independent of the solar model, it depends on the equation of state. This He abundance corresponds to $A_{\rm He} = 10.93$ i.e. $N_{\rm He}/N_{\rm H} = 8.5\%$.

**Table 1** Element abundances in the present-day solar photosphere and in meteorites (C1 chondrites). Indirect solar estimates are marked with [..]. For He, see text

| | Elem. | Photosphere | Meteorites | | Elem. | Photosphere | Meteorites |
|---|---|---|---|---|---|---|---|
| 1 | H | 12.00 | 8.25 ± 0.05 | 44 | Ru | 1.84 ± 0.07 | 1.77 ± 0.08 |
| 2 | He | [10.93±0.01] | 1.29 | 45 | Rh | 1.12 ± 0.12 | 1.07 ± 0.02 |
| 3 | Li | 1.05 ± 0.10 | 3.25 ± 0.06 | 46 | Pd | 1.66 ± 0.04 | 1.67 ± 0.02 |
| 4 | Be | 1.38 ± 0.09 | 1.38 ± 0.08 | 47 | Ag | 0.94 ± 0.25 | 1.20 ± 0.06 |
| 5 | B | 2.70 ± 0.20 | 2.75 ± 0.04 | 48 | Cd | 1.77 ± 0.11 | 1.71 ± 0.03 |
| 6 | C | 8.39 ± 0.05 | 7.40 ± 0.06 | 49 | In | 1.60 ± 0.20 | 0.80 ± 0.03 |
| 7 | N | 7.78 ± 0.06 | 6.25 ± 0.07 | 50 | Sn | 2.00 ± 0.30 | 2.08 ± 0.04 |
| 8 | O | 8.66 ± 0.05 | 8.39 ± 0.02 | 51 | Sb | 1.00 ± 0.30 | 1.03 ± 0.07 |
| 9 | F | 4.56 ± 0.30 | 4.43 ± 0.06 | 52 | Te | | 2.19 ± 0.04 |
| 10 | Ne | [7.84±0.06] | −1.06 | 53 | I | | 1.51 ± 0.12 |
| 11 | Na | 6.17 ± 0.04 | 6.27 ± 0.03 | 54 | Xe | [2.24±0.02] | −1.97 |
| 12 | Mg | 7.53 ± 0.09 | 7.53 ± 0.03 | 55 | Cs | | 1.07 ± 0.03 |
| 13 | Al | 6.37 ± 0.06 | 6.43 ± 0.02 | 56 | Ba | 2.17 ± 0.07 | 2.16 ± 0.03 |
| 14 | Si | 7.51 ± 0.04 | 7.51 ± 0.02 | 57 | La | 1.13 ± 0.05 | 1.15 ± 0.06 |
| 15 | P | 5.36 ± 0.04 | 5.40 ± 0.04 | 58 | Ce | 1.70 ± 0.10 | 1.58 ± 0.02 |
| 16 | S | 7.14 ± 0.05 | 7.16 ± 0.04 | 59 | Pr | 0.58 ± 0.10 | 0.75 ± 0.03 |
| 17 | Cl | 5.50 ± 0.30 | 5.23 ± 0.06 | 60 | Nd | 1.45 ± 0.05 | 1.43 ± 0.03 |
| 18 | Ar | [6.18±0.08] | −0.45 | 62 | Sm | 1.00 ± 0.03 | 0.92 ± 0.04 |
| 19 | K | 5.08 ± 0.07 | 5.06 ± 0.05 | 63 | Eu | 0.52 ± 0.06 | 0.49 ± 0.04 |
| 20 | Ca | 6.31 ± 0.04 | 6.29 ± 0.03 | 64 | Gd | 1.11 ± 0.03 | 1.03 ± 0.02 |
| 21 | Sc | 3.17 ± 0.10 | 3.04 ± 0.04 | 65 | Tb | 0.28 ± 0.30 | 0.28 ± 0.03 |
| 22 | Ti | 4.90 ± 0.06 | 4.89 ± 0.03 | 66 | Dy | 1.14 ± 0.08 | 1.10 ± 0.04 |
| 23 | V | 4.00 ± 0.02 | 3.97 ± 0.03 | 67 | Ho | 0.51 ± 0.10 | 0.46 ± 0.02 |
| 24 | Cr | 5.64 ± 0.10 | 5.63 ± 0.05 | 68 | Er | 0.93 ± 0.06 | 0.92 ± 0.03 |
| 25 | Mn | 5.39 ± 0.03 | 5.47 ± 0.03 | 69 | Tm | 0.00 ± 0.15 | 0.08 ± 0.06 |
| 26 | Fe | 7.45 ± 0.05 | 7.45 ± 0.03 | 70 | Yb | 1.08 ± 0.15 | 0.91 ± 0.03 |
| 27 | Co | 4.92 ± 0.08 | 4.86 ± 0.03 | 71 | Lu | 0.06 ± 0.10 | 0.06 ± 0.06 |
| 28 | Ni | 6.23 ± 0.04 | 6.19 ± 0.03 | 72 | Hf | 0.88 ± 0.08 | 0.74 ± 0.04 |
| 29 | Cu | 4.21 ± 0.04 | 4.23 ± 0.06 | 73 | Ta | | −0.17 ± 0.03 |
| 30 | Zn | 4.60 ± 0.03 | 4.61 ± 0.04 | 74 | W | 1.11 ± 0.15 | 0.62 ± 0.03 |
| 31 | Ga | 2.88 ± 0.10 | 3.07 ± 0.06 | 75 | Re | | 0.23 ± 0.04 |
| 32 | Ge | 3.58 ± 0.05 | 3.59 ± 0.05 | 76 | Os | 1.25 ± 0.11 | 1.34 ± 0.03 |
| 33 | As | | 2.29 ± 0.05 | 77 | Ir | 1.38 ± 0.05 | 1.32 ± 0.03 |
| 34 | Se | | 3.33 ± 0.04 | 78 | Pt | | 1.64 ± 0.03 |
| 35 | Br | | 2.56 ± 0.09 | 79 | Au | 1.01 ± 0.15 | 0.80 ± 0.06 |
| 36 | Kr | [3.25±0.08] | −2.27 | 80 | Hg | | 1.13 ± 0.18 |
| 37 | Rb | 2.60 ± 0.15 | 2.33 ± 0.06 | 81 | Tl | 0.90 ± 0.20 | 0.78 ± 0.04 |
| 38 | Sr | 2.92 ± 0.05 | 2.88 ± 0.04 | 82 | Pb | 2.00 ± 0.06 | 2.02 ± 0.04 |
| 39 | Y | 2.21 ± 0.02 | 2.17 ± 0.04 | 83 | Bi | | 0.65 ± 0.03 |
| 40 | Zr | 2.58 ± 0.02 | 2.57 ± 0.02 | 90 | Th | | 0.06 ± 0.04 |
| 41 | Nb | 1.42 ± 0.06 | 1.39 ± 0.03 | 92 | U | $<-0.47$ | −0.52 ± 0.04 |
| 42 | Mo | 1.92 ± 0.05 | 1.96 ± 0.04 | | | | |

Our new results only concern the elements C, N, O, Na to Ca, and Fe, as well as Ne and Ar. The abundances of all the elements not directly reconsidered here have been taken from our most recent compilation (Grevesse et al. 2005) updated with very recent works for Zr (Ljung et al. 2006), Pd (Xu et al. 2006), Sm (Lawler et al. 2006), Gd (Den Hartog et al. 2006) and Os (Quinet et al. 2006). All of these results do not use 3D models but rather the classical 1D models. However, they use improved atomic data, in particular transition probabilities.

As already mentioned the new solar analyses for C, N, O, Na to Ca and Fe have been carried out using the 3D hydrodynamic model discussed in Sect. 2. In each case, and whenever possible, we used as many indicators as possible of the abundance—forbidden and permitted atomic lines as well as molecular lines—in order to minimize systematic errors. A special effort has also been made to utilize only the very best available solar lines and line data. It is better to retain only a small number of best-quality abundance indicators rather than using larger samples of less reliable lines. In several incidences, detailed non-LTE calculations have been carried out after the required atomic data had become available.

### 3.1 Carbon, Nitrogen and Oxygen

Detailed accounts of our new analyses of these very important elements have recently been published (Asplund et al. 2005b; Asplund et al. 2004; Scott et al. 2006); the N results are currently being prepared for publication. They are also discussed in our recent reviews (Asplund et al. 2005a; Grevesse et al. 2005).

Table 2 summarizes the results obtained for the three elements with the 3D model and with the 1D model of Holweger and Müller (1974), which has been widely used for solar studies. We used a large number of abundance indicators: atomic and molecular lines covering the wide wavelength range from the visible to the infrared, and formed in different layers of the photosphere and having different sensitivities to temperature. For C and O, we see from Table 2 that, in sharp contrast with the analysis using the 1D model where the spread of the results is very large (0.31 dex for C, 0.23 dex for O), excellent agreement is found

**Table 2** C, N, O abundances as implied from a variety of different atomic and molecular indicators using a 3D hydrodynamic model of the solar atmosphere (Asplund et al. 2005a). Results from the semi-empirical model of Holweger and Müller (1974) are given for comparison

| Lines | $A_{C,N,O}$ | |
|---|---|---|
| | 3D | HM |
| [CI] | 8.39 | 8.45 |
| CI | $8.36 \pm 0.03$ | $8.39 \pm 0.03$ |
| CH $\Delta v = 1$ | $8.38 \pm 0.04$ | $8.53 \pm 0.04$ |
| $C_2$ Swan | $8.44 \pm 0.03$ | $8.53 \pm 0.03$ |
| CH A-X | $8.45 \pm 0.04$ | $8.59 \pm 0.04$ |
| CO $\Delta v = 1$ | $8.41 \pm 0.02$ | $8.62 \pm 0.02$ |
| CO $\Delta v = 2$ | $8.38 \pm 0.02$ | $8.70 \pm 0.03$ |
| NI | $7.85 \pm 0.08$ | $7.97 \pm 0.08$ |
| NH $\Delta v = 1$ | $7.73 \pm 0.05$ | $7.95 \pm 0.05$ |
| [OI] | $8.68 \pm 0.01$ | $8.76 \pm 0.02$ |
| OI | $8.64 \pm 0.02$ | $8.64 \pm 0.08$ |
| OH $\Delta v = 0$ | $8.61 \pm 0.03$ | $8.82 \pm 0.01$ |
| OH $\Delta v = 1$ | $8.61 \pm 0.03$ | $8.87 \pm 0.03$ |

between all abundance indicators when employing the 3D model. This excellent agreement between transitions of very different formation depths and temperature and pressure sensitivities is a very strong argument in favour of our new abundances as well as for the realism of the 3D model. In particular, we note with satisfaction that consistent results are now finally provided by the infrared vibration-rotation CO lines which have previously caused a great deal of troubles when analysed with a 1D model (Grevesse et al. 1995; Ayres 2002; Scott et al. 2006).

Nitrogen has only a few very faint NI lines, many of them blended with CN lines, and faint vibration-rotation lines of NH in the infrared, to offer as direct indicators of its abundance.

The new solar abundances of C, N, and O are much lower than those recommended in the widely used compilation of Anders and Grevesse (Anders and Grevesse 1989): $-0.17$ dex (C), $-0.27$ dex (N) and $-0.27$ dex (O) respectively. They are also much lower than the values recommended by Grevesse and Noels (1993) and Grevesse and Sauval (1998) in more recent compilations: $-0.13$ dex (C), $-0.14$ dex (N) and $-0.17$ dex (O) respectively.

### 3.2 Neon and Argon

It is well known that no spectral lines of Ne and Ar are present in the photospheric spectrum. The "photospheric" abundances of these elements, Ne being the most important one because of its high abundance, have therefore to be estimated from measurements made in coronal matter of various types. Ratios Ne/O and Ar/O are generally measured using X-ray and vacuum ultraviolet spectroscopy of various types of active centers and in the quiet corona, in the solar wind (SW) and in solar energetic particles (SEP). As these ratios refer to oxygen, the Ne and Ar abundances are directly affected by the new solar abundance of oxygen.

Measurements of the coronal Ne/O abundance ratio show however a very broad scatter by a factor of about 5, from very low, $\mathrm{Ne/O} \sim 0.1$, to very high values, $\mathrm{Ne/O} \sim 0.5$. In spite of these variations, a large number of analyses, using the various techniques mentioned above, applied to quite different types of coronal matter, lead to values around $\mathrm{Ne/O} = 0.15$ and $\mathrm{Ar/O} = 0.033$. We shall adopt the SEP values published by Reames (1999). Therefore the solar abundances of Ne and Ar become $A_{\mathrm{Ne}} = 7.84$ and $A_{\mathrm{Ar}} = 6.18$, much lower than older values around 8.1 and 6.4, respectively.

Since increasing Ne might possibly help reconciling the standard solar model and the observations of helioseismology (see Sect. 4.7), many new analyses have been devoted to the abundance of Ne: some confirm the low Ne/O ratio adopted here but others lead to higher values of this ratio.

Recently, Schmelz et al. (2005), analysing X-ray spectra of different coronal features and of the full-sun, and Young (2005a), from XUV spectra of the quiet sun, confirmed the low Ne/O value.

Drake and Testa (2005) however suggested, from X-ray analyses of a sample of highly active stars, that the Sun should have a much higher ratio, $\mathrm{Ne/O} = 0.41$. Recently, Liefke and Schmitt (2006) showed that these highly active stars are not at all representative of solar-type stars because, as Güdel (2004) already found, those stars exhibit the inverse FIP effect: elements with high first ionization potential (FIP) are enhanced rather than low FIP elements (see Sect. 4.5). Neon, having a very high FIP (21.6 eV), much larger than O (13.6 eV), is therefore enhanced and large Ne/O ratios are observed. We suggest, with Güdel (2004), that the solar corona, in some highly magnetically active centers, might perhaps also exhibit such an inverse FIP effect. This could maybe explain some of the high Ne/O ratios observed in various coronal structures (see also Sect. 4.5).

More puzzling are the solar results of Feldman and Widing (2007 and references therein) and Bochsler (2007 and references therein). Feldman and Widing derived the Ne/O and Mg/Ne ratios in different types of solar matter. Ne/O remains rather constant, around 0.15, whereas Mg/Ne varies tremendously because of the FIP effect (Sect. 4.5). Adopting the lowest value for the ratio Mg/Ne, corresponding to no FIP effect, they derive a high value of the abundance of Ne and, from Ne/O, a high value of O, in agreement with the older values of both Ne, $A_{\mathrm{Ne}} = 8.1$, and O, $A_{\mathrm{O}} = 8.9$. The only way to reconcile those results with ours is to suspect that FIP effects smaller than 1 can occur for the low FIP element Mg (see Sect. 4.5).

On the other hand, Bochsler (2007) derived the abundances of Ne and O from solar wind data. We note that the pioneer aluminum foil experiment by Johannes Geiss and collaborators, during the Apollo missions on the Moon, are still considered to give the best solar wind data ever obtained. From the very variable He/Ne and $^{4}\mathrm{He}/^{3}\mathrm{He}$ ratios, Bochsler finally derived a ratio He/Ne independent of any fractionation. Combined with the value for He in the solar convection zone reported in Table 1, he found a high value of the abundance of Ne, $A_{\mathrm{Ne}} = 8.10$. With the rather stable Ne/O ratio observed in the solar wind, he also found a high value for O, $A_{\mathrm{O}} = 8.92$.

Also very puzzling is the recent analysis of Cunha et al. (2006) of a sample of B stars. While they confirm the low solar abundances of CNO described here, they, on the contrary, find a high abundance of Ne, $A_{\mathrm{Ne}} = 8.11$, in agreement with the old solar value.

Although we still believe the low Ne/O ratio and the low Ne abundance reported in Table 1, to be representative of the present Sun, we have to keep these last results in mind and wait for additional analyses of the Sun, stars and interstellar medium.

### 3.3 Intermediate Elements: Na to Ca, Fe

3D analyses of Na, Mg, Al, Si, P, S, K, Ca and Fe have also been performed. Detailed results have been published (Si: Asplund 2000; Fe: Asplund et al. 2000b; Na, Mg, Al, P, S, K, Ca: Asplund et al. 2005a). When possible, departures from LTE have also been taken into account. As for CNO, the 3D-based abundances are lower than the 1D-based results but the impact of the 3D model atmosphere is smaller than for CNO, mainly since the abundances are based on atomic transitions rather than on very temperature-sensitive molecular lines. The results reported in Table 1 for these elements are actually 0.05 to 0.10 dex lower than those recommended by Anders and Grevesse (1989) and Grevesse and Sauval (1998). Once again, as for CNO, the difference 3D−1D is much larger for the most sensitive indicators of temperature, like NaI or CaI, which are minor species compared to NaII or CaII, or to PI and SI, which are also major species.

## 4 Implications of the New Results and Comments

### 4.1 Solar Metallicity

Because we decreased by rather large amounts the abundances of elements which contribute much to the metallicity, $Z$ will decrease accordingly. With the solar composition given in Table 1, the new present day solar metallicity is $Z = 0.0122$.

This metallicity is much lower than the previously recommended and widely used values, $Z = 0.0189$ (Anders and Grevesse 1989) and $Z = 0.017$ (Grevesse and Noels 1993; Grevesse and Sauval 1998).

## 4.2 Protosolar Chemical Composition

The well known diffusion of the elements at the bottom of the solar convection zone slowly depletes this reservoir as well as the photosphere on top of it (see e.g. Turcotte et al. 1998; Turcotte and Wimmer-Schweingruber 2002). Corrections to the present day abundances as in Table 1 in order to obtain the protosolar abundances can be estimated: the values in Table 1 have to be increased by 0.05 dex for all the so-called metals and the abundance of He has to be increased by 0.057 dex (Grevesse et al. 2005).

The protosolar values for $Y$, the abundance by mass of He, and the metallicity, $Z$, become $Y_0 = 0.2735$ and $Z_0 = 0.0132$.

## 4.3 The Sun is a Sunlike Star

Previous studies suggested that the Sun had too high abundances of O and C compared to the solar neighbourhood. The new lower solar abundances of C and O show that the Sun is now in agreement with the surrounding interstellar medium and nearby B-stars (Turck-Chièze et al. 2004; Asplund et al. 2005a, 2005b).

## 4.4 Comparison with Meteorites

Since many years, it is well known that the agreement between photospheric and meteoritic abundances is very good. The mean difference is $0.01 \pm 0.06$ dex, when ignoring the obvious, known cases where the elements are depleted in the Sun (Li) or in meteorites or where the photospheric abundances are doubtful because of the lack of unperturbed lines, the lack of accurate transition probabilities and/or the problem of departure from LTE impossible to handle.

## 4.5 Photospheric Versus Coronal Abundances: the FIP Effect

Abundance measurements in various types of coronal matter, like different coronal structures, solar wind (SW), solar energetic particles (SEP), show the well known FIP (First Ionization Potential) effect. Elements with low FIP ($<$10 eV) are overabundant in the corona relative to the photosphere whereas higher FIP elements ($>$10 eV) have photospheric-type abundances. This highly variable phenomenon is discussed at large in different papers of this Volume. In spite of the rather large variations of this FIP effect, canonical mean enhancement factor values (or FIP factors, i.e. ratios of photospheric versus "coronal" abundances) can be derived: they amount to ~2.7 (slow SW) and ~1.8 (rapid SW, Bochsler 2006), ~3.25 (SEP, Reames 1999) and 1.25 to 1.66 for the quiet corona (Young 2005b). With our new photospheric abundances, these values are reduced to ~2.0 (slow SW), ~1.4 (rapid SW), ~2.4 (SEP) and 0.8 to 1.1 (quiet sun). The last numbers show that there exists coronal matter which shows no FIP effect i.e. which has photospheric-type composition. This is true as well for the polar plumes (Del Zanna et al. 2003). Possibly also FIP factors smaller than 1 could exist: this has also been suggested during this symposium.

A further remark needs to be made. The low FIP elements are well represented, by 14 elements from Na to Zn with FIP ranging from 5.1 to 9.4 eV, but the high FIP side is really underpopulated. Actually, only 3 elements have accurate photospheric as well as coronal abundances: C (FIP: 11.3 eV), O (13.6 eV) and N (14.5 eV). The uncertainties of the abundances of Cl (13.5 eV) and F (17.5 eV) both in the photosphere and corona are much too large and these two elements cannot be considered among the high FIP elements. Furthermore, Ar (15.8 eV) and Ne, with a very high FIP of 21.6 eV, cannot be taken into account because their photospheric abundances are not directly derived from the photosphere itself. Therefore we do not really know if neon does not behave differently from the other high FIP elements, such as C, N and O, which do however have a much lower FIP than Ne.

### 4.6 Miscellaneous

As the stellar abundances are generally referred to the solar values, the new solar scale will alter the cosmic yardstick. This has important impacts in various fields of astrophysics like stellar modeling as a whole, giant planets, T-Tauri models, Herbig Ae/Be, gas/dust ratio in dense interstellar clouds, ...

We have also to be very cautious when comparing our new 3D-based solar results with stellar abundance results for stars having outer convection zones. These stellar abundances could be severely biased because of the use of theoretical 1D models instead of 3D models (Asplund 2005). We have to keep in mind that stellar element abundances are not observed but interpreted based on models of the stellar atmospheres and line formation.

### 4.7 Problems with the Standard Solar Model

While the new abundances have positive implications as described hereabove, they introduce at least one new problem. Solar interior models computed with our new abundances completely disagree with the extremely precise measurements of the sound speed profile, the convection zone helium abundance ($Y = 0.2485$) and the depth of the convective envelope ($r_e = 0.713\ \mathrm{R}_\odot$) inferred from helioseismology while the same models computed with older solar abundances agree very well with these measurements. Since O, C and Ne, by order of decreasing importance, are very important contributors to the opacity in the layers just below the convection zone where the problems arise, the revised solar abundances of these elements decrease the opacity and this significantly alters the structure of these layers.

A flurry of papers, too numerous to be cited here, have appeared where solar scientists and others like J.N. Bahcall, S. Basu, H.M. Antia, M.H. Pinsonneault, J.A. Guzik, J. Montalban, A. Miglio, A. Noels, M. Seaton, S. Turck-Chièze, M. Castro, S. Vauclair, and their collaborators, have been reexamining systematically all the ingredients that enter the models and trying to find a solution. Indeed, no real solution has yet been found: only ad hoc measures such as artificially increasing the opacity in the region below the convection zone, increasing the diffusion velocities, increasing the neon abundance, accreting low $Z$ matter, ..., currently exist.

## 5 Conclusions

The new solar abundances presented here are systematically lower, for the elements Na to Ca and Fe, to much lower, for C, N, O, as well as Ne and Ar, than previously recommended values.

It should be noted that not the whole difference with previous models is attributed to the use of a 3D model atmosphere over classical 1D models since the adoption of more recent transition probabilities, more realistic non-LTE procedures when possible, better observations (infrared transitions not observed from ground-based facilities) and a proper accounting of blends play also an important role in this respect.

The use of a 3D hydrodynamic model represents however a real step forward in the modelling of the very inhomogeneous ever changing solar atmosphere. For the first time, we can reproduce the observed photospheric line profiles including their widths, shifts and asymmetries, the atomic- and molecular-based abundances now finally agree, there are no more significant trends with line strengths or excitation potentials in the derived abundances, very different lines, probing very different layers of the solar atmosphere with greatly different sensitivities to the physical conditions, now lead to the same results.

Although efforts should be continued in refining the 3D models, in the non-LTE calculations, in improving the number and accuracy of the atomic and molecular data, in extending the work to many more elements, we believe that the arguments hereabove mentioned are very strong in favour of our new results as well as for the realism of the 3D model.

**Acknowledgements** We would like to thank various collaborators, including Carlos Allende-Prieto, Paul Barklem, Mats Carlsson, Dan Kiselman, David Lambert, Åke Nordlund, Pat Scott, Bob Stein, and Regner Trampedach.

We also thank Manuel Güdel for fruitful informations on the solar Ne abundance problem, Andrea Miglio, Josefina Montalban, Arlette Noels-Grötsch, Gregor Rauw, Uri Feldman, Peter Bochsler, Don Reames, Sylvie Théado and Sylvaine Turck-Chièze for helpful discussions and the referee for very constructive suggestions.

NG thanks the organizers and Dick Mewaldt for their invitation, and Léo Houziaux, Secrétaire Perpétuel of the Belgian Royal Academy of Sciences and the Foundation Ochs-Lefèbvre for financial support.

## References

E. Anders, N. Grevesse, Geochim. Cosmochim. Acta **53**, 197 (1989)

M. Asplund, Astron. Astrophys. **359**, 755 (2000)

M. Asplund, Astron. Astrophys. **417**, 769 (2004)

M. Asplund, Annu. Rev. Astron. Astrophys. **43**, 481 (2005)

M. Asplund, A. Nordlund, R. Trampedach, C. Allende Prieto, R.F. Stein, Astron. Astrophys. **359**, 729 (2000a)

M. Asplund, A. Nordlund, R. Trampedach, R.F. Stein, Astron. Astrophys. **359**, 743 (2000b)

M. Asplund, N. Grevesse, A.J. Sauval, C. Allende Prieto, D. Kiselman, Astron. Astrophys. **417**, 751 (2004)

M. Asplund, N. Grevesse, A.J. Sauval, in *Cosmic Abundances as Records of Stellar Evolution and Nucleosynthesis*, ed. by T.G. Barnes III, F.N. Bash. ASP Conf. Ser., vol. 336 (2005a), p. 25

M. Asplund, N. Grevesse, A.J. Sauval, C. Allende Prieto, R. Blomme, Astron. Astrophys. **431**, 693 (2005b)

T.R. Ayres, Astrophys. J. **575**, 1104 (2002)

S. Basu, H.M. Antia, Astrophys. J. **606**, L85 (2004)

P. Bochsler (2007, personal communication)

P. Bochsler, Astron. Astrophys. Rev. (2006, in press)

K. Cunha, I. Hubeny, Th. Lanz, Astrophys. J. **647**, L143 (2006)

G. Del Zanna, B.J.I. Bromage, H.E. Mason, Astron. Astrophys. **398**, 743 (2003)

E.A. Den Hartog, J.E. Lawler, C. Sneden, J.J. Cowan, Astrophys. J. Suppl. Ser. **167**, 292 (2006)

J.J. Drake, P. Testa, Nature **436**, 525 (2005)

U. Feldman, K.G. Widing (2007, this volume). doi:10.1007/s11214-007-9157-7

N. Grevesse, A. Noels, in *Origin and Evolution of the Elements*, ed. by N. Prantzos, E. Vangioni-Flam, M. Cassé (Cambridge University Press, Cambridge, 1993), p. 14

N. Grevesse, A.J. Sauval, Space Sci. Rev. **85**, 161 (1998)

N. Grevesse, A. Noels, A.J. Sauval, in *Laboratory and Astronomical High Resolution Spectra*, ed. by A.J. Sauval, R. Blomme, N. Grevesse. ASP Conf. Ser., vol. 81 (1995), p. 174

N. Grevesse, M. Asplund, A.J. Sauval, in *Elements Stratification in Stars, 40 years of Atomic Diffusion*, ed. by G. Alecian, O. Richard, S. Vauclair. EAS Publ. Series, vol. 17 (2005), p. 21

M. Güdel, Astron. Astrophys. Rev. **12**, 71 (2004)

H. Holweger, E.A. Müller, Sol. Phys. **39**, 19 (1974)

E.A. Lawler, J.E. Den Hartog, C. Sneden, J.J. Cowan, Astrophys. J. Suppl. Ser. **162**, 227 (2006)

C. Liefke, J.H.M.M. Schmitt, Astron. Astrophys. **458**, L1 (2006)

G. Ljung, H. Nilsson, M. Asplund, S. Johansson, Astron. Astrophys. **456**, 1181 (2006)

K. Lodders, Astrophys. J. **591**, 1220 (2003)

P. Quinet, P. Palmeri, E. Biémont, A. Jorissen, S. van Eck, S. Svanberg, H.L. Xu, B. Plez, Astron. Astrophys. **448**, 1207 (2006)
D.V. Reames, Space Sci. Rev. **90**, 413 (1999)
J.T. Schmelz, K. Nasraoui, J.K. Roames, L.A. Lippner, J.W. Garst, Astrophys. J. **634**, L197 (2005)
P. Scott, M. Asplund, N. Grevesse, A.J. Sauval, Astron. Astrophys. **456**, 675 (2006)
S. Turck-Chièze, S. Couvidat, L. Piau, J. Ferguson, P. Lambert, J. Ballot, R.A. Garcia, P. Nghiem, Phys. Rev. Lett. **93**, 211102 (2004)
S. Turcotte, J. Richer, G. Michaud, C.A. Iglesias, F.J. Rogers, Astrophys. J. **504**, 539 (1998)
S. Turcotte, R.F. Wimmer-Schweingruber, J. Geophys. Res. **107**(A12), SSH5-1, 1442 (2002)
P.R. Young, Astron. Astrophys. **444**, L45 (2005a)
P.R. Young, Astron. Astrophys. **439**, 361 (2005b)
H.L. Xu, Z.W. Sun, Z.W. Dai, Z.K. Jiang, P. Palmeri, P. Quinet, E. Biémont, Astron. Astrophys. **452**, 357 (2006)

Space Sci Rev (2007) 130: 115–126
DOI 10.1007/s11214-007-9157-7

# Spectroscopic Measurement of Coronal Compositions

**U. Feldman · K.G. Widing**

Received: 18 December 2006 / Accepted: 2 February 2007 /
Published online: 27 April 2007


**Abstract** Although the elemental composition in all parts of the solar photosphere appears to be the same this is clearly not the case with the solar upper atmosphere (SUA). Spectroscopic studies show that in the corona elemental composition along solar equatorial regions is usually different from polar regions; composition in quiet Sun regions is often different from coronal hole and active region compositions and the transition region composition is frequently different from the coronal composition along the same line of sight. In the following two issues are discussed. The first involves abundance ratios between the high-FIP O and Ne and the low-FIP Mg and Fe that are important for meaningful comparisons between photospheric and SUA compositions and the second involves a review of composition and time variability of SUA plasmas at heights of $1.0 \le h \le 1.5R_{\odot}$.

**Keywords** Sun · Solar wind

## 1 Introduction

The 15 most abundant elements in the solar photosphere are composed of two distinct groups. The first includes the non-volatile Na, Mg, Al, Si, S, Ca, Fe, Ni and the second the volatiles H, N, C, O and the noble gases He, Ne, Ar. The elemental composition of the non-volatiles in C I carbonaceous chondrite meteorites, which are believed to be formed together with the rest of the solar system and did not undergo processing, happen to agree

U. Feldman (✉)
Artep Inc., 2922 Excelsior Spring Circle, Ellicott City, Columbia, MD 21042, USA
e-mail: uri.feldman@nrl.navy.mil

U. Feldman · K.G. Widing
E.O. Hulburt Center for Space Research, Naval Research Laboratory, Washington, DC 20375-5352, USA

K.G. Widing
Computational Physics Inc., Fairfax, VA 22031, USA

very well with the photospheric composition. Thanks to the refined capabilities of laboratory measurements, C I carbonaceous chondrite abundances are well established.

Photospheric abundances of volatile elements, which are not faithfully retained in meteoritic material, could not be derived in the laboratory, and thus are not as well established. Neutral H, C, N, and O, which have relatively low excitation energies, are well represented in photospheric spectra and in principle their abundances could be derived from the spectral features. Although attempts to determine their abundances began long ago, they are still being modified. A case in point is the abundance of O, which when measured on the logarithmic scale of $\log \mathrm{H} \equiv 12$, changed from 8.93 (Anders and Grevesse 1989) to $8.83 \pm 0.035$ (Grevesse and Sauval 1998), to $8.69 \pm 0.05$ (Allende Prieto et al. 2001), and to $8.66 \pm 0.05$ (Asplund et al. 2004), a factor of 1.9. Since the excitation energies of neutral He, Ne and Ar are very high, they cannot be observed in photospheric spectra and their abundances are usually derived by association with other elements. The Ne abundance is assumed to be 0.15 of the O abundance. Thus a decrease in the O abundance implies a similar decrease in the Ne abundance.

In the solar upper atmosphere (SUA) abundances, most often are derived relative to the well established Mg, Si or Fe. Thus, uncertainties introduced by the O abundance are avoided. In a few special cases SUA abundances are directly compared with the abundance of H derived from an H I emission line or from the free–free continuum.

Observational studies indicate that in all photospheric regions the elemental composition is the same; however; in the SUA this often is not the case. Pottasch (1963, 1964) was the first to present spectroscopic evidence of composition differences between the SUA and the photosphere. In scaling the ultraviolet abundance to H he found that Mg, Al, and Si were three times more abundant in the SUA compared to the then standard photospheric abundances (Goldberg et al. 1960). Unfortunately, his finding did not attract attention. The first conclusive findings of differences between SUA and photospheric composition resulted from in-situ solar wind (SW) and solar energetic particle observations (Hovestadt 1974; Geiss 1982; Breneman and Stone 1985). When compared with the photosphere, elements with first ionization potential (FIP) $\leq 10$ eV (low-FIP) were often overabundant by about a factor 4–5 while elements with FIP $\geq 11$ eV (high-FIP) stayed unchanged. Inspired by SW and Solar Energetic Particles findings, Meyer (1985) reexamined the literature and found that, relative to the photosphere, the SUA spectroscopic results are consistent with a factor of 4–6 under-abundance by the high-FIP when compared with the low-FIP elements. Later, Meyer (1996) concluded that the low-FIP elements were overabundant by a factor of 4–5, while high-FIP elements stayed nearly photospheric. Following Meyer's 1985 publication, patterns in composition modification of SUA plasmas from the photosphere began to emerge. For details see review articles by Feldman (1992), Saba (1995), Fludra et al. (1999), Feldman and Laming (2000), Feldman and Widing (2003) and Raymond (2004).

## 2 Photospheric Abundance of Neon and Oxygen

The lowest excited energy levels of neutral He, Ne and Ar are too energetic to be produced under photospheric conditions; however, in the much hotter SUA plasmas they are easily ionized and excited. Using spectra emitted by SUA plasmas Ne abundance was derived independent of the O abundance by two methods. In the first, using SUA spectra emitted by plasmas known to have photospheric composition, the Ne abundance was compared with well-established Mg, or Fe abundances. Such is the case when photospheric material is violently ejected into the corona, in flaring plasmas that erupt in close proximity to the photosphere, in bright active regions shortly after emerging, in prominences or in coronal hole (CH) plasmas. In the second method the Ne line intensities are directly compared with the H line intensities or with the free–free continuum. Below we evaluate published results that were based on measurements by different instruments spanning wide wavelength ranges.

### 2.1 The Neon Photospheric Abundance Derived from SUA Plasmas

Recently Lodders (2003), Asplund et al. (2005) and Grevesse et al. (2005) published modified photospheric abundance values. The photospheric abundances of Mg and Fe in the Lodders (2003) publication are $\log \mathrm{Mg} = 7.55 \pm 0.02$ and $\log \mathrm{Fe} = 7.47 \pm 0.03$. These values are 0.02 higher than values published by Asplund et al. (2005) and Grevesse et al. (2005). In a flare, believed to occur in or near the photosphere, a Ne/Mg abundance ratio of $3.6 \pm 0.3$ was determined (Feldman and Widing 1990). By combining the derived ratio with the Mg abundance a log Ne abundance of $8.11 \pm 0.09$ is obtained. Widing (1997) measured the Ne/Mg abundance ratio in newly emerging active regions. The average Ne/Mg abundance ratio of the three regions that were measured very shortly after emerging was $3.30 \pm 0.09$, implying a log Ne abundance of $8.07 \pm 0.06$. Young and Mason (1998) derived for a newly emerging flux region an abundance ratio of $\mathrm{Mg/Ne} = 0.26 \pm 0.05$ ($\mathrm{Ne/Mg} = 3.85 \pm 0.75$) corresponding to a $\log \mathrm{Ne} = 8.13 \pm 0.08$. The transition region is known to have a FIP bias of $\sim$1.5. Young (2005a) studied the Mg/Ne line ratios in the transition region network and found a $\mathrm{Mg/Ne} = 0.316$ ($\mathrm{Ne/Mg} = 3.16$). Using the derived value as a lower limit of the Ne photospheric abundance a $\log \mathrm{Ne} = 8.05 \pm 0.07$ is derived. McKenzie and Feldman (1992) found over a factor of 4 variation when deriving the Fe/Ne abundance ratio in a large number of flares. The smallest Fe/Ne values assumed to arise from flares having nearly photospheric abundances were $\mathrm{Fe/Ne} \sim 0.27$ ($\mathrm{Ne/Fe} = 3.7$). In combining this value with the photospheric Fe abundance a $\log \mathrm{Ne} = 8.04 \pm 0.20$ is obtained. Using spectra from high temperature quiescent prominences expected to have photospheric abundances an average $\mathrm{Ne/Mg} = 3.1$ is derived resulting in a $\log \mathrm{Ne} = 8.04$ (Spicer et al. 1998).

Using the intensity ratio between the He-like 1s2s–1s2p Ne IX line and the nearby continuum emitted by a moderate flare Landi et al. (2007) derived an absolute Ne abundance of $\log \mathrm{Ne} = 8.10 \pm 0.10$.

### 2.2 The Oxygen Photospheric Abundance

Measurements from radiation emitted by a variety of astrophysical sources indicated that the abundance value is $\sim$0.15. Recent derivations of the Ne/O abundance ratio in the SUA support the above ratio. Young (2005b) derived in the QS transition region a ratio of 0.17. Using Active region data in the X-rays Schmelz et al. (2005) derived an average ratio of 0.15 which is similar to an earlier result by McKenzie and Feldman (1992). Widing et al. (1986) derived in an eruptive prominence (surge), expected to have photospheric or nearly photospheric abundances, ratios of $\mathrm{Ne/O} = 0.12$ and $\mathrm{O/Mg} = 17.8$. Adopting a $\mathrm{Ne/O} = 0.15$ abundance ratio and combining it with the Ne abundance discussed in Sect. 2.1 log O abundance values vary between 8.80 and 8.95 (Table 1).

### 2.3 Summary of the Photospheric O and Ne Abundance Ratios

The Ne abundances as derived from the Ne/Mg, Ne/Fe and the Ne/free–free ratios discussed in Sect. 2.1 is $\log \mathrm{Ne} = 8.08 \pm 0.08$. By accepting a $\mathrm{Ne/O} = 0.15$ ratio an O abun-

**Table 1** Abundance ratios in plasmas expected to have photospheric or nearly photospheric compositions

| Wavelength | Ne/Mg | Ne/Fe[a] | O/Mg | log Ne[b] | log O[c] |
|---|---|---|---|---|---|
| 400 Å Flare[d] | $3.6 \pm 0.3$ | | | $8.11 \pm 0.09$ | $8.93 \pm 0.09$ |
| 400 Å Emerg. AR[e] | $3.30 \pm 0.09$ | | | $8.07 \pm 0.06$ | $8.89 \pm 0.07$ |
| 350–560 Å Emerg. AR[f] | $3.85 \pm 0.75$ | | | $8.13 \pm 0.08$ | $8.95 \pm 0.09$ |
| 308–633 Å Network[g] | $\geq 3.16 \pm 0.101$ | | | $\geq 8.05 \pm 0.10$ | $\geq 8.87 \pm 0.10$ |
| 9–15 Å Flares[h] | | 3.7 | | $8.04 \pm 0.20$ | $8.86 \pm 0.20$ |
| ~400 Å Prominence[i] | 3.1 | | | 8.04 | 8.86 |
| 300–600 Å Surge[j] | | | 17.8 | | 8.80 |
| 1248 Å Ne IX[k] | | | | $8.10 \pm 0.10$ | $8.92 \pm 0.08$ |

[a]It is assumed that $\log \text{Mg} = 7.55 \pm 0.02$, $\log \text{Fe} = 7.47 \pm 0.03$ (Lodders 2003)

[b]It is assumed that $\log \text{Mg} = 7.55 \pm 0.02$, $\log \text{Fe} = 7.47 \pm 0.03$ (Lodders 2003)

[c]A photospheric abundance ratio of $\text{Ne/O} = 0.15$, is assumed (Sect. 2.2)

[d]Feldman and Widing (1990)

[e]Widing (1997)

[f]Young and Mason (1998)

[g]Young (2005a)

[h]McKenzie and Feldman (1992)

[i]Spicer et al. (1998)

[j]Widing et al. (1986)

[k]Landi et al. (2007)

dance of $\log \text{O} = 8.90 \pm 0.08$ is obtained. Clearly, the O abundance derived from measurements in selected solar upper atmosphere plasmas, believed to have photospheric compositions, are at odds with the recently published values of Allende Prieto et al. (2001), and Asplund et al. (2004). The Asplund et al. (2004) values in which the O and Ne photospheric abundances are reduced by a factor of ~1.9 while the Mg abundance by only a factor of 1.1 result in a photospheric abundance ratio of $\text{Ne/Mg} = 2.0$. Results from various SUA plasmas indicate values of Ne/Mg~3.6 (Table 1), implying that by accepting a ratio of $\text{Ne/Mg} = 2$ Ne needs to either be enriched or Mg depleted in the discussed SUA regions; a conclusion that appears to be at odds with earlier inferences. As a result in the following comparisons between SUA and photospheric abundances we will assume the Feldman and Widing (2003) Ne and O abundance values.

## 3 The Composition of the SUA

*Definition 1* The FIP bias is defined as the ratio of the low-FIP elemental abundance in the SUA to its value in the photosphere, i.e., FIP bias $= A_{\text{El}}(\text{SUA})/A_{\text{El}}(\text{Ph})$

In the last two decades a large body of evidence regarding compositions of SUA plasmas have been accumulated. Below we present a selected set of observations elaborating the following main observational conclusions:

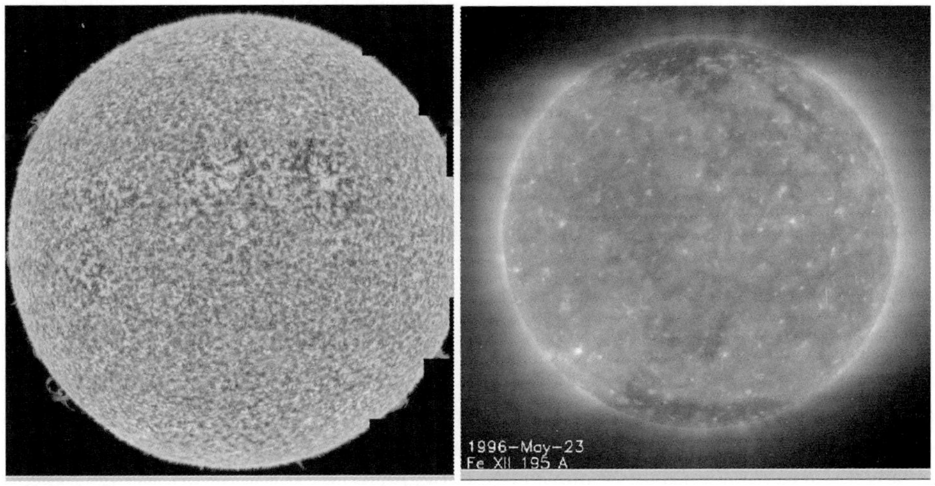


**Fig. 1** The QS and CH morphology in the lines of (a) C III ($T_e \sim 6 \times 10^4$ K) and (b) Fe XII ($\sim 1.4 \times 10^6$ K)

- Compositions appear to be modified only in plasmas confined in loop like structures.
- In the corona over quiet Sun (QS) regions where only average values where derived the FIP bias = 3–4, while in the open magnetic field coronal hole (CH) regions the FIP bias $\sim 1$.
- The magnitude of FIP bias is time dependent and in old active regions (AR) values as high as 15 were measured.
- FIP bias modification times are measured in hours or days and not in seconds or minutes.

### 3.1 The Average Elemental Composition Above CH, and QS Regions

In its 11-year cycle the SUA gradually changes from *solar minimum*, where only occasional small AR appear, to *solar maximum*, where large and complex areas of activity are present. When viewed through narrow temperature band filters in the $3 \times 10^4 \le T_e \le 7 \times 10^5$ K range polar CH plasmas and equatorial QS plasmas look fairly similar (Fig. 1a). However, at $T_e \ge 8 \times 10^5$ K the QS equatorial regions morphology becomes distinctly different from that of polar CH regions (Fig. 1b). While the maximum temperature of polar CH plasmas is $\le 1 \times 10^6$ K in the QS corona it is $\sim 1.4 \times 10^6$ K.

The $3 \times 10^4 \le T_e \le 7 \times 10^5$ K plasmas in both QS and CH regions appear to be confined in small loop-like structures. Also, in the hotter QS regions the plasma is confined in loops but in QS the loop like structures are much larger and may reach lengths in excess of $5 \times 10^5$ km. The morphology of polar CH, $T_e \sim 8 \times 10^5$ K, is much simpler and other than sporadic *polar plumes* that are present in the region, the plasma does not appear to be confined by closed magnetic structures.

The number of loop-like structures per unit area in QS and CH regions is too large to enable a modern instrument to resolve individual structures and follow their development over extended times. Thus in QS and CH regions information is available only on average values of the abundances close to the solar surface $h \le 1.5 R_\odot$; the composition changes with temperature, and the composition changes with height.

### *3.1.1 The Composition of the* $3 \times 10^4 \leq T_e \leq 7 \times 10^5$ *K Plasmas During Solar Minimum in CH and in QS Regions*

Using the Mg VI/Ne VI intensity ratios Feldman and Widing (1993) derived a FIP bias of 1.5–2 for transition region ($3 \times 10^4 \leq T_e \leq 7 \times 10^5$ K) structures in CH and QS regions. Young and Mason (1998) derived in transition region cell interiors and cell boundaries a FIP bias of $\sim$2. In a recent publication Young (2005b) derived a FIP bias of 1.25 for the network and 1.66 for the cell centers. Laming et al. (1995) analyzed the transition region spectra of the Sun as a star and found a FIP bias of slightly larger than 1. Based on the above we conclude that the $3 \times 10^4 \leq T_e \leq 7 \times 10^5$ K CH and QS plasmas have a FIP bias of $1.5 \pm 0.5$.

### *3.1.2 The Composition of the* $\sim 1.4 \times 10^6$ *K QS Corona*

Coronal QS plasmas during solar minimum are nearly isothermal. Using theoretical emissivities of lines from high- and low-FIP elements the FIP bias vs. temperature could be obtained. An example of the FIP bias in equatorial region plasma is shown in Fig. 2 (Feldman et al. 1998). As indicated by the figure both high- and low-FIP lines intersect each other at a temperature of $\log T_e \cong 6.13$ ($T_e \cong 1.35 \times 10^6$ K), the high FIP lines intersect at a FIP value of $\sim$1 while the low FIP lines at a value of 3–4. Landi et al. (2006) studied the coronal composition early in the rise of the activity cycle over 36 locations distributed in a rectangular grid bounded in the east-west direction by $1R_\odot$ to $1.5R_\odot$ and in the north–south directions by $-0.5R_\odot$ and $+0.5R_\odot$. As it turned out, with the exception of the extreme southwest corner, which resembled CH plasma properties, the region exhibited typical QS properties, i.e., the derived FIP bias at heights $h < 1.4R$ was $\sim$3.5 with a tendency for somewhat increased values at larger heights. Warren (1999) and Widing et al. (2005) studied QS regions in southeast and northwest quadrants midway on the rise of the next solar cycle at heights of $h < 1.1R_\odot$. The analysis indicated isothermal conditions in both regions and FIP bias values of 2.3 and 1.4 respectively.

QS coronal loops, which may exceed heights of a solar radius, survive for days. Since in such loops the electron density varies dramatically between the bases and tops, and the plasma turbulence is fairly small, elemental settling, which results in a FIP bias modification as a function of height, occurs. Figure 3 shows the normalized intensities of Ne VIII, Mg X, Si XI and Si XII lines vs. height; all are elements with similar atomic weights. Since their intensities decrease with height at the same rate it is safe to conclude that the plasma temperature does not vary with height. More important, the fact that the intensity ratios of the high-FIP to low-FIP lines do not vary with height shows that the FIP bias also does not vary with height, at least for elements with similar atomic weights. The decrease of the Fe line intensities with height in Fig. 3, which are formed in about the same temperature, is faster than for the lighter element lines. Between 1.05–$1.45R_\odot$ the Fe lines decrease three times faster than the Ne, Mg, and Si lines. A similar result is obtained when plotting the normalized Fe XII/S XI intensity ratios over the western hemisphere. Again, other than in the extreme southern region, elemental settling can account for the decrease of the Fe XII/S XI intensity ratio by a factor of 2–3 at $1.5R_\odot$.

### *3.1.3 The Composition of the* $\sim 8 \times 10^5$ *K CH Plasma*

Studies of the $T_e \cong 8 \times 10^5$ K polar CH plasmas indicate that in such regions the FIP bias is very close to 1 (Fig. 2b). Evidently CH plasmas undergo little, if any, composition modification once emerging from under the photosphere.

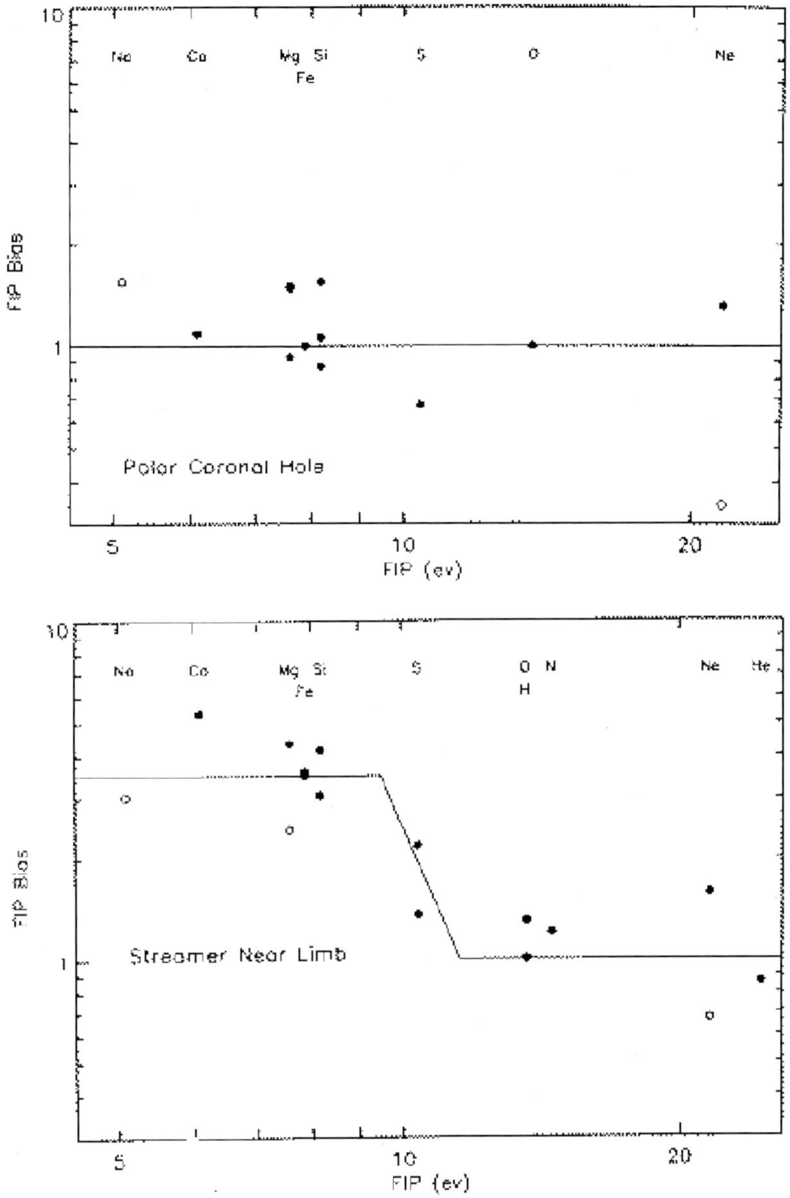

**Fig. 2** Plot of the effective FIP bias in the corona, (a) above the QS equatorial streamer, (b) above a polar CH region

## 3.2 FIP Bias in Distinct Solar Structures

It is expected that plasmas confined in loop like structures at the instant of emergence from under the photosphere will have photospheric composition. However, once above the photosphere the composition will start modifying. The amount of modification has been found to depend on the age of the structure.

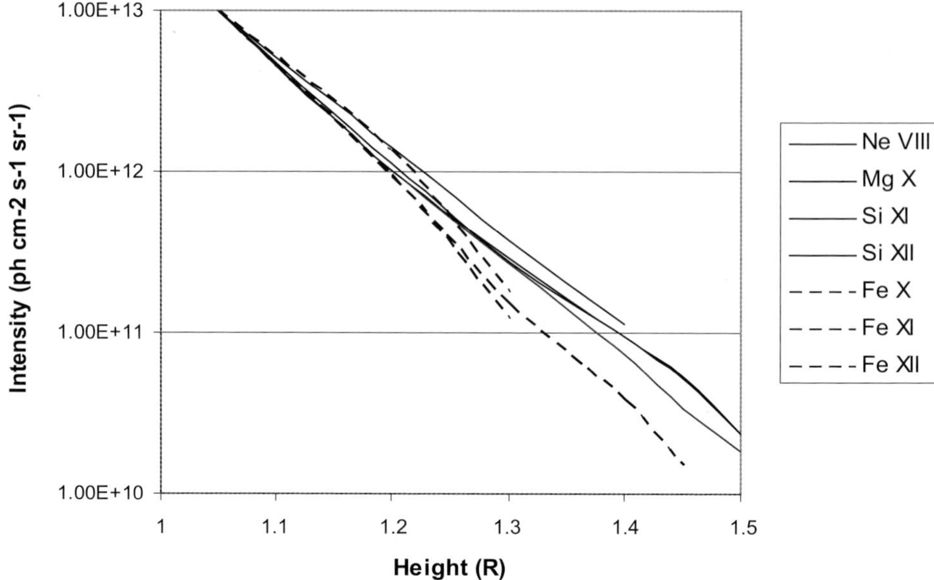


**Fig. 3** Relative line intensities as a function of height above the quiet equatorial region of a number of prominent spectral lines. The Ne VIII, Mg X 609 Å, Si IX and Si XII lines are plotted as solid curves. The Fe X, Fe XI and Fe XII lines are plotted as dashed curves

### *3.2.1 An Eruptive Prominence*

Properties of a plasma volume ~10 minutes after it was violently ejected from under the photosphere at a speed of ~400 km $s^{-1}$ and reached temperatures of $1 \times 10^6$ K were studied by Widing et al. (1986). Typically coronal O/Mg abundance ratios are ~6 but in the ejected plasma it was 17.8, quite similar to the photospheric ratio. The eruptive prominence is an indication that ejected plasmas while undergoing intense heating maintain for at least the first 10 minutes its photospheric composition.

### *3.2.2 Active Regions*

Shortly after emerging from under the photosphere plasma, trapped in rising AR magnetic field is expected and indeed have photospheric composition (Sheeley 1995, 1996; Widing 1997). The FIP bias of established hot (2–5 $\times 10^6$ K) AR plasmas were studied in the soft X-rays (i.e., Strong et al. 1988; McKenzie and Feldman 1992). While low-FIP intensity ratios of two low FIP elements (Fe XVIII to Mg XI) show minimal variations and are in close agreement with photospheric values the abundance ratio between low- to high-FIP elements (Fe XVII to Ne IX) vary by factors of 1–4 (McKenzie and Feldman 1992). In using the same line ratios Strong et al. (1991a, 1991b) and 1992 found in AR variations as high as 7 in Fe/Ne abundance ratios.

Widing and Feldman (1995) studied the O/Ne and Mg/Ne abundance ratios in a sample of AR using XUV lines. For three of the AR the FIP bias factor given by Mg VI to Ne VI lines was 4.8, 5.4 and 5.9: a testimony that for AR observed at high spatial resolution the measured FIP bias could exceed 4.

#### 3.2.3 Flares

Flares are transient phenomena. They can last from tens of minutes to tens of hours; so that in shorter events they do not last long enough to alter their composition. Thus, it is reasonable to expect that if the plasma source of a flare is local its composition should resemble the composition of the regions in which it erupted.

The Skylab 1973 December 2 flare erupted in the photosphere or very close to it and as expected its composition as derived from EUV lines was nearly photospheric (Feldman and Widing 1990). Sylwester et al. (1990), Fludra et al. (1990), Sterling et al. (1993), Strong et al. (1991a, 1991b), McKenzie and Feldman (1992), and Phillips et al. (2003) studied large populations of flares from their soft X-ray spectra and also found FIP bias values of 1–4.

Feldman et al. (2004) used the SUMER high spatial resolution observations to derive the composition at a height of $\sim 1.1 R_{\odot}$ in a recurrent flare. By comparing the intensity of the free–free continuum with the intensities of highly ionized Ca and Fe lines they obtained for the low-FIP elements a FIP bias of 8–10, a value significantly larger than the earlier measurements.

### 3.3 The FIP Bias Rate of Change in SUA

A most interesting issue associated with elemental abundances in the SUA is its rate of change and the maximum FIP bias values a plasma volume can have. At birth SUA plasmas have a FIP bias of 1, and within the following few hours it is nearly unchanged. However, in the average QS corona and in the slow speed SW FIP bias values of 4–5 are detected. Furthermore, in some old AR and in some flaring plasmas the FIP bias values can be significantly higher (8–16) (Widing and Feldman 1995; Young and Mason 1998; Dwivedi et al. 1999a; Feldman et al. 2004). Clearly a fairly slow FIP bias rate of change exists in the SUA.

The composition of AR plasmas, which are born near the east limb, could be monitored from birth until they decay or rotate beyond the west limb, usually for a week to 10 days. In searching the Skylab spectroheliograph records Widing and Feldman (1995) identified 6 fairly bright AR that were born near the east limb and lasted for at least 3 days. The composition of each of the regions was determined for as long as the active region features were visible. Figure 4 displays, for four of the regions, plots of the FIP bias values vs. time since emergence. As seen from the figure, soon after emergence the composition of the AR plasma was photospheric, but soon after it began to change. After 2–3 days it reaches FIP bias values of 4–5 and after 4.6 and 5.5 days the values were 8 and 9 respectively. The trend of increasing bias was similar in all AR. In order to establish the rate of FIP bias change in various types of SUA structures one needs to identify well-defined regions immediately after "birth" and follow them for a long time. Unfortunately, the Skylab spectroheliograph was the only instrument where this was thus far accomplished. Assuming that the rate of change derived for most magnetically confined SUA plasmas is similar to the one derived for the Skylab small AR we could conclude the following:

- Since in QS coronal plasmas average FIP bias values are $\sim 4$, QS coronal structures last on average 2–3 days.
- The colder QS and CH transition region structures which have a FIP bias of $\sim 1.5$ on average last much less, most likely only hours.
- The fact that in AR FIP bias of $\sim 15$ is often encountered is an indication that in some regions plasmas may become confined in the SUA for well over a week.

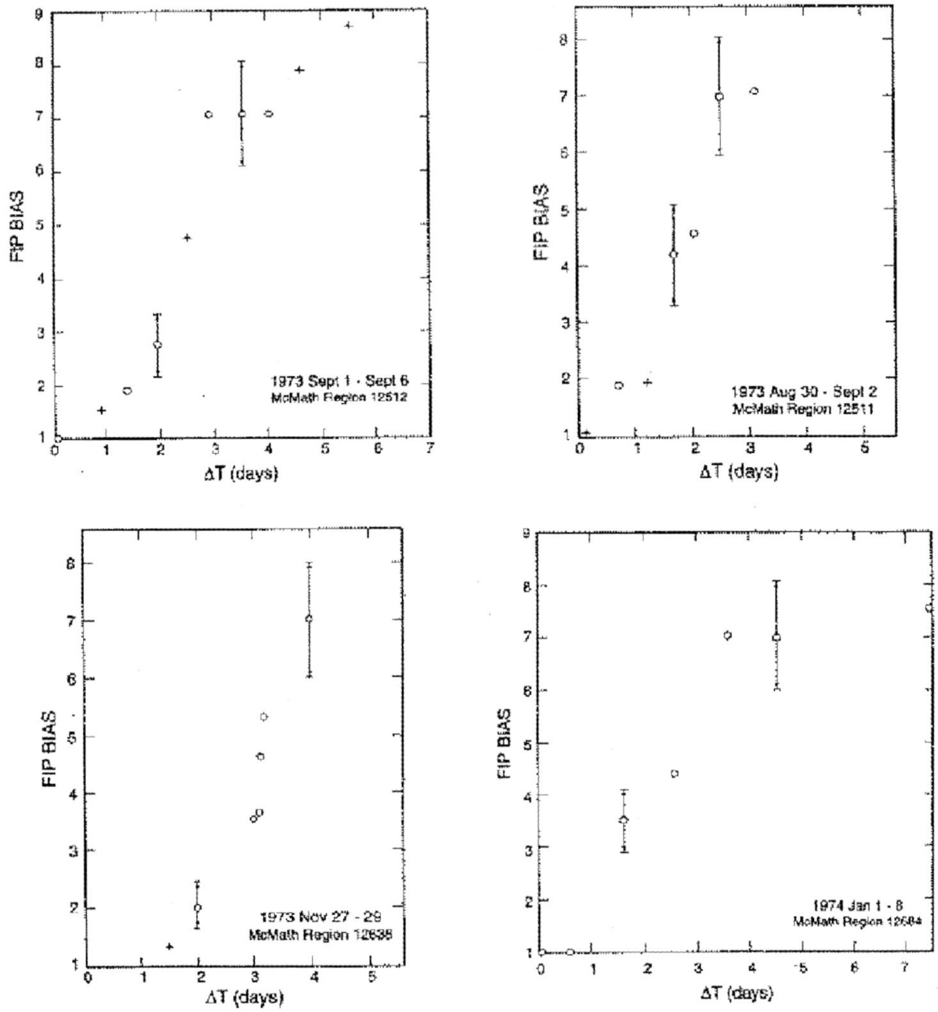

**Fig. 4** FIP bias values plotted against the time interval since emergence of the new regions. Open circles represent bias values determined from visual estimates for which a diagnostic line ratio equals 1. Crosses give the bias values determined from photometric calibration of diagnostic intensity ratios

## 4 Concluding Remarks

One major conclusion gained from composition surveys in the SUA is the close relationship between the lengths of plasma confinement and the magnitude of the FIP bias. Clearly no FIP enhancement is present in coronal holes where loop like structures that may confine the coronal plasma are rare. Similarly in the $3 \times 10^4$ K–$6 \times 10^5$ K plasma regions where loop-like structures are short lived, the FIP increase is small. From a SUMER/SOHO raster study of activity above the limb Dwivedi et al. (1999b) concluded that FIP fractionation strongly depends on plasma structures.

There is evidence that the FIP-biased composition of the slow SW is fed from loops in the AR belts. In a review of SW composition Zurbuchen et al. (1998) stated that "abundances

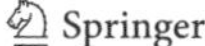

in the SW are observed to be highly variable in time and they exhibit structures on very small spatial scales". They give an example of the SW observed with SWICS/ACE in which the Fe/O abundance ratio and the $O^{7+}/O^{6+}$ charge-state ratio show large fluctuations with time. There are correlated changes of the two parameters on time scales of days or fraction of days. This type of behavior is difficult to reconcile with the time-stationary, diffusion models feeding the corona and the wind. The observations are easier to understand if the coronal composition is derived from a collection of AR loops of different ages and temperatures which are opened (possibly by reconnection) to release their FIP-biased material produced by a dynamic fractionation mechanism operating in the loop legs.

## References


C. Allende Prieto, D.L. Lambert, M. Asplund, Astrophys. J. Lett. **556**, L63 (2001)
E. Anders, N. Grevesse, Geochim. Cosmochim. Acta **53**, 197 (1989)
M. Asplund, N. Grevesse, A.J. Sauval, C. Allende Prieto, D. Kiselman, Astron. Astrophys. **417**, 751 (2004)
M. Asplund, N. Grevesse, A.J. Sauval, F.N., in *Cosmic Abundances as Records of Stellar Evolution and Nucleosynthesis, in Honor of David L. Lambert*, ed. by T.G. Barnes III, F.N. Bash. Astronomical Society of the Pacific Conference Series, vol. 336 (2005), p. 25
H.H. Breneman, E.C. Stone, Astrophys. J. Lett. **299**, L57 (1985)
B.N. Dwivedi, W. Curdt, K. Wilhelm, Astrophys. J. **517**, 516 (1999a)
B.N. Dwivedi, W. Curdt, K. Wilhelm, in *Proc. 8th SOHO Workshop, Plasma Dynamic and Diagnostics in Solar Transition Region and Corona*, ed. by J.-C. Vial (ESA, Paris, 1999b), p. 293.
U. Feldman, Phys. Scr. **46**, 202 (1992)
U. Feldman, J.M. Laming, Phys. Scr. **61**, 222 (2000)
U. Feldman, K.G. Widing, Astrophys. J. **363**, 292 (1990)
U. Feldman, K.G. Widing, Astrophys. J. **414**, 381 (1993)
U. Feldman, K.G. Widing, Space Sci. Rev. **107**, 665 (2003)
U. Feldman, U. Schühle, K.G. Widing, J.M. Laming, Astrophys. J. **505**, 999 (1998)
U. Feldman, I. Dammasch, E. Landi, G.A. Doschek, Astrophys. J. **609**, 439 (2004)
A. Fludra, R.D. Bentley, J.L. Culhane, J. Jakimiec, J.R. Lemen et al., in *Proceedings of the EPS 6th European Solar Meeting* (1990), p. 266
A. Fludra, J.L.R. Saba, J.-C. Hénoux, R.J. Murphy, D.V. Reams et al., in *The Many Faces of the Sun*, ed. by K.T. Strong, J.L.R. Saba, B.M. Haisch, J.T. Schmelz (1999), p. 89
J. Geiss, Space Sci. Rev. **33**, 201 (1982)
L. Goldberg, E. Müller, L.H. Aller, Astrophys. J. Suppl. **4**, 1 (1960)
N. Grevesse, A.J. Sauval, Space Sci. Rev. **85**, 161 (1998)
N. Grevesse, M. Asplund, A.J. Sauval, in *Element Stratification in Stars: 40 Years of Atomic Diffusion*, ed. by G. Alecian, O. Richard, S. Vauclair. EAS Publications Series, vol. 17 (2005), p. 21
D. Hovestadt, in *Solar Wind III*, ed. by C.T. Russel (Univ. of California Press, Los Angeles, 1974), p. 2
E. Landi, U. Feldman, G. Doschek, Astrophys. J. **643**, 1258 (2006)
E. Landi, U. Feldman, G. Doschek, Astrophys. J. **659**, 743 (2007)
J.M. Laming, J.J. Drake, K.G. Widing, Astrophys. J. **443**, 416 (1995)
K. Lodders, Astrophys. J. **591**, 1220 (2003)
D.L. McKenzie, U. Feldman, Astrophys. J. **389**, 764 (1992)
J.P. Meyer, Astrophys. J. Suppl. **57**, 173 (1985)
J.P. Meyer, in *Cosmic Abundances*, ed. by S.S. Holt, G. Sonneborn. ASP Conf. Series (Astronomical Society of the Pacific, 1996), p. 127
K.J.H. Phillps, J. Sylwester, B. Sylwester, E. Landi, Astrophys. J. **589**, L113 (2003)
S.R. Pottasch, Astrophys. J. **137**, 945 (1963)
S.R. Pottasch, Space Sci. Rev. **3**, 816 (1964)
J.C. Raymond, in *The Sun and the Heliosphere as an Integrated System*, ed. by G. Poletto, S.T. Suess (Kluwer, Dordrecht, 2004), p. 353
J.L.R. Saba, Adv. Space Res. **15**(7), 13 (1995)
J.L.R. Saba, K.T. Strong, in *Proc. 1st SOHO Workshop, Coronal Streamers, Coronal Loops, and Coronal and Solar Wind Composition*, ed. by V. Domingo (ESA, Noordwijk, 1992), p. 347
N.R. Sheeley Jr., Astrophys. J. **440**, 884 (1995)
N.R. Sheeley Jr., Astrophys. J. **469**, 423 (1996)

J.T. Schmelz, K. Nasraoui, J.K. Roames, L.A. Lippner, J.W. Garst, Astrophys. J. **634**, L197 (2005)
D.S. Spicer, U. Feldman, K.G. Widing, M. Rilee, Astrophys. J. **494**, 450 (1998)
A.C. Sterling, G.A. Doschek, U. Feldman, Astrophys. J. **404**, 394 (1993)
K.T. Strong, E.S. Clafin, J.R. Leman, G.A. Linford, Adv. Space Res. **8**(11), 167 (1988)
K.T. Strong, E.S. Clafin, J.R. Leman, G.A. Linford, Astrophys. J. **8**(11), 167 (1991a)
K.T. Strong, J.R. Leman, G.A. Linford, Adv. Space Res. **11**(1), 151 (1991b)
B. Sylwester, J. Sylwester, R.D. Bentley, A. Fludra, Sol. Phys. **126**, 177 (1990)
H.P. Warren, Sol. Phys. **190**, 363 (1999)
K.G. Widing, Astrophys. J. **480**, 400 (1997)
K.G. Widing, U. Feldman, Astrophys. J. **442**, 446 (1995)
K.G. Widing, U. Feldman, A.K. Bhatia, Astrophys. J. **308**, 982 (1986)
K.G. Widing, E. Landi, U. Feldman, Astrophys. J. **622**, 1211 (2005)
P.R. Young, Astron. Astrophys. **439**, 361 (2005a)
P.R. Young, Astron. Astrophys. **444**, L45 (2005b)
P.R. Young, H.E. Mason, Space Sci. Rev. **85**, 315 (1998)
T.H. Zurbuchen, L.A. Fisk, G. Gloeckler, N.A. Schwadron, Space Sci. Rev. **85**, 397 (1998)

Space Sci Rev (2007) 130: 127–138
DOI 10.1007/s11214-007-9197-z

# Solar Gamma-Ray Spectroscopy

**R.J. Murphy**

Received: 22 January 2007 / Accepted: 16 April 2007 /
Published online: 22 May 2007


**Abstract** Interactions of ions accelerated in solar flares produce gamma-ray lines and continuum and neutrons. These emissions contain a rich set of observables that provides information about both the accelerated ions and the environment where the ions are transported and interact. Ion interactions with the various nuclei present in the ambient medium produce gamma-ray lines at unique energies. How abundance information is extracted from the measurements is discussed and results from analyses of a number of solar flares are presented. The analyses indicate that the composition of the ambient gas where the ions interact (typically at chromospheric densities) is different from that of the photosphere and more like the composition of the corona, exhibiting low-FIP elemental enhancements that may vary from flare to flare. Evidence for increased Ne/O and the photospheric $^3$He abundance is also discussed.

**Keywords** Sun: abundances · Chromosphere · Photosphere · Flares · X-rays · Gamma rays

## 1 Introduction

The composition of the various regions of the solar atmosphere have been studied using a variety of techniques (see, e.g., Anders and Grevesse 1989) using data from X-ray and optical spectroscopy, meteorites, and solar energetic particles (SEP). These studies have revealed considerable abundance variations; for example, based on SEP and other data it has been shown (e.g., Myer 1985) that, compared to photospheric abundances, the coronal abundances of elements with low first ionization potential (FIP), such as Mg, Si and Fe, are enhanced relative to those with high FIP, such as C, N and O.

Gamma-ray line spectroscopy offers another technique for determining abundances and has some significant advantages: (1) the cross sections for production of the strongest

R.J. Murphy (✉)
Naval Research Laboratory, E.O. Hulburt Center for Astrophysics, Washington, DC, WA 20375, USA
e-mail: ronald.murphy@nrl.navy.mil

gamma-ray lines are well-measured with uncertainties as low as 10%; (2) because the interactions involve the nucleus only, they are not dependent on the temperature and density of the environment which affect the ionization states of the ambient ions; and (3) since the gamma rays are very penetrating, consideration of radiation transfer is generally not required.

Abundances determined with gamma-ray spectroscopy are useful in several ways. They can provide information about species whose abundances cannot be directly determined by other means, such as Ne (which does not produce atomic lines in the lower solar atmosphere) and photospheric $^3$He. By determining abundances in the chromosphere where the gamma rays are produced, gamma-ray spectroscopy provides an important constraint on understanding the FIP composition bias of the corona relative to the photosphere. Abundances of the accelerated ions producing the gamma-ray lines can be compared with directly-measured abundances of SEPs to determine the relationship of the two populations

In a solar flare, ions can be accelerated to GeV energies. Nuclear interactions of these ions with the solar atmosphere produce excited and radioactive nuclei, neutrons, and pi-mesons. Ion acceleration probably occurs in the corona where low energy losses at coronal densities allow efficient acceleration, but the ion interactions generally occur at higher densities, probably similar to those of the chromosphere or upper photosphere (see below), where interaction rates are greater (see Hua et al. 1989). All of the nuclear interaction products subsequently produce observable gamma-ray emission via secondary processes, and the neutrons that escape may be observed directly in space and on Earth and indirectly in space via their decay protons.

The composition of the ambient medium where the accelerated ions interact can be derived by gamma-ray spectroscopy by comparing fluences of nuclear deexcitation lines, mostly occurring in the range from ∼1 to 8 MeV. When accelerated ions interact with ambient elements, nuclei are excited and then de-excite on very short time scales (typically less than $10^{-6}$ s). As an excited nucleus de-excites through its allowed levels, the transition energies appear as gamma rays, resulting in gamma-ray lines whose central energies are unique to the element. When accelerated proton and alpha particles react with ambient He and heavier nuclei, the interactions are referred to as "direct" reactions. Lines produced by direct reactions are "narrow" with fractional FWHM ∼2%, due to the relatively low recoil velocity of the heavy nucleus. "Inverse" reactions, when accelerated ions heavier than He interact with ambient H and He, produce "broad" lines with fractional FWHM ∼20% due to the relatively high recoil velocity of the nucleus. Lines produced by heavy–heavy interactions are not significant because of the low abundances.

The yield of a deexcitation line from a direct reaction of an accelerated proton or alpha particle with an ambient element is linearly proportional to the abundance of that ambient element. Comparison of the various narrow line fluences therefore provides information on the relative abundances of the medium where the interactions take place. This is complicated by the fact that a given excited nucleus can be produced not only via inelastic interactions with that nucleus but additionally via spallation reactions with heavier nuclei. The elements that produce deexcitation lines of sufficient strength to allow reliable abundance determinations (and their corresponding strongest line energies) are He (∼0.45 MeV), C (4.439 MeV), N (2.313 MeV), O (6.129 MeV), Ne (1.634 MeV), Mg (1.369 MeV), Si (1.789 MeV) and Fe (0.847 MeV). We emphasize again that gamma-ray deexcitation lines are typically produced in the chromosphere or upper photosphere and so it is this region whose composition is revealed by gamma-ray spectroscopy.

The photospheric $^3$He/H abundance ratio can be determined from the neutron-capture line. Neutrons initially produced moving downward either decay, react with $^3$He, or are captured on H. Capture on other nuclei is less important due to their smaller relative abundances. The capture on H results in the formation of deuterium with the binding energy appearing as a 2.223 MeV neutron-capture line photon. The reaction with $^3$He is charge exchange, $^3\mathrm{He}(n,p)^3\mathrm{H}$, without the emission of radiation. Since the probability for elastic scattering is much larger than the probability for either of these reactions, most of the neutrons thermalize first, causing a delay in the formation of the capture photon relative to neutron formation. The cross sections for the H and $^3$He reactions are $2.2 \times 10^{-30}\beta^{-1}$ cm$^2$ and $3.7 \times 10^{-26}\beta^{-1}$ cm$^2$, respectively, where $\beta$ is neutron velocity in units of the speed of light. Therefore, if the $^3$He/H ratio at the capture site is $\sim 5 \times 10^{-5}$(which is comparable to that observed in the solar wind), nearly equal numbers of neutrons will be captured on H as react with $^3$He. The ambient $^3$He/H abundance ratio therefore affects the total yield of the capture line and also the delay of its formation. Because effective capture on H or reaction with $^3$He requires high density, these reactions occur deep in the photosphere. Study of the 2.223 MeV neutron-capture line therefore provides information on the photospheric $^3$He/H abundance ratio.

While gamma-ray line spectroscopy can also provide information on the composition of the accelerated ions in solar flares via the broad deexcitation lines, in this paper we only discuss narrow deexcitation lines and the neutron-capture line, providing information on the composition of the ambient chromosphere and photosphere. The exception is the accelerated alpha/proton ratio which must be known to correctly interpret the narrow line fluences in terms of abundances. In Sect. 2, we discuss the transport and interaction model used in the calculation of gamma-ray line spectra, noting those parameters of the model other than the ambient abundances themselves that affect deexcitation line fluences. In Sect. 3, we discuss the application of the calculations to solar-flare gamma-ray data obtained with spectrometers on the *Solar Maximum Mission* (*SMM*), the *Compton Gamma Ray Observatory* (*CGRO*), *Yohkoh*, *Integral*, and the *Reuven Ramaty High Energy Solar Spectroscopic Imager* (*RHESSI*).

## 2 Accelerated Ion Transport and Interaction Model and Analysis Technique

### 2.1 Accelerated Ion Transport and Interaction Model

Extracting information about ambient abundances from gamma-ray line observations requires a transport and interaction model. A transport model was developed by Hua et al. (1989) in which a Monte Carlo simulation follows individual energetic ions through a flare loop until they either undergo a nuclear reaction or their energy falls below nuclear reaction thresholds. The most important aspects of ion transport are included in the model: energy losses due to Coulomb collisions, removal by nuclear reactions, magnetic mirroring in the convergent flux tube, and MHD pitch angle scattering (PAS).

The model consists of a semicircular coronal portion of length $L$ and two straight portions extending vertically from the transition region through the chromosphere into the photosphere. Below the transition region, the magnetic field strength is assumed proportional to a power $\delta$ of the pressure (Zweibel and Haber 1983). Pitch angle scattering is characterized by the mean free path $\Lambda$, assumed to be independent of particle energy (see Hua et al. 1989 for discussion). The level of PAS is then characterized by $\lambda$, the ratio of $\Lambda$ to the loop half-length $L_c$ $(= L/2)$. A particular height-density profile for the solar atmosphere (e.g., that of Avrett 1981) is assumed. The accelerated ions are released isotropically at the top of the loop with an assumed kinetic energy spectral shape. All species are assumed to share the same spectral shape. The ions are followed until they interact according to the cross section for the reaction being considered (typically near the loop footpoints in the chromosphere or upper photosphere) or thermalize. The effect of the model parameters on the observable quantities of solar flare gamma-ray emission was discussed by Murphy et al. (2007).

When an interaction occurs, the time after the initial release of the energetic ion, the location along the loop, and the direction of motion of the ion are recorded. For deexcitation line reactions, the resulting interacting-particle angular distribution can then be used with the gamma-ray line production code (see below) to calculate line shapes and shifts. The attenuation of the gamma ray along the line of sight is calculated from the location of the interaction along the loop. For neutron-producing reactions, the Monte Carlo simulation follows the neutron until it escapes into space, is captured on H to contribute to the neutron-capture line, reacts with $^3$He without production of radiation, or decays. If a neutron-capture gamma ray is produced, the time of its production is recorded and its attenuation along the line of site is calculated.

In addition to the relative ambient elemental abundances, deexcitation line yields also depend on the shape of the accelerated ion energy spectrum since the various reaction cross sections have different dependences on the accelerated ion energies. Since both accelerated proton and alpha interactions contribute to a narrow line, relative line yields are also affected by the accelerated alpha/proton ratio. On the other hand, Hua et al. (1989) found that nuclear excitation reactions typically occur at relatively shallow depths. Unless the flare is located at or beyond the limb of the Sun, attenuation of the escaping deexcitation gamma rays is not significant and deexcitation line yields do not depend on where the flare occurs on the solar disk. Also, because the ions are confined to the magnetic loop and produce the nuclear interactions as they slow down and thermalize rather than as they escape, the total deexcitation line emission is "thick target". Line yields therefore do not depend on the loop parameters $L$, $\lambda$, and $\delta$ or on the ambient density.

This is not true of the neutron-capture line. Because the captures occur at photospheric depths resulting in significant attenuation of this line, both the yield and time history of the line depend on the flare location. In addition, because the fate of the neutron depends on its direction of motion, the yield and time history also depend on $\lambda$ and $\delta$ since these model parameters determine the direction of motion of the accelerated ion when it interacts which is then reflected in the neutron direction of motion. The loop size $L$ also affects the time history of the capture line. The neutron-capture line yield and time history, then, not only depend on the photospheric $^3$He/H ratio, but also on the accelerated ion spectral shape and alpha/proton ratio, and on the loop parameters $L$, $\lambda$ and $\delta$. Unless the accelerated ion spectrum is very steep, most neutrons are produced by interactions of accelerated protons and alpha particles with ambient H and He and by interactions of heavy accelerated ions with ambient H and He if the heavy elements are enhanced as in impulsive SEPs. Abundances of ambient elements heavier than He therefore do not significantly affect the neutron-capture line yield. Murphy et al. (2007) found that the density-height profile of the solar atmosphere also does not significantly affect the capture line yield or time history.

Table 1 summarizes the parameters of the loop transport and interaction model that significantly affect gamma-ray line yields. It is seen that deriving well-constrained values for relative ambient abundances requires knowledge of several parameters. Early abundance determinations using gamma-ray lines often ignored this requirement, fixing some parameters at assumed values without acknowledging the additional abundance uncertainties that necessarily result.

**Table 1** Parameter dependences of gamma-ray line yields

| Gamma-ray line | Parameters |
|---|---|
| Deexcitation line | ambient abundances, $\alpha$/p, ion spectrum |
| Neutron-capture line | $L$, $\delta$, $\lambda$, $\alpha$/p, ion spectrum |

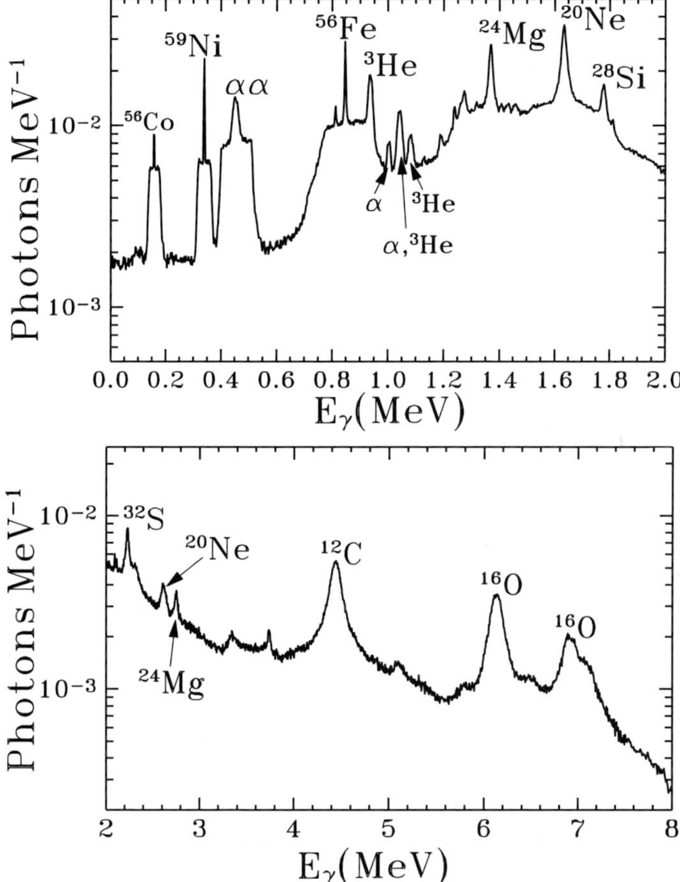


**Fig. 1** Calculated gamma-ray deexcitation line spectrum. The elements responsible for the strongest lines are noted. See text for full discussion

## 2.2 Analysis Technique

A comprehensive treatment of nuclear deexcitation gamma-ray line emission was given by Ramaty et al. (1979). The resulting gamma-ray line production code has been continuously updated (most recently by Kozlovsky et al. 2002) to incorporate new cross-section and nuclear kinematic data and includes more than 100 explicit lines. This code calculates a complete gamma-ray line spectrum along any observing direction from the flare site for any assumed accelerated-ion spectral shape, accelerated and ambient composition, and interacting-particle angular distribution. An example of such a calculated spectrum is shown in Fig. 1 for an isotropic accelerated-particle angular distribution, impulsive flare

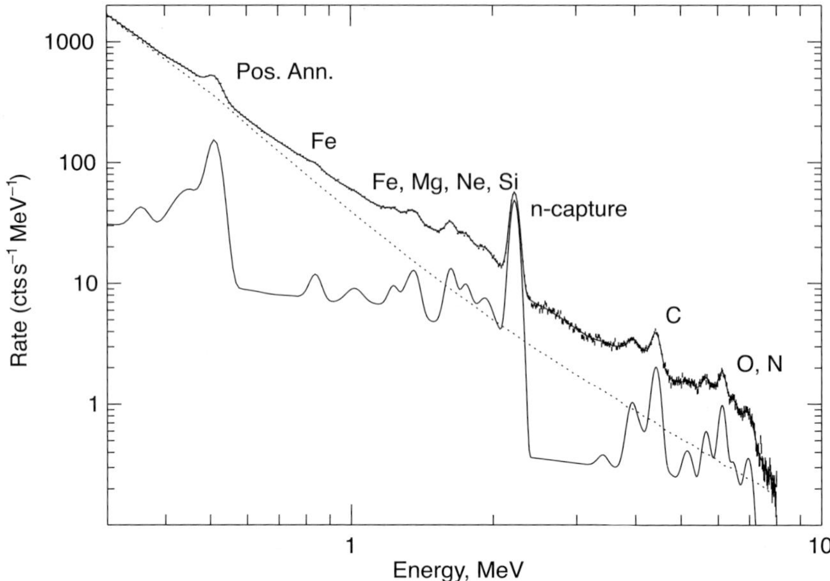


**Fig. 2** Count spectrum derived from the sum of 19 flares observed with *SMM*. The *solid line* running through the data is the best-fitting count spectrum. The *dotted line* is the fitted electron bremsstrahlung component and the *other solid line* is the fitted narrow nuclear deexcitation line component

(alpha/proton = 0.5) accelerated ion and coronal ambient compositions (see Ramaty et al. 1996), and a power-law accelerated-ion spectrum with an index $s$ of 4. The nuclei responsible for the strongest lines are indicated. The narrow lines are superposed on a continuum composed of the broad lines and a quasi-continuum due to the large number of relatively weak and overlapping lines from deexcitation of heavy elements. An observed flare spectrum would also include continuum electron bremsstrahlung emission. An example is shown in Fig. 2 where the summed count spectrum from 19 flares observed with *SMM* is shown.

The gamma-ray line code can be used in two ways to determine ambient abundances. One is to calculate gamma-ray spectra which can then be compared with observed flare spectra. The other is to calculate expected fluences for the various gamma-ray lines which then can be compared with measured fluences. In either case, the parameters (including the abundances) are adjusted until the measured and calculated data optimally agree. Both approaches have been used in past research.

## 3 Ambient Abundance Determinations

### 3.1 Accelerated Ion Spectral Shape and the Chromospheric Ne/O Abundance

Since the yields of the various deexcitation lines depend on the shape of the accelerated ion spectrum in addition to the ambient abundances, this shape must be known before reliable abundance determinations can be made. So, before discussing the results of ambient abundance determinations from gamma-ray spectroscopy, we discuss determination of the accelerated ion spectrum which first involves discussing the ambient Ne/O abundance ratio. Even if the ambient composition were known, deexcitation-line ratios alone cannot establish the shape of the spectrum over a broad energy range because the cross sections for the various reactions are significant only over a relatively narrow range of ion energies (from a few to several tens of MeV nucleon$^{-1}$). Because the neutrons responsible for the neutron-capture line are produced by ions with energies higher than those producing excitation reactions (see Murphy et al. 2007), comparing the ratio of the neutron-capture line to a deexcitation line with the ratio of two deexcitation lines gives information over a broader energy range (up to ~50 MeV nucleon$^{-1}$). The positron annihilation line and, when present, pion-decay emission can also be useful since their production is again due to accelerated ions of energies higher than those for deexcitation line reactions.

Of the deexcitation line ratios, the 6.129 MeV $^{16}$O – 1.634 MeV $^{20}$Ne line ratio is most sensitive to the steepness of the ion spectrum because of the very different production cross section threshold energies for the two lines (see Kozlovsky et al. 2002). This ratio is also useful because both lines are typically quite strong in flares and because it is only weakly dependent on the composition of ambient Mg, Si and Fe. It does depend on the accelerated alpha/proton ratio and, of course, on the Ne/O abundance.

Early efforts at gamma-ray abundance determinations (e.g., Murphy et al. 1985a, 1985b, 1991) simply assumed a Bessel function shape (see Forman et al. 1986). In these studies, the Ne/O abundance ratio was found to be surprisingly high: $0.38 \pm 0.08$ (1-$\sigma$ error) compared to the accepted photospheric and coronal value of ~0.15. While these analyses also showed enhancements of low-FIP elements (see Sect. 3.2), the Ne result was surprising since Ne has a high FIP. After Share and Murphy (1995) pointed out that this could be the result of the assumed accelerated ion spectral shape, Ramaty et al. (1995, 1996) used a power law spectral shape for the accelerated ions and analyzed nine *SMM* flares for which measured deexcitation line and neutron-capture line fluence data were available (Share and Murphy 1995). In order to obtain the same spectral indexes from both the 6.129 MeV $^{16}$O – 1.634 MeV $^{20}$Ne line ratio and the neutron-capture—4.439 MeV $^{12}$C ratio (required if the ion spectrum is an unbroken power law), an ambient Ne/O abundance ratio of ~0.25 was required. While this is less than that obtained in the earlier studies, it is still higher than the accepted value of 0.15.

Similar Ne/O enhancements have been seen in other measurements. For example, Schmelz (1993) found Ne/O $= 0.32 \pm 0.02$ using observations of a solar flare obtained with the *SMM* Flat Crystal Spectrometer (FCS). Drake and Testa (2005) found Ne/O $= 0.41 \pm 0.02$ for a sample of 21 stars located within 100 pc of the Sun observed with the High Energy Transmission Grating on *Chandra*. But the Ne/O abundance ratio derived in a recent analysis of FCS data excluding data taken during flares and long-duration events (Schmelz et al. 2005) was consistent with the accepted 0.15 value. Schmelz et al. (2005) also reconsidered older full-Sun observations and again found Ne/O consistent with the accepted value. Considering that the Ne/O enhancements could be flare-related, Shemi (1991) suggested that pre-flare soft X-ray radiation preferentially ionizes Ne which could then be transported to higher levels of the atmosphere. Since the *Chandra* spectra analyzed by Drake and Testa (2005) are most likely dominated by photons from very intense flares, their enhanced neon abundances might result from a similar mechanism. However, Schmelz et al. (2005) point out that this is not likely since low-FIP elements in active stellar coronae are generally depleted rather than enhanced, as they are in the solar corona. Finally, Bahcall et al. (2005) have suggested an increased Ne abundance to explain the discrepancy of recent solar abundance determinations (e.g., Lodders 2003) with helioseismological measurements.

In the Ramaty et al. (1995, 1996) studies mentioned above, assuming Ne/O $= 0.15$ resulted in a softer deduced ion spectrum than that deduced using the neutron-capture line.

However, the neutron-capture line yield also depends on the loop parameters $\lambda$ and $\delta$ (Table 1) since these parameters affect the angular distribution of the interacting ions and thus the angular distribution of the neutrons. In the Ramaty et al. studies, although the accelerated alpha/proton ratio was varied, only two angular distributions were considered: downward isotropic and "fan beam" (where the ions move parallel to the solar surface). But distributions more downward directed than downward isotropic would increase the neutron-capture line yield and result in a softer derived ion spectrum which could be consistent with that derived from the O/Ne line ratio. In a recent analysis of gamma-ray data obtained with *RHESSI* and *Integral* from the 2003 October 28 flare, Murphy (private communication 2007) used gamma-ray line shifts and delays of the deexcitation line time history to independently determine $\lambda$ and $\delta$. Using these values, they found that the power law spectral index derived from the O/Ne line ratio was consistent with the index derived using the neutron-capture line for assumed Ne/O $= 0.15$ but not for 0.25. It is clear, then, that the Ramaty et al. (1995, 1996) *SMM* studies must be revisited, exploring a wider range of the parameters, before the ambient Ne/O abundance ratio can be confidently determined.

### 3.2 FIP Effect in the Chromosphere

Narrow gamma-ray lines from solar flares were first observed with the gamma-ray spectrometer on *SMM*. Using *SMM* data for the 27 April 1981 flare, Forrest (1983) showed that a calculated gamma-ray spectrum (Ramaty et al. 1979) for which both the ambient material and the accelerated ions have photospheric abundances provided an unacceptable fit to the data, but that a better fit resulted by enhancing the abundances of ambient Ne, Mg, Si and Fe relative to C and O. Using the same data, Murphy et al. (1985a, 1985b) derived the best fitting ambient abundances. In this study, the accelerated particle abundances were held fixed at large proton flare values and the ambient abundances were varied to achieve the best fit to the data. The results again suggested that ambient Mg/C, Si/C and Fe/C are enhanced relative to the corresponding photospheric ratios, consistent with coronal values; that Ne/C and Ne/O are larger than the coronal values; and that O/C is consistent with both the photospheric and coronal values. Also using these same data, Murphy et al. (1991) relaxed the constraint that the accelerated particle composition is fixed, and determined the abundances of both these particles and the ambient gas, again confirming the previous ambient abundance results. The resulting abundances relative to the photospheric abundances (Anders and Grevesse 1989) are shown Fig. 3, renormalized so that the C ratio is unity. The O abundance is seen to be photospheric but the Mg, Si and Fe abundances are enhanced, similar to the coronal abundances also shown in the figure; i.e., enhanced by about a factor of three. (The Ne/O abundance was discussed in Sect. 3.1.)

The gamma-ray data from this flare imply that the FIP bias, known to exist in the corona (e.g. Feldman 1992; Reames 1998; Geiss 1998), already exists in the chromosphere, where the bulk of gamma rays are probably produced. This finding is consistent with results from atomic spectroscopy that suggest a chromospheric origin for the effect (McKenzie and Feldman 1992; Athay 1994), and supports some models of the FIP effect (Judge and Peter 1998; Mullan and Arge 2000; Laming 2004).

Share and Murphy (1995) measured the fluxes of 10 narrow gamma-ray lines in 19 X-class solar flares observed with *SMM*/GRS. Their analysis did not determine ambient abundances directly, but comparison of the flare-to-flare line flux variations provided insight into the corresponding abundances. The variations suggested that the ambient abundances are grouped with respect to FIP: line fluxes from elements with similar FIP correlate well with

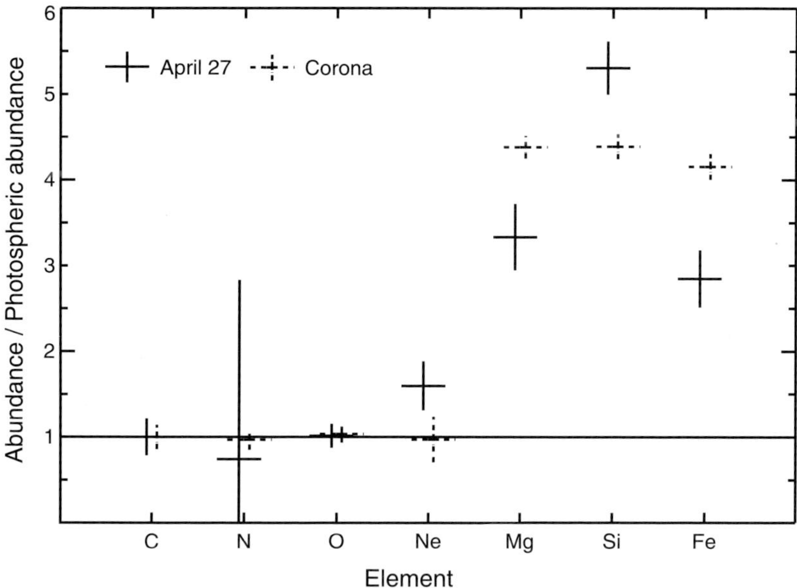


**Fig. 3** Derived ambient abundances relative to the photosphere with the C ratio normalized to 1 (*solid*). Also shown are coronal abundances relative to the photosphere (*dotted*)

one another. In contrast, the low-FIP to high-FIP line ratios were not consistent with a common value, varying by about a factor of four from flare to flare. The authors showed that there was little correlation of this ratio with accelerated ion spectral indexes, suggesting that the line flux variation was due to abundance variations rather than a spectral effect. The magnitude of the variation was similar to that of the low-FIP enhancement of the corona over the photosphere, implying that the flare ambient abundances spanned a range from photospheric to coronal.

Ramaty et al. (1995) used these *SMM* line fluxes with calculated line yields to directly determine the abundances. They found that the flare-averaged composition of the plasma in which the ions interact is similar to coronal and that the low FIP-to-high FIP ratio for individual flares was always higher than photospheric. The authors suggested that variability, if any, was less than that obtained with atomic spectroscopy; however, this is in part due to limited statistics. The factor of four variability was still present in the best-fitting abundance values (e.g., for Mg/O).

In their analysis of the 1991 June 4 flare observed with *CGRO*/OSSE, Murphy et al. (1997) also did not directly determine abundances, using measured line fluences to infer abundance information. The flare-averaged deexcitation line fluence ratios were similar to the averages measured for the 19 *SMM* flares (Share and Murphy 1995). The June 4 ratios therefore suggested that the ambient low-FIP elements at the interaction site were enhanced relative to high-FIP elements by about a factor of $\sim$2.5 compared to photospheric abundances. The improved statistics available with the OSSE detectors allowed a search for variability within the June 4 flare. The flux ratios measured for the first and second orbits of emission were different (5.6 $\sigma$), increasing from more photospheric to more coronal values. The first-orbit ratio corresponded to about a factor of two enhancement of the low-FIP/high-FIP abundance ratio relative to the photosphere and the second-orbit ratio to about a factor of 3.5, or similar to the corona. Evidence for an increasing ratio was also provided by *RHESSI*

observations of the 2003 October 28 and November 2 flares (Share et al. 2004). Such variability could possibly be due to time-dependent compositional changes at the flare site, but the time scale is probably too short for this to be a reasonable explanation. Alternatively, the location of the gamma-ray production site could have changed with time, progressing from deeper in the chromosphere-photosphere toward the corona. This could happen if the height of the magnetic field mirroring point increased as the flare progressed.

### 3.3 Chromospheric $^4$He/H

The $^4$He abundance is not measured spectroscopically in the photosphere or chromosphere because no significant atomic excitations are possible at those temperatures, but $[\mathrm{He/H}]_{\mathrm{ph}} = 8.4\%$ has been inferred from helioseismology (Richard et al. 1998). This value is higher than the average coronal value of 3–5% inferred from solar wind data (Coplan et al. 1990) and gradual SEP events (Mazur et al. 1993). The alpha–alpha gamma-ray line feature at $\sim$0.45 MeV is a combination of two Doppler broadened lines produced by accelerated alpha-particles interacting with ambient $^4$He to produce $^7$Li and $^7$Be (Kozlovsky and Ramaty 1974). The intensity of this feature depends on both the ambient and accelerated He abundances.

The high flux in the alpha-alpha feature relative to other deexcitation lines in the *SMM* flares and in the 1991 June 4 flare observed by *CGRO*/OSSE led to the conclusion (Share and Murphy 1997) that the accelerated alpha/proton ratio typically had to be large, $\sim$0.5, for an assumed ambient $^4$He/H abundance ratio of 0.1. Mandzhavidze et al. (1997) instead suggested that the ambient ratio might be higher in some flares and described a test which could distinguish between the two explanations: the measurement of lines which only result from interactions of accelerated alpha particles; e.g., at 0.339 MeV. Although such lines are relatively weak and narrow and would be more easily observed with high spectral resolution spectrometers such as on *RHESSI*, Share and Murphy (1998) used the moderate resolution *SMM* detectors and concluded that on average the ambient $^4$He abundance is consistent with the accepted photospheric value and therefore flare-accelerated ions have an enhanced alpha/proton ratio. However, owing to the weakness of the 0.339 MeV line, average ambient $^4$He/H abundance ratios up to 0.2 to 0.3 were still acceptable. Mandzhavidze et al. (1999) made detailed studies of the line intensities from individual *SMM* flares and concluded that a high ambient $^4$He/H abundance may be required in some of them.

### 3.4 Photospheric $^3$He/H

Several investigators have determined the photospheric $^3$He/H ratio using neutron-capture line time-history measurements obtained with *SMM*, *GRANAT* and *CGRO*. Chupp et al. (1981), Prince et al. (1983), Trottet et al. (1993), Murphy et al. (1997) and Rank et al. (2001) approximated the expected line time history from instantaneous neutron production as the sum of exponentials and made simplifying assumptions about the density structure of the solar atmosphere and the interacting particle angular distribution. Hua and Lingenfelter (1987) used a Monte Carlo technique to calculate neutron-capture line production. The derived $^3$He/H ratios from these analyses ranged from 0 to $5 \times 10^{-5}$, with the smallest uncertainties ($1\text{–}2 \times 10^{-5}$) obtained by Hua and Lingenfelter (1987) and Rank et al. (2001). In these latter two analyses, a fixed interacting ion angular distribution was used and the uncertainties of the measured time history used to represent the neutron-production time profile were neglected. Because of these various simplifying assumptions, the photospheric $^3$He/H abundance derived from gamma-ray spectroscopy was not reliably determined in these early studies.

Murphy et al. (2003) improved the analysis technique with their study of the neutron-capture line in the 2002 July 23 flare observed with *RHESSI*. Capture line time profiles were calculated for various values of PAS scattering ($\lambda$) and the effect of uncertainties of the deexcitation line time history used to represent the nuclear interaction time history were included. However, they fixed $\delta$ at 0.2 and $L$ at 11,500 km and did not derive the accelerated ion spectral index independently from line ratios. Even though they ignored the impact of these assumptions on the derived $^3$He/H uncertainty, their photospheric $^3$He/H abundance was still not constraining, with a $\pm 1\sigma$ range of 0.5–10 $\times$ $10^{-5}$.

Murphy et al. (2007) attempted to address all of the parameters affecting the neutron-capture line in their analysis of the 1991 June 4 flare observed by *CGRO*/OSSE. $L$ was determined from optical images of foot point emission, the spectral index and the accelerated alpha/proton ratio was determined from the O/Ne and alpha-alpha/C line ratios, and $\delta$ and $\lambda$ were determined from deexcitation line shifts and time history delays. However, the combined effect of the allowed ranges for the various derived parameters resulted in only an upper limit of 4.5 $\times$ $10^{-5}$ (3-$\sigma$) for the photospheric $^3$He/H abundance.

## 4 Summary

Gamma-ray spectroscopy of solar flares offers a unique method for determining abundances of the solar atmosphere. One of its advantage over other techniques is that interpretation of the gamma-ray data is independent of the temperature and density of the medium. Gamma-ray spectroscopy reveals the composition of the chromosphere and upper photosphere since that is where most of the interactions resulting in deexcitation lines occur. Analyses of data from a number of solar flares has shown that the enhancement of low-FIP elements known to exist in the corona is already present in the region where the gamma rays are produced. The degree of enhancement varies from flare to flare and can also vary with time within flares. The analyses also suggest that the chromospheric Ne/O abundance is higher than the accepted value of 0.15, but this result is uncertain, pending more extensive re-analyses of the data. Studies of the neutron-capture line can reveal the $^3$He/H abundance ratio in the photosphere. While the results are consistent with estimates made by other techniques, uncertainties associated with the various parameters affecting calculations of the capture line flux make the determination very uncertain.

**Acknowledgement** This work was supported by NASA DPR W19,977 and the Office of Naval Research.

## References

E. Anders, N. Grevesse, Geochim. Cosmochim. Acta **53**, 197 (1989)
R.G. Athay, Astrophys. J. **423**, 516 (1994)
J.N. Bahcall, S. Basu, A.M. Serenelli, Astrophys. J. **631**, 1281 (2005)
E.L. Chupp et al., Astrophys. J. **244**, L171 (1981)
M.A. Coplan et al., Sol. Phys. **128**, 195 (1990)
J.J. Drake, P. Testa, Nature **436**, 525 (2005)
U. Feldman, Phys. Scr. **46**, 202 (1992)
M.A. Forman, R. Ramaty, E.G. Zweibel, in *The Physics of the Sun*, vol. II, ed. by P.A. Sturrock et al. (1986), p. 249
D.J. Forrest, in *Positron Electron Pairs in Astrophysics*, ed. by M.L. Burns, A.K. Harding, R. Ramaty (AIP, New York, 1983), p. 3
J. Geiss, Space Sci. Rev. **85**, 241 (1998)
X.-M. Hua, R.E. Lingenfelter, Astrophys. J. **319**, 555 (1987)

X.-M. Hua, R. Ramaty, R.E. Lingenfelter, Astrophys. J. **341**, 516 (1989)
P.G. Judge, H. Peter, Space Sci. Rev. **85**, 187 (1998)
B. Kozlovsky, R. Ramaty, Astrophys. J. **191**, L43 (1974)
B. Kozlovsky, R.J. Murphy, R. Ramaty, Astrophys. J. Suppl. Ser. **141**, 523 (2002)
M. Laming, Astrophys. J. **614**, 1063 (2004)
K. Lodders, Astrophys. J. **591**, 1220 (2003)
N. Mandzhavidze, R. Ramaty, B. Kozlovsky, Astrophys. J. **489**, L99 (1997)
N. Mandzhavidze, R. Ramaty, B. Kozlovsky, Astrophys. J. **518**, 918 (1999)
J.E. Mazur, G.M. Mason, B. Klecker, R.E. McGuire, Astrophys. J. **404**, 810 (1993)
D.L. McKenzie, U. Feldman, Astrophys. J. **389**, 764 (1992)
D.J. Mullan, C.N. Arge, in *High Energy Solar Physics – Anticipating HESSI*, ed. by R. Ramaty, N. Mandzhavidze. Proc. Astron. Soc. of the Pacific, vol. 206 (2000), p. 71
R.J. Murphy, D.J. Forrest, R. Ramaty, B. Kozlovsky, *19th ICRC*, vol. 4 (1985a), p. 253
R.J. Murphy, R. Ramaty, D.J. Forrest, B. Kozlovsky, *19th ICRC*, vol. 4 (1985b), p. 249
R.J. Murphy, R. Ramaty, B. Kozlovsky, D.V. Reames, Astrophys. J. **437**, 793 (1991)
R.J. Murphy, G.H. Share, J.E. Grove, W.N. Johnson, R.L. Kinzer, J.D. Kurfess, M.S. Strickman, G.V. Jung, Astrophys. J. **490**, 883 (1997)
R.J. Murphy, G.H. Share, X.-M. Hua, R.P. Lin, D.M. Smith, R.A. Schwartz, Astrophys. J. **595**, L93 (2003)
R.J. Murphy, B. Kozlovsky, G.H. Share, X.-M. Hua, R.E. Lingenfelter, Astrophys. J. **168** (2007, in press)
J.P. Myer, Astrophys. J. Suppl. Ser. **57**, 151 (1985)
T.A. Prince, D.J. Forrest, E.L. Chupp, G. Kanbach, G.H. Share, *18th ICRC*, vol. 4 (1983), p. 79
R. Ramaty, B. Kozlovsky, R.E. Lingenfelter, Astrophys. J. Suppl. Ser. **40**, 487 (1979)
R. Ramaty, N. Mandzhavidze, B. Kozlovsky, R.J. Murphy, Astrophys. J. **455**, L193 (1995)
R. Ramaty, N. Mandzhavidze, B. Kozlovsky, in *High Energy Solar Physics*, ed. by R. Ramaty, N. Mandzhavidze, X.-M. Hua (Am. Inst. Phys., New York, 1996), pp. 172, 374
G. Rank, J. Ryan, H. Debrunner, M. McConnell, V. Schonfelder, Astron. Astrophys. **378**, 1046 (2001)
D.V. Reames, Space Sci. Rev. **85**, 327 (1998)
O. Richard, W.A. Dziembowski, R. Sienkiewicz, P.R. Goode, Astron. Astrophys. **338**, 756 (1998)
J.T. Schmelz, Astrophys. J. **408**, 373 (1993)
J.T. Schmelz, K. Nasraoui, J.K. Roames, L.A. Lippner, J.W. Garst, Astrophys. J. **634**, L197 (2005)
G.H. Share, R.J. Murphy, Astrophys. J. **452**, 933 (1995)
G.H. Share, R.J. Murphy, Astrophys. J. **485**, 409 (1997)
G.H. Share, R.J. Murphy, Astrophys. J. **508**, 876 (1998)
G.H. Share, R.J. Murphy, D.M. Smith, R.A. Schwartz, R.P. Lin, Astrophys. J. **615**, L169 (2004)
A. Shemi, Mon. Not. Roy. Astron. Soc. **251**, 221 (1991)
G. Trottet et al., Astron. Astrophys. Suppl. Ser. **97**, 337 (1993)
E.G. Zweibel, D. Haber, Astrophys. J. **264**, 648 (1983)

Space Sci Rev (2007) 130: 139–152
DOI 10.1007/s11214-007-9189-z

# The Composition of the Solar Wind in Polar Coronal Holes

**George Gloeckler · Johannes Geiss**

Received: 16 January 2007 / Accepted: 9 April 2007 /
Published online: 30 May 2007


**Abstract** The solar wind charge state and elemental compositions have been measured with the Solar Wind Ion Composition Spectrometers (SWICS) on Ulysses and ACE for a combined period of about 25 years. This most extensive data set includes all varieties of solar wind flows and extends over more than one solar cycle. With SWICS the abundances of all charge states of He, C, N, O, Ne, Mg, Si, S, Ar and Fe can be reliably determined (when averaged over sufficiently long time periods) under any solar wind flow conditions. Here we report on results of our detailed analysis of the elemental composition and ionization states of the most unbiased solar wind from the polar coronal holes during solar minimum in 1994–1996, which includes new values for the abundance S, Ca and Ar and a more accurate determination of the $^{20}$Ne abundance. We find that in the solar minimum polar coronal hole solar wind the average freezing-in temperature is $\sim 1.1 \times 10^6$ K, increasing slightly with the mass of the ion. Using an extrapolation method we derive photospheric abundances from solar wind composition measurements. We suggest that our solar-wind-derived values should be used for the photospheric ratios of Ne/Fe $= 1.26 \pm 0.28$ and Ar/Fe $= 0.030 \pm 0.007$.

**Keywords** Sun: solar wind · Sun: elemental composition · Sun: isotopic abundance ratios

## 1 Introduction

The first precise measurements of the composition of noble gases in the solar wind were made using the Solar Wind Composition Experiment on the five Apollo missions to the

G. Gloeckler (✉)
Department of Oceanic, Atmospheric and Space Sciences, University of Michigan, Ann Arbor, MI 48109-2143, USA
e-mail: gglo@umich.edu

G. Gloeckler
Department of Physics, University of Maryland, College Park, MD 20742-0001, USA

J. Geiss
International Space Science Institute, Hallerstrasse 6, 3012 Bern, Switzerland
e-mail: geiss@issibern.ch

moon starting in 1969 (Geiss et al. 1970a). The first measurements of the charge state as well as the mass and charge of solar wind ions were made with a time-of-flight (TOF) mass spectrometer (CHEM) on the AMPTE spacecraft in 1982 in the magnetosheath of the Earth (Gloeckler and Geiss 1989). Both of these observations were made in the slow in-ecliptic solar wind during short time intervals of the declining phase of solar cycle 20 and 21 respectively, close to solar maximum. With the launch in 1990 of the TOF Solar Wind Ion Composition Spectrometer (SWICS) on Ulysses (Gloeckler et al. 1992) it was finally possible to observe the solar wind and continuously measure its elemental and charge state composition with a time resolution of ~15 minutes for tens of years under all solar wind flow conditions. The unique solar polar orbit of Ulysses allowed for the first time high-latitudes observations of the fast polar coronal hole solar wind. Geiss et al. (1995) presented the first observations of solar wind charge states (from which the freezing-in coronal temperature was determined) and composition in the fast stream from the south polar coronal hole.

A number of reports have been published that used SWICS Ulysses and ACE data to analyze solar wind composition in a variety of solar wind flows and changing solar wind conditions, employing various analysis techniques. For example, in their comprehensive study, von Steiger et al. (2000) contrasted the solar wind composition and charge state distributions in the fast, polar coronal hole solar wind with that of the slow, in-ecliptic wind. The forward model they developed estimates charge states of various elements based on probability distributions of TOF and energy values, and is best suited for studies of the variations of the composition of the more abundant elements (He, C, O, Mg, Si and Fe) and their charge states over relatively short time periods. In this paper we use a new method described below, different from that of von Steiger et al. (2000), and apply it to determine the composition and charge states in the fast solar wind at high latitude during a ~3 year time period starting at the end of 1993.

## 2 Instrumentation and Data Analysis Technique

The Solar Wind Ion Composition Spectrometer (SWICS) launched on Ulysses in 1990 uses a combination of energy per charge ($\varepsilon$), post-acceleration ($V$), followed by a time-of-flight ($\tau$) and energy measurement ($E$) to determine the speed ($u$), mass ($m$) and charge ($q$) of solar wind and suprathermal ions with energies between 0.6 to 60 keV/charge (see Gloeckler et al. 1992 for a full description of the SWICS instrument). It then follows that $m = 2(E/\alpha) \cdot (\tau/d)^2$, $q = (E/\alpha)/(V + \varepsilon)$ and $u = (2\varepsilon \cdot q/m)^{1/2}$, where $d$ is the time-of-flight distance (~10 cm) and $\alpha(m, E)$ is the energy defect, the fraction of the total energy of the ion that a solid-state detector measures. For solar wind energies, $q \approx (E/\alpha)/V$ and $m/q \approx 2V \cdot (\tau/d)^2$. Using this technique, the background noise is extremely low due to the triple coincidence measurements.

First, we accumulate a mass ($m$) versus mass/charge ($m/q$) matrix (Fig. 1), where $m$ and $m/q$ are calculated from measured ($\tau$, $E$) pulse-height pairs using preflight calibrations of $\alpha(m, E)$, and select our data such that $u$ is within a few percent of the simultaneously measured solar wind bulk speed, $V_{SW}$. For the present study we further selected data when $V_{SW}$ was between 700 and 800 km/s. The rest of our analysis is performed using the $m - m/q$ matrix data accumulated over a three year time period and selected according to the criteria just described.

We construct a mass histogram (Fig. 2) for some selected $m/q$ value (e.g. $2.370 \pm 0.036$, blue stripe in Fig. 1). Using the least squares method, we then find best fits to the most

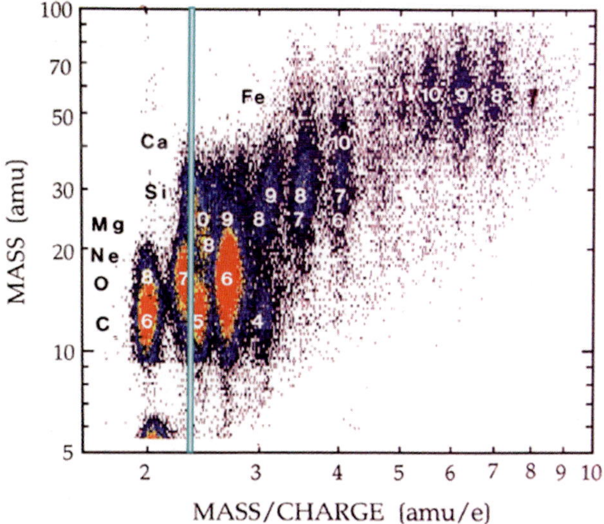


**Fig. 1** Typical mass – mass/charge ($m - m/q$) matrix derived from measured time-of-flight and energy pulse-height signal pairs using preflight calibrations of the energy defect. (See text)

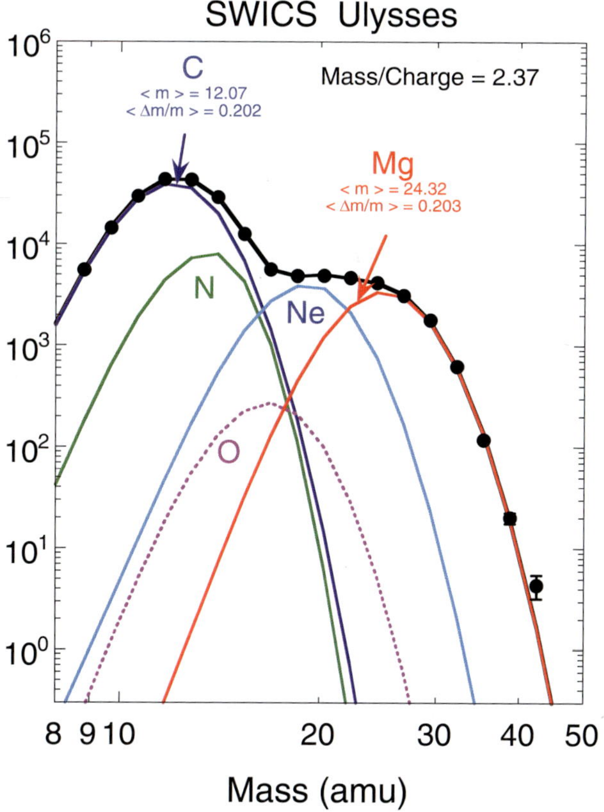


**Fig. 2** Mass histogram for $m/q = 2.370 \pm 0.036$ showing the most prominent element, C, and contributions from four other elements (N, Ne, Mg and O)

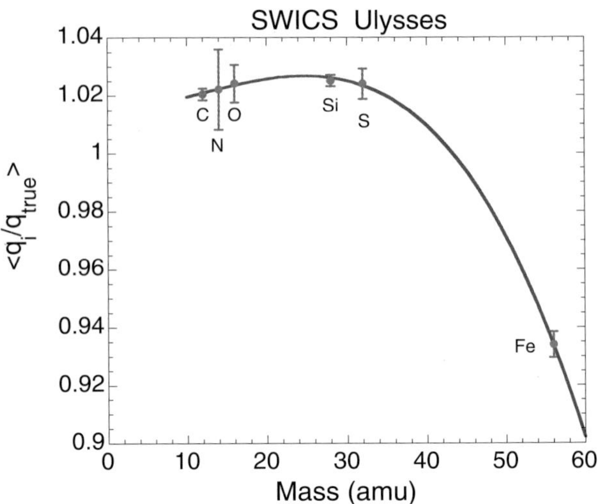


**Fig. 3** Ratio of the derived to the true charge state, $\langle q_j \rangle / q_{\text{true}}$, for C, N, O, Si, S and Fe. The curve is a polynomial fit to the five values with the least uncertainties

prominent peaks (i.e. C and Mg in Fig. 2) with Gaussian distributions for each of the individual elements, $j$, where each distribution is characterized by $c_j$ (integrated counts), $\langle m_j \rangle$ (mean mass) and $\langle (\Delta m/m)_j \rangle$ (width).

We repeat this procedure for all other mass histograms at different $m/q$ values, $k$, and obtain $c_{jk}$, $\langle m_{jk} \rangle$ and $\langle (\Delta m/m)_{jk} \rangle$ for as many of elements and from as many mass histograms as possible. Finally, we find the best estimates for the ratio of the derived to the true charge state of element $j$ $\langle q_j \rangle / q_{\text{true}} = (1/k) \cdot \sum q_{jk}/q_{\text{true}}$, where $q_{jk} = \langle m_{jk} \rangle / (m/q)_k$. Plotting $\langle q_j \rangle / q_{\text{true}}$ versus $j$, or equivalent mass $m$, as shown in Fig. 3, and fitting a polynomial curve to the best determined $\langle q_j \rangle / q_{\text{true}}$ values (i.e. C, O, Si, S and Fe) allows us to obtain from the polynomial fit estimates of $\langle q_j \rangle / q_{\text{true}}$ (and from these $\langle m_{jk} \rangle = \langle q_j \rangle \cdot (m/q)_k$) for the other, less abundant elements such as N, Ne, Mg, Ar and Ca. A similar procedure is employed to estimate best values for $\langle (\Delta m/m)_{jk} \rangle$. We then repeat the entire sequence enough times until all values for $\langle q_j \rangle / q_{\text{true}}$ and $\langle (\Delta m/m)_{jk} \rangle$ converge.

Figures 4, 5 and 6 show mass per charge distributions of individual elements. These distributions were derived using the integrated counts ($c_{jk}$) obtained in the final fits from mass histograms such as those shown in Fig. 2. With this technique clear separation of individual charge states of each element is obtained, and detection of rare charge states, such as $C^{3+}$, $N^{4+}$, $O^{5+}$ and $Ne^{7+}$, is possible.

## 3 Results

### 3.1 Ionization Fractions

We obtain ion fraction distributions for each element by dividing the total counts for each charge state by the detection efficiency (typically 0.4 to 0.6) of that charge state. Detection efficiencies were determined from preflight calibrations using a limited number of elements (H, He, C, N, O, Ne, Ar and Kr), and contribute most to the systematic uncertainties of our measurements. We used the efficiency model of Christina Cohen (private communication).

In Figs. 7 and 8 we show the average ion fraction distributions for $^{12}C$, $^{14}N$, $^{16}O$, $^{20}Ne$, $^{24}Mg$, $^{28}Si$, $^{32}S$ and $^{56}Fe$ in the 700–800 km/s high-latitude solar wind from polar coronal

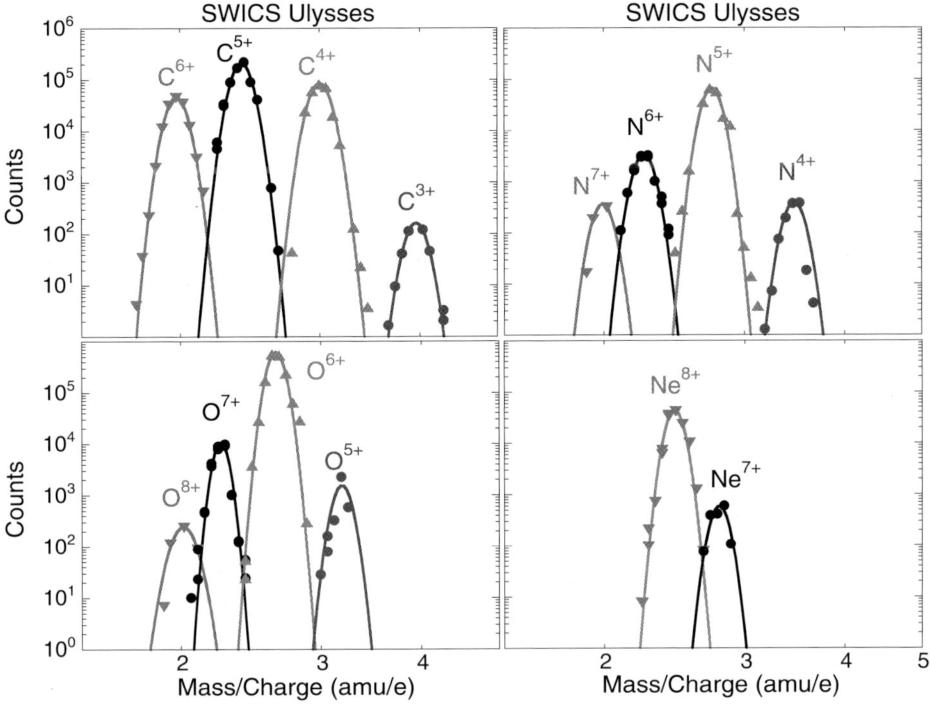


**Fig. 4** Mass per charge distributions for C, N, O and Ne. Multiple data points at a given $m/q$ value are from different estimates of integrated counts and indicate roughly the uncertainties in these estimates. *Curves* are Gaussian fits to the individual distributions

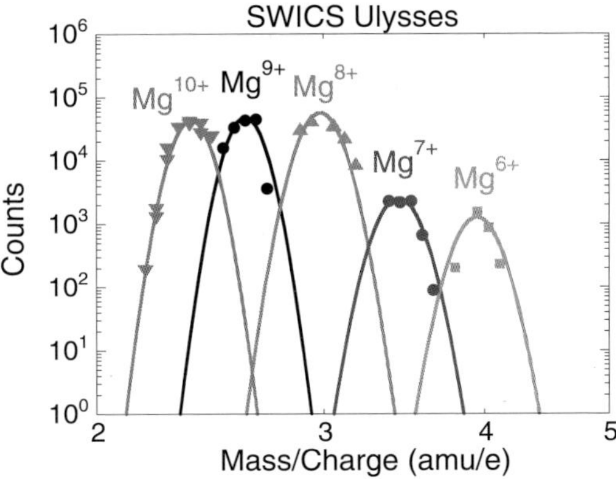


**Fig. 5** Same as Fig. 4 but for Mg

holes during solar minimum. At this stage of our analysis we could only obtain the most dominant charge state for Ar and Ca, 8 and 10 respectively. The ion fraction distributions for all of the elements shown, except for carbon, are well described by an equilibrium charge state distribution at a single freezing-in temperature. The most probable freezing-in temperature for each element is given in each respective panel of Figs. 7 and 8 and in Table 1.

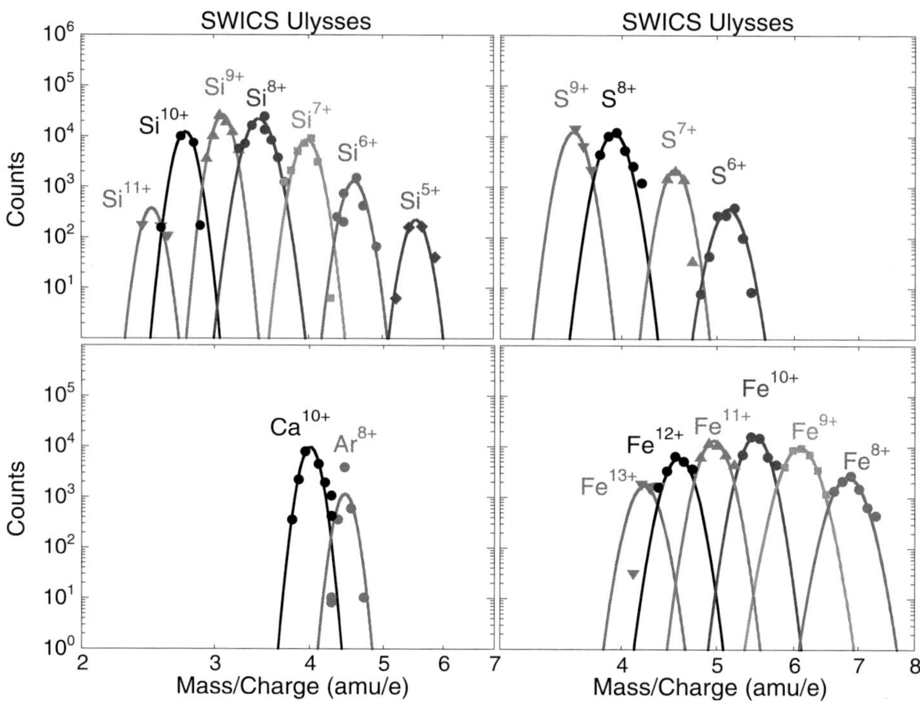


**Fig. 6** Same as Fig. 4 but for Si, S, Ar, Ca and Fe

The observed ion fraction distribution for carbon is narrower than that of any of the equilibrium charge state model distributions at a single temperature. No combination of equilibrium model distributions with different temperatures can match the observed distribution. However, the wave driven, two-fluid coronal hole model of Bürgi and Geiss (1986) predicts an ion fraction distribution for carbon that fits the measured distribution extremely well, as shown in upper left-hand panel of Fig. 7. Bürgi and Geiss (1986) have shown that due to the low electron density, carbon ions depart from local equilibrium at low solar altitude and are frozen-in below the temperature maximum. Their coronal hole model also gives good fits to the observed ion fraction distributions for N, O and Ne but not for heavier elements, for which their model predicts a lower freezing-in temperature for higher charge states and a lower abundance of higher charge states compared to those observed. As the authors point out, this discrepancy could be eliminated, if coronal electron velocity distributions had suprathermal tails, instead of being maxwellian as they assumed, and/or if differential velocities between charge states played an important role.

The freezing-in temperatures and errors are listed in the last column of Table 1. The dependence on ion mass of the freezing-in temperature, $T$, of C, N, O, Ne, Mg, Si, S and Fe are shown in Fig. 9. The average freezing-in temperature is $1.09 \times 10^6$ K and appears to increase slightly with increasing mass, a trend also seen in the results of the coronal hole model of Bürgi and Geiss (1986). Fitting a power law to our best-determined temperatures (O, Si, S and Fe), we find that $T = 0.84M^{0.084}$. We point out that the freezing-in temperature of nitrogen is well below that of the other ions. While at this stage of our analysis we could only determine the abundance of a single charge state of Ar and Ca respectively, the identified charge states are compatible with a freezing-in temperature of $\sim 10^6$ K.

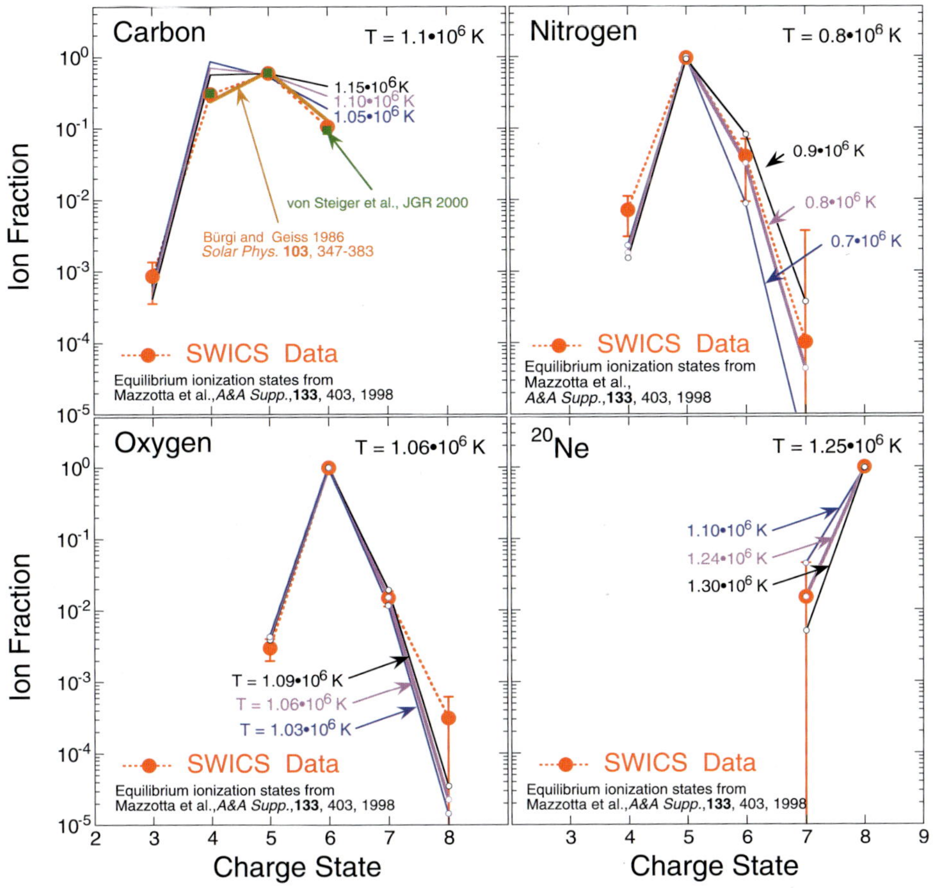

**Fig. 7** Ion fraction distributions for $^{12}C$, $^{14}N$, $^{16}O$ and $^{20}Ne$ measured in the high-speed stream solar wind from polar coronal holes during solar minimum (*large red filled circles*). *Error bars* (1-$\sigma$) include statistical and estimated systematic uncertainties. Equilibrium charge state model distributions (Mazzotta et al. 1998) for the average and $\pm 1$-$\sigma$ values of the freezing-in temperature are shown next to the temperature for which they are computed. For carbon, ion fraction measurements of von Steiger et al. (2000) for the polar coronal hole solar wind are indicated by filled *green squares*

## 3.2 Elemental Abundance

We obtain the elemental abundances by summing the efficiency-corrected counts for each charge state of a given element. The abundance ratios of $^{12}C$, $^{14}N$, $^{20}Ne$, $^{24}Mg$, $^{28}Si$, $^{32}S$, $^{36}Ar$, $^{40}Ca$ and $^{56}Fe$, relative to $^{16}O$, are given in column 2 of Table 1. The errors are due mostly to the estimated systematic uncertainties of the efficiencies for detecting a given ion.

# 4 Discussion

We now compare the elemental abundance in the high-speed solar wind from polar coronal holes during solar minimum (Table 1, column 2) to photospheric abundance ratios (Table 1, column 3; Asplund et al. 2005). Plotted in the left-hand panel of Fig. 10 as a function of

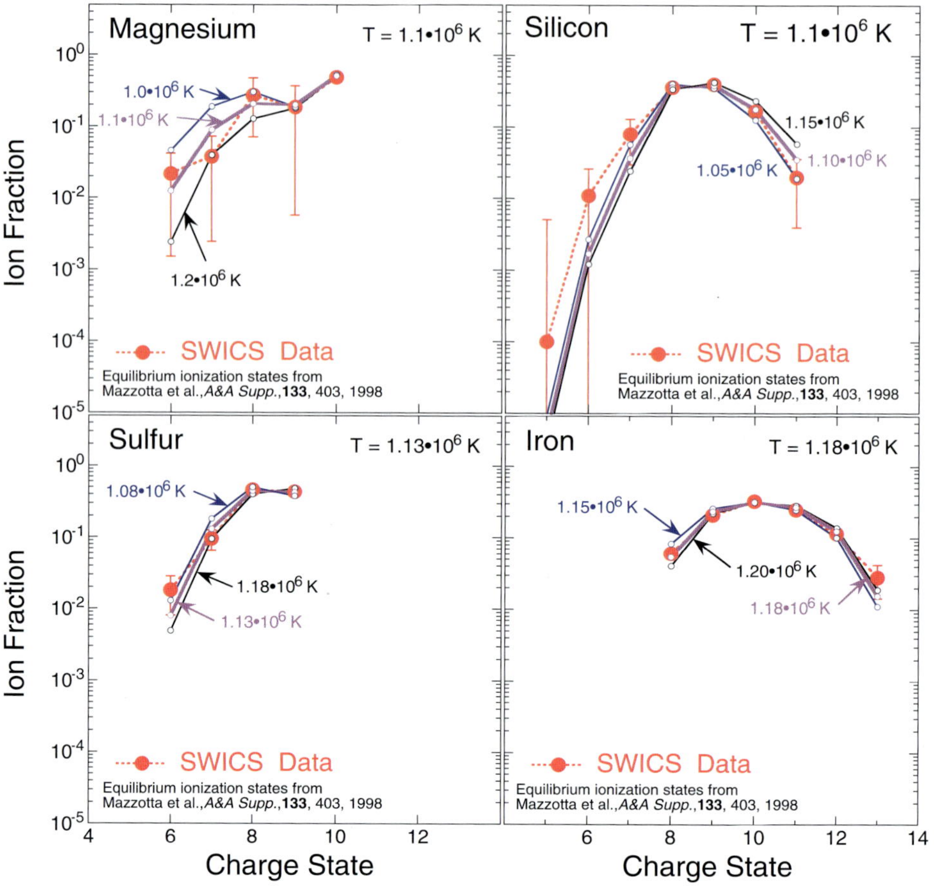

Magnesium T = 1.1•10^6 K
1.0•10^6 K
1.1•10^6 K
1.2•10^6 K
SWICS Data
Equilibrium ionization states from Mazzotta et al., *A&A Supp.*, **133**, 403, 1998

Silicon T = 1.1•10^6 K
1.15•10^6 K
1.05•10^6 K
1.10•10^6 K
SWICS Data
Equilibrium ionization states from Mazzotta et al., *A&A Supp.*, **133**, 403, 1998

Sulfur T = 1.13•10^6 K
1.08•10^6 K
1.18•10^6 K
1.13•10^6 K
SWICS Data
Equilibrium ionization states from Mazzotta et al., *A&A Supp.*, **133**, 403, 1998

Iron T = 1.18•10^6 K
1.15•10^6 K
1.20•10^6 K
1.18•10^6 K
SWICS Data
Equilibrium ionization states from Mazzotta et al., *A&A Supp.*, **133**, 403, 1998

Ion Fraction
Charge State

**Fig. 8** Same as Fig. 7 except for Mg, Si, S and Fe

the first ionization potential (FIP) is $R$, the abundance ratio, relative to Fe, in the solar wind divided by the respective abundance ratio, also relative to Fe, in the photosphere. We have chosen to normalize to iron rather than to oxygen, as is usually done, because the solar wind Fe can now be measured accurately and the value of the photospheric abundance of iron has remained fairly stable for many years, unlike that of oxygen. Except for Ar and Ne, all ratios are, within uncertainties, equal to one. Neon and argon are about a factor of two below one.

In the right-hand panel of Fig. 10 we show the same solar wind to photospheric ratios, $R$, as in the left-hand panel but now as a function of standard ionization time (SIT), introduced by Geiss and Bochsler (1985) (see also von Steiger and Geiss 1989; Geiss 1998) as a more physical parameter than FIP. Neon and argon are a factor of two below N and O, even though all these elements have about same standard ionization time.

While composition of Ne and Ar can be measured in the corona using spectroscopic observations (e.g. Feldman and Widing 2003), the photospheric abundances of these noble gases are not available directly from spectroscopy due to lack of suitable spectral lines, and in the past have been estimated from solar energetic particle measurements (Reames 1999). We believe that a far better method to derive the abundance of Ne and Ar in the photosphere is to use solar wind measurements of Ne and Ar in both the high-speed stream

**Table 1** Elemental abundances and freezing-in temperature of the high-speed solar wind from the polar coronal holes during solar minimum, and photospheric abundance ratios derived from solar wind and spectroscopic measurements

| Element | Coronal hole abundance ratio[a] | Photospheric abundance ratio[b] | Photospheric abundance ratio | Freezing-in temperature (K) |
|---|---|---|---|---|
| $^{12}$C | $0.710 \pm 0.080$ | $0.537 \pm 0.09$ | $0.715 \pm 0.09$[c] | $(1.1 \pm 0.08) \times 10^6$ |
| $^{14}$N | $0.143 \pm 0.016$ | $0.132 \pm 0.025$ | $0.145 \pm 0.019$[c] | $(0.8 \pm 0.1) \times 10^6$ |
| $^{16}$O | $\equiv 1.000$ | $\equiv 1.000$ | $\equiv 1.000$ | $(1.06 \pm 0.03) \times 10^6$ |
| $^{20}$Ne | $0.071 \pm 0.010$ | $[0.151 \pm 0.028]$ | $0.078 \pm 0.017$[c] | $(1.25 \pm 0.15) \times 10^6$ |
| $^{24}$Mg | $0.108 \pm 0.022$ | $0.074 \pm 0.018$ | $0.077 \pm 0.028$[c] | $(1.1 \pm 0.15) \times 10^6$ |
| $^{28}$Si | $0.088 \pm 0.014$ | $0.071 \pm 0.011$ | $0.066 \pm 0.026$[c] | $(1.1 \pm 0.05) \times 10^6$ |
| $^{32}$S | $0.035 \pm 0.004$ | $0.030 \pm 0.005$ | $0.035 \pm 0.004$[d] | $(1.13 \pm 0.05) \times 10^6$ |
| $^{36}$Ar | $0.0018 \pm 0.0004$ | $[0.0033 \pm 0.0008]$ | $0.0018 \pm 0.0004$[d] | – |
| $^{40}$Ca | $0.007 \pm 0.003$ | $0.0045 \pm 0.0007$ | – | – |
| $^{56}$Fe | $0.067 \pm 0.007$ | $0.062 \pm 0.010$ | – | $(1.175 \pm 0.025) \times 10^6$ |

[a]Errors are mostly systematic

[b]From Asplund et al. (2005). Ne/O and Ar/O ratios based on extrapolations from solar energetic particle abundances (Reames 1999)

[c]Extrapolated from solar wind measurements (see Fig. 11)

[d]Polar coronal hole value from column 2

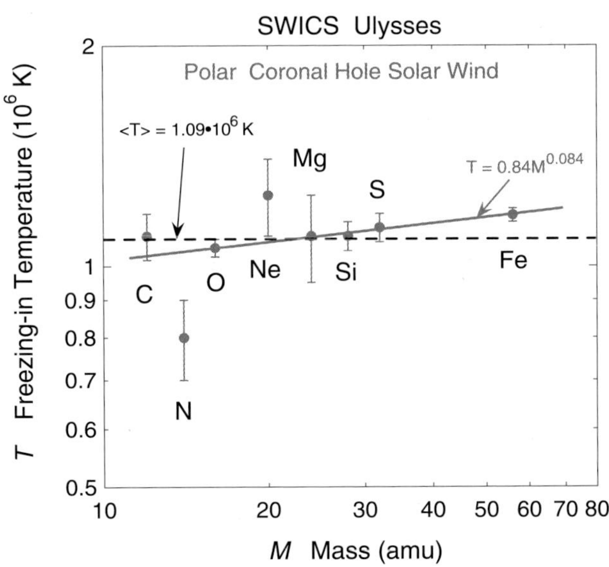


**Fig. 9** Freezing-in temperature of C, N, O, Ne Mg, Si, S and Fe as a function of ion mass. Error limits are freezing-in temperatures of the $\pm 1$-$\sigma$ equilibrium model fits to the ion fraction distributions of the respective ions from Figs. 7 and 8

and in the slow wind. This method was first use by Gloeckler and Geiss (2000) to determine the photospheric (or outer convective zone) abundance of $^3$He/$^4$He (also see Fig. 6 of Gloeckler and Fisk 2007), and is illustrated in Fig. 11 for the case of Si/O and Ne/O. The photospheric values of these two ratios using this solar wind extrapolation method are listed in column 4 of Table 1 along with those of C, N, Mg, S and Ar. Except for Ne and Ar,

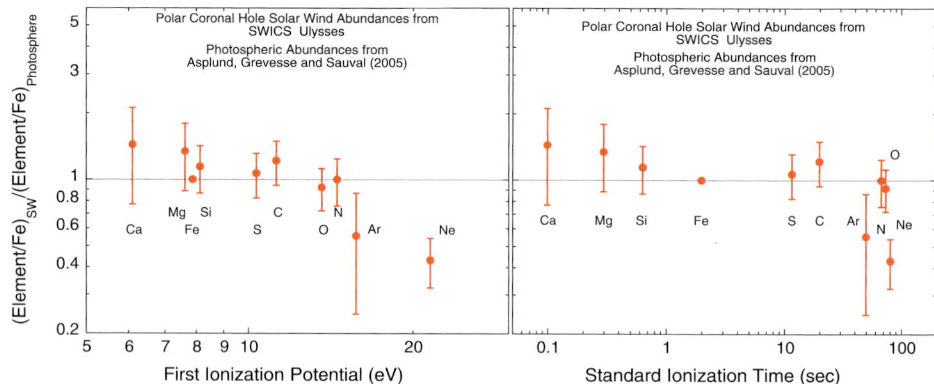


**Fig. 10** *R*, the abundance ratio, relative to Fe, of the high-speed solar wind from polar coronal holes at solar minimum divided by the respective abundance ratio, also relative to Fe, of the photosphere, versus the first ionization potential or FIP (*left-hand panel*), and versus the standard ionization time or SIT (*right-hand panel*). *Error bars* include uncertainties of both the solar wind and photospheric ratios

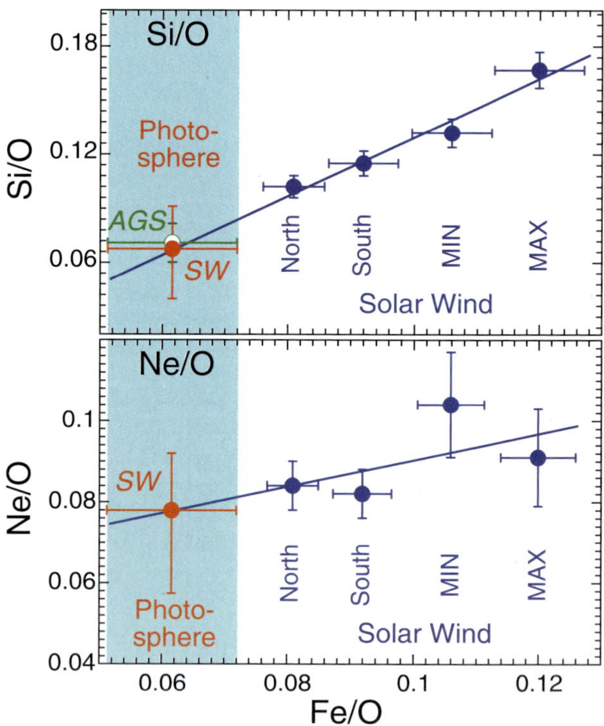


**Fig. 11** Solar wind Si/O (*top panel*) and Ne/O (*bottom panel*) abundance ratios versus Fe/O ratios measured with SWICS on Ulysses (von Steiger et al. 2000) during various solar wind flow conditions (*blue filled circles*; see von Steiger et al. 2000 for description of the four time periods: North, South, MIN and MAX). The *blue lines* represent least squares fits to the four respective solar wind data points. Extrapolation of the best-fit lines to the photospheric Fe/O abundance ratio (Asplund et al. 2005) gives an estimate of the Si/O (*top panel*) and Ne/O (*bottom panel*) abundance ratios in the photosphere (*red filled circles*). Errors of the solar wind ratios are mostly systematic and will be reduced in the future. The well-established photospheric Si/O ratio derived from spectroscopic measurements (Asplund et al. 2005) is indicated by the green *unfilled circle* labeled *AGS* in the *top panel*

whose photospheric abundance cannot be obtained from spectroscopic measurements, and for C, the agreement between the solar-wind-derived photospheric values (column 4) and the corresponding ratios of Asplund et al. (2005), listed in column 2 of Table 1, are quite remarkable. We therefore suggest that for the photospheric values of neon and argon our solar-wind-derived ratios, Ne/O $= 0.078 \pm 0.017$ and Ar/O $= 0.0018 \pm 0.0004$, be adopted.

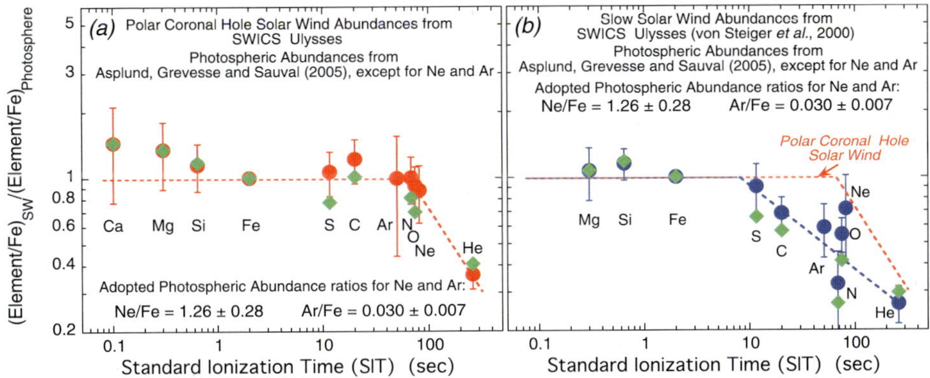


**Fig. 12** (**a**) Same as right-hand panel of Fig. 10, but now using values of $1.26 \pm 0.28$ and $0.030 \pm 0.007$ for the photospheric abundance of Ne and Ar relative to Fe, respectively (*filled red circles*). The fast solar wind He/Fe ratio is the average of the southern and northern polar coronal hole values of von Steiger et al. (2000). (**b**) Same as (**a**), but for the slow solar wind. The solar wind abundances for all elements shown (except Ar) are the average of the MIN and MAX data of von Steiger et al. (2000). The solar wind Ar/Fe ratio is derived from the Geiss et al. (2004) Apollo $^{20}$Ne/$^{36}$Ar ratio ($49 \pm 7$) and the von Steiger et al. (2000) solar maximum (MAX) period Ne/Fe ratio of $0.76 \pm 0.32$. The photospheric Ne/Fe and Ar/Fe ratios used were $1.26 \pm 0.28$ and $0.03 \pm 0.007$, respectively. The slow solar wind He/Fe ratio is the MAX value of von Steiger et al. (2000). Date points represented by *filled green diamonds in both panels* are computed using Grevesse and Sauval (1998) photospheric abundance ratios instead of the Asplund et al. (2005) values (Ne and Ar points are not shown). With these older photospheric ratios the SIT patterns remains basically the same, but the onset of progressively reduced abundance of high-SIT elements would move to ~2 to 3 times lower SIT values. Our basic conclusions remain, however, unchanged

The solar wind C/O and Fe/O abundance ratios are especially well determined from SWICS measurements. It is therefore surprising that both the solar-wind-derived photospheric C/O and the measured C/O ratio in the solar minimum polar coronal solar wind are about 30% higher than the Asplund et al. (2005) C/O ratio. We hope that this discrepancy will be resolved with further improvements of models of the solar atmosphere and better efficiency models for the SWICS instruments.

In Fig. 12a we use our solar-wind-derived photospheric values for Ne/Fe $= 1.26 \pm 0.28$ and Ar/Fe $= 0.030 \pm 0.007$ (obtained by dividing the corresponding ratios in column 4 of Table 1 by the photospheric Fe/O $= 0.062$ ratio). With this adjustment the solar wind composition is, within errors, the same as the photospheric composition for all elements with standard ionization times (SIT) below ~70.

The ratio $R$ for the slow, mostly in-ecliptic solar wind as a function of SIT is shown in Fig. 12b. We used the average of the two slow solar wind (MIN and MAX) composition measurements of von Steiger et al. (2000) for C, N, O, Ne, Mg, Si, S and Fe and the Apollo Foil Experiment value for the Ar/Ne ratio (Geiss et al. 2004). For the photospheric composition we used the Asplund et al. (2005) value, except for Ne and Ar for which we used our adopted photospheric abundance of Ne/Fe $= 1.26 \pm 0.28$ and Ar/Fe $= 0.030 \pm 0.007$. It is clear that in this slow solar wind, unlike in the fast polar coronal hole solar wind, the ratios $R$ begin to decrease from one at SIT of about 7 seconds. Thus, whatever process causes the neutral gas in the chromosphere to become ionized, it is at least an order of magnitude faster in the region of origin of the slow wind compared to polar coronal hole regions. What this process is remains an open question.

In Fig. 12 we also plotted the He/Fe ratios for both the fast and slow solar wind. In both cases this ratio is far below its photospheric value. When helium was first measured in the solar wind (Neugebauer and Snyder 1966), the He/H ratio was found to be typically $\leq 0.04$, much lower than the expected solar ratio. The helium deficit was attributed to a lack of acceleration in the corona. Geiss et al. (1970b) derived a drag factor $\Gamma = Q^2/(2A - Q - 1)$ that governs the collisional acceleration of minor ions with $Q$ and atomic weight $A$ in a proton–electron plasma, and they confirmed by numerical integration that in such a two-component solar wind, momentum transfer is lower for $^4He^{++}$ than it is for all heavier ions up to and beyond iron. Geiss et al. (1970b) emphasized that their numerical results and their drag factor $\Gamma$ neglect thermal diffusion, wave-particle interactions, and the momentum transfer of helium to the other components of the plasma in the corona and solar wind. The momentum equation integrated by F. Bürgi included these three factors, and showed that they have a major influence on the acceleration, charge state distribution and abundance of individual ion species in the solar wind (Bürgi and Geiss 1986). While the factor $\Gamma$ or similar drag factors remain useful for roughly comparing the importance of coulomb drag between different ion species, numerical models are needed for quantitatively assessing ion abundances or charge state distributions.

In 1977 Bame et al. (1977) discovered that the He/H ratio in the high speed streams coming out of coronal holes is very steady at 0.048, but still considerably lower than in the outer convective zone where its value is now found to be 0.084 (Perez Hernandez and Christensen-Dalsgaard 1994). Following these observations, Geiss (1982) argued that the persistent depletion of the He/H ratio by a factor of 1.6 cannot be produced in the corona, but could result from a separation process operating below the corona at temperatures $T \approx 10^4$ K, i.e. in the chromosphere (and perhaps the transition region), that level in the solar atmosphere, where the so called FIP effect, which is the overabundance of Mg, Si and Fe relative to C, N and O, is produced. Consequently, FIP effect models that included helium were developed for chromospheric conditions (Geiss and Bochsler 1985; von Steiger and Geiss 1989). Shortly afterwards, it was discovered what nobody had predicted: The FIP effect was very much lower in the fast, coronal hole solar wind than in the slow solar wind (Gloeckler et al. 1989; von Steiger et al. 1992).

In Figs. 10 and 12 we normalize abundances to the Mg–Si–Fe plateau (we use Fe, the best determined element in this group). This normalization makes sense, because the proposed FIP mechanisms operate in competition between ionization time and a time constant (SIT) that characterizes the separation of ions from atoms. The existence of the plateau of low FIP elements implies that these elements were fully ionized at the time of separation. On the other hand, elements with FIP $> \sim 10$ eV or SIT $> \sim 10$ seconds are depleted in the slow solar wind, because they were not yet fully ionized at the time of ion–atom separation.

Our best estimates of the photospheric abundance of isotopes and those elements for which no suitable spectral lines exist now come from solar wind measurements in the slow wind and high-speed streams as illustrated in Fig. 11. Still, it would be good to underpin this empirical method by a theory of the FIP or SIT effect that explains the difference in depletion of high FIP elements, including helium, between the slow solar wind and the coronal-hole high-speed streams.

## 5 Conclusions

We have determined the average abundances and ion fraction distributions of $^{12}C$, $^{14}N$, $^{16}O$, $^{20}Ne$, $^{24}Mg$, $^{28}Si$, $^{32}S$, $^{36}Ar$, $^{40}Ca$ and $^{56}Fe$ in the high-speed solar wind from polar coronal holes during solar minimum using a new analysis technique. We find and conclude that:

1. Except for carbon all other elements either have, or are consistent with having, a single equilibrium freezing-in temperature.
2. The average freezing-in temperature is $\sim 1.1 \times 10^6$ K and shows a slight increase with increasing mass.
3. Carbon alone has an observed ion fraction distribution that is significantly narrower than that predicted by equilibrium charge state models at any single freezing-in temperature. Its ion fraction distribution, however, agrees well with predictions of the multi-fluid coronal hole model of Bürgi and Geiss (1986).
4. Except for He and probably Ne, the composition of the high-speed solar wind from polar coronal holes during solar minimum has no FIP effect, i.e. the elemental ratios relative to Fe, in this solar wind and in the photosphere are, within errors, the same.
5. We used an extrapolation method to derive photospheric abundances from solar wind composition measurements in various types of solar wind flows, ranging from the polar coronal hole flow at solar minimum to the solar maximum slow solar wind. These solar-wind-derived photospheric ratios are in excellent agreement (except for C) with corresponding photospheric ratios that can be obtained from spectral lines. We therefore recommend that for the best estimates of the photospheric abundance of neon and argon are our solar-wind-derived values of $\text{Ne/Fe} = 1.26 \pm 0.28$ and $\text{Ar/Fe} = 0.030 \pm 0.007$.
6. The slow solar wind, unlike the fast polar coronal hole solar wind, has a SIT effect that causes the abundance of elements in the solar wind with standard ionization times larger than about 7 sec to be below that of corresponding elements in the photosphere.
7. Whatever process causes the ionized gas in the chromosphere to separate from the neutral gas and then escape, is at least an order of magnitude faster in the region of origin of the slow solar wind than it is in polar coronal hole regions.

In our future work we will apply this technique to obtain more accurate estimates of solar wind composition and charge states using SWICS Ulysses and ACE data in a variety of solar wind flow conditions, and in particular as a function solar wind speed.

**Acknowledgements** This work was supported, in part, by NASA contract NAGR-10975, and by JPL contract 1237843.

## References


M. Asplund, N. Grevesse, A.J. Sauval, in *Cosmic Abundances as Records of Stellar Evolution and Nucleosynthesis in honor of David L. Lambert*, ed. by T.G. Barnes III, F.N. Bash, vol. 336 (Astronomical Society of the Pacific, 2005), p. 25

S.J. Bame, J.R. Asbridge, W.C. Feldman, J.T. Gosling, J. Geophys. Res. **82**, 1487 (1977)

A. Bürgi, J. Geiss, Sol. Phys. **103**, 347–383 (1986)

U. Feldman, K.G. Widing, Space Sci. Rev. **107**, 665–720 (2003)

J. Geiss, Space Sci. Rev. **33**, 201 (1982)

J. Geiss, Space Sci. Rev. **85**, 241–252 (1998)

J. Geiss, P. Bochsler, in *Proc. Rapports Isotopiques dans le Système Solaire* (Cepadues Editions, Paris, 1985), pp. 213–228

J. Geiss, P. Eberhardt, F. Bühler, J. Meister, P. Signer, Geophys. Res. **75**, 5972–5979 (1970a)

J. Geiss, P. Hirt, H. Leutwyler, Sol. Phys. **12**, 458 (1970b)

J. Geiss, F. Bühler, H. Cerutti, P. Eberhardt, C.H. Filleux, J. Meister, P. Signer, Space Sci. Rev. **110**, 307–335 (2004)

J. Geiss, G. Gloeckler, R. von Steiger, H. Balsiger, L.A. Fisk, A.B. Galvin, F.M. Ipavich, S. Livi, J.F. McKenzie, K.W. Ogilvie, B. Wilken, Science **268**, 1033–1036 (1995)

G. Gloeckler, L.A. Fisk (2007), this volume

G. Gloeckler, J. Geiss, in *Cosmic Abundances of Matter*, ed. by C.J. Waddington. AIP Conf. Proc., vol. 183 (1989), p. 49

G. Gloeckler, J. Geiss, in *The Light Elements and Their Evolution*, ed. by L. Da Silva, M. Spite, J.R. De Medeiros. AIU Symposium, vol. 198 (2000), p. 224
G. Gloeckler, J. Geiss, H. Balsiger, P. Bedini, J.C. Cain, J. Fischer et al., Astron. Astrophys. Suppl. Ser. **92**, 267 (1992)
G. Gloeckler, F.M. Ipavich, D.C. Hamilton, B. Wilken, G. Kremser, EOS Trans. AGU **70**, 424 (1989)
N. Grevesse, A.J. Sauval, Space Sci. Rev. **85**, 161–174 (1998)
P. Mazzotta, G. Mazzitelli, S. Colafrancesco, N. Vittorio, Astron. Astrophys. Suppl. Ser. **133**, 403–409 (1998)
M. Neugebauer, C.W. Snyder, J. Geophys. Res. **71**, 4469 (1966)
F. Perez Hernandez, J. Christensen-Dalsgaard, Mon. Not. Roy. Astron. Soc. **269**, 475 (1994)
D.V. Reames, Space Sci. Rev. **90**, 413 (1999)
R. von Steiger, J. Geiss, Astron. Astrophys. **225**, 222–238 (1989)
R. von Steiger, S.P. Christon, G. Gloeckler, F.M. Ipavich, Astrophys. J. **389**, 791–799 (1992)
R. von Steiger, N.A. Schwadron, L.A. Fisk, J. Geiss, G. Gloeckler, S. Hefti, B. Wilken, R.F. Wimmer-Schweingruber, T.H. Zurbuchen, J. Geophys. Res. **105**, 27,217 (2000)

Space Sci Rev (2007) 130: 153–160
DOI 10.1007/s11214-007-9180-8

# Acceleration and Composition of Solar Wind Suprathermal Tails

**L.A. Fisk · G. Gloeckler**

Received: 17 January 2007 / Accepted: 27 March 2007 /
Published online: 17 May 2007


**Abstract** Observations in the solar wind have revealed important insights into how energetic particles are accelerated in astrophysical plasmas. In circumstances where stochastic acceleration is expected, a suprathermal tail on the distribution function is formed with a common spectral shape: the spectrum is a power law in particle speed with a spectral index of $-5$. Recent theories for this phenomenon, in which thermodynamic constraints are applied to explain the common spectral shape, are reviewed. As an example of potential extensions of this theoretical work, consideration is given to the acceleration of Anomalous Cosmic Rays in the heliosheath.

**Keywords** Solar wind · Stochastic acceleration · Cosmic rays

## 1 Introduction

The composition of energetic particles is determined, in large part, by the processes by which they are accelerated. It follows, therefore, that to understand and interpret the composition of various species—cosmic rays, solar energetic particles, etc.—we must understand their acceleration.

The heliosphere remains a most useful laboratory for understanding particle acceleration. We can observe both the accelerated particles and the plasma conditions and processes responsible for the acceleration. Moreover, there are numerous examples of acceleration—stochastic acceleration in turbulence, propagating and standing shocks, large-scale and small-scale shocks.

L.A. Fisk (✉) · G. Gloeckler
Department of Atmospheric, Oceanic and Space Sciences, University of Michigan, Ann Arbor, MI 48109-2143, USA
e-mail: lafisk@umich.edu

G. Gloeckler
Department of Physics and IPST, University of Maryland, College Park, MD 20742, USA

Currently, the heliosphere is providing us with extraordinary insight into particle acceleration. There are recent observations that reveal that under many different circumstances the spectrum of suprathermal particles is always the same—it is a power law with a spectral index of $-1.5$ when expressed as differential intensity, or $-5$ when expressed as a distribution function in velocity space (Gloeckler et al. 2000; Gloeckler 2003; Simunac and Armstrong 2004; Fisk and Gloeckler 2006). This common spectrum occurs in the quiet solar wind in the absence of shocks, in disturbed conditions downstream from shocks, and in particular throughout the heliosheath currently being explored by Voyager 1 (Decker et al. 2005).

There has been a great deal of work done on the acceleration of energetic particles, dating back to Fermi in the 1940s. Stochastic acceleration in the solar wind, sometimes known as second-order Fermi acceleration, has been studied by, e.g., Fisk (1976), Schwadron et al. (1996), le Roux et al. (2001, 2002), Giacalone et al. (2002), and Webb et al. (2003). Acceleration at shocks, known as diffusive shock acceleration, developed into useable theories with the work of Krymsky (1977), Axford et al. (1978), Blandford and Ostriker (1978), and Bell (1978).

Not one of these theories for particle acceleration, either stochastic acceleration or diffusive shock acceleration, naturally yields a spectrum that is a power law with a unique spectral index of $-5$. Diffusive shock acceleration naturally yields power-law spectra, but the spectral index depends sensitively on the jump in flow speed across the shock, which clearly varies from shock to shock. Stochastic acceleration can yield power-law spectra, although exponential spectra are more common. Moreover, the spectral shape of the stochastic spectra depends sensitively on the properties of the turbulence.

The message then from the heliospheric observations is that there is some additional process at work, beyond the simple stochastic and shock acceleration theories developed to date, which is forcing the spectra of the accelerated particles into both a unique shape and one that occurs commonly in many different plasma conditions. Clearly, if we understand this additional process, and the conditions under which it occurs, we should be able to predict the spectral shape of accelerated particles in many other astrophysical settings, well beyond the heliosphere. The universality of the common spectral shape in the heliosphere indicates a much broader universality.

Fisk and Gloeckler (2006, 2007) have introduced a theory for the common spectral shape. In Fisk and Gloeckler (2007) the theory is based on thermodynamic constraints. The thermodynamic approach is promising for the simple reason that anything that occurs so commonly, in so many different circumstances, must have its roots in a fundamental property of the system.

In this paper, we review briefly the observations of the common spectral shape. We then summarize the theoretical arguments of Fisk and Gloeckler (2007); for details on these arguments refer to the published work. Finally, we provide an example of how this theory for the acceleration of energetic particles can be applied in other settings by speculating on how it might apply to the acceleration of Anomalous Cosmic Rays (ACRs).

## 2 Observations

Shown in Fig. 1 is a typical proton distribution function, observed in the slow, quiet solar wind by the *SWICS* instrument on Ulysses (Gloeckler et al. 1992). The distribution function is plotted versus the ion speed divided by the solar wind flow speed. The spectrum consists

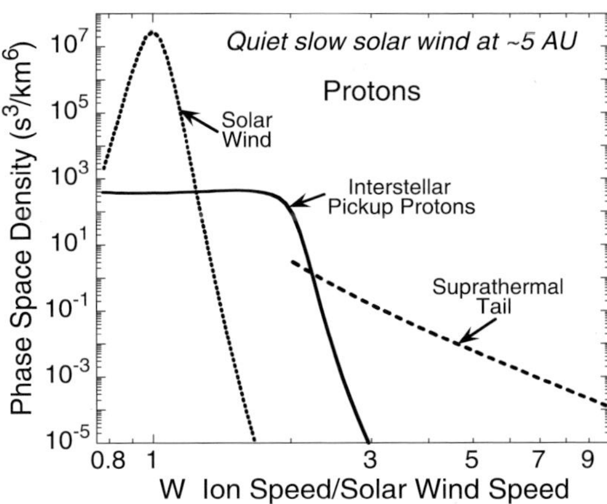


**Fig. 1** Typical proton velocity distribution function observed during times of quiet (least disturbed) solar wind conditions at about 5 AU from the Sun (after Fisk and Gloeckler 2007). The spectrum is made up of three distinct proton populations: (1) The thermal bulk solar wind particles that carry most of the mass (*dotted curve*), (2) the interstellar pickup protons (*solid curve*) that have a flat spectrum that drops sharply at about twice the solar wind speed (in the spacecraft frame), and (3) the suprathermal tail (*dashed curve*) that has a power-law spectrum in the frame of the solar wind with a unique spectral index of −5

of three parts: the thermal distribution of the solar wind, interstellar pickup ions (interstellar neutral gas that is swept into the heliosphere, ionized by photo-ionization and charge exchange, and picked up by the solar wind); and the suprathermal tail.

The suprathermal tail is a power law with spectral index of −5 in the frame of the solar wind. In the spacecraft frame, as is shown in the figure, the spectrum is not a simple power law; transformation to the solar wind frame is necessary to reveal the power law. The power law extends up to higher energies than can be observed by *SWICS*, typically reaching energies of up to a few MeV/nucleon, particularly in the outer heliosphere.

Figure 1 is for the quiet, slow solar wind. This same spectral shape occurs in disturbed conditions downstream from shocks in the solar wind. Most striking, the common spectral shape occurs throughout the heliosheath currently being explored by Voyager 1 (Decker et al. 2005). Indeed, the spectral shape of low-energy energetic particles in the heliosheath is the most constant spectrum ever observed by space instrumentation exploring our solar system.

## 3 A Theoretical Explanation

In this section, we summarize the theoretical explanation for the common spectral shape of Fisk and Gloeckler (2007); for details refer to the published work. The acceleration of low-energy energetic particles in the solar wind is considered to result from a stochastic process. In Fig. 1, the common spectral shape occurs in the quiet solar wind far from shocks, which requires some form of stochastic acceleration. Even in the presence of shocks, stochastic acceleration in the accompanying turbulence can be important, and in the case of the heliosheath, large-scale turbulent motions are present that yield a natural stochastic acceleration (Burlaga et al. 2006).

The basic argument of Fisk and Gloeckler (2007) is that it is possible to place thermodynamic constraints on stochastic acceleration in compressional turbulence that naturally yields the observed common spectral shape. The summary of these arguments is:

- The turbulence is assumed to be compressional, i.e., there are random compressions and expansions. Stochastic acceleration in compressional turbulence should naturally tend to yield power-law spectra, since in a compression and/or expansion the rate of change of energy of a particle is proportional to the particle energy.

- Present in the turbulence are two distinct sets of particles. (1) Very low-energy particles that are not very mobile and which are simply alternatively expanded and compressed. These are referred to as core particles. (2) At some higher energy, the particles are sufficiently mobile along the magnetic field so that they can experience a statistically significant number of random compressions and expansions. These particles, referred to as tail particles, will diffuse upward in energy and create a suprathermal tail on the distribution. This is a classic stochastic acceleration process.
- The upward diffusion of the particles is bounded. At some high energy, the particle gyroradii exceed the scale size of the compressions and expansions, and the particles no longer undergo the stochastic acceleration.
- From the point of view of the turbulence, all particles, both core and tail, are simply being adiabatically compressed and expanded. There is no dissipation of the turbulence. The function of the turbulence is only to redistribute energy from the core to the tail.
- The system of core and tail particles is thus thermally isolated. There is no dissipation of the turbulence to increase the total energy in the particles. The tail is bounded at high energies and there is no escape and loss of energy. (Below we consider the consequences of some slow escape when the particles reach high energies.)
- The diffusion in energy space—the stochastic acceleration—is an irreversible process that increases the entropy of the core and tail particles. It is analogous to a Joule expansion of a gas into a vacuum, in which the volume of a thermally isolated gas increases irreversibly without performing any work.
- The equilibrium state of a thermally isolated system is a state of maximum entropy, i.e., in equilibrium the entropy of the tail must obtain a maximum. Moreover, in a thermally isolated system, no process can decrease the entropy, so upon attaining maximum entropy, the entropy must be constant.
- The tail particles undergo random compressions and expansions, and so in equilibrium, when the entropy is constant, the random compressions and expansions must be isentropic.
- The behavior of the tail particles is governed by the standard transport equation for energetic particles (Parker 1965; Gleeson and Axford 1967). In the solar wind, the stochastic acceleration should be more rapid than adiabatic deceleration. Also, spatial diffusion can be ignored. For the problem considered here, the system is assumed to be thermally isolated, and thus escape by diffusion is not permitted.
- In order for the standard transport equation, with the above assumptions about adiabatic cooling and spatial diffusion, to yield an isentropic compression or expansion in the compression turbulence, Fisk and Gloeckler (2007) show that the only allowable spectral index for a power-law spectrum is $-5$ (when expressed as a distribution function).
- *A power-law spectrum with spectral index of $-5$ is unique. It is the one spectrum where the pressure in the tail, by itself, undergoes isentropic compressions and expansions.*

Fisk and Gloeckler (2007) also use the thermodynamic arguments to specify the pressure that is possible in the tail:

- A compression of the core particles increases their temperature, and raises their speeds to be above the threshold speed required for the particles to enter into the stochastic acceleration process.
- Equilibrium should occur when the energy that can flow from the core to the tail is equal to the average energy or equivalently the pressure in the tail, including the pressure resulting from the compression of the tail particles.

- *With these constraints*, Fisk and Gloeckler (2007) *show that the pressure in the tail particles in equilibrium must be*

$$P_t = \frac{2}{5} \frac{\beta}{(\beta + 1)} P_c. \tag{1}$$

*Here, $P_t$ and $P_c$ are, respectively, the pressure in the tail and the core particles; $\beta = \delta P_c / P_c = \delta P_t / P_t$ is the maximum relative increase in the pressure of the core or tail particles ($\beta$ can be shown to be the same for both core and tail particles).*

Fisk and Gloeckler (2007) also reconcile the thermodynamic arguments with the arguments presented in Fisk and Gloeckler (2006) that show that the observed common spectral shape can arise from a cascade in energy analogous to a turbulent cascade.

- Suppose that at high energies, when the particle gyroradii exceed the scale size of the turbulence and the particles no longer experience the random compressions and expansions, some small fraction of the particles escape. The fraction needs to be sufficiently small so that the thermodynamic constraints apply to the particles below the high-energy threshold.
- Fisk and Gloeckler (2007) recast the transport equation into an energy equation and show that with the common spectral shape, a power law with spectral index of $-5$, there is no dissipation of energy in the tail. There can be flow of energy through the tail, but not dissipation.
- The situation is analogous to a turbulent cascade. There is energy in the core particles, which flows upward through the tail. The energy does not dissipate in the tail, and a small fraction escapes at high energy. The tail is thus the equivalent of the inertial range in a turbulent cascade.
- *Arguments analogous to those of* Kolmogorov (1941), *which yields the $-5/3$ spectral index of the inertial range of incompressible hydrodynamic turbulence, can be applied to demonstrate that the spectral index for the suprathermal tails in the solar wind is the observed value of $-5$.*

## 4 Application to Anomalous Cosmic Rays

Consider the application of these principles of stochastic acceleration in compressional turbulence to Anomalous Cosmic Rays (ACRs). ACRs are considered to originate from interstellar pickup ions, since they exhibit the same composition; ACRs contain only those elements that are expected to be neutral in the interstellar medium (Fisk et al. 1974). Pickup ions are injected into the solar wind with energies of ~1 keV/nucleon. ACRs have energies in excess of ~10 MeV/nucleon, thus requiring a more than four-orders-of-magnitude acceleration somewhere in the solar wind. It has been expected that the likely site for this acceleration would be the termination shock of the solar wind, where the supersonic solar wind goes subsonic to begin the process of merging with the local interstellar medium (Pesses et al. 1981; Jokipii 1990, and references therein; Zank 1999, and references therein).

However, one of the principal surprises of the Voyager 1 crossing of the termination shock was that ACRs are not being accelerated at this location. Stone et al. (2005) and Decker et al. (2005) report that the intensity of ACRs did not peak at the shock, indicating that the source of the ACRs was at least not in the region of the termination shock observed by Voyager 1. Gloeckler and Fisk (2006) have argued from an analysis of the upstream beams seen by Voyager 1 that Voyager observed representative conditions. It is likely then that ACRs are not accelerated at the termination shock and another explanation needs to be found.

The heliosheath exhibits large-scale compressional turbulence (Burlaga et al. 2006). It is thus an ideal site for stochastic acceleration of energetic particles. Moreover, the subsonic flow of the solar wind in the heliosheath should not experience appreciable adiabatic deceleration, as occurs in the expanding supersonic solar wind, and thus there is no competing deceleration.

The observed spectrum of low-energy ions in the heliosheath, with the common spectral index of $-5$ when expressed as a distribution function (Decker et al. 2005), is testimony that fully evolved stochastic acceleration in the compressional turbulence is occurring in the heliosheath. The thermodynamic constraints of Fisk and Gloeckler (2007) apply.

Indeed, it appears as if stochastic acceleration in compressional turbulence is the dominant acceleration process occurring throughout the outer heliosphere. Gloeckler and Fisk (2006) demonstrate that underlying the anisotropic beams observed upstream from the termination shock is an ambient spectrum with the common spectral index of $-5$, with a cutoff at about 3 MeV/nucleon. The composition of these particles is that of the pickup ions in that they are depleted in carbon. The particles are accelerated crossing the shock and the spectrum remains with the common spectral index of $-5$ throughout the downstream heliosheath. The anisotropic beams upstream can be understood simply as downstream particles leaking upstream (Gloeckler and Fisk 2006). The beams have highly variable spectra due to velocity dispersion, but the spectra averaged over time exhibit the common spectral shape of the downstream spectrum. Stochastic acceleration in compressional turbulence is thus generating the common spectral shape in the solar wind upstream and downstream from the termination shock, and appears even to be responsible for the thermalization of the particles occurring at the termination shock (Fisk et al. 2006).

It makes sense then to consider that the traditional high-energy ACRs are also accelerated by stochastic acceleration in compressional turbulence, this time in the distant heliosheath. The stochastic acceleration mechanism described here has a high-energy cutoff, which occurs where the particle gyroradii exceed the scale size of the turbulence. We require then that the scale size of the compressional turbulence increases with increasing distance downstream from the termination shock. The suprathermal tails seen at the termination shock can then be extended to the high energies of the ACRs. We should expect that the cutoff energy will ultimately not depend on the scale size of the turbulence, but rather on the diffusion coefficient for spatial diffusion in the heliosheath. When the particles reach energies where they can readily diffuse, they will propagate back into the heliosphere to be seen as ACRs.

One issue that needs to be addressed in this explanation for the ACRs is their composition. The particles accelerated at the termination shock have a higher abundance of hydrogen relative to other species than do the ACRs (Stone et al. 2005). An explanation for this difference is possible if stochastic acceleration in compressional turbulence is also responsible for the thermalization and thus the acceleration of energetic particles at the termination shock (Fisk et al. 2006).

Consider first the composition in equilibrium conditions. The role of the compressional turbulence is to redistribute the energy from the core to the tail particles and thus (1) should hold separately for each species. The threshold between the core and the tail for pickup ions should depend only on particle speed. The pressure in the tail, or equivalently the intensity of the tail particles since the spectral shape is constant, is thus proportional to the pressure in the core, which in turn is proportional to the core density. We expect then in equilibrium that the composition of the tail and the core particles is the same.

Equilibrium conditions should prevail in the ambient solar wind upstream from the termination shock, or in the distant heliosheath downstream. However, immediately behind the termination shock we should not expect equilibrium conditions. Core particles are heated in crossing the shock, non-adiabatically, and raised in energy above the threshold between the core and tail. They can flow into the tail, and an enhanced tail will result. If the heating raises the temperature of the core particles to the same value for all species, a true thermalization, then more hydrogen will be raised above the threshold energy than will the heavier species. The resulting enhanced tails will be biased in favor of hydrogen.

Further into the heliosheath, the pressure in the tail should return to the equilibrium condition described in (1). The tail composition should be that of the core particles, that is, the composition of interstellar pickup ions in the core. However, in the case of hydrogen, the pressure in the core should be reduced. Part of the core pressure was used to create an enhanced tail immediately downstream from the termination shock. In returning to equilibrium conditions in the distant heliosheath, this excess pressure may escape when the particles reach sufficiently high energies to be mobile. With a reduced core pressure for hydrogen, there is a reduced tail pressure according to (1), and the equilibrium composition in the distant heliosheath is biased against hydrogen. As the scale size of the turbulence in the heliosheath increases with distance behind the termination shock, the particles in equilibrium are stochastically accelerated to high energies and result in the ACRs.

In this model, then, there is a systematic variation in the composition of accelerated particles. Upstream from the termination shock, in the ambient solar wind, we expect that the composition will be that of interstellar pickup ions. Immediately behind the termination shock, the composition should be biased in favor of the lighter elements, e.g., hydrogen. In the distance heliosheath, where the particles are accelerated to sufficiently high energies to become ACRs, the composition should be biased against the lighter elements.

## 5 Concluding Remarks

Observations in the heliosphere are providing profound insights into how energetic particles are accelerated in astrophysical plasmas. A common spectral shape is observed—a power law with spectral index of $-5$, when the spectrum is expressed as a distribution function in velocity space (the spectral index is $-1.5$ when the spectrum is expressed as differential intensity). The acceleration mechanism is likely to be stochastic acceleration in compressional turbulence. Fisk and Gloeckler (2007) have demonstrated that simple thermodynamic arguments can be applied to this acceleration mechanism to yield the required common spectral shape.

With this knowledge that stochastic acceleration in compressional turbulence has a unique output, we can search other astrophysical settings for application of this result. The acceleration of Anomalous Cosmic Rays in the distant heliosheath is one such possibility, but there are others as well. It may be, for example, that compressional turbulence exists in the solar corona, and thus the unique spectrum should occur for particles accelerated in coronal loops or in the open magnetic flux present in the corona. The interstellar medium is also a likely site for compressional turbulence, and could yield a suprathermal low-energy particle population with the unique spectral shape. This, in fact, would provide an explanation, as suggested by Gloeckler et al. (1997), for the apparent thermal imbalance in the local interstellar medium discussed by Bowyer et al. (1995).

**Acknowledgements** This work was supported, in part, by NSF grants ATM 03-18590 and ATM 06-32471, by NASA contact NAGR-10975, and by JPL contract 1237843.

## References


W.I. Axford, E. Leer, G. Skadron, in *Proc. 15th Int. Cosmic Ray Conf.*, vol. 11 (Plovdiv, 1978), p. 132.
A.R. Bell, MNRAS **182**, 147 (1978)
R.D. Blandford, J.P. Ostriker, Astrophys. J. **221**, L29 (1978)
S. Bowyer, R. Lieu, S.D. Sidher, M. Lampton, J. Kunde, Nature **375**, 212 (1995)
L.L. Burlaga, N.F. Ness, M.H. Acuna, J. Geophys. Res. **111**, A09112 (2006). doi: 10.1029/2006JA011651
R.B. Decker, S.M. Krimigis, E.C. Roelof, M.E. Hill, T.P. Armstrong, G. Gloeckler, D.C. Hamilton, L.J. Lanzerotti, Science **309**, 2020 (2005)
L.A. Fisk, J. Geophys. Res. **81**, 4633 (1976)
L.A. Fisk, G. Gloeckler, Astrophys. J. Lett. **640**, L79 (2006)
L.A. Fisk, G. Gloeckler, PNAS **104**, 5749 (2007)
L.A. Fisk, B. Kozlovsky, R. Ramaty, Astrophys. J. Lett. **190**, L35 (1974)
L.A. Fisk, G. Gloeckler, T.H. Zurbuchen, Astrophys. J. **664**, 631 (2006)
J. Giacalone, J.R. Jokipii, J. Kota, Astrophys. J. **573**, 845 (2002)
L.J. Gleeson, W.I. Axford, Astrophys. J. Lett. **149**, L115 (1967)
G. Gloeckler, in *AIP Conf. Proc. 679, Solar Wind Ten*, ed. by M. Velli, R. Bruno, F. Malara (2003), p. 583
G. Gloeckler, L.A. Fisk, J. Geiss, Nature **386**, 374 (1997)
G. Gloeckler, L.A. Fisk, Astrophys. J. Lett. **648**, L63 (2006)
G. Gloeckler et al., Astron. Astrophys. Suppl. Ser. **92**(2), 267 (1992)
G. Gloeckler, L.A. Fisk, T.H. Zurbuchen, N.A. Schwadron, in *AIP Conf. Proc. 528: Acceleration and Transport of Energetic Particles Observed in the Heliosphere: Proc. of the ACE-2000 Symp.*, ed. by R.A. Mewaldt, J.R. Jokipii, M.A. Lee, E. Möbius, T.H. Zurbuchen (2000), p. 221
J.R. Jokipii, in *Physics of the Outer Heliosphere*, ed. by S. Grzedielshki, D.E. Page (1990), p. 169
A.N. Kolmogorov, Dokl. Akad. Nauk SSSR **30**, 301 (1941)
G.F. Krymsky, Dokl. Akad. Nauk SSSR **234**, 1306 (1977)
J.A. le Roux, W.H. Matthaeus, G.P. Zank, Geophys. Res. Lett. **28**, 3831 (2001)
J.A. le Roux, G.P. Zank, W.H. Matthaeus, J. Geophys. Res. **107**, SSH 9-1 (2002). doi: 10.1029/2001JA000285
E.N. Parker, Planet. Space Sci. **13**, 9 (1965)
M.E. Pesses, J.R. Jokipii, D. Eicher, Astrophys. J. Lett. **246**, L85 (1981)
N.A. Schwadron, L.A. Fisk, G. Gloeckler, Geophys. Res. Lett. **23**, 2871 (1996)
K.D.C. Simunac, T.P. Armstrong, J. Geophys. Res. **109**, A10101 (2004). doi: 10.1029/2003JA010194
E.C. Stone, A.C. Cummings, F.B. McDonald, B.C. Heikkila, N. Lal, W.R. Webber, Science **309**, 2017 (2005)
G.M. Webb, C.M. Ko, G.P. Zank, J.R. Jokipii, Astrophys. J. **595**, 195 (2003)
G. Zank, Space Sci. Rev. **89**, 413 (1999)

Space Sci Rev (2007) 130: 161–171
DOI 10.1007/s11214-007-9227-x

# Solar and Solar-Wind Composition Results from the Genesis Mission

**R.C. Wiens · D.S. Burnett · C.M. Hohenberg · A. Meshik · V. Heber · A. Grimberg · R. Wieler · D.B. Reisenfeld**

Received: 21 February 2007 / Accepted: 29 May 2007 / Published online: 3 August 2007


**Abstract** The Genesis mission returned samples of solar wind to Earth in September 2004 for ground-based analyses of solar-wind composition, particularly for isotope ratios. Substrates, consisting mostly of high-purity semiconductor materials, were exposed to the solar wind at L1 from December 2001 to April 2004. In addition to a bulk sample of the solar wind, separate samples of coronal hole (CH), interstream (IS), and coronal mass ejection material were obtained. Although many substrates were broken upon landing due to the failure to deploy the parachute, a number of results have been obtained, and most of the primary science objectives will likely be met. These objectives include He, Ne, Ar, Kr, and Xe isotope ratios in the bulk solar wind and in different solar-wind regimes, and $^{15}N/^{14}N$ and $^{18}O/^{17}O/^{16}O$ to high precision. The greatest successes to date have been with the noble gases. Light noble gases from bulk solar wind and separate solar-wind regime samples have now been analyzed. Helium results show clear evidence of isotopic fractionation between CH and IS samples, consistent with simplistic Coulomb drag theory predictions of fractionation between the photosphere and different solar-wind regimes, though fractionation by wave heating is also a possible explanation. Neon results from closed system stepped etching of bulk metallic glass have revealed the nature of isotopic fractionation as a function of depth, which in lunar samples have for years deceptively suggested the presence of

R.C. Wiens (✉)
Los Alamos National Laboratory, MS D466, Los Alamos, NM 87545, USA
e-mail: rwiens@lanl.gov

D.S. Burnett
Geological & Planetary Sciences, MS 100-23, Caltech, Pasadena, CA 91125, USA

C.M. Hohenberg · A. Meshik
Physics Department, Washington University, St. Louis, MO 63130, USA

V. Heber · A. Grimberg · R. Wieler
Isotope Geology NW C, ETH, 8092 Zurich, Switzerland

D.B. Reisenfeld
Department of Physics & Astronomy, U. of Montana, MS 1080, 32 Campus Dr., Missoula, MT 59812, USA

an additional, energetic component in solar wind trapped in lunar grains and meteorites. Isotope ratios of the heavy noble gases, nitrogen, and oxygen are in the process of being measured.

**Keywords** Composition: solar-wind · Composition: solar · Noble gases: solar

## 1 Introduction

Our understanding of the processes involved in solar-system formation comes mostly from two sources: Cosmochemistry studies of primitive materials left from the period of solar-system formation and, in recent years, observations of other stellar systems in various stages of planet formation. The photosphere of the Sun can be considered one of the most important primitive material reservoirs in the solar system because the Sun contains more than 99.8% of all the known material in the solar system. The photosphere is considered primitive because it was not subject to large-scale volatile-solid differentiation such as experienced by the planets. Rather, only a modest gravitational settling is inferred to have occurred (e.g., Cox et al. 1989).

The solar isotopic composition has historically been much less well known than its elemental composition, which has been studied over nearly the last 200 years, starting with the discovery by Frauenhofer in 1814 of absorption lines in the solar spectrum. Although greater elemental precision is desirable, modeling of the observed absorption lines has led to relatively accurate (on the order of 10%) abundances for most of the elements, with the main area of controversy relegated to a very few elements, such as Ne, whose presence in the region near the photosphere does not produce absorption lines. For isotopes, however, the solar composition had to be inferred based on the isotopic composition of primitive meteorites (e.g., Suess and Urey 1956 and references therein). However, this inference does not extend to volatile elements.

The first definitive isotopic measurements of solar material were provided by the Solar Wind Composition (SWC) experiments provided by the U. Bern during the Apollo missions, summarized recently by Geiss et al. (2004). These provided He, Ne, and Ar isotopic compositions unsurpassed in accuracy until now. In the past dozen years, compositions of the most abundant isotopes of the major elements from N through Fe have been cataloged by the MASS sensor on the SMS instrument on the Wind spacecraft (Gloeckler et al. 1995), the MTOF sensor on the CELIAS instrument on SOHO (Hovestadt et al. 1995), and the SWIMS instrument on ACE (Gloeckler et al. 1998). Notable isotope ratio measurements include those of Mg (e.g., Bochsler et al. 1997; Kucharek et al. 1997, 2001), $^{16}O/^{18}O$ in fast (Wimmer-Schweingruber et al. 2001) and slow (Collier et al. 1998) solar wind, $^{15}N/^{14}N$ (Kallenbach 2003), and isotope ratios of Si (Kallenbach et al. 1998a; Wimmer-Schweingruber et al. 1998), S (Wimmer-Schweingruber 2002), Ca (Kallenbach et al. 1998b), and Fe (Ipavich et al. 2001). These measurements, summarized by Wimmer-Schweingruber (2002), Kallenbach (2003), and Wiens et al. (2004) have verified that in all cases the solar isotopic composition is identical to Earth's, except for He and Ne, within uncertainties generally between $\pm 8$ and $\pm 30\%$.

Besides solar-wind (SW) measurements, the only other isotopic measurements of the Sun are from the vibration-rotation modes of CO molecules forming in the solar atmosphere over sunspot regions (e.g., Hall 1973; Harris et al. 1987; Asplund et al. 2005; Ayres et al. 2006). These measurements, most summarized by Wiens et al. (2004), have provided data on the most abundant isotopes of C and O to a best accuracy of roughly $\pm 5\%$. The latest study gives

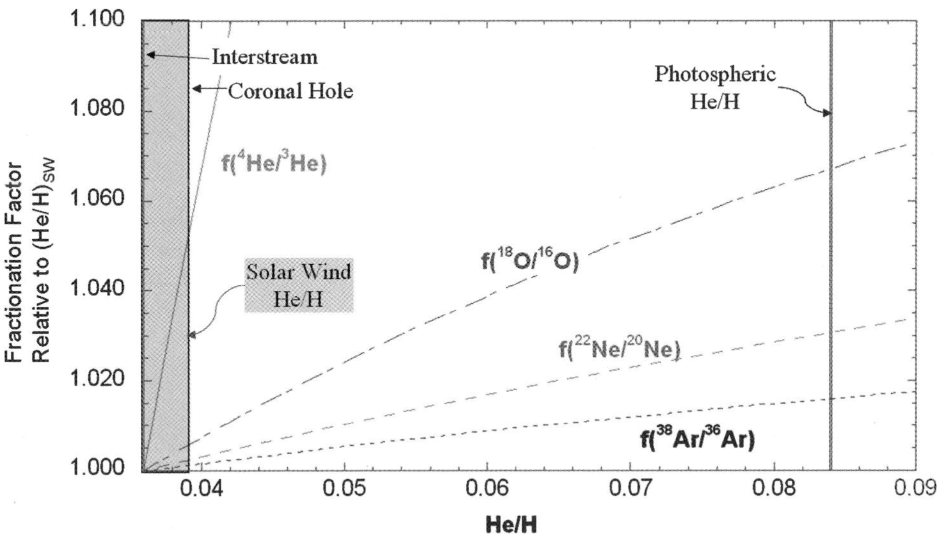


**Fig. 1** Possible fractionation predicted between solar-wind isotopic compositions in the Genesis samples and the photosphere. The *curves* are based on the Coulomb drag theory proposed by Bodmer and Bochsler (1998), and provide upper limits to fractionation, given that much of the He/H fractionation is in the supply of gases to the corona rather than by Coulomb drag acceleration (Geiss 1982; von Steiger and Geiss 1989). *Curves* show fractionation relative to the He/H ratio of the Genesis interstream sample at the far left side. The *box on the left* shows the range of He/H ratios between the Genesis interstream sample (0.0358) and the Genesis coronal-hole sample (0.0389). The *line at the right* shows the photospheric He/H ratio determined by helioseismology

significantly lower uncertainties ($1-\sigma$), with $^{18}O/^{16}O = 440 \pm 6$, $^{17}O/^{16}O = 1700 \pm 220$, and $^{13}C/^{12}C = 80 \pm 1$ (Ayres et al. 2006).

More precise isotopic measurements of the solar composition are important to answer a number of questions across a range of disciplines. If SW data are to be used to answer these questions, a fundamental issue that must be addressed is whether the solar wind is isotopically fractionated from the photosphere. Geiss et al. (1970) extended Parker's theory of solar wind expansion to include ion species other than protons. Using assumptions of a polytropic, isothermal steady state with spherical homogeneity and constant charge state, and ignoring wave acceleration, they integrated the non-linear momentum equation from the low corona into the supersonic SW regime, concluding that isotopic fractionation of the solar wind was likely. Later works (e.g., Bürgi and Geiss 1986 and thereafter) treated He as a major ion, included thermal diffusion and non-resonant wave–particle interactions. A relatively simple implementation of Coulomb drag theory has been used by Bodmer and Bochsler (1998) and Bochsler (2000) to predict isotopic fractionation of various species in the solar wind relative to the photosphere using a linear collision term and ignoring wave heating. In this model the isotopic fractionation is correlated to the He/H ratio of solar wind relative to the photospheric value, as determined by helioseismology (e.g., Basu and Antia 2004). Figure 1 shows the photospheric fractionation factors, relative to the SW composition, for several species of interest, as a function of He/H. A mass-dependent fractionation of up to 6% is expected between the low-speed SW $^{18}O/^{16}O$ ratio and that in the photosphere. However, it is clear that Coulomb drag does not account for the bulk of the elemental fractionation in the low-speed wind, which is related to the first ionization potential (FIP), with related theoretical explanations (e.g., Geiss 1982; von Steiger and Geiss

1989). If ionization efficiency, an atomic property, alone is the total cause of fractionations, then essentially no fractionation of *isotopes* would be expected in the solar wind relative to the photosphere. To the extent that some FIP fractionation is involved in producing the SW He/H depletion, the fractionations predicted by Coulomb Drag, e.g., in Fig. 1, are upper limits.

A second possible mass-dependent fractionation mechanism is ion cyclotron heating. Ion cyclotron heating tends to be preferential to heavy ions, resulting in greater thermal diffusion of heavy isotopes in the source region. This would result in the same general trend shown in Fig. 1, but there are no quantitative predictions of the magnitude of the fractionation. Resonant ion cyclotron heating could conceivably also result in non-mass dependent fractionation (e.g., Bochsler and Kallenbach 1994), but this is quite unlikely to be detected in long-term samples of solar wind.

Although the photospheric isotopic composition is not accurately known for comparison, isotopic fractionation between the photosphere and the solar wind by either Coulomb drag or wave heating may be discernable by a much smaller fractionation between interstream (low-speed; hereafter IS) and coronal-hole (CH) wind, for example, as shown along the left side of Fig. 1. Spacecraft data on solar wind isotopes, particularly of Mg and Ne (e.g., Kallenbach et al. 1997, 1998a), are consistent with the magnitude of fractionations shown in Fig. 1, though uncertainties have also allowed the possibility of no isotopic fractionation. Comparison of He isotopes from ecliptic and polar flows encountered by the Ulysses spacecraft, which differ significantly in He/H and Si/O, has been interpreted over the last several years as strongly suggestive of isotopic fractionation (e.g., Geiss and Gloeckler 2003), though below the upper limits suggested by Fig. 1.

If the issue of fractionation between the solar wind and photosphere can be solved such that the photospheric isotopic composition can be precisely determined from SW measurements, a number of questions can be solved for planetary science and cosmochemistry. Arguably the issue of greatest interest is the $^{16,17,18}O$ composition of the Sun. Oxygen is unique among non-noble elements with more than two isotopes in that it displays a large variability of up to 8% in $^{18}O/^{16}O$ among early solar-system objects. This heterogeneity has been attributed to nebular mixing (Clayton et al. 1977), to non-mass-dependent fractionation during condensation of silicates from the gas phase in the hot solar nebula (Thiemens and Heidenreich 1983), and to a self-shielding effect in which $^{16}O$-containing molecules in the inner solar system extinguished ultra-violet light (Clayton 2002). A relatively precise measurement (e.g., to $\pm 0.5\%$) of the solar $^{18}O/^{17}O/^{16}O$ would indicate which of the above theories is correct, revealing a completely new aspect of the formation process of the solar system. Attempts to determine the composition of SW O trapped in the surface layers of metallic lunar grains have been frustrating, as measurements by two different groups yielded opposing results and conclusions (Hashizume and Chaussidon 2005; Ireland et al. 2006).

Precise isotopic determinations of other elements in the Sun would yield new insights in other areas of planetary science. The isotopic ratio of solar N has been in dispute ever since the first analyses of SW-bearing lunar soil due to the very large ($>30\%$) range of values measured in different temperature steps of the gas extraction process (e.g., Marty et al. 2003). There currently appear to be at least two initial N reservoirs in the solar system, evidenced by measurements of the Jovian system and meteorites (e.g., Owen and Bar-Nun 2001). The solar isotopic composition will, we hope, elucidate the nature of these reservoirs and their relationship to each other. Other high-priority solar isotopic ratios include the $^{13}C/^{12}C$, the heavy noble gas composition, and $^{7}Li/^{6}Li$.

## 2 The Genesis Mission

The Genesis SW sample return mission was conceived to address the above issues as well as to return elemental abundances of as many elements as possible. The idea was to extend the successful Solar Wind Composition (SWC) experiments (Geiss et al. 2004) in which aluminum and platinum foils were exposed to the solar wind on the surface of the Moon between 1 and 45 hours during the Apollo lunar missions. The NASA Discovery class of missions allowed a low-cost dedicated spacecraft to be used to collect solar wind outside the magnetosphere. The mission was selected in 1997 and was launched in 2001 (Burnett et al. 2003).

The spacecraft carried a re-entry capsule which housed a payload canister holding five arrays of SW collectors and a SW concentrator. Two of the five arrays were designed to collect solar wind continuously, while the remaining three arrays collected specific SW regimes. The regimes were determined by on-board SW ion and electron monitors (Barraclough et al. 2003). The key factor in distinguishing IS from CH material was proton velocity, while coronal mass ejections (CMEs) were determined by a range of factors including bi-directional electron streaming, proton temperature as a function of speed, and He/H ratio. An on-board algorithm (Neugebauer et al. 2003) commanded the regime-specific arrays in real time with a latency of about five minutes once a decision was made. The CME detection was weighted such that if there was uncertainty, the CME collector would be deployed so as to avoid contaminating either the IS or CH material with CME material. The exposure times for each regime array were recorded. The Genesis-based regime selections have been compared with other spacecraft data available, and SW statistics from other spacecraft instruments have been studied for the integrated regime exposure times (e.g., Reisenfeld et al. 2007, this volume).

The SW collector arrays consisted of a number of high-purity materials, including float-zone and Czochralski-grown silicon, detector-grade germanium, diamond-like carbon, aluminum, gold, and sapphire (Jurewicz et al. 2003). During exposure, SW ions were implanted at mean depths of ∼40 nm within the substrates. Extreme care was taken to minimize surface contamination in light of the shallow depths of implantation. The SW concentrator was a 40 cm diameter electrostatic ion telescope that used high-transparency grids to steer the ions, first by rejecting the majority of the protons, then by accelerating and reflecting the heavier ions onto a 6 cm diameter target (Nordholt et al. 2003; Wiens et al. 2003). It was built primarily to enhance the SW O fluence above the ubiquitous O background in terrestrial materials.

Exposure of the substrates occurred as the spacecraft orbited the L1 point, sunward of the Earth. The exposure commenced in November, 2001, and was completed in April, 2004, covering a period starting just after solar maximum. The bulk collector arrays were exposed for 853 days. During this time, the Genesis ion monitor recorded a total fluence of 1.84e16 protons/cm$^2$, and 7.30e14 alphas/cm$^2$. The relative percentages of alpha particles collected by the three regime-specific collectors was 41.5% for the interstream collector array, 31.9% for the CH array, and 26.7% for the CME array. The fraction of CH material was larger than expected, with most of the high-speed streams occurring during the second half of the exposure period. The CME collection fraction was also relatively high, boosted slightly by the default collection in periods of uncertainty.

Re-entry of the capsule occurred in late 2004 over the Dugway Proving Grounds in Utah. It was the first NASA capsule to re-enter since the end of the Apollo program in the 1970s. The Genesis capsule failed to deploy its parachute, and experienced a hard landing. Nearly all of the 100 mm diameter hexagonal collectors were broken. There were approximately

400 fragments recovered >25 mm, 1,700 additional fragments >10 mm, and more than 7,200 fragments <10 mm. Most of the larger fragments were sapphire-backed samples of either gold or aluminum films, or exposed sapphire, with silicon comprising most of the smaller fragments (Allton et al. 2005). The SW concentrator target was almost completely intact, a very promising outcome allowing eventual $^{18}O/^{17}O/^{16}O$ analyses. All in all, a relatively large amount of SW collector materials are available for analysis as long as small samples can be used.

A less immediately obvious problem was in-flight contamination of the wafer surfaces. Although great care was taken to avoid organic materials in the sample canister, ellipsometry and X-ray photo-electron spectroscopy (XPS) revealed the presence of a thin contamination film containing Si, C, O, and F (Allton et al. 2006). The average thickness is <5 nm, and in >95% of the samples the thickness is <10 nm, so that it did not significantly affect the implantation of solar wind. However, it has caused difficulty for analysis of a number of elements susceptible to surface contamination.

## 3 Results

This section summarizes the isotopic results to date. A separate paper (Reisenfeld et al. 2007, this volume) reports on the elemental abundances determined from Genesis samples, and their relationship to other elemental abundance measurements. Additionally, papers by Grimberg et al. (2007, this volume) and Heber et al. (2007, this volume) report in greater depth on the solar energetic particle (SEP) experiment and on Ne isotopic verification of the SW concentrator operation, respectively.

### 3.1 Helium and Neon Isotopes and Photospheric Fractionation

The Genesis mission's greatest successes to date have been with the noble gases, which are relatively immune to contamination issues. Helium, Ne, and Ar isotopic abundances have been measured by mass spectrometry in bulk collector material as well as in the separate SW regimes. The $^4He/^3He$ ratio from Genesis (Mabry et al. 2007; Heber et al. 2007) is very close to the weighted average of $2350 \pm 120$ from the Apollo SWC measurements (Geiss et al. 2004), though measurements to date have stayed on the low side of this value. The composition of the bulk solar wind Ne (Grimberg et al. 2006; Hohenberg et al. 2006; Heber et al. 2007) is within uncertainty of accepted SW values from lunar samples (e.g., $^{20}Ne/^{22}Ne = 13.86$; Palma et al. 2002), which itself is nominally slightly heavier but within uncertainty of the Apollo SWC foil average of $13.7 \pm 0.3$ (Geiss et al. 1972; Geiss et al. 2004). For both He and Ne, individual measurements have varied somewhat, and questions of whether the materials quantitatively retained all noble gases in all cases, as well as measurement questions such as the extent of hydride interferences (Mabry et al. 2007), are still being determined.

A significant issue is how well the current Genesis measurements can constrain the SW fractionation relative to the photosphere. In the Coulomb drag theory, the fractionation relative to H of a species of mass number $A$ and atomic charge $q$ scales as $(2A - q - 1) \times [(1 + A)/A]^{1/2} \times q^{-2}$ (e.g., Bodmer and Bochsler 1998). A simple model which assumes all (rather than ~2/3; J. Raines, T. Zurbuchen, personal communication) of the Ne charge is +8, and uses a polytropic index $\alpha$ of 1.4 for distances far from the sun (Geiss et al. 1970), predicts a maximum $^{22}Ne/^{20}Ne$ fractionation of 2.4 permil (parts per thousand) between the IS and CH samples, with the CH sample enriched in the heavy isotope. The same treatment

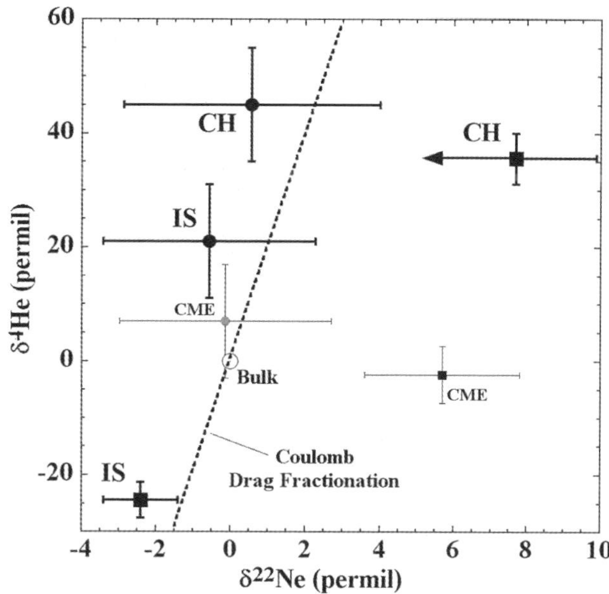


**Fig. 2** Deviations of $^4$He/$^3$He vs. $^{22}$Ne/$^{20}$Ne in Genesis solar-wind regime samples relative to bulk solar wind measurements that were made at the respective laboratories (*circles* = Mabry et al. 2007; *squares* = Heber et al. 2007 and this work) displayed in permil with $1-\sigma$ uncertainties. IS = interstream, CH = coronal hole, and CME = coronal mass ejection. The *dashed curve* represents the fractionation line predicted by the simplistic Coulomb Drag model for CH and IS solar wind, as mentioned in the text

predicts a maximum fractionation of 53 permil for $^4$He/$^3$He, some 20 times stronger than for $^{22}$Ne/$^{20}$Ne (cf. Fig. 1).

Figure 2 shows the regime data in permil deviations relative to $^4$He/$^3$He and $^{22}$Ne/$^{20}$Ne in the bulk solar wind. Correction of ion backscatter out of the collectors was made for He. The difference in backscatter correction between regimes is negligible for Ne. The bulk SW measurements are shown for reference at the origin. The $^{22}$Ne/$^{20}$Ne plotted from the Zurich CH regime is an upper limit. CME samples are plotted in Fig. 2 along with CH and IS samples, though CME material is not expected to behave similarly to the other regimes in terms of Coulomb drag fractionation, given its transitory nature and elemental ratio variability observed by other instruments.

The fractionation trend predicted for CH and IS by the simplistic Coulomb drag model (Bodmer and Bochsler 1998) is indicated in Fig. 2 by a dashed line. The Ne isotopic data are consistent with, but certainly not diagnostic of Coulomb drag, given the uncertainties. For the He data, the difference between CH and IS data points from the Washington University data (Fig. 2, filled circles) suggests a slight fractionation at the $1-\sigma$ level, though the fact that all three regimes lie above the bulk datum defies mass balance rules. The difference between CH and IS regime samples in the Zurich $^4$He/$^3$He data is much more significant, with a nominal separation of 6.0%, consistent with the maximum predicted by the Coulomb drag model. These results lead us to conclude that long-term averages of SW regime samples are isotopically fractionated from one another, and more significantly from the photosphere. Differences between measurements still need to be reconciled to determine the magnitude of the fractionation. Future Genesis He and Ne analyses should confirm the anticorrelation between $^3$He/$^4$He and $^4$He/$^{20}$Ne observed in the Apollo SWC measurements, which showed that the variations between solar-wind samples was not merely due to a mass-dependent effect (Geiss et al. 1972, 2004).

## 3.2 Search for SEP

Solar noble gases implanted in the lunar regolith appear to consist of two components: a normal SW component and an energetic particle ("SEP") component implanted at greater depth. This phenomenon was noted in essentially all noble gases, and was best documented for Ne. Although the SW component was identified with a $^{20}Ne/^{22}Ne$ ratio of ~13.8, the "SEP" component appeared to have $^{20}Ne/^{22}Ne$ ~11.2 (e.g., Benkert et al. 1993). The composition and source of the "SEP" component was discussed in numerous papers. Most puzzling was that the "SEP" abundance, approximately 20–30% of the SW abundances, was orders of magnitude higher than observed for solar particles of this energy range by spacecraft. Also, the SWC foil experiment did not observe "SEP" contributions in the solar material collected during the Apollo missions (Geiss et al. 2004). These observations suggested that "SEP" must be variable in time or from a non-solar source. The identification of the "SEP" component in lunar soils and meteorites was complicated by the presence, particularly in Ne, of spallation nuclei produced by galactic cosmic rays.

The Genesis mission carried a target specifically to test for "SEP" noble gases. It was a bulk metallic glass that would allow the solar wind to be released in multiple steps (Jurewicz et al. 2003). Spallation contributions were essentially nonexistent in the Genesis samples, potentially allowing unhindered observation of the "SEP" component. This target was returned to earth unbroken.

Analysis of the bulk metallic glass sample (Grimberg et al. 2006; Grimberg et al. 2007, this volume) revealed a smooth isotopic trend as a function of the fraction of gas released, mimicking the trend predicted purely from SW implantation. Re-analysis of the trends produced by Ne released from lunar grains showed that these could also be modeled simply from SW implantation combined with spallation. What appeared to be a genuine component at $^{20}Ne/^{22}Ne$ ~11.2 was merely a point at which the implantation trend was overtaken by the trend caused by addition of spallation gases.

## 3.3 Argon Isotopes

In retrospect, the isotopic trending with depth resulting from normal SW implantation also clearly led to greater uncertainty in the derived solar isotopic compositions of the heavier noble gases and may have contributed to the reports of isotopic variability of surface-correlated N in lunar grains (e.g., Kerridge 1993). Especially for Ar, the reported solar wind value was biased toward the lighter end of the possible range, well separated from the supposedly isotopically heavy "SEP" (e.g., Palma et al. 2002). Figure 3 shows historical SW Ar isotopic measurements (e.g. Geiss et al. 1972, 2004; Benkert et al. 1993; Becker et al. 1998; Palma et al. 2002), along with the Genesis-derived ratio (Mabry et al. 2007). The Genesis uncertainty of $\pm 0.009$, within the size of the data point, represents the standard deviation of a number of measurements. Ratios for the atmospheres of Earth (Lee et al. 2006), Venus (Istomin et al. 1983), and Jupiter (Mahaffy et al. 1998) are shown for comparison. The Genesis measurement is nearly identical to the nominal ratio from SOHO (Weygand et al. 2001) and the Benkert et al. (1993) ratio from lunar soils, is and within error of the Apollo SWC results (Geiss et al. 2004).

The Ar isotopes in separate SW regime samples have also been measured. Argon is clearly less diagnostic than the lighter gases in distinguishing Coulomb drag fractionation, with a predicted maximum difference between regimes of only 1.2 permil. As expected, the regime samples are all within uncertainty of each other and of the bulk SW measurement.

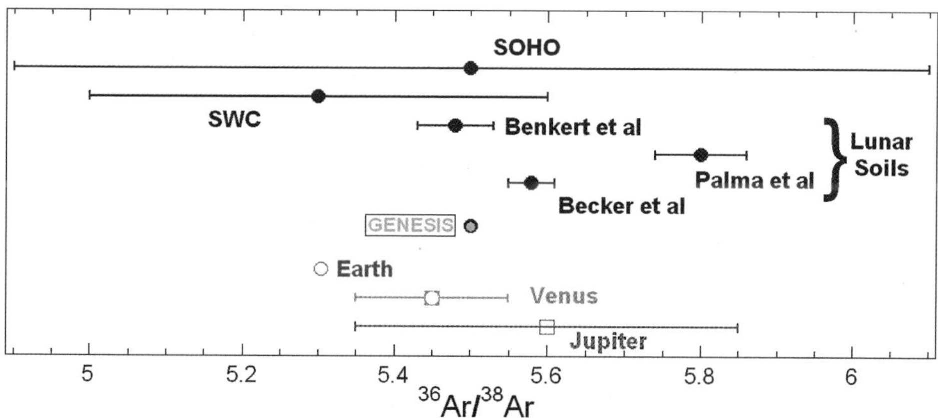


**Fig. 3** Genesis bulk solar wind composition (Mabry et al. 2007) in comparison with previous solar-wind measurements, and with planetary atmosphere compositions. The Genesis $1-\sigma$ measurement uncertainty is within the size of the data point

## 3.4 Prognosis for Future Measurements

Several of the highest priority measurements will, we hope, be made within the next year. These include heavy noble gas elemental and isotopic compositions, and solar $^{18}O/^{17}O/^{16}O$ and $^{15}N/^{14}N$. Here we elaborate on the prognosis for the N and O isotopes, which should be obtainable to significantly better than $\pm 1\%$ accuracy.

The O isotopic composition measurement is expected to use the SW concentrator targets, which received between a few and up to 30 times higher fluences than the passive collectors. A fractionation, varying as a function of the target radius, was introduced by concentrating the ions. While the fractionation was modeled (Wiens et al. 2003), verification of the fractionation by measuring Ne along the radius of the target is an important step in confirming an accurate result. Numerous Ne analyses have now been made on a gold-plated retainer piece of the target assembly. The results confirm proper operation of the concentrator, but give a somewhat different fractionation pattern than the model (Heber et al. 2007, this volume). Efforts are currently underway to understand the differences between model and observation for Ne. Pending a resolution, the O measurements are expected to be made either by gas mass spectrometry or by secondary ion mass spectrometry using a specially modified instrument incorporating an accelerator sector.

For $^{15}N/^{14}N$ a gold foil was expected to be used, applying a vacuum mercury amalgamation technique to release the gas. However, the foil was found to have significant N contamination in the near-surface region containing the solar wind (R.O. Pepin, personal communication). Fortunately, a number of gold-on-sapphire collectors were also flown, and these are now being used for the N analysis (Marty et al. 2007), expected to be done shortly.

**Acknowledgements** We thank the NASA Discovery Mission program, the NASA Jet Propulsion Laboratory, and Lockheed-Martin Astronautics, for their support of the Genesis mission. Thousands of people contributed to the mission, and our thanks also go to all of them. We also thank the organizers of this ISSI symposium.

## References

J.H. Allton, E.K. Stansbery, K.M. McNamara, Lunar Planet. Sci. **XXXVI**, 2083 (2005)

J.H. Allton, M.J. Calaway, M.C. Rodriguez, J.D. Hittle, S.J. Wentworth, E.K. Stansbery, K.M. McNamara, Lunar Planet. Sci. **XXXVII**, 1611 (2006)

M. Asplund, N. Grevesse, A.J. Sauval, in *Cosmic Abundances as Records of Stellar Evolution and Nucleosynthesis*, ed. by F.N. Bash, T.G. Barnes, ASP Conf. Ser., vol. XXX, 2005

T.R. Ayres, C. Plymate, C.U. Keller, Astrophys. J. **S165**, 618 (2006)

B.L. Barraclough, E.E. Dors et al., Space Sci. Rev. **105**, 627 (2003)

S. Basu, H.M. Antia, Astrophys. J. **606**, L85 (2004)

R.H. Becker, D.J. Schlutter, P.E. Rider, R.O. Pepin, Meteorit. Planet. Sci. **33**, 109 (1998)

J.-P. Benkert, H. Baur, P. Signer, R. Wieler, J. Geophys. Res. **98**, 13147 (1993)

P. Bochsler, Rev. Geophys. **38**, 247 (2000)

P. Bochsler, R. Kallenbach, Meteoritics **29**, 653 (1994)

P. Bochsler, H. Balsiger, R. Bodmer, O. Kern, T. Zurbuchen, G. Gloeckler, D.C. Hamilton, M.R. Collier, D. Hovestadt, Phys. Chem. Earth **22**, 401 (1997)

R. Bodmer, P. Bochsler, Astron. Astrophys. **337**, 921 (1998)

A. Bürgi, J. Geiss, Sol. Phys. **1986**, 347 (1986)

D.S. Burnett et al., Space Sci. Rev. **105**, 509 (2003)

R.N. Clayton, Nature **415**, 860 (2002)

R.N. Clayton, N. Onuma, L. Grossman, T.K. Mayeda, Earth Planet. Sci. Lett. **34**, 209 (1977)

M.R. Collier, D.C. Hamilton, G. Gloeckler, G. Ho, P. Bochsler, R. Bodmer, R. Sheldon, J. Geophys. Res. **103**, 7 (1998)

A.N. Cox, J.A. Guzik, R.B. Kidman, Astrophys. J. **342**, 1187 (1989)

J. Geiss, Space Sci. Rev. **33**, 201 (1982)

J. Geiss, G. Gloeckler, Space Sci. Rev. **106**, 3 (2003)

J. Geiss, P. Hirt, H. Leutwyler, Sol. Phys. **12**, 458 (1970)

J. Geiss, F. Bühler, H. Cerutti, P. Eberhardt, C. Filleux, in *Apollo 16 Prelim. Sci. Rep.* NASA SP-315, 14-1, 1972

J. Geiss, F. Bühler, H. Cerutti, P. Eberhardt, C. Filleux, J. Meister, P. Signer, Space Sci. Rev. **110**, 307 (2004)

G. Gloeckler et al., Space Sci. Rev. **71**, 79 (1995)

G. Gloeckler et al., Space Sci. Rev. **86**, 497 (1998)

A. Grimberg et al., Science **314**, 1133 (2006)

A. Grimberg et al., Space Sci. Rev. (2007, this issue), doi: 10.1007/s11214-007-9150-1

D.N.B. Hall, Astrophys. J. **182**, 977 (1973)

K. Hashizume, M. Chaussidon, Nature **434**, 619 (2005)

M.J. Harris, D.L. Lambert, A. Goldman, Mon. Not. Roy. Astr. Soc. **224**, 237 (1987)

V. Heber et al., Space Sci. Rev. (2007, this issue), doi: 10.1007/s11214-007-9179-1

V. Heber, H. Baur, D.S. Burnett, R. Wieler, Lunar Planet. Sci. **XXXVIII**, 1894 (2007)

C.M. Hohenberg, A.P. Meshik, Y. Marrocchi, J.C. Mabry, O.V. Pravdivtseva, J.H. Allton, D.S. Burnett, Lunar Planet. Sci. **XXXVII**, 2439 (2006)

D. Hovestadt et al., Sol. Phys. **162**, 441 (1995)

F.M. Ipavich, J.A. Paquette, P. Bochsler, S.E. Lasley, P. Wurz, in *Solar and Galactic Composition*, ed. by R.F. Wimmer-Schweingruber, AIP Conf. Proc., Melville, NY, 2001, p. 121

V.G. Istomin, K.V. Gechnev, V.A. Kochnev, Cosmic Res. **21**, 329 (1983)

T.R. Ireland, P. Holden, M.D. Norman, J. Clarke, Nature **440**, 776 (2006)

A.J.G. Jurewicz et al., Space Sci. Rev. **105**, 535 (2003)

R. Kallenbach, Space Sci. Rev. **106**, 305 (2003)

R. Kallenbach, F.M. Ipavich, P. Bochsler, S. Hefti, D. Hovestadt, H. Grünwald, M. Hilchenbach et al., J. Geophys. Res. **102**, 26895 (1997)

R. Kallenbach, F.M. Ipavich, H. Kucharek et al., Space Sci. Rev. **85**, 357 (1998a)

R. Kallenbach, F.M. Ipavich, P. Bochsler, S. Hefti, P. Wurz et al., Astrophys. J. **498**, L75 (1998b)

J.F. Kerridge, Rev. Geophys. **31**, 423 (1993)

H. Kucharek, F.M. Ipavich, R. Kallenbach, P. Bochsler, H. Hovestadt, D. Grünwald, M. Hilchenbach, W.I. Axford et al., ESA SP-404, 1997, p. 473

H. Kucharek, B. Klecker, F.M. Ipavich, R. Kallenbach, H. Grünwald, M.R. Aellig, P. Bochsler, in *Recent Insights int the Physics of the Sun and Heliosphere: Highlights from SOHO and Other Space Missions*, ed. by P. Brekke, B. Fleck, J.B. Gurman, IAU Symp., 2001, pp. 203, 562

J.-Y. Lee, K. Marti, J.P. Severinghaus, K. Kawamura, H.-S. Yoo, J.B. Lee, J.S. Kim, Geochim. Cosmochim. Acta **70**, 4507 (2006)

J.C. Mabry, A.P. Meshik, C.M. Hohenberg, Y. Marrocchi, O.V. Pravdivtseva, R.C. Wiens, C. Olinger, D.B. Reisenfeld, J. Allton, R. Bastien, K. McNamara, E. Stansbery, D.S. Burnett, Lunar Planet. Sci. **XXXVIII**, 2412 (2007)

P.R. Mahaffy, T.M. Donahue, S.K. Atreya, T.C. Owen, H.B. Niemann, Space Sci. Rev. **84**, 251 (1998)

B. Marty, K. Hashizume, M. Chaussidon, R. Wieler, Space Sci. Rev. **106**, 175 (2003)
B. Marty, L. Zimmermann, P. Burnard, Lunar Planet. Sci. **XXXVIII**, 1704 (2007)
M. Neugebauer, J.T. Steinberg, R.L. Tokar, B.L. Barraclough, E.E. Dors, R.C. Wiens, D.E. Gingerich, D. Luckey, D.B. Whiteaker, Space Sci. Rev. **105**, 661 (2003)
J.E. Nordholt, R.C. Wiens et al., Space Sci. Rev. **105**, 561 (2003)
T.C. Owen, A. Bar-Nun, Orig. Life Evol. Biosphere **31**, 435 (2001)
R.L. Palma, R.H. Becker, R.O. Pepin, D.J. Schlutter, Geochim. Cosmochim. Acta **66**, 2929 (2002)
D.B. Reisenfeld et al., Space Sci. Rev. (2007, this issue), doi: 10.1007/s11214-007-9215-1
H.E. Suess, H.C. Urey, Rev. Mod. Phys. **28**, 53 (1956)
M.H. Thiemens, J.E. Heidenreich, III, Science **219**, 1073 (1983)
R.H. von Steiger, J. Geiss, Astron. Astrophys. **225**, 222 (1989)
J.M. Weygand, F.M. Ipavich, P. Wurz, J.A. Paquette, P. Bochsler, Geochim. Cosmochim. Acta **65**, 4589 (2001)
R.C. Wiens, M. Neugebauer, D.B. Reisenfeld, R.W. Moses, Jr., J.E. Nordholt, Space Sci. Rev. **105**, 601 (2003)
R.C. Wiens, P. Bochsler, D.S. Burnett, R.F. Wimmer-Schweingruber, Earth Planet. Sci. Lett. **222**, 697 (2004)
R.F. Wimmer-Schweingruber, Adv. Space Res. **30**, 23 (2002)
R.F. Wimmer-Schweingruber, P. Bochsler, O. Kern, G. Gloeckler, D.C. Hamilton, J. Geophys. Res. **103**, 20621 (1998)
R.F. Wimmer-Schweingruber, P. Bochsler, G. Gloeckler, Geophys. Res. Lett. **28**, 2763 (2001)

Space Sci Rev (2007) 130: 173–182
DOI 10.1007/s11214-007-9216-0

# Isotopic Composition of the Solar Wind Inferred from In-Situ Spacecraft Measurements

**R. Kallenbach · K. Bamert · M. Hilchenbach**

Received: 11 February 2007 / Accepted: 10 May 2007 / Published online: 19 July 2007


**Abstract** The Sun is the largest reservoir of matter in the solar system, which formed 4.6 Gyr ago from the protosolar nebula. Data from space missions and theoretical models indicate that the solar wind carries a nearly unfractionated sample of heavy isotopes at energies of about 1 keV/amu from the Sun into interplanetary space. In anticipation of results from the Genesis mission's solar-wind implanted samples, we revisit solar wind isotopic abundance data from the high-resolution CELIAS/MTOF spectrometer on board SOHO. In particular, we evaluate the isotopic abundance ratios $^{15}N/^{14}N$, $^{17}O/^{16}O$, and $^{18}O/^{16}O$ in the solar wind, which are reference values for isotopic fractionation processes during the formation of terrestrial planets as well as for the Galactic chemical evolution. We also give isotopic abundance ratios for He, Ne, Ar, Mg, Si, Ca, and Fe measured in situ in the solar wind.

**Keywords** Isotopic composition · Solar system · Solar wind

## 1 Introduction

Solar system samples are usually compared to terrestrial material as a reference. However, because the Sun contains more than 99% of solar system matter, any non-nuclear fractionation process during the formation of the inner solar system from the early solar nebula could have changed the terrestrial isotopic composition, but not the solar isotopic composition. Consequently, solar matter is a better witness of the composition of the protosolar nebula than terrestrial material.

R. Kallenbach (✉) · M. Hilchenbach
Max Planck Institute for Solar System Research, Max-Planck-Strasse 2, 37191 Katlenburg-Lindau, Germany
e-mail: kallenbach@mps.mpg.de

R. Kallenbach · K. Bamert
Physikalisches Institut, University of Bern, Sidlerstrasse 5, 3012 Bern, Switzerland

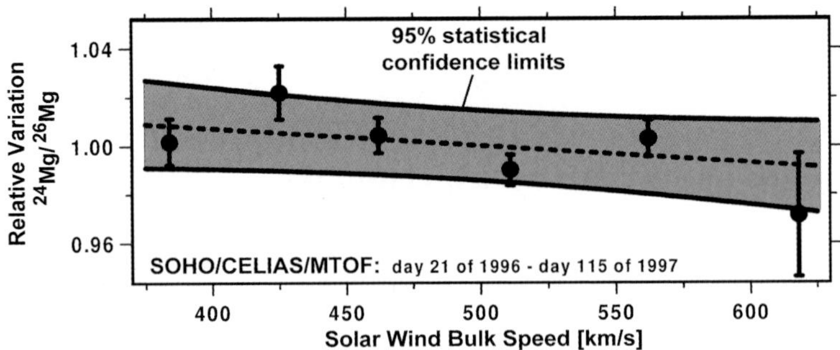


**Fig. 1** Relative variation of the solar wind $^{24}$Mg/$^{26}$Mg ratio versus the solar wind bulk speed. Data from Kallenbach et al. (1998a)

Kallenbach (2003) concluded from data of spaceborne instrumentation and from theoretical models that the solar wind carries a sample of heavy solar isotopes that is fractionated by, at most, about one percent per atomic mass unit (Fig. 1).

This review gives the most recent summary of solar wind isotopic abundance data from spaceborne sensors in order to give a reference for results soon to be expected from solar-wind implanted samples of the Genesis mission. The primary scientific objective of the Genesis mission is to determine the nitrogen and oxygen isotopic abundance ratio in the solar wind. These ratios are difficult to measure with spaceborne instrumentation, as will be discussed. Nonetheless, we give our best effort to constrain the values on the ratios $^{15}$N/$^{14}$N, $^{18}$O/$^{16}$O, and $^{17}$O/$^{16}$O in the solar wind. These ratios are derived from data taken during the first two years of the SOHO mission. A more precise analysis may still be possible when evaluating all data of the SOHO mission which has operated for more than ten years.

## 2 Nitrogen Isotopes in the Solar System

The sources and reservoirs of nitrogen in the solar nebula from which planetary bodies formed are still under debate. The processes in the early solar nebula are usually understood in terms of the chemistry and physics of interstellar molecular clouds (Herbst 2003). Ion-molecule reactions at temperatures of 10–30 K enrich $^{15}$N in the HCN and $NH_3$ molecules, while the dominant (>90%) form of nitrogen in interstellar clouds, molecular $N_2$, is relatively depleted in $^{15}$N.

This overall picture matches qualitatively the measurements of the $^{15}$N/$^{14}$N ratio in different solar system samples. The gas envelope of Jupiter presumably had been accreted from the volatile component of the solar nebula which mainly consisted of molecular $N_2$. The $^{15}$N/$^{14}$N ratio of $(2.3 \pm 0.3) \times 10^{-3}$ in Jupiter's atmosphere (Owen et al. 2001; Abbas et al. 2004; Fouchet et al. 2004) is indeed lower than the $^{15}$N/$^{14}$N ratio of $(3.10 \pm 0.45) \times 10^{-3}$ observed in HCN (Jewitt et al. 1997) in Comet Hale-Bopp (C/1995 O1). The CN radicals observed in comets de Vico (122P/1995 S1) and Ikeya-Zhang (153P/2002 C1) have even larger $^{15}$N enrichments, $^{15}\mathrm{N}/^{14}\mathrm{N} \approx (6.5 \pm 2.0) \times 10^{-3}$ (Jehin et al. 2004). This points toward the existence of (an)other unknown parent(s) of CN besides HCN, with an even higher $^{15}$N excess. These could be organic compounds like those found in interplanetary dust particles (Messenger et al. 2003).

Observations with the ion neutral mass spectrometer on board Cassini during the first flyby of Titan yield $^{15}\mathrm{N}/^{14}\mathrm{N} \approx (5.20 \pm 0.55) \times 10^{-3}$ in $N_2$ of Titan's atmosphere (Niemann et al. 2005), while the observations in HCN yield even higher $^{15}$N, $^{15}\mathrm{N}/^{14}\mathrm{N} \approx 0.017$

(Lammer and Bauer 2003). Lammer and Bauer (2003) suggested that the $^{15}$N enrichment in HCN indicates fractionation by gravitational escape from an early massive atmosphere, while Niemann et al. (2005) proposed that the $^{15}$N enrichment in HCN relative to the $^{15}$N in $N_2$ is caused by a photochemical process.

Most recently, Liang et al. (2007) measured the photoabsorption cross-section of $NH_3$ in the wavelength range 140 to 220 nm and concluded that $NH_3$ in Jupiter's atmosphere at the pressure level of 400 mbar, where the Galileo Probe measurements took place, should be depleted in $^{15}$N with respect to nitrogen in the solar nebula. This leaves the solar wind as the only reliable witness for the protosolar isotopic composition of nitrogen.

The isotopic composition of nitrogen in the solar wind has been searched in lunar surface samples. Nitrogen that is implanted in the first tens of nanometers of lunar regolith grains and associated with the deuterium-free solar wind hydrogen is depleted in $^{15}$N by at least 24% (Hashizume et al. 2000) with respect to nitrogen in the terrestrial atmosphere ($^{15}\mathrm{N}/^{14}\mathrm{N} \approx 0.00368$). In contrast, a component associated with deuterium-rich hydrogen in silicon-bearing coatings at the surface of ilmenite grains has been found to be enriched in $^{15}$N with respect to terrestrial nitrogen. The strong enrichments in $^{15}$N may be hard to explain with isotopic fractionation in only nebular or planetary environments. Possibly, the compounds strongly enriched in $^{15}$N are of interstellar origin and never equilibrated in the protosolar nebula. These compounds may carry signatures of nucleosynthetic processes (Marty et al. 2003). About 90% of nitrogen in lunar regolith has been identified as non-solar (Wieler et al. 1999).

The measurement of the $^{15}$N/$^{14}$N has been regarded as not feasible with the MTOF (mass time-of-flight) sensor of the Charge, Element, and Isotope Analysis System (CELIAS) onboard SOHO (Solar and Heliospheric Observatory). However, efforts have been made to retrieve the $^{15}$N abundance in the solar wind by Kallenbach et al. (1998b), which we refer to as K98 hereafter. The main problems arise for principal instrumental reasons. In the spectra of the MTOF sensor, the $^{15}\mathrm{N}^{+}$ counts have to be corrected for interference with $^{30}\mathrm{Si}^{++}$, which has the same mass-to-charge ratio (Fig. 2). In K98, the $^{30}\mathrm{Si}^{++}$ counts interfering with $^{15}\mathrm{N}^{+}$ have been estimated from the $^{29}\mathrm{Si}^{++}$ counts which are in principal isolated but hard to determine on top of the 'wing' of the $^{14}\mathrm{N}^{+}$ & $^{28}\mathrm{Si}^{++}$ main peak. In this reassessment, the $^{30}\mathrm{Si}^{++}$ counts are estimated from the Si/N elemental abundance ratio which in turn is estimated from the Mg/O ratio. For Mg and Si the FIP (First-Ionization Potential) enrichment (von Steiger and Geiss 1989) over their photospheric abundance is approximately identical, while N and O are both high-FIP elements with very little FIP fractionation in the solar wind. During the measurement period for the data shown in Fig. 2 the FIP enrichment of Mg is approximately 1.4 (Kallenbach 2001) including instrumental fractionation. The MTOF sensor detects high-speed—that is, low-FIP-fractionation—solar wind more efficiently, which results in a lower count rate in the slow solar wind and a larger uncertainty for the high-FIP data in Fig. 3. In K98, a fit to the line shapes by analytical functions yielded a FIP enrichment of about 2.9. This value would match expectations of FIP fractionation in the slow solar wind. However, the elemental instrument efficiencies of the MTOF sensor have to be included for the analysis of the data in Fig. 2.

The discrepancy in the Si abundance implied by Kallenbach (2001) and that implied by K98 has motivated a more detailed analysis of the line shapes in the MTOF spectra, which is described in the Appendix for reproducability. The analysis using the current technique to describe the line shapes yields $^{14}\mathrm{N}/^{15}\mathrm{N} \approx 320$ or $^{15}\mathrm{N}/^{14}\mathrm{N} \approx (3.1 \pm 0.7) \times 10^{-3}$ ($1\sigma$-error). However, considering the uncertainty of the data analysis with such large background, we basically repeat the result of Kallenbach (2003) that $^{14}\mathrm{N}/^{15}\mathrm{N} > 200$ in the solar wind.

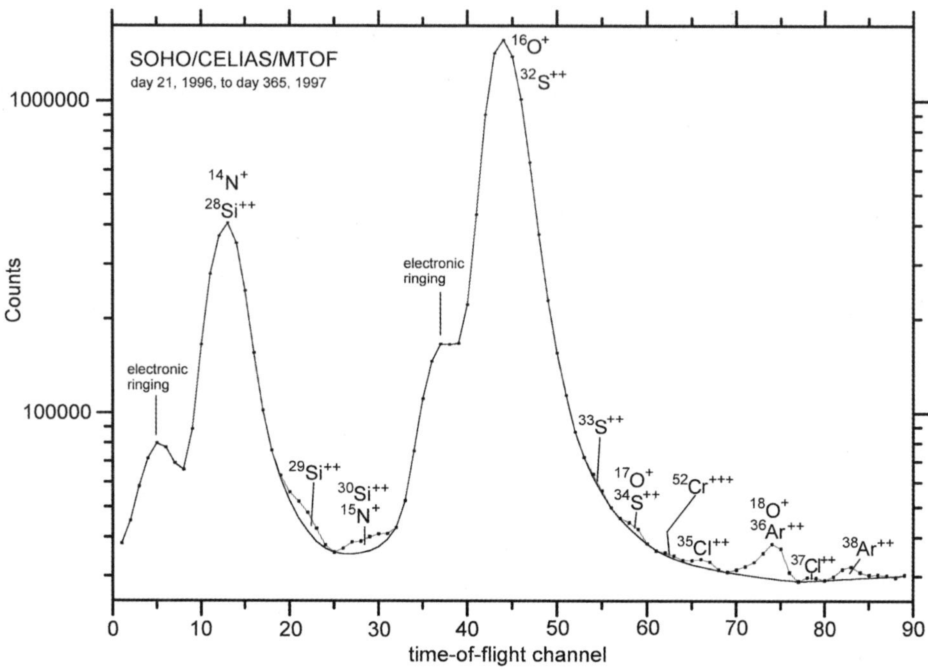


**Fig. 2** The SOHO/CELIAS/MTOF spectrum in the atomic mass-per-charge range from about 13 to 20 for the time period 21 January 1996 to 31 December 1997. The data are identical to those published in Kallenbach et al. (1998b)

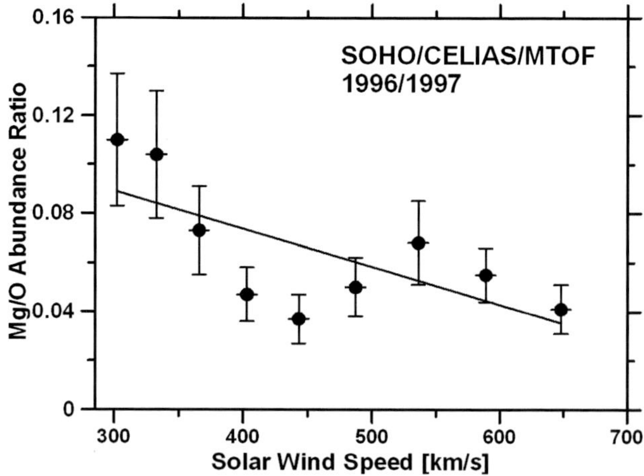


**Fig. 3** The Mg/O ratio in the solar wind as measured with the MTOF sensor (Kallenbach 2001). The larger *error bars* at the high Mg/O ratio—that is, at high-FIP fractionation—are due to the lower detection efficiency of MTOF in the slow solar wind; that is, the count rates are lower in the slow solar wind

## 3 Oxygen Isotopic Ratios in the Solar System

Since the pioneering work of Thiemens and Heidenreich (1983), the process of photo-selfshielding is discussed in the context of isotopic ratios as tracers for planetary formation processes in the early solar nebula. The basic idea is that fractionation occurs through selective photo-dissociation of molecules such as $C^{17,18}O$, HD, or $^{14}N^{15}N$. The more abundant molecules such as pure $H_2$, or $^{14}N_2$ or $C^{16}O$, absorb the respective ultraviolet (UV) line that dissociates them within a much smaller depth in the accretion disk. That means, the medium is optically thicker for these UV lines than for the shifted UV lines that are absorbed by the rare molecules. Therefore, relatively more $C^{17,18}O$ (HD, $^{14}N^{15}N$) molecules are dissociated so that relatively more $^{17,18}O$ (D, $^{15}N$) radicals are available for chemical reactions. At present, a chemist divides the approximately cylindrical protoplanetary disk into three types of vertical layers: (1) the upper and lower photo-dissociation region (PDR), (2) the upper and lower molecular layer, and (3) the central condensation layer. The PDR is the outermost layer of the proto-planetary disk and is exposed to the UV, X-ray, and particle radiation of the young Sun, because the accretion disk is warped and becomes thicker at large distances from the Sun. The PDR is also exposed to radiation from the interstellar medium or from other stars. Closer to the central layer, molecules such as HCN or $NH_3$ are formed from the radicals. Oxygen may be incorporated into rocky and icy materials by the reactions (Clayton 2002)

$$\begin{aligned} Mg_{(g)} + SiO_{(g)} &\Rightarrow MgSiO_{3(s)} \\ H_{2(g)} + O_{(g)} &\Rightarrow H_2O_{(g)}. \end{aligned} \tag{1}$$

In the central condensation layer, molecules accrete on the surface of ices and dust grains. An important point of this view is that at almost any heliocentric distance the disk is exposed to optical and particle radiation from both the interstellar medium and the central star. Therefore, planetary material formed from ices and rocks that also have trapped HCN or $NH_3$ tends to be enriched in the rare isotopes with respect to solar material.

Recent work by Hashizume and Chaussidon (2005) indicates that the self-shielding of CO molecules is reflected in the oxygen isotopic ratios in lunar samples and the solar wind implanted into these samples (Fig. 4).

The data analysis for the MTOF data is again shown in the Appendix. We find $^{16}O/^{18}O \geq 500$. Although the ratio is rather uncertain it would be concordant with the conclusions of Hashizume and Chaussidon (2005). In a similar way, we find $^{16}O/^{17}O \geq 2000$ if there is no contribution of $^{34}S^{++}$ to the peak of $^{17}O^{+}$. However, this seems unlikely. Therefore, a depletion of $^{17}O$ in the solar wind with respect to terrestrial $^{17}O$ does not seem unlikely although it cannot be strongly supported by MTOF data because of the measurement's large uncertainty.

## 4 Summary

Table 1 summarizes the isotopic ratios measured in the solar wind by spacecraft instrumentation and compares these ratios to meteoritic values or to values determined with the Apollo Solar Wind Collection (SWC) Foil Experiment (Geiss et al. 1972).

Regarding the nitrogen and oxygen isotopic abundance ratios we can state only that the MTOF data do not deliver any contradiction to the results of Owen et al. (2001) and Hashizume and Chaussidon (2005) that the main isotopes $^{14}N$ and $^{16}O$ are enriched in the solar wind with respect to their abundance in terrestrial samples. This enrichment could indicate that the terrestrial isotopic composition of nitrogen and oxygen had been fractionated with respect to the main reservoir, the solar material, by the process of photo-selfshielding in the protosolar nebula. However, the MTOF data on the isotopic abundance ratios of oxygen and nitrogen in the solar wind are consistent with the terrestrial isotopic ratios as well.

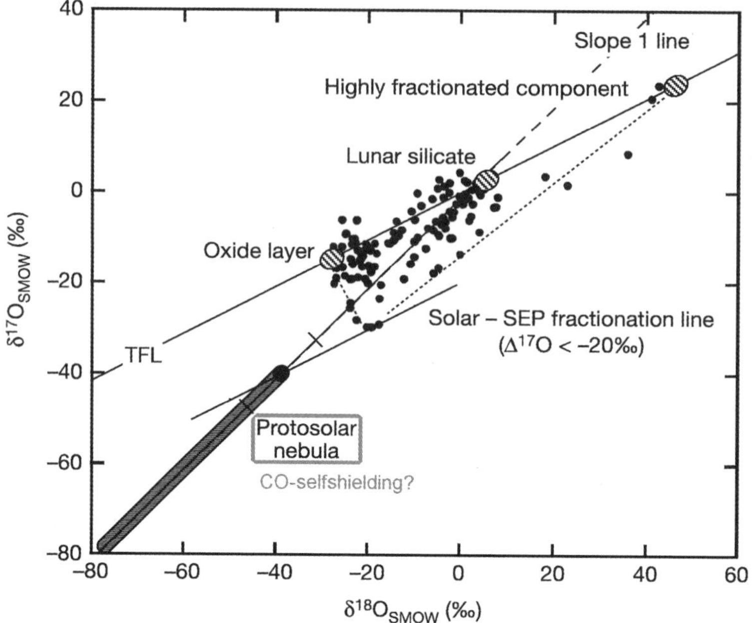


**Fig. 4** Theoretical prediction of the solar-nebula fractionation line by Hashizume and Chaussidon (2005). The upper limit for the composition of the protosolar nebular gas is predicted to be $\delta^{17}O \approx \delta^{18}O = -4 \pm 0.8\%$ (1-$\sigma$ error). The abbreviation TFL labels the terrestrial fractionation line

**Table 1** Mean isotopic abundance ratios in the solar wind from spaceborne sensors

| Volatile isotopes | | | Refractory isotopes | | |
|---|---|---|---|---|---|
| Ratio | Spacecraft data | Apollo SWC/ meteorites | Ratio | Solar wind | Meteoritic values |
| $^3He/^4He^{(s)}$ | $0.00041 \pm 0.000025$[1)] | 0.000142[2)] | $^{25}Mg/^{24}Mg$ | $0.1260 \pm 0.0014$[3)] | 0.12658[2)] |
| $^3He/^4He^{(f)}$ | $0.00033 \pm 0.000027$[1)] | 0.000142[2)] | $^{26}Mg/^{24}Mg$ | $0.1380 \pm 0.0031$[3)] | 0.13947[2)] |
| $^{15}N/^{14}N$ | $\leq 0.005$[4)] | 0.0023(3)[5)] | $^{29}Si/^{28}Si$ | $0.05012 \pm 0.00072$[3)] | 0.050634[2)] |
| $^{17}O/^{16}O$ | $\leq 0.0005$[4)] | 0.00038[2)] | $^{30}Si/^{28}Si$ | $0.03344 \pm 0.00024$[3)] | 0.033612[2)] |
| $^{18}O/^{16}O$ | $\leq 0.002$[4,6)] | 0.0020[2)] | $^{42}Ca/^{40}Ca$ | $0.00657 \pm 0.00017$[3)] | 0.006621[2)] |
| $^{21}Ne/^{20}Ne$ | $0.0023 \pm 0.0006$[3)] | 0.0024(3)[7)] | $^{44}Ca/^{40}Ca$ | $0.0209 \pm 0.0011$[3)] | 0.021208[2)] |
| $^{22}Ne/^{20}Ne$ | $0.0728 \pm 0.0013$[3)] | 0.0726(3)[7)] | $^{54}Fe/^{56}Fe$ | $0.068 \pm 0.004$[9)] | 0.06327[2)] |
| $^{38}Ar/^{36}Ar$ | $0.183 \pm 0.018$[8)] | 0.1880[2)] | $^{57}Fe/^{56}Fe$ | $0.025 \pm 0.005$[9)] | 0.02339[2)] |

1) Gloeckler and Geiss (1998); 2) Anders and Grevesse (1989); 3) Kallenbach (2000; 2001); 4) this work; 5) Owen et al. (2001); 6) Collier et al. (1998), Wimmer-Schweingruber et al. (2001); 7) mean of solar wind ratios from Apollo SWC (Geiss et al. 1972) and from Palma et al. (2002); 8) Weygand et al. (2001); 9) Ipavich et al. (2001); (f): 'fast' solar wind; (s): 'slow' solar wind

It will be the great challenge of the Genesis team to deliver more precise results. A future analysis of the full measurement period of MTOF onboard SOHO of more than ten years may yield results with a higher precision due to better counting statistics. Principal difficulties will remain because of the interference of doubly charged Si and S isotopes in the time-of-flight spectra. However, when Genesis data are available, a re-evaluation of MTOF data may still be worthwhile.

## Appendix

As mentioned in Sect. 2, the main difficulty in determining the abundance of the rare $^{15}$N isotopes in MTOF data comes from the interference of $^{30}Si^{++}$ with $^{15}N^{+}$ in the time-of-flight spectra. Additionally, the tail of the peak measured for the more abundant $^{14}N^{+}$&$^{28}Si^{++}$ needs to be subtracted from the peaks of the less abundant $^{29}Si^{++}$ and $^{15}N^{+}$&$^{30}Si^{++}$ that lie on top of it. This subtraction is very critical and presumably the main source of possible systematic uncertainties. However, the technique presented here reproduces the correct $^{29}Si^{++}$ and $^{30}Si^{++}$ count numbers that has been derived independently from the determination of the FIP-fractionation in the solar wind during the measurement period.

In K98 the line shapes have been described by analytical functions. These functions may not be correct in the far ends of the tails of the peaks. In this analysis, we have verified that the various peaks in the MTOF spectra can be described by a series of ratios $R_{\mathrm{down};i} = \Delta N_{i+1}/\Delta N_i$, $\Delta N_i = N_{i+1} - N_i$, where $N_i$ are the counts in channel $i$. The ratios typically range from $R_{\mathrm{down};i} \approx 0.5$ at a few channels above the peak center to $R_{\mathrm{down};i} \approx 0.8$ in the far end of the peak. This kind of numerical fit has most precisely been applied to the $^{16}O^{+}$ peak which is assumed to be similar to the $^{14}N^{+}$&$^{28}Si^{++}$ peak. This is justified because all peaks in the MTOF data have a similar form and in particular because of the fact that the mass-per-charge ratios of $^{16}O^{+}$, $^{14}N^{+}$, and $^{28}Si^{++}$ are similar. In the following, we assume the technique being precise. However, in the final result we include the difference to the result obtained with the original method in K98 as an upper limit for a possible systematic error.

### Data Analysis for the Nitrogen Isotopes

#### *Determination of the $^{29}Si^{++}$ and $^{15}N^{+}$&$^{30}Si^{++}$ Counts from Line Shape Analysis*

As stated earlier, we assume that the shape of the $^{14}N^{+}$&$^{28}Si^{++}$ peak extending to the $^{29}Si^{++}$ and $^{15}N^{+}$&$^{30}Si^{++}$ mass range and the shape at the right-hand side of the $^{16}O^{+}$ peak are described by identical ratios $R_{\mathrm{down};i} = \Delta N_{i+1}/\Delta N_i$, $\Delta N_i = N_{i+1} - N_i$, where $N_i$ are the counts in channel $i$. This yields the black line in Fig. 2. The actual counts for the different channels are listed in Table 2.

Beyond channel 25 the sequence includes $^{17}O^{+}$ for channels 58 through 63. However, the average number $R_{\mathrm{down}} = 0.80 \pm 0.08$ reproduces the black line below the $^{17}O^{+}$ peak. This sequence converges to the background level of 30,000 apparent from the convergence of the right wing of the $^{16}O^{+}$ peak. This background is also used in Table 2.

The background level of 30,000 defines a mean $R_{\mathrm{up}}$ number for channels 26 through 31. Extrapolating the left-hand side of $^{16}O^{+}$ leads to a total difference of $(52{,}289 - 42{,}997) \times R_{\mathrm{up}}/(1 - R_{\mathrm{up}}) = 42{,}997 - 30{,}000 = 12{,}997$ and thus $R_{\mathrm{up}} \approx 0.583$. From that the $^{16}O^{+}$ "wing" below $^{15}N^{+}$&$^{30}Si^{++}$ is determined. In our view, there is no better explicit way to estimate the line shapes below the $^{15}N^{+}$&$^{30}Si^{++}$ counts. In particular, the "electronic ringing" peak hampers the analysis of the line shape. "Ringing" is an instrumental effect that occurs in the time-of-flight electronics. It means that a fraction of 0.12 of TOF measurements for any ion species is systematically shifted by a fixed channel number to shorter times. This is best visible at the left-hand side of the $^{16}O^{+}$ peak but also to the left of the $^{14}N^{+}$/$^{28}Si^{++}$

**Table 2** Overview of data analysis

| Ch # | Counts | $R_{down}$* | $^{14}N^+$ & $^{28}Si^{++}$ | $R_{up}$* | $^{16}O^+$ | $^{14}N^+$& $^{28}Si^{++}$ + $^{16}O^+$ a) |
|---|---|---|---|---|---|---|
| 17 | 102,137 | 0.579 | 102,137 | – | – | 102,137 |
| 18 | 75,731 | 0.490 | 75,731 | – | – | 75,731 |
| 19 | 63,007 | 0.605 | 59,769 | – | – | 59,769 |
| 20 | 55,757 | 0.595 | 50,271 | – | – | 50,271 |
| 21 | 52,034 | 0.593 | 44,639 | – | – | 44,639 |
| 22 | 47,966 | 0.630 | 41,094 | – | – | 41,094 |
| 23 | 42,615 | 0.670 | 38,718 | – | – | 38,718 |
| 24 | 37,713 | 0.710 | 37,032 | – | – | 37,032 |
| 25 | 35,757 | 0.759 | 35,757 | – | – | 35,757 |
| 26 | 36,851 | 0.800 | 34,612 | 0.583 | 30,517 | 35,129 |
| 27 | 38,566 | 0.800 | 33,696 | 0.583 | 30,882 | 34,578 |
| 28 | 38,816 | 0.800 | 32,963 | 0.583 | 31,508 | 34,471 |
| 29 | 40,124 | 0.800 | 32,377 | 0.583 | 32,581 | 34,958 |
| 30 | 40,972 | 0.800 | 31,908 | 0.583 | 34,422 | 36,330 |
| 31 | 41,085 | 0.800 | 31,533 | 0.583 | 37,580 | 39,113 |
| 32 | 42,997 | – | – | 0.401 | 42,997 | 42,997 |
| 33 | 52,289 | – | – | 0.645 | 52,289 | 52,289 |

$R_{down}$ belongs to the $^{14}N^+$&$^{28}Si^{++}$ peak, $R_{up}$ to the $^{16}O^+$ peak

a) Sum of $^{14}N^+$&$^{28}Si^{++}$ and $^{16}O^+$ based on a background level of 30,000 in the channels between the $^{14}N^+$&$^{28}Si^{++}$ and $^{16}O^+$ lines

*Averaged over four channels of the $^{16}O^+$ peak to reduce statistical variations and multiplied by a factor 1.0425 in order to connect the count number of channel 18 to the count number of channel 25

peak. It has been checked that the used line shapes—that is, the numbers $R_{up/down}$—are similar to the numbers found for instance at the $^{28}Si^+$ peak and the $^{56}Fe^+$ peak. The uncertainty of the numbers $R_{up/down}$ is typically 15%.

The procedure yields 27,569 counts for $^{29}Si^{++}$. The background below $^{29}Si^{++}$ is 271,523 counts and yields an uncertainty of 521 counts. For $^{15}N^+$&$^{30}Si^{++}$ we get 21,835 counts with a background of 214,579; that is, with an uncertainty of 463 counts. We now have to correct the $^{29}Si^{++}$ by $0.12 \times 21,835 \approx 2,620$ counts which spill over from the $^{15}N^+$&$^{30}Si^{++}$ peak to the $^{29}Si^{++}$ peak; that is, $^{29}Si^{++} = 24,949 \pm 521$. The $^{30}Si/^{29}Si$ ratio in the solar wind is about $0.68 \pm 0.03$ (Kallenbach 2003), consistent with the meteoritic value (Anders and Grevesse 1989), and $^{30}Si^{++}$ is detected more efficiently than $^{29}Si^{++}$ by a factor 1.04. Therefore, we must multiply $^{29}Si^{++}$ by a factor 0.7072 to get the number $^{30}Si^{++} = 17,644$. This method finally yields $^{15}N^+ = 4,191$.

### *$^{29}Si^{++}$ Counts from Determination of the FIP Fractionation*

We have verified the Si counts by determining the FIP fractionation and thus the elemental abundance ratio Si/N in the solar wind during the measurement time period. In the slow solar wind Si can be enriched by the FIP fractionation process up to a factor of 3 to 4 with respect to its photospheric abundance, however, from Fig. 3 it follows that MTOF *counts* an average enrichment of Mg by the FIP effect of only about 1.4. From the photospheric

**Table 3** Data analysis: Oxygen isotopes

| Ch # | Counts | $R_{down}$ | $^{16}O^{+}$ | Ch # | Counts |
|---|---|---|---|---|---|
| 67 | 33,314 | | | 78 | 29,688 |
| 68 | 31,431 | | | 79 | 29,601 |
| 69 | 30,867 | 0.8 | 30,867 | 80 | 29,090 |
| 70 | 31,409 | 0.8 | 31,409 | 81 | 29,848 |
| 71 | 32,180 | 0.8 | 32,180 | 82 | 31,491 |
| 72 | 33,262 | 0.8 | 33,262 | 83 | 32,091 |
| 73 | 35,603 | 0.8 | 35,603 | 84 | 30,792 |
| 74 | 38,125 | 0.8 | 38,125 | 85 | 30,222 |
| 75 | 36,787 | 0.8 | 36,787 | 86 | 30,300 |
| 76 | 30,799 | 0.8 | 30,799 | 87 | 30,077 |
| 77 | 28,850 | 0.8 | 28,850 | 88 | 29,642 |

ratio $^{28}Si/^{14}N = 0.296$ (Anders and Grevesse 1989) it follows that $^{28}Si/^{14}N \approx 0.415$ in the solar wind during the measurement period. A detailed analysis of calibrated instrument functions (Gonin et al. 1994; Kallenbach et al. 1995) gives a higher detection efficiency of $^{14}N^{+}$ than for $^{28}Si^{+}$ by a factor 1.18. Therefore, we have a fraction of 26% of the 1,635,988 counts within 6 channels around the peak center of $^{14}N^{+}$&$^{28}Si^{++}$ as $^{28}Si^{++}$ counts; that is, $^{28}Si^{++} \approx 425{,}357$. The $^{28}Si/^{29}Si$ ratio is about 19.7 (Anders and Grevesse 1989), as confirmed by MTOF measurements (Kallenbach 2003), and $^{29}Si^{++}$ is detected more efficiently than $^{28}Si^{++}$ by a factor 1.07 (K98). This yields $^{29}Si^{++} \approx 23{,}103$. Correcting the 2,620 counts of spill-over from $^{15}N^{+}$&$^{30}Si^{++}$ through "ringing" this yields $^{29}Si^{++} \approx 20{,}483$. Comparison to $^{29}Si^{++} \approx 24{,}949$ suggests that the line shape method may lead to a background level that gives 4,516 counts more for $^{29}Si^{++}$. This would mean that the line shape method also could overestimate the $^{15}N^{+}$&$^{30}Si^{++}$ counts by as much as 4,516 counts; that is, $^{15}N^{+}$&$^{30}Si^{++}$ could be as low as 17,319. Multiplying $^{29}Si^{++}$ by 0.7072 as above, we get $^{30}Si^{++} \approx 14{,}486$. The method finally yields $^{15}N^{+} = 3{,}433$.

*Determination of $^{15}N/^{14}N$*

With this information we now have to determine a ratio $^{15}N/^{14}N$. The two methods above yield an average value $^{15}N^{+} \approx 3{,}812 \pm 536$. However, we have to add the statistical uncertainty introduced by the background of roughly 271,523 below $^{29}Si^{++}$ and 214,579 below $^{15}N^{+}$&$^{30}Si^{++}$. The total statistical uncertainty is thus about 591 counts. This yields $^{15}N^{+} \approx 3{,}812 \pm 798$.

The number of $^{14}N^{+}$ counts is $1{,}635{,}988 - 425{,}357 = 1{,}210{,}631$. Considering the fact that $^{15}N^{+}$ is detected more efficiently than $^{14}N^{+}$ by a factor 1.055 (K98), this yields $^{14}N/^{15}N \approx 320$ or $^{15}N/^{14}N \approx (3.1 \pm 0.7) \times 10^{-3}$. This is marginally consistent with the value $^{15}N/^{14}N \approx (2.3 \pm 0.3) \times 10^{-3}$ measured with the GPMS in Jupiter's atmosphere (Owen et al. 2001), but also consistent with the terrestrial ratio. Considering the uncertainty of the analysis of data with such large background, we basically repeat the result of Kallenbach (2003) that $^{14}N/^{15}N > 200$ in the solar wind.

Data Analysis for the Oxygen Isotopes

The CELIAS/MTOF data give about 7,158,000 counts within seven channels for $^{16}O^{+}$ and about $31{,}400 \pm 500$ counts for $^{18}O^{+}$ & $^{36}Ar^{++}$ (Table 3). The contribution of $^{36}Ar^{++}$ can be estimated from the $^{38}Ar^{++}$ peak in which the contribution of $^{57}Fe^{+++}$ is negligible. Taking a background of $29,803 \pm 394$ from channels 78–80 and 85–88, that is, for a maximum background of 30,197 we get at least 3,911 counts for $^{38}Ar^{++}$. With $^{36}Ar/^{38}Ar$ of about 5.3 in the solar wind and a more efficient detection of $^{38}Ar^{++}$ by about 20% as a typical maximum number for MTOF, we find at most about 14,600 counts of $^{18}O^{+}$ and thus $^{16}O/^{18}O \geq 500$. In a similar way, we find $^{16}O/^{17}O \geq 2{,}000$ if there is no contribution of $^{34}S^{++}$ to the peak of $^{17}O^{+}$. However, the meteoritic $^{34}S/^{17}O$ ratio is about 2.4 and the ratio of the detection efficiencies of $^{34}S^{++}$ and $^{17}O^{+}$ is about 0.4 (Kallenbach et al. 1995) so that $^{34}S^{++}/^{17}O^{+}$ may be of order unity in the MTOF spectrum. Therefore, a depletion of $^{17}O$ in the solar wind with respect to terrestrial $^{17}O$ does not seem unlikely.

## References

M.M. Abbas et al., Astrophys. J. **602**, 1063 (2004)
E. Anders, N. Grevesse, Geochim. Cosmochim. Acta **53**, 197 (1989)
R.N. Clayton, LPSI 33 # 1326, 2002
M.R. Collier et al., J. Geophys. Res. **103**, 7 (1998)
T. Fouchet et al., Icarus **172**, 50 (2004)
J. Geiss et al., NASA SP **315**, 14.1 (1972)
G. Gloeckler, J. Geiss, Space Sci. Rev. **84**, 275 (1998)
M. Gonin et al., NIM B **94**, 15 (1994)
K. Hashizume et al., 31st Lun. Planet. Sci. Conf., #1565, 2000
K. Hashizume, M. Chaussidon, Nature **434**, 619 (2005)
E. Herbst, Space Sci. Rev. **106**, 293 (2003)
F.M. Ipavich et al., AIP CP **598**, 121 (2001)
E. Jehin et al., Astrophys. J. **613**, L161 (2004)
D.C. Jewitt et al., Science **278**, 90 (1997)
R. Kallenbach et al., NIM B **103**, 111 (1995)
R. Kallenbach et al., ESA SP **415**, 33 (1998a)
R. Kallenbach et al., Astrophys. J. **507**, L185 (1998b)
R. Kallenbach, *Isotopic Composition of the Solar Wind* (University of Bern, Bern, 2000)
R. Kallenbach, AIP CP **598**, 113 (2001)
R. Kallenbach, Space Sci. Rev. **106**, 305–316 (2003)
H. Lammer, S.J. Bauer, Space Sci. Rev. **106**, 281 (2003)
M.-C. Liang et al., Astrophys. J. **657**, L117 (2007)
B. Marty et al., Space Sci. Rev. **106**, 175 (2003)
S. Messenger et al., Space Sci. Rev. **106**, 155 (2003)
H.B. Niemann et al., Nature **438**, 779 (2005)
T. Owen et al., Astrophys. J. **553**, L77 (2001)
R.L. Palma et al., Geochim. Cosmochim. Acta **66**, 2929 (2002)
M.H. Thiemens, J.E. Heidenreich III, Science **219**, 1073 (1983)
R. von Steiger, J. Geiss, Astron. Astrophys. **225**, 222–238 (1989)
J.M. Weygand et al., Geochim. Cosmochim. Acta **65**, 4589 (2001)
R. Wieler et al., EPSL **167**, 47 (1999)
R.F. Wimmer-Schweingruber et al., Geophys. Res. Lett. **28**, 2763 (2001)

Space Sci Rev (2007) 130: 183–194
DOI 10.1007/s11214-007-9218-y

# Solar Elemental Composition Based on Studies of Solar Energetic Particles

**C.M.S. Cohen · R.A. Mewaldt · R.A. Leske · A.C. Cummings · E.C. Stone · M.E. Wiedenbeck · T.T. von Rosenvinge · G.M. Mason**

Received: 7 February 2007 / Accepted: 19 May 2007 / Published online: 21 July 2007


**Abstract** Solar abundances can be derived from the composition of the solar wind and solar energetic particles (SEPs) as well as obtained through spectroscopic means. Past comparisons have suggested that all three samples agree well, when rigidity-related fractionation effects on the SEPs were accounted for. It has been known that such effects vary from one event to the next and should be addressed on an event-by-event basis. This paper examines event variability more closely, particularly in terms of energy-dependent SEP abundances. This is now possible using detailed SEP measurements spanning several decades in energy from the Ultra Low Energy Isotope Spectrometer (ULEIS) and the Solar Isotope Spectrometer (SIS) on the ACE spacecraft. We present examples of the variability of the elemental composition with energy and suggest they can be understood in terms of diffusion from the acceleration region near the interplanetary shock. By means of a spectral scaling procedure, we obtain energy-independent abundance ratios for 14 large SEP events and compare them to reported solar wind and coronal abundances as well as to previous surveys of SEP events.

**Keywords** Sun: abundances · Sun: coronal mass ejections (CMEs) · Sun: flares

## 1 Introduction

The Sun provides us with the unique opportunity to study stellar phenomena through the examination of solar composition and activity. Two samples of solar material are directly

C.M.S. Cohen (✉) · R.A. Mewaldt · R.A. Leske · A.C. Cummings · E.C. Stone
California Institute of Technology, Pasadena, CA 91125, USA
e-mail: cohen@srl.caltech.edu

M.E. Wiedenbeck
Jet Propulsion Laboratory, Pasadena, CA 91109, USA

T.T. von Rosenvinge
NASA/Goddard Space Flight Center, Greenbelt, MD 20771, USA

G.M. Mason
Johns Hopkins Applied Physics Laboratory, Laurel, MD 20723, USA

available to space-based instrumentation: solar wind and solar energetic particles (SEPs). Additionally, the solar composition can be measured spectroscopically through absorption and emission lines observed at different wavelengths by ground- and space-based telescopes. Analysis of each of the three samples provides unique information, but is subject to individual challenges due to limitations in the measurement techniques and to physical processes that can alter the composition during the formation of the sample. Comparison of composition results from solar wind, SEPs, and spectroscopy yields more robust values for the composition of the Sun as well as insights into the physical processes involved in heating, accelerating, and transporting solar material.

Although previous comparisons between SEP, solar wind, and spectroscopic abundances indicated reasonable agreement (Meyer 1985a), the more recent data indicate substantial differences between the SEP and solar wind abundances (Mewaldt et al. 2001, 2007). As noted by Meyer (1985b) and Breneman and Stone (1985), fractionation of SEPs related to the ionic charge to mass ratios ($Q/M$) needs to be considered when making such comparisons. The magnitude of such effects can differ from one SEP event to the next, however, causing substantial variability in the measured SEP abundances. As discussed in the following, there are other effects which can result in composition variability within a single SEP event as well. In this paper we discuss some of these effects, and using data from the Ultra-Low Energy Isotope Spectrometer (ULEIS; Mason et al. 1998) and the Solar Isotope Spectrometer (SIS; Stone et al. 1998) on the ACE spacecraft we attempt to correct for them to obtain values for the average SEP composition.

## 2 SEP Variability

Before examining how SEP composition varies from one event to the next, variations within a given event should be understood. Evidence for changes in the composition as a function of time have been attributed to (1) differences in transport times for individual elements, typically assumed to be organized by ion rigidity (Ng et al. 1999) and (2) the superposition of two source populations created by distinct acceleration mechanisms (Cane et al. 2003). Most reported SEP abundances are obtained by averaging measured intensities over the duration of each SEP event to obtain fluences; if transport is the cause of any time-dependent composition, then calculating abundances from the ratios of measured fluences should better reflect the source composition. Alternatively, if the variation is due to separate components (which themselves may vary in composition and intensity from one SEP event to the next), using such event-averaged abundances is inappropriate. In this study, event-averaged abundances are used, however, some effort has been made to select events that are not likely to have two significant source populations (as discussed in the Analysis section).

It was shown by Breneman and Stone (1985) that the (event-averaged) SEP elemental abundances often vary systematically with $Q/M$ for a given SEP event as compared to the abundances obtained by averaging over many SEP events. An example of this using data from two SEP events as measured by SIS is shown in Fig. 1. The event of October 16, 2002, is rich in low $Q/M$ elements, such as Fe, when compared to the average large SEP event abundances reported by Reames (1995), while the September 12, 2004, event is systematically depleted in these elements. The fact that $Q/M$ provides a reasonable organization of the enhancements/depletions suggests that the physical process responsible is governed by the particle rigidity.

In a related effect, a number of SEP events exhibit elemental abundances that vary with energy. Most often for large events the pattern is one in which ratios of lower $Q/M$ to higher

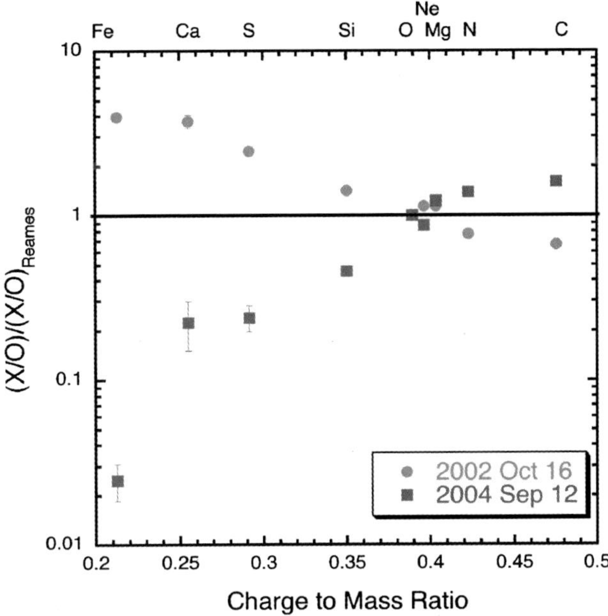


**Fig. 1** An example of how the observed abundance ratios can depend on the charge-to-mass ratio of the individual elements. A temperature of 1.6 MK was assumed for the determination of the charge states (Mazzotta et al. 1998). Measured abundances (relative to oxygen and integrated from 12 to 60 MeV/nucleon) for both SEP events were normalized by the average SEP abundances reported by Reames (1995)

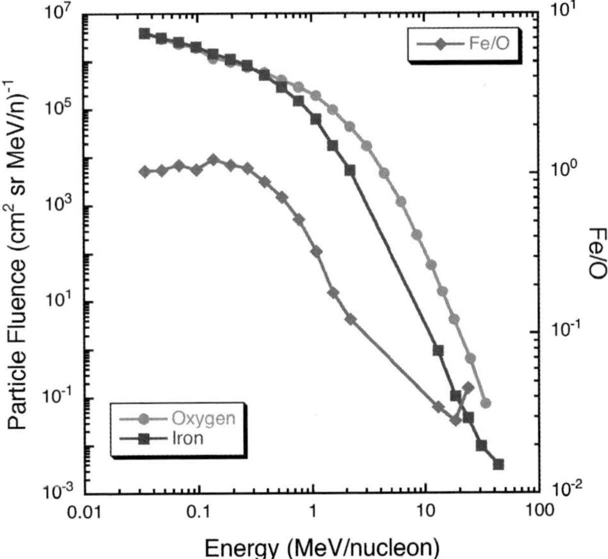


**Fig. 2** Oxygen and iron energy spectra from ULEIS and SIS averaged over the SEP event of December 2, 2003 (*left y-axis*; *circles* and *squares*, respectively). Note that the spectra roll-off at different energies. The calculated Fe/O abundance (*right y-axis*; *diamonds*) strongly decreases with increasing energy as a result

$Q/M$ particles (e.g., Fe/O) decrease with increasing energy, sometimes by more than an order of magnitude over the energy range of 1 to 10 MeV/nucleon (see, e.g., Fig. 2). In such cases, the reported composition of the event is highly dependent on the energy range over which the spectra are integrated. As illustrated in Fig. 2, the decrease in the Fe/O ratio with increasing energy is a result of the different energies at which the oxygen and iron spectra roll-off (steepen). Below the roll-off energy of Fe ($E_r$(Fe)) the Fe/O ratio is approximately independent of energy; however, the value of $E_r$ (for each element) varies from one SEP

event to the next. Here, we strive to understand and remove the energy dependence of the abundances in each event before calculating an average SEP composition.

## 3 Understanding Energy Variations

Acceleration of SEPs by interplanetary shocks is a mature theory with numerical and analytical models that produce results qualitatively in agreement with the observational data (Lee 2005; Li et al. 2005). One of the hallmarks of diffusive shock acceleration has been that the resulting spectra of the accelerated ions should be a power law with an index that depends only on the shock parameters and not on the species being accelerated. Although power law spectra are observed at low energies in the SEP spectra, it is typical to see roll-offs at higher energies (see, e.g., Fig. 2 and Cohen et al. 2003). These roll-offs were discussed by Ellison and Ramaty (1985) and could be a result of adiabatic deceleration, limited particle acceleration times, or diffusive escape from the shock. Diffusive escape would likely be a rigidity-dependent process and so would be expected to set in at different energies for species with different $Q/M$ values.

Similar arguments have been used to explain the roll-overs observed at low energies in spectra of anomalous cosmic rays (ACRs) (Cummings et al. 1984). By expressing the diffusion coefficient, $\kappa$, as a function of the particle's mean free path ($\lambda$; $\kappa = 1/3v\lambda$) and assuming that $\lambda$ can be expressed as a power law in rigidity ($\lambda \sim R^{\alpha}$; $R = Mv/Q$), we have:

$$\kappa \sim (M/Q)^{\alpha}\, E^{(\alpha+1)/2}, \tag{1}$$

where $v$ is the particle's velocity and $E$ is its energy/mass in MeV/nucleon.

If diffusive escape is the cause of the spectral roll-offs, we would expect the roll-offs to occur at the same value of the diffusion coefficient for each species. Equation (1) can be equated for two different particles (X and Y) to yield a general scaling law for the $E_{\mathrm{r}}$ values in terms of $Q/M$ (see also Cohen et al. 2005):

$$E_{\mathrm{r}}(\mathrm{X})/E_{\mathrm{r}}(\mathrm{Y}) = \left[(Q/M)_{\mathrm{X}}/(Q/M)_{\mathrm{Y}}\right]^{2\alpha/(\alpha+1)}. \tag{2}$$

Unfortunately, in the vast majority of SEP events, the charge states of the ions are not measured; thus the $Q/M$ values cannot be determined (particularly at energies much above 1 MeV/nucleon). Although values can be assumed, it is known that ionic charge states (especially for heavy elements such as Fe) vary substantially from one SEP event to the next (Labrador et al. 2005) and often assumptions of a single ionization temperature are not valid (Klecker et al. 2006). A further complication arises from the observation that large events can exhibit energy-dependent charge states (Klecker et al. 2006; Labrador et al. 2005; Mazur et al. 1999) making it difficult to select a single charge state to use in (2). For this work we are primarily interested in scaling the individual elemental spectra relative to that of oxygen (as a reference element) to remove as much of the energy dependence as possible. To maintain the composition resulting from the shock-accelerated power-law portion of the spectrum (i.e., below $E_{\mathrm{r}}$), this scaling must be done for both the energy/nucleon ($E$) and the particle fluence ($f$), i.e.,

$$E' = E \times S \quad \text{and} \quad f' = f \times S^{\delta} \tag{3}$$

(from $f = A \times E^{\delta} = A \times (E'/S)^{\delta}$), where $S$ is the scaling factor and the primed values are the results of scaling. In this manner an energy-independent abundance ratio can be determined empirically without a priori calculating a specific scaling factor for each element. More details of our method are given in the next section.

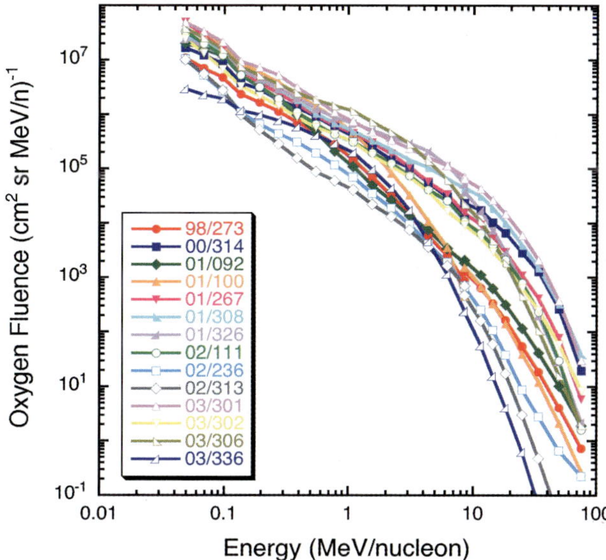


**Fig. 3** Event averaged ULEIS + SIS oxygen spectra for all the events analyzed in this study. The different roll-off energies are readily apparent. The legend gives the starting year/DOY of each event in the SIS data

## 4 Analysis

As our analysis technique requires roll-offs in the spectra to be clearly present, the dominant criterion for the selection of SEP events was the shape of the event-integrated oxygen spectrum. The individual SEP-event spectra were constructed from the particle fluences as a function of energy as measured by the ULEIS and SIS instruments. The spectral shape requirement, combined with the need for statistically accurate spectra for major elements from C to Fe over the energy range of <0.5 to >30 MeV/nucleon, resulted in a selection of 14 large SEP events. The oxygen spectra for the selected events are plotted in Fig. 3 and the year/day of year for the onsets of the events, as measured by SIS, are given in the figure legend. Although the spectral roll-offs are apparent in each event, it is also clear that the slope of the spectra above $E_r(\mathrm{O})$ is quite variable, as is the value of $E_r(\mathrm{O})$.

As discussed previously, interpreting event-integrated spectra can be complicated by the presence of multiple components. We believe this is not a great concern for the events examined in this study for a number of reasons. All the events except for one (2002/236) were classified as either "shock dominated" or "other" by Cane et al. (2006), suggesting that any flare-related component is small or negligible. Additionally, 11 of the 14 events had shock transit speeds in excess of 1000 km/s indicating the shock-accelerated component is likely to be strong and probably dominant. Finally, the $^3\mathrm{He}/^4\mathrm{He}$ values at 0.5–2 MeV/nucleon reported by Desai et al. (2006) are merely upper limits for 11 of the 14 events and 0.4%, 0.2%, and 0.09% for events 2001/092, 2002/236, and 2003/336, respectively (high $^3\mathrm{He}/^4\mathrm{He}$ values are a signature of flare-related material).

There is one event included in this study which does show possible evidence for two components. Event 2002/236, as studied by Tylka et al. (2005), has an Fe/O ratio that strongly increases with energy above 10 MeV/nucleon, unlike the other events analyzed here. As the event satisfies the initial selection criteria, it has been retained for the analysis but will be discussed more explicitly in the next section.

For each SEP event, the individual element spectra are compared to that of oxygen and the X/O ratio is calculated as a function of energy, where X denotes the particular element

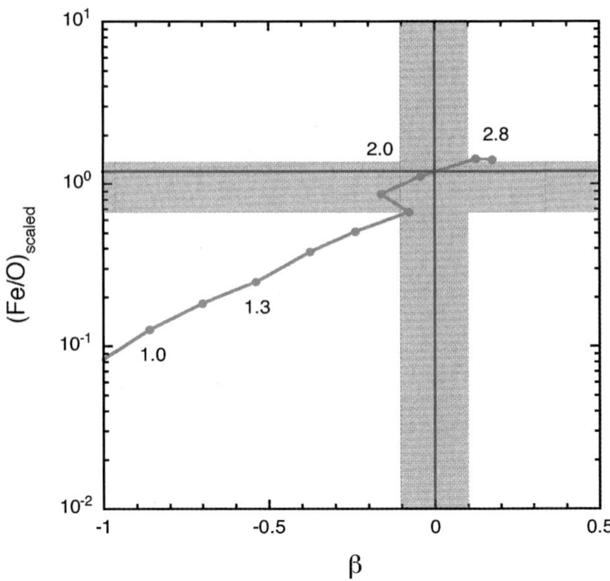


**Fig. 4** A plot of $(\mathrm{Fe/O})_{\mathrm{scaled}}$ values as a function of $\beta$ (see text for definition) for different amounts of scaling (values of $S$ are given next to some of the points for reference) for the same event shown in Figs. 2 and 5. The *horizontal line* show the adopted energy-independent $(\mathrm{Fe/O})_{\mathrm{scaled}}$ value obtained from where the *curve* crosses the *vertical line* at $\beta = 0$ (*shaded regions* give the uncertainty)

under consideration (see Fig. 2 for Fe). A scaling factor is applied to the spectrum of element X (per (3), with a $\delta$ value determined from fitting the eight lowest energy points of the oxygen spectrum) and the X/O values are recalculated using the scaled X and unscaled O spectra. The amount of energy dependence in the scaled ratio is roughly determined by fitting $(\mathrm{X/O})_{\mathrm{scaled}} \sim (E')^{\beta}$ from ~0.5 to ~50 MeV/nucleon. This is repeated for a range of scaling factors; for each factor, $S$, the power law index ($\beta$) and the average $(\mathrm{X/O})_{\mathrm{scaled}}$ value is recorded. From a plot of $(\mathrm{X/O})_{\mathrm{scaled}}$ versus $\beta$, the adopted $(\mathrm{X/O})_{\mathrm{scaled}}$ value is taken from where the curve crosses $\beta = 0$ and the uncertainties from where the curve crosses $\beta = \pm 0.1$ (Fig. 4). The scaling of the iron spectrum (using $S = 2.5$) is illustrated in Fig. 5 for the same SEP event as shown (with no scaling) in Fig. 2. With appropriate scaling the Fe/O ratio is approximately independent of energy (Fig. 5) as compared to decreasing by over an order of magnitude (Fig. 2). Not surprisingly, Fe required the largest and most varied values of S, ranging from 0.75 to 2.7. The final energy-independent $(\mathrm{Fe/O})_{\mathrm{scaled}}$ ratios for each SEP event are shown in Fig. 6 (left panel).

## 5 Results and Discussion

Some of the remaining variability in the $(\mathrm{Fe/O})_{\mathrm{scaled}}$ ratios is undoubtedly due to differing amounts of fractionation in each event resulting from the first-ionization-potential (FIP) effect. This effect causes the corona to be enriched in elements with low FIP values (e.g., Fe) relative to those with FIP values greater than ~10 eV (e.g., O). It is known that the degree of FIP fractionation varies between SEP events (Garrard and Stone 1994; Mewaldt et al. 2000, 2002; Slocum et al. 2003). To better examine how our scaling procedure has affected the event-to-event variability, we have renormalized the abundances to Si (see Fig. 6, right panel, for Fe/Si). The ratios between the two low-FIP elements Fe and Si are significantly less variable than those between Fe and O, which has a high FIP.

Figure 7 illustrates the overall effect of scaling the spectra before calculating abundances. In the left panel the individual abundance ratios (relative to Si) as measured

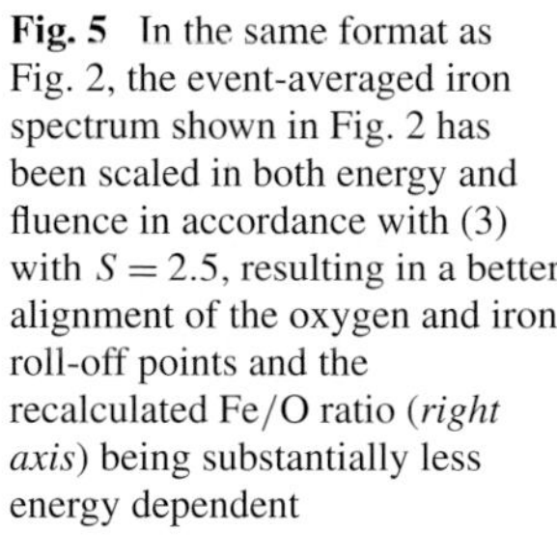

**Fig. 5** In the same format as Fig. 2, the event-averaged iron spectrum shown in Fig. 2 has been scaled in both energy and fluence in accordance with (3) with $S = 2.5$, resulting in a better alignment of the oxygen and iron roll-off points and the recalculated Fe/O ratio (*right axis*) being substantially less energy dependent

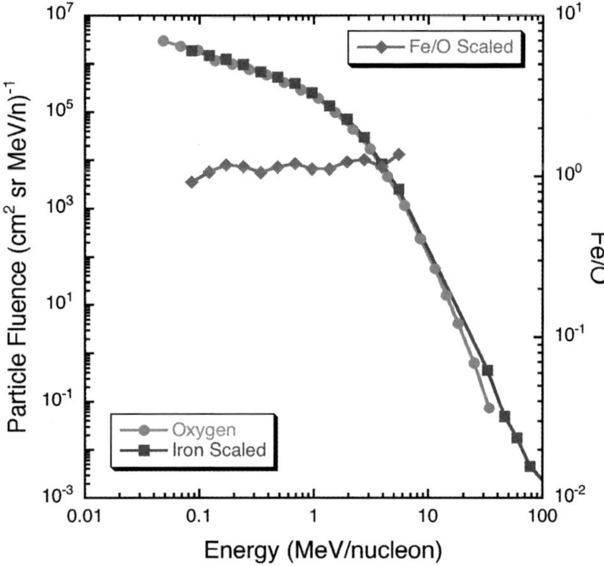


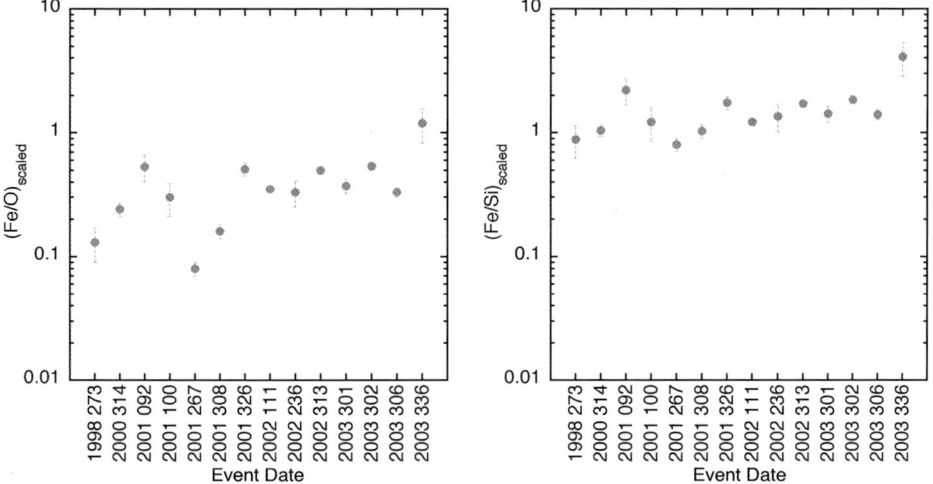


**Fig. 6** The energy-independent Fe/O (*left panel*) and Fe/Si (*right panel*) abundance for each of the SEP events used in this study. Event dates are given as the year and day of year of the onset of the event as measured by SIS. The larger spread of the Fe/O abundances is partly due to variation in the FIP factor from event to event, which is removed when comparing the two low FIP elements Fe and Si

at 50 MeV/nucleon are normalized to the (unweighted) average composition (at 50 MeV/nucleon) and plotted as a function of nuclear charge. The observed variation in Fe/Si for this group of 14 events is a factor of 200. This is reduced to less than a factor of 5 variation when the energy-independent ratios of the individual events are compared. The unweighted average of the scaled abundance ratios for all 14 SEP events is given in Table 1.

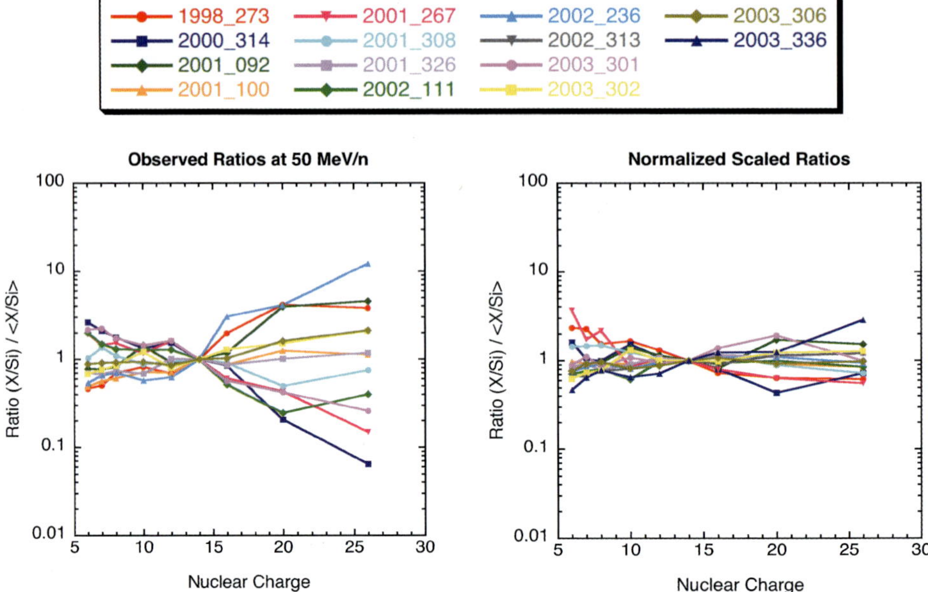


**Fig. 7** The variation of the observed abundance ratios (relative to Si) for each event (normalized to the average composition of all the events) as a function of nuclear charge as calculated at 50 MeV/nucleon with no spectral scaling (*left panel*) and after scaling the spectra to minimize the energy dependence (*right panel*)

**Table 1** Derived abundance ratios

| Element | X/O scaled | X/Si scaled |
|---|---|---|
| C | $0.40 \pm 0.03$ | $1.76 \pm 0.16$ |
| N | $0.15 \pm 0.01$ | $0.65 \pm 0.04$ |
| O | 1.00 | 4.41 |
| Ne | $0.15 \pm 0.01$ | $0.68 \pm 0.04$ |
| Mg | $0.22 \pm 0.01$ | $0.95 \pm 0.05$ |
| Si | $0.23 \pm 0.01$ | $1.00 \pm 0.05$ |
| S | $0.056 \pm 0.003$ | $0.25 \pm 0.01$ |
| Ca | $0.021 \pm 0.001$ | $0.091 \pm 0.005$ |
| Fe | $0.32 \pm 0.03$ | $1.43 \pm 0.11$ |

The 2002/236 event stands out in the left panel of Fig. 7 as having the highest Fe/Si ratio at 50 MeV/nucleon. In fact, the Fe/O ratio for this event is not a monotonic function of energy and thus the analysis procedure does not remove all the energy dependence from the Fe/O ratio in this event (particularly at energies above 10 MeV/nucleon). However, the energy dependence is reduced for other ratios (e.g., Si/O) and the resulting set of scaled ratios (relative to Si) are not very different than those from other events as can be seen in the right panel of Fig. 7. If the results from event 2002/236 were discarded the average abundances in Table 1 would change by $<3\%$. Given such a small change, we continue to include this event in the remaining discussion.

Previous studies have dealt with apparent FIP- and $Q/M$-related fractionation on an event-by-event basis by assuming $Q/M$ values and fitting the data with a function that

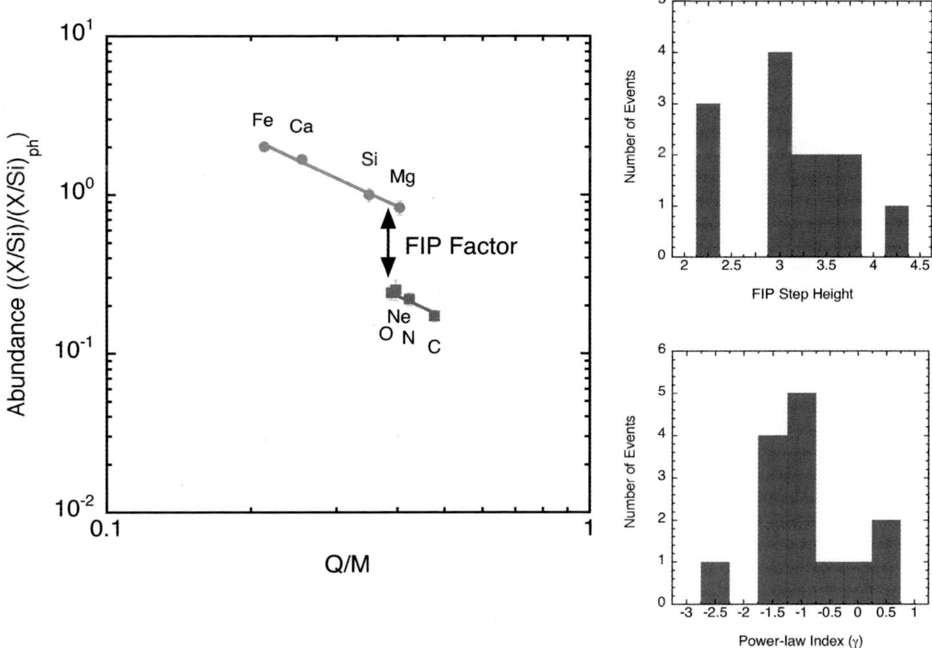


**Fig. 8** Example of fitting the abundances to determine the $Q/M$ and FIP fractionation (*left panel*; event 2002/313). The distributions of the FIP step height (*upper right panel*) and the power law index of the $Q/M$ dependence (*lower right panel*) of the scaled abundances as obtained by such fits to the individual event abundance ratios

combined a $Q/M$ power-law dependence with a FIP step (Garrard and Stone 1994; Mewaldt et al. 2000; Slocum et al. 2003). In an effort to examine such effects in the sample of events studied here (in particular the degree of FIP variation), we have assumed charge states corresponding to an ionization temperature of 1.6 MK using the tables of Mazzotta et al. (1998), plotted the $(X/Si)_{scaled}$ abundances (normalized to the photospheric abundances of Lodders 2003) versus $Q/M$, and fit two power law curves of equal index, vertically displaced by a constant factor, to the low- and high-FIP elements (Fig. 8, left panel). The resulting index ($\gamma$) of the power law curves gives the amount of $Q/M$ fractionation, while the separation of the two curves gives the magnitude of the FIP fractionation. The resulting fit values for the events in this study are shown as histograms in the right panels of Fig. 8.

It is clear from these results the FIP fractionation varies by nearly a factor of 2 for our events consistent with the results of Garrard and Stone (1994), Mewaldt et al. (2000, 2002), and Slocum et al. (2003). Additionally, there appears to be some remaining $Q/M$ fractionation in our scaled abundances. The degree to which this changes from event to event would be more accurately known if we were able to use the true $Q/M$ values for each event. Before correcting for these effects in individual events, it would be preferable to understand their cause; in particular the $Q/M$ fractionation could be due to relative efficiencies of injecting ions of different rigidities into the acceleration process or it could be an effect present in the source population itself. As to the FIP effect, it is known that the degree of FIP fractionation varies in the corona itself (Feldman 1998).

The average $(X/Si)_{scaled}$ abundances (without correcting for any remaining $Q/M$ fractionation) obtained in this work are compared to other measurements in Fig. 9. In addition

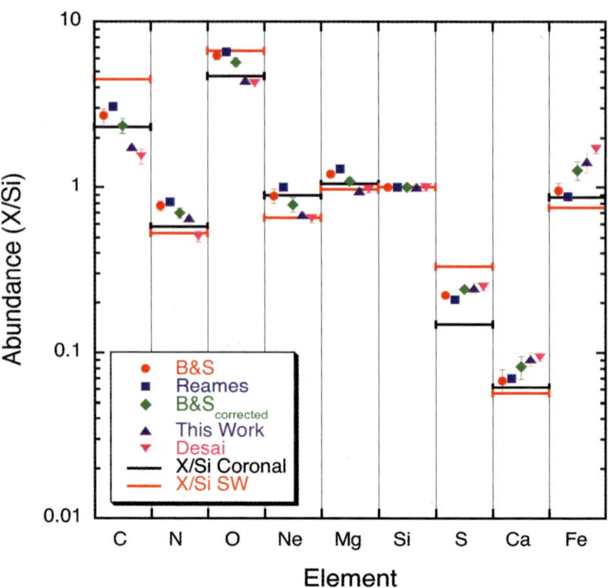


**Fig. 9** Comparison of the average composition of the studied events after spectral scaling to those obtained from previous studies of SEP events (Breneman and Stone 1985; Desai et al. 2006; Reames 1995), the quiet corona (Feldman and Widing 2003), and the slow solar wind (von Steiger et al. 2000). Note that both the coronal and solar wind abundances are variable, 10–40% for the solar wind

to the previous studies of large SEP events by Breneman and Stone (1985) (B&S), Reames (1995), and Desai et al. (2006), composition from slow solar wind (von Steiger et al. 2000) and spectroscopic coronal measurements (Feldman and Widing 2003) are also plotted. The older studies of B&S and Reames were based on data obtained near 5 MeV/nucleon (3.5–50 MeV/nucleon for B&S and 5–12 MeV/nucleon for Reames), producing consistent abundances for SEP events of the 1970s (B&S) and 1980s (Reames). More recently, Desai et al. examined data from events of the last 10 years (1997–2005), at significantly lower energies, 0.38 MeV/nucleon. These abundances are markedly different than those of the older studies, with elements heavier than Si having higher abundances (particularly Fe) and lighter elements having lower abundances. This is not surprising in light of our current analysis. Figure 2 shows that many spectral roll-offs occur near or below the energy at which the B&S and Reames studies were conducted, suggesting their abundances could be affected.

Such considerations prompted B&S to correct their average SEP abundances according to a power law in $Q/M$ in order to obtain a source composition. These values, shown in Fig. 9 as $\mathrm{B\&S}_{\mathrm{corrected}}$, fall between the B&S and Desai et al. values as expected. B&S made the correction on the average abundances, rather than for individual events, because of the limited statistical accuracy of their data. In addition to the excellent statistical accuracy of the spectra studied here, we also have spectra spanning three orders of magnitude in energy, allowing the energy dependence to be precisely determined and corrected for on an event-by-event basis. That the Desai et al. average abundance for Fe/Si is slightly higher than our average scaled abundance can be attributed to the broader SEP event sample (and therefore the inclusion of more Fe-rich events) of the Desai et al. study.

Finally the SEP results can be compared to abundances as measured in the slow solar wind and corona. These are indicated as horizontal bars in Fig. 9. Although for some elements our scaled results are in good agreement with the coronal abundances, for other elements the results are in better agreement with the solar wind. Some of our abundances agree with neither the coronal nor solar wind abundances. Assuming that the corona is source of the seed populations for the SEP events and the solar wind, it is surprising that the average coronal composition does not agree well with that of the average SEP or the average slow

solar wind. This is partly a consequence of the fact that there is substantial variability in the FIP fractionation present in the corona, solar wind, and SEP events. Although correcting our SEP results for residual $Q/M$ fractionation (Fig. 8) might improve the agreement with the coronal abundances, it is unlikely to improve the agreement with the solar wind values as the differences do not appear to be organized by $Z$ or $Q/M$. To quantitatively explain the compositional differences apparent in Fig. 9 one must understand, in detail, how the solar wind and SEP events are heated, accelerated, and transported from their coronal source(s). This paper has examined one aspect of this for SEP events and quantified that effect on the observed SEP composition.

## 6 Summary

We analyzed the spectra of heavy ions (C–Fe) in 14 large SEP events and evaluated the composition as a function of energy. Using arguments related to diffusive escape from a shock-associated acceleration region, we attempted to remove the energy dependence on an event-by-event basis. From the scaled individual results we obtained an average SEP composition which indicates the older studies of data near 5 MeV/nucleon were probably affected, somewhat, by the presence of spectral roll-offs. We also compared the SEP abundances to average slow solar wind and coronal abundances. Some of the apparent differences are undoubtedly due to the variability present in the composition of the solar wind and of different samples of the corona as well as from one SEP event to another. Our removal of the energy dependence of the SEP abundances did not fully eliminate the event-to-event compositional differences; some of which can be related to changes in the degree of FIP-fractionation, an aspect that affects the solar wind and coronal composition as well.

It should be noted that merely due to the nature of our analysis, our selection of events was not democratic and, perhaps, is not a representative sample of SEP events. There remain a number of events with energy spectra which are power laws over a large range of energies and others which are better described as broken power laws (Mewaldt et al. 2005). Some of these events have abundances relatively independent of energy (Mason et al. 1999), while others have abundances that actually increase with increasing energy (Tylka et al. 2002). Accurately interpreting and, in some cases, correcting for these characteristics to obtain a table of SEP source abundances is a large, and ongoing, task.

**Acknowledgements** This work was supported by NASA at Caltech (under grants NNG04GB55G, NNG04088G, NAG5-12929), JPL, and GSFC.

## References

H.H. Breneman, E.C. Stone, Astrophys. J. **299**, L57 (1985)
H.V. Cane, T.T. von Rosenvinge, C.M.S. Cohen, R.A. Mewaldt, Geophys. Res. Lett. **30**, SEP 5 (2003)
H.V. Cane, R.A. Mewaldt, C.M.S. Cohen, T.T. von Rosenvinge, J. Geophys. Res. Atmos. **111**, A06590 (2006). doi:10.1029/2005JA011071
C.M.S. Cohen et al., Adv. Space Res. **32**, 2649 (2003). doi:10.1016/S0273-1177(03)00901-3
C.M.S. Cohen et al., J. Geophys. Res. **110**, A09S16 (2005). doi:10.1029/2005JA011004
A.C. Cummings, E.C. Stone, W.R. Webber, Astrophys. J. **287**, L99 (1984)
M.I. Desai et al., Astrophys. J. **649**, 470 (2006). doi:10.1086/505649
D.C. Ellison, R. Ramaty, Astrophys. J. **298**, 400 (1985)
U. Feldman, Space Sci. Rev. **85**, 227 (1998). doi:10.1023/A:1005146332450
U. Feldman, K.G. Widing, Space Sci. Rev. **107**, 665 (2003). doi:10.1023/A:1026103726147
T.L. Garrard, E.C. Stone, Adv. Space Res. **14**, 589 (1994). doi:10.1016/0273-1177(94)90514-2

B. Klecker, E. Mobius, M.A. Popecki, Space Sci. Rev. **124**, 289–301 (2006). doi:10.1007/s11214-006-9111-0
A.W. Labrador, R.A. Leske, R.A. Mewaldt, E.C. Stone, T.T. von Rosenvinge, in *International Cosmic Ray Conference*, vol. 1 (2005), p. 99
M.A. Lee, Astrophys. J. Suppl. Ser. **158**, 38 (2005)
G. Li, G.P. Zank, W.K.M. Rice, J. Geophys. Res. Atmos. **110**, 06104 (2005). doi:10.1029/2004JA010600
K. Lodders, Astrophys. J. **591**, 1220 (2003). doi:10.1086/375492
G.M. Mason et al., Space Sci. Rev. **86**, 409 (1998)
G.M. Mason et al., Geophys. Res. Lett. **26**, 141 (1999)
J.E. Mazur, G.M. Mason, M.D. Looper, R.A. Leske, R.A. Mewaldt, Geophys. Res. Lett. **26**, 173 (1999)
P. Mazzotta, G. Mazzitelli, S. Colafrancesco, N. Vittorio, Astron. Astrophys. Suppl. Ser. **133**, 403 (1998)
R.A. Mewaldt et al., in *Acceleration and Transport of Energetic Particles Observed in the Heliosphere*. AIP Conf. Proc., vol. 528 (2000), p. 123
R.A. Mewaldt et al., in *International Cosmic Ray Conference*, vol. 8 (2001), p. 3132
R.A. Mewaldt et al., Adv. Space Res. **30**, 79 (2002)
R.A. Mewaldt et al., in *4th IGPP Conference*. Physics of Collisionless Shocks, vol. 781 (2005), pp. 227
R.A. Mewaldt et al., Space Sci. Rev. (2007). doi:10.1007/s11214-007-9187-1
J.P. Meyer, Astrophys. J. Suppl. Ser. **57**, 173 (1985a). doi:10.1086/191001
J.P. Meyer, Astrophys. J. Suppl. Ser. **57**, 151 (1985b). doi:10.1086/191000
C.K. Ng, D.V. Reames, A.J. Tylka, Geophys. Res. Lett. **26**, 2145 (1999)
D.V. Reames, Adv. Space Res. **15**, 41 (1995)
P.L. Slocum et al., Astrophys. J. **594**, 592 (2003)
E.C. Stone et al., Space Sci. Rev. **86**, 357 (1998)
A.J. Tylka et al., Astrophys. J. **581**, L119 (2002)
A.J. Tylka et al., Astrophys. J. **625**, 474 (2005)
R. von Steiger et al., J. Geophys. Res. **105**, 27217 (2000). doi:10.1029/1999JA000358

Space Sci Rev (2007) 130: 195–205
DOI 10.1007/s11214-007-9185-3

# Solar Isotopic Composition as Determined Using Solar Energetic Particles

**R.A. Leske · R.A. Mewaldt · C.M.S. Cohen · A.C. Cummings · E.C. Stone · M.E. Wiedenbeck · T.T. von Rosenvinge**

Received: 2 February 2007 / Accepted: 3 April 2007 /
Published online: 25 May 2007


**Abstract** Solar energetic particles (SEPs) provide a sample of the Sun from which solar composition may be determined. Using high-resolution measurements from the Solar Isotope Spectrometer (SIS) onboard NASA's Advanced Composition Explorer (ACE) spacecraft, we have studied the isotopic composition of SEPs at energies ≥20 MeV/nucleon in large SEP events. We present SEP isotope measurements of C, O, Ne, Mg, Si, S, Ar, Ca, Fe, and Ni made in 49 large events from late 1997 to the present. The isotopic composition is highly variable from one SEP event to another due to variations in seed particle composition or due to mass fractionation that occurs during the acceleration and/or transport of these particles. We show that various isotopic and elemental enhancements are correlated with each other, discuss the empirical corrections used to account for the compositional variability, and obtain estimated solar isotopic abundances. We compare the solar values and their uncertainties inferred from SEPs with solar wind and other solar system abundances and find generally good agreement.

**Keywords** Sun: abundances · Sun: particle emission · Sun: coronal mass ejections (CMEs) · Sun: flares

## 1 Introduction

The Sun's composition may be determined using samples of solar material in the form of either the solar wind or solar energetic particles (SEPs), and each approach has its own challenges. In situ solar wind composition measurements typically require detailed knowledge

R.A. Leske (✉) · R.A. Mewaldt · C.M.S. Cohen · A.C. Cummings · E.C. Stone
Space Radiation Laboratory, California Institute of Technology, Pasadena, CA 91125, USA
e-mail: ral@srl.caltech.edu

M.E. Wiedenbeck
Jet Propulsion Laboratory, California Institute of Technology, Pasadena, CA 91109, USA

T.T. von Rosenvinge
NASA/Goddard Space Flight Center, Greenbelt, MD 20771, USA

of instrument response functions, efficiencies, and backgrounds, whereas SEP abundances are often distorted by fractionation processes during particle acceleration or transport. It is therefore valuable to compare solar abundances obtained from both types of studies.

The seed material for gradual SEP events is thought to be suprathermal solar wind or coronal material, possibly supplemented by flare particles or remnant material from earlier impulsive events (Tylka et al. 2005; Cane et al. 2003), accelerated by large shocks driven by fast coronal mass ejections (Reames 1995a). After correcting measured SEP elemental abundances for fractionation associated with acceleration and transport (Breneman and Stone 1985; Garrard and Stone 1993) or averaging over many events (Reames 1995b), estimates of the underlying solar elemental composition have been obtained. Similar analyses have been used in the past to determine the solar isotopic composition for a few elements (Mewaldt and Stone 1989; Williams et al. 1998).

Our previous studies using the Solar Isotope Spectrometer (SIS) on the Advanced Composition Explorer (ACE) have found large enhancements and event-to-event variability in SEP isotopic abundance ratios (see, e.g., Leske et al. 2001b, 2003, and references therein) with good correlations between various elemental and isotopic abundances. In the present work, we report ACE/SIS isotopic abundance measurements for C, O, Ne, Mg, Si, S, Ar, Ca, Fe, and Ni in as many as 49 individual SEP events. Using the observed correlations we empirically correct for the compositional variability and derive solar isotopic abundances from SEPs.

## 2 Observations and Analysis

The SIS instrument uses the $dE/dx$ versus residual energy technique in a pair of silicon solid-state detector telescopes to obtain the nuclear charge, $Z$, mass, $M$, and total kinetic energy, $E$, for particles with energies of ~10 to ~100 MeV/nucleon (Stone et al. 1998). For this study, we examined all SEP events with high-energy heavy ion intensities large enough to yield statistically meaningful isotope abundances for at least $^{22}$Ne/$^{20}$Ne and $^{26}$Mg/$^{24}$Mg. This selection resulted in 49 large SEP events, ~50% more than were used in our earlier analysis (Leske et al. 2003). During the very highest rate periods, mass resolution is degraded by chance coincidences between heavy ions and low energy protons. Therefore, time periods near the peaks of the 10 largest events (with ~20–60 MeV/nucleon oxygen intensities exceeding ~$10^{-3}$ (cm$^2$ sr s MeV/nucleon)$^{-1}$) were not used for the isotopic analysis. Mass resolution depends on $Z$ and $E$; for the species and energies studied here it ranges from ~0.15 to ~0.3 amu, as illustrated by the well-resolved mass peaks in Fig. 1. Obtaining event-integrated isotope abundance ratios from these data is straightforward; further analysis details are given elsewhere (Leske et al. 1999a, 1999b). At present we do not attempt to isolate $^{15}$N, $^{17}$O, or $^{21}$Ne, but these species may be measurable at high energies in some events with further work.

Deriving the source composition is complicated by the event-to-event variability of SEP isotopic abundances, as is evident from the two neon mass histograms shown in Fig. 2. Note that the $^{22}$Ne/$^{20}$Ne abundance ratio, relating two species that differ by only 10% in mass, can vary by a factor of ~4 from event to event.

If there is any variation in an isotopic abundance ratio with energy, it is generally small compared with statistical uncertainties or compared with the deviation of that ratio from the solar wind value, at least over the energy interval of our measurements. Examples for the case of the $^{22}$Ne/$^{20}$Ne ratio in the two extreme events shown in Fig. 2 are illustrated in Fig. 3. Within statistical uncertainties, no energy dependence is evident in either case.

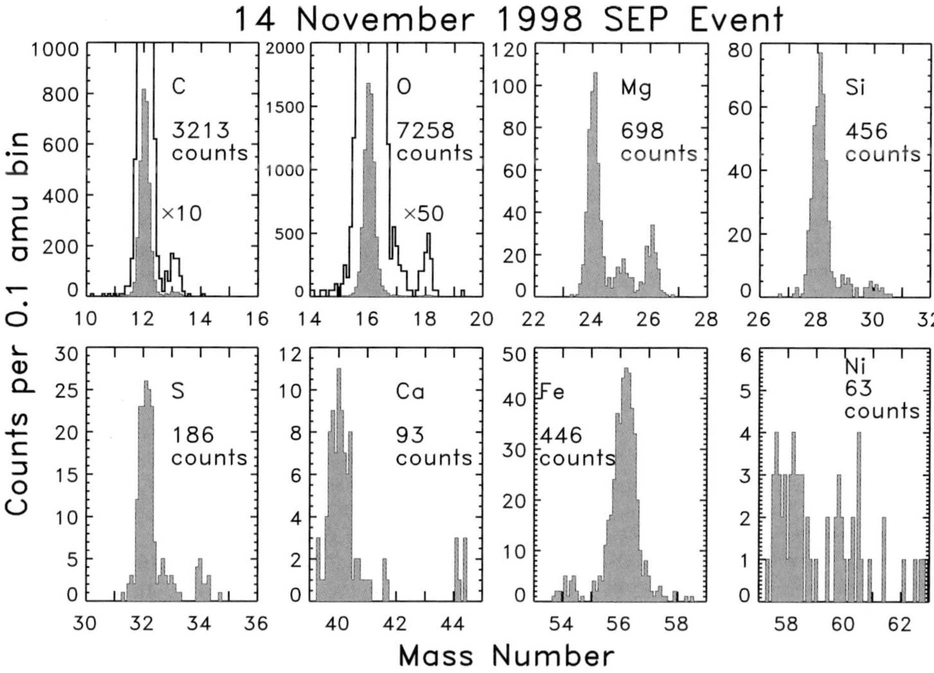


**Fig. 1** Selected mass histograms from ACE/SIS for $E \gtrsim 20$ MeV/nucleon during the 14 November 1998 SEP event. Expanded views by the factors indicated are used to improve the visibility of the $^{13}$C and $^{18}$O peaks

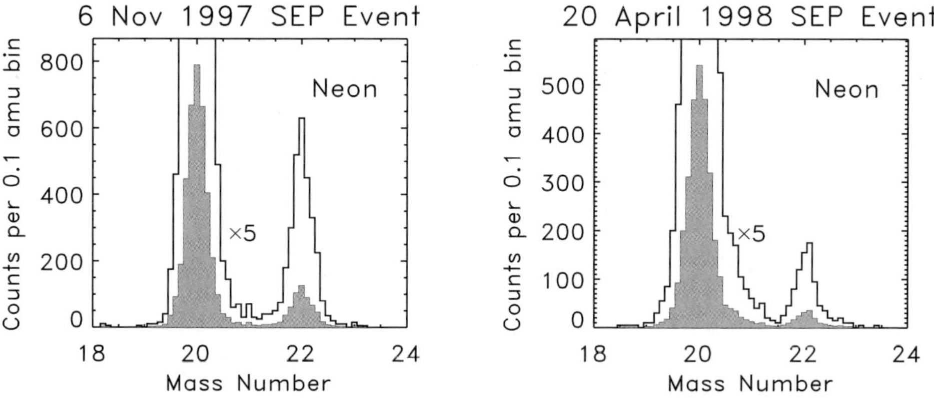


**Fig. 2** Neon mass histograms from the 6 November 1997 SEP event (*left*) and 20 April 1998 event (*right*), plotted with the $^{20}$Ne peaks at the same height. The $^{22}$Ne/$^{20}$Ne ratio is clearly different in the two events, as seen more easily by the factor of 5 expanded views

The $^{22}$Ne/$^{20}$Ne ratio for the 6 November 1997 event is significantly enhanced over the solar wind value throughout the SIS energy interval, while it is consistently somewhat depleted in the 20 April 1998 event. The Fe/O ratio in this latter event varies strongly with energy, dropping by a factor of ~50 between 10 and 40 MeV/nucleon (Tylka et al. 2000). Both the Fe and O spectra are essentially exponentials (rather than power laws) in energy per

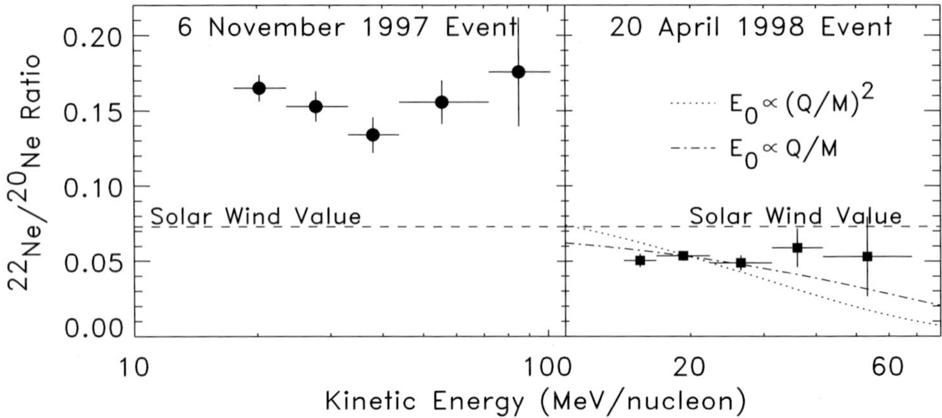


**Fig. 3** The SEP $^{22}$Ne/$^{20}$Ne isotopic abundance ratio plotted versus energy for the 6 November 1997 event (*left*) and the 20 April 1998 event (*right*). Also shown is the expected energy dependence in the 20 April 1998 event for spectra that are exponential in energy per nucleon with e-folding energies $E_0$ scaling as $(Q/M)^2$ (*dotted curve*) or $Q/M$ (*dot-dashed curve*). The curves were normalized at 20 MeV/nucleon, using a fitted $E_0$ for $^{20}$Ne of 6.3 MeV/nucleon and assuming the mean $Q$ for $^{20}$Ne is the same for $^{22}$Ne. The *horizontal dashed* line marks the solar wind $^{22}$Ne/$^{20}$Ne value

nucleon. The large energy dependence of the abundance ratios observed in such events is often due to spectra with e-folding energies which scale with the ratio of the mean ionic charge, $Q$, to mass, $M$, and $Q/M$ is lower for Fe than for O. Shock acceleration theory (Li et al. 2005) suggests these e-folding energies should scale as $(Q/M)^2$. However in this particular event, analysis of elemental spectra (Tylka et al. 2000) shows the scaling goes more nearly as $Q/M$. As shown in Fig. 3, e-folding energies scaling as $(Q/M)^2$ would result in a larger energy dependence in the $^{22}$Ne/$^{20}$Ne ratio than can be accommodated by the data, and while $Q/M$ scaling yields more energy dependence than necessary to fit the data, it is not inconsistent with the 1-$\sigma$ uncertainties. Since any energy dependence is expected to be quite small over our energy interval, for the present study we assume all isotopic abundance ratios are independent of energy.

Earlier studies (Breneman and Stone 1985) have shown that SEP heavy ion elemental abundance ratios are also quite variable from event to event and scale approximately as a power law in $Q/M$, with a power law index which itself varies from event to event. Any process which fractionates elements based on $Q/M$ must also affect isotopes with different $M$, and there should be a predictable correlation between elemental and isotopic abundances. Power-law fractionation in $Q/M$ implies that the enhancement or depletion of any measured SEP abundance ratio $X_1/X_2$, when compared with the abundance ratio of any two reference species $R_1/R_2$, should be:

$$\frac{(X_1/X_2)_{\rm SEP}}{(X_1/X_2)_{\rm solar}} = \left(\frac{(R_1/R_2)_{\rm SEP}}{(R_1/R_2)_{\rm solar}}\right)^{\frac{\ln[(Q/M)_{X_1}/(Q/M)_{X_2}]}{\ln[(Q/M)_{R_1}/(Q/M)_{R_2}]}}. \tag{1}$$

We find that isotopic abundances are indeed correlated with elemental abundances, as illustrated in Fig. 4. To compare these correlations with those expected from (1) the mean ionic charge states $Q$ of the species involved must be known. It is reasonable to assume that $\langle Q(^{22}\mathrm{Ne})\rangle = \langle Q(^{20}\mathrm{Ne})\rangle$, so the values of $Q(X_1)$ and $Q(X_2)$ in (1) factor out, but this will not be true for the charge states of Fe and O or Na and Mg used in Fig. 4. For most

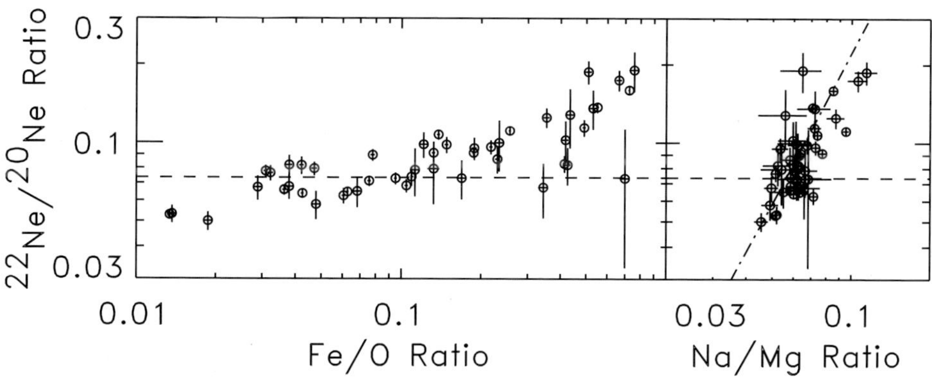


**Fig. 4** The SEP $^{22}Ne/^{20}Ne$ isotopic ratio plotted versus the Fe/O (*left*) and Na/Mg (*right*) elemental abundance ratios. The *diagonal line* shows the correlation expected from (1), assuming $\langle Q(\text{Na})\rangle = 9$ and $\langle Q(\text{Mg})\rangle = 10$. The *horizontal dashed line* marks the solar wind $^{22}Ne/^{20}Ne$ value

SEP events $Q$ is not measured at energies of tens of MeV/nucleon. Lower energy charge-state measurements may not apply as heavy elements often exhibit energy-dependent charge states (Oetliker et al. 1997; Klecker et al. 2006; Popecki 2006), while those higher energy charge state measurements that do exist show considerable event-to-event variability (see, e.g., Leske et al. 2001a and references therein; Labrador et al. 2005). Also, in the derivation of (1) it was implicitly assumed that any elemental fractionation associated with the first ionization potential (FIP) is the same magnitude in the SEP event as it is in the solar source material. However, the size of the FIP effect in SEP events (Garrard and Stone 1994; Mewaldt et al. 2000; Slocum et al. 2003), in the corona (Widing and Feldman 2001), and in solar wind (von Steiger et al. 2000) is variable, which can affect certain elemental but not isotopic ratios and may blur the expected correlations.

The $^{22}Ne/^{20}Ne$ ratio generally correlates better with Na/Mg than with Fe/O. Note that those events in Fig. 4 with $^{22}Ne/^{20}Ne$ values near that of the solar wind exhibit Fe/O ratios which span nearly 2 orders of magnitude, while nearly all their Na/Mg ratios are tightly clustered around the standard solar system value of 0.056 (Lodders 2003). Na and Mg have similar FIP values, and their charge states are much less variable than those of Fe. As noted in Cohen et al. (1999), both Na and Mg ions should have 2 electrons attached over a broad range of coronal temperatures (Arnaud and Rothenflug 1985). The dot-dashed line in Fig. 4 shows the expected correlation if $\langle Q(\text{Na})\rangle = 9$ and $\langle Q(\text{Mg})\rangle = 10$. This very simple model provides a good first order fit to most of the data.

The predicted correlations are very sensitive to $Q/M$ for reference ratios involving similar values of $Q/M$; changing $Q(\text{Na})/Q(\text{Mg})$ or the mean mass of Mg by only 1% changes the expected slope by ~20% (Leske et al. 2001b). Even if $Q$ could be measured to this accuracy at SIS energies, it is $Q$ at the time of fractionation that matters, which may differ from that at 1 AU if fractionation happens early and if stripping occurs in acceleration through the corona (see, e.g., Barghouty and Mewaldt 2000). The sensitivity of the correlations to variability in the Na and Mg charge states or mean Mg mass may account for the scatter remaining in the right panel of Fig. 4.

Since the mean $Q$ should be the same for isotopes of the same element, if we use an isotope ratio as the reference value in (1) the charge states factor out and also the mass is fixed, so we might expect a tighter correlation. This is illustrated by the plot of $^{22}Ne/^{20}Ne$ versus $^{26}Mg/^{24}Mg$ in Fig. 5. Most of the events with small uncertainties agree very well with

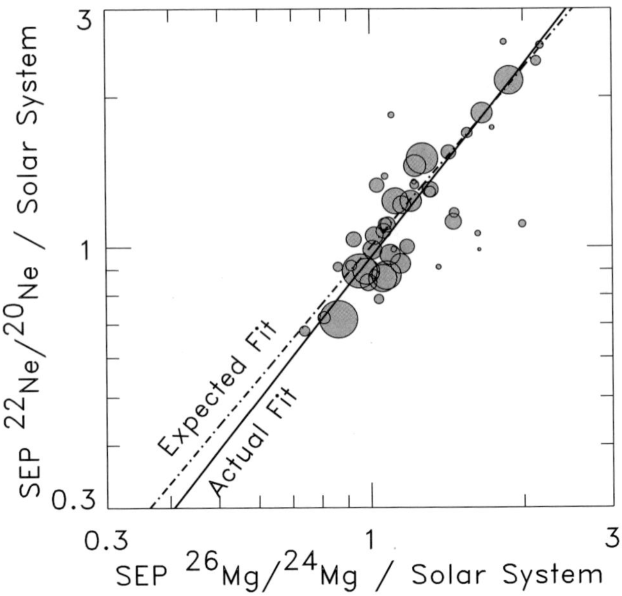

**Fig. 5** The $^{22}Ne/^{20}Ne$ versus $^{26}Mg/^{24}Mg$ isotopic ratios in each of the 49 SEP events, normalized to standard solar system values (Lodders 2003). The correlation expected using (1) and the actual fit are also shown. The *symbol* diameters scale inversely with the size of the uncertainties (so larger points have a greater significance)

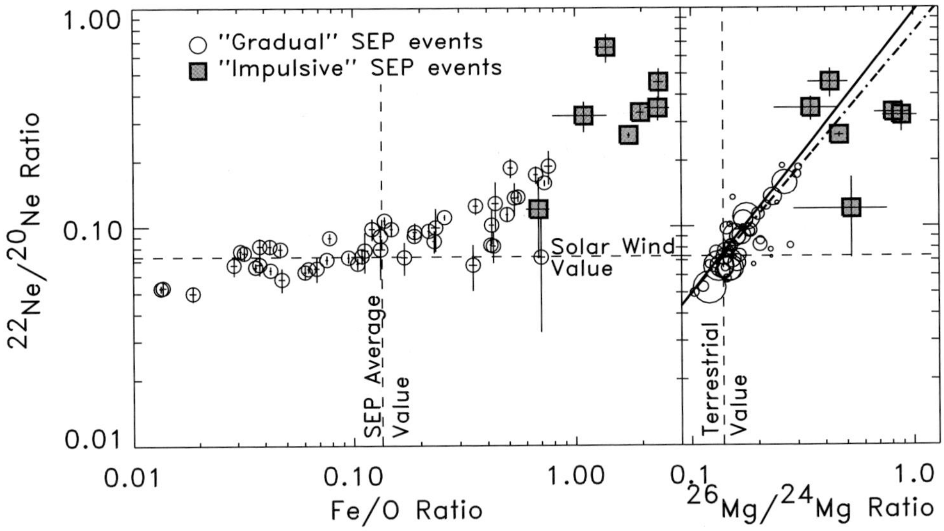

**Fig. 6** The SEP $^{22}Ne/^{20}Ne$ isotopic ratio plotted versus the Fe/O (*left*) and $^{26}Mg/^{24}Mg$ (*right*) abundance ratios for both gradual and impulsive SEP events. *Diagonal lines* are the same as in Fig. 5

the expected line, and the actual fit differs from the expected correlation by no more than ~7% over the entire range of measured values.

The few impulsive (i.e., $^{3}He$-rich) events we have observed at energies of tens of MeV/nucleon with good heavy-ion statistics tend to be extremely fractionated and variable, with $^{22}Ne/^{20}Ne$ as much as 9 times the solar wind value and $^{26}Mg/^{24}Mg$ enhanced by more than 6 times the terrestrial value. Such extreme isotopic fractionation in impulsive events has also been reported at lower energies (Mason et al. 1994; Dwyer et al. 2001). In Fig. 6, nearly all the impulsive events shown have an Fe/O ratio higher than in gradual events, as has long been known (e.g., Reames et al. 1994). The very tight correlation between $^{22}$Ne/$^{20}$Ne and $^{26}$Mg/$^{24}$Mg (Fig. 5) seen in gradual events seems absent from impulsive events, with individual events scattered about, but not on, the correlation line. Peculiar abundances in individual impulsive events that are not simply correlated with $Q/M$ may be due to resonant plasma heating of the source material (see, e.g., Mason et al. 2002 and references therein). However, since Fe-rich (and therefore from Fig. 4 $^{22}$Ne/$^{20}$Ne-rich) gradual events are thought to be contaminated by or even dominated by impulsive flare material (Tylka et al. 2005; Cane et al. 2003), one might expect both event classes to show similar fractionation patterns. Perhaps the very large impulsive events we observe with SIS are compositionally peculiar compared to a typical impulsive event. Or the scatter in individual impulsive event composition may suggest that the impulsive contribution to gradual events represents some sort of average impulsive material—either a temporal average over several impulsive events (Tylka et al. 2005) or a spatial average resulting from a large shock passing through spatially inhomogeneous flare material.

## 3 Results

The SEP abundance values for 11 isotope ratios for elements from C to Ni are shown plotted versus $^{26}$Mg/$^{24}$Mg in Fig. 7, with the correlations expected from (1) indicated by diagonal lines. The data roughly agree with the expected trends for Ne, Mg, and Si. Elements heavier than Si tend to have fewer data points and lower statistical significance, and the correlation

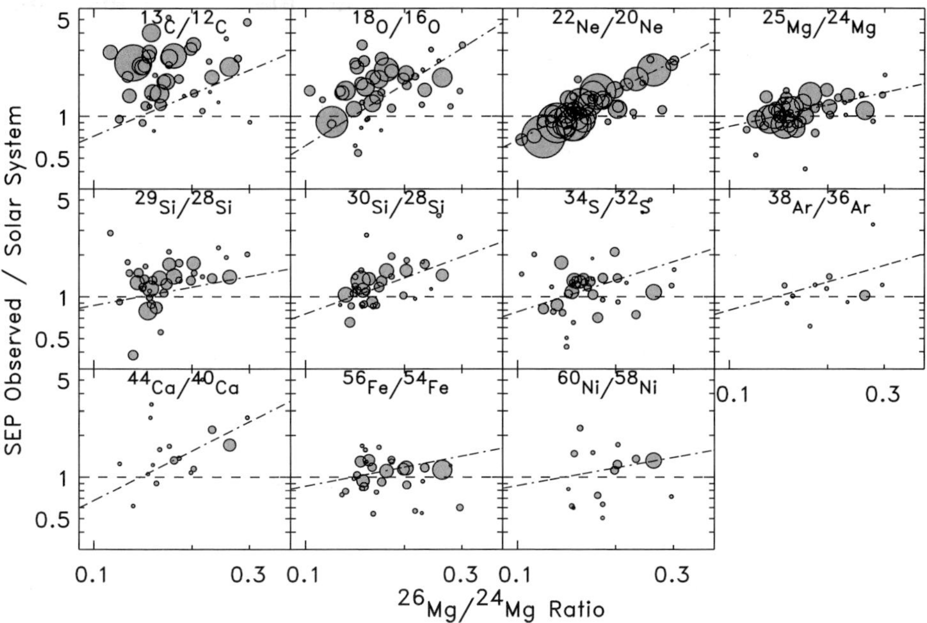


**Fig. 7** Eleven SEP isotope abundance ratios (normalized to standard solar system abundances, Lodders, 2003) in up to 49 SEP events plotted versus the $^{26}$Mg/$^{24}$Mg ratio. *Diagonal lines* show the correlations expected using (1), where the different slopes arise from the different mass ratios. *Symbol* diameters scale inversely with the size of the uncertainties

between the actual fractionation and the expected trends is less clear. The correlations may break down for species with $Q/M$ far from that of the Mg reference ratio if the actual dependence on $Q/M$ is not a simple power law as we assumed. At lower $Z$, $^{13}$C is often enhanced or $^{12}$C is depleted in SEP events relative to terrestrial abundances, but the reason for this is not understood. Higher $^{13}$C values do not appear to be due to spillover from $^{12}$C, as the two mass peaks are generally well separated (Leske et al. 1999a).

Using $^{26}$Mg/$^{24}$Mg as the reference ratio in (1) with its solar value taken to be the terrestrial value (Lodders 2003), and assuming $Q$ is the same for different isotopes of the same element, we obtain the solar isotope ratios for each SEP event. Averaging over all the SIS measurements for each isotope ratio results in the SEP-derived solar values shown in Fig. 8. We also show the weighted average without correcting for fractiona-

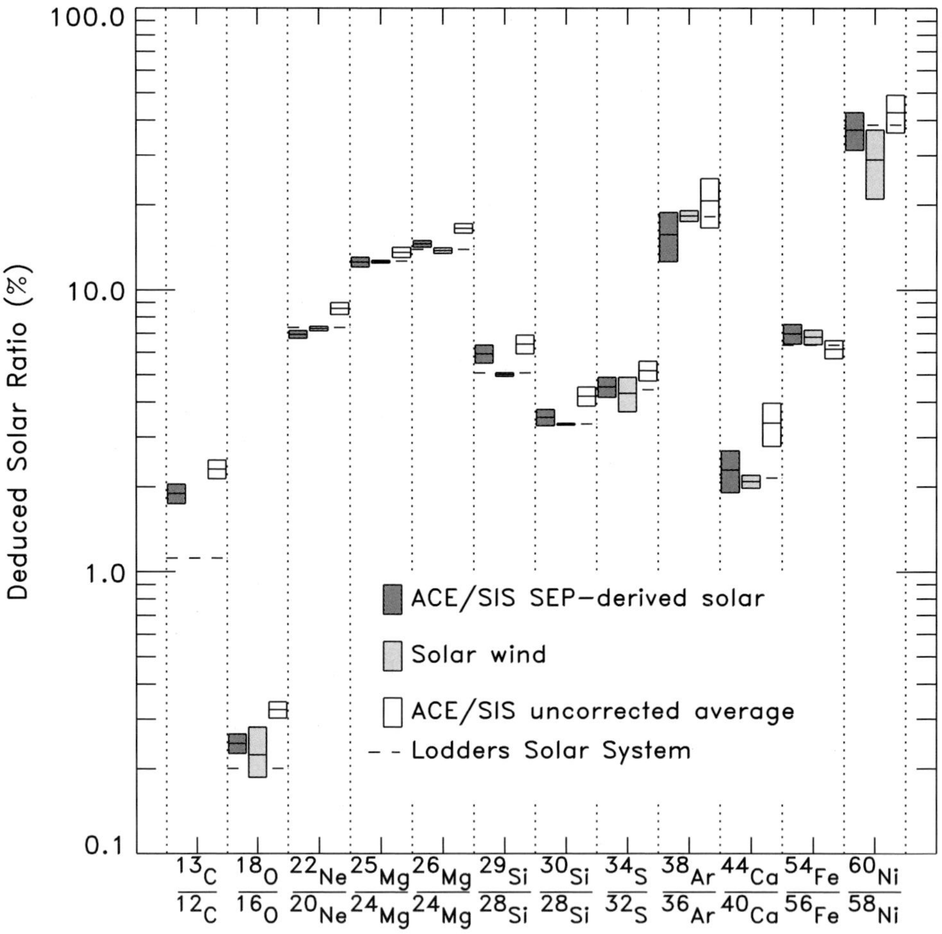

**Fig. 8** Average solar isotopic abundance ratios from SIS SEP measurements after correcting for fractionation (*dark grey boxes*). The $^{26}$Mg/$^{24}$Mg ratio served as the reference value in (1) for everything other than $^{26}$Mg/$^{24}$Mg, for which $^{22}$Ne/$^{20}$Ne was used. For comparison, averages without fractionation corrections (*open boxes*), standard solar system values (*dashed lines*; Lodders 2003) and measured solar wind values (*light grey boxes*; Kallenbach 2001; Wimmer-Schweingruber et al. 2001; Ipavich et al. 2001; Wimmer-Schweingruber 2002; Karrer et al. 2007) are shown

tion, which may be more appropriate for cases where the data may not follow the expected fractionation correlations or for studies more interested in arriving fluences of various species rather than source composition. For comparison Fig. 8 also shows standard solar system values (Lodders 2003) and existing solar wind values (Kallenbach 2001; Wimmer-Schweingruber et al. 2001; Ipavich et al. 2001; Wimmer-Schweingruber 2002; Karrer et al. 2007). As new results from Genesis become available the solar wind values may change, although preliminary results for the $^{22}Ne/^{20}Ne$ ratio (Grimberg et al. 2007) appear to be consistent with the solar wind values plotted here. Enhancement factors over standard abundances for our SEP-derived solar values and for solar wind values are shown in Fig. 9.

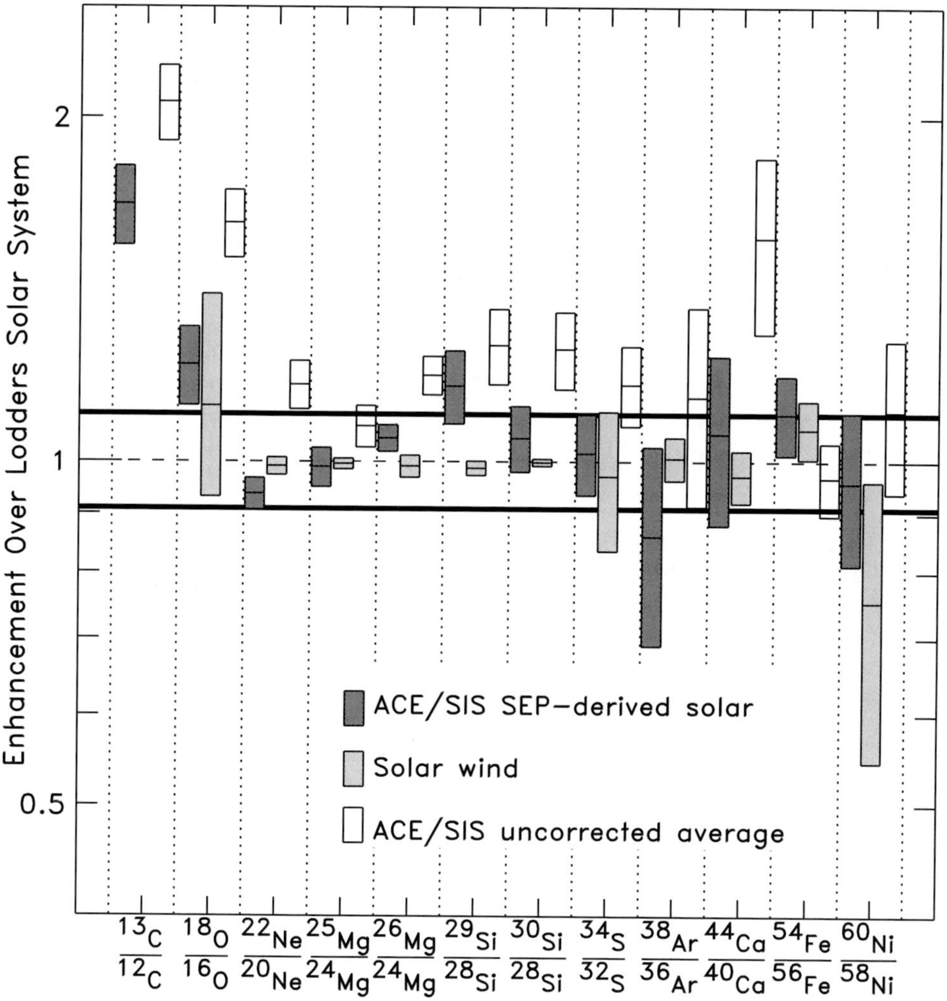

**Fig. 9** Ratio of deduced source values (see Fig. 8) to the standard solar system values (Lodders 2003). SEP measurements after correcting for fractionation (*dark grey boxes*), SEP averages without fractionation corrections (*open boxes*), and measured solar wind values (*light grey boxes*; Kallenbach 2001; Wimmer-Schweingruber et al. 2001; Ipavich et al. 2001; Wimmer-Schweingruber 2002; Karrer et al. 2007) are shown. *Solid horizontal lines* indicate ±10% deviation from the standard solar system values

Although our fractionation correction makes simplifying assumptions, the resulting solar abundances mostly appear to be quite reasonable. All but 2 values are within $2\sigma$ of the standard "solar system" values (Lodders 2003) (which are actually terrestrial values except for Ne and Ar, for which solar wind values were used) and most are within 10% of the standard values (Fig. 9). Our results also agree well with the average values shown for the solar wind, which can be compositionally quite variable (Raymond et al. 2001). Performing our analysis on subsets of the SEP events selected on the basis of iron enrichment or spectral shape (as done for elemental abundances in Cohen et al. 2007) naturally increases the uncertainties on our derived solar abundances, but does not significantly change their values. In particular, our $^{13}C$ measurements suggest that some additional process affects fractionation of this isotope or that fractionation is more than just a simple power law for all $Q/M$. For several ratios, uncertainties on our SEP-derived solar abundances are comparable to or better than those presently obtained from the solar wind. Additional SEP events may reduce the uncertainties for the heaviest species where there are still only a few measurements, but a better theoretical understanding of the mass fractionation process would allow much further progress. Continued study of puzzles such as the frequent enhancement of $^{13}C$ may provide clues to the nature of the fractionation process.

**Acknowledgements** This work was supported by NASA at the California Institute of Technology (grant NAG5-12929), the Jet Propulsion Laboratory, and the Goddard Space Flight Center.

## References

M. Arnaud, R. Rothenflug, Astron. Astrophys. Suppl. **60**, 425 (1985)

A.F. Barghouty, R.A. Mewaldt, in *AIP Conf. Proc. 528: Acceleration and Transport of Energetic Particles Observed in the Heliosphere*, ed. by R.A. Mewaldt et al. (2000), p. 71

H.H. Breneman, E.C. Stone, Astrophys. J. Lett. **299**, L57 (1985)

H.V. Cane, T.T. von Rosenvinge, C.M.S. Cohen, R.A. Mewaldt, Geophys. Res. Lett. **30** (2003). doi:10.1029/2002GL016580

C.M.S. Cohen, R.A. Mewaldt, R.A. Leske, A.C. Cummings, E.C. Stone, M.E. Wiedenbeck, E.R. Christian, T.T. von Rosenvinge, Geophys. Res. Lett. **26**, 2697 (1999)

C.M.S. Cohen, R.A. Mewaldt, R.A. Leske, A.C. Cummings, E.C. Stone, M.E. Wiedenbeck, T.T. von Rosenvinge, G.M. Mason, Space Sci. Rev. (2007, this volume)

J.R. Dwyer, G.M. Mason, J.E. Mazur, R.E. Gold, S.M. Krimigis, E. Möbius, M. Popecki, Astrophys. J. **563**, 403 (2001)

T.L. Garrard, E.C. Stone, in *Proc. 23rd Internat. Cosmic Ray Conf.* (Calgary), vol. 3 (1993), p. 384

T.L. Garrard, E.C. Stone, Adv. Space Res. **14**(10), 589 (1994)

A. Grimberg, D.S. Burnett, F. Bühler, P. Bochsler, V.S. Heber, H. Baur, R. Wieler, Space Sci. Rev. (2007, this volume)

F.M. Ipavich, J.A. Paquette, P. Bochsler, S.E. Lasley, P. Wurz, in *AIP Conf. Proc. 598: Solar and Galactic Composition*, ed. by R.F. Wimmer-Schweingruber (2001), p. 121

R. Kallenbach, in *AIP Conf. Proc. 598: Solar and Galactic Composition*, ed. by R.F. Wimmer-Schweingruber (2001), p. 113

R. Karrer, P. Bochsler, C. Giammanco, F. Ipavich, J. Paquette, P. Wurz, Space Sci. Rev. (2007, this volume)

B. Klecker, E. Möbius, M.A. Popecki, L.M. Kistler, H. Kucharek, M. Hilchenbach, Adv. Space Res. **38**, 493 (2006)

A.W. Labrador, R.A. Leske, R.A. Mewaldt, E.C. Stone, T.T. von Rosenvinge, *Proc. 29th Internat. Cosmic Ray Conf.* (Pune), vol. 1 (2005), p. 99

R.A. Leske, C.M.S. Cohen, A.C. Cummings, R.A. Mewaldt, E.C. Stone, B.L. Dougherty, M.E. Wiedenbeck, E.R. Christian, T.T. von Rosenvinge, Geophys. Res. Lett. **26**, 153 (1999a)

R.A. Leske, R.A. Mewaldt, C.M.S. Cohen, A.C. Cummings, E.C. Stone, M.E. Wiedenbeck, E.R. Christian, T.T. von Rosenvinge, Geophys. Res. Lett. **26**, 2693 (1999b)

R.A. Leske, R.A. Mewaldt, A.C. Cummings, E.C. Stone, T.T. von Rosenvinge, in *AIP Conf. Proc. 598: Solar and Galactic Composition*, ed. by R.F. Wimmer-Schweingruber (2001a), p. 171

R.A. Leske, R.A. Mewaldt, C.M.S. Cohen, E.R. Christian, A.C. Cummings, P.L. Slocum, E.C. Stone, T.T. von Rosenvinge, M.E. Wiedenbeck, in *AIP Conf. Proc. 598: Solar and Galactic Composition*, ed. by R.F. Wimmer-Schweingruber (2001b), p. 127

R.A. Leske, R.A. Mewaldt, C.M.S. Cohen, E.R. Christian, A.C. Cummings, P.L. Slocum, E.C. Stone, T.T. von Rosenvinge, M.E. Wiedenbeck, in *AIP Conf. Proc. 679: Solar Wind Ten*, ed. by M. Velli, R. Bruno, F. Malara (2003), p. 616

G. Li, G.P. Zank, W.K.M. Rice, J. Geophys. Res. **110**, A06104 (2005). doi:10.1029/2004JA010600

K. Lodders, Astrophys. J. **591**, 1220 (2003)

G.M. Mason, J.E. Mazur, D.C. Hamilton, Astrophys. J. **425**, 843 (1994)

G.M. Mason, J.E. Mazur, J.R. Dwyer, Astrophys. J. Lett. **565**, L51 (2002)

R.A. Mewaldt, E.C. Stone, Astrophys. J. **337**, 959 (1989)

R.A. Mewaldt, C.M.S. Cohen, R.A. Leske, E.R. Christian, A.C. Cummings, P.L. Slocum, E.C. Stone, T.T. von Rosenvinge, M.E. Wiedenbeck, in *AIP Conf. Proc. 528: Acceleration and Transport of Energetic Particles Observed in the Heliosphere*, ed. by R.A. Mewaldt et al. (2000), p. 123

M. Oetliker, B. Klecker, D. Hovestadt, G.M. Mason, J.E. Mazur, R.A. Leske, R.A. Mewaldt, J.B. Blake, M.D. Looper, Astrophys. J. **477**, 495 (1997)

M.A. Popecki, in *Geophys. Mono. 165: Solar Eruptions and Energetic Particles*, ed. by N. Gopalswamy, R.A. Mewaldt, J. Torsti (2006), p. 127

J.C. Raymond, J.E. Mazur, F. Allegrini, E. Antonucci, G. Del Zanna, S. Giordano, G. Ho, Y.-K. Ko, E. Landi, A. Lazarus et. al, in *AIP Conf. Proc. 598: Solar and Galactic Composition*, ed. by R.F. Wimmer-Schweingruber (2001), p. 49

D.V. Reames, Rev. Geophys. **33**, 585 (1995a)

D.V. Reames, Adv. Space Res. **15**(7), 41 (1995b)

D.V. Reames, J.P. Meyer, T.T. von Rosenvinge, Astrophys. J. Suppl. **90**, 649 (1994)

P.L. Slocum, E.C. Stone, R.A. Leske, E.R. Christian, C.M.S. Cohen, A.C. Cummings, M.I. Desai, J.R. Dwyer, G.M. Mason, J.E. Mazur, R.A. Mewaldt, T.T. von Rosenvinge, M.E. Wiedenbeck, Astrophys. J. **594**, 592 (2003)

E.C. Stone et al., Space Sci. Rev. **86**, 357 (1998)

A.J. Tylka, P.R. Boberg, R.E. McGuire, C.K. Ng, D.V. Reames, in *AIP Conf. Proc. 528: Acceleration and Transport of Energetic Particles Observed in the Heliosphere*, ed. by R.A. Mewaldt et al. (2000), p. 147

A.J. Tylka, C.M.S. Cohen, W.F. Dietrich, M.A. Lee, C.G. Maclennan, R.A. Mewaldt, C.K. Ng, D.V. Reames, Astrophys. J. **625**, 474 (2005)

R. von Steiger, N.A. Schwadron, L.A. Fisk, J. Geiss, G. Gloeckler, S. Hefti, B. Wilken, R.F. Wimmer-Schweingruber, T.H. Zurbuchen, J. Geophys. Res. **105**, 27217 (2000)

K.G. Widing, U. Feldman, Astrophys. J. **555**, 426 (2001)

D.L. Williams, R.A. Leske, R.A. Mewaldt, E.C. Stone, Space Sci. Rev. **85**, 379 (1998)

R.F. Wimmer-Schweingruber, Adv. Space Res. **30**(1), 23 (2002)

R.F. Wimmer-Schweingruber, P. Bochsler, G. Gloeckler, Geophys. Res. Lett. **28**, 2763 (2001)

Space Sci Rev (2007) 130: 207–219
DOI 10.1007/s11214-007-9187-1

# On the Differences in Composition between Solar Energetic Particles and Solar Wind

**R.A. Mewaldt · C.M.S. Cohen · G.M. Mason · A.C. Cummings · M.I. Desai · R.A. Leske · J. Raines · E.C. Stone · M.E. Wiedenbeck · T.T. von Rosenvinge · T.H. Zurbuchen**

Received: 5 February 2007 / Accepted: 4 April 2007 /
Published online: 26 May 2007


**Abstract** Although the average composition of solar energetic particles (SEPs) and the bulk solar wind are similar in a number of ways, there are key differences which imply that solar wind is not the principal seed population for SEPs accelerated by coronal mass ejection (CME) driven shocks. This paper reviews these composition differences and considers the composition of other possible seed populations, including coronal material, impulsive flare material, and interplanetary CME material.

**Keywords** Sun: particle emission · Sun: solar wind · Sun: coronal mass ejections (CMEs) · Sun: abundances · Sun: flares · Acceleration of particles

## 1 Introduction

As solar cycle 23 began, two classes of solar energetic particle (SEP) events were recognized that could be distinguished by their composition, spatial, and temporal characteristics,

R.A. Mewaldt (✉) · C.M.S. Cohen · A.C. Cummings · R.A. Leske · E.C. Stone
California Institute of Technology, Pasadena, CA 91125, USA
e-mail: rmewaldt@srl.caltech.edu

G.M. Mason
Applied Physics Laboratory, Johns Hopkins University, Laurel, MD 20723, USA

M.I. Desai
Southwest Research Institute, San Antonio, TX 78238, USA

J. Raines · T.H. Zurbuchen
University of Michigan, Ann Arbor, MI 48109, USA

M.E. Wiedenbeck
Jet Propulsion Laboratory, Pasadena, CA 91109, USA

T.T. von Rosenvinge
NASA/Goddard Space Flight Center, Greenbelt, MD 20771, USA

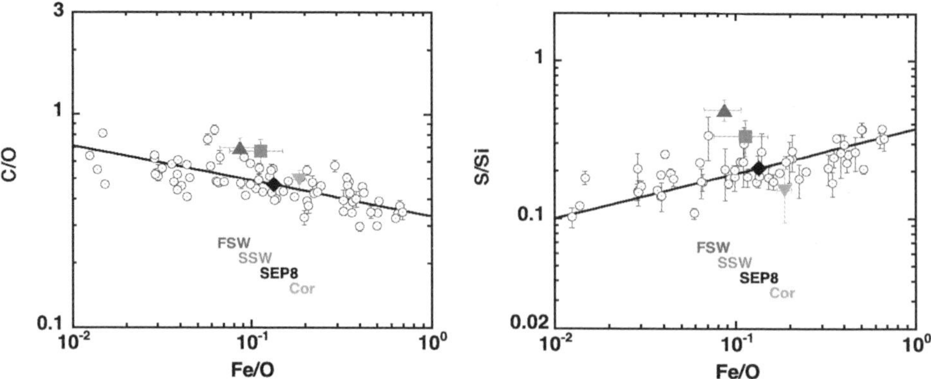


**Fig. 1** Plot of C/O versus Fe/O (*left*) and S/Si versus Fe/O (*right*) illustrating how $Q/M$-dependent fractionation leads to correlated abundance variations. Data are 12–30 MeV/nuc SEP abundances observed in gradual SEP events by the ACE/SIS instrument. Also shown are average abundances for the slow and fast solar wind (SSW and FSW; von Steiger et al. 2000), 5–12 MeV/nuc SEPs (SEP8; Reames 1995), and the corona (Cor, Feldman and Widing 2003). The *solid black lines* show the expected trends if $Q/M$-dependent fractionation follows a power-law in $Q/M$, as discussed in the text

and by associated solar phenomena (e.g., Reames 1999). Impulsive events were generally smaller events associated with impulsive X-ray flares that were enriched in $^3$He, heavy elements such as Fe, and electrons. The mean charge states were high (e.g., $Fe^{20+}$) suggesting that flare-heated material was being accelerated or that electron stripping was taking place (Klecker et al. 2006). Gradual events were larger events with a more normal composition thought to be due to the acceleration of coronal material and solar wind by shocks driven by fast coronal mass ejections (CMEs). The mean charge states in gradual events (e.g., $Fe^{10+}$ to $Fe^{15+}$) indicated coronal temperatures.

Although it has long been known that there is considerable variation in the composition of SEP events, two organizing principles were identified by Breneman and Stone (1985). They found that the most important factor relating the SEP abundance variations was fractionation that depended on the charge-to-mass ratio ($Q/M$). This is illustrated in Fig. 1, which shows abundance data for SEP events observed by the Solar Isotope Spectrometer (SIS) on ACE from 1997–2005. Note that variations in both C/O and S/Si are correlated with the Fe/O variations. Breneman and Stone showed that $Q/M$-dependent fractionation effects could be represented as a power-law in $Q/M$. Assuming average 5–12 MeV/nuc SEP abundances from Reames (1995) and mean SEP charge states measured by Luhn et al. (1984) at $\sim$1 MeV/nuc, the allowed C/O, Si/S, and Fe/O combinations can be found by varying the power-law index, which leads to the solid lines in Fig. 1. A similar fractionation pattern results using 20 MeV/nuc charge-state measurements (Leske et al. 1995). Some of the scatter in Fig. 1 is undoubtedly because the mean charge states often differ from these average values (see, e.g., Labrador et al. 2005; Popecki 2006). Note that while the C/O and S/Si ratios in some SEP events overlap with the solar wind values, for SEP events with Fe/O ratios similar to those in the average solar wind the C/O and S/Si ratios are significantly lower than the corresponding solar wind values.

Breneman and Stone's second organizing factor was the "FIP fractionation factor" first recognized by Hovestadt (1974). When compared to photospheric abundances, the SEP abundances of elements with first ionization potential (FIP) $>$ 10 eV are depleted relative to those with lower FIP values. When compared to earlier photospheric compilations (e.g.,

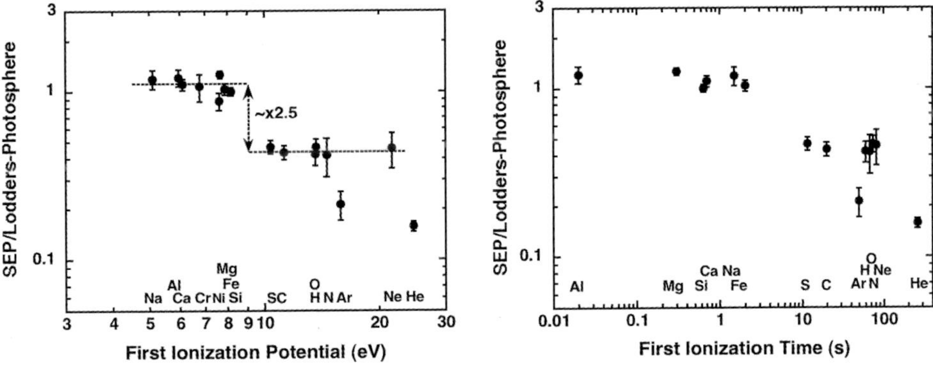


**Fig. 2** Ratio of the average 5 to 12 MeV/nuc SEP abundances of Reames (1995) to the photospheric abundances of Lodders (2003), plotted versus first ionization potential (*left panel*; adapted from Mewaldt et al. 2006) and against first ionization time (*right panel*). The FIT values are from Geiss (1998), supplemented by values from Marsch et al. (1995)

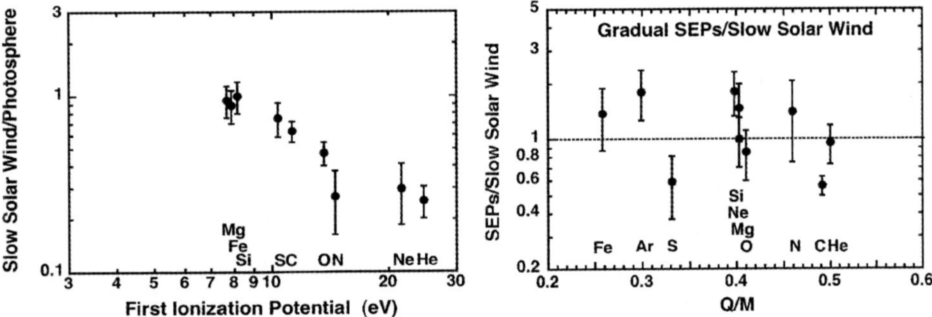


**Fig. 3** *Left panel*: Ratio of the average slow solar wind abundances (von Steiger et al. 2000) to the photospheric abundances of Lodders (2003), plotted versus first ionization potential. *Right panel*: Ratio of average 5–12 MeV/nuc SEP abundances (Reames 1995) to slow-wind abundances plotted versus $Q/M$. Uncertainties on the SEP, solar wind, and photospheric abundances are all included (figures from Mewaldt et al. 2006)

Anders and Grevesse 1989) the depletion factor is ~3.5 to 4 (e.g., Mewaldt et al. 2002); with the revised photospheric abundances of C, N, O (e.g., Lodders 2003) this factor is reduced to ~2.5 (see Fig. 2). Also shown are these abundance ratios plotted versus "first ionization time" (FIT), which may be a more relevant atomic parameter (Geiss and Bochsler 1985; Geiss 1998).

Solar wind abundances also show a fractionation pattern organized by FIP (von Steiger et al. 2000), or possibly by FIT. However, the solar wind fractionation pattern (Fig. 3) differs from the SEP pattern. There is considerable scatter in the ratio of the SEP and SW abundances (Fig. 3) that does not seem to be organized by either FIP or $Q/M$, or any other simple combination of nuclear charge ($Z$), mass ($M$), or charge state ($Q$) (Mewaldt et al. 2001a, 2006). In particular, there are several element ratios that are distinctly different in the solar wind and in SEPs, including C/O and Ne/O (Table 1). On the basis of this comparison Mewaldt et al. (2001a, 2002) concluded that SEPs in gradual events are not just an accelerated sample of bulk solar wind—they must routinely also contain major contributions from other seed-particle populations.

**Table 1** Solar wind and SEP abundance ratios

| Ratio | Slow solar wind (von Steiger et al. 2000) | Solar particles (Reames 1995) |
|---|---|---|
| C/O | $0.67 \pm 0.07$ | $0.46 \pm 0.01$ |
| Ne/O | $0.10 \pm 0.03$ | $0.15 \pm 0.01$ |

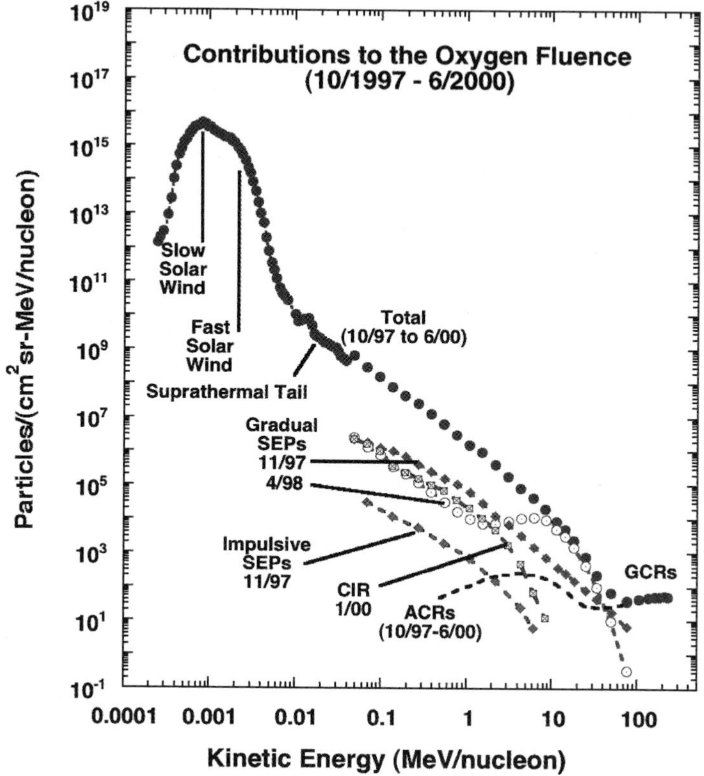


**Fig. 4** Fluence of energetic nuclei measured from October 1997 through June 2000 by four instruments on ACE (Mewaldt et al. 2001b). Also shown are the contributions from typical solar and interplanetary events and from anomalous and galactic cosmic rays

It is likely that most particles accelerated to high energy in SEP events come from a pool of suprathermal particles originating from a variety of sources, including small impulsive SEP events, previous gradual events, corotating interactions regions (CIRs), and suprathermal tails on the solar wind (e.g., Mason 2001). Figure 4 shows the fluence of energetic oxygen nuclei from solar wind to cosmic ray energies observed over a 2.75-year period. Also indicated are the contributions of fast and slow solar wind, examples of SEP and other interplanetary events, and anomalous and galactic cosmic rays. There is considerable controversy about the relative contribution of these seed populations to SEPs, and how they are accelerated. In this paper we review solar wind and SEP composition differences and discuss how the suggested seed populations might explain these differences. We restrict our attention to SEP abundances based on event fluences in order to minimize differences that occur early in some events due to particle transport.

## 2 Assessing the Contribution of Flare Material

With the advent of higher resolution SEP observations during solar cycle 23 it became apparent that many events did not fit the simple two-class picture described above. First, many large events that appeared to be of gradual origin had $^3$He/$^4$He ratios well in excess of the solar wind abundance ratio, along with overabundances of heavy elements such as Fe (Cohen et al. 1999; Mason et al. 1999; Mewaldt 2000). In addition, in some events the mean ionic charge states increased with energy (Mazur et al. 1999; Moebius et al. 1999; Labrador et al. 2005). These events appeared to contain a (sometimes energy-dependent) mixture of flare and shock-accelerated material, prompting the name "hybrid events". Cliver (1996) used this term in a similar context.

There are several suggested explanations for the presence of flare material in gradual events. Mason et al. (1999) suggested that hybrid events result when the CME-driven shock accelerates remnant suprathermal ions from earlier impulsive events (see also Desai et al. 2006a, 2007). Such ions should be preferentially accelerated over thermal ions because of their higher energy. Tylka et al. (2005) extended this idea by suggesting that shock geometry plays a key role in defining the seed population—in their model the required injection energy threshold is higher at quasi-perpendicular shocks, favoring suprathermal ions, while at quasi-parallel shocks a lower injection threshold allows more solar-wind ions to be accelerated (see also Lee and Tylka 2007). Cane et al. (2003, 2006) suggested that $^3$He, Fe, and high-charge-state enrichments occur in events in which the intensity of flare-accelerated ions escaping the Sun is comparable to or exceeds the intensity of shock-accelerated ions. This occurs most easily in magnetically well-connected events. It is also possible that CME-driven shocks can reaccelerate flare material from the same event (Kocharov and Torsti 2003; Li and Zank 2005).

The isotope $^3$He is the best indicator of flare material. Enrichments of $^3$He in gradual events by $> \times 10$ are common at 10 MeV/nuc and above (Cohen et al. 1999; Wiedenbeck et al. 2000; Torsti et al. 2003) and also below 1 MeV/nuc (Mason et al. 1999; Desai et al. 2006a). Assuming the $^3$He and $^4$He shock-acceleration efficiencies are similar, the $^3$He/$^4$He ratio can provide a measure of the contribution of flare-material. During moderately quiet times the typical $^3$He/$^4$He ratio in the suprathermal energy range (presumably due to flare material) is 2–5% (Desai et al. 2006b). The left panel of Fig. 5 shows the $\sim$10 MeV/nuc $^3$He/$^4$He ratio in 18 gradual events along with upper limits when a finite $^3$He fluence was not identified. Only a small fraction of the events are consistent with the quiet-time suprathermal ratio of 0.02 to 0.05, suggesting that in most gradual events the majority of $\sim$10 MeV/nuc He comes from a source other than remnant suprathermals. Taking into account the upper limits, we find that $\sim$20% of $\sim$10 MeV/nuc $^4$He originates from remnant flare material, on average. The same restriction does not apply to directly-accelerated flare particles or to material from the associated flare that is accelerated by the shock because it is known that the average $^3$He/$^4$He ratio is reduced in larger impulsive events (Reames et al. 1994).

Another indicator of flare-accelerated particles is the Ne/O ratio (Reames et al. 1994). Figure 5 (right panel) shows Ne/O versus Fe/O for both impulsive and gradual events with 10–30 MeV/nuc. The average for these impulsive events is Ne/O = 0.48, compared to an average of 0.18 in the gradual events. Note that for events enriched in Fe (e.g., Fe/O > 0.4) Ne is typically also enriched. The solid line shows a simple calculation in which material

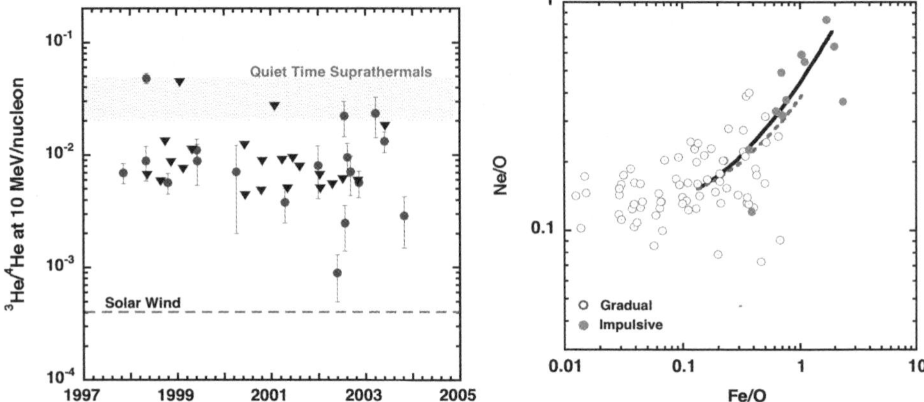


**Fig. 5** The *left panel* shows $^3$He/$^4$He ratios (*circles*) and upper limits (*triangles*) for 5 to 14 MeV/nuc He measured in gradual SEP events (see Cohen et al. 1999; Wiedenbeck et al. 2000; and Cane et al. 2003). Also indicated are the average solar-wind $^3$He/$^4$He ratio (Gloeckler and Geiss 1998) and the typical range of suprathermal $^3$He/$^4$He during solar quiet times (Desai et al. 2006b). The *right panel* shows 10 to 30 MeV/nuc Ne/O vs Fe/O in gradual and impulsive SEP events (separated on the basis of $^3$He). The *solid and dashed lines* indicate mixtures of gradual and flare material as described in the text

with Fe/O = 0.134 and Ne/O = 0.152 (Reames 1995) is mixed with "flare" material having Fe/O = 1.8 and Ne/O = 0.8 (selected to match the extreme Ne/O values). This example gives a reasonable representation of the trend of the data for Fe/O > 0.134. Mixing material with Ne/O = 0.152 and Fe/O = 0.134 with "flare" material having Ne/O = 0.40 and Fe/O = 1.07 (Reames 1999) gives a similar mixing line (dashed), and leads to an estimated average flare contribution of ~15% to the gradual events, with only 9 of 88 having flare contributions >50%.

Overabundances of heavy elements such as Fe have also been used as indicators of remnant flare material (Mason et al. 1999; Tylka et al. 2005; Desai et al. 2006a) or of direct contributions from flare-accelerated particles (Cane et al. 2003, 2006). Mewaldt et al. (2006) found that in the suprathermal energy range ~95% of large SEP events have Fe/O > 0.134 (the average SEP value found by Reames 1995) while only ~50% of the 10 to 40 MeV/nuc Fe/O ratios are >0.134 (see Fig. 6). The data in Fig. 6 led Mewaldt et al. (2006) to conclude that either (1) the vast majority of SEP events accelerate an Fe-rich seed population or (2) there must be preferential injection or acceleration of heavy elements like Fe at CME-driven shocks near the Sun. However, interplanetary shocks at 1 AU do not accelerate Fe as efficiently as lighter ions (Desai et al. 2003), leading Desai et al. (2006a, 2007) to conclude that the majority of SEP events contain an accelerated Fe-rich seed population, most likely remnant flare material.

## 3 Obtaining an Unbiased SEP Composition

In obtaining average SEP abundances previous studies have often taken the mean abundances of large samples of SEP events, assuming that $Q/M$-dependent fractionation effects are negligible or will average out (e.g., Reames 1995; Desai et al. 2006a). However, there are $Q/M$-dependent spectral breaks in most large SEP events (typically at >1 MeV/nuc) in which the low-energy power-law shape "rolls over" at higher energies for lighter species

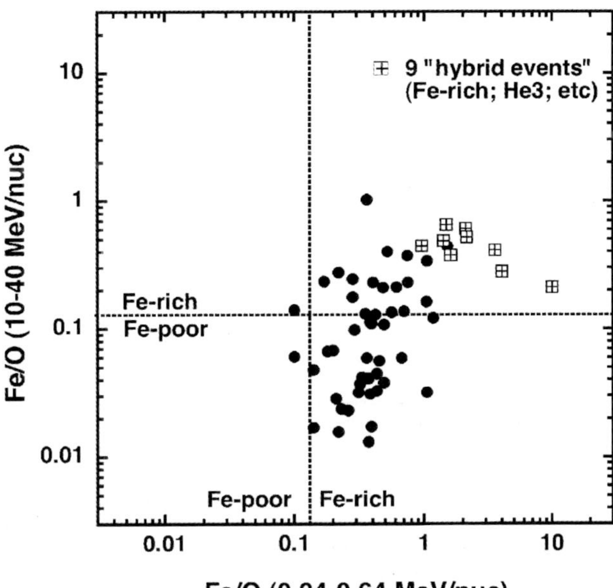


**Fig. 6** Plot of 10–40 MeV/nuc Fe/O versus 0.04–0.64 MeV/nuc Fe/O in large SEP events (from Mewaldt et al. 2006). Also indicated are nine "hybrid" events identified by their Fe/O ratios and by enriched $^{3}He/^{4}He$ and/or highly-ionized Fe. Data are from the SIS and ULEIS instruments on ACE

such as C or O than it does for heavier species like Fe (see, e.g., Tylka et al. 2000; Cohen et al. 2005; Mewaldt et al. 2005). Including events with these spectral characteristics might bias the elemental composition against heavy elements at energies >1 MeV/nuc (Desai et al. 2006a). Breneman and Stone (1985) corrected their average abundances for the effects of $Q/M$ bias.

A second possible approach is to select only those SEP events where there is relatively little $Q/M$-dependent fractionation (Meyer 1985; McGuire et al. 1986). We did this by requiring that the fluence ratio for Fe/(Mg + Si) be within ±45% of the photospheric ratio for these elements. Since the FIP of Fe is between that of Mg and Si, this criterion is largely independent of any changes in the degree of FIP-fractionation (Mewaldt et al. 2002). It was also required that the spectral slopes of 12 to 40 MeV/nuc C, O, Ne, Mg, Si, and Fe be consistent with the mean slope of these species to within 10%. Applied to the 1997–2002 SIS database, these criteria resulted in 14 SEP events.

The summed 12–40 MeV/nuc fluences from these 14 events provided relative abundances for 14 species. Additional species were added using abundance ratios of rare-species to abundant-neighbors (e.g., K/Ca and Cr/Fe) measured by Cohen et al. (2001) in the same instrument. Figure 7 shows that the FIP-fractionation pattern for this high-energy compilation extends below Si down to ∼8 eV or less. In addition, the magnitude of the step between high-FIP and the lowest-FIP species is ∼2 times greater than in Fig. 2 for the 5–12 MeV/nuc abundances of Reames (1995). Figure 8 shows the ratio of this "unbiased" sample of SEPs to the coronal spectroscopic abundances of Feldman and Widing (2003). Note that these abundances are consistent to within 20% for all ten species. Evidently SEP abundances agree better with the coronal composition than with the solar wind composition if $Q/M$-dependent fractionation and spectral effects are minimized (see also Breneman and Stone 1985; Meyer 1985; and McGuire et al. 1986).

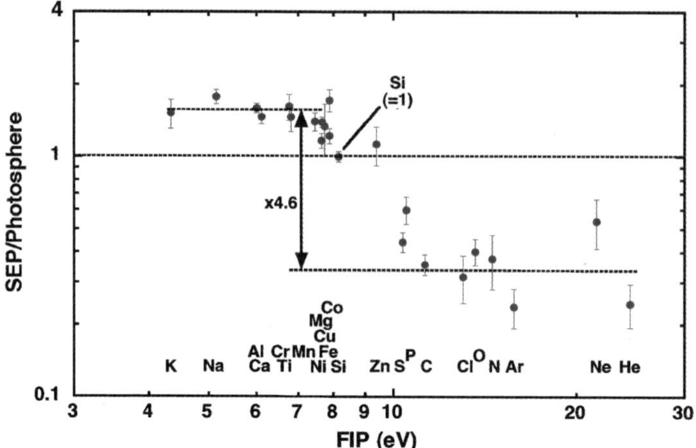


**Fig. 7** Ratio of the 12–40 MeV/nuc "unbiased" SEP abundances from the ACE/SIS instrument to the photospheric abundances (Lodders 2003), plotted versus FIP. Comparing this FIP-fractionation pattern to that in Fig. 2, the depletion of high-FIP species is greater, and it begins below Si (i.e., Si is depleted by ∼35–40%), rather than above Si as suggested by the data in Fig. 2

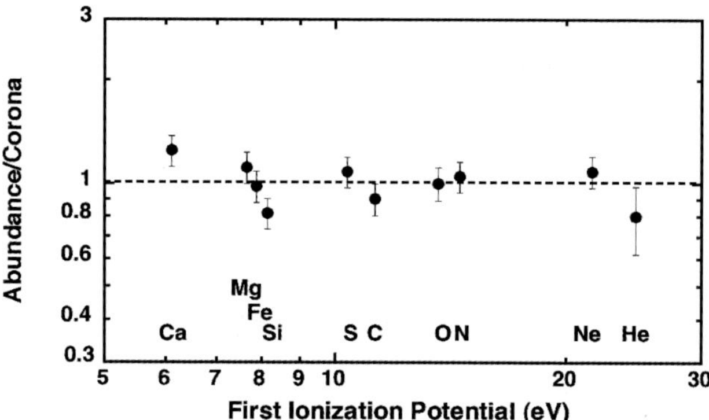


**Fig. 8** Ratio of the "unbiased" average SEP abundances shown in Fig. 7 to the coronal abundances of Feldman and Widing (2003). The plotted uncertainties include the SEP uncertainties added in quadrature with an assumed 10% uncertainty in the coronal abundances

## 4 CME Material as a Seed Population for SEPs

Zurbuchen and von Steiger (2003) suggested that the differences between SEP and SW abundances are smaller if one compares to solar wind with a higher freeze-in temperature. Using Ulysses data from 2000, Zurbuchen et al. (2002) found that the FIP fractionation factor (FFF) in the solar wind increases with the freeze-in temperature, with a FIP factor ∼1.7 for slow solar wind (normalized to Lodders' photospheric abundances), ∼1.3 for fast wind, and ∼2.0 for oxygen freeze-in temperatures >2 MK. Here FFF is the abundance ratio of low-FIP to high-FIP species $[(\mathrm{Mg} + \mathrm{Si} + \mathrm{S} + \mathrm{Fe})/(\mathrm{C} + \mathrm{N} + \mathrm{O} + \mathrm{Ne})]$ divided by the photospheric value for this abundance ratio (von Steiger et al. 2000).

To investigate this possibility further ACE/SWICS data from 2001 and 2003 were used to derive the relative abundances of key species in the solar wind as a function of freeze-in temperature. Figure 9 shows 2-hr averages of the C/O and Fe/O ratios versus the $O^{7+}/O^{6+}$ ratio (which depends on freeze-in temperature as indicated). Note that $O^{7+}/O^{6+}$ ratios >1 correspond mainly to interplanetary CME (ICME) material (Zurbuchen et al. 2002). Although

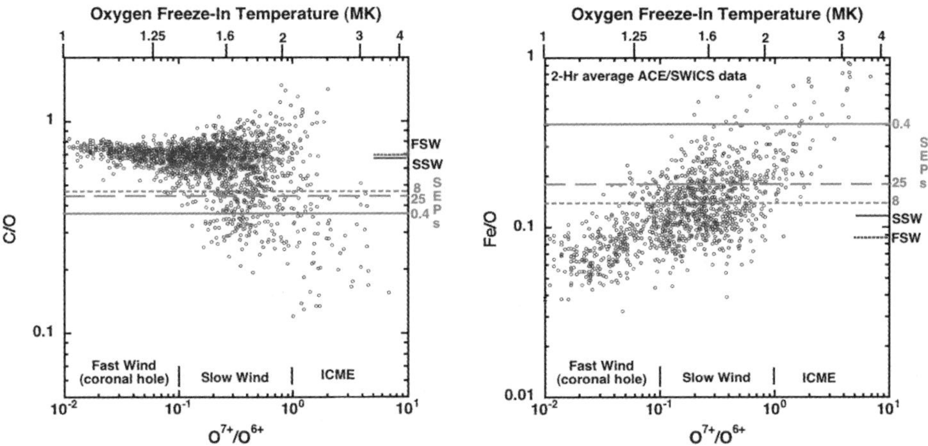


**Fig. 9** Two-hour averages of C/O (*left*) and Fe/O (*right*) from ACE/SWICS during 2001 and 2003 are plotted versus the $O^{7+}/O^{6+}$ ratio. The average abundance ratios for fast and slow solar wind are also indicated (von Steiger et al. 2000), along with the average ratios for 0.4, 8, and 25 MeV/nuc SEPs (Desai et al. 2006a; Reames 1995; Fig. 6 in this paper respectively)

there is considerable scatter in these 2-hr averages (especially for high-temperature C/O), as the freeze-in temperature increases, the solar wind composition becomes more iron-rich (see also Ipavich et al. 1986) and carbon-poor, in better agreement with typical SEP abundances. We find similar trends in Mg/O (see also Richardson and Cane 2004) and Si/O.

Using data from all of 2001 and 2003, and estimates of FFF based only on O, Mg, Si, and Fe (von Steiger et al. 2000), we find $\mathrm{FFF} = 5.0 \pm 0.9$ for solar wind with a freeze-in temperature >2 MK, considerably greater than for either fast or slow solar wind, and also greater than the ICME FIP factor of $2.0 \pm 0.2$ found by Zurbuchen et al. (2002). However, their measurements were made during 2000, while Ulysses was over the southern pole of the Sun. An explanation for this latitude dependence is suggested below.

Richardson and Cane (2004) have previously reported an elevated Mg/O ratio in ICMEs (which they interpreted as increased FIP fractionation) and they found a "universal" relationship between Mg/O and $O^{7+}/O^{6+}$ in both the ambient wind and ICMEs. We find a similar correlation between Fe/O and $O^{7+}/O^{6+}$ (Fig. 9), as well as a broader increase in Si/O when $O^{7+}/O^{6+}$ exceeds $\sim 0.8$. As expected, the mean Fe charge state also increases for $O^{7+}/O^{6+} > 0.6$ (Reinard 2005). Von Steiger et al. (2005) found that composition signatures of ICMEs were much weaker at high latitude. Reinard (2005) showed that high-charge-state material in ICMEs is concentrated near the center of the ejecta, so that in situ crossings of an ICME flank miss the enhanced charge states. Taken together, these observations explain why our FIP factor for in-ecliptic ICMEs is much greater than Zurbuchen et al. (2002) found at high latitudes. ACE observations in the ecliptic would have seen a much greater fraction of ICME center crossings than would have been observed at heliographic latitudes of $-42°$ to $-80°$, causing a latitude dependence in the ICME FIP factor.

The FIP factor that we find for ICME material is similar to that for the average SEP compositions of Desai et al. (2006a, 2006b) based on ACE/ULEIS data and of Cohen et al. (2007), who combined data from ACE/SIS and ACE/ULEIS. It is also interesting that ICMEs often contain high charge states of Fe (Lepri et al. 2001; Richardson and Cane 2004), similar to those observed in many high-energy SEP events (Oetliker et al. 1997; Mazur et al. 1999; Cohen et al. 1999; Labrador et al. 2005).

## 5 Discussion

In Fig. 10, we compare the FIP fractionation factors for several SEP and solar wind samples, for the corona, and for a loop model. All but the ICME point are based on the ratio [(Mg + Si + S + Fe)/(C + N + O + Ne)]. Note that S behaves as a low-FIP species in solar wind (Fig. 3), but more like a high-FIP species in SEPs (Figs. 2 and 7). However, including S or not makes a difference of $<10\%$. All SEP samples have a considerably greater FIP factor than those for either fast or slow solar wind.

In the model of Fisk (2003) the slow solar wind originates from the reconnection of open field lines with coronal loops, which releases material stored in the loops. It is interesting that spectroscopic studies show that the FIP effect in coronal loops grows with time—larger loops have a greater FIP effect than smaller loops (Feldman 1998). Perhaps SEPs originate from material in larger loops than the material making up the bulk solar wind. Also shown in Fig. 10 is the FFF for a coronal loop model (Schwadron et al. 1999) in which the FIP fractionation depends on wave-heating of minor ions.

From a comparison of Figs. 3 and 8 it is clear that if $Q/M$-dependent fractionation and related spectral differences can be minimized, then the resulting SEP composition agrees better with coronal abundances based on spectroscopy than with the solar wind composition. This provides renewed support for the idea that SEPs can provide coronal abundance data for species not readily measured by other means. The FIP-pattern for this higher-energy SEP sample (Fig. 7), including 21 species, has a break below 8 eV, somewhat less than indicated by earlier measurements. It is interesting that Feldman and Widing (2003) suggested that the break in the FIP fractionation pattern might be reduced to FIP $\leq$7 eV in very low-temperature ($\sim$3000 K) photospheric material near sunspots. The more gradual slope between the high and low-FIP species in Fig. 7 suggests that material from regions with a range of temperatures has been sampled.

It is interesting that the ICME composition is more similar to SEPs than to the bulk solar wind. Although ICME charge states are also similar to those in SEPs, it is hard to see how ICME material can be a viable seed population for most SEP events. While ICME material is nearly always ejected in large SEP events, it is located behind the shock (Richardson and Cane 2004), and it is unclear how it might be injected into the shock acceleration process. On the other hand, Gopalswamy et al. (2004) have found that most of the largest SEP events occur when there is a second fast CME from the same active region as an earlier CME.

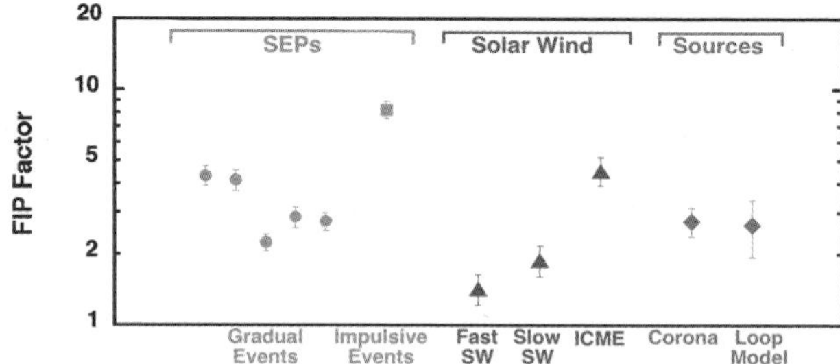


**Fig. 10** A comparison of FIP fractionation factors. Included are five gradual SEP compilations ordered, left to right, from lower to higher mean energy (Desai et al. 2006a; Cohen et al. 2007; Reames 1995; Breneman and Stone 1985; Fig. 6 of this paper); the impulsive SEP abundances of Mason et al. (2004); slow and fast solar wind (von Steiger et al. 2000) and ICME data from this paper; corona spectroscopic data (Feldman and Widing 2003), and a coronal loop model (Schwadron et al. 1999). All use the 8-species FIP fractionation factor except the ICME data, which is based on O, Mg, Si and Fe

In such events the shock does pass through the ICME material from the first event. There could also be other conditioning that occurs (e.g., elevated turbulence) that facilitates the acceleration of remnant ICME material. It is also possible that mechanisms other than CME-driven shock acceleration are able to accelerate ICME ejecta, or that CME material and SEP seed particles somehow share a common origin.

The presence of $^3$He in a large fraction of gradual SEP events provides strong evidence that flare material from either earlier flares or the associated flare is often incorporated into gradual SEP events, and enrichments in heavy elements and high charge states support this conclusion. While there is no question that suprathermal ions (either remnant flare or other suprathermals) encountered by an interplanetary shock will be further accelerated to higher energy (e.g., Desai et al. 2003), there is a question as to whether there is a sufficient density of these ions available to account for the observed fluence of particles in very large SEP events (Mewaldt et al. 2006). Comparing the $^3$He/$^4$He ratio in gradual events (Fig. 5) to that in the suprathermal region (Desai et al. 2006b) suggests that an average of ~20% of the $^4$He at ~10 MeV/nuc originated from flare material, indicating that in most gradual events there must be another source of $^4$He seed particles that contributes more than remnant material (see also Mewaldt et al. 2006), unless $^3$He is much less efficiently accelerated. Similarly, in most large SEP events the 10–30 MeV/nuc Ne/O ratio agrees with the average "gradual" value of ~0.15; only in Fe-rich events does Ne/O approach values typical of impulsive events. We find that on average less than 20% of the O originates from flare material, again indicating that there must be a more important seed population than flare material. Note that the relative contributions of various seed populations undoubtedly vary with energy.

Several additional seed populations should be considered. Mewaldt et al. (2006) suggested that previous gradual events account for much of the remnant suprathermal material encountered by CME-driven shocks. Note that this material is Fe-rich (see Fig. 6) since most particles are at low-energy. Re-accelerating gradual material helps supply sufficient suprathermal seed particles, but does not answer why gradual events are Fe-rich in the first place.

Gloeckler et al. (2000) have shown that there are ubiquitous suprathermal tails on the solar wind with characteristic $E^{-1.5}$ energy spectra starting at several times the solar wind speed and extending to >1 MeV/nuc (see Fig. 4). The origin of these tails is not known, but they apparently result from an interplanetary acceleration process operating well inside 1 AU (e.g., Fisk and Gloeckler 2006). Perhaps such a process is responsible for the pre-acceleration required to elevate thermal solar wind, ICME, and/or other material above the injection threshold for shock acceleration. It is important to measure the composition of these tails accurately to see if they can explain the composition differences between solar wind and SEPs. Also warranting investigation is the question of whether $Q/M$-dependent injection and/or acceleration of coronal material, solar wind and suprathermals can explain the Fe-rich character of almost all gradual SEP events below 1 MeV/nuc (Fig. 6 and Desai et al. 2006a). Although Fe is not as efficiently accelerated in shock acceleration processes observed near 1 AU, we know very little about conditions and processes in the high corona.

## 6 Summary

The differences in composition between SEPs and solar wind show that bulk solar wind cannot be the main seed population for large gradual events. In the energy range from ~10 to ~50 MeV/nuc, $^3$He/$^4$He and Ne/O data show that either remnant or directly accelerated flare material is a key contributor to Fe-rich events, but not the most important contributor in most gradual events. The results of our attempt to obtain an SEP sample unbiased by $Q/M$-fractionation or spectral effects agrees better with the coronal composition based on spectroscopic data, than with slow solar wind. A survey of in-ecliptic ICME material shows that it shares several properties with SEPs, including greater FIP fractionation and a greater degree of ionization than either slow or fast solar wind. However, it is not easy to see how ICME material can be injected and accelerated in the standard picture of CME-driven shock acceleration. The ubiquitous suprathermal tails on the solar wind represent an additional promising seed population if their composition can help explain the SEP composition.

**Acknowledgements** We are grateful to the local organizing committee at ISSI for their contributions to a very successful and enjoyable Symposium. This work was supported by NASA under grants NNG04GB55G, NNG04088G, and NAG5-12929.

## References

E. Anders, N. Grevesse, Geochim. Cosmochim. Acta **53**, 197–214 (1989)
H.H. Breneman, E.C. Stone, Astrophys. J. Lett. **299**, L57–L61 (1985)
H.V. Cane, T.T. von Rosenvinge, C.M.S. Cohen, R.A. Mewaldt, Geophys. Res. Lett. **30**, 5–1 (2003). doi: 10.1029/2002GL016580
H.V. Cane, R.A. Mewaldt, C.M.S. Cohen, T.T. von Rosenvinge, J. Geophys. Res. **111** (2006). doi: 10.1029/2005JA011071
E.W. Cliver, in *High Energy Solar Physics*, ed. by R. Ramaty et al., AIP Conf. Proc., vol. 374 (1996), pp. 45–60
C.M.S. Cohen et al., Geophys. Res. Lett. **26**, 2697–2700 (1999)
C.M.S. Cohen et al., in *Solar and Galactic Composition*. AIP Conf. Proc., vol. 598 (2001), p. 107
C.M.S. Cohen et al., J. Geophys. Res. **110**, (2005). doi: 10.1029/2004JA011004
C.M.S. Cohen et al. (2007), this volume. doi: 10.1007/s11214-007-9218-y
M.I. Desai et al., Astrophys. J. **588**, 1149–1162 (2003)
M.I. Desai et al., Astrophys. J. **649**, 470–489 (2006a)
M.I. Desai et al., Astrophys. J. Lett. **645**, L81–L84 (2006b)
M.I. Desai et al. (2007), this volume
U. Feldman, Space Sci. Rev. **85**, 227–240 (1998)
U. Feldman, K.G. Widing, Space Sci. Rev. **107**, 665–720 (2003)
L.A. Fisk, J. Geophys. Res. **108**, (2003). doi: 10.1029/2002JA009284
L.A. Fisk, G. Gloeckler, Astrophys. J. **640**, L79–L82 (2006)
J. Geiss, Space Sci. Rev. **85**, 241–252 (1998)
J. Geiss, P. Bochsler, in *Proc. Rapports Isotopiques dans le Systeme Solaire* (Cepadues-Editions, Paris, 1985), pp. 213–228.
G. Gloeckler, J. Geiss, Space Sci. Rev. **84**, 275 (1998)
G. Gloeckler, L.A. Fisk, T.H. Zurbuchen, N.A. Schwadron, in *Acceleration and Transport of Energetic Particles in the Heliosphere*. AIP Conf. Proc., vol. 528 (2000), p. 229
N. Gopalswamy, S. Yashiro, S. Krucker, G. Stenborg, R.A. Howard, J. Geophys. Res. **109**, (2004). doi: 10.10029/2004JA010602
D. Hovestadt, in *Solar Wind Three*, ed. by C.T. Russell (1974), pp. 2–25
F.M. Ipavich et al., J. Geophys. Res. **91**, 4133 (1986)
B. Klecker et al., Adv. Space Res. **38**(3), 493–497 (2006)
L. Kocharov, J. Torsti, Astrophys. J. **586**, 1430–1435 (2003)
A.W. Labrador, R.A. Leske, R.A. Mewaldt, E.C. Stone, T.T. von Rosenvinge, in *Proc. 29th Internat. Cosmic Ray Conf.*, vol. 1 (2005), pp. 99–102
M.A. Lee, A.J. Tylka (2007), this volume
S. Lepri, T.H. Zurbuchen, L.A. Fisk, I.G. Richardson, H.V. Cane, G. Gloeckler, J. Geophys. Res. **106**, 29231 (2001)
R.A. Leske, J.R. Cummings, R.A. Mewaldt, E.C. Stone, T.T. von Rosenvinge, Astrophys. J. Lett. **452**, L149–L152 (1995)
G. Li, G.P. Zank, Geophys. Res. Lett. **32** (2005). doi: 10.1029/2004GL021250
K. Lodders, Astrophys. J., **591**, 1220–1247 (2003)
A. Luhn et al., Adv. Space Res. **4**(2), 161–166 (1984)

E. Marsch, R. von Steiger, P. Bochsler, Astron. Astrophys. **301**, 261–276 (1995)
G.M. Mason et al., Astrophys. J. **606**, 555 (2004)
G.M. Mason, Space Sci. Rev. **99**, 119 (2001)
G.M. Mason, J.E. Mazur, J.R. Dwyer, Astrophys. J. Lett. **525**, L133–L136 (1999)
J.E. Mazur, G.M. Mason, M.D. Looper, R.A. Leske, R.A. Mewaldt, Geophys. Res. Lett. **26**, 173–176 (1999)
R.E. McGuire, T.T. von Rosenvinge, F.B. McDonald, Astrophys. J. **301**, 938–961 (1986)
R.A. Mewaldt, in *Proc. 26th Internat. Cosmic Ray Conference: Invited, Rapporteur, and Highlight Papers*. AIP Conf. Proc., vol. 516 (2000), pp. 265–270
R.A. Mewaldt et al., in *Proc. 27th Internat. Cosmic Ray Conf.*, vol. 8 (2001a), pp. 3132–3135
R.A. Mewaldt et al., in *Solar and Galactic Composition*. AIP Conf. Proc., vol. 598 (2001b), pp. 165–170
R.A. Mewaldt et al., Adv. Space Res. **30**(1), 79–84 (2002)
R.A. Mewaldt et al., in *Physics of Collisionless Shocks*. AIP Conf. Proc., vol. 781 (2005), pp. 227–232
R.A. Mewaldt, C.M.S. Cohen, G.M. Mason, in *Solar Eruptions and Energetic Particles*, ed. by N. Gopalswamy, et al., AGU Monograph, vol. 165 (2006), pp. 115–126
J.P. Meyer, Astrophys. J. Suppl. **57**, 151–171 (1985)
E. Moebius et al., Geophys. Res. Lett. **226**, 145–148 (1999)
M. Oetliker et al., Astrophys. J. **477**, 495–501 (1997)
M.A. Popecki, in *Solar Eruptions and Energetic Particles*, ed. by N. Gopolswamy, R.A. Mewaldt, J. Torsti. AGU Monograph, vol. 165 (2006), pp. 127–136
D.V. Reames, Adv. Space Res. **15**(7), 41–51 (1995)
D.V. Reames, Space Sci. Rev. **90**, 413 (1999)
D.V. Reames, J.P. Meyer, T.T. von Rosenvinge, Astrophys. J. **90**, 649–667 (1994)
A. Reinard, Astrophys. J. **620**, 501–505 (2005)
I.G. Richardson, H.V. Cane, J. Geophys. Res. **109** (2004). doi: 10.1029/2004JA010598
N.A. Schwadron, L.A. Fisk, T.H. Zurbuchen, Astrophys. J. **521**, 859–867 (1999)
J. Torsti, J. Laivola, L. Kocharov, Astron. Astrophys. **408**, L1–L4 (2003)
A.J. Tylka, P.R. Boberg, R.E. McGuire, C.K. Ng, D.V. Reames, in *Acceleration and Transport of Energetic Particles Observed in the Heliosphere*. AIP Conf. Proc., vol. 528 (2000), pp. 147–152
A.J. Tylka et al., Astrophys. J. **625**, 474–495 (2005)
R. von Steiger et al., J. Geophys. Res. **105**, 27217–27238 (2000)
R. von Steiger, T.H. Zurbuchen, A. Kilchenmann, in *Proc. Solar Wind 11 – SOHO 16 "Connecting Sun and Heliosphere"*. ESA SP-592 (2005), pp. 317–323
M.E. Wiedenbeck et al., in *Acceleration and Transport of Energetic Particles Observed in the Heliosphere*. AIP Conf. Proc., vol. 528 (2000), pp. 107–110
T. Zurbuchen, L.A. Fisk, G. Gloeckler, R. von Steiger, Geophys. Res. Lett. **29** (2002). doi: 10.1029/2001GL013946
T.H. Zurbuchen, R. von Steiger, EOS Trans. AGU **84**(46) (2003), Fall Meet. Suppl., Abstract SH11A-03

Space Sci Rev (2007) 130: 221–229
DOI 10.1007/s11214-007-9188-0

# What Determines the Composition of SEPs in Gradual Events?

**Martin A. Lee**

Received: 30 December 2006 / Accepted: 6 April 2007 /
Published online: 8 June 2007


**Abstract** Gradual solar energetic particle (SEP) events are evidently accelerated by coronal/interplanetary shocks driven by coronal mass ejections. This talk addresses the different factors which determine the composition of the accelerated ions. The first factor is the set of available seed populations including the solar wind core and suprathermal tail, remnant impulsive events from preceding solar flares, and remnant gradual events. The second factor is the fractionation of the seed ions by the injection process, that is, what fraction of the ions are extracted by the shock to participate in diffusive shock acceleration. Injection is a controversial topic since it depends on the detailed electromagnetic structure of the shock transition and the transport of ions in these structured fields, both of which are not well understood or determined theoretically. The third factor is fractionation during the acceleration process, due to the dependence of ion transport in the turbulent electromagnetic fields adjacent to the shock on the mass/charge ratio. Of crucial importance in the last two factors is the magnetic obliquity of the shock. The form of the proton-excited hydromagnetic wave spectrum is also important. Finally, more subtle effects on ion composition arise from the superposition of ion contributions over the time history of the shock along the observer's magnetic flux tube, and the sequence of flux tubes sampled by the observer.

**Keywords** Solar energetic particles (SEPs) · Gradual SEP events · Shock acceleration · Ion composition

## 1 Introduction

I hope my viewgraphs are a relief from the professionalism of Power Point! Composition is a crucial aspect of the acceleration and transport of solar energetic particles (SEPs). On the one hand, it provides an important tool for distinguishing the origins of the particles and probing the basic processes of acceleration and transport. On the other hand, it provides

M.A. Lee (✉)
Space Science Center, Institute for the Study of Earth, Oceans and Space, University of New Hampshire, Durham, NH 03824, USA
e-mail: marty.lee@unh.edu

a challenge for theory to understand in one framework the behavior of the electrons and the large number of ion species measured by spacecraft such as ACE, Wind and Ulysses as functions of time and energy. We have heard from Cohen (2007), Leske (2007) and Mewaldt (2007) about the large compositional variations within a SEP event and between events. It is the purpose of this talk to describe the aspects of, or the different processes operating in, the large gradual SEP events, which control their composition. Gradual events are those likely to be accelerated primarily by coronal/interplanetary shocks driven by coronal mass ejections (CMEs). I shall not address the composition of the smaller impulsive events. In spite of a general consensus on the nature of gradual events, predictive theory is challenging for reasons which I shall try to elucidate.

The nature of gradual events is as follows: A rapid CME erupts from a solar active region into the corona with a speed greater than the magnetohydrodynamic "fast" speed, which in the corona is approximately equal to the Alfvén speed. The resulting pressure enhancement ahead of the CME steepens into a shock wave which drapes around the CME and propagates ahead of it. The shock extracts ions (and some electrons) from the plasma through which it propagates and accelerates them by the first-order Fermi process. The accelerating protons upstream of the shock stream relative to the upstream plasma and excite hydromagnetic waves, preferentially propagating nearly parallel to the upstream magnetic field, which attempt to outrun the shock in the antisunward direction. The excited waves are essential for acceleration of ions to the observed energies since the ambient magnetic fluctuations are too weak to provide the required rapid scattering of the ions back and forth across the shock. Further upstream of the shock, where the excited wave intensity decays to ambient levels, ions escape the shock upstream by virtue of the magnetic mirror force in the weakening interplanetary magnetic field. These escaping particles stream past Earth and beyond with streaming anisotropies that are initially very large. As the shock, with its accelerating particles and excited waves, propagates outwards, a spacecraft observes first the escaping particles, then a particle intensity increase to its maximum value at or near the shock, and finally a decay after shock passage as the accelerated particles adiabatically cool in the divergent solar wind. I shall now describe the different phases of the acceleration process and their influence on ion composition.

## 2 Available Seed Populations

The first important feature of gradual events that determines event composition is the number of available seed populations. These constitute the pool of particles that are available for acceleration. Each seed population has its own distinctive composition, spatial location and distribution function: The solar wind is always present with a relatively large number density. The solar wind suprathermal particles inside Earth orbit originate primarily from the thermal solar wind and the "inner source" of pickup ions (Fisk and Gloeckler 2006; Gloeckler 2007; Geiss et al. 1995). Particularly during solar maximum conditions, when SEPs are plentiful, we expect remnant particles from previous "impulsive" and "gradual" events to be present. One important issue for the remnant particles is how long they remain in the inner heliosphere where acceleration by the shock wave to the highest energies is possible. The remnant particles are eventually swept into the outer heliosphere by a combination of solar wind advection and spatial diffusion, and cooled by adiabatic deceleration. During solar minimum conditions, it may be more likely to observe SEP events without large contributions from remnant particles.

## 3 Direct Flare Component

Usually accompanying the eruption of a CME is a flare at the site of magnetic reconnection behind the CME. Particles are accelerated at the reconnection site either by direct acceleration in the reconnection electric field or by turbulent/stochastic acceleration in the turbulence generated at the flare site. These accelerated particles form an impulsive event which may have access to open field lines and escape before or during the formation of the shock wave ahead of the CME. In principle these particles can contribute to a gradual event without being re-accelerated by the shock; if they are re-accelerated by the shock they assume the role of remnant impulsive particles. Although the presence of a direct flare component is not exactly a phase of the gradual event acceleration process, its importance is a possibility and a subject of current controversy (Cane et al. 2003; Tylka et al. 2005).

## 4 Injection Fractionation

Not all ions in a seed population are able to participate in the first-order Fermi acceleration process at the shock. Consider first a parallel shock in which the upstream magnetic field is parallel to the shock normal. The shock has large-amplitude magnetic fluctuations downstream of the shock transition. An upstream ion approaching the shock can only participate in the Fermi process if it is sufficiently mobile while interacting with the downstream turbulence that it can in principle return to the upstream plasma. The mobility is determined primarily by the particle speed $v$ and to a lesser extent by the particle mass-to-charge ratio, $A/Q$, where $A$ and $Q$ are expressed relative to the proton mass and charge. At Earth's quasi-parallel bow shock (Lee 1982) and at large interplanetary traveling shocks (Lee 1983) only $\sim$1% of incident solar wind ions with $v \approx V$, where $V$ is the normal component of the upstream plasma velocity relative to the shock, are able to scatter back upstream and participate in diffusive shock acceleration. Since the diffuse ions accelerated at Earth's bow shock have approximately solar wind composition (Ipavich et al. 1984; Desai et al. 2000), we may infer that the injection process is insensitive to $A/Q$ in these cases. [However, recent observations of diffuse ions over a solar cycle (Desai et al. 2006) reveal an admixture of SEPs during solar maximum conditions.]

In contrast, incident ions with $v \gg V$ are all able to participate in the acceleration process. Thus, we may define an injection fraction $\xi(v)$ at a parallel shock which varies from $\xi(V) \approx 0.01$ to $\xi(v \gg V) = 1$ as shown by the dashed curve in Fig. 1. The transition

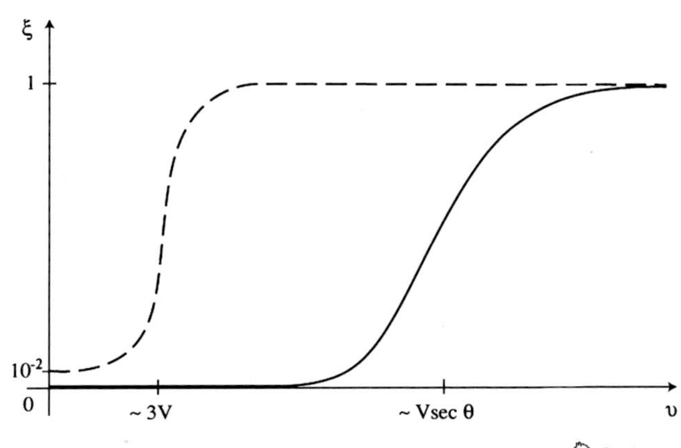


**Fig. 1** The injection fraction $\xi$ versus $v$ for a parallel shock (*dashed curve*) and an oblique shock (*solid curve*). See text for definitions

speed is probably a few times $V$ as indicated. The injection rate of species $i$ is then given by $n_i V \xi_i$ where $n_i$ is the upstream density of species $i$.

Shock magnetic obliquity plays a crucial role in determining the injection fraction, $\xi(v, A/Q, \theta)$, where $\theta$ is the angle between the upstream magnetic field and the shock normal. Unless the upstream magnetic field is turbulent, a particle can only participate in the acceleration process if $v \geq V \sec\theta$ so that the particle is able to propagate upstream against the plasma flow into the shock. Otherwise the particle is simply swept downstream. Thus, the parallel-shock injection fraction must be multiplied by a function $\eta(v)$ which increases from $\eta(v \ll V\sec\theta) = 0$ to $\eta(v \gg V\sec\theta) = 1$. If $\sec\theta \gg 1$, the injection fraction has the form shown by the solid line in Fig. 1.

The injection fraction can depend more sensitively on $A/Q$ at quasi-perpendicular shocks where the electrostatic shock potential plays an important role in shock dissipation. Although the large gyroradius of ions with large $A/Q$ provides mobility, these ions pass the potential more easily and penetrate deeper into the downstream plasma from which location escape back upstream is more difficult. In addition the shock-surfing or multiply-reflected-ion mechanism (Lee et al. 1996; Zank et al. 1996) favors low $A/Q$ ions which are more easily reflected from the shock potential before skipping along the shock surface and gaining energy in the motional electric field. However, other more subtle effects may modify this trend (Zank et al. 2001).

It is important to remember that the theory governing the injection fraction is rudimentary. Inferring injection fractions from observations of energetic particles at shocks of known strength and magnetic obliquity would be very worthwhile.

## 5 Acceleration

At first consideration we do not expect further fractionation of the ions during the acceleration process. The defining characteristic of the theory of diffusive shock acceleration is the power-law energy spectrum downstream of a planar stationary shock, at which particles are injected at low energies:

$$f_i(v, z < 0) \propto n_i \xi_i (v/v_{0i})^{-\beta}, \tag{1}$$

where $f_i(v, z)$ is the omnidirectional distribution function of ion species $i$, $z$ is distance upstream of the shock along the shock normal, $v_{0i}$ is the characteristic injection speed of the ions, $\beta = 3X(X-1)^{-1}$, and $X$ is the shock compression ratio. If $v_{0i} \sim V$, then the composition of the downstream ions at such a shock is determined by seed-particle composition ($n_i$) and injection fraction ($\xi_i$), with no additional fractionation during the acceleration process. This initial impression is false, however, since the upstream ions are fractionated and shocks in the interplanetary medium are neither planar nor stationary.

The transport of nonrelativistic ions in the vicinity of the shock is governed by (Parker 1965)

$$V_z \frac{\partial f_i}{\partial z} - \frac{\partial}{\partial z}\left(K_{i,zz} \frac{\partial f_i}{\partial z}\right) - \frac{1}{3}\frac{dV_z}{dz} v \frac{\partial f_i}{\partial v} = 0, \tag{2}$$

where we assume that the shock is stationary and, initially, that the shock is planar. Here $\mathbf{V}$ is the plasma velocity in the shock frame and $\mathbf{K}_i$ is the spatial diffusion tensor. Equation (2) is valid as long as the distribution function is nearly isotropic in the frame of the shock, which is generally satisfied for particles in the vicinity of the shock with $v \gg |\mathbf{V}|$. The injected population is included by specifying $f_i(v, z \to \infty) = f_{i\infty}(v)$. We take $V_z = -VS(z) - V_d S(-z)$, where $S(z)$ is the Heaviside step function, and $V$ and $V_d$ are constant and positive.

Since $K_{i,zz}$ depends on the magnetic power spectrum, which depends on distance $z$ upstream of the shock, the natural variable to use in place of $z$ is

$$\zeta(z) = \int_0^z [V/K_{zz}(z')]dz', \tag{3}$$

where the subscript $i$ and dependence on $v$ are suppressed. Lee (2005a) found that if $K_{zz}$ is replaced by $K_{zz}(1 + z/r_S)^n$, where $r_S$ is the heliocentric distance of the shock and $n > 1$, then the solution includes effectively the magnetic mirror force on the ions in the spherical geometry of the solar wind with approximately radial magnetic field. This replacement is crucial since it models the extraction of ions from the scattering sheath adjacent to the shock by the mirror force. With that replacement we have $\zeta(z \to \infty) = \zeta_\infty < \infty$. The solution of (2) is then

$$f(z > 0, v) = f_0 - (f_0 - f_\infty)(1 - \mathrm{e}^{-\zeta})(1 - \mathrm{e}^{-\zeta_\infty})^{-1}, \tag{4}$$

$$f_0 \equiv \beta \int_0^v \frac{dv'}{v'} f_\infty(v')(1 - \mathrm{e}^{-\zeta'_\infty})^{-1} \left(\frac{v}{v'}\right)^{-\beta} \exp\left[-\beta \int_{v'}^v \frac{dv''}{v''} \mathrm{e}^{-\zeta''_\infty}(1 - \mathrm{e}^{-\zeta''_\infty})^{-1}\right], \tag{5}$$

where $\zeta'_\infty \equiv \zeta_\infty(v')$, $\zeta''_\infty \equiv \zeta_\infty(v'')$ and $f(z \le 0) = f_0$. Note that if $f_\infty \propto \delta(v - v_{0i})$ and $\zeta_\infty \to \infty$ (characteristic of a planar geometry), then we recover the standard power-law spectrum in speed shown in (1).

Equations (4) and (5) reveal three types of fractionation in gradual events. First we note with Lee (2005a) that, if $K_{zz}$ is dominated by diffusion parallel to the magnetic field as expected, then $(K_{i,zz})^{-1} = K_{i0}^{-1} \langle I(z, k = \Omega_i/v\mu)\rangle$, where $K_{i0} \propto (A_i/Q_i)^2 v^3$, $I(z,k)$ is the wave intensity, $k$ is wavenumber parallel to the magnetic field $\mathbf{B}$, $\Omega_i$ is the ion gyrofrequency $[\Omega_i = Q_i e_p B/(A_i m_p c)]$, $e_p$ and $m_p$ are the proton charge and mass, and $\mu$ is the cosine of the ion pitch angle. The average is a particular weighted average over $\mu$ (Earl 1974).

The first type of fractionation occurs in the foreshock ($z > 0$) where the distribution is determined by a balance between advection of particles into the shock with the plasma velocity and diffusion away from the shock. At lower energies for which $\zeta_\infty \gg 1$ we have

$$f - f_\infty = (f_0 - f_\infty) \exp\left[-\int_0^z V K_{i0}^{-1} \langle I(z', \Omega_i/v\mu)\rangle (1 + z'/r_S)^{-n} dz'\right]. \tag{6}$$

Thus, each ion species decays exponentially with increasing distance from the shock with its characteristic lengthscale $L_i$. If $I \propto k^{-\lambda}$, then $L_i \propto (A_i/Q_i)^{2-\lambda} v^{3-\lambda}$.

Lee (2005b) has shown that wave excitation upstream of the shock by the accelerated protons injected out of the solar wind results in a wave intensity with a "knee" connecting two power laws. We have recently extended that work to include $f_\infty \propto v^{-\gamma} \neq 0$, as presented above. For the generalized treatment the wavenumber of the knee is proportional to $z^{-1/(\alpha-3)}$, where $\alpha = \min(\beta, \gamma)$. Since both $\beta$ and $\gamma$ are greater than 4, $\alpha > 4$. [Since $\beta = 3X(X-1)^{-1}$ and $X < 4$ for a nonrelativistic ideal gas, $\beta > 4$. Also $\gamma > 4$ since otherwise the energy density of the distribution $f_\infty(v)$ is unbounded. Formally a nonrelativistic version of this theory, as presented here, requires $\gamma > 5$; if $4 < \gamma \le 5$ relativistic effects are essential.] Therefore, the knee moves to lower wavenumber as $z$ increases. Assuming that the ambient wave intensity is negligible, below the knee $\lambda = 6 - \alpha$ and above the knee $\lambda = 2$. Thus, near the shock and at lower frequencies $\lambda = 6 - \alpha$, and further from the shock and at higher frequencies $\lambda = 2$. Within the foreshock, therefore, we expect fractionation as $L_i \propto (A_i/Q_i)^{\alpha-4}v^{\alpha-3}$ near the shock and $L_i \propto v$ further from the shock. As a result, we expect large-$A/Q$ ions to become enhanced in the foreshock with increasing distance from the shock up to a certain distance, beyond which fractionation ceases. Some events show this pattern of fractionation, but others do not as clearly (Tylka et al. 1999; Reames 1999). I should also mention that, since $K_{i0} \propto \cos^2\theta$ and $I(z,k)$ also depends on $\theta$ (Lee 2005a), $L_i$ depends on $\theta$ such that $L_i$ is smaller for quasi-perpendicular shocks than for quasi-parallel shocks.

The second type of fractionation occurs for the ions which escape the foreshock. From (4) their flux is given by

$$-K_{zz}\frac{\partial f}{\partial z}\bigg|_{z\to\infty} = V(f_0 - f_\infty)\mathrm{e}^{-\zeta_\infty}(1-\mathrm{e}^{-\zeta_\infty})^{-1}. \quad (7)$$

For large streaming anisotropy (7) becomes

$$f(z\to\infty) - f_\infty \sim v^{-1}[-K_{zz}(\partial f/\partial z)]_{z\to\infty}. \quad (8)$$

At higher energies ($\zeta_\infty \le 1$) (8) becomes

$$f(z\to\infty) - f_\infty \sim (V/v)(f_0 - f_\infty)\zeta_\infty^{-1}. \quad (9)$$

The factor $\zeta_\infty^{-1} \propto (A_i/Q_i)^{2-\lambda}v^{3-\lambda}$, where now $\lambda$ is a weighted average over $\mu$ and $z$ of $6-\alpha, 2$ and $5/3$ (characteristic of the ambient fluctuations in the solar wind) according to the integral in (3) with $z\to\infty$. Again, since $\alpha > 4$, $\zeta_\infty^{-1}$ favors ions with larger $A/Q$, and larger $v$. Equation (9) describes the escaping ions which are the first to arrive at Earth orbit. They are anisotropic, richer in heavy ions, and have a harder energy spectrum than at the shock. They also account for the "streaming limit" identified by Reames and coworkers (Reames 1990; Ng and Reames 1994; Reames and Ng 1998; Lee 2005a).

The third type of fractionation arises from the escaping ions, which introduce the exponential rollover in the energy spectrum apparent in (5). For the higher energy ions where the rollover becomes important ($\zeta_\infty \le 1$) we have a rollover of the form

$$\exp\left[-\beta\int^{v} dv'(v'\zeta'_\infty)^{-1}\right]. \quad (10)$$

As for the escaping ions, we have $(\zeta'_\infty)^{-1} \propto (A_i/Q_i)^{2-\lambda}v^{3-\lambda}$, where $\lambda$ is the weighted average which is less than 2. Note that if $\lambda = 1$ we obtain an exponential rollover in energy per charge, which is often found in observed spectra (Ellison and Ramaty 1985; Tylka et al. 2001). In any case the energy spectral rollovers are fractionated in the sense that larger-$A/Q$ ions rollover at a lower speed. This fractionation is due to the fact that the larger-$A/Q$ ions escape more easily from the shock and are therefore not available for further acceleration.

These basic types of fractionation can combine with those due to injection and seed populations during the course of a given time-dependent event to yield complex patterns of fractionation. For example, consider the time history of the event shown in Fig. 2. The observer (indicated by a schematic spacecraft) measures the composition of an event which occurs when a spherical shock encounters and propagates along the observer's field line **B**. If the shock originates a distance $L$ from the field line and propagates with constant speed $V_0$, then $V_0 t\sin\theta = L$. Differentiating this expression we obtain the incremental time interval

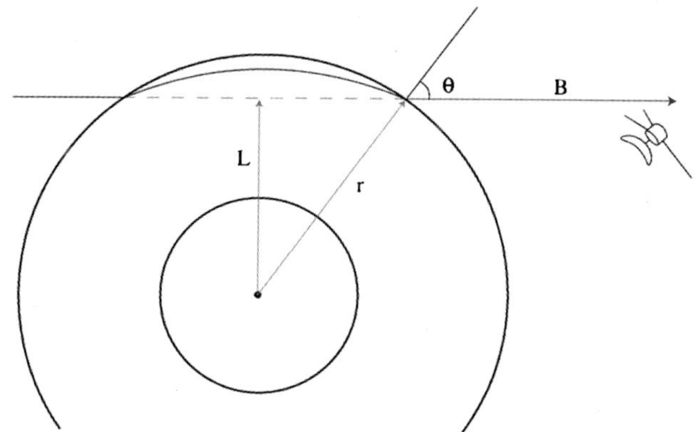


**Fig. 2** Schematic diagram of a spherical shock (*dark solid concentric circles*) intercepting the field line **B** (*light solid line*) of the observing spacecraft. The shock originates a distance $L$ from the original straight field line. The radius of the shock is $r = V_0 t$, and the angle between the shock normal and the upstream **B** on the chosen field line is $\theta$

during which the magnetic obliquity of the shock is within the range $(\omega, \omega + d\omega)$ where $\omega = \cos\theta$:

$$dt = (L/V_0)(1 - \omega^2)^{-3/2}\omega\, d\omega. \tag{11}$$

The fluence of the energetic particle event is then given by the time (or $\omega$) integration over the event weighted by the distribution function characteristic of the shock at each time (or $\omega$) along the field line. The fluence could be represented, in the simplest case in which the distribution depends only on the parameter $\omega$, by

$$\int_0^1 d\omega g(\omega) f_i(\omega, v), \tag{12}$$

where $g(\omega)$ is a function which represents the kind of weighting described by (11).

Tylka and Lee (2006) have explored the compositional consequences of a further simplified model in which at each time the distribution is a sum of two components arising from the injection of solar wind and remnant impulsive event ions. The distribution of each is a power-law times an exponential rollover in energy $E$ with an e-folding energy of the form $E_0(Q_i/A_i)/\omega$. The injection fraction for the remnant energetic ions is constant. However, the injection fraction for the solar wind ions is taken to be proportional to $\omega$ to represent suppressed injection of solar wind ions at quasi-perpendicular shocks. The time integration gives equal weighting to all values of $\omega$ between $\omega_1$ and $\omega_2$. The resulting fluence is then proportional to

$$\int_{\omega_1}^{\omega_2} d\omega\big[n_i^{\mathrm{rem}} + \xi_0 n_i^{\mathrm{SW}}\omega\big]E^{-\Gamma}\exp\big[-\omega(E/E_0)(A_i/Q_i)\big], \tag{13}$$

where $n_i^{\mathrm{rem}}$ and $n_i^{\mathrm{SW}}$ are the number densities of remnant impulsive event ions and solar wind ions, respectively, and $\xi_0$ is the injection fraction of solar wind ions at a parallel shock. Based on (13), Fig. 3 shows the fluence values of Fe/O relative to its value in the corona as a function of $E$ for various values of $R[= n_O^{\mathrm{rem}}/(\xi_O n_O^{\mathrm{SW}})]$ with $\omega_1 = 0$ and $\omega_2 = 1$, and in the lowest curve for $R = 0$ with $\omega_1 = 0.5$ and $\omega_2 = 1$. Remnant impulsive event and solar wind abundances and charge states are taken to be the standard values. See Tylka and Lee (2006) for details. The point is that even this simple model can yield a diverse array of abundance ratios as a function of energy. Note that the spread in Fe/O can increase at higher energies by up to two orders of magnitude, as observed between some events (Tylka

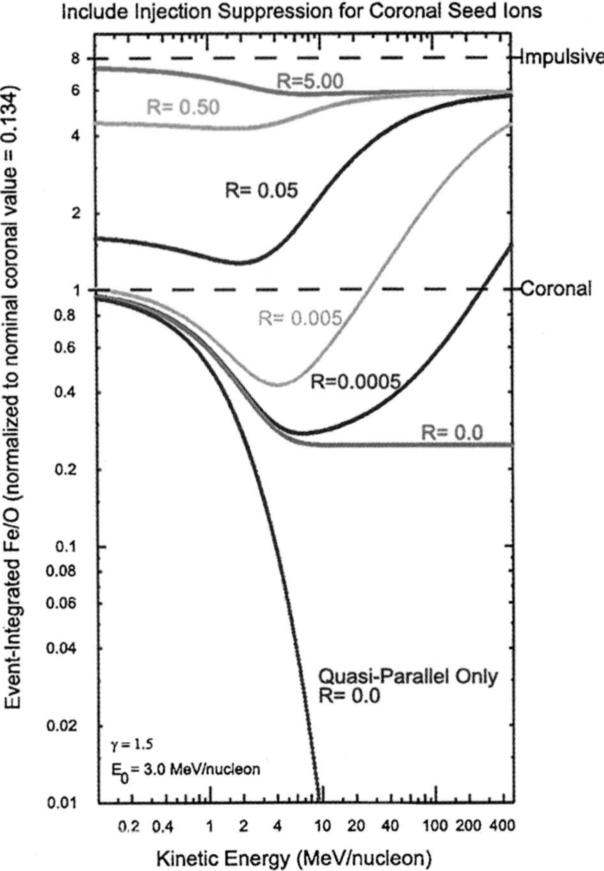


**Fig. 3** Model calculations of Fe/O (normalized to the nominal coronal value) versus energy based on (13). $E_0$ is taken to be 3.0 MeV/nucleon. The *lowest curve* with $R[=n_o^{\mathrm{rem}}/(\xi_0 n_o^{\mathrm{SW}})]=0$ assumes $\omega_1=0.5$ and $\omega_2=1$. *All other curves* assume $\omega_1=0$ and $\omega_2=1$. The abundances and charge states of the solar wind and remnant impulsive events are taken to have nominal values. The equation $\gamma=1.5$ should refer to the symbol $\Gamma$ used in (13). This figure is adapted from Fig. 5 of Tylka and Lee (2006)

et al. 2005). The compositional variations evident in Fig. 3 arise from the interplay between the $\theta$-dependence of the rollover, the relative composition of impulsive events and the solar wind, and the suppression of solar wind injection at quasi-perpendicular shocks. We should keep in mind that this simple model ignores the contributions of the ions in the foreshock (decreasing with increasing $z$ upstream of the shock) and of the escaping ions. A predictive model must include all these phases of a gradual event.

## 6 Conclusions

The ion compositional variations within a SEP event and between events present an exciting challenge for research into the origin and behavior of gradual SEP events. At the same time these variations provide important clues on the nature of gradual events: the importance of magnetic obliquity and the upstream escape of ions, the important role of the proton-excited waves upstream of the shock, and the conditions for acceleration close to the Sun where the highest energy ions are produced and where the ions probe an environment not accessible to direct observation. An understanding of SEP composition will increase our ability to develop predictive models for these potentially disruptive events.

**Acknowledgements** I wish to thank ISSI for organizing this excellent interdisciplinary conference in honor of Johannes Geiss on the occasion of his 80th birthday in such a beautiful location and with such warm hospitality. I also wish to thank a very helpful reviewer for a critical reading of the original manuscript. This work was supported, in part, by NASA Living With a Star Targeted Research and Technology Grants DPR NNH05AB581 and NNX06AG92G.

## References

H.V. Cane, T.T. von Rosenvinge, C.M.S. Cohen, R.A. Mewaldt, Geophys. Res. Lett. **30**(12), 8017 (2003). doi:10.1029/2002GL016580
C.M.S. Cohen, R.A. Mewaldt, R.A. Leske, A.C. Cummings, E.C. Stone, M.E. Wiedenbeck et al., this volume (2007). doi:10.1007/s11214-007-9218-y
M.I. Desai, G.M. Mason, J.R. Dwyer, J.E. Mazur, T.T. von Rosenvinge, R.P. Lepping, J. Geophys. Res. **105**, 61–78 (2000)
M.I. Desai, G.M. Mason, J.E. Mazur, J.R. Dwyer, Geophys. Res. Lett. **33**, L18104 (2006). doi:10.1029/2006GL027277
J.A. Earl, Astrophys. J. **193**, 231–242 (1974)
D.C. Ellison, R. Ramaty, Astrophys. J. **298**, 400–408 (1985)
L.A. Fisk, G. Gloeckler, Astrophys. J. **640**, L79–L82 (2006)
J. Geiss, G. Gloeckler, L.A. Fisk, R. von Steiger, J. Geophys. Res. **100**, 23373–23378 (1995)
G. Gloeckler, L. Fisk, this volume (2007). doi:10.1007/s11214-007-9226-y
F.M. Ipavich, J.T. Gosling, M. Scholer, J. Geophys. Res. **89**, 1501–1507 (1984)
M.A. Lee, J. Geophys. Res. **87**, 5063 (1982)
M.A. Lee, J. Geophys. Res. **88**, 6109 (1983)
M.A. Lee, Astrophys. J. Suppl. **158**, 38–67 (2005a)
M.A. Lee, in *The Physics of Collisionless Shocks*, ed. by G. Li, G.P. Zank, C.T. Russell (AIP, Melville, 2005b), pp. 240–245
M.A. Lee, V.D. Shapiro, R.Z. Sagdeev, J. Geophys. Res. **101**, 4777–4789 (1996)
R.A. Leske, R.A. Mewaldt, C.M.S. Cohen, A.C. Cummings, E.C. Stone, W.E. Wiedenbeck et al., this volume (2007). doi:10.1007/s11214-007-9185-3
R.A. Mewaldt, C.M.S. Cohen, G.M. Mason, A.C. Cummings, M.I. Desai, R.A. Leske et al., this volume (2007). doi:10.1007/s11214-007-9187-1
C.K. Ng, D.V. Reames, Astrophys. J. **424**, 1032–1048 (1994)
E.N. Parker, Planet. Space Sci. **13**, 9 (1965)
D.V. Reames, Astrophys. J. **358**, L63–L67 (1990)
D.V. Reames, Space Sci. Rev. **90**, 413–491 (1999)
D.V. Reames, C.K. Ng, Astrophys. J. **504**, 1002–1005 (1998)
A.J. Tylka, D.V. Reames, C.K. Ng, Geophys. Res. Lett. **26**(14), 2141–2144 (1999)
A.J. Tylka, C.M.S. Cohen, W.F. Dietrich, C.G. Maclennan, R.E. McGuire, C.K. Ng, D.V. Reames, Astrophys. J. **558**, L59–L63 (2001)
A.J. Tylka, C.M.S. Cohen, W.F. Dietrich, M.A. Lee, C.G. Maclennan, R.A. Mewaldt, C.K. Ng, D.V. Reames, Astrophys. J. **625**, 474–495 (2005)
A.J. Tylka, M.A. Lee, Astrophys. J. **646**(2), 1319–1334 (2006)
G.P. Zank, H.L. Pauls, I.H. Cairns, G.M. Webb, J. Geophys. Res. **101**, 457–478 (1996)
G.P. Zank, W.K.M. Rice, J.A. le Roux, W.H. Matthaeus, Astrophys. J. **556**(1), 494–500 (2001)

Space Sci Rev (2007) 130: 231–242
DOI 10.1007/s11214-007-9156-8

# $^3$He-Rich Solar Energetic Particle Events

**G.M. Mason**

Received: 16 November 2006 / Accepted: 2 February 2007 /
Published online: 28 April 2007


**Abstract** $^3$He-rich solar energetic particle (SEP) events show huge enrichments of $^3$He and association with kilovolt electrons and Type-III radio bursts. Observations from a new generation of high resolution instruments launched on the Wind, ACE, Yohkoh, SOHO, TRACE, and RHESSI spacecraft have revealed many new properties of these events: the particle energy spectra are found to be either power-law or curved in shape, with the $^3$He spectrum often being distinctly different from other species. Ultra-heavy nuclei up to $>200$ amu are found to be routinely present at average enrichments of $>200$ times solar-system abundances. The high ionization states previously observed near $\sim 1$ MeV/nucleon have been found to decrease towards normal solar coronal values in these events. The source regions have been identified for many events, and are associated with X-ray jets and EUV flares that are associated with magnetic reconnection sites near active regions. This paper reviews the current experimental picture and theoretical models, with emphasis on the new insights found in the last few years.

**Keywords** Acceleration of particles · Sun: flares · Sun: activity · Sun: coronal mass ejections (CMEs) · Sun: particle emission

## 1 Introduction

Small solar energetic particle (SEP) events with $^3$He/$^4$He ratios greatly enhanced over solar system abundances were discovered in the 1970s and immediately became the focus of experimental and theoretical studies which sought to explain the mechanisms that could lead to the observed $10^3$–$10^4$-fold enhancements. Figure 1 shows an example of an event with $^3$He/$^4$He $> 1$ that also had only upper limits of $^2$H and $^3$H, thereby ruling out production by nuclear reactions. By the mid 1980s it had been established that these events were associated with kilovolt electrons, Type-III radio bursts, and sometimes small X-ray flares. In addition

G.M. Mason (✉)
Applied Physics Laboratory, Johns Hopkins University, Laurel, MD 20723, USA
e-mail: glenn.mason@jhuapl.edu

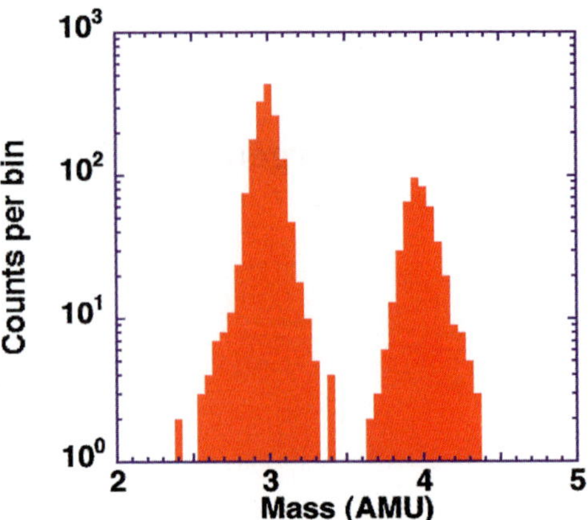


**Fig. 1** He mass histogram for event of March 21, 1999, where $^3$He/$^4$He $\gg 1$ at 750 keV/nucleon (from Mason et al. 2002)

they showed enrichments of heavier nuclei such as Fe by a factor of $\sim$10 over solar abundances, and the ionization states of these heavier species were significantly larger than in the corona or in large shock-associated SEP events. Theoretical investigations showed that the most promising mechanism to explain the $^3$He enrichment was plasma resonance heating that could single out the rare isotope due to its unique charge to mass ratio (see e.g., reviews by Kocharov and Kocharov 1984; Reames 1999).

## 2 New Observations

### 2.1 Spectral Forms

Figure 2 shows differential energy spectra of two $^3$He-rich events observed with high resolution spectrometers on the ACE spacecraft (ULEIS, Mason et al. 1998; and SIS, Stone et al. 1998). Figure 2(a) shows "Class-1" spectra which are power-laws or broken power laws for all species. The abundance ratios of $^3$He/$^4$He $\sim 0.2$ and Fe/O $\sim 1.8$ are typical, and are nearly constant over the entire energy range. Figure 1(b) shows an event with "Class-2" spectra which are curved for $^3$He and Fe, with significantly different shape from $^4$He. Indeed, in this event the $^3$He/$^4$He ratio is $<1$ below 200 keV/nucleon, increasing steadily to $^3$He/$^4$He $= 4$ at 800 keV/nucleon, then approaching $\sim$1 near 10 MeV/nucleon.

Figure 3 shows $^3$He/$^4$He ratios for three events that illustrate the range of variability observed. In addition to the two events shown in Fig. 2, there is a third, Sept. 27, 2000, where all the spectra were power laws (Class-1) but the $^3$He spectrum was significantly flatter than the $^4$He, O, and Fe spectra. This led to a factor of $\sim$30 increase in the $^3$He/$^4$He between $\sim$150 keV/nucleon and 10 MeV/nucleon. These spectra show that in many cases there is no single $^3$He/$^4$He ratio that can be used to characterize an event; rather, the energy of the ratio measurement must be specified. The peaking in the $^3$He/$^4$He ratio usually occurs near $\sim$1 MeV/nucleon (see also Möbius et al. 1982).

### 2.2 Heavy Ion Abundance Enhancements

Heavy ion enrichments in the mass range up to Fe have long been known to be a property of $^3$He-rich events (e.g., Mason et al. 1986), and the ACE instruments have sufficient sensitivity at low energies to measure the correlation between the heavy ion enhancements

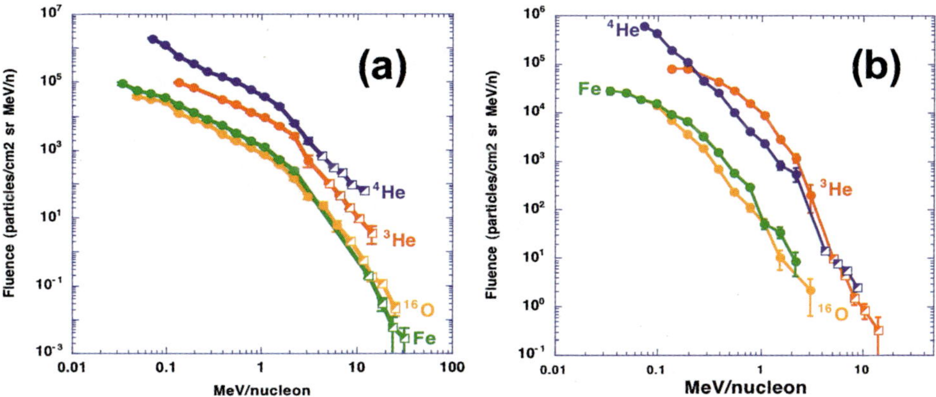


**Fig. 2** Examples of different types of differential energy spectra in $^3$He-rich events. **a** broken power law spectra (Sept. 9, 1998 event), and **b** curved $^3$He and Fe spectra (March 21, 1999 event) where the $^3$He/$^4$He ratio changes significantly over the range ~0.2–10 MeV/nucleon (from Mason et al. 2002)

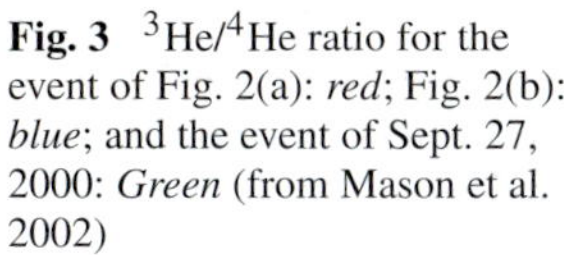

**Fig. 3** $^3$He/$^4$He ratio for the event of Fig. 2(a): *red*; Fig. 2(b): *blue*; and the event of Sept. 27, 2000: *Green* (from Mason et al. 2002)

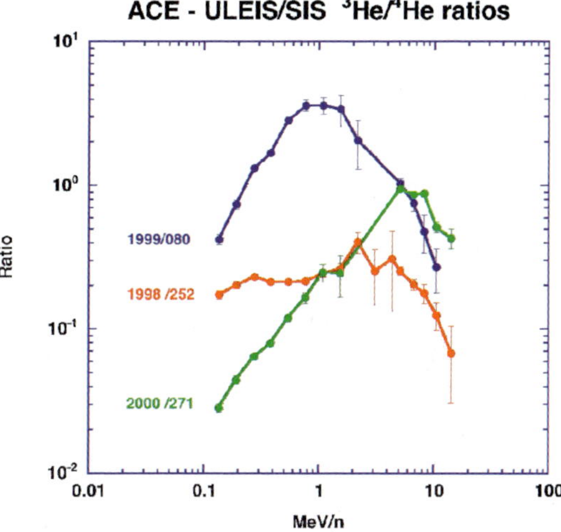


for several elements in individual events, as shown in Fig. 4. In the figure, each element's ratio compared to O is plotted against the event's Fe/C ratio. The enhancements are obviously highly significant: over the range Ne–Ca, the regression coefficients range from 0.65 to 0.91, and thus each has a <0.1% probability of coming from a random distribution. This contrasts with the situation at several MeV/nucleon, where the heavy ion enhancements are uncorrelated (Reames et al. 1994). Since most of the material accelerated in these events is at lower energies, the correlations shown in Fig. 4 must be reproduced in any model that explains the bulk heavy ion abundances in $^3$He-rich events.

## 2.3 Ionization States

A critical clue to the origin of $^3$He-rich events comes from the ionization state of the heavy ions, which are higher than coronal values, or values from large SEP events (Klecker et al. 1984). Two advanced instruments (STOF, Hovestadt et al. 1995; SEPICA, Möbius et

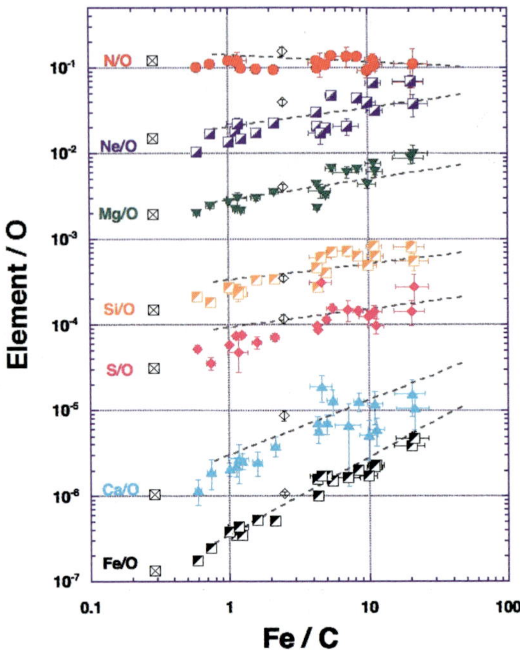


**Fig. 4** Correlation between heavy ion enhancements in 20 $^3$He-rich events: the enhancements compared to O are plotted vs. each event's Fe/O ratio. Each event contributes a single point to each element plot. The correlations are statistically highly significant (see text) (Mason et al. 2004)

al. 1998) on the SOHO and ACE spacecraft have allowed study of Fe ionization states in individual $^3$He-rich SEP events and have extended the energy range of the measurements almost an order of magnitude below previous work. Figure 5 shows ionization states from 180–500 keV/nucleon from 4 events observed with SEPICA, and a sum of STOF for events 2–4 for 10–100 keV/nucleon. The dashed lines show ionization states calculated with the model of Kocharov et al. (2000) which includes effects of radiation as well as impact ionization by protons and electrons. At higher energies, the ionization states reflect stripping in the ambient plasma, while the low energy curves reflect the ambient temperature. It may be that the disagreement between the energy range of the calculated ionization increase vs. the observed increase is due to adiabatic deceleration of the particles near the Sun (Kartavykh et al. 2005) although other possibilities also exist (Klecker et al. 2006). If the energy dependence of the ionization state shown in Fig. 5 is due to stripping, then the most likely source region is low in the corona (e.g., Dröge et al. 2006). This is consistent with electron and radio observations that suggest a low coronal source (Klein and Posner 2005; Wang et al. 2006a). However, we must note that recent timing studies with suprathermal electrons compared to the ions suggest ion acceleration higher in the corona (Wang et al. 2005) so the observational evidence for the ion source location is mixed.

### 2.4 Ultra-Heavy Nuclei Abundances

Ultra-heavy nuclei (i.e. mass range well above Fe) were first detected in SEP events by Shirk and Price (1973) who examined a command window module from the Apollo program. Reames (2000) surveyed the period Nov. 1994 to May 2000 using a high sensitivity instrument (LEMT, von Rosenvinge et al. 1995) on Wind and found 9 cases of impulsive events with large enrichments of nuclei up to $Z = 82$; 6 of the events were observed in Sept. 1998, and one each in Sept. 1997, Aug. 1999, and May 2000.

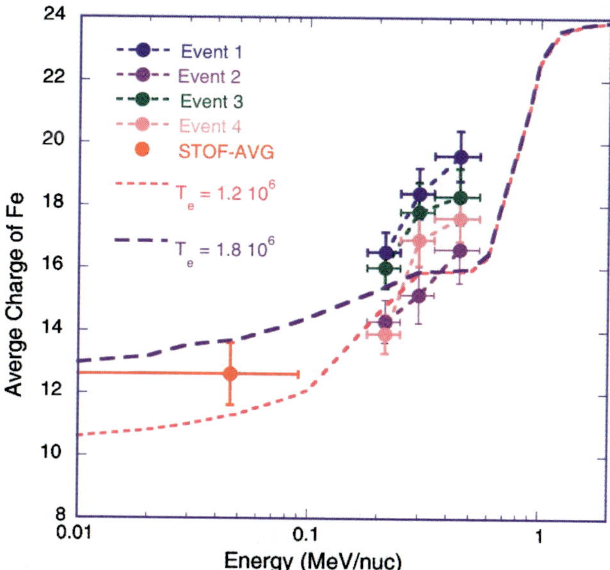


**Fig. 5** Ionization states of Fe measured by the ACE/SEPICA instrument in 4 $^3$He-rich SEP events on (1) Sept. 9, 1998; (2) July 3, 1999; (3) July 20, 1999; (4) and May 1, 2000. The low energy *red* point is the average of events 2–4 from the SOHO/STOF instrument (from Klecker et al. 2006)

The ULEIS instrument on ACE, whose low energy threshold results in greater statistical accuracy than earlier instruments, surveyed the period Sept. 1997 through April 2003 and found that UH nuclei were commonly seen in impulsive events, as shown, e.g., in Fig. 6. Note from the figure that when the $^3$He (*red line*) pops up, generally there are $m > 70$ amu ions observed, while in periods with no $^3$He enrichment, the ions are not seen. Summing up about 297 days of observing time when intensities were low and Fe/O $\sim 1$ or greater, Mason et al. (2004) constructed mass histograms over the range 50–250 amu shown in Fig. 7. Although individual spikes in the histogram are not significant, the data shows that masses are present up to the top of the periodic table, and that the broad enhancements seen in solar system abundances around 85, 130, and 190 amu are seen in the impulsive SEP histogram. Figure 7(b) compares smoothed data from the upper panel with smoothed solar system abundances, showing that the relative enhancement of SEPs increases with mass.

The high ionization states of heavy ions in $^3$He-rich SEP events poses a significant problem in models of preferential acceleration, since these models require different $Q/M$ ratios to be effective, while the elements will have similar or identical $Q/M$ ratios if fully stripped (e.g., Luhn et al. 1987). Reames et al. (1994) suggested that if the original acceleration took place in a $\sim$3–5 MK plasma then the different $Q/M$ ratios of heavy elements might make an enhancement mechanism feasible (after acceleration the ions were assumed to be stripped to the values observed at 1 AU). To explore the enhancement for UH ions, we assume a 3.2 MK plasma, and show in Fig. 8 the enhancements as a function of $Q/M$ (see also discussion in Mason et al. 2004). The *average* enhancement is close to a factor of 200. In events with very large Fe/C ratios (see Fig. 4) the UH nuclei are enhanced by up to a factor of $\sim$1000, greater than the $^3$He/$^4$He enrichment in the same events (Mason et al. 2004). Since the UH nuclei at temperatures of a few MK have only a fairly narrow range of ionization states, together with a large range of masses, the $Q/M$ ratios for these nuclei can be expected to cover an nearly continuous range, making it difficult to see how a specific resonance mechanism such as that invoked for $^3$He could also explain the UH enhancements (Mason et al. 2004). However, detailed exploration of this issue has not yet

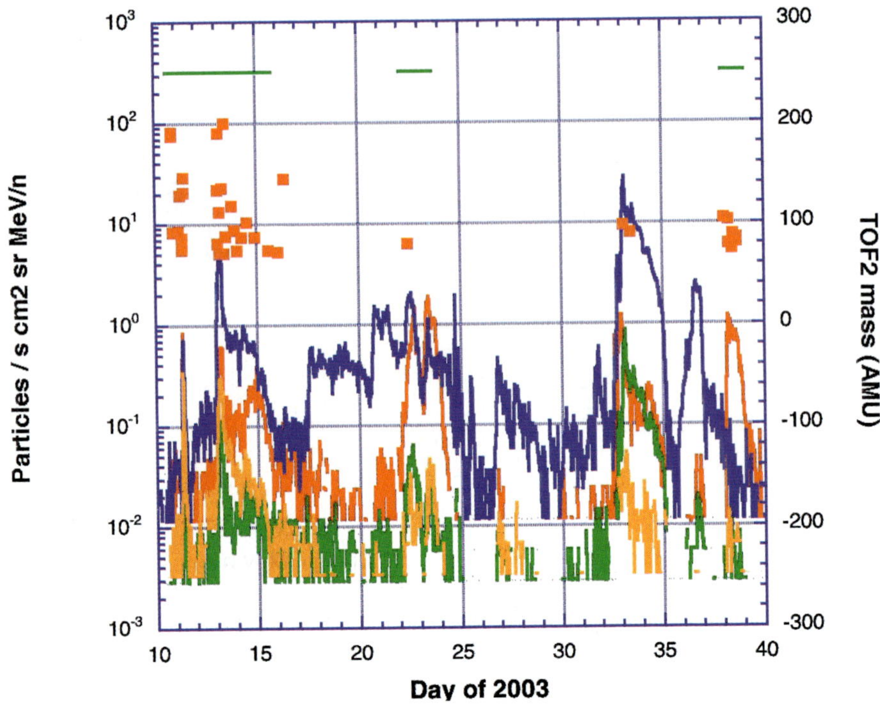


**Fig. 6** *Solid lines*: 1-hr average intensity of 225–320 keV/nucleon $^3$He (*red*), $^4$He (*blue*), O (*green*) and Fe (*orange*) for a 30 day period in 2003. *Red squares*: arrival times and mass of ions above 70 amu (right scale). *Green horizontal lines*: impulsive periods for sum

been done. A summary of the relative abundances in $^3$He-rich flares is given in Table 1 (from Mason et al. 2004).

## 2.5 Sources of $^3$He-rich SEP Events

Less than one-half of $^3$He-rich events have well identified sources on the Sun (e.g., Reames 1999) which may in part be due to the small size of the events. Recent searches for $^3$He-rich event sources have made exciting progress by using multiple sensors of higher sensitivity than available in the past. Krucker et al. (2003) used RHESSI, TRACE, Wind, and ACE measurements to produce movies of $^3$He-rich events that were characterized by jets and magnetic field topologies associated with reconnection. Wang et al. (2006b) surveyed the origin of 25 $^3$He-rich SEP events observed with ACE/ULEIS, as illustrated in Fig. 9. During a period when ACE/ULEIS observed multiple injections of $^3$He (Fig. 9 left), SOHO EUV observations (Fig. 9 center) showed a flaring active region in the western hemisphere, which was also identified in imaging radio observations of Type-III bursts (Pick et al. 2006). The white light images showed jet-like ejections around the particle injection times. Potential field source surface (PFSS) calculations (Fig. 9 right) showed that the source region had magnetic field lines that were connected to the ecliptic. The source regions found by Wang et al. (2006b) were generally active regions with nearby coronal holes. They interpreted the jets as signatures of magnetic reconnection ("footpoint exchange") between closed and open field lines. $^3$He enrichments would be expected whenever Earth-connected field lines

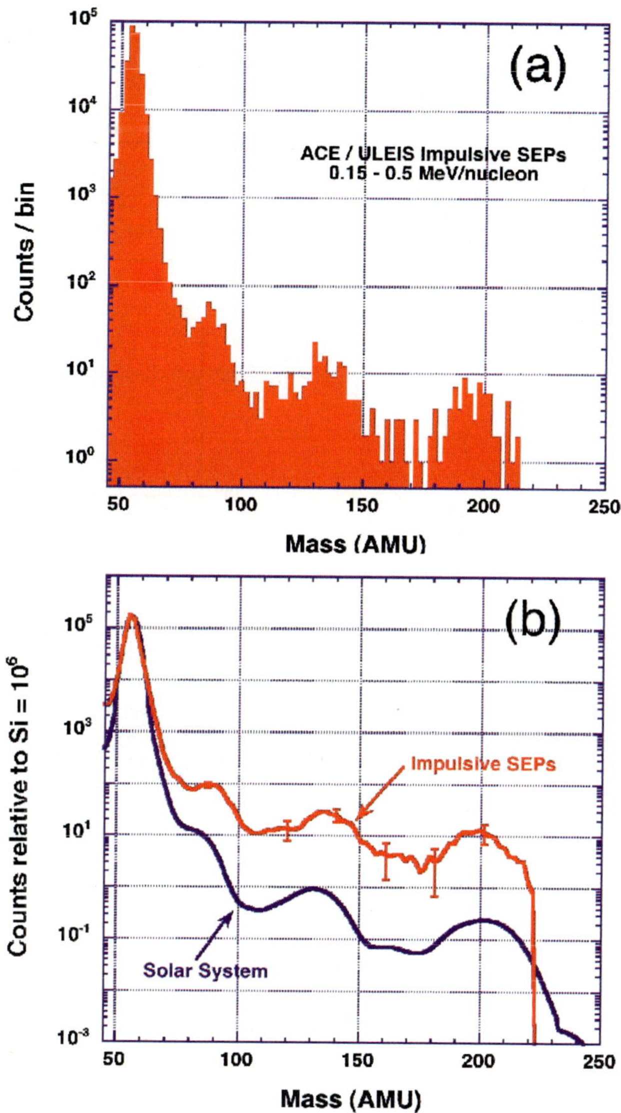


**Fig. 7** **a** mass histogram of 0.15–0.5 MeV/nucleon Fe peak and heavy nuclei during all impulsive periods. **b** *red curve*: 5-bin smoothed data from top panel normalized at the Fe peak; *blue curve*: solar system abundances smoothed by ULEIS mass resolution (from Mason et al. 2004)

undergo footpoint exchanges with nearby source regions (for a discussion of the association of EUV jets and reconnection, see Shimojo and Shibata 2000; see also Wang and Sheeley 2002).

## 2.6 Periods of Continuous $^3$He Emission

A qualitatively new feature of $^3$He-rich SEPs discovered on ACE has been the existence of periods of nearly continuous emission of suprathermal $^3$He. Wiedenbeck et al. (2003, 2005) surveyed all time periods from the 1997 launch of ACE through solar maximum, and showed that $^3$He was present the majority of the days over the energy range up to several MeV/nucleon. Among these periods are times when the suprathermal $^3$He inten-

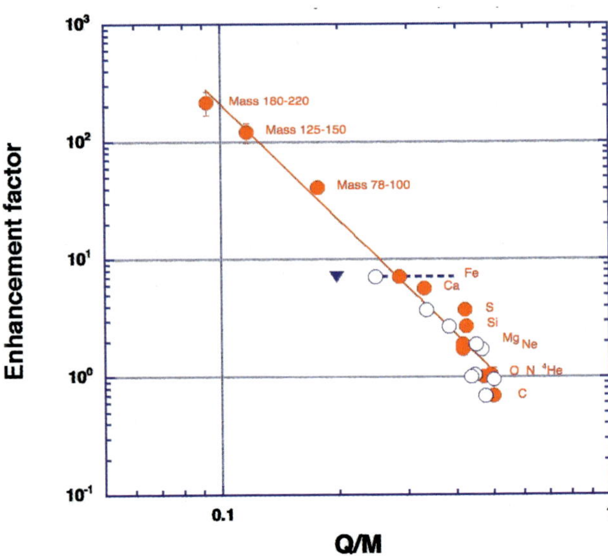


**Fig. 8** Abundance enhancement factor for impulsive event survey plotted vs. $Q/M$ ratio, with charge state calculated assuming 3.2 MK equilibrium plasma (*red filled circles*), and direct measurement of energetic particles in several large SPE events (*blue symbols*). The *red line* is a power law fit to the *red points* and has a slope of $-3.26$. Energy range of the particle data is 320–450 keV/nucleon, Mason et al. (2004)

**Table 1** $^3$He-rich solar energetic particle event abundances

| Species | Mass | $^3$He-rich SEP abundance[a] | Gradual SEP abundance[b] | Solar system abundance[c] | $^3$He-rich SEP vs. reference |
|---|---|---|---|---|---|
| $^4$He | 4 | $54 \pm 14$ | $57 \pm 3$[d] | | $0.95 \pm 0.24$[d] |
| C | 12 | $0.322 \pm 0.003$ | $0.465 \pm 0.009$ | | $0.69 \pm 0.01$ |
| N | 14 | $0.129 \pm 0.002$ | $0.124 \pm 0.003$ | | $1.04 \pm 0.03$ |
| O | 16 | $=1.000 \pm 0.006$ | $=1 \pm 0.01$ | | $1.00 \pm 0.01$ |
| Ne | 20 | $0.261 \pm 0.003$ | $0.152 \pm 0.004$ | | $1.72 \pm 0.05$ |
| Mg | 24 | $0.370 \pm 0.003$ | $0.196 \pm 0.004$ | | $1.89 \pm 0.04$ |
| Si | 28 | $0.409 \pm 0.004$ | $0.152 \pm 0.004$ | | $2.69 \pm 0.07$ |
| S | 32 | $0.118 \pm 0.015$ | $0.0318 \pm 0.0007$ | | $3.70 \pm 0.47$ |
| Ca | 40 | $0.060 \pm 0.003$ | $0.0106 \pm 0.0004$ | | $5.66 \pm 0.38$ |
| Fe | 56 | $0.950 \pm 0.005$[e] | $0.134 \pm 0.004$ | =1 | $7.09 \pm 0.21$ |
| | 78–100 | $(9.6 \pm 1.1)$e–04 | | 1.7e–04 | $40.7 \pm 4.5$ |
| | 125–150 | $(3.1 \pm 0.6)$e–04 | | 1.95e–05 | $119.6 \pm 23.0$ |
| | 180–220 | $(2.2 \pm 0.5)$e–04 | | 7.63e–06 | $215.4 \pm 49.5$ |

[a] 385 keV nucleon$^{-1}$, this work

[b] 5–12 MeV nucleon$^{-1}$, Reames (1995)

[c] Meteoritic, Anders and Grevesse (1989)

[d] Uncertainty primarily due to instrumental efficiency (see discussion in Desai et al. 2003)

[e] Multiplied by a factor of 0.92 to correct for larger mass range summed in Fe group peak

sity is far above background. In some cases (e.g. Fig. 9) these multi-day periods contain individual injections that stand out over a lower but continuously present $^3$He presence.

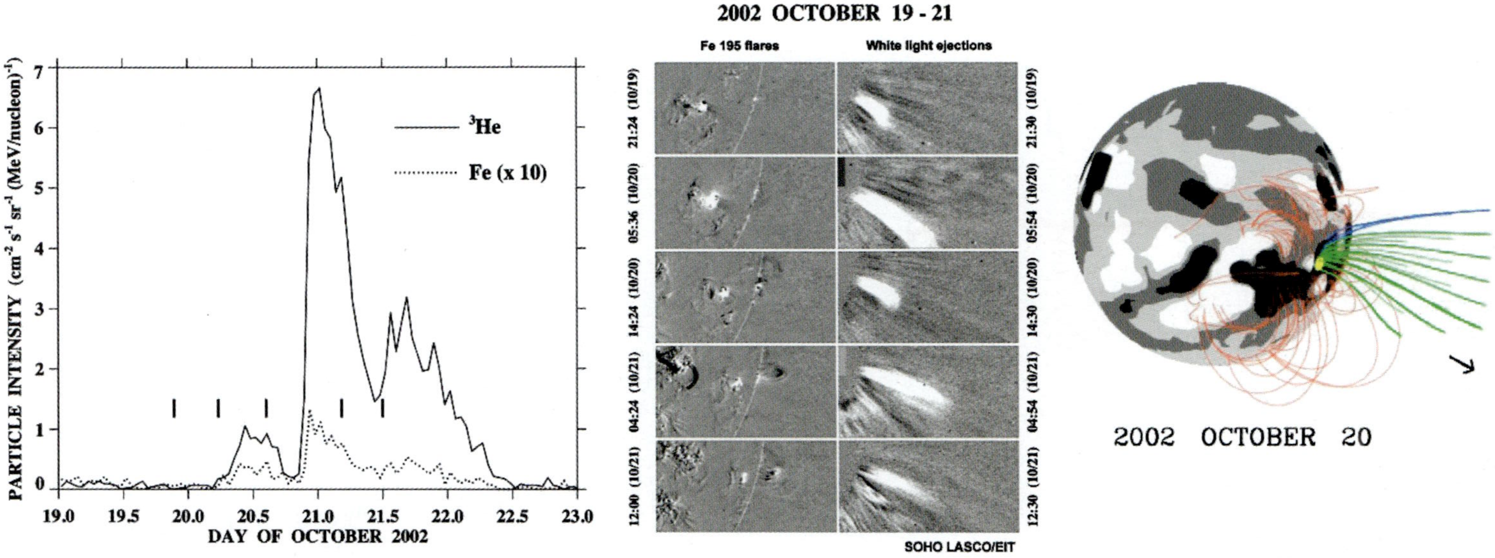

**Fig. 9** *Left*: extended $^3$He-rich period 19–22 Oct. 2002 showing high $^3$He abundance over several days. *Center*: successive jets from associated flaring active region, with each flare marked by vertical line in left panel; *Right*: PFSS calculation showing source location (*yellow dot*), closed (*red*) and open field lines (*green*), and open field lines to ecliptic (*blue*). Adapted from Wang et al. (2006b)

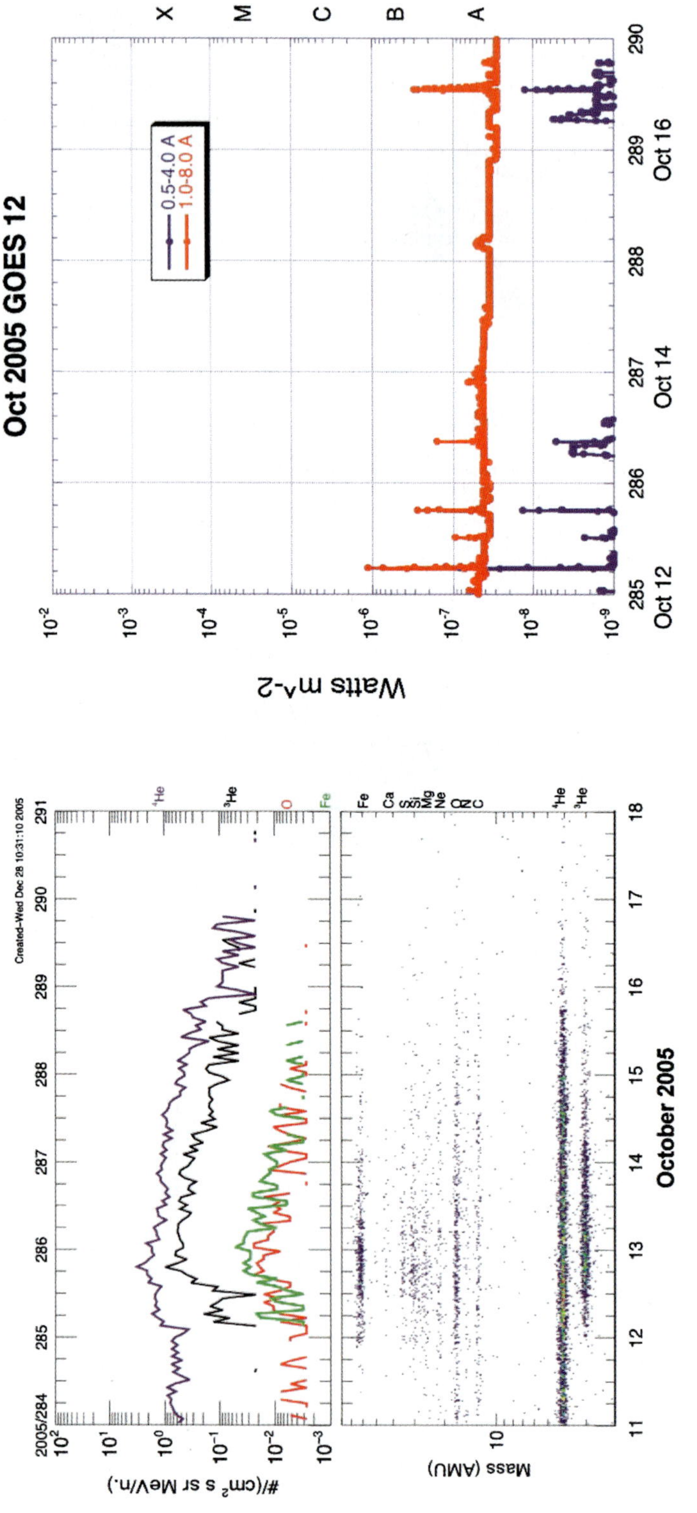

**Fig. 10** *Left*: example of a multiday period in where ~300 keV/nucleon $^3$He is present without clear ion injections, and X-ray activity (*right*) is quite low

A different situation is illustrated in Fig. 10, which shows a multi-day $^3$He-rich period with ion injections late on Oct. 11 and early on Oct. 12, and none thereafter event though the $^3$He emission continues. During this period, roughly Oct. 12–16, 2005, AR10814 was in the western hemisphere (at $\sim$S10), and passed over the limb on the 16th, in coincidence with the tapering off of the $^3$He. Although there is evidence for ion injection early in the period, for most of the time the $^3$He appears to be continuously present, perhaps suggesting that a more nearly steady process associated with reconnection is associated with the $^3$He acceleration. There were $\sim$15 such periods observed on ACE during the recent solar activity maximum.

## 3 Progress in Theoretical Models

Fisk (1978) was the first to propose that $^3$He was preferentially heated by electrostatic ion cyclotron (EIC) waves due to its unique charge to mass ratio. Since that time a number of theories have been proposed that use EIC or other plasma waves (e.g., Möbius et al. 1982; e.g., Temerin and Roth 1992; Roth and Temerin 1997; Zhang 1999; Paesold et al. 2003; Liu et al. 2004). Some of the recent models have included predictions for the spectral forms of $^3$He and $^4$He, and the model of Liu et al. (2006) has produced reasonable fits to curved spectra such as that shown in Fig. 2(b). Some of these models include preferential heating of heavy ions by a higher harmonic as qualitatively suggested by Fisk, while others have used cascading MHD turbulence (e.g., Miller 1997, 1998). While much progress has been made, challenges remain: for example, none of the published theories addresses the enhancements of UH nuclei, and given the large range of $Q/M$ likely present in this mass range, it is not clear that the current ideas will extrapolate into that mass range.

## 4 Summary

New observational capabilities have greatly expanded our knowledge of $^3$He-rich SEP events: the $^3$He spectral shape has been found to often differ from other ions, perhaps indicating that a key component of its energization is different from the other ions; UH nuclei have been discovered to be routinely present in these events, with enhancements sometimes larger than the $^3$He enrichment, reaching values of $\sim$1000 over solar system material; emission of energetic $^3$He from the Sun is often observed nearly continuously over periods of several days, and at a lower level, $^3$He is present in the interplanetary medium near Earth a majority of the days during the peak of the sunspot cycle; source identification of regions of $^3$He emission have been achieved using ion, electron, imaging radio, EUV, and hard and soft X-rays, revealing an association with EUV and white light jets previously identified as sites of magnetic reconnection from newly emerging flux; ionization states of the heavy ions show a strong energy dependence which suggests stripping, most likely in a low coronal source; however, impulsive electron timing studies seem to imply a high coronal source (several $R_s$) and so may contradict the ionization data, or point to a second source. Further progress is anticipated with new theoretical studies, along with improved observations from the recently launched STEREO mission, and future missions to the inner heliosphere such as Solar Orbiter and Inner Heliospheric Sentinels.

**Acknowledgements** For permission to use copyrighted material we thank the American Astronomical Society (Figs. 1–4, 7–9). This work was supported by NASA grant NNG04GJ51G at the Johns Hopkins University Applied Physics Laboratory. The GOES X-ray data in Fig. 10 was obtained from the NOAA/SEC web site.

## References

E. Anders, N. Grevesse, Geochim. Cosmochim. Acta **53**, 197 (1989)
M.I. Desai et al., Astrophys. J. **588**, 1149 (2003)
W. Dröge, J.J. Kartavykh, B. Klecker, G.M. Mason, Astrophys. J. **645**, 1516 (2006)
L.A. Fisk, Astrophys. J. **224**, 1048 (1978)
D. Hovestadt et al., Solar Phys. **162**, 441 (1995)
J.J. Kartavykh, W. Dröge, V.M. Ostryakov, G.A. Kovaltsov, Solar Phys. **227**, 123 (2005)
B. Klecker et al., Astrophys. J. **281**, 458 (1984)
B. Klecker et al., Adv. Space Res. **38**, 493 (2006)
K.-L. Klein, A. Posner, Astron. Astrophys. **438**, 1029 (2005)
L.G. Kocharov, G.E. Kocharov, Space Sci. Rev. **38**, 89 (1984)
L.G. Kocharov, G.A. Kovaltsov, J. Torsti, V.M. Ostryakov, Astron. Astrophys. **357**, 716 (2000)
S. Krucker, R.P. Lin, E.P. Kontar, G.M. Mason, M.E. Wiedenbeck, EOS, Trans. Am. Geophys. U 2003 Fall Meeting Suppl., paper SH11D, 2003
S. Liu, V. Petrosian, G.M. Mason, Astrophys. J. Lett. **613**, L81 (2004)
S. Liu, V. Petrosian, G.M. Mason, Astrophys. J. **636**, 462 (2006)
A. Luhn, B. Klecker, D. Hovestadt, E. Möbius, Astrophys. J. **317**, 951 (1987)
G.M. Mason, D.V. Reames, B. Klecker, D. Hovestadt, T.T. von Rosenvinge, Astrophys. J. **303**, 849 (1986)
G.M. Mason et al., Space Sci. Rev. **86**, 409 (1998)
G.M. Mason et al., Astrophys. J. **574**, 1039 (2002)
G.M. Mason et al., Astrophys. J. **606**, 555 (2004)
J.A. Miller, Astrophys. J. **491**, 939 (1997)
J.A. Miller, Space Sci. Rev. **86**, 79 (1998)
E. Möbius et al., Space Sci. Rev. **86**, 449 (1998)
E. Möbius, M. Scholer, D. Hovestadt, B. Klecker, G. Gloeckler, Astrophys. J. **259**, 397 (1982)
G. Paesold, R. Kallenbach, A.B. Benz, Astrophys. J. **582**, 495 (2003)
M. Pick, G.M. Mason, Y.-M. Wang, C. Tan, L. Wang, Astrophys. J. **648**, 1247 (2006)
D.V. Reames, Adv. Space Res. **15**, 41 (1995)
D.V. Reames, Space Sci. Rev. **90**, 413 (1999)
D.V. Reames, Astrophys. J. Lett. **540**, L111 (2000)
D.V. Reames, J.P. Meyer, T.T. von Rosenvinge, Astrophys. J. Suppl. **90**, 649 (1994)
I. Roth, M. Temerin, Ap. J. **477**, 940 (1997)
M. Shimojo, K. Shibata, Astrophys. J. **542**, 1100 (2000)
E.K. Shirk, P.B. Price, in *Proc. 13th ICRC*, vol. 2, Denver (1973), p. 1474
E.C. Stone et al., Space Sci. Rev. **86**, 357 (1998)
M. Temerin, I. Roth, Astrophys. J. Lett. **391**, L105 (1992)
T.T. von Rosenvinge et al., Space. Sci. Rev. **71**, 155 (1995)
Y.-M. Wang, N.R. Sheeley Jr., Astrophys. J. **575**, 542 (2002)
L. Wang, R.P. Lin, S. Krucker, G.M. Mason, in: *Solar Wind 11/SOHO 16, A Study of the Solar Injection for Eleven Impulsive Electron/*$^3$*He-rich SEP Events*. Whistler, Canada, 2005
L. Wang, R.P. Lin, S. Krucker, J.T. Gosling, Geophys. Res. Let. **33**, L03106 (2006a) doi: 10.1029/2005GL024434
Y.-M. Wang, M. Pick, G.M. Mason, Astrophys. J. **639**, 495 (2006b)
M.E. Wiedenbeck et al., in *AIP Conf. Proc. #679, How Common are Flare Suprathermals in the Inner Heliosphere?* ed. by M. Velli, R. Bruno, F. Malara (AIP, New York, 2003), p. 652
M.E. Wiedenbeck et al., in *ICRC 29, The Time Variation of Energetic* $^3$*He in Interplanetary Space from 1997 to 2005*, Pune, India, 2005
T.X. Zhang, Astrophys. J. **518**, 954 (1999)

Space Sci Rev (2007) 130: 243–253
DOI 10.1007/s11214-007-9219-x

# Evidence for a Two-Stage Acceleration Process in Large Solar Energetic Particle Events

**M.I. Desai · G.M. Mason · R.E. Gold · S.M. Krimigis · C.M.S. Cohen · R.A. Mewaldt · J.E. Mazur · J.R. Dwyer**

Received: 7 December 2006 / Accepted: 19 May 2007 / Published online: 18 July 2007


**Abstract** Using high-resolution mass spectrometers on board the Advanced Composition Explorer (ACE), we surveyed the event-averaged ∼0.1–60 MeV/nuc heavy ion elemental composition in 64 large solar energetic particle (LSEP) events of cycle 23. Our results show the following: (1) The Fe/O ratio decreases with increasing energy up to ∼10 MeV/nuc in ∼92% of the events and up to ∼60 MeV/nuc in ∼64% of the events. (2) The rare isotope $^3$He is greatly enhanced over the corona or the solar wind values in 46% of the events. (3) The heavy ion abundances are not systematically organized by the ion's $M/Q$ ratio when compared with the solar wind values. (4) Heavy ion abundances from C–Fe exhibit systematic $M/Q$-dependent enhancements that are remarkably similar to those seen in $^3$He-rich SEP events and CME-driven interplanetary (IP) shock events. Taken together, these results confirm the role of shocks in energizing particles up to ∼60 MeV/nuc in the majority of large SEP events of cycle 23, but also show that the seed population is not dominated by ions originating from the ambient corona or the thermal solar wind, as previously believed. Rather, it appears that the source material for CME-associated large SEP events originates predominantly from a suprathermal population with a heavy ion enrichment pattern that is organized according to the ion's mass-per-charge ratio. These new results indicate that current LSEP models must include the routine production of this dynamic suprathermal seed population as a critical pre-cursor to the CME shock acceleration process.

M.I. Desai (✉)
Southwest Research Institute, 6220 Culebra Road, San Antonio, TX 78238, USA
e-mail: mdesai@swri.edu

G.M. Mason · R.E. Gold · S.M. Krimigis
The Johns Hopkins University, Applied Physics Laboratory, Laurel, MD 20723, USA

C.M.S. Cohen · R.A. Mewaldt
California Institute of Technology, Pasadena, CA 91125, USA

J.E. Mazur
The Aerospace Corporation, 15049 Conference Center Drive, CH3/210, Chantilly, VA 20151, USA

J.R. Dwyer
Florida Institute of Technology, Melbourne, FL 32901, USA

**Keywords** Sun: energetic particles · Sun: coronal mass ejections · Sun: flares

## 1 Introduction

Early observations of solar energetic particle or SEP events extending up to GeV energies were made with ground-based ionization chambers and neutron monitors (Forbush 1946; Meyer et al. 1956). Such events, also known as ground level events or GLEs, were closely associated with the maximum of H$\alpha$ flares on the Sun. Consequently, it was presumed that there was a causal relationship between the flare and the energetic particles observed at 1 AU.

Later, however, on the basis of a close association between SEP events and slow-drifting Type II and various kinds of Type IV radio bursts, Wild et al. (1963) proposed that the energetic particles might be accelerated at magnetohydrodynamic shock waves that typically accompanied the flares. In addition, Lin (1970) reported that the SEP events observed at 1 AU could essentially be grouped into two types. First, "pure" electron events that were closely associated with flares and metric Type III emission, and second, "mixed" events where protons, relativistic electrons, and flares were associated with Type II/IV radio events. Lin (1970) proposed two distinct acceleration processes for the pure and the mixed SEP events.

Using Skylab observations, Kahler et al. (1978) were the first to report a close association between coronal mass ejections (CMEs) and large solar proton events. They suggested that the CME could either create open field lines for flare particles to escape into the interplanetary medium or that the protons could be accelerated near a region above or around the outward moving ejecta far above the flare site. Subsequently, detailed analyses of flare durations, longitudinal distributions from multi-spacecraft observations, high resolution ionic charge state and elemental composition measurements, and clearer associations with radio bursts led most researchers in the 1990s to accept the view that the SEP events observed at 1 AU belong to two classes, namely impulsive and gradual (e.g., Kahler et al. 1978, 1984; Cliver et al. 1982; Kocharov 1983; Luhn et al. 1984; Mason et al. 1984; Cane et al. 1986; Reames 1988).

In this two-class picture the gradual events occurred as a result of diffusive acceleration of ambient coronal or solar wind material at CME-driven coronal and interplanetary (IP) shocks, while the impulsive events were attributed to stochastic acceleration of coronal material heated up to $\sim$10 MK during magnetic reconnection in solar flares (e.g., Reames 1999). The gradual or CME shock-accelerated events lasted several days and had larger fluences, while the impulsive or flare-accelerated events lasted a few hours and had smaller fluences. Impulsive events were observed when the observer was magnetically connected to the flare site, while ions accelerated at the expanding large-scale CME-driven shocks populate magnetic field lines over a broad range of longitudes (Cane et al. 1988).

The impulsive SEP events were electron-rich and associated with Type III radio bursts. These events also had $^3$He/$^4$He ratios enhanced between factors of $10^3$–$10^4$ and Fe/O ratios enhanced by up to a factor of 10 over the corresponding solar wind values, and had Fe with ionization states up to $\sim$20. In contrast, the gradual events were proton-rich, were associated with Type II bursts, had average Fe/O ratios of $\sim$0.1 with Fe ionization states of $\sim$14, and were assumed to have $^3$He/$^4$He ratios similar to those measured in the solar wind (Reames 1999; Cliver 2000).

Since those earlier studies, instruments with greater sensitivity and resolution on board the Advanced Composition Explorer (ACE) (Stone et al. 1998a) have provided three major observational advances, making it possible to re-examine questions about the origin of the seed populations and probe details of the acceleration mechanisms for individual large SEP events. First, the newly developed capability of directly measuring the solar wind ion composition and its variations (Gloeckler et al. 1992; Geiss et al. 1996) has provided definitive solar wind abundances that can be compared with the SEPs. Second, improved ionization state measurements over a broader energy range have enabled us to investigate the energy dependence and the event-to-event variability of the ionic charge states (e.g., Oetliker et al. 1997; Mazur et al. 1999; Möbius et al. 1999, 2000; Klecker et al. 2006). And third, sophisticated mass spectrometers have enabled us to identify many SEP elements over a broad energy range, making it possible to quantify the affects of particle energy spectra on the average abundances.

## 2 Observations

In this paper we use data from the Ultra-Low-Energy Isotope Spectrometer (ULEIS: Mason et al. 1998) and the Solar Isotope Spectrometer (SIS: Stone et al. 1998b) on board ACE. We selected 64 large SEP events from NOAA's list of 85 solar proton events (SPEs) that affected Earth's environment between November 1997 and January 2005. Hereafter we refer to these 64 events as large SEP events or LSEP events. Details of the selection of events and their sampling intervals were provided by Desai et al. (2006a).

### 2.1 $^3$He Enhancements

Figure 1a shows the 0.5–2.0 MeV/nuc $^3$He and $^4$He time-intensity profiles in a large SEP event that occurred on June 4, 1999 (from Mason et al. 1999). The temporal profiles of the two species are remarkably similar, which indicates that they probably have the same acceleration and transport history. In this event, the $^3$He is enriched by a factor of $16 \pm 3$ while the Fe/O ratio (not shown) is enhanced by about a factor of 10 over the corresponding solar wind values.

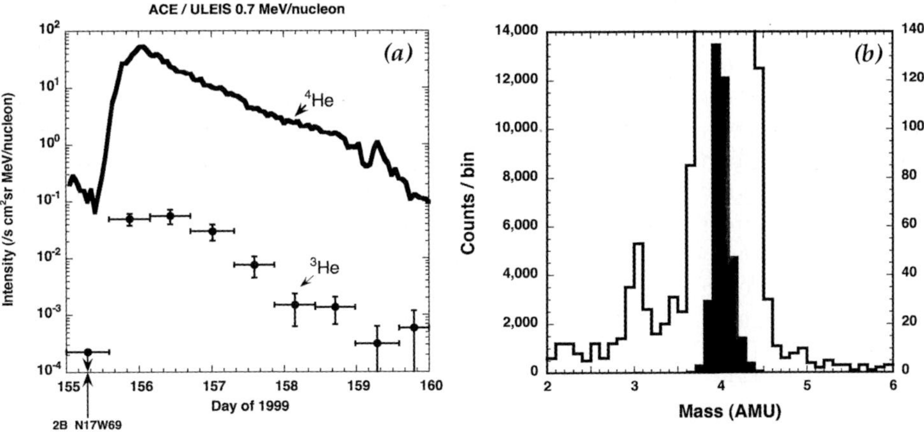


**Fig. 1** (**a**) Temporal profiles of ~0.7 MeV/nuc $^3$He and $^4$He ions in a large SEP event. (**b**) 0.5–2.0 MeV/nuc He mass histogram obtained during several large SEP events. The *right scale* corresponds to the open histogram (taken from Mason et al. 1999)

In this survey, we find that the 0.5–2.0 MeV/nuc $^3He/^4He$ ratio in 29 of the 64 events (~46%) is enhanced between factors ~4–150 over the corresponding solar wind value. Large enrichments in the $^3He/^4He$ ratio are also observed above ~10 MeV/nuc (e.g., Cohen et al. 1999; Wiedenbeck et al. 2000). Figure 1b shows the ~1 MeV/nuc He mass histogram from several such events. Notice that the $^3He$ is clearly resolved from $^4He$ and the background. In summary, our results (see Desai et al. 2006a) indicate the presence of flare-accelerated material during a sizable fraction of large SEP events of cycle 23.

### 2.2 Heavy Ion Abundances

Table 1 provides the 0.32–0.45 MeV/nuc heavy ion abundances averaged over 64 large SEP (LSEP) events and those measured in a variety of heliospheric populations. The LSEP averages are the arithmetic mean values for the 64 event sample; the abundances for each event are obtained by taking ratios of the fluences integrated over the duration of the event. Figure 2 compares the large SEP abundances with the solar wind values (von Steiger et al. 2000), as a function of $M/Q$ ratio. The charge states are average values measured in the slow solar wind (from von Steiger et al. 1997). These are: $He^{2+}$, $C^{5.38+}$, $N^{5.47+}$, $O^{6.05+}$, $Ne^{7.97+}$, $Mg^{9.5+}$, $Si^{8.57+}$, $S^{8.75+}$, $Ca^{9.02+}$, and $Fe^{9.84+}$.

From Fig. 2 we note the following: (1) The LSEP abundances are poorly correlated with the solar wind values. (2) C, N, O, Ne, and Mg have similar $M/Q$ values but exhibit highly unsystematic enhancements and depletions relative to the solar wind values, with C in particular being depleted by about a factor of 2. (3) The abundances of Si, Ca, and Fe are enhanced relative to the solar wind values.

Figure 3 shows the energy-dependent behavior of Fe/O in 37 of the 64 large SEP events in this survey. The Fe/O ratios are 0.11–0.14 MeV/nuc from ULEIS (Desai et al. 2006a); 3.3–10 MeV/nuc from LEMT (Reames and Ng 2004); and 12–60 MeV/nuc from SIS (Desai et al. 2006a). Note that the Fe/O ratio either decreases or remains constant with increasing

**Table 1** Heavy ion abundances in 64 Large SEP events at ACE, compared with those measured in other solar and heliospheric populations

| Species | Large SEPs[a] | Slow SW[b] | ISEPs[c] | Photosphere[d] | Corona[e] |
|---|---|---|---|---|---|
| $^4$He | 75.0±23.6 | 95.9±28.8 | 54±14 | 162±14 | 126±11 |
| C | 0.361±0.012 | 0.670±0.067 | 0.322±0.003 | 0.501±0.058 | 0.490±0.056 |
| N | 0.119±0.003 | 0.069±0.021 | 0.129±0.002 | 0.138±0.022 | 0.123±0.020 |
| O | ≡ 1.0±0.02 | ≡1 | ≡ 1.0±0.06 | ≡ 1.0±0.161 | ≡ 1.0±0.161 |
| Ne | 0.152±0.005 | 0.091±0.027 | 0.261±0.003 | 0.151±0.021 | 0.191±0.026 |
| Mg | 0.229±0.007 | 0.147±0.030 | 0.370±0.003 | 0.072±0.009 | 0.224±0.026 |
| Si | 0.235±0.011 | 0.167±0.034 | 0.409±0.004 | 0.071±0.007 | 0.214±0.022 |
| S | 0.059±0.004 | 0.049±0.010 | 0.118±0.015 | 0.032±0.008 | 0.032±0.008 |
| Ca | 0.022±0.002 | 0.017±0.003 | 0.060±0.003 | 0.005±0.0001 | 0.013±0.0002 |
| Fe | 0.404±0.047 | 0.120±0.024 | 0.950±0.005 | 0.061±0.006 | 0.186±0.017 |

[a] 0.32–0.45 MeV/nuc, from Desai et al. (2006a)

[b] From von Steiger et al. (2000); Ca/O ratio is from Wurz et al. (2003)

[c] ~0.385 MeV/nuc, from Mason et al. (2004)

[d] From Lodders et al. (2003)

[e] ~1.4 MK quiet corona, from Feldman and Widing (2003)

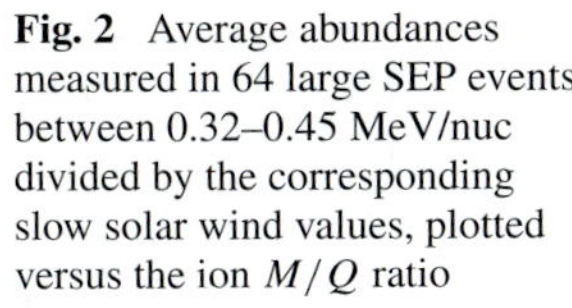

**Fig. 2** Average abundances measured in 64 large SEP events between 0.32–0.45 MeV/nuc divided by the corresponding slow solar wind values, plotted versus the ion $M/Q$ ratio

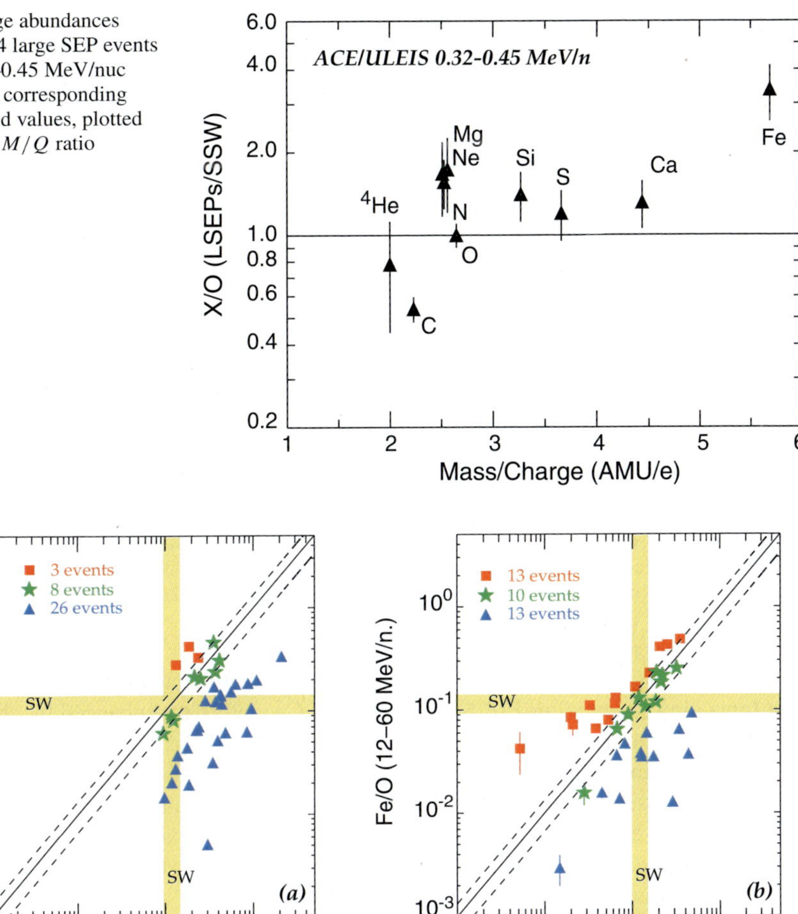


**Fig. 3** (**a**) Fe/O ratio measured at 0.11–0.14 MeV/nuc versus that measured at 3.3–10 MeV/nuc for 37 large SEP events. (**b**) Fe/O ratio measured at 3.3–10 MeV/nuc versus that measured at 12–60 MeV/nuc for 36 SEP events. *Yellow bands* represent error limits for the Fe/O ratio measured in the slow solar wind (adapted from Desai et al. 2006a)

energy in 34 of the 37 events (~92%) up to ~10 MeV/nuc and in 26 of the 36 events (~64%) up to ~60 MeV/nuc. However, an unexpected and puzzling feature of Fig. 3a is that the low-energy Fe/O ratio between 0.11–0.14 MeV/nuc in most of the LSEP events (55 of the 64 events; see Desai et al. 2006a) is enhanced when compared with the average solar wind value.

## 2.3 Mass-per-Charge Dependent Fractionation

An important question arises: What mechanisms are responsible for the large enhancements and the event-to-event variability seen in the low-energy Fe/O during the LSEP events of cycle 23? This section investigates whether these enhancements are systematically organized by the ion's mass-per-charge ($M/Q$) ratio. Other surveys of large SEP abundances above

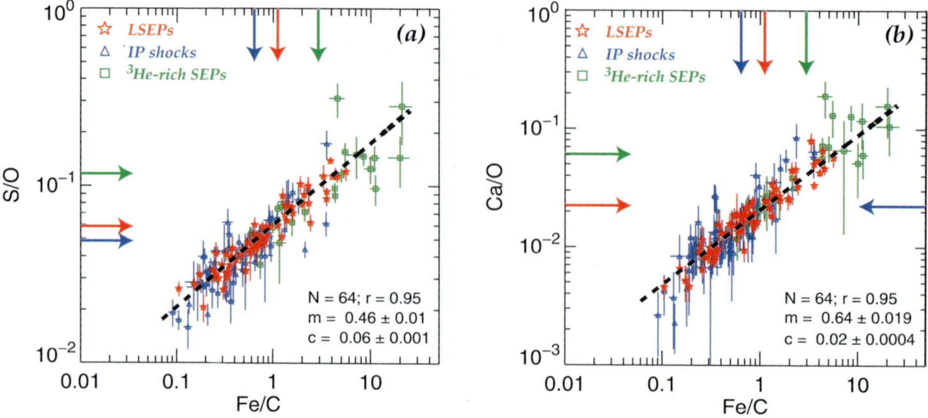


**Fig. 4** (**a**) S/O and (**b**) Ca/O plotted versus Fe/C ratio at 0.38 MeV/nuc for three different types of events. *Blue triangles* = IP shock events of Desai et al. (2003); *red asterisks* = SEP events in this survey; and *green squares* = $^3$He-rich SEP events of Mason et al. (2004). *Colored arrows along the axes* represent the average values of the respective populations. *Dashed black* line represents the linear fit to the large SEP data. The quantities $N$ and $r$ represent the number of SEP events and the correlation coefficient. The quantities $m$ and $c$ represent the slopes and intercepts of the linear fits (taken from Desai et al. 2006a)

~5 MeV/nuc (e.g., Mewaldt et al. 2007) have shown that the abundances of elements with First Ionization Potential or FIP $\lesssim$10 eV like Fe, Ca, and S are enhanced by about a factor of ~2.5 relative to those that have higher FIP values like C and O.

Thus, to examine the $M/Q$-dependent fractionation of ions with higher $M/Q$ values compared to those with lower $M/Q$ values, we first need to minimize the FIP fractionation effect discussed earlier. We do this in Fig. 4 by plotting the event-averaged abundances of S, Ca, and Fe (i.e., low FIP elements) relative to C and O (i.e., high FIP elements) and then plot (a) the S/O ratio and (b) the Ca/O ratio versus the Fe/C ratio at ~0.38 MeV/nuc for the 64 LSEP events in our survey.

Figure 4 also compares the LSEP abundances with those seen in IP shock events of Desai et al. (2003) and $^3$He-rich SEP events of Mason et al. (2004). The dashed line represents a linear fit to the LSEP data with slope $m$ and intercept $c$. We remark that the slopes and intercepts of the linear fits obtained independently by fitting the data for IP shocks (blue) and $^3$He-rich events (green) are well within the respective $1\sigma$ error limits of the fit parameters for the three types of events.

Figure 4 shows the following: (1) Enhancements in the S/O and Ca/O ratios are accompanied by simultaneous enhancements in the Fe/C ratio. (2) The fits to the data provide an excellent representation of the event-to-event variations and the enhancement pattern of the heavy ion abundances in large SEP events. (3) For each element, the power-law dependence in large SEP events provides remarkably good fits to the corresponding event-to-event variations seen in IP shock events and in $^3$He-rich SEP events.

## 3 Discussion

Our survey of the ~0.1–60 MeV/nuc heavy ion abundances in 64 large SEP events of cycle 23 shows the following:

- The Fe/O ratio decreases with increasing energy in ~92% of the events up to ~10 MeV/nuc and in ~64% of the events up to ~60 MeV/nuc.
- The rare isotope $^3$He is greatly enhanced over the corona or the solar wind values in 46% of the events.
- The heavy ion abundances show large, variable, and unsystematic enhancements and depletions relative to the solar wind values.
- Event-to-event variations in the heavy ion abundances in LSEP events are organized according to the $M/Q$ ratio and are remarkably similar to those seen in $^3$He-rich SEP events and IP shock events.

### 3.1 Two-Stage Acceleration in Large SEP events

We now discuss the implications of these results for the acceleration mechanisms operating in large SEP events. Rigidity-dependent shock acceleration processes tend to deplete the abundances of heavier ions when compared with the lighter ions (e.g., Klecker et al. 1981; Cane et al. 1991; Desai et al. 2003, 2004). Figure 3 shows that the Fe/O ratio decreases with increasing energy in the majority of the LSEP events in our survey. Such systematic energy dependence of the SEP Fe/O ratios were also reported by a number of previous studies (e.g., Mazur et al. 1992; Tylka et al. 2005; Cohen et al. 2005; Mewaldt et al. 2006). Likewise, the Fe/O ratios in individual CME-driven IP shock events at ACE also exhibited similar spectral properties (see Desai et al. 2003, 2004), and are consistent with the behavior predicted by shock acceleration models (e.g., Lee 2005) in which ions with higher $M/Q$ ratios are accelerated less efficiently than those with lower $M/Q$ ratios.

However, the $^3$He/$^4$He ratio (Fig. 1b) and the heavy ion abundances in large SEP events (Figs. 3a and 4) are significantly enhanced relative to the slow solar wind or the ambient coronal abundances (see Mason et al. 1999; Cohen et al. 1999; Desai et al. 2006a). Since $^3$He has a smaller $M/Q$ ratio than $^4$He while ions such as Fe have a larger $M/Q$ ratio than O, such large simultaneous enrichments are difficult to reconcile with rigidity-dependent shock acceleration of solar wind material. This implies that the dominant seed population for the majority of the large SEP events could not have originated either from the thermal or suprathermal solar wind or from the ambient corona (see also Mewaldt et al. 2002, 2006).

In fact, taken together Figs. 3 and 4 indicate the occurrence of *two* distinct $M/Q$-dependent fractionation processes in the same large SEP event. The first process results in producing $M/Q$-dependent enhancements, similar to those seen in the $^3$He-rich SEP events, while the second process—probably due to shock acceleration—causes the Fe/O ratio to decrease with increasing energy.

Previously, Desai et al. (2003) found that the heavy ion abundances in CME-driven IP shocks were related to those measured in the ambient suprathermal population measured prior to the arrival of the shocks at 1 AU according to the simple relation $\log(\Gamma_X) = c + m[(M_X Q_O)/(M_O Q_X)]$, where $c = 0.59 \pm 0.06$, and $m = -0.62 \pm 0.06$, and $(M_X/Q_X)$ is the $M/Q$ ratio of element X. Here $\Gamma_X = (X/O)_S/(X/O)_A$ is the enhancement or depletion of the X/O ratio in IP shocks, $(X/O)_S$ relative to that in the suprathermal population, $(X/O)_A$ (see Fig. 12 of Desai et al. 2003).

We infer average abundances for the "source" population by applying this function to the survey-averaged large SEP abundances provided in Table 1. Since CME-driven IP shocks depleted the abundances of heavier ions more than those of the lighter ions, using this function results in the inferred source population for large SEP events to be even more abundant in heavier ions than the average abundances reported here (see Desai et al. 2006a).

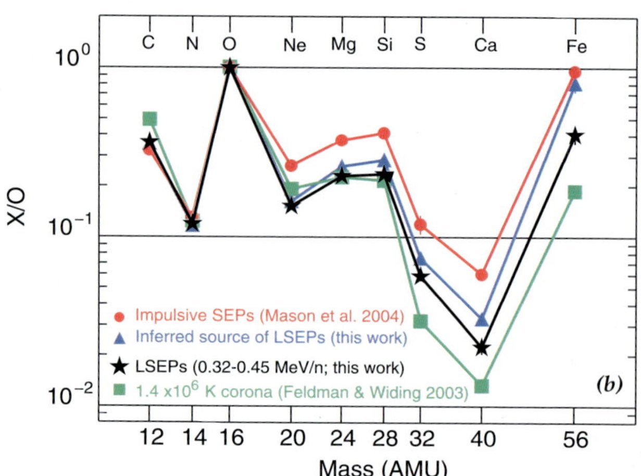


**Fig. 5** Average abundances measured in 64 large SEP events, compared with those measured in $^3$He-rich SEP events (from Mason et al. 2004) and in the ~1.4 MK quiet corona (from Feldman and Widing 2003). Also shown are the inferred source abundances for large SEP events of cycle 23

Figure 5 compares the inferred source population with spectroscopic measurements of the ~1.4 MK quiet corona (Feldman and Widing 2003), average values measured in $^3$He-rich SEP events (Mason et al. 2004), and large SEP events as a function of atomic mass. Clearly, the heavy ions in the inferred source population are greatly enhanced over the quiet coronal values. Assuming that the inferred source abundances are a simple linear mixture of ambient coronal and heavy ion enriched impulsive flare-like material with no energy dependence, we find that more than 75% of the seed population must comprise material that is already enriched in heavy ions.

### 3.2 Role of Flares Versus CME-Driven Shocks

Figure 3 shows that the Fe/O decreases or remains constant with increasing energy up to ~60 MeV/nuc in ~64% of the events, and is therefore consistent with the rigidity-dependent effects of the CME-driven shock acceleration processes observed near 1 AU (e.g., Desai et al. 2004). This result taken together with observations of species-dependent spectral breaks (e.g., Cohen et al. 2005; Mewaldt et al. 2005) therefore points to acceleration at CME-driven shocks as the dominant process in the majority of LSEP events of cycle 23.

However, for ~36% of the events there is a decrease in Fe/O around ~3–10 MeV/nuc followed by an increase at the higher energies. This behavior is puzzling and could point to a separate mechanism operating at these high energies e.g., a direct Fe-rich acceleration event at or near the Sun (e.g., Cane et al. 2003, 2006), or perhaps the re-acceleration of a flare seed population with Fe/O ratios that increase with energy (e.g., Tylka et al. 2005). No doubt other possibilities exist as well, and we believe that there is insufficient data to resolve the issue at this time.

### 3.3 Origin of the Seed Population for Large SEP Events

An important question is what is the origin of the seed population for CME-driven shocks that produce large SEP events? Since tracer ion species like $^3$He and interstellar pickup $He^+$ are extremely rare in the solar wind (Gloeckler and Geiss 1998), their mere presence during IP shock associated energetic particle events (see e.g., Desai et al. 2001; Kucharek et al. 2003) points to a seed population comprising suprathermal ions with speeds more than twice the bulk solar wind speed.

Such suprathermal ions probably have at least two advantages over the thermal solar wind ions. First, they have higher speeds that presumably exceed the injection threshold speeds typically associated with IP shocks. Second, they have broader velocity and nearly isotropic angular distributions when compared with the narrower velocity and anisotropic distributions of the thermal solar wind ions (Giacalone et al. 1994; Scholer et al. 1999).

In fact, Gloeckler (2003) reported the presence of a suprathermal ion population between ~2–10 times the solar wind speed even in the absence of solar activity and interplanetary shocks. Presently, however, the origin of this suprathermal material is not clear. Indeed, many sources such as interstellar and inner source pickup ions, heated solar wind ions, and ions accelerated in prior solar and interplanetary activity are known to contribute to this energy regime (e.g., Mason 2000; Mewaldt et al. 2001; Mason et al. 2005).

One possibility is that flares provide much of the suprathermal seed population for CME-driven shocks near the Sun and in interplanetary space. Support for this is bolstered by the fact that $^3$He enrichments are seen in the interplanetary medium for most of the time during solar active periods (e.g., Wiedenbeck et al. 2003; Desai et al. 2006b), perhaps indicating that a quasi-continuous mechanism (e.g., micro-flares) rather than short-lived episodes replenishes the solar corona and the interplanetary medium with suprathermal ions. Further evidence for this scenario was provided by Desai et al. (2006b) who found that the 80–100 keV/nucleon suprathermal heavy ion population during quiet times was remarkably similar to that measured in SEP events during solar maximum from 1998–2004, but similar to solar wind/CIR values during solar minimum conditions of 1994–1997, and 2005.

Another possibility is that the heavy ion enriched suprathermal seed population originates from flares that typically accompany the CMEs (e.g., Li and Zank 2005 and the references therein). However, even though the $^3$He enrichments indicate the presence of impulsive-flare-accelerated material in the seed population, neither scenario can satisfactorily account for the relatively low $^3$He/$^4$He ratios ($<10\%$) that are accompanied by the substantially large Fe/O ratios ($\sim 1$) in many of the large SEP events of cycle 23 (see Mewaldt et al. 2006; Desai et al. 2006a).

Other possibilities are (1) a common, unidentified mechanism produces the heavy-ion enriched suprathermal seed population, (2) heavy ions from the ambient corona or the solar wind are heated preferentially to produce a heavy-ion-enriched suprathermal tail, and (3) heavy ions are preferentially injected into the shock-acceleration process. New measurements from future missions like the Solar Orbiter and the Inner Heliospheric Sentinels will enable us to better understand the origin of this suprathermal seed population.

## 4 Conclusions

The new composition results obtained over solar cycle 23 can no longer be reconciled with the 1990s viewpoint that the large SEP events are produced by the acceleration of ambient coronal or solar wind material at CME-driven shocks near the Sun and in interplanetary space. Instead, any new comprehensive model for most of the large SEP events of cycle 23 must involve at least two steps: first, the routine production of a heavy-ion enriched suprathermal seed population, and second, CME shock acceleration of this material resulting in species-dependent spectral breaks and the decrease of Fe/O with increasing energy.

**Acknowledgements** Work at SwRI was partially supported by NASA grant NNG05GQ94G and NSF grant ATM-0555878. Work at Caltech was supported by NASA grant NAG5-12929.

## References


H.V. Cane, T.T. von Rosenvinge, C.M.S. Cohen, R.A. Mewaldt, J. Geophys. Res. **111**, A06S90 (2006). doi:10.1029/2005JA011071

H.V. Cane, T.T. von Rosenvinge, C.M.S. Cohen, R.A. Mewaldt, Geophys. Res. Lett. **30**(12), 8017 (2003). doi:10.1029/2002GL016580

H.V. Cane, D.V. Reames, T.T. von Rosenvinge, Astrophys. J. **373**, 675 (1991)

H.V. Cane, D.V. Reames, T.T. von Rosenvinge, Astrophys. J. **93**, 9555 (1988)

H.V. Cane, R.E. McGuire, T.T. von Rosenvinge, Astrophys. J. **301**, 448 (1986)

E.W. Cliver, *Acceleration and Transport of Energetic Particles Observed in the Heliosphere*. AIP Conference Proceedings, vol. 528 (2000), p. 21

E.W. Cliver, S.W. Kahler, M.A. Shea, D.F. Smart, Astrophys. J. **260**, 362 (1982)

C.M.S. Cohen et al., Geophys. Res. Lett. **26**, 2697 (1999)

C.M.S. Cohen et al., J. Geophys. Res. **110**, A09S16 (2005)

M.I. Desai et al., Astrophys. J. **649**, 470 (2006a)

M.I. Desai, G.M. Mason, J.R. Dwyer, J.E. Mazur, Astrophys. J. **645**, L81 (2006b)

M.I. Desai et al., Astrophys. J. **611**, 1156 (2004)

M.I. Desai et al., Astrophys. J. **558**, 1149 (2003)

M.I. Desai, G.M. Mason, J.R. Dwyer, J.E. Mazur, C.W. Smith, R.M. Skoug, Astrophys. J. **553**, L89 (2001)

U. Feldman, K.G. Widing, Space Sci. Rev. **107**, 665 (2003)

S.E. Forbush, Phys. Rev. **70**, 771 (1946)

J. Geiss, G. Gloeckler, R. von Steiger, Space Sci. Rev. **78**, 43 (1996)

J. Giacalone, J.R. Jokipii, J. Kota, J. Geophys. Res. **99**, 19351 (1994)

G. Gloeckler, in *Solar Wind Ten*, ed. by M. Velli, R. Bruno, F. Malara, P.B. Bucci. AIP Conference Proceedings, vol. 679 (Melville, 2003), pp. 583

G. Gloeckler, J. Geiss, Space Sci. Rev. **84**, 275 (1998)

G. Gloeckler et al., Astron. Astrophys. Suppl. Ser. **92**, 267 (1992)

S.W. Kahler et al., J. Geophys. Res. **89**, 9683 (1984)

S.W. Kahler, E. Hildner, M.A.I. Van Hollebeke, Solar Phys. **57**, 429 (1978)

B. Klecker et al., Space Sci. Rev. **124**(1–4), 289 (2006)

B. Klecker et al., Astrophys. J. **251**, 391 (1981)

G.E. Kocharov, *Proceedings of the 18th International Cosmic Ray Conference, Invited and Rapporteur Papers*, vol. 12, (1983), p. 235

H. Kucharek et al., J. Geophys. Res. **108**, A10 (2003). doi:10.1029/2003JA009938

M.A. Lee, Astrophys. J. **158**, 38 (2005)

G. Li, G.P. Zank, Geophys. Res. Lett. **32**, 2101 (2005)

R.P. Lin, Sol. Phys. **12**, 266 (1970)

K. Lodders et al., Astrophys. J. **591**, 1220 (2003)

A. Luhn et al., Adv. Space Res. **4**, 161 (1984)

G.M. Mason, M.I. Desai, J.E. Mazur, J.R. Dwyer, *The Physics of Collisionless Shocks*. AIP Conference Proceedings, vol. 781 (2005), p. 219

G.M. Mason et al., Astrophys. J. **606**, 555 (2004)

G.M. Mason, *Acceleration and Transport of Energetic Particles Observed in the Heliosphere*. AIP Conference Proceedings, vol. 528 (2000), p. 234

G.M. Mason, J.E. Mazur, J.R. Dwyer, Astrophys. J. Lett. **525**, L133 (1999)

G.M. Mason et al., Space Sci. Rev. **86**, 409 (1998)

G.M. Mason, G. Gloeckler, D. Hovestadt, Astrophys. J. **280**, 902 (1984)

J.E. Mazur et al., Astrophys. J. **401**, 398 (1992)

J.E. Mazur et al., Geophys. Res. Lett. **26**, 173 (1999)

R.A. Mewaldt et al., Space Sci. Rev. (2007), this issue. doi:10.1007/s11214-007-9187-1

R.A. Mewaldt, C.M.S. Cohen, G.M. Mason, in *Geophys. Monograph Series*, vol. 165 (2006), p. 115

R.A. Mewaldt et al., J. Geophys. Res. **110**, A09S18 (2005)

R.A. Mewaldt et al., Adv. Space Res. **30**, 79 (2002)

R.A. Mewaldt et al., in *Solar and Galactic Composition*. AIP Conference Proceedings, vol. 598 (2001), p. 165

P. Meyer, E.N. Parker, J.A. Simpson, Phys. Rev. **104**, 768 (1956)

Möbius et al., Geophys. Res. Lett. **26**, 145 (1999)

Möbius, et al.: *Acceleration and Transport of Energetic Particles Observed in the Heliosphere*. AIP Conference Proceedings, vol. 528 (2000), p. 131

M. Oetliker et al., Astrophys. J. **447**, 495 (1997)

D.V. Reames, C.K. Ng, Astrophys. J. **610**, 510 (2004)

D.V. Reames, Space Sci. Rev. **90**, 413 (1999)
D.V. Reames, Astrophys. J. **220**, L71 (1988)
M. Scholer et al., Geophys. Res. Lett. **26**, 29 (1999)
R. von Steiger et al., J. Geophys. Res. **105**, 27217 (2000)
R. von Steiger, J. Geiss, G. Gloeckler, Cosmic Winds and the Heliosphere (1997), p. 581
E.C. Stone, A.M. Frandsen, R.A. Mewaldt, E.R. Christian, D. Margolies, J.F. Ormes, F. Snow, Space Sci. Rev. **86**, 1 (1998a)
E.C. Stone et al., Space Sci. Rev. **86**, 285–356 (1998b)
A.J. Tylka et al., Astrophys. J. **625**, 474 (2005)
M.E. Wiedenbeck et al., *Solar Wind Ten*. AIP Conference Proceedings, vol. 679 (2003), p. 652
M.E. Wiedenbeck et al., *Acceleration and Transport of Energetic Particles Observed in the Heliosphere*. AIP Conference Proceedings, vol. 528 (2000), p. 107
J.P. Wild, S.F. Smerd, A.A. Weiss, Annu. Rev. Astron. Astrophys. **1**, 291 (1963)
P. Wurz et al., Astrophys. J. **583**, 489 (2003)

Space Sci Rev (2007) 130: 255–272
DOI 10.1007/s11214-007-9214-2

# Particle Acceleration at Interplanetary Shocks

**G.P. Zank · Gang Li · Olga Verkhoglyadova**

Received: 5 April 2007 / Accepted: 14 May 2007 / Published online: 30 June 2007


**Abstract** This paper briefly reviews proton acceleration at interplanetary shocks. This is key to describing the acceleration of heavy ions at interplanetary shocks because wave excitation—and hence particle scattering—at oblique shocks is controlled by the protons and not the heavy ions. Heavy ions behave as test particles, and their acceleration characteristics are controlled by the properties of proton-excited turbulence. As a result, the resonance condition for heavy ions introduces distinctly different signatures in abundance, spectra, and intensity profiles, depending on ion mass and charge. Self-consistent models of heavy-ion acceleration and the resulting fractionation are discussed. This includes discussion of the injection problem and the acceleration characteristics of quasi-parallel and quasi-perpendicular shocks.

**Keywords** Solar energetic particles · Coronal mass ejections · Particle acceleration

## 1 Introduction

Understanding the problem of particle acceleration at interplanetary shocks is increasingly important, especially in the context of understanding the space environment. The basic physics is thought to have been established in the late 1970s and 1980s with the seminal papers of Axford et al. (1977) and Bell (1978a, 1978b), but detailed interplanetary observations are not easily interpreted in terms of the simple original models of particle acceleration at shock waves. Three fundamental aspects make the interplanetary problem more complicated than the typical astrophysical problem: the time dependence of the acceleration and the

G.P. Zank (✉) · G. Li · O. Verkhoglyadova
Institute of Geophysics and Planetary Physics (IGPP), University of California, Riverside, CA 92521, USA
e-mail: zank@ucr.edu

G. Li
e-mail: gang.li@ucr.edu

O. Verkhoglyadova
e-mail: olgav@ucr.edu

solar wind background; the geometry of the shock; and the long mean free path for particle transport away from the shock. Consequently, the shock itself introduces a multiplicity of time scales, ranging from shock propagation time scales to particle acceleration time scales at parallel and perpendicular shocks, and many of these time scales feed into other time scales (such as determining maximum particle energy scalings and escape time scales).

Solar energetic particle (SEP) events are historically classified into two classes, "impulsive" and "gradual". Historically, flares have been thought to be the sites of both impulsive and gradual SEPs. A clear association of coronal mass ejections (CMEs) with observed gradual SEP events was not made until Kahler et al. (1984), who found that large SEPs and CMEs were very highly correlated, as were SEP intensities and the size and speed of the CME. Some earlier charge state measurements (Luhn et al. 1984, 1987; Leske et al. 1995; Mason et al. 1995; Tylka et al. 1995; Oetliker et al. 1997) of energetic ions in gradual SEP events also suggested that these ions originated in regions having $T \simeq 2 \times 10^6$ K, corresponding to coronal material, rather than at a much higher temperature ($T \simeq 2 \times 10^7$ K) flare site. However, puzzling observations—especially after the launch of the ACE spacecraft in the past decade—suggest that high-energy (above tens of MeV) and low-energy particles may result from different seed and acceleration mechanisms dominating different energy regimes (Mason et al. 1999). Using observations from ACE, Cohen et al. (1999) and Cliver and Cane (2000) presented examples of large SEP events associated with impulsive soft X-ray events. These SEP events also had some characteristics of impulsive events, such as high Fe/O ratios. More recently, Cane et al. (2003) studied 29 intense SEP events and found that there are four mixed events which possess both flare- and shock-accelerated characteristics. Cane et al. (2003) considered 30–50 MeV/nuc. energies in Fe and O. These events usually have a time–intensity profile that looks like those due to shock acceleration but have a flare-like composition. It is possible that these are the events in which CMEs and the accompanying flares occur temporally close to each other and are both well connected magnetically to the Earth. For such cases, some of the solar flare material will undergo shock re-acceleration (see Li and Zank 2005) and thus have a time–intensity profile similar to shock accelerated particles.

To better understand in situ measurements (time–intensity profiles, particle spectra, relative heavy ion abundances, etc.) of gradual SEP events and to help clarify the ambiguous relationship between flares and CMEs, a detailed model of ion acceleration and transport at CME-driven shocks is necessary. Earlier works by Heras et al. (1992, 1995), Ruffolo (1995), Kallenrode and Wibberenz (1997), Kallenrode and Hatzky (1999), Lario et al. (1998), and Ng et al. (1999, 2003) adopted a "black-box" approach to particle acceleration at propagating CME-driven shocks, simply injecting ion spectra at the shock, thus neglecting all the physics of diffusive shock acceleration, and focused on the transport of energetic particles in the interplanetary medium by solving the Boltzmann–Vlasov equation numerically. Ng et al. (2003) extended his earlier models (Ng and Reames 1995; Ng et al. 1999) by allowing energetic particles and the generated Alfvén waves to interact self-consistently. The amplification of the Alfvén waves is determined by the particle anisotropy. Particles escaping from the shock at earlier times generate the necessary waves for accelerating particles at a later time. Nontrivial SEP time intensity profiles can result (Ng et al. 2003). Although the Ng et al. (2003) approach marked a significant step toward a better understanding of gradual SEP events, it assumed a prescribed particle spectrum (power law at low energies and an exponential roll-over at high energies) at a CME shock propagating at a constant speed with an assumed radial magnetic field. This approach therefore neglects much of the basic physics of time-dependent diffusive particle acceleration. Furthermore, the assumption that the shock propagates at a constant speed in the solar wind is clearly not reasonable.

Li et al. (2005) presented a model that calculates the acceleration of heavy ions at a CME-driven or an interplanetary shock wave and their subsequent transport in the interplanetary medium. This work is based on a series of papers by Zank et al. (2000), Rice et al. (2003), and Li et al. (2003). These models assume particle acceleration at a quasi-parallel shock. The extension of these ideas to quasi-perpendicular shocks has been presented in Zank et al. (2006). The basic physics underlying these models is discussed in the following.

We give a brief overview of shock acceleration at quasi-parallel and quasi-perpendicular shocks, including our approach to particle transport in the interplanetary medium. These ideas will be applied to the acceleration of heavy ions at interplanetary shocks, and we conclude by modeling some real SEP and ESP events.

## 2 Particle Acceleration Models and Transport

### 2.1 Acceleration at Quasi-Parallel Shocks

Zank et al. (2000) modeled the evolution of a CME-driven shock wave. At the shock front, the accelerated particle spectrum is produced by diffusive shock acceleration. For diffusive shock acceleration to be applicable, magnetic turbulence near the shock is necessary to scatter particles repeatedly across the shock. For quasi-parallel shocks, the turbulence (Alfvén waves) is generated by the anisotropic energetic protons at the shock (Bell 1978a; Lee 1983). Furthermore, since the number density of heavy ions is negligible compared with that of protons, the turbulence is due solely to the streaming protons. Heavy ions can therefore be treated as test particles, experiencing the turbulence but not contributing to its generation. We consider two species of heavy ions, specifically, CNO particles and iron, with charge $Q = 6$ and mass $A = 14$ corresponding to CNO particles and $Q = 14$ and $A = 56$ corresponding to iron.

The resonance condition for particles and waves is given by the Doppler condition,

$$\omega - \Omega = k_{\parallel} \mu p / \gamma m, \tag{1}$$

where $\omega$ is the resonant wave frequency, $\Omega$ is the local ion gyrofrequency, $k_{\parallel}$ is the wave vector along the magnetic field, $\mu$ the particle pitch angle cosine, $p$ the particle momentum, $m$ ($m_p$) the particle (proton) mass, and $\gamma$ the Lorentz factor. Higher order resonances are not considered. The ion gyrofrequency is $\Omega = (Q/A)eB/\gamma m_p c$ (in Gaussian units), where $Q$ is the particle charge, $A$ the ion mass number, and $c$ the speed of light. In the context of particle acceleration of heavy ions at a quasi-parallel shock, the resonant wave number for an ion of mass $A$ and charge $Q$ is

$$k \simeq \frac{Q}{A} \frac{eB}{\tilde{p}c}, \tag{2}$$

where $\tilde{p} = p/A$ denotes particle momentum per nucleon. (See Gordon et al. 1999 and Li et al. 2005 for a discussion of the resonance-broadening assumption implicit in (2).) Physically, the Doppler condition means the phase of the Alfvén wave seen by the particle is unchanged after a gyration. The diffusion coefficient for particles in resonance with the Alfvén wave depends on the wave intensity $I(k)$ through

$$\kappa(p) = \frac{vp^2c^2}{8\pi Q^2 e^2} \frac{1}{I(k = \Omega/v)}. \tag{3}$$

The wave intensity $I(k,t)$ can be obtained from the wave kinetic equation (Lee 1983; Gordon et al. 1999; Rice et al. 2003), and depends on the obliquity of the shock going to zero as the shock becomes increasingly perpendicular. For strong shocks, the wave intensity can be approximated by $B^2/k$, and the Bohm approximation used in Zank et al. (2000) is recovered.

The propagation of the shock in the solar wind is modeled numerically and the post shock flow is determined completely consistently. We use an MHD model for the shock and solar wind. In the shell model of Zank et al. (2000), the region over which the shock sweeps is approximated by a set of concentric shells. At some time $t$, the shock is located at $R(t)$. A new shell is then created with the outer edge and inner edge initially coincident with the shock front at $R(t)$. After a time $\Delta t$ later, the outer edge of the shell, which is attached to the shock, propagates to $R(t+\Delta t)$ whereas the inner edge of the shell propagates to $R(t)+v_{sw}\Delta t$. Thus, the shell is created and will decouple from the shock. After $t+\Delta t$, the shock continues to propagate and convect in the supersonic solar wind according to the self-consistent MHD model, and a new shell is created. Shock properties, such as shock speed and compression ratio, are followed numerically as the shock wave expands. At the shock, the accelerated particle spectrum is obtained by adopting the diffusive shock-acceleration solution for a planar shock, subject to the assumption that the time dependent parameters of the shock (i.e., the shock Mach number and compression ratio) can be regarded as locally constant. This leads to,

$$f(t_i,p)=\frac{\beta N}{4\pi p_{\mathrm{inj}}^3 u_{\mathrm{up}}}(p/p_{\mathrm{inj}})^{-\beta(t_i)}\{H[p-p_{\mathrm{inj}}(t_i)]-H[p-p_{\max}(t_i)]\}. \tag{4}$$

Here, $\beta(t_i)$ [$=3s_i/(s_i-1)$, $s_i$ the shock compression ratio] is the spectral index of the shock-accelerated ions, the variable $t_i$ represents the $i$th time step in the code, $N$ the injection rate, $H$ the Heaviside step function, $u_{\mathrm{up}}$ the upstream flow speed, and $p_{\mathrm{inj}}$ and $p_{\max}$ are the injection and maximum momentum of the accelerated protons, and correspond to the maximum turbulence wave number $k_{\max}=eB/p_{\mathrm{inj}}^p c$ and the minimum turbulence wave number $k_{\min}=eB/p_{\max}^p c$. The superscript $p$ indicates proton. For wave numbers smaller (larger) than the minimum (maximum) wave number, the corresponding wave intensity will be that of the ambient solar wind.

We choose $p_{\mathrm{inj}}$ so that $p_{\mathrm{inj}}^2/2m_p$ corresponds to the downstream thermal energy per nucleon, and assume that the injection momentum per nucleon is the same for protons and heavy ions, $\tilde{p}_{\mathrm{inj}}^p=\tilde{p}_{\mathrm{inj}}^{\mathrm{CNO}}=\tilde{p}_{\mathrm{inj}}^{\mathrm{Fe}}$. We therefore assume that at a quasi-parallel shock, the injection mechanism for diffusive shock acceleration does not distinguish protons from heavy ions. Furthermore, because of the weak dependence of $\tilde{p}_{\max}$ on $\tilde{p}_{\mathrm{inj}}$, we expect this to be a reasonable assumption.

As the interplanetary shock propagates outward, it slows, its strength decreases, and its area increases, and the steady-state solution (4) holds only approximately (and locally). The maximum momentum for protons can be determined by equating the dynamical timescale of the shock and the acceleration time scale (Drury 1983; Zank et al. 2000),

$$\frac{R(t)-R_0}{\dot{R}(t)}\simeq\frac{\beta(t)}{u_{\mathrm{up}}^2}\int_{p_{\mathrm{inj}}}^{p_{\max}}\kappa(p')\,\mathrm{d}(\ln(p')). \tag{5}$$

In (5), $R(t)$ is the shock speed in the spacecraft frame, $R_0$ the position of shock formation ($\sim 5\ \mathrm{R}_\odot$). The solution of (5) yields $p_{\max}$ for protons, and hence the lower limit of the excited

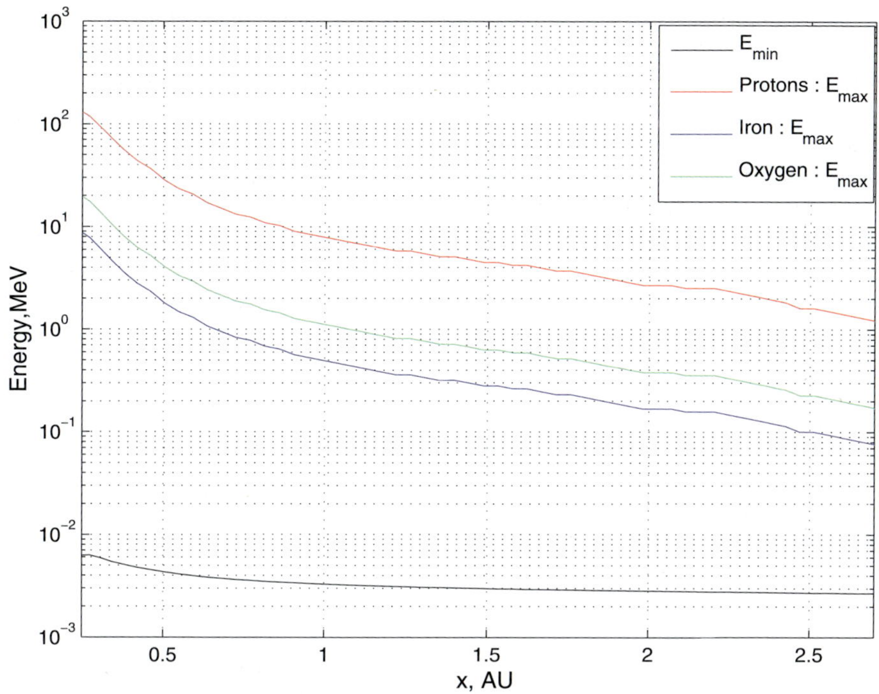


**Fig. 1** Plot of the maximum and minimum proton, O, and Fe energies per nucleon as a function of heliocentric distance for the September 29, 2001, SEP event. Note that the ordering of maximum energies is inversely dependent on particle mass, and that the maximum energy decreases with distance i.e., both $\dot{R}(t)$ and $B$ compensate for $R(t)$. Verkhoglyadova et al. (2007)

resonant wave number, as (Zank et al. 2000)

$$p_{\max} \simeq \left\{ \left[ \frac{M^2(t)+3}{5M^2(t)+3} \frac{(R(t)-R_0)}{\beta(t)\kappa_0} \frac{B}{B_0} + \sqrt{\left(\frac{m_p c}{p_0}\right)^2 + \left(\frac{p_{\text{inj}}}{p_0}\right)^2} \right]^2 - \left(\frac{m_p c}{p_0}\right)^2 \right\}^{1/2}. \quad (6)$$

Expression (6) illustrates explicitly that the maximum energy to which a particle can be accelerated depends on the age of the shock $R(t)$, the strength of the shock $\dot{R}(t)$, and the interplanetary magnetic field strength $B$. Heavy ions experience only the turbulence generated by the streaming protons. Thus the maximum achievable momentum for heavy ions is subject to the constraint (2), which shows that the maximum heavy ion momentum $\tilde{p}^i_{\max}$ and the maximum proton momentum $\tilde{p}^p_{\max}$ are related through the minimum turbulence wave number $k_{\min}$ by

$$\frac{eB}{\tilde{p}^p c} = k_{\min} = \frac{Q^i}{A^i} \frac{eB}{\tilde{p}^i c}, \quad (7)$$

and $i$ indicates either CNO or Fe particles. The $(A/Q)$ dependence of $\tilde{p}$ is evident. Figure 1 illustrates the maximum particle energies for protons, O, and Fe ions as a function of radial distance. In most of the plots that follow, we use the results of Verkhoglyadova et al. (2007) who modeled several observed gradual SEP events on the basis of the theory described here.

Downstream of the shock, energetic particles convect with the solar wind, adiabatically cool, and diffuse. The shell model allows us to follow the evolution of the particle phase

space distribution function $f$ downstream of the shock quite straightforwardly. The convection, adiabatic cooling, and diffusion of particles are treated using an operator-splitting technique. First, the convection of particles with the solar wind is followed by updating the locations of the front and back boundaries of each shell after each time step. As the shell convects outward, it also expands radially. From Liouville's theorem, the radial expansion leads to a cooling in $p$ space. Finally, particles can diffuse to adjacent shells depending on their diffusion coefficient, and also leak out from the shock complex if during time step they reach some distance $l(r, p, t)$ in front of the shock (see Li et al. 2005).

### 2.2 Quasi-Perpendicular Shocks

Although the theory of particle acceleration at a quasi-parallel shock appears to be reasonably well understood, no similar theory exists for perpendicular shocks. Zank et al. (2006) developed an approach for diffusive shock acceleration at perpendicular shocks. Particle acceleration via the first-order Fermi mechanism at a perpendicular shock wave remains an outstanding problem for two essential reasons. The first is that, unlike the quasi-parallel shocks discussed earlier, accelerated particles at a perpendicular shock cannot excite the [Alfvén] wave field that is responsible for scattering the particles repeatedly across the shock. This is because the growth term in the wave equation is proportional to $\cos\theta_{bn}$, where $\theta_{bn}$ is the angle between the upstream magnetic field and the shock normal direction, and is thus 90° for a strictly perpendicular shock (e.g., Gordon et al. 1999; Li et al. 2003; Rice et al. 2003). Unlike the particle-acceleration model described in Sect. 2.1, this therefore requires that particle scattering at a perpendicular shock be the result of in situ upstream turbulence that is convected into the shock. At the heart of the problem for perpendicular shocks is the need for a viable model of the perpendicular component of the diffusion tensor. The second problem is that since an accelerated particle is essentially tied to a magnetic field line, its ability to cross the shock repeatedly is limited to the time it takes a magnetic field line to cross from upstream to downstream, assuming that the magnetic field line experiences some wandering to make the transmission time nonzero. Thus, a fast-moving particle is necessary if it is to experience multiple crossings of the perpendicular shock (see Fig. 2) so that it can be diffusively accelerated. Consequently, diffusive shock acceleration of particles at a perpendicular shock is effective for particles that are already energetic. This is referred to as the "injection problem" for perpendicular shocks (Jokipii 1987; Zank et al. 1996).

It is generally thought that the acceleration time at a quasi-parallel shock is much larger than that at a quasi-perpendicular shock. This result is, however, predicated on the assumption that the intensity of the upstream turbulence is the same at a parallel and perpendicular shock, which, as we discussed earlier, is incorrect because waves are excited at a quasi-parallel shock and not at a quasi-perpendicular shock. A physically meaningful comparison of acceleration time scales must take this into account.

On the basis of a recently developed nonlinear guiding center theory for the perpendicular spatial diffusion coefficient $\kappa_\perp$ used to describe the transport of energetic particles (Matthaeus et al. 2003; Zank et al. 2004), Zank et al. (2006) constructed a model for diffusive particle acceleration at highly perpendicular shocks—i.e., shocks whose upstream magnetic field is almost orthogonal to the shock normal. They used $\kappa_\perp$ to investigate energetic particle anisotropy and injection energy at shocks of all obliquities, finding that at 1 AU, for example, parallel and perfectly perpendicular shocks (i.e., the shock normal is exactly 90° to the upstream magnetic field) can inject protons with equal facility. However, at highly oblique shocks, high injection energies are necessary, as illustrated in Fig. 3. As

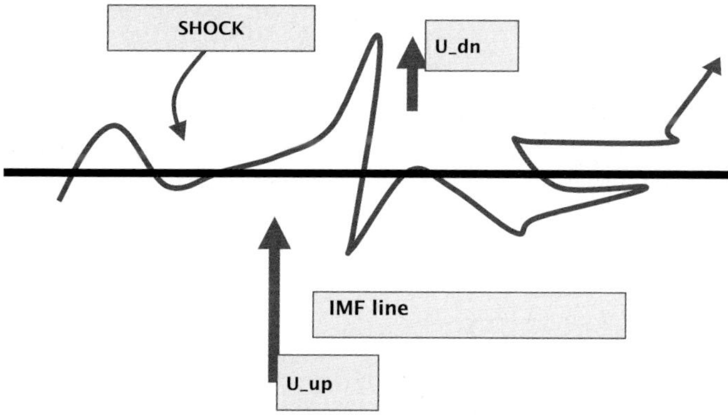


**Fig. 2** Schematic showing an interplanetary magnetic field (IMF) line experiencing random walking as it is convected through a shock. A particle attached to the field line can make multiple excursions up- and downstream as the field line is carried completely through the shock, provided the particle is moving sufficiently fast. This requires that particles already be energetic if they are to experience diffusive shock acceleration at a perpendicular shock. $U_{\mathrm{up}}$ and $U_{\mathrm{dn}}$ refer to the upstream and downstream flow speeds (Zank et al. 2006)

discussed by Zank et al. (2006), since interplanetary shocks are typically highly oblique (rather than perfectly perpendicular), injection is likely to be sporadic, suggesting, as found by van Nes et al. (1984), that the low-energy event intensity decreases as a shock becomes more perpendicular.

An important point not hitherto recognized is that the inclusion of self-consistent wave excitation at quasi-parallel shocks in evaluating the particle acceleration timescale ensures that it is significantly smaller than that for highly perpendicular shocks at low to intermediate energies and comparable at high energies. Thus, higher proton energies are achieved at quasi-parallel rather than highly perpendicular interplanetary shocks within 1 AU. However, both injection energy and the acceleration timescale at highly perpendicular shocks are sensitive to assumptions about the ratio of the 2D correlation length scale $\lambda_{\mathrm{2D}}$ to the slab correlation length scale $\lambda_{\mathrm{slab}}$ (Fig. 3).

By adapting the approach proposed by Zank et al. (2000) and Li et al. (2003), Zank et al. (2006) developed a model for particle acceleration at a perpendicular interplanetary shock. Just as at a quasi-parallel shock, the usual stationary cosmic ray transport equation can be solved at a perpendicular shock, yielding the same spectral dependence on the shock compression ratio in both cases. The differences between the two cases are the scaling of the upstream exponential decay of particle intensity, the injection energy, and the maximum energy, since all of these depend on the spatial diffusion coefficient.

As discussed earlier, the local injection momentum or velocity for particles to be accelerated diffusively at highly oblique shocks is much more stringent than at quasi-parallel shocks, and is given by (Zank et al. 2006)

$$v_{\mathrm{inj}} = 3u\left[\frac{1}{(s-1)^2} + \frac{r_L^2 + \lambda_\parallel^2\cos^2\theta_{bn}}{(\lambda_\perp + \lambda_\parallel\cos^2\theta_{bn})^2}\right]^{1/2}. \tag{8}$$

As with the quasi-parallel case, a similar approach, but now using the NLGC model for the perpendicular diffusion coefficient, can be followed to estimate the maximum particle momentum $p_{\mathrm{max}}$. This is now very complicated analytically within the NLGC framework, and Zank et al. (2006) solved the corresponding integro-differential equation numerically.

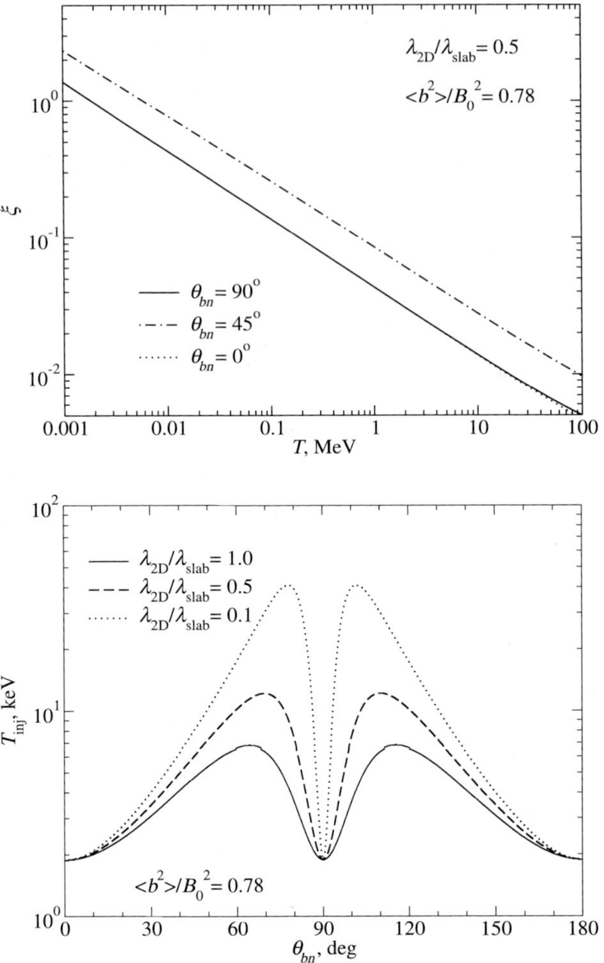


**Fig. 3** *Top*: A plot of the anisotropy $\xi$ ahead of a shock as a function of energy for parallel, oblique, and perpendicular shocks. A compression ratio of $s = 3$ is assumed. Note that at low energies, the perpendicular and parallel shocks have almost identical values of $\xi$. *Bottom*: Plot of the injection energy as a function of the shock obliquity under the assumption that $\xi = 1$ defines the threshold energy for injecting particles into the diffusive shock-acceleration mechanism. Three possible ratios of $\lambda_{2D}/\lambda_{slab}$ are considered. In both the *left* and *right panels*, parameters corresponding to an interplanetary shock located at 1 AU are used. Note that the $\theta_{Bn} = 0°$ curve is almost completely obscured by the $\theta_{Bn} = 90°$ curve (Zank et al. 2006)

However, estimates for $p_{\max}$ can be determined for shocks within several AU, i.e., for shock-accelerated particles with gyroradii that exceed the correlation length scale. For arbitrary mass and charge ions, Zank et al. (2006) computed the maximum particle momentum as

$$\tilde{p}_{\max} \simeq \left(\frac{Q}{A}\right)^{2/3} \left(\frac{e}{m_p}\right)^{2/3} \frac{R}{\dot{R}} \frac{(s-1)}{s} \frac{V_{sh}^2}{1.17\alpha} \lambda_c^{\text{slab}} \langle b_{\text{slab}}^2 \rangle, \tag{9}$$

where $\tilde{p}$ is the momentum per nucleon, and $\langle b_{\text{slab}}^2\rangle$ is the energy density of magnetic field fluctuations. The dimensional constant $\alpha$ has units cm$^{4/3}$ s$^{-2/3}$ (see Zank et al. 2006). The maximum momentum expression reveals several interesting points. A fundamental difference between the perpendicular and quasi-parallel expressions is that the former assumes only pre-existing turbulence in the solar wind, whereas the latter results from solving the coupled wave energy and cosmic ray streaming equation explicitly; i.e., in the perpendicular case, the energy density in slab turbulence corresponds to that in the ambient solar wind whereas in the case of quasi-parallel shocks it is determined instead by the self-consistent excitation of waves by the accelerated particles themselves. From another perspective, unlike the quasi-parallel case, the resonance condition does not enter into the evaluation of $p_{\max}$ given in (9). As a result, the diffusion coefficient is fundamentally different in each case, and hence the maximum attainable energy is different for a parallel or perpendicular shock.

Some comments regarding the determination of $p_{\max}$ for different shock configurations and ionic species are useful because three approaches have been identified (Zank et al. 2000; Li et al. 2003; Zank et al. 2006):

1. For protons accelerated at a quasi-parallel shock, $p_{\max}$ is determined purely on the basis of balancing the particle acceleration time resulting from resonant scattering with the dynamical timescale of the shock. The wave–turbulence spectrum excited by the streaming energized protons extends in wave number as far as the available dynamical time allows.
2. For heavy ions at a quasi-parallel shock, the maximum energy is also computed on the basis of a resonance condition but only up to the minimum $k$ excited by the energetic streaming protons, which control the development of the wave spectrum. Thus, maximum energies for heavy ions are controlled by the accelerated protons and their self-excited wave spectrum. This implies a $(Q/A)^2$ dependence of the maximum attainable particle energy for heavy ions.
3. For protons at a highly perpendicular shock, the maximum energy is independent of the resonance condition, depending only on the shock parameters and upstream turbulence levels. For heavy ions, this implies either a $(Q/A)^{1/2}$ or a $(Q/A)^{4/3}$ dependence of the maximum attainable particle energy, depending on the relationship of the maximum energy particle gyroradius compared to the turbulence correlation length scale.

The summary presented here suggests that it may be possible to extract observational signatures related to the mass–charge ratio that distinguish particle acceleration at quasi-parallel and highly perpendicular shocks.

Zank et al. (2006) considered a CME-driven shock propagating from 0.1 AU to $\sim$1 AU, making the somewhat simplistic assumption that the shock remains either parallel or that $\theta_{bn} = 85°$. Figure 4 illustrates, for a perpendicular shock, the dependence of injection energy and the maximum energy to which a particle can be accelerated as a function of radial distance. The figure compares the corresponding energies for an otherwise identical parallel shock where the energetic particles excite the upstream wave spectrum (a subtlety that has been neglected in previous comparisons of particle acceleration at perpendicular and parallel shocks), assuming that the shock remains parallel for the duration of its propagation to 1 AU. Since we assume wave excitation in the parallel shock calculation, this results in maximum energies that can be as much as an order of magnitude larger at a parallel shock than at a perpendicular shock close to the sun. The maximum energies accelerated at parallel and perpendicular shocks begin to converge towards 1 AU. The decay in maximum energy is slower for the perpendicular shock than the parallel because of the slower WKB-like decrease in energy in turbulent fluctuations and because of the $1/r^2$ dependence of $B$ within 1 AU for the parallel case. We stress that in the absence of the self-consistent wave excitation, the perpendicular shock would accelerate particles to higher energies than a parallel shock.

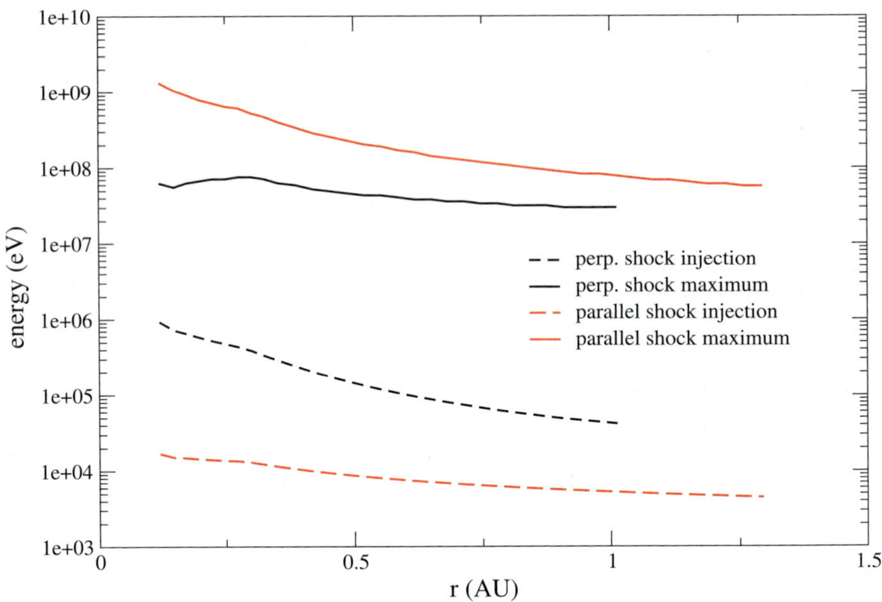


**Fig. 4** Maximum particle energy and injection energy for a highly perpendicular shock (85°, *black curves*) as a function of heliocentric distance. For comparison, we plot the corresponding results for a parallel shock (*red curves*) (Zank et al. 2006)

## 2.3 Particle transport in the interplanetary medium

The Boltzmann–Vlasov equation describes the evolution of the distribution function $f(r, p, t)$ for energetic particles escaping from the shock front into the interplanetary medium,

$$\frac{\partial f}{\partial t} + \frac{p}{m} \cdot \nabla f + \mathbf{F} \cdot \nabla_p f = \left.\frac{\partial f}{\partial t}\right|_{\text{coll}}, \tag{10}$$

where the right-hand side describes particle collisions; in this case, the scattering of charged particle by fluctuations in the IMF. Ignoring the collision term, the characteristics of the Boltzmann equation are simply the equations of motion for a single particle. Thus, solving the Boltzmann equation can be achieved by following individual particle motion, which consists of an interplay between collisionless motion and occasional pitch angle scattering. A Monte Carlo code was developed by Li et al. (2003) to follow charged particle motion in the IMF. The frequency of scattering can be parameterized by the mean free path $\lambda_{\|}$. Following Zank et al. (1998) and Li et al. (2003, 2005), we can express the particle mean free path as

$$\lambda_{\|} = \lambda_0 \left(\frac{\tilde{p}c}{1\ \text{GeV}}\right)^{1/3} \left(\frac{A}{Q}\right)^{1/3} \left(\frac{r}{1\ \text{AU}}\right)^{2/3}, \tag{11}$$

where $\lambda_0$ is a normalizing mfp (typically 0.1–1 AU, from observations). Note the presence of an $(A/Q)^{1/3}$ dependence in (11). For Fe and CNO particles with the same momentum per nucleon $\tilde{p}$, Fe will have a larger mean free path than CNO particles because of the larger $A/Q$ ratio.

Between consecutive pitch angle scatterings, charged particles gyrate along the IMF. As the magnetic field expands radially, particles experience adiabatic cooling and focusing effects as a consequence of adiabatic invariants of the motion. The geometry of the magnetic field $B$ is given by the usual Parker spiral (Parker 1958),

$$B = B_0\left(\frac{R_0}{r}\right)^2\left[1+\left(\frac{\Omega_0 R_0}{u}\right)\left(\frac{r}{R_0}-1\right)^2 \sin^2\theta\right]^{1/2}, \tag{12}$$

where $\theta$ is the colatitude of the solar wind with respect to the solar rotation axis; $\Omega_0$ the solar rotation rate; $u$ the radial solar wind speed, and $B_0$ the interplanetary magnetic field (IMF) at the corotation radius $R_0$ (typically, $R_0 = 10\ \mathrm{R}_\odot$, $B_0 = 1.83 \times 10^{-6}$ T, $u = 400$ km/s, and $\Omega_0 = 2\pi/25.4$ days).

After the initial condition $(t, p)$ of a test particle escaping from the shock complex is decided, Li et al. (2003, 2005) then followed its subsequent motion, i.e., "free" motion and pitch angle scattering until the shock passes 1 AU. The scattering is assumed to be isotropic and Markovian; thus, the new pitch angle after a scattering has no memory of the previous pitch angle. Since the shock is expanding outward, particles are subject to possible "absorptions" by the shock before it reaches 1 AU. This could happen, for example, if a particle is moving inward towards the sun due to pitch angle scattering or if the particle speed along the magnetic field is so small that the shock can catch it from behind. Once a particle is absorbed, it can be re-accelerated if its energy is smaller than the current maximum energy associated with the shock. The code contains adiabatic cooling, focusing, and mirroring. See Li et al. (2003) for details.

## 3 Modeling Observed Events

### 3.1 Differences between Particle Acceleration at Perpendicular and Parallel Shocks

The differences in particle injection and maximum particle momenta between quasi-parallel and perpendicular shocks were illustrated in Fig. 4. Figure 5 shows the evolution of the accelerated energetic particle relative number density at 1 AU as a function of time for the example of a quasi-perpendicular and parallel shock. The solid curves correspond to the quasi-perpendicular shock $\theta_{bn} = 85^\circ$ and the dotted curves to the parallel shock-accelerated protons. The total number of injected particles in the parallel shock case is assumed to be 20 times greater than that of the perpendicular shock. This ratio is in agreement with results obtained by Mewaldt et al. (2001); see Zank et al. (2006). Since the Zank et al. (2006) model is crude, they did not follow the time evolution of injected particles at a perpendicular shock, but instead assumed that the total number of injected particles at the quasi-perpendicular shock remains 20 times smaller than that at the parallel shock.

At early times, the quasi-perpendicular shock has a much clearer power law extending from lower energies than the parallel case. This is a consequence of quasi-perpendicular shocks not being as effective at trapping particles at the shock front as parallel shocks, which ensures that relatively more particles escape at lower energies from the highly perpendicular shock than at the parallel shock. Consequently, a power law spectrum is more likely to be seen at earlier times. The parallel shock can accelerate particles to higher energies and this is revealed clearly in Fig. 5. The parallel shock spectra tend to fill out into a power law over time until at 1 AU it almost resembles the perpendicular shock example, except that it is shifted out to higher energies.

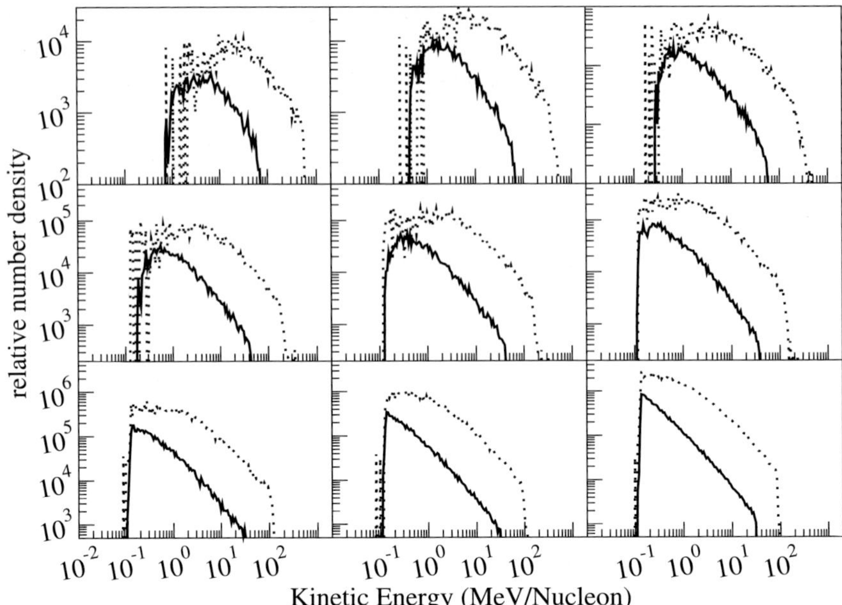


**Fig. 5** The time interval spectra for a perpendicular (*solid line*) and a parallel shock (*dotted line*). From left to right and top to bottom, the panels correspond to the time intervals $t = (1 - 8/9)T$, $t = (1 - 7/9)T, \ldots, t = (1 - 1/9)T$, where $T$ is the time taken for the shock to reach 1 AU. Note particularly the hardening of the spectrum with increasing time for the perpendicular shock example (Zank et al. 2006)

Figure 6 shows the corresponding intensity profiles for a quasi-perpendicular (top panel) and parallel (bottom panel) shock. Three energies are illustrated. Clear differences between the two models are evident. Thanks to wave excitation by the streaming instability at the parallel shock, the parallel diffusion coefficient is smaller at these energies than the diffusion coefficient at the quasi-perpendicular shock. Consequently, particle trapping is more efficient at the parallel shock, thus limiting particle escape at the energies illustrated. This is reflected in the intensity profiles, which show a very rapid rise time and formation of a plateau in the quasi-perpendicular shock case. A much slower rise time is exhibited in the parallel shock example and only the $T = 50$ MeV protons reach a plateau phase (by contrast, the 50 MeV particles accelerated at the quasi-perpendicular shock are released rapidly enough that they are no longer observed at 1 AU after $\sim$1 day). Like the spectra illustrated in Fig. 5, distinctive differences are present in the intensity profiles associated with quasi-perpendicular and parallel shocks. These will, of course, not be revealed as clearly in observations as they are in our models since we have deliberately neglected the changing obliquity of the shock with heliocentric radius and the spacecraft connection to different magnetic field lines with time.

### 3.2 Modeling a Quasi-Parallel Event

The theory discussed in Sects. 2.1, 2.2, 2.3 is now becoming reasonably well established, but very few attempts have been made to directly model particle acceleration and transport of specific events (see, e.g., Li et al. 2005; Tylka et al. 2005). In part, this is due to (1) the difficulty in modeling the physical conditions pertaining to the initial conditions in the solar

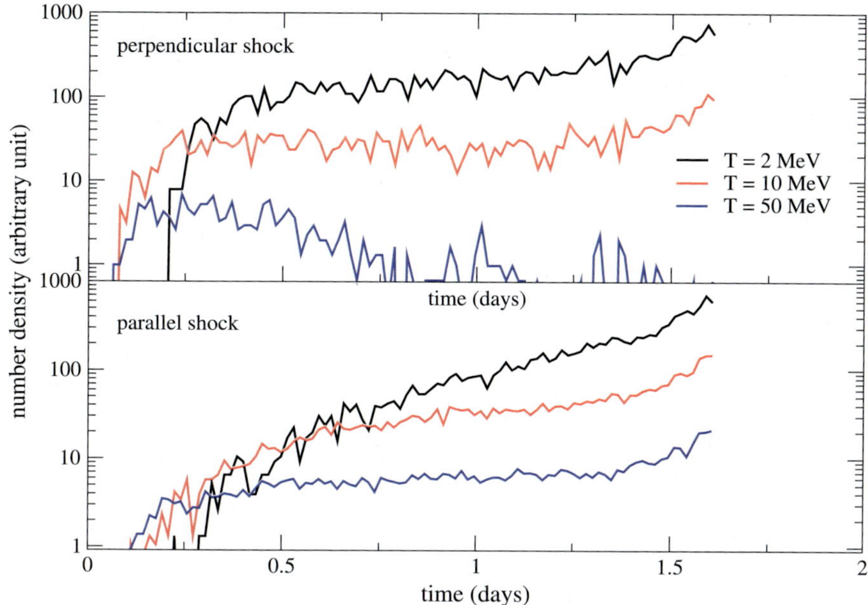


**Fig. 6** Time–intensity profiles for the quasi-perpendicular and parallel shock cases. The *upper panel* is for the perpendicular shock case and the *lower panel* is for the parallel shock case. The shock arrived at 1 AU, about 1.6 days after initiation. The basic details of the model and the related physics were discussed by Zank et al. (2000), Li et al. (2003), and Rice et al. (2003). The intensity profiles emphasize the important role of the time-dependent maximum energy to which protons are accelerated at a shock and the subsequent efficiency of trapping these particles in the vicinity of the shock. Compared to the parallel shock case, the particle intensity reaches the plateau phase faster for the quasi-perpendicular shock example. This is because $\kappa_\perp$ at a highly perpendicular shock is larger than the stimulated $\kappa_\parallel$ at a parallel shock, so particles (especially at low energies) find it easier to escape from the quasi-perpendicular shock than the parallel shock (Zank et al. 2006)

wind, including shock initiation, formation, and location; the initial shock speed, obliquity, upstream solar wind plasma conditions; the pre-turbulence levels at the shock; the ambient seed population, and so on, and (2) the temporal variation in the magnetic connection of the spacecraft to the propagating shock.

To address the second problem, a fully 2D model is needed. In particular, the model should include a treatment of particle acceleration at perpendicular shocks similar to that outlined in Sect. 2.2. Although a 2D model is important and perhaps necessary to understand most SEP events, a model that focuses on quasi-parallel shocks can nevertheless still provide a reasonable description of some SEP events. Needless to say, these events are not common, since they must maintain a quasi-parallel configuration during their propagation from the Sun to 1 AU. This section discusses a very detailed modeling effort by Verkhoglyadova et al. (2007) for a particular quasi-parallel event.

As a first step, Verkhoglyadova et al. (2007) identified seven quasi-parallel shocks at 1 AU from the ACE List of Disturbances and Transients (maintained by C.W. Smith) and the MIT shock database (maintained by J. Kasper). Of course, these shocks may not have been quasi-parallel near the Sun, in which case we expect that our 1D model simulation and the observations may differ, especially in the early time–intensity profiles. In view of the distinctions between quasi-parallel and quasi-perpendicular shock time–intensity profiles and the dependence of the break point on $(Q/A)$, we can expect that if the observed time–intensity profiles and the modeled calculation agree reasonably well, then the shock being modeled was quasi-parallel throughout its propagation. More accurately, the shock associated with the connection point of the interplanetary magnetic field and the spacecraft as the shock traversed the interplanetary medium to 1 AU remained in a quasi-parallel configuration. Were this the case, we can further examine the $(Q/A)$ dependence of the break point of the energy spectra for heavy ions to better support our assumption of quasi-parallel shock propagation from the Sun to 1 AU.

Having identified a potentially suitable quasi-parallel shock, we next determine if SEP signatures at 1 AU can be associated with it. This step of the analysis is frequently complicated by the possibility that an observed SEP event at 1 AU may often be the result of particle acceleration at and/or interaction with two or more CME shocks. Ideally, the identification of "clean" SEP events with pre-event backgrounds that are clearly free of "contamination" from earlier (small or large) SEP events or even previous impulsive events is preferred. However, besides the difficulty in identifying such infrequent events, a persuasive argument can be made that such "stirred up" conditions are conducive to particle acceleration at interplanetary shocks (enhanced levels of both pre-energized particles—for injection—and low-frequency magnetic turbulence—for particle scattering). Consequently, we do not impose the requirement that a clean pre-event background is necessary when selecting an event. Instead, a quasi-parallel shock is included in our study if careful examination of particle spectra and time intensity suggests the possibility of diffusive shock acceleration.

The numerical model of Sects. 2.1 and 2.3 developed at the IGPP, University of California at Riverside, is called the **P**article **A**cceleration and **T**ransport in the **H**eliosphere model (hereafter PATH). The model consists of two major parts. The first part includes modeling of an evolving shock and particle acceleration, and the second part treats energetic particle transport throughout the heliosphere (from 0.1 to $>$1 AU). The core of the PATH model was described by Zank et al. (2000); Rice et al. (2003); and Li et al. (2003, 2005). As described in Sect. 2.1, the spatial diffusion coefficient for the quasi-parallel shock is calculated self-consistently (Gordon et al. 1999; Rice et al. 2003). The injection momentum is a parameter that can be adjusted, and is currently taken as $\sim$10 keV.

After verifying the appropriate SEP signatures at 1 AU, we then numerically model the solar wind background and shock, tuning the model to yield as accurately as possible the pre-event conditions. This includes using real-time solar wind parameters prior to the event as the background input and adjusting the initial CME-driven shock speed, density, and temperature jumps at the inner boundary to obtain results that closely mimic the observed shock speed and compression ratio at 1 AU. After obtaining a realistic solar wind, the PATH model then follows the shock evolution, the associated particle acceleration, and finally the transport of accelerated energetic particles throughout the heliosphere (from $\sim$0.1 to beyond 1 AU).

A case identified by Verkhoglyadova et al. (2007) as an appropriate candidate to model was an event that occurred on September 29, 2001. From the MIT shock database maintained by J. Kasper, the average angle between the shock normal and the background magnetic field is estimated as $\sim 19°$ (another estimate gives 25.7°) which justifies the approximation of a quasi-parallel shock. The shock reached 1 AU with a speed of $\sim$700 km/sec at 9:10 on September 29, 2001. The compression ratio was $s \sim 2.3$. Prior to the arrival of the shock, a partial halo CME with a speed of 1,109 km/s was observed by SOHO/LASCO at 04:54:05 on September 27, 2001 (from the LASCO CME catalogue). From 04:32 to 04:38 on September 27, 2001, X-rays were also observed in the active region 9628, located at S20W27, by GOES.

As documented by Desai et al. (2003), there was a large SEP event on September 24, 2001. The example considered by Verkhoglyadova et al. (2007) occurred during the decay

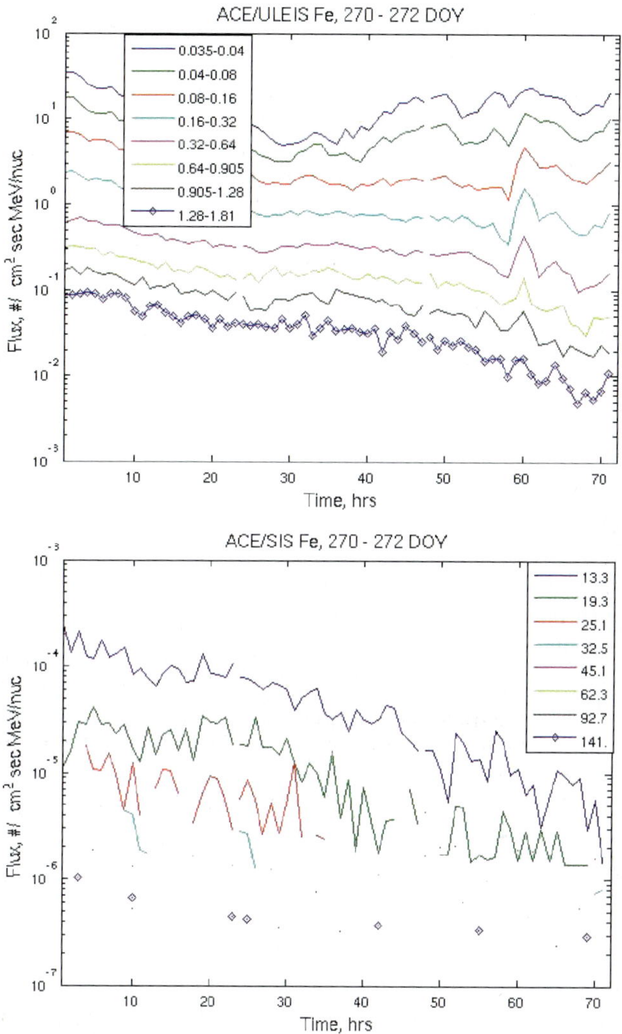


**Fig. 7** Time–intensity profiles drawn from ACE ULEIS (*top*) and SIS data (*bottom*) at a range of energies, given in MeV/nucleon (Verkhoglyadova et al. 2007)

phase of this large SEP event. A careful examination of the Fe time–intensity profile, for example, shows quite convincingly that particle acceleration occurred at the September 29, 2001, shock. Figure 7 shows plots at various energies of the Fe time–intensity profile using ULEIS and SIS energies. At energies drawn from ULEIS (Fig. 7, top panel), a local bump in the time intensity is clearly seen close to $t = 60$ hours (with 0 starting from day 270) when the September 29, 2001, shock arrived, a clear signature of diffusive shock acceleration. At higher SIS energies (Fig. 7, bottom panel), although the gradually decaying background from the previous event is high, we can still infer particle acceleration associated with the September 29, 2001, event through, for example, the increase in the $E = 19.3$ MeV/nucleon profile. Between $t = 10$–35 hours (from day 270), there is a relatively abrupt enhancement over the background followed by an equally abrupt decrease. As we discuss in the following,

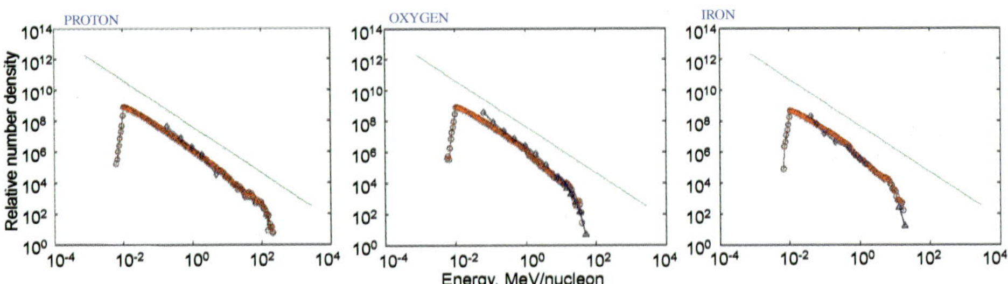


**Fig. 8** Event-integrated proton, iron, and oxygen spectra at 1 AU determined by the PATH model (*red circles*) and compared to ACE SIS (*blue triangles*) and ULEIS (*blue diamonds*) measured spectra (Verkhoglyadova et al. 2007)

we may anticipate that this enhancement is associated with the early acceleration of Fe ions by the September 29 shock for a short period. (It is possible that the time–intensity profiles of these two events could be separated using a detailed simulation that includes both shocks. This is left for future work.)

As suggested earlier, a prior shock primes the interplanetary medium by seeding it with energetic particles that the second shock can inject and (re)accelerate. Diffusive shock acceleration may well therefore be easier or more efficient for a shock that follows another. A careful injection model for pre-existing energetic particles should include their spectral form, abundances, and so on, but for the present, Verkhoglyadova et al. (2007) assumed that these particles are simply assimilated and re-accelerated at the second shock. This is reasonable since the re-acceleration of a pre-existing spectrum produces a new spectrum with characteristics determined by the local shock compression ratio (unless the energies are higher than the shock is capable of accelerating particles to, in which case the spectral index is unchanged although the distribution experiences compression). Because of the pre-existing energetic particle population, we focus on event-integrated spectra rather than the evolution of the accelerated spectrum at 1 AU. The former is largely independent of the pre-existing energetic particle population whereas the latter can be contaminated by features of the initial spectrum. The event-integrated spectra will reflect fully the characteristics of the later shock as it propagates to 1 AU.

Verkhoglyadova et al. (2007) found that the shock reaches 1 AU in $\sim$50 hours. They modeled the acceleration and transport of protons and Fe and (CN)O ions, with a charge-to-mass ratio of 14 : 56 and 6 : 16, respectively, in the vicinity of the quasi-parallel shock. As the shock propagates it slows and the maximum energies to which the particle species can be accelerated decreases (Fig. 1), and is inversely proportional to the $(Q/A)$ of the ion species. Accelerated particles leak/escape from the shock and propagate to 1 AU and beyond. Figure 8 shows the integrated proton, iron, and oxygen spectra at 1 AU. A power law with theoretical limit of $\sim(s+2)/(s-1)/2$ (shown by the green line) is plotted to guide the eye for $s = 2.3$. Corresponding data taken from ACE measurements are also plotted. We use ULEIS (Mason et al. 1998) and SIS (Stone et al. 1998) particle detectors to estimate the integrated fluxes and spectra. The low-energy part of the spectrum is fitted to ULEIS measurements (shown by diamonds) and the high-energy part is fitted to SIS measurements (shown by triangles). Note that the modeled fluxes are presented in arbitrary units (see Verkhoglyadova et al. 2007 for details).

The spectral shapes are interesting in that they possess a "double power-law" structure with the break in energy ordered approximately by $(Q/A)^2$. Although the underlying diffusive shock acceleration theory predicts only strict power laws at any given time at the shock front, that the event-integrated spectra show a break or roll-over at high energies can be understood by recalling that the event-integrated spectrum results from overlaying a series of time-dependent spectra as the shock front propagates to 1 AU. As the maximum particle energy accelerated at the shock front decreases with time and the turbulence responsible for trapping particles in the shock complex weakens at the corresponding wave numbers, the event-integrated spectra no longer receive a contribution at high energies, which results in a "broken power law" or gradual roll-over. That the proton, Fe, and O modeled spectra simultaneously agree so well with observations is remarkable. Finally, we note that by comparing the modeled spectra with observations for different heavy ion species, we can examine the validity of our assumption that this event is quasi-parallel. If acceleration at the shock were dominated by perpendicular diffusion, it would be impossible to fit the modeled spectra to the observed O and Fe spectra since the maximum energy has a different $(Q/A)$ dependence at a quasi-perpendicular shock, than at a quasi-parallel shock; see (6–9).

## 4 Conclusions

For the case of quasi-parallel shocks, we have described a time-dependent model of shock wave propagation in the solar wind, at which particles are accelerated by the diffusive shock-acceleration mechanism. The model, called PATH for quasi-parallel shocks, includes local particle injection, Fermi acceleration at the shock, self-consistent excitation of the waves responsible for scattering, particle trapping and escape at the shock complex, and non-diffusive transport in the interplanetary medium, and does remarkably well in describing observed SEP events. This includes spectra, intensity profiles, and particle anisotropies. We modeled both proton and heavy ion acceleration and transport in gradual events, and can fit simultaneously event-integrated spectra for protons, Fe, and O. This allows us to understand the Fe/O ratios, for example. We have begun to model mixed events to explore the consequences of a pre-accelerated particle population (from flares, for example) and have also related this to the timing of flare-CME events. These results are not discussed here (Li and Zank 2005), nor did we discuss preliminary results for multi-D shocks and particle acceleration.

Secondly, we discussed a basic theory for particle acceleration at highly perpendicular shocks based on the convection of in situ solar wind turbulence into the shock. We found that the highest injection energies were needed for quasi-perpendicular shocks rather than for properly perpendicular shocks, making the 90° shock a "singular" example. Unlike the quasi-parallel case, the determination of the perpendicular diffusion coefficient is not based on a resonance condition but on the Nonlinear Guiding Center theory instead. Maximum energies at quasi-perpendicular shocks are smaller than those achieved at quasi-parallel shocks near the sun when self-consistent wave excitation is included. The injection energy threshold is much higher for quasi-perpendicular shocks than for quasi-parallel shocks, and we can therefore expect distinctive compositional differences for the two cases. Finally, although not discussed here, observations support the notion that diffusive particle acceleration at shocks can occur in the absence of stimulated wave activity (Zank et al. 2006).

**Acknowledgements** The authors acknowledge the partial support of NASA grants NNG04GF83G, NNG05GH38G, NNG05GM62G, a Cluster University of Delaware subcontract BART372159/SG, and NSF grants ATM0317509, and ATM0428880.

## References

W.I. Axford, E. Leer, G. Skadron, *Proc.15th Int. Cosmic Ray Conf. (Plovdiv)*, vol. 11 (1977), p. 132
A.R. Bell, Mon. Not. Roy. Astron. Soc. **182**, 147–156 (1978) 443–455
H.V. Cane et al., Geophys. Res. Lett. **30**(12), 8017 (2003). doi:10.1029/2002GL016580
E.W. Cliver, H.V. Cane, *AIP Conf. Proc.*, vol. 516 (Am. Inst. of Phys., College Park, 2000)
C.M.S. Cohen, R.A. Mewaldt, R.A. Leske, A.C. Cummings, E.C. Stone, M.E. Wiedenbeck, E.R. Christian, T.T. von Rosenvinge, Geophys. Res. Lett. **26**, 2697 (1999)
M.I. Desai, G.M. Mason, J.R. Dwyer, J.E. Mazur, R.E. Gold, S.M. Krimigis, C.W. Smith, R.M. Skoug, Astrophys. J. **588**(2), 1149–1162 (2003)
L.O. Drury, Rep. Progr. Phys. **46**, 973–1027 (1983)
B.E. Gordon, A.M. Lee, E. Möbius, K.J. Trattner, J. Geophys. Res. **104**, 28–263 (1999)
A.M. Heras, B. Sanahuja, Z.K. Smith, T. Detman, M. Dryer, Astrophys. J. **391**, 359–369 (1992)
A.M. Heras, B. Sanahuja, D. Lario, Z.K. Smith, T. Detman, M. Dryer, Astrophys. J. **445**, 497–508 (1995)
J.R. Jokipii, Astrophys. J. **313**, 842 (1987)
S.W. Kahler, N.R. Sheeley, R.A. Howard, M.J. Koomen, D.J. Michels, R.E. McGuire, T.T. von Rosenvinge, D.V. Reames, J. Geophys. Res. **89**, 9683–9693 (1984)
M.-B. Kallenrode, G. Wibberenz, J. Geophys. Res. **102**, 22,311–22,334 (1997)
M.-B. Kallenrode, R. Hatzky, *Proc. Int. Conf. Cosmic Rays 26th*, vol. 6 (1999), pp. 324–327
D. Lario, B. Sanahuja, A.M. Heras, Astrophys. J. **509**, 415–434 (1998)
M.A. Lee, J. Geophys. Res. **88**, 6109 (1983)
R.A. Leske, J.R. Cummings, R.A. Mewaldt, E.C. Stone, T.T. von Rosenvinge, Astrophys. J. **452**, L149–L152 (1995)
G. Li, G.P. Zank, W.K.M. Rice, J. Geophys. Res. **108**(A2), 1082 (2003). doi:10.1029/2002JA009666
G. Li, G.P. Zank, Geophys. Res. Lett. **32**, L02101 (2005). doi:10.1029/2004GL021250
G. Li, G.P. Zank, W.K.M. Rice, J. Geophys. Res. A06104 (2005)
G. Li, G.P. Zank, M. Desai, G.M. Mason, W. Rice, *Particle Acceleration in Astrophysical Plasmas in Geospace and Beyond*. Geophys. Monogr. Ser. (AGU, Washington, 2005), pp. 51–58
A. Luhn, B. Klecker, D. Hovestadt, M. Scholer, G. Gloeckler, F.M. Ipavich, C.Y. Fan, L.A. Fisk, Adv. Space Res. **4**, 161 (1984)
A. Luhn, B. Klecker, D. Hovestadt, E. Mobius, Astrophys. J. **317**, 951–955 (1987)
G.M. Mason, J.E. Mazur, M.D. Looper, R.A. Mewaldt, Astrophys. J. **452**, 901–911 (1995)
G.M. Mason, R.E. Gold, S.M. Krimigis, J.E. Mazur, G.B. Andrews, K.A. Daley, J.R. Dwyer, K.F. Heuerman, T.L. James, M.J. Kennedy, T. Lefevere, H. Malcolm, B. Tossman, P.H. Walpole, Space Sci. Rev. **86**(1/4), 409–448 (1998)
G.M. Mason et al., Geophys. Res. Lett. **26**, 141–144 (1999)
W.H. Matthaeus, G. Qin, J.W. Bieber, G.P. Zank, Astrophys. J. **590**, L53 (2003)
R.A. Mewaldt et al., *Proc. Int. Conf. Cosmic Rays 27th*, vol. 1 (2001), p. 3984
C.K. Ng, D.V. Reames, Astrophys. J. **453**, 890 (1995)
C.K. Ng, D.V. Reames, A.J. Tylka, Geophys. Res. Lett. **26**, 2145–2148 (1999)
C.K. Ng, D.V. Reames, A.J. Tylka, Astrophys. J. **591**, 461–485 (2003)
M. Oetliker, B. Klecker, D. Hovestadt, G.M. Mason, J.E. Mazur, R.A. Leske, R.A. Mewaldt, J.B. Blake, M.D. Looper, Astrophys. J. **477**, 495–501 (1997)
E.N. Parker, Astrophys. J. **123**, 664–676 (1958)
W.K.M. Rice, G.P. Zank, G. Li, J. Geophys. Res. **108**, 1369 (2003). doi:10.1029/2002JA009756
D. Ruffolo, Astrophys. J. **442**, 861–874 (1995)
E.C. Stone, C.M.S. Cohen, W.R. Cook, A.C. Cummings, B. Gauld, B. Kecman, R.A. Leske, R.A. Mewaldt, M.R. Thayer, B.L. Dougherty, R.L. Grumm, B.D. Milliken, R.G. Radocinski, M.E. Wiedenbeck, E.R. Christian, S. Shuman, T.T. von Rosenvinge, Space Sci. Rev. **86**(1/4), 357–408 (1998)
A.J. Tylka, P.R. Boberg, J.H. Adams, L.P. Beahm, W.F. Dietrich, T. Kleis, Astrophys. J. **444**, L109–L113 (1995)
A.J. Tylka, C.M.S. Cohen, W.F. Dietrich, M.A. Lee, C.G. Maclennan, R.A. Mewaldt, C.K. Ng, D.V. Reames, Astrophys. J. **625**(1), 474–495 (2005)
P. van Nes, R. Reinhard, T.R. Sanderson, K.-P. Wenzel, R.D. Zwickl, J. Geophys. Res. **89**, 2122 (1984)
O.P. Verkhoglyadova, G. Li, G.P. Zank, Q. Hu, Geophys. Res. Lett. (2007, in press)
G.P. Zank, H.L. Pauls, I.H. Cairns, G.M. Webb, J. Geophys. Res. **101**, 457 (1996)
G.P. Zank, W.H. Matthaeus, J.W. Bieber, H. Moraal, J. Geophys. Res. (Space) **103**, 2085–2097 (1998)
G.P. Zank, W.K.M. Rice, C.C. Wu, J. Geophys. Res. (Space) **105**, 25079–25095 (2000)
G.P. Zank, G. Li, V. Florinski, W.H. Matthaeus, G.M. Webb, J.A. le Roux, J. Geophys. Res. **109**, A04107 (2004). doi:10.1029/2003JA010301
G.P. Zank, L. Gang, V. Florinski, Q. Hu, D. Lario, C.W. Smith, J. Geophys. Res. **1**(A6), (2006). CiteID A06108

Space Sci Rev (2007) 130: 273–282
DOI 10.1007/s11214-007-9207-1

# Ionic Charge States of Solar Energetic Particles: A Clue to the Source

**B. Klecker · E. Möbius · M.A. Popecki**

Received: 8 February 2007 / Accepted: 17 April 2007 /
Published online: 7 June 2007


**Abstract** The ionic charge of solar energetic particles (SEP) as observed in interplanetary space is an important parameter for the diagnostic of the plasma conditions at the source region and provides fundamental information about the acceleration and propagation processes at the Sun and in interplanetary space. In this paper we review the new measurements of ionic charge states with advanced instrumentation onboard the SAMPEX, SOHO, and ACE spacecraft that provide for the first time ionic charge measurements over the wide energy range of ~0.01 to 70 MeV/nuc (for Fe), and for many individual SEP events. These new measurements show a strong energy dependence of the mean ionic charge of heavy ions, most pronounced for iron, indicating that the previous interpretation of the mean ionic charge being solely related to the ambient plasma temperature was too simplistic. This energy dependence, in combination with models on acceleration, charge stripping, and solar and interplanetary propagation, provides constraints for the temperature, density, and acceleration time scales in the acceleration region. The comparison of the measurements with model calculations shows that for *impulsive* events with a large increase of $Q_{\mathrm{Fe}}(E)$ at energies $\leq 1$ MeV/nuc the acceleration occurs low in the corona, typically at altitudes $\leq 0.2\ R_S$.

**Keywords** Sun: solar energetic particles · Ionic charge states

## 1 Introduction

The ionic charge of solar energetic particles (SEPs) is an important parameter for the diagnostic of the plasma conditions at the source region. Furthermore, the acceleration and transport processes depend generally on velocity and rigidity, i.e., on the mass and ionic charge of the ions. The large variations of elemental and isotopic abundances by up to several orders

B. Klecker (✉)
Max-Planck-Institut für extraterrestrische Physik, 85741 Garching, Germany
e-mail: berndt.klecker@mpe.mpg.de

E. Möbius · M.A. Popecki
University of New Hampshire, Durham, NH 03824, USA

of magnitude and variations of the mean ionic charge of heavy ions, in particular of Fe, at ~1 MeV/nuc have been used for dividing SEP events into two classes, *gradual* and *impulsive* (see, e.g., Reames 1999 for an early review), following the classification of flares based on the duration of their soft X-ray emission (Pallavicini et al. 1977): (1) *Gradual* events show large interplanetary ion intensities, small electron to proton ratios, on average elemental abundances similar to coronal abundances, and ionic charge states consistent with source temperatures of (1.5–2) $\times$ $10^6$ K, characteristic for the solar corona. These events show long-duration soft X-ray emission and are associated with interplanetary shocks, driven by coronal mass ejections (CMEs). Particles are accelerated at the shock front over a wide range of solar longitudes, and then propagate along the interplanetary magnetic field before reaching the spacecraft. For a recent review on energetic particle observations related to CMEs and coronal and interplanetary shocks see Klecker et al. (2006a). (2) *Impulsive* events show small interplanetary ion intensities, a high electron to proton intensity ratio, enhanced abundances of heavy elements (e.g., by a factor ~10 for Fe relative to O), and enhancements of $^3$He relative to $^4$He by up to a factor of $10^4$. Because of limited sensitivity the heavy ion ionic charge states could only be determined previously as an average over several $^3$He-rich time periods. This event average showed high mean ionic charge states for Si ($Q \sim 14$) and Fe ($Q \sim 20$), which were interpreted as being due to a high temperature of ~$10^7$ K in the source region (Klecker et al. 1984; Luhn et al. 1987). These events show short-duration soft X-ray emission and the acceleration process is thought to be related to the flare. These "flare particles" can reach the spacecraft only from a narrow range of solar longitudes connecting the acceleration site with the spacecraft.

However, new measurements with improved sensitivity onboard the SAMPEX, SOHO, and ACE spacecraft demonstrate that energetic ions in coronal and interplanetary shock related (*gradual*) events, in particular at energies $\geq$10 s of MeV/nuc, often show signatures usually associated with *impulsive* events, including high ionic charge states, with $Q \sim$ 15–20 for Fe (e.g., Leske et al. 1995; Mazur et al. 1999) and enrichments in heavy ions and $^3$He (e.g., Cohen et al. 1999; Mason et al. 1999; Torsti et al. 2002; Desai et al. 2003; von Rosenvinge and Cane 2006, and references therein). Therefore, the two-class paradigm is now in question. It is, however, not questioned that SEPs originate in (at least) two different ways: (1) Coronal and interplanetary shocks related to CMEs can accelerate particles and (2) particles are accelerated in the flare process, possibly related to reconnection, and the high charge states and the large enrichments of $^3$He and heavy ions are related to this acceleration process.

In this paper we will review recent developments on ionic charge state measurements. We first present a short summary of the different techniques to determine the ionic charge of solar energetic particles, and then discuss recent results and their implications for the acceleration processes.

## 2 Measurement Techniques

Over the past ~30 years basically three methods have been developed to determine or infer the ionic charge of energetic ions: (1) direct in situ determination of the particle parameters mass ($M$) and/or nuclear charge ($Z$), kinetic energy ($E$), and ionic charge ($Q$); (2) in situ measurements of particle mass and kinetic energy, and determination of $Q$ from the rigidity-dependent cutoff of the magnetic field of the Earth; and (3) indirect methods, inferring the ionic charge from rigidity-dependent acceleration or propagation processes. For a detailed discussion of these methods see the recent reviews by Popecki (2006) and Klecker et al. (2006b).

The direct measurements are the most accurate; however, they are technically limited to energies less than a few MeV/nuc. At higher energies, novel instrumentation with large collecting power onboard the polar orbiting SAMPEX (*Solar, Anomalous and Magnetospheric Particle Explorer*) spacecraft (Baker et al. 1993) provided for the first time ionic charge measurements for many elements in the mass range C to Fe over the extended energy range of ~0.3–70 MeV/nuc, utilizing the magnetic field of the Earth as a magnetic spectrometer (Mason et al. 1995; Leske et al. 1995; Oetliker et al. 1997). In recent years, the energy range was also extended to lower energies. At the same time sensitivity and resolution were improved with the SEPICA (Solar Energetic Particle Ionic Charge Analyzer) instrument (Möbius et al. 1998) onboard the *Advanced Composition Explorer* (ACE), and with the STOF sensor of the CELIAS experiment (Hovestadt et al. 1995) onboard the *Solar and Heliospheric Observatory* (SOHO). Over the past ~10 years these new instruments have now extended the measurement of ionic charge states to many SEP events with low interplanetary particle intensity, in particular, also to many individual *impulsive* events.

## 3 Ionic Charge State Measurements: New Results

### 3.1 New Results in Gradual SEP Events

With the new measurements that provide for the first time ionic charge measurements in large (*gradual*) events over an extended energy range, a significant variation with energy and a large variability of the energy dependence of heavy ions have been found in many events, most notably for iron ions. At low energies ($\leq$200 keV/nuc) the mean ionic charge of Fe is usually ~9–11 (Bogdanov et al. 2000), which is similar to solar wind charge states (Ko et al. 1999). At higher energies a large variability is observed. The mean ionic charge of Fe at energies $\leq$1 MeV/nuc is either constant or increases with energy, in a few cases by up to 4 charge units (Möbius et al. 1999; Mazur et al. 1999; Möbius et al. 2002). At energies above ~10 MeV/nuc, however, the mean ionic charge is often observed to be significantly larger than at low energies, with $Q_{\rm Fe} \sim 15$–20 (Leske et al. 1995; Oetliker et al. 1997; Labrador et al. 2005). The variation of $Q_{\rm Fe}$ with energy is illustrated in Fig. 1 (left-hand side), which shows $Q_{\rm Fe}$ event averages in three energy ranges between 0.18 and 65 MeV/nuc for several large events. These results indicate that the previous interpretation of heavy ion charge states being solely related to the plasma temperature was too simplistic. The compilation of Fe/O ratios and heavy ion charge states in Fig. 1 (right-hand side) shows that the observed variability of $Q_{\rm Fe}$ at $E \geq 10$ MeV/nuc is strongly correlated with the relative abundance of Fe (Labrador et al. 2005).

### 3.2 New Results in Impulsive SEP Events

The high-sensitivity ionic charge measurement with the SEPICA experiment onboard ACE provided ionic charge measurements for a large number of *gradual* and *impulsive* SEP events. Figure 2 shows the distributions of the mean ionic charge of Fe at 0.18–0.25 MeV/nuc in ~40 impulsive events compiled by DiFabio et al. (2006) and, for comparison, in ~40 interplanetary shock-related time periods (Klecker et al. 2006c). Figure 2 shows that the distributions of the mean charge of Fe are completely different for the two classes of events: Those correlated with interplanetary shocks show in this energy range a

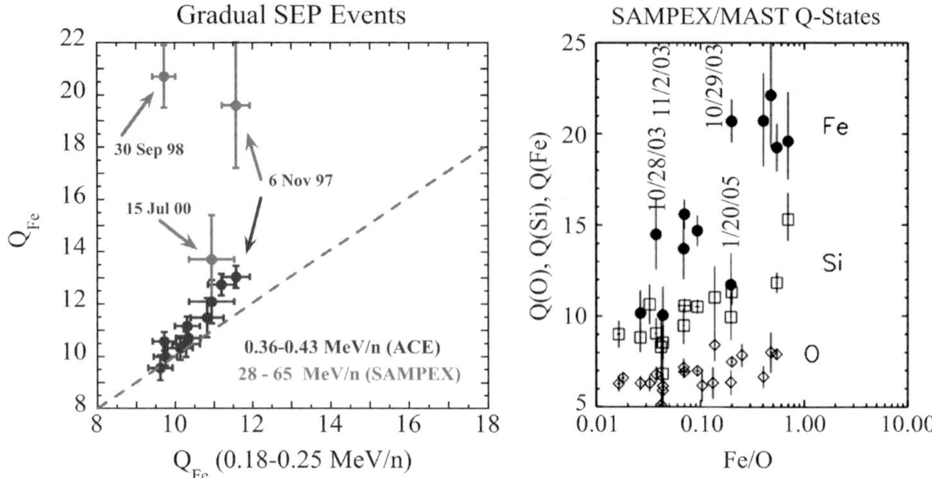


**Fig. 1** *Left*: Ionic charge $Q_{\mathrm{Fe}}$ at 0.18–0.25 MeV/nuc versus 0.36–0.43 and 28–65 MeV/nuc (compilation from Möbius et al. 1999; Labrador et al. 2005; Popecki 2006). *Right*: Correlation of heavy ion abundances and charge states at $\geq$10 MeV/nuc (Labrador et al. 2005)

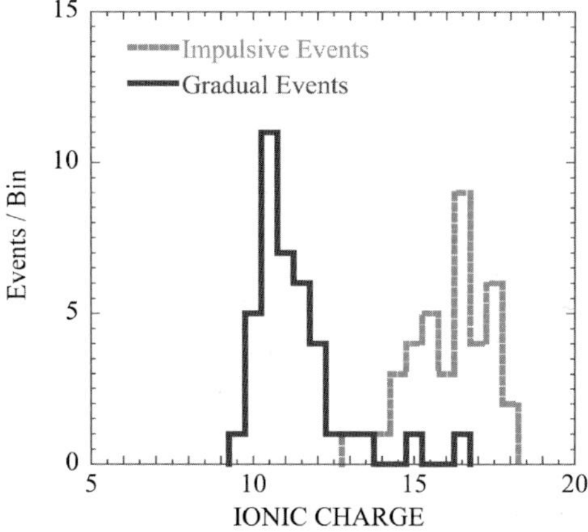


**Fig. 2** Average ionic charge of Fe in the energy range 0.18–0.24 MeV/nuc in ~40 impulsive and ~40 gradual SEP events; for details see the text

distribution with a peak around $Q_{\mathrm{Fe}} \sim 10$, whereas the corresponding distribution of average charge states in impulsive events is considerably wider and ranges from ~14 to 18. Furthermore, all *impulsive* events observed with SEPICA onboard ACE during 1997–2000 show a strong increase of the mean ionic charge of Fe in the narrow energy range of ~0.1–0.55 MeV/nuc, on average by $\sim 4.5 \pm 1$ charge units (Möbius et al. 2003; Klecker et al. 2006d; Popecki 2006; DiFabio et al. 2006).

Figure 3 illustrates this energy dependence with three typical impulsive events observed during 1998–2000. The observations show a monotonic increase of $Q$ with energy in the measurement range of the instrument, with $Q_{\mathrm{Fe}} \sim 13$–15 at ~0.1 MeV/nuc and $Q_{\mathrm{Fe}} \sim 18$–20 at ~0.55 MeV/nuc (Popecki 2006, and references therein), at the high-energy end, consistent with the early measurements at ~1 MeV/nuc.

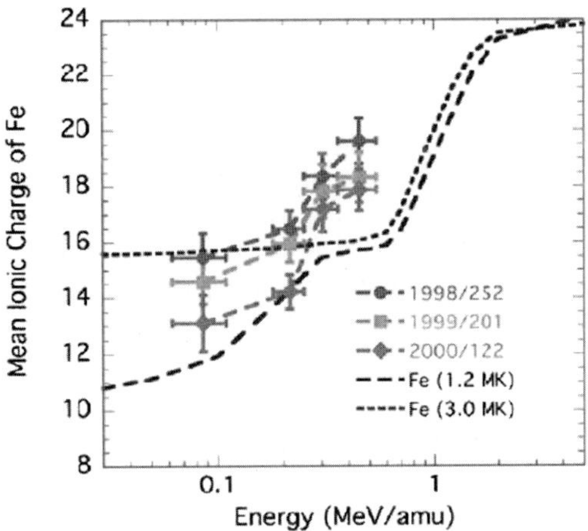


**Fig. 3** The energy dependence of the ionic charge of Fe in three representative impulsive events (Klecker et al. 2006d; DiFabio et al. 2006). Also shown is the energy dependence of the equilibrium charge from ionization effects in a dense environment for $1.2 \times 10^6$ and $3 \times 10^6$ K (Klecker et al. 2006d)

## 4 The Energy Dependence of the Ionic Charge

Several mechanisms resulting in an increase of the mean charge states with energy have been discussed recently (for a more detailed discussion see also Popecki 2006, Klecker et al. 2006b and references therein): (1) charge changing effects resulting from ionization by thermal electrons and ions, (2) mixing of sources with different ionic charge distributions, and (3) $M/Q$-dependent energy spectra. However, a large increase of $Q_{\text{Fe}}$ at energies $\leq 1.0$ MeV/nuc, as systematically observed in *impulsive* events, could so far only be explained by additional ionization of more energetic ions in a dense environment. If the particles propagate in a sufficiently dense environment in the lower corona during or after the acceleration, a large increase of the mean ionic charge at energies of $\sim 0.2$–1 MeV/nuc is a natural consequence of the cross sections for ionization by thermal electrons and protons (Kocharov et al. 2000). Calculations of the equilibrium ionic charge of energetic heavy ions, including the effects of radiative and dielectronic recombination and ionization by thermal electrons and ions ($p$, $^4$He), show that the mean ionic charge of an ion with speed $V$ increases monotonically as a function of $N\tau$, where $N$ is the plasma density and $\tau$ is the acceleration time scale. The ionic charge approaches asymptotically an upper limit, the equilibrium mean charge $Q_{\text{eq}}$, where $(N\tau)_{\text{eq}}$ depends on ion speed and plasma temperature and is in the range of $10^{10}$–$10^{11}$ s cm$^{-3}$ (Kocharov et al. 2000; Kovaltsov et al. 2001). Figure 3 shows as an example the equilibrium mean ionic charge of Fe as a function of energy for two temperatures ($1.2 \times 10^6$ and $3 \times 10^6$ K) in the surrounding plasma, demonstrating the strong increase of $Q_{\text{eq}}$, for Fe at energies $\geq 0.2$ MeV/nuc. The simple equilibrium model also demonstrates that the new measurements of $Q_{\text{Fe}}$ at low energies are not compatible with a high coronal temperature of $\sim 10^7$ K, as previously inferred from the ionic charge measurements at $\sim 1$ MeV/nuc.

Figure 3 also shows that the steep increase of $Q_{\text{Fe}}$ with energy is typically observed at somewhat lower energies than predicted by the equilibrium stripping model. It has been demonstrated recently that this difference between the observed and predicted energy dependence of $Q$ can be explained by propagation effects (Kartavykh et al. 2005): On their way from the acceleration site at the Sun to 1 AU low-energy particles can lose a significant fraction of their energy by adiabatic deceleration. This results, for an average ionic charge $Q(E)$ at the Sun, in the same mean ionic charge at a lower energy $E$ at 1 AU, as seen qualitatively in Fig. 3. Furthermore, the equilibrium mean charge is an upper limit that may never be reached under realistic conditions.

Therefore, a quantitative comparison of the observations with calculations requires more realistic models that include the effects of acceleration, ionization, and recombination, Coulomb energy losses at the Sun, and propagation at the Sun and in interplanetary space. Also, the details of the dependence of $Q$ on $E$ will depend on the type of model, i.e., whether acceleration and charge changing effects are concurrent as in the leaky-box models with stochastic or shock acceleration, or whether there is charge stripping also after acceleration, as would be the case for acceleration by a shock low in the corona with subsequent escape of the particles into interplanetary space. For a recent review on the modeling of energy-dependent charge states of SEPs see Kocharov (2006) and references therein.

## 5 Implications for the Source Location

### 5.1 Impulsive Events

The mean ionic charge of Fe and the energy spectra in impulsive events have recently been modeled by a leaky-box model including stochastic acceleration and charge changing processes. In these models, the intensity–time profiles and anisotropies of particles with different mass per charge ratio (e.g., $H^+$, $He^{2+}$, electrons) are used to infer the injection profile at the Sun and the propagation characteristics in interplanetary space. Then, the observed energy spectra and charge spectra of heavy ions are used to infer the model parameters for the acceleration and the plasma parameters of the acceleration environment. In these models stochastic acceleration is assumed with the model parameters $N\tau_A$, $\tau_A/\tau_D$, $\gamma$, and $T_e$, where $N$ and $T_e$ are plasma density and temperature, $\tau_A$ and $\tau_D$ are the time scales for acceleration and diffusion in the source region at the Sun, and $\gamma$ is the power-law index of the power spectrum of the wave turbulence. The model calculations show that a steep increase of $Q_{\mathrm{Fe}}$ with energy at $E \leq 1$ MeV/nuc as observed in all *impulsive* events can be explained with this type of model and both the energy spectra and the observed energy dependence of the ionic charge can be reproduced satisfactorily, if interplanetary propagation is included (Kartavykh et al. 2006; Dröge et al. 2006). Typical values of the model parameters are $N\tau_A \sim 10^{10}$–$10^{11}$ s cm$^{-3}$, $\tau_A/\tau_D \sim 0.1$, and $T_e \sim 10^6$ K. However, some of the charge spectra could only be reproduced by assuming two acceleration regions with different plasma parameters with significantly larger $T_e$ ($\sim 10^7$ K) in the second region (Dröge et al. 2006). If we assume acceleration time scales in the range of $\sim$1 to 10 s this corresponds to densities of $\sim 10^9$–$10^{11}$ cm$^{-3}$. This is similar to the density range of (0.6–10)$\times 10^9$ cm$^{-3}$ inferred from radio and electron measurements for the density of the acceleration region of electrons (e.g., Aschwanden 2002 and references therein); that is, it indicates acceleration in the low corona, at altitudes $\leq 0.2\ R_S$.

### 5.2 Gradual Events

In many gradual events the mean ionic charge is approximately constant up to energies of a few MeV/nuc. This is consistent with acceleration by a shock high in the corona or in interplanetary space, because charge stripping by the processes discussed thus far will not change the ionic charge states as a function of energy significantly for values of $N\tau_A \leq 10^8$ cm$^{-3}$ s, as shown by nonequilibrium calculations (Kovaltsov et al. 2001). These nonequilibrium models also show, for higher values of $N\tau_A$, a significant increase of $Q_{\mathrm{Fe}}$ at energies $\geq 10$ MeV/nuc, as often observed. However, there are also alternative explanations for the increase of heavy ion charge states at high energies as discussed in the following non-equilibrium calculations.

For the cases where a significant increase of $Q_{\mathrm{Fe}}$ is observed below $\sim$1 MeV/nuc, models including charge stripping and shock acceleration can qualitatively reproduce the observed energy dependence, if the acceleration starts in the low corona (e.g., Barghouty and Mewaldt 1999; Ostryakov and Stovpyuk 1999; Ostryakov and Stovpyuk 2003; Lytova and Kocharov 2005). In these models the charge state variation with energy depends on the coronal density profile $N(r)$, $T_e$, shock speed, compression ratio, and the absolute value and rigidity dependence of the spatial diffusion coefficients, upstream and downstream of the shock. It also depends on the type of model, i.e., whether stripping after acceleration is included. Typical forms of $Q_{\mathrm{Fe}}(E)$ are presented in Fig. 4 (Kocharov 2006), which shows the results of a numerical model for iron, including charge stripping, Coulomb losses, diffusion, convection, adiabatic deceleration, and shock acceleration, assuming an initial altitude of $\sim$0.2 $R_S$ for the shock, a turbulent layer with thickness 0.5 $R_S$, a shock speed of 600 km/s, and a coronal density profile adopted from Guhathakurta et al. (1996). Figure 4 demonstrates that there is a significant difference in $Q_{\mathrm{Fe}}$ for particles at the shock (plus signs), at a fixed distance from the shock (histogram), and for escaping particles with additional stripping after acceleration (diamonds).

Limits of the altitude range where the acceleration by an outward traveling shock needs to start to reproduce a large increase of $Q_{\mathrm{Fe}}$ at $\leq$1 MeV/nuc may be estimated from the requirement that the local fast magnetosonic speed $V_F$ is sufficiently small for traveling disturbances to produce fast-mode shocks (e.g., Mann et al. 1999; Gopalswamy et al. 2001) and from the acceleration model. However, there was so far no attempt to systematically explore the altitude range compatible with the observed variation of $Q_{\mathrm{Fe}}$ with energy. Uncertainties in the coronal density distribution, shock geometry, and turbulence will result in large uncertainties in the altitude estimate. In a recent review an estimate of $\sim$0.5–1 $R_S$ (with a

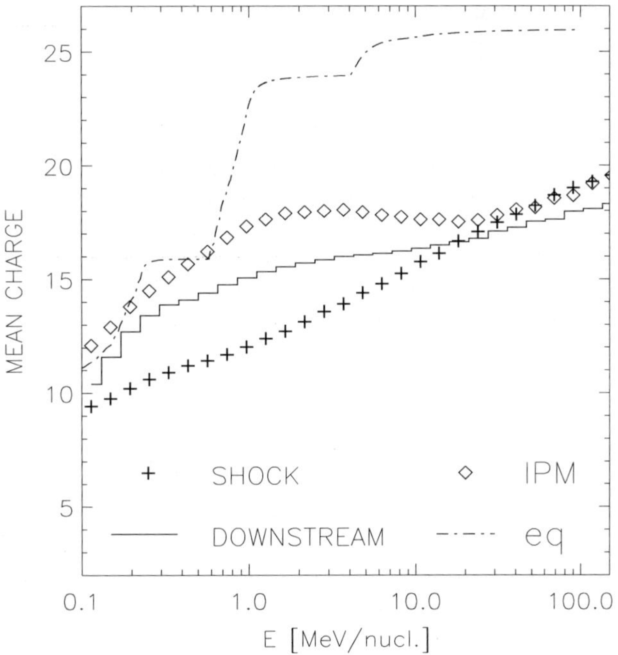


**Fig. 4** Mean charge states of Fe computed with a shock acceleration model. The symbols correspond to particles at the shock front (*plus signs*), escaping particles at a fixed distance from the shock (*histogram*), escaping particles with additional stripping during escape (*diamonds*), and equilibrium charge states (*dash-dotted line*) (Kocharov 2006)

large uncertainty of ~50%) was provided for the altitude range compatible with the energy dependence of $Q_{\mathrm{Fe}}$ in these events (Kocharov 2006).

A different approach, primarily intended to explain the large variability of compositional variations and spectral features of heavy ions in large interplanetary shock related SEP events, also predicts an increase of the mean ionic charge of heavy ions at high energies (Tylka et al. 2005; Tylka and Lee 2006). In this model, compositional (and ionic charge) variations are caused by the interplay of two factors: shock geometry and the mixture of two seed populations with composition and charge states similar to solar wind and flare particles, respectively. Key elements in this scenario are (1) a power-law energy spectra with an exponential rollover at high energies,

$$J(E) \sim E^{-\gamma} \exp(-E/E_0), \tag{1}$$

with $E_0$ depending on $M/Q$ (Ellison and Ramaty 1985; Tylka et al. 2000) and on the angle $\theta_{BN}$ between the magnetic field and the shock normal (Lee 2005) as

$$E_0 = E_{0p}(Q/M) f(\theta_{BN}), \tag{2}$$

(2) a higher injection threshold for large $\theta_{BN}$, and (3) averaging the energy spectra over $\theta_{BN}$ (by assuming contributions from the parallel and perpendicular regions of the evolving shock). In this scenario shock geometry determines via the injection threshold which of the two components dominate and pre-accelerated flare particles are preferentially injected at the perpendicular shock. However, whether the injection threshold at parallel and perpendicular shocks is in fact drastically different is presently disputed (Giacalone 2005) and needs further investigation.

In an alternative two-component model one assumes a direct flare component with high Fe charge states and high Fe/O ratio, which dominates at high energies, with a shock-accelerated component with charge states and heavy ion abundances similar to solar wind values dominating at low energies (e.g., Klein and Trottet 2001; Cane et al. 2003).

## 6 Summary

1. In all *impulsive* events observed during 1997–2000, the mean ionic charge of iron increases significantly, on average by $4.5 \pm 1$ charge units in the rather small energy range of ~0.1–0.55 MeV/nuc. At ~0.01–0.1 MeV/nuc an average value of $Q_{\mathrm{Fe}} \sim 12$ was observed for 3 events (Klecker et al. 2006d) and at ~0.1 MeV/nuc $Q_{\mathrm{Fe}} \sim 11$–15 was observed for 14 events (DiFabio et al. 2006).
2. In *gradual* events, the mean ionic charge at low energies of ~0.1 MeV/nuc is mostly compatible with solar wind charge states. At higher energies a large event-to-event variability is observed, with $Q_{\mathrm{Fe}} \sim 15$–20 at energies above ~10 MeV/nuc in several events.

The energy dependence of the ionic charge in *impulsive* events can be used to infer the location of the source and acceleration region. With models including the effect of stochastic acceleration, propagation, charge stripping, and Coulomb losses at the Sun and interplanetary propagation, $Q(E)$, energy spectra, and intensity–time and anisotropy–time profiles can be reproduced. These models show that the ionic charge states are determined by a combination of the parameters temperature, density, acceleration, and propagation time scales at the Sun, and interplanetary propagation conditions, with $N\tau_A \sim 10^{10}$–$10^{11}$ s cm$^{-3}$, placing the acceleration low in the corona, at altitudes $\leq 0.2\ R_S$. To further constrain the model parameters, the particle measurements need to be complemented by the measurements of, for example, X rays, electrons, and $\gamma$ rays to independently determine the acceleration and propagation time scales at the Sun. Furthermore, the present stripping models do not reproduce the heavy ion abundance enhancements observed in impulsive events.

In *gradual* events, the energy dependence of the ionic charge provides information on the various sources contributing to the accelerated particle population and on the source location. A large increase of $Q_{\mathrm{Fe}}$ at $\leq 1$ MeV/nuc can be explained by shock acceleration and impact ionization low in the corona, with a shock starting at heliocentric distances of $\sim$1.5–2 $R_S$ (Kocharov 2006). Events with $Q_{\mathrm{Fe}} \sim$ const. at low energies and with a large increase of the mean ionic charge at energies of tens of MeV/nuc, accompanied by an increase of the Fe abundance at high energies, suggest injection of ions from two sources: (1) a source with heavy ion abundances and charge states similar to solar wind or corona and (2) a flare component with heavy ion enrichment and charge states determined by charge stripping low in the corona. Whether this flare component is directly injected or further accelerated at a coronal shock is presently under debate and needs further investigation.

**Acknowledgements** BK thanks the organizing committee for the invitation to the Symposium on the Composition of Matter at Grindelwald. The paper also benefited greatly from the discussions at the ISSI working group on impulsive SEP events. The work was partially supported by NASA under NAG 5-12929.

## References

M.J. Aschwanden, Space Sci. Rev. **101**, 1 (2002)
D.N. Baker, G.M. Mason, O. Figueroa, G. Colon, J.G. Watzin, R.M. Aleman, IEEE Trans. Geosci. Remote Sensing **31**, 531 (1993)
A.F. Barghouty, R.A. Mewaldt, Astrophys. J. Lett. **520**, L127 (1999)
A.T. Bogdanov, B. Klecker, E. Möbius et al., *Acceleration and Transport of Energetic Particles Observed in the Heliosphere: ACE 2000 Symposium*, ed. by R.A. Mewaldt et al. (2000), pp. 143–146
H.V. Cane, T.T. von Rosenvinge, C.M.S. Cohen, R.A. Mewaldt, Geophys. Res. Lett. **30**(Sept.), 1 (2003)
C.M.S. Cohen, R.A. Mewaldt, R.A. Leske, A.C. Cummings, E.C. Stone, M.E. Wiedenbeck, E.R. Christian, T.T. von Rosenvinge, Geophys. Res. Lett. **26**, 2697 (1999)
M.I. Desai, G.M. Mason, J.R. Dwyer, J.E. Mazur, R.E. Gold, S.M. Krimigis, C.W. Smith, R.M. Skoug, Astrophys. J. **588**, 1149 (2003)
R. DiFabio, E. Möbius, M.A. Popecki, G.M. Mason, B. Klecker, H. Kucharek, *Astrophys. J.* (2006, submitted)
W. Dröge, Y.Y. Kartavykh, B. Klecker, G.M. Mason, Astrophys. J. **645**, 1516 (2006)
D.C. Ellison, R. Ramaty, Astrophys. J. **298**, 400 (1985)
J. Giacalone, Astrophys. J. Lett. **628**, L37 (2005)
N. Gopalswamy, A. Lara, M.L. Kaiser, J.-L. Bougeret, J. Geophys. Res. **106**, 25261 (2001)
M. Guhathakurta, T.E. Holzer, R.M. MacQueen, Astrophys. J. **458**, 817 (1996)
D. Hovestadt, M. Hilchenbach, A. Bürgi, B. Klecker et al., Sol. Phys. **162**, 441 (1995)
J.J. Kartavykh, W. Dröge, G.A. Kovaltsov, V.M. Ostryakov, Sol. Phys. **227**, 123 (2005)
Y.Y. Kartavykh, W. Dröge, G.A. Kovaltsov, V.M. Ostryakov, Adv. Space Res. **38**, 516 (2006)
B. Klecker, D. Hovestadt, M. Scholer, G. Gloeckler, F.M. Ipavich, C.Y. Fan, L.A. Fisk, Astrophys. J. **281**, 458 (1984)
B. Klecker, H. Kunow, H.V. Cane et al., Space Sci. Rev. **123**, 217 (2006a)
B. Klecker, E. Möbius, M.A. Popecki, Space Sci. Rev. **124**, 289 (2006b)
B. Klecker, E. Möbius, M.A. Popecki, L.M. Kistler, AGU Fall Meeting, Paper SH 41B-07, B7 (2006c)
B. Klecker, E. Möbius, M.A. Popecki, L.M. Kistler, H. Kucharek, M. Hilchenbach, Adv. Space Res. **38**, 493 (2006d)
K. Klein, G. Trottet, Space Sci. Rev. **95**, 215 (2001)
Y.-K. Ko, G. Gloeckler, C.M.S. Cohen, A.B. Galvin, J. Geophys. Res. **104**, 17005 (1999)
L. Kocharov, in *Solar Eruptions and Energetic Particles*, ed. by N. Gopalswamy, R.A. Mewaldt, J. Torsti. AGU Geophysical Monograph, vol. 165 (2006), pp. 137–145
L. Kocharov, G.A. Kovaltsov, J. Torsti, V.M. Ostryakov, Astron. Astrophys. **357**, 716 (2000)

G.A. Kovaltsov, A.F. Barghouty, L. Kocharov, V.M. Ostryakov, J. Torsti, Astron. Astrophys. **375**, 1075 (2001)
A.W. Labrador et al., in *Proc. 29th Int. Cosmic Ray Conf., Pune, India*, vol. 1 (2005), pp. 99–102
M.A. Lee, Astrophys. J. Suppl. **158**, 38 (2005)
R.A. Leske, J.R. Cummings, R.A. Mewaldt, E.C. Stone, T.T. von Rosenvinge, Astrophys. J. Lett. **452**, L149 (1995)
A. Luhn, B. Klecker, D. Hovestadt, E. Möbius, Astrophys. J. **317**, 951 (1987)
M. Lytova, L. Kocharov, Astrophys. J. Lett. **620**, L55 (2005)
G. Mann, A. Klassen, C. Estel, B.J. Thompson, in *ESA SP-446: 8th SOHO Workshop: Plasma Dynamics and Diagnostics in the Solar Transition Region and Corona*, ed. by J.-C. Vial, B. Kaldeich-Schümann (1999), p. 477
G.M. Mason, J.E. Mazur, J.R. Dwyer, Astrophys. J. Lett. **525**, L133 (1999)
G.M. Mason, J.E. Mazur, M.D. Looper, R.A. Mewaldt, Astrophys. J. **452**, 901 (1995)
J.E. Mazur, G.M. Mason, M.D. Looper, R.A. Leske, R.A. Mewaldt, Geophys. Res. Lett. **26**, 173 (1999)
E. Möbius, L.M. Kistler, M.A. Popecki et al., Space Sci. Rev. **86**, 449 (1998)
E. Möbius, M. Popecki, B. Klecker, L.M. Kistler, A. Bogdanov, A.B. Galvin, D. Heirtzler, D. Hovestadt, E.J. Lund, D. Morris, W.K.H. Schmidt, Geophys. Res. Lett. **26**, 145 (1999)
E. Möbius, M. Popecki, B. Klecker, L.M. Kistler, A. Bogdanov, A.B. Galvin, D. Heirtzler, D. Hovestadt, D. Morris, Adv. Space Res. **29**, 1501 (2002)
E. Möbius, Y. Cao, M. Popecki, L. Kistler, H. Kucharek, D. Morris, B. Klecker, in *Proc. 28th Int. Cosmic Ray Conf., Tsukuba, Japan*, vol. 6 (2003), pp. 3273–3276
M. Oetliker, B. Klecker, D. Hovestadt, G.M. Mason, J.E. Mazur, R.A. Leske, R.A. Mewaldt, J.B. Blake, M.D. Looper, Astrophys. J. **477**, 495 (1997)
V.M. Ostryakov, M.F. Stovpyuk, Solar Phys. **189**, 357 (1999)
V.M. Ostryakov, M.F. Stovpyuk, Solar Phys. **217**, 281 (2003)
R. Pallavicini, S. Serio, G.S. Vaiana, Astrophys. J. **216**, 108 (1977)
M.A. Popecki, in *Solar Eruptions and Energetic Particles*, ed. by N. Gopalswamy, R.A. Mewaldt, J. Torsti. AGU Geophysical Monograph, vol. 165 (2006), pp. 127–135
D.V. Reames, Space Sci. Rev. **90**, 413 (1999)
J. Torsti, L. Kocharov, J. Laivola, N. Lehtinen, M.L. Kaiser, M.J. Reiner, Astrophys. J. Lett. **573**, L59 (2002)
A.J. Tylka, P.R. Boberg, R.E. McGuire, C.K. Ng, D.V. Reames, in *Acceleration and Transport of Energetic Particles Observed in the Heliosphere: ACE 2000 Symposium*, ed. by R.A.M. et al. (2000), pp. 147–152
A.J. Tylka, C.M.S. Cohen, W.F. Dietrich, M.A. Lee, C.G. Maclennan, R.A. Mewaldt, C.K. Ng, D.V. Reames, Astrophys. J. **625**, 474 (2005)
A.J. Tylka, M.A. Lee, Astrophys. J. **646**, 1319 (2006)
T.T. von Rosenvinge, H.V. Cane, in *Solar Eruptions and Energetic Particles*, ed. by N. Gopalswamy, R.A. Mewaldt, J. Torsti. AGU Geophysical Monograph, vol. 165 (2006), pp. 103–114

Space Sci Rev (2007) 130: 283–291
DOI 10.1007/s11214-007-9166-6

# Pickup Ions and Cosmic Rays from Dust in the Heliosphere

**N.A. Schwadron · G. Gloeckler**

Received: 31 December 2006 / Accepted: 13 February 2007 /
Published online: 28 April 2007


**Abstract** The combination of recent observational and theoretical work has completed the catalog of the sources of heliospheric Pickup Ions (PUIs). These PUIs are the seed population for Anomalous Cosmic Rays (ACRs), which are accelerated to high energies at or beyond the Termination Shock (TS). For elements with high First Ionization Potentials (high-FIP atoms: e.g., H, He, Ne, etc.), the dominant source of PUIs and ACRs is from neutral atoms that drift into the heliosphere from the Local Interstellar Medium (LISM) and, prior to ionization, are influenced primarily by solar gravitation and radiation pressure (for H). After ionization, these interstellar ions are pickup up by the solar wind, swept out, and are either accelerated near the TS or beyond it. Elements with low first ionization potentials (low-FIP atoms: e.g., C, Si, Mg, Fe, etc.) are also observed as PUIs by Ulysses and as ACRs by Wind and Voyager. But the low-FIP composition of this additional component reveals a very different origin. Low-FIP interstellar atoms are predominantly ionized in the LISM and therefore excluded from the heliosphere by the solar wind. Remarkably, a low-FIP component of PUIs was hypothesized by Banks (J. Geophys. Res. **76**, 4341, 1971) over twenty years prior to its direct detection by Ulysses/SWICS (Geiss et al., J. Geophys. Res. **100**(23), 373, 1995) The leading concept for the generation of Inner Source PUIs involves an effective recycling of solar wind on grains near the Sun, as originally suggested by Banks. Voyager and Wind also observe low-FIP ACRs, and a grain-related source appears likely and necessary. Two concepts have been proposed to explain these low-FIP ACRs: the first concept involves the acceleration of the Inner Source of PUIs, and the second involves a so-called Outer Source of PUIs generated from solar wind interaction with the large population of grains in the Kuiper Belt. We review here the observational and theoretical work over the last decade that shows how solar wind and heliospheric grains interact to produce pickup

N.A. Schwadron (✉)
Department of Astronomy, Center for Spaceweather Modeling, Center for Space Physics,
Boston University, Commonwealth Ave 725, Boston, MA 02215, USA
e-mail: nathanas@bu.edu

G. Gloeckler
Department of Atmospheric, Oceanic and Space Science, University of Michigan,
Hayward St 2455, Ann Arbor, MI 48109-2143, USA

ions, and, in turn, anomalous cosmic rays. The inner and outer sources of pickup ions and anomalous cosmic rays exemplify dusty plasma interactions that are fundamental throughout the cosmos for the production of energetic particles and the formation of stellar systems.

**Keywords** Heliosphere: anomalous cosmic rays · Heliosphere: pickup ions · Sun: inner source · Sun: dust populations

## 1 Critical Roles and Sources of Pickup Ions

Pickup ions play a critical role in astrophysical plasmas by linking the neutral and ionized particle populations. As illustrated in Fig. 1, pickup ions catalyze a host of complex physical interactions that bring about mass loading (Szegö et al. 2000), excite electromagnetic waves, which in turn modify pickup ion transport (Lee ande Ip 1987) and pickup ion acceleration due to wave–particle interactions (Schwadron et al. 1996; Fisk and Gloeckler 2006) and at shocks (Pesses et al. 1981).

ACR populations reflect the complex pickup ion interactions indicated in Fig. 1. Most ACRs arise from neutral atoms that penetrate the heliosphere from the interstellar medium (Fisk et al. 1974) are subsequently ionized, picked up and swept out by the solar wind, pre-accelerated by shocks and waves in the solar wind (Schwadron et al. 1996; Fisk and Gloeckler 2006), and then accelerated at the solar wind termination shock to hundreds of MeV. This last step cannot occur by diffusive acceleration alone because there is a lower

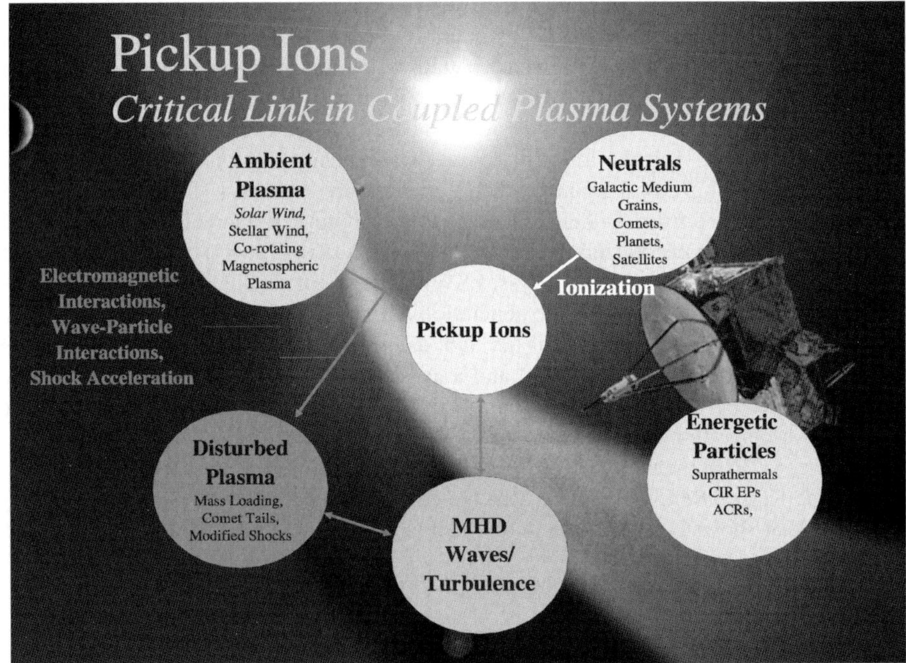


**Fig. 1** An illustration showing PUI interactions, which link a plasma with neutral particle populations. The physics of pickup ion linkages are quite general and apply in diverse plasma environments including the solar wind, stellar winds, and co-rotating magnetospheric plasmas (e.g., at Saturn). In addition to interstellar neutrals, sources that produce neutral particles by various processes include grains, comets, planets and satellites

energy threshold, or injection energy, above which particles are accelerated through diffusive acceleration, and below which the acceleration may be due to a second-order Fermi process and shock-drift acceleration (shock surfing). Easily ionized elements such as C, Si, and Fe, should be strongly depleted in ACRs since such elements are not predominantly neutral in the interstellar medium and therefore cannot drift into the heliosphere.

Instruments like SWICS on Ulysses have been able to detect pickup ions directly and ongoing measurements by cosmic ray detectors on spacecraft such as Voyager and Wind have been able to detect non-traditional components of ACRs (Reames 1999; Mazur et al. 2000; Cummings et al. 2002). These observations clearly show that the traditional interstellar source is, by itself, not adequate to explain the detailed ACR composition. Rather, grain-produced pickup ions must also be an important source of ACRs. Grains near the Sun produce an "Inner Source" of pickup ions (Sect. 2); these distributions cool substantially as they are swept outward from the Sun, making injection of inner source PUIs very difficult at the TS. While these PUIs are not the most likely source for the missing ACR source, a recently proposed "Outer Source" of pickup ions and anomalous cosmic rays (Sect. 3) from grains from the Kuiper Belt (Schwadron et al. 2002) can help fill this gap.

Comets are also a source of pickup ions the heliosphere. Until recently it was thought that the only way to observe cometary pickup ions is to send spacecraft to sample cometary matter directly. However, Ulysses has now had two serendipitous crossings of distant cometary tails (Gloeckler et al. 2000; Gloeckler et al. 2004). Comets were formed during the birth of our solar system and provide important primordial samples of matter.

We review here the observational and theoretical work which has led to our understanding of the generation of pickup ions from grains and comets in the heliosphere. Although many puzzles remain, the inner, outer and cometary sources of pickup ions are the most likely sources of newly observed components of ACRs with low first ionization potentials that cannot be generated from interstellar neutral atoms.

## 2 The Interstellar and Inner Source of Pickup Ions

The tell-tale signature of interstellar pickup ions, as seen for $H^+$ in Fig. 2, is a cutoff in the distribution at ion speeds twice that of the solar wind in the spacecraft reference frame. This cutoff is exactly what we would expect since interstellar ions are initially nearly stationary in the spacecraft reference frame and subsequently change direction, but not energy, in the solar wind reference frame due to wave–particle interactions and gyration, thereby causing them to be distributed over a range of speeds between zero and twice the solar wind speed in the spacecraft frame.

By comparison to the interstellar pickup ion distributions, the $C^+$ and $O^+$ distributions in Fig. 2 are puzzling. These ions were discovered and attributed to an Inner Source by Geiss et al. (1995). As opposed to being flat for ion speeds near the solar wind speed, there is a peak. The fact that the ions are singly charged precludes them from having been emitted directly by the Sun; if this were the case, they would be much more highly charged, e.g., $C^{5+}$ and $O^{6+}$ each common solar wind heavy ions. The $C^+$ and $O^+$ distributions in Fig. 2 are consistent with a pickup ion source close to the Sun. As the ions travel out in the solar wind, they cool adiabatically in the solar wind reference frame, and as opposed to having a distribution that cuts off at twice the solar wind speed in the spacecraft reference frame, they have a peak near the solar wind speed. The solid curve that goes through the $C^+$ and $O^+$ data points is based on a transport model (Schwadron 1998; Schwadron et al. 2000) for ions picked up close to the Sun (at $10R_\odot$), thereby proving the source.

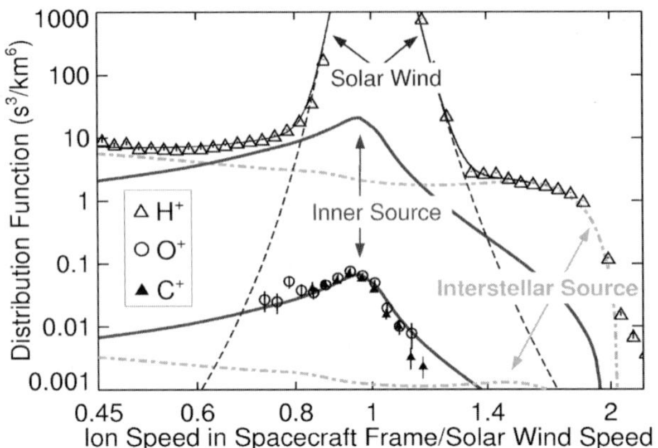


**Fig. 2** Observed distributions from Ulysses/SWICS of $C^+$ (*solid triangles*), $O^+$ (*open circles*), and $H^+$ (*open triangles*) vs. ion speed in the spacecraft frame normalized by the solar wind speed (Schwadron et al. 2000). During the observation period, all of 1994, Ulysses moved between 48°S latitude (at 3.8 AU) to 80°S and then back to 45°S (at 1.6 AU). The observations are compared with simulated distributions of solar wind $H^+$ (*dashed black curve*), interstellar $H^+$ (*upper grey dash-dot curve*), inner source $H^+$ (*upper thick black line*), inner source $C^+$, $O^+$ (*lower thick black line*), and interstellar $O^+$ (*lower grey dash-dot curve*)

Based on data like those presented in Fig. 2, Geiss et al. (1995), Gloeckler and Geiss (1998), Gloeckler et al. (2000) and Schwadron et al. (2000) concluded that there was evidence for other pickup ion sources in addition to interstellar neutrals. Since the source was peaked near the Sun like interplanetary grains, it was concluded that grains were associated with the source.

Based on a grain source, a naive expectation was that the inner source composition would resemble that of grains, i.e., enhancements of carbon and oxygen and strong depletions of noble elements such as neon were expected. However, the composition is almost identical to the solar wind (Gloeckler et al. 2000), as shown in Fig. 3. This conundrum was resolved by assuming a production mechanism whereby solar wind ions become embedded within grains and subsequently reemitted as neutrals. Remarkably, the existence of the inner source produced through the embedding and re-emission of solar wind particles was hypothesized (Banks 1971) many years prior to its discovery by Geiss et al. (1995).

The concept that the inner source is generated due to the embedding of re-emission of solar wind particles requires that sputtered atoms do not strongly contribute to the source since the composition would then resemble grains. If sputtering were dominant, elements such as Ne and He would be strongly depleted compared to the solar wind since these volatile species are so strongly depleted in comets and grains. Wimmer-Schweingruber and Bochsler (2003) raised a problem, however, by arguing that grains larger than a micron or so have sputtering yields larger than the yields from the embedding and re-emission of solar wind particles. To avoid this problem, it was suggested that the grains that give rise to the inner source are extremely small (hundreds of angstroms). A small grain population generated through catastrophic collisions of larger interplanetary grains would also yield a very large filling factor[1] which is consistent with observations of the inner source that require

[1]The filling factor is the net area of the sky filled in by grains divided by the net area of the sky over which they are distributed; so, for example, a large solid object would have a filling factor of 1, whereas a finite set of points would have a filling factor of 0.

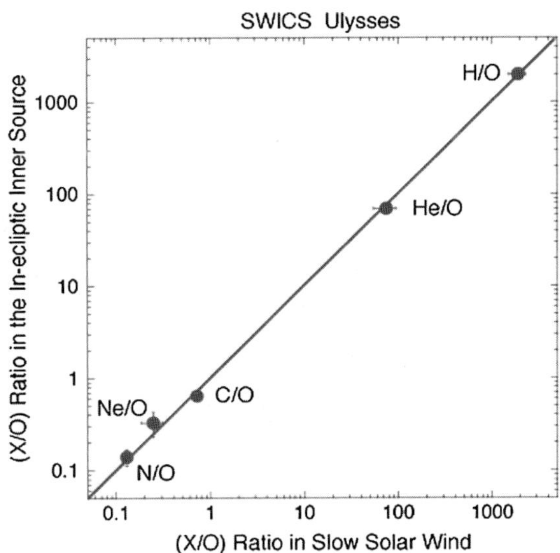


**Fig. 3** Since the H/O, He/O, C/O, N/O, and Ne/O ratios in the in-ecliptic Inner Source are nearly identical to the corresponding ratios in the solar wind, it is likely that most of the Inner Source pickup ions originate as dust-desorbed atoms and molecules. Because the in-ecliptic Inner Source has He and Ne, it is extremely unlikely that most of the Inner Source pickup ions come from cometary material

a net geometric cross-section typically 100 times larger than that inferred from zodiacal light observations (Schwadron et al. 2000). The very small grains act effectively as ultrathin foils for neutralizing solar wind (Funsten et al. 1993), but are not effective for scattering light owing to their very small size.

Our understanding of the inner source is not yet definitive. The observed flux of inner source ions is much larger than from estimates based on the zodiacal light observations (Schwadron et al. 2000). Does this imply a much larger population of grains near the Sun, or a population of very small grains produced by catastrophic collisions? In either case, we are strongly motivated to observe pickup ions and heliospheric grains in regions much closer to the peak production of the inner source at $\sim 10 R_{\odot}$.

## 3 Distant Cometary Tails

Direct sampling (e.g., von Rosenvinge et al. 1986 and references therein) and remote sensing observations (Huebner and Benkhoff 1999) of material from a few comets have established the characteristic composition of cometary gas. Direct detection of cometary matter requires spacecraft on trajectories that take them close to targeted comets (von Rosenvinge et al. 1986). Two unplanned crossings of cometary tails by Ulysses (Gloeckler et al. 2000, 2004) have shown that it is also possible to directly sample cometary ions with spacecraft that are not only far away but also at relatively large angular separation from the comet.

As a comet approaches the Sun it emits at an increasing rate volatile material consisting mainly of water group molecules. This neutral gas, evaporated from the comet's surface, moves out in roughly all directions at typical speeds of $\sim 1$ km/s and is then quickly dissociated and ionized by solar radiation and the solar wind. Immediately after being ionized the predominantly singly-charged cometary pickup ions are picked up and swept away from the Sun by the solar wind, forming a thin, long ion tail that extends radially outward to distances of at least several AU. Pickup ions have a characteristic velocity spectrum with a sharp drop in density at a well defined cutoff speed which is related to the radial distance from where the gas was first ionized to the location where the pickup ions were observed.

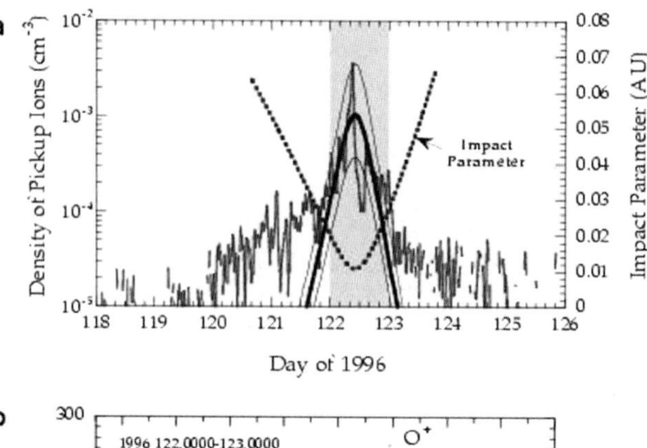


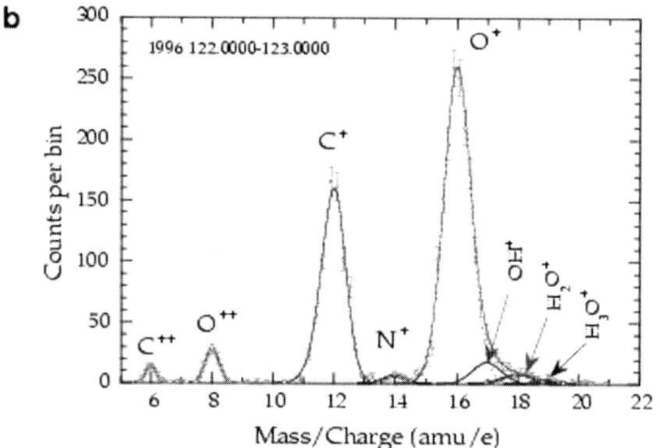


**Fig. 4** **a** Pickup $O^+$ density observed during the Hyakutake event shown in *red*. The *blue dashed curve* represents the distance of Ulysses from the Sun-comet line. The *three black curves* represent pickup ion distributions from a simple propagation model. **b** Abundances of various pickup ion species are derived from fitting counts binned according to the pickup ion mass-per-charge. These figures are reproduced from Gloeckler et al. (2000)

The first unplanned comet tail crossing occurred in 1996 when comet Hyakutake was at 0.35 AU and Ulysses was at 3.7 AU (Gloeckler et al. 2000). In this case there was a nearly precise radial alignment between the comet and Ulysses. Figure 4 shows measurements from the SWICS instrument on Ulysses. The top panel shows the density of pickup $O^+$ as a function of the distance between Ulysses and the Sun-comet line. The second panel shows abundances inferred from the pickup ion observations.

The second unexpected comet tail crossing by Ulysses (Gloeckler et al. 2004) involved detection of ions from comet C/1999 T1, named McNaught–Hartley (IAU Circ., 1999, 7273), and perhaps comet C/2000 S5 (SOHO). Unlike the case of comet Hyakutake (Gloeckler et al. 2000; Jones et al. 2000), these comets were at large angular separations from the Ulysses spacecraft. While both Ulysses and McNaught–Hartley had almost the same latitude ($\sim$70° south) they were separated in longitude by $\sim$35°. The minimum distance of McNaught–Hartley (at 1.492 AU from the Sun) to a line connecting the Sun and Ulysses position (at 2.55 AU) $\sim$4.5 days later (the impact parameter) reached its lowest value of $\sim$0.318 AU or $4.77 \times 10^7$ km on day 293, when the maximum rate of $O^+$ was observed. The detection of these cometary ions was possible due to a coronal mass ejection (CME) that mixed and distorted the heliospheric magnetic field sufficiently to guide pickup ions produced from near ($\lesssim$1 million km) the comets to the Ulysses spacecraft located more than 150 million km from the comets. The ability of CMEs to carry cometary ions far from their radial paths significantly increases the probability of detecting these ions.

In many respects, the first unplanned comet-tail crossing by Ulysses was seen as an anomaly. However, the second crossing shows that chance detection of comet tails is more likely than we thought. Clearly, the presence of a CME increases the odds of an event. More importantly, however, these events suggest that a mission equipped with improved pickup ion instrumentation (e.g., with factor $\sim$100 times the observing power of Ulysses/SWICS) would be capable of observing many cometary tails, and may significantly enhance our knowledge of cometary composition. It would also enhance the prospects of detecting pickup ions from a sungrazing comet destroyed near the Sun. The observation of such an event may be very valuable since the observed pickup ions would provide a measure of the net compositional inventory of the destroyed comet.

There are many ways to advance our understanding of cometary abundances and the distribution and properties of small comets. Future planned experimentation should optimize geometric factors, folding in expected ion production from smaller comets. Further modeling would help to infer comet composition from pickup ion measurements, and results should be compared with the abundances inferred from optical measurements. In the case of the inner source dust, ground-based composition measurements focusing on regions inside of 50 solar radii would help identify the spatial distribution of the inner source.

## 4 Outer Source of Pickup Ions and Anomalous Cosmic Rays

Recent observations from the Voyager and Wind spacecraft have resolved ACR components comprised of easily ionized elements such as Si, C, Mg, S, and Fe (Reames 1999; Mazur et al. 2000; Cummings et al. 2002). An interstellar source for these "additional" ACRs, other than a possible interstellar contribution to C, is not possible (Cummings et al. 2002). Thus, the source for these ACRs must reside within the heliosphere. There are a number of potential ACRs sources within the heliosphere (Schwadron et al. 2002). The solar wind particles are easily ruled out as a source since they are highly ionized, whereas ACRs are predominantly singly charged (Klecker et al. 1995). Discrete sources such as planets are also easily ruled out since their source rate is not sufficient to generate the needed amounts of pickup ions. Another potential source is from comets. The net cometary source rate is sufficiently large only inside 1.5 AU, a location so close to the Sun that the generated pickup ions are likely to be strongly cooled by the time they reach the termination shock, making injection into diffusive shock acceleration extremely difficult. Moreover, a cometary source would naturally be rich in C, which is not consistent with the compositional observations of easily ionized ACRs.

The inner source is another possibility, but again, adiabatic cooling poses a significant problem since these particles are picked up so close to the Sun. It has been suggested that the inner source may be substantially pre-accelerated in the inner heliosphere, thereby overcoming the effects of adiabatic cooling. Contradicting this suggestion, however, inner source observations in slow solar wind show clearly the pronounced effects of adiabatic cooling (Schwadron et al. 1999) and charge-state observations of energetic particles near 1 AU indicate no evidence for acceleration of the inner source (Mazur et al. 2002). *It appears that the additional population of ACRs requires a large source of pickup ions inside the heliosphere that is produced beyond 1 AU.*

Schwadron et al. (2002) suggest that there exists a strong outer source of heliospheric pickup ions that explains the presence of easily ionized ACRs. The source is material extracted from the Kuiper belt through a series of processes (shown schematically by the blue lines in Fig. 5): First, micron-sized grains are produced due to collisions of objects within the Kuiper Belt; grains spiral in toward the Sun due to the Poynting–Robertson effect; neutral atoms are produced by sputtering and are converted into pickup ions when they become ionized; the pickup ions are transported by the solar wind to the termination shock and, as they

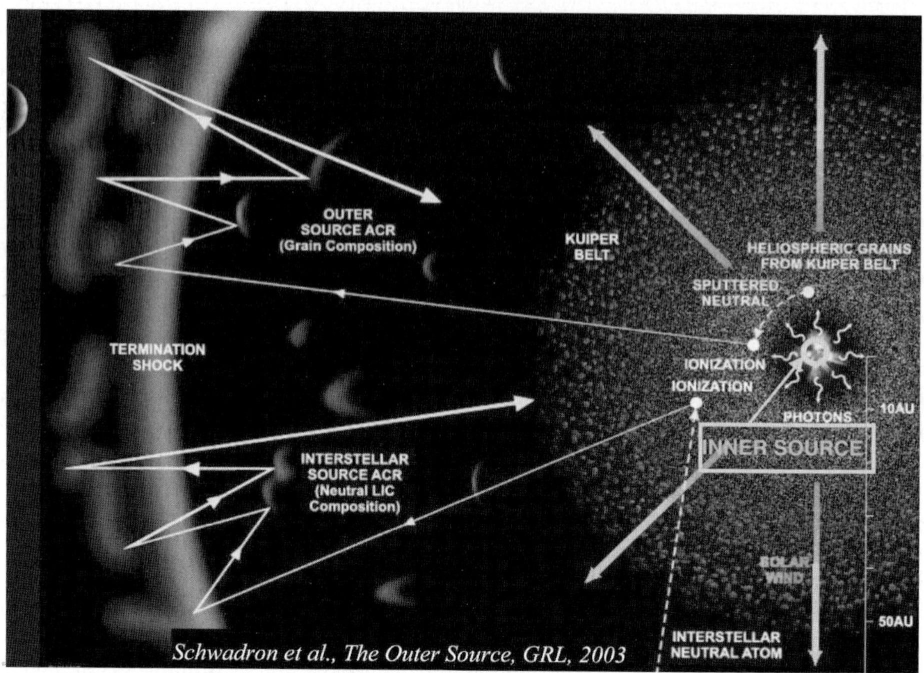


**Fig. 5** An illustration of ACR production (Schwadron et al. 2002). *Yellow curves* apply to the known interstellar source ACRs, as adapted from Jokipii and McDonald (1995), while *blue curves* apply to the outer source. Also shown by the *blue circle* close to the Sun is the location of the inner source of pickup ions

are convected, are pre-accelerated due to interaction with shocks and due to wave–particle interactions (e.g., Fisk and Gloeckler, 2006); finally, they are injected into an acceleration process at the termination shock to achieve ACR energies. The predicted abundances of this outer source are all within a factor of two of observed values.

## 5 Conclusions

We have briefly reviewed recent findings concerning the sources of pickup ions from grains in the heliosphere:

- **The inner source** of pickup ions has been observed and is caused by interactions between a large grain population near the Sun and the solar wind (Geiss et al. 1995; Gloeckler and Geiss 1998; Schwadron et al. 2000; Wimmer-Schweingruber and Bochsler 2003). The discovery of the inner source demonstrates the importance of grains in the production of heliospheric pickup ion populations.
- **Distant comet tails** have now been observed at least two times by the Ulysses: the first tail extended from comet Hyakutake (Gloeckler et al. 2000), and the second from comet McNaught–Hartley (Gloeckler et al. 2004). The odds of detecting distant comet tails are much higher than once thought, suggesting that a properly designed pickup ion instrument may be able to sample and study cometary material through detection of tails from a many different comets. It also raises the possibility of detecting pickup ions from a sun-grazing comet destroyed during its approach close to the Sun; such pickup ions would provide a measure of the net compositional inventory of the destroyed sun-grazing comet.

- **The outer source** of pickup ions is generated due to sputtering of grains produced in the Kuiper Belt (Schwadron et al. 2002). The pickup ions are generated far from the Sun in regions where pre-acceleration within the heliosphere is natural (i.e., at co-rotating shocks and in the slow solar wind). Hence, outer source pickup ions are likely to have high energy tails (Fisk and Gloeckler 2006, and references therein) at the termination shock and therefore may be efficiently injected into diffusive acceleration. The abundance and source location of outer source pickup ions suggest that they may contribute substantially to the anomalous cosmic rays, thereby explaining an additional population of easily ionized ACRs that cannot be accounted for with traditional heliospheric populations.

The discoveries of the inner and outer sources of pickup ions and anomalous cosmic rays exemplify dusty plasma interactions in our heliosphere. Solar wind interacts with dust to produce neutral atoms both through sputtering and the embedding and re-emission of solar wind particles. Photo-ionization or charge-exchange then converts the neutrals into pickup ions, which are subsequently accelerated at shocks and through stochastic acceleration. When these particles are accelerated at or beyond the termination shock, they generate a new population of anomalous cosmic rays composed of elements such Fe, Mg, Si, and C with relatively low first ionization potentials.

## References


P.M. Banks, J. Geophys. Res. **76**, 4341 (1971)
A.C. Cummings, E.C. Stone, C.D. Steenberg, Astrophys. J. **578**, 194 (2002)
L.A. Fisk, G. Gloeckler, Astrophys. J. **640**, L79 (2006)
L.A. Fisk, B. Kozlovsky, R. Ramaty, Astrophys. J. **190**, L35 (1974)
H.O. Funsten, B.L. Barraclough, D.J. McComas, Nucl. Instrum. Meth. Phys. Res. B **80/81**, 49 (1993)
J. Geiss, G. Gloeckler, L.A. Fisk, R. von Steiger, J. Geophys. Res. **100**(23), 373 (1995)
G. Gloeckler, J. Geiss, Space Sci. Rev. **86**, 127 (1998)
G. Gloeckler, L.A. Fisk, J. Geiss, N.A. Schwadron, T.H. Zurbuchen, J. Geophys. Res. **105**, 7459 (2000)
G. Gloeckler, J. Geiss, N.A. Schwadron et al., Nature **404**, 576 (2000)
G. Gloeckler, F. Allegrini, H.A. Elliott et al., Astrophys. J. Lett. **604**, L121–L124 (2004)
W.F. Huebner, J. Benkhoff, Space Sci. Rev. **90**, 117 (1999)
J.R. Jokipii, F.B. McDonald, Sci. Am. **272**, 58 (1995)
G.H. Jones, A. Balogh, T.S. Horbury, Nature **404**, 574 (2000)
B. Klecker, M.C. McNab, J.B. Blake et al., Astrophys. J. Lett. **442**, L69 (1995)
M.A. Lee, W. Ip, J. Geophys. Res. **92**, 11104 (1987)
J. Mazur, G. Mason, J. Blake et al., J. Geophys. Res. **105**, 21015 (2000)
J. Mazur, G. Mason, R. Mewaldt, Astrophys. J. **566**, 555 (2002)
M.E. Pesses, J.R. Jokipii, D. Eichler, Astrophys. J. **246**, L85 (1981)
D.V. Reames, Astrophys. J. **518**, 473 (1999)
N.A. Schwadron, J. Geophys. Res. **103**, 20643 (1998)
N.A. Schwadron, L.A. Fisk, G. Gloeckler, Geophys. Res. Lett. **23**, 2871 (1996)
N.A. Schwadron, G. Gloeckler, L.A. Fisk, J. Geiss, T.H. Zurbuchen, in *AIP Conference Proceedings*, vol. 471 (Am. Inst. Phys., New York, 1999) p. 487.
N.A. Schwadron, J. Geiss, L.A. Fisk et al., J. Geophys. Res. **105**, 7465 (2000)
N.A. Schwadron, M. Combi, W. Huebner, D.J. McComas, Geophys. Res. Lett. **29**(54), 1 (2002)
K. Szegö, K. Glassmeier, R. Bingham et al., Space Sci. Rev. **94**, 429 (2000)
T.T. von Rosenvinge, J.C. Brandt, R.W. Farquhar, Science **232**, 353 (1986)
R.F. Wimmer-Schweingruber, P. Bochsler, Geophys. Res. Lett. **30**, 49 (2003)

Space Sci Rev (2007) 130: 293–300
DOI 10.1007/s11214-007-9150-1

# Composition of Light Solar Wind Noble Gases in the Bulk Metallic Glass flown on the Genesis Mission

**A. Grimberg · D.S. Burnett · P. Bochsler · H. Baur · R. Wieler**

Received: 20 December 2006 / Accepted: 1 February 2007 /
Published online: 22 March 2007


**Abstract** We discuss data of light noble gases from the solar wind implanted into a metallic glass target flown on the Genesis mission. Helium and neon isotopic compositions of the bulk solar wind trapped in this target during 887 days of exposure to the solar wind do not deviate significantly from the values in foils of the Apollo Solar Wind Composition experiments, which have been exposed for hours to days. In general, the depth profile of the Ne isotopic composition is similar to those often found in lunar soils, and essentially very well reproduced by ion-implantation modelling, adopting the measured velocity distribution of solar particles during the Genesis exposure and assuming a uniform isotopic composition of solar wind neon. The results confirm that contributions from high-energy particles to the solar wind fluence are negligible, which is consistent with in-situ observations. This makes the enigmatic "SEP-Ne" component, apparently present in lunar grains at relatively large depth, obsolete. $^{20}$Ne/$^{22}$Ne ratios in gas trapped very near the metallic glass surface are up to 10% higher than predicted by ion implantation simulations. We attribute this superficially trapped gas to very low-speed, current-sheet-related solar wind, which has been fractionated in the corona due to inefficient Coulomb drag.

**Keywords** Sun: solar wind · Isotopic abundance ratios · Methods: laboratory

## 1 Introduction

To infer the isotopic and elemental composition of the solar photosphere from that of the solar wind (SW), it is essential to have precise information on the fractionation mechanisms

A. Grimberg (✉) · H. Baur · R. Wieler
Isotope Geology, ETH Zürich, CH-8092 Zürich, Switzerland
e-mail: Grimberg@erdw.ethz.ch

D.S. Burnett
Calif. Inst. of Technology, GPS MC 100-23, Pasadena, CA 91125, USA

P. Bochsler
Physikalisches Institut, University of Bern, CH-3012 Bern, Switzerland

operating at the source, i.e. the Sun and its outer convective zone, and in the solar wind itself. This is one of the major goals of NASA's Genesis mission (Burnett et al. 2003), which collected SW particles for more than 2 ½ years at the Lagrangian point L1 for high-precision analysis on earth. Three SW regime collector panels sampled the low-speed interstream wind, high-speed wind from coronal holes, and SW related to coronal mass ejections, respectively (Reisenfeld et al. 2007). Additional collectors sampled all SW regimes, including high-energy particles, to provide compositional information about the bulk SW and potential variations as a function of energy.

In the following we present noble gas data from a bulk metallic glass (BMG) target (Jurewicz et al. 2003) exposed to the solar corpuscular radiation during the entire exposure period. The main purpose of this experiment was to determine the dependence of the isotopic and elemental composition of light noble gases on implantation depth in order to study possible variations in their isotopic composition as a function of solar particle energy. We compare the BMG data with those from the Apollo Solar Wind Composition (SWC) experiments (Geiss et al. 2004), in situ measurements (Kallenbach et al., 1997a, 1997b, 1998) and depth profiles of the Ne isotopic composition measured in lunar samples (Black and Pepin 1969; Etique et al. 1981; Wieler et al. 1986; Benkert et al. 1993; Wieler 1998).

Noble gases are among the key elements to be studied in Genesis targets, since their composition in the Sun and the solar nebula cannot be deduced from meteorites, where they are heavily depleted. Moreover, light noble gases are relatively easy to measure in Genesis targets, and therefore are particularly well suited to study possible fractionation processes in different solar wind regimes. Precise measurements of He, Ne and Ar isotopes in the SW have already been obtained from the SWC-foils (Geiss et al. 2004). Their short exposure duration of two days at most restricts the interpretation of the obtained data to specific SW conditions, especially to low-speed solar wind. In situ measurements, e.g. with instruments onboard the Advanced Composition Explorer (ACE) (Gloeckler et al. 1998; Mason et al. 1998; McComas et al. 1998; Smith et al. 1998) or the Solar and Heliospheric Observatory (SOHO/CELIAS) (Hovestadt et al. 1995), provide the full range of energy and regime-dependent data (Kallenbach et al. 1998; Wimmer-Schweingruber et al. 1998), but often suffer from lack of precision, which is necessary to constrain theories on isotopic fractionation in the SW. Noble gases measured in lunar regolith samples that were exposed to the SW for several million years include all SW regimes. The complicated sample history, variable losses of solar wind gases (Wielet et al. 1986, 1998; Pepin et al. 1999, 2000; Burnett et al. 2003) and a non-ideal behaviour during stepwise noble gas extraction make it difficult to deduce the bulk SW composition from such samples and to obtain reliable constraints on further parameters such as the dependence of trapped noble gas composition on implantation depth and thus particle energy.

The BMG on Genesis is well suited to deduce the composition of SW noble gases as a function of implantation depth, and hence, to possibly provide information on fractionation effects depending on the implantation energy. On the one hand, this is because the BMG etches very homogeneously in nitric acid (Heber 2002). A sufficient number of solar wind Ne ions were collected to analyse released gases in several tens of steps, yielding a high depth resolution. At the same time, the exposure period was short enough to prevent loss of near-surface-sited gas due to surface sputtering, as is often the case in natural (meteoritic or lunar) samples, and to minimise the production of spallogenic isotopes by galactic cosmic rays (GCR). On the other hand, the trapping behaviour for He and Ne irradiation in the metallic glass and other target materials was carefully tested experimentally (Grimberg et al.

2005) and simulated with the SRIM-code (Ziegler 2004). Furthermore, precise input parameters on SW conditions during the Genesis exposure for SRIM modelling were provided by instruments onboard ACE and the Genesis spacecraft itself (Reisenfeld et al. 2007).

## 2 Bulk Solar Wind He and Ne Data

Although the BMG was primarily designed to study the dependence of trapped noble gas composition on implantation depth, we also report data on the composition of the bulk solar wind trapped by this target during the maximum activity of solar cycle 23. Such data are important for comparison with other target materials on Genesis as well as with earlier SW collection experiments, e.g. the SWC foils or with gases trapped in lunar samples. The bulk BMG data (Table 1) were obtained by total noble gas extraction via melting of the glass or as the sum of the closed-system etching steps described below. The measured bulk isotopic ratios of He and Ne from both methods are in good agreement. To compare these data with those obtained on other target materials and to deduce absolute solar wind fluxes, one has to consider that the BMG mainly consists of transition metals ($Zr_{58.5}Nb_{2.8}Cu_{15.6}Ni_{12.8}Al_{10.3}$, subscripts in atomic-%) with a relatively high atomic number. Since backscattering of impinging ions, the relevant process controlling the trapping efficiency of a material, depends on the atomic masses, of both, the projectile and the target, care has to be taken when comparing results from different target materials. Backscatter losses from the BMG are larger than for targets consisting of lighter elements, hence the correction is more important for the heavy BMG target than for, e.g., the SWC Al-foils. Correction factors for He and Ne isotopes listed in Table 1 have been determined

**Table 1** Bulk isotopic and elemental composition from the Genesis BMG and from SWC foils reported by Geiss et al. (2004)

| Sample | $^3He/^4He$ ($\times 10^{-4}$) | $\pm 2\sigma$ ($\times 10^{-4}$) | $^{20}Ne/^{22}Ne$ | $\pm 2\sigma$ | $^4He/^{20}Ne$ | $\pm 2\sigma$ |
|---|---|---|---|---|---|---|
| **Uncorr. Genesis BMG**[a] | | | | | | |
| CSSE, mean[b] | 4.34 | 0.04 | 13.64 | 0.11 | 466 | 20 |
| Total Extraction, mean | 4.31 | 0.01 | 13.56 | 0.05 | 504 | 21 |
| **Corr. Genesis BMG**[c] | | | | | | |
| CSSE, mean | 4.44 | 0.04 | 13.83 | 0.11 | 479 | 20 |
| Total Extraction, mean | 4.41 | 0.01 | 13.75 | 0.05 | 518 | 21 |
| **SWC Foils**[d] | | | | | | |
| Apollo 11 | 5.38 | 0.40 | 13.50 | 1.00 | 430 | 90 |
| Apollo 12 | 4.08 | 0.17 | 13.25 | 0.50 | 620 | 70 |
| Apollo 14 | 4.48 | 0.28 | 13.65 | 0.40 | 550 | 70 |
| Apollo 15 | 4.33 | 0.22 | 13.70 | 0.40 | 550 | 50 |
| Apollo 16 | 4.42 | 0.20 | 13.80 | 0.40 | 570 | 50 |
| SWC time weighted average | 4.26 | 0.22 | 13.70 | 0.30 | 570 | 70 |

[a]This work; mean ratios of three samples for both, CSSE and total extraction

[b]Ratio of the integrated isotope abundances for all steps in a certain sample (12, 31, and 21 steps respectively)

[c]Backscatter correction factors for the BMG (SWC foils) are 1.024 (1.02) for $^3He/^4He$, 1.014 (1.00) for $^{20}Ne/^{22}Ne$ and 1.029 (1.12) for $^4He/^{20}Ne$

[d]Backscatter corrected values from Geiss et al. (2004), see [c] for correction factors

with implantation experiments at the CASYMS facility in Bern (Ghielmetti et al. 1983; Grimberg et al. 2005) and SRIM. For Ne, the experimental and the modelled value are consistent and we choose the SRIM value because of its smaller uncertainty. For He we adopted the CASMYS-deduced correction factor noting that SRIM simulations for the BMG seem to overestimate backscattering of very light elements in the SW energy range.

Corrected He and Ne isotopic ratios in the BMG agree well with the results of the SWC foils reported by Geiss et al. (2004). We note that the bulk SW isotopic and elemental composition of He and Ne does not vary significantly between the daily scale recorded by the SWC experiments and the 2.5 year average recorded by the Genesis BMG target. The ratios of $^{4}$He/$^{20}$Ne and $^{4}$He/$^{3}$He measured in different SWC experiments correlate with each other (Geiss et al. 2004), which can be explained by fractionation caused by Coulomb drag, affecting both, the elemental and the isotopic ratio simultaneously (Geiss et al. 1970; Bodmer and Bochsler 1998). The BMG data point falls right on the Coulomb drag fractionation line which is consistent with the SWC foil data (Bochsler 2007). This also confirms that even the more mobile He has been quantitatively retained in the BMG despite the estimated temperatures of ~180°C during exposure.

## 3 Depth Dependence of Trapped Solar Ne Composition in the BMG

The mean penetration distance of ions into a target material scales with their implantation energy (Ziegler 2004). Therefore, the depth-dependent distribution of SW isotopes and elements in a material contains information about the energy-dependence of the SW composition. This information is smeared out by two effects, however. On the one hand, particles of a given implantation energy have a relatively wide depth distribution. Changes in SW velocity over a collection period result in an overlap of different depth distributions for a given particle. On the other hand, at a given moment, all SW species have similar velocities, causing heavier particles to have higher kinetic energies and thus greater penetration depths. This leads to mass fractionation upon implantation.

The depth-dependent isotopic composition of neon obtained by closed-system stepwise etching (CSSE) analyses (Heber 2002; Grimberg et al. 2005) in three BMG samples is shown in Fig. 1. In the first steps from very close to the surface, $^{20}$Ne/$^{22}$Ne-ratios are distinctly higher than the measured bulk SW average in the BMG of $13.75 \pm 0.05$ (Table 1) or any value reported from depth-dependent analyses of lunar grains (Black and Pepin 1969; Etique et al. 1981; Wieler et al. 1986; Benkert et al. 1993). With progressive etching the $^{20}$Ne/$^{22}$Ne-ratio slowly decreases and the final etching steps, releasing deeply sited particles, display $^{20}$Ne/$^{22}$Ne-ratios as low as $10.5 \pm 0.3$.

Overall, the depth distribution found in the BMG is in very good agreement with SRIM simulations (Fig. 1) assuming a uniform SW composition, using the backscatter-corrected average of $^{20}$Ne/$^{22}$Ne ratios determined in the total extraction analyses of $13.75 \pm 0.05$ (Table 1) and taking the velocity histogram as obtained with ACE instruments for the exposure period (Reisenfeld et al. 2007). Minor deviations of the measured data from the SRIM model predictions, showing-up in elevated $^{20}$Ne/$^{22}$Ne-ratios released in small gas fractions from very shallow depths in the BMG and a less steep gradient towards heavier composition with increasing depth, will be discussed in the following chapter.

The generally very good agreement of the measured depth distribution of Ne isotopes with SRIM predictions corroborates mass-dependent fractionation of SW ions upon implantation to be the responsible process for the observed Ne data pattern in the BMG (Grimberg et al. 2006). This process has been discussed earlier (Tamhane and Agrawal 1979;

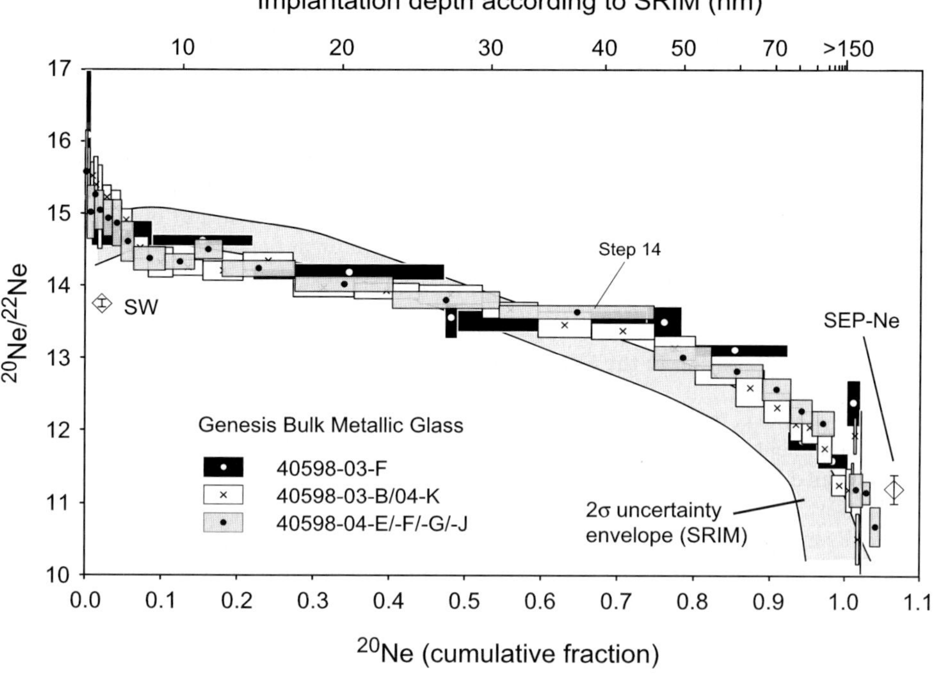


**Fig. 1** Depth-dependent neon isotopic composition of implanted SW derived from three Genesis BMG samples (*white, grey and black boxes*) by CSSE together with SRIM simulation data (*light grey shaded area in between the two solid black lines* displaying the $2\sigma$ uncertainty envelope including the statistical error and the error of the bulk measurement by total extraction, assuming a velocity-independent $^{20}$Ne/$^{22}$Ne ratio of $13.75 \pm 0.05$), versus the cumulative $^{20}$Ne fraction. Length of boxes in horizontal direction indicates gas amounts released per etching step, vertical extension of boxes indicates $2\sigma$ uncertainties of $^{20}$Ne/$^{22}$Ne ratios including ion statistics, extraction blank variability, interferences and mass discrimination. Step 14 of sample 40598-04-EFGJ had to be corrected for ~23% of Ne contribution from a gas inclusion in the BMG opened during the etching process. $^{20}$Ne/$^{22}$Ne ratios of all three BMG samples follow a trend that gets progressively heavier with depth. The implantation depth of the measured gas given on the upper abscissa is estimated from the simulated depth according to SRIM. Simulated and measured $^{20}$Ne/$^{22}$Ne profiles agree very well with each other. Thus the data pattern of the BMG can be explained by a fractionation of an isotopically uniform solar wind (SW = BMG bulk average from total extraction) upon implantation. Remarkably, ratios in the first 8% of the gas from very close to the surface show clearly higher $^{20}$Ne/$^{22}$Ne ratios than predicted by the SRIM simulations. This suggests this fraction to be very low-speed, current-sheet-related SW. The putative "SEP-Ne" data point is only for reference and should not be considered any longer.

Pepin et al. 2000; Mewaldt et al. 2001) but the solar Ne in mineral grains from lunar soils has widely been interpreted to be a mixture of two isotopically distinct components in the solar corpuscular radiation, SW-Ne near the surface, and "SEP-Ne" with a $^{20}$Ne/$^{22}$Ne ratio of $11.2 \pm 0.2$ (Etique et al. 1981; Wieler et al. 1986; Benkert et al. 1993) at larger depth. The main problem with this interpretation has always been the very high required fluence of solar energetic particles (Wieler 1998; Mewaldt et al., 2001; Wimmer-Schweingruber and Bochsler 2001) since in situ analyses measured SEP/SW ratios $<0.001$, being orders of magnitude smaller than those reported from lunar samples (SEP-Ne/SW-Ne ~0.1–0.4). The BMG data now show unambiguously that the putative "SEP-Ne" component is not needed to explain the lunar soil data. The interpretation of the BMG data here is also fully consistent with the solar energetic particles fluences measured during the exposure period of Genesis with the ACE instruments SIS, CRIS and ULEIS (Mason et al. 1998; McComas et al. 1998; Stone et al. 1998a, 1998b). The integrated Ne fluence ($1\pi$) of suprathermal ions in the range from 5.0 keV/nucleon to 1000 keV/nucleon is four orders of magnitude lower than the SW fluence (SEP/SW $\sim$ 0.0001) for the Genesis period (Mewaldt R.A., private communication). The corresponding Ne concentrations in the BMG are much too low to be resolved from the SW contribution. The data reported here show that fractionation upon implantation is capable to produce $^{20}$Ne/$^{22}$Ne ratios as low and even lower than the putative "SEP-Ne" value formerly inferred from lunar data. The "SEP"-noble gas component is thus an artefact and should not be considered any longer.

## 4 Isotopic Fractionation of the Solar Wind

As discussed above, the depth distribution of noble gas isotopes also contains, in principle, information about the fractionation operating on the SW prior to implantation, though this is complicated by implantation-induced fractionation overprinting the original SW signature. However, the initial etch steps, representing the first 8% of the Ne from very close to the BMG surface (Fig. 1), show $^{20}$Ne/$^{22}$Ne ratios up to 10% higher than implied by SRIM simulations (Fig. 2). We note that losses of superficially implanted neon by diffusion would

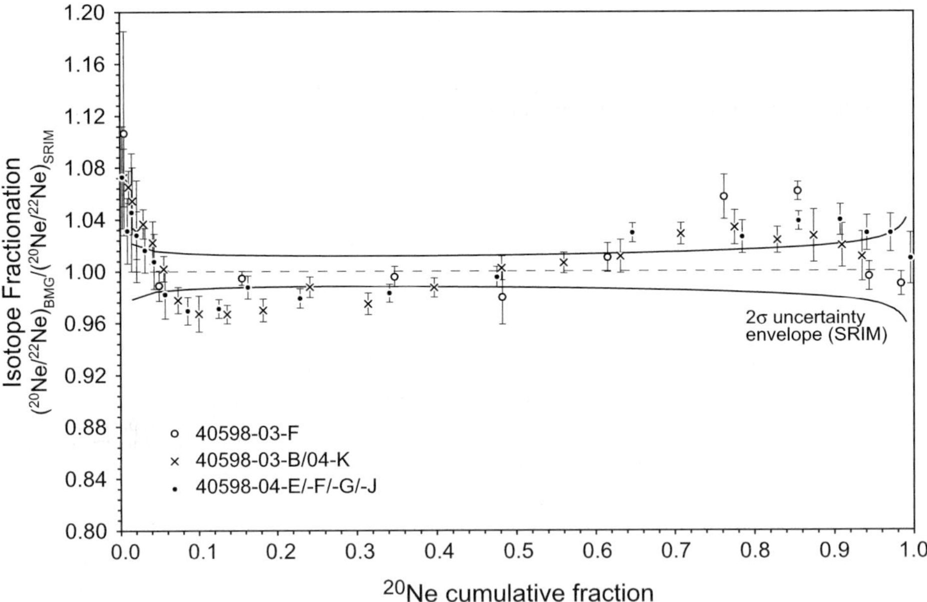


**Fig. 2** Deviations of $^{20}$Ne/$^{22}$Ne ratios measured in the BMG from values predicted according to SRIM (*horizontal dashed line* with $2\sigma$ uncertainty envelope including the statistical error and the error of the bulk measurement by total extraction) for a uniform SW with a $^{20}$Ne/$^{22}$Ne ratio of 13.75 as a function implantation depth represented by the cumulative $^{20}$Ne fraction. A depletion of the heavy isotope up to a $^{20}$Ne fraction of 8% can be explained by inefficient Coulomb drag according to the model of Bodmer and Bochsler (1998). The weak trend in gas released in later steps towards an enrichment of the light isotope is contrary to the Coulomb drag model as well as the findings by Kallenbach et al. (1997a, 1997b, 1998). It might indicate a not perfectly homogeneous etching rather than a fractionation of the SW at higher speed. Hence, its significance is questionable

produce the opposite trend. The same statement applies for the effect of a possible superficial contamination of the BMG with air neon. The depletion of the heavy isotope relative to SW implanted with higher energies supports the model of isotopic fractionation due to inefficient Coulomb drag in low-speed, current-sheet-related solar wind (e.g. Wimmer-Schweingruber 1994; Bodmer and Bochsler 1998). The inefficiency of Coulomb drag in low-speed wind is corroborated by a coincident depletion of $\alpha$-particles represented by small He/H ratios $<0.015$ (cf. Geiss et al. 1970) at low SW velocities as measured with the Genesis ion monitor. Adopting the correlation of $^{4}He/^{20}Ne$ and $^{20}Ne/^{22}Ne$ observed with the Apollo Foil experiments the Coulomb drag model (Bochsler 2007) predicts a fractionation of Ne towards a lighter isotopic composition in the low-speed SW relative to the SW average. Since the surface-near neon composition in the BMG and the Coulomb drag model are, in general, consistent we attribute the superficially trapped gas to very low-speed, current-sheet-related solar wind.

In greater depths ($^{20}Ne$ fraction $>0.1$), however, the BMG data show a slightly less steep gradient towards heavier Ne composition than predicted by the SRIM model (Fig. 1). This pattern might be explained by an enhancement of the light isotope with increasing SW energy relative to the bulk SW (Fig. 2). However, this is in contrast to in situ measurements from SOHO/CELIAS/MTOF reported by Kallenbach et al. (1997a, 1997b, 1998), that display a marginally significant enrichment of the heavy isotope by $\sim 2\%$ over the range of 350 km/s to 650 km/s, and it is also contradicted by any solar wind fractionation concept. The reproducibility of the three different CSSE measurements at a given implantation depth in turn limits the determination of SW fractionation prior to implantation. Hence, the significance of the trend found in greater depths ($^{20}Ne$ fraction $>0.1$) of the BMG needs further investigation; for instance some smearing of the isotopic distribution due to somewhat inhomogeneous etching cannot be ruled out at this point.

## 5 Conclusions

The isotopic composition of Ne as well as the He/Ne ratio of trapped solar wind in Genesis' Bulk Metallic Glass (BMG) target is fully consistent with data from the previous Apollo Solar Wind Composition experiment (Geiss et al. 2004). This shows that the BMG target quantitatively trapped and retained light noble gases from the solar wind, which is also encouraging for other Genesis targets. Thanks to the homogeneous etching of the BMG, allowing high resolution depth profiling of the Ne isotopic composition, and the detailed monitoring of SW conditions during the Genesis exposure, we have been able to show that the observed depth distribution of Ne isotopes is, in general, consistent with fractionation of a SW with uniform isotopic composition upon implantation. Since the measured depth profile is also similar to the distribution of solar Ne in lunar soils, an important consequence of this finding is that the putative "SEP-Ne" component is obsolete. While our data are thus consistent with a uniform SW Ne isotopic composition over most of the SW energy range, isotopically very light Ne representing the most shallowly implanted gas indicates that very low-speed, current-sheet related SW is fractionated relative to the bulk SW composition due to inefficient Coulomb drag. Further measurements of the He and Ar isotopic depth distribution in this target have the potential to examine the dependence of the elemental composition on SW energy.

**Acknowledgements** We would like to thank R.A. Mewaldt, G.M. Mason, C.M.S. Cohen, R.A. Leske, and M.E. Wiedenbeck for providing suprathermal fluence data from ACE, F. Bühler for valuable discussions, and S. Tosatti and O.J. Homan for XPS and plasma cleaning of the BMG. We also would like to thank the entire Genesis team. This work was supported by the Swiss National Science Foundation and NASA.

## References

J.P. Benkert, H. Baur, P. Signer, R. Wieler, J. Geophys. Res. – Planets **98**, 13147 (1993)
D.C. Black, R.O. Pepin, Earth Planet. Sci. Lett. **6**, 395 (1969)
R. Bodmer, P. Bochsler, Astron. Astrophys. **337**, 921 (1998)
P. Bochsler, Space Sci. Rev. this volume (2007)
D.S. Burnett et al., Space Sci. Rev. **105**, 509 (2003)
P. Etique, P. Signer, R. Wieler, Lunar Planet. Sci. Conf. **XII** (1981), pp. 265–267
J. Geiss, P. Hirt, H. Leutwyler, Sol. Phys. **12**, 458 (1970)
J. Geiss et al., Space Sci. Rev. **110**, 307 (2004)
A.G. Ghielmetti, H. Balsiger, R. Bänninger, P. Eberhardt, J. Geiss, D.T. Young, Rev. Sci. Instruments **54**, 425 (1983)
G. Gloeckler et al., Space Sci. Rev. **86**, 497 (1998)
A. Grimberg, F. Bühler, P. Bochsler, V.S. Heber, H. Baur, R. Wieler, Lunar Planet. Sci. **XXXVI**, abstract #1355 (2005)
A. Grimberg et al., Science **314**, 1133 (2006)
V. Heber, PhD Thesis, ETH Zürich, Nr. 14579, 2002, p. 157
D. Hovestadt et al., Sol. Phys. **162**, 441 (1995)
A.J.G. Jurewicz et al., Space Sci. Rev. **105**, 535 (2003)
R. Kallenbach et al., J. Geophys. Res. – Space Phys. **102**, 26895 (1997a)
R. Kallenbach et al., in *31st ESALAB Symposium, Correlated Phenomena at the Sun, in the Heliosphere and in Geospace* (1997b), pp. 33
R. Kallenbach et al., Space Sci. Rev. **85**, 357 (1998)
G.M. Mason et al., Space Sci. Rev. **86**, 409 (1998)
D.J. McComas et al., Space Sci. Rev. **86**, 563 (1998)
R.A. Mewaldt, R.C. Ogliore, G. Gloeckler, G.M. Mason, in R. Wimmer-Schweingruber (ed.), Solar and galactic composition. AIP Conf. Proc. **598**, 393–398 (2001)
R.O. Pepin, R.H. Becker, D.J. Schlutter, Geochimica et Cosmochimica Acta **63**, 2145 (1999)
R.O. Pepin, R.L. Palma, D.J. Schlutter, Meteorit. Planet. Sci. **35**, 495 (2000)
D.B. Reisenfeld et al., Space Sci. Rev. this volume (2007)
C.W. Smith et al., Space Sci. Rev. **86**, 613 (1998)
E.C. Stone et al., Space Sci. Rev. **86**, 357 (1998a)
E.C. Stone et al., Space Sci. Rev. **96**, 285 (1998b)
A.S. Tamhane, J.K. Agrawal, Earth Planet. Sci. Lett. **42**, 243 (1979)
R. Wieler, Space Sci. Rev. **85**, 303 (1998)
R. Wieler, H. Baur, P. Signer, Geochimica et Cosmochimica Acta **50**, 1997 (1986)
R.F. Wimmer-Schweingruber, PhD Thesis, University Bern, 1994
R.F. Wimmer-Schweingruber, P. Bochsler, O. Kern, G. Gloeckler, J. Geophys. Res. **103**(A9), 20621 (1998)
R.F. Wimmer-Schweingruber, P. Bochsler, in *Solar and Galactic Composition. AIP Conf. Proc.*, ed. by R. Wimmer-Schweingruber, vol. 598 (2001), p. 399
J.F. Ziegler, Nucl. Instruments Methods Phys. Res. Sect. B **219**(20), 1027 (2004)

Space Sci Rev (2007) 130: 301–307
DOI 10.1007/s11214-007-9201-7

# Fe/O Ratios in Interplanetary Shock Accelerated Particles

**H.V. Cane · I.G. Richardson · T.T. von Rosenvinge**

Received: 29 December 2006 / Accepted: 22 April 2007 /
Published online: 25 May 2007


**Abstract** It is widely accepted that diffusive shock acceleration is an important process in the heliosphere, in particular in producing the energetic particles associated with interplanetary shocks driven by coronal mass ejections. In its simplest formulation shock acceleration is expected to accelerate ions with higher mass to charge ratios less efficiently than those with lower mass to charge. Thus it is anticipated that the Fe/O ratio in shock-accelerated ion populations will decrease with increasing energy above some energy. We examine the circumstances of five interplanetary shocks that have been reported to have associated populations in which Fe/O increases with increasing energy. In each event, the situation is complex, with particle contributions from other sources in addition to the shock. Furthermore, we show that the Fe/O ratio in shock-accelerated ions can decrease even when the shock is traveling through an Fe-rich ambient ion population. Thus, although shock acceleration of an Fe-rich suprathermal population has been proposed to explain large Fe-rich solar particle events, we find no support for this proposal in these observations.

**Keywords** Acceleration of particles · Shock waves · Sun: flares

## 1 Introduction

In recent studies, Desai et al. (2003, 2004) have examined the composition of energetic ions observed by the ULEIS instrument on the ACE spacecraft in the vicinity of 72 interplanetary shocks. Such observations provide important clues about the source population(s) of

H.V. Cane (✉) · I.G. Richardson · T.T. von Rosenvinge
Astroparticle Physics Laboratory, NASA/Goddard Space Flight Center, Greenbelt, MD 20771, USA
e-mail: hilary.cane@utas.edu.au

H.V. Cane
School of Mathematics and Physics, University of Tasmania, Hobart, Tasmania, Australia

I.G. Richardson
Department of Astronomy, University of Maryland, College Park, MD 20742, USA

the accelerated ions and about the acceleration process. Desai et al. (2004) found that the majority of the events (93%) had Fe/O ratios either constant or decreasing with energy, as expected for the rigidity-dependent shock acceleration process. However, five of the shocks were found to have Fe/O that increased with energy. These events also had Fe/O ratios well above values measured in the ambient corona and solar wind; two shocks had Fe/O $>1$. Such high values of Fe/O are not consistent with the paradigm proposed by Reames (1999) in which shocks in large solar particle events accelerate ambient material. Desai et al. (2006) provide further strong evidence that the source population (at least at energies below a few MeV/nuc) is likely to be from solar flares.

Tylka et al. (2005) have noted that the five unusual shocks in the Desai et al. (2004) study were all quasi-perpendicular. This observation has been used by Tylka et al. (2005) to support their proposal that the enhanced Fe/O occasionally seen at high energies in event-averaged abundances of large solar particle events occurs because quasi-perpendicular shocks preferentially accelerate Fe-rich flare suprathermals. Tylka et al. (2005) suggest that these suprathermals originate in small flares that occur prior to the large flare that accompanies each large particle event. This contrasts with the proposal by Cane et al. (2006, 2003) that high energy, Fe-rich, ions seen in some large events are accelerated by the concomitant flare.

In this paper, we examine the circumstances of the five shocks studied by Desai et al. (2003, 2004) in which the Fe/O ratio is reported to increase with energy. We find that in two cases, there is clear evidence for the onset of an unrelated major solar particle event close to or within the time interval over which Desai et al. (2003, 2004) determined their Fe/O ratios. In a third case, the shock is bathed in an Fe-rich particle population but a localized particle "spike" at the shock has a lower Fe/O ratio. The remaining two shock periods are more complicated and do not conform to the simple picture of a shock-accelerated particle population that reaches maximum intensity in the vicinity of the shock. We conclude that the observational foundation for the scenario of Tylka et al. (2005) from the studies of Desai et al. (2003, 2004) is weak. Note that we are not calling into question the main results of the Desai et al. (2003, 2004) studies and are only focusing on the 5 specific events with Fe/O reported to increase with energy. In the next section, we discuss these events. Our conclusions are summarized in Sect. 3.

## 2 Observations

### 2.1 April 24, 2000 and July 11, 2000

We begin with the minor shock ($\delta V \sim 50$ km/s; compression $\sim 1.6$; $\theta_{Bn} \sim 70°$) of April 24, 2000. Figure 1a shows energetic particle intensity–time profiles from several near-Earth spacecraft during April 21–24. These include 38–53 and 175–315 keV electrons from the EPAM instrument on ACE, Fe and O in several energy ranges from the ULEIS experiment on ACE, proton data at 1.5 to 67 MeV from the ERNE instrument on SOHO, and 65–580 keV ion data from ACE/EPAM. The bottom panels show the solar wind magnetic field intensity, density and speed at ACE. The shaded region is the interval over which Desai et al. (2003) consider the particles present to be shock-accelerated and hence integrated to obtain their Fe/O ratios. It is immediately evident, considering the EPAM and ERNE observations, that a solar energetic particle (SEP) event commenced during this interval, at $\sim$14 UT on April 23. This was associated with a fast (1187 km/s) halo CME predominantly above the west limb first observed by the LASCO coronagraph on the SOHO spacecraft at 1254 UT.

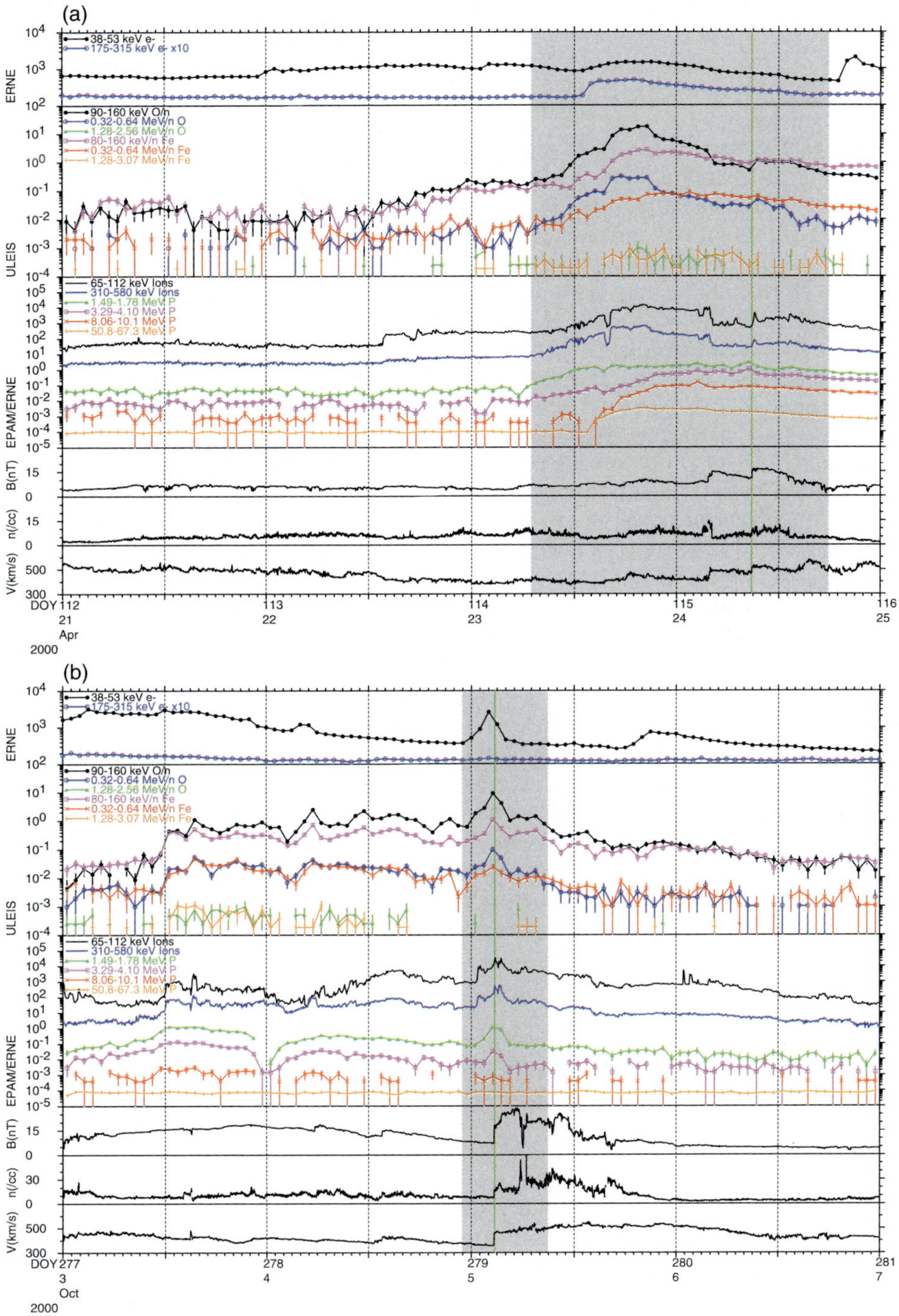

**Fig. 1** (**a**) Energetic particle and solar wind observations in the vicinity of the shock of April 24, 2000. Note the solar energetic particle event onset during the "shock" integration period (indicated by *gray shading*) of Desai et al. (2003). (**b**) Similar observations for the shock of October 5, 2000

Cane et al. (2002) conclude that the weak flare signatures (C2.7 in soft X-rays) arise because the solar event was associated with an active region behind the west limb. The higher energy particles at Earth, extending down to ~1 MeV, in the vicinity of the shock are clearly dominated by this event. At lower energies, there is an additional preceding particle enhancement that is particularly clear in the ULEIS Fe and O data. We cannot identify a specific source for this enhancement. Note the absence of velocity dispersion, one of the criteria used by Desai et al. to identify a shock-accelerated population. In view of this, and the absence of clear flare signatures in the shock integration interval, it is perhaps not surprising that the unrelated SEP event was evidently overlooked by Desai et al. Nevertheless, when a wide range of particle energies is considered, it is clear that the SEP event contributes to the particles observed in the vicinity of the shock and to the particles detected by ULEIS. It is also interesting that the Fe and O time profiles are dissimilar. In particular, the "square-topped" feature associated with peak 0.32–0.64 MeV/nuc oxygen is not seen in the Fe profile at the same energy. In summary, the averaging interval around the shock of April 24, 2000 includes a contribution from an unrelated solar particle event which could account for the high Fe/O ratio, increasing with energy, reported by Desai et al. (2003).

A similar situation prevails for the minor shock ($\delta V \sim 50$ km/s; compression ~2.3; $\theta_{Bn} \sim 84°$) at 1123 UT on July 11. During the Desai et al. (2003) shock particle integration period of July 11 0204 UT to July 12 0555 UT, $\geq$1 MeV particle populations are dominated by particles from an SEP event commencing at the end of July 10. The probable origin is an M6 flare at N18°E49°; a type III-*l* radio burst was also detected from 2105 UT (Cane et al. 2002) together with a fast (1352 km/s) partial halo CME observed by LASCO at 2150 UT on July 10. Though apparently poorly connected to the Earth, particle observations similar to those in Fig. 1 (not shown here) clearly show particles arriving at Earth from this event. In the ULEIS observations, this new particle event is less easily identified because it merges with the decay of a preceding particle event accompanied by a shock on July 10. However, the presence of particles from the July 10 event is unambiguous when the ULEIS data are considered in conjunction with data from a wider range of energies, and these particles show higher Fe/O ratios (~1) than particles from the preceding shock-associated increase. In summary, we conclude that Fe/O ratios during the integration period of the July 11, 2000 shock are also compromised by the occurrence of Fe-rich SEP particles. Note that for both events discussed in this section, low energy charge state measurements from the SEPICA instrument on ACE also indicate the presence of flare particles (Popecki et al. 2001).

### 2.2 October 5, 2000

Observations in the vicinity of this shock ($\delta V \sim 100$ km/s; compression ~2.4; $\theta_{Bn} \sim 66°$) are shown in Fig. 1b. Here, there is a clear particle "spike" centered on shock passage that likely results from shock drift acceleration at the quasi-perpendicular shock. The spike is superimposed on an extended particle increase that is Fe-rich throughout and more Fe-rich at higher energies; Fe/O ~ 0.3 at 0.09–0.16 MeV/n and ~0.9 at 0.32–0.64 MeV/n. At the time the shock passes the Fe/O ratio decreases but to values that increase with energy. However if background levels are subtracted then the shock spike has Fe/O ~ 0.1 in both energy ranges. Regardless of which particles are attributed to local shock acceleration it is clear that the background population has increasing Fe/O with energy for a period of about 30 hours before, to about 12 hours after, shock passage and that Fe/O decreases relative to the background values right at shock passage. This is particularly clear in the 0.32–0.64 MeV/n data (second pair of curves in the second panel) for which the O and Fe profiles follow each other before the shock passes and then there is a clear spike in the O profile (blue curve) but a barely significant rise in the Fe profile (red curve).

### 2.3 February 20, 2000

This shock ($\delta V \sim 80$ km/s; compression ~2.2; $\theta_{Bn} \sim 85°$) passed the Earth following particle events on February 17 and 18, associated with solar events at S29°E07° and W120° respectively (Cane et al. 2002). The February 17 event is the probable origin of the shock, although the far-western event may have been an important contributor to the particles observed at Earth. At the lowest energies a brief shock spike is evident (Fig. 2a) and at all but the highest energies there is a local enhancement commencing several hours following the shock. This feature is relatively Fe-rich, with Fe/O nearly 1, and more distinct in heavy ions than in protons at the same energy. The feature extends to the leading edge of an ICME following the shock (Cane and Richardson 2003) where the intensity falls abruptly. Note that the Desai et al. integration period is dominated by this Fe-rich feature at higher energies and there is no increase at shock passage in the intermediate energy Fe or O. Since the particles during the day ahead of shock passage are relatively Fe-poor it seems difficult to argue that the post shock increase is a signature of acceleration local to the shock. Another possibility is that it is a trapped population of either previously accelerated ions, in which case the Fe/O richness needs to be accounted for, or flare-accelerated ions. We note that the ions are not trapped in the ICME, and hence are not carried out from the Sun inside the ICME, but are in the upstream sheath. An interesting point is that the enhancement appears to commence in the vicinity of a prominent decrease in magnetic field intensity and increase in plasma density in the sheath. We might envisage particles being trapped in this low field region, but clearly the particle enhancement extends well beyond it. While we do not fully understand the origin of this interesting Fe-rich population, we suggest that it is not simply a consequence of the local interplanetary shock.

### 2.4 October 21, 1999

This shock ($\delta V \sim 90$ km/s; compression ~2.5; $\theta_{Bn} \sim 77°$) lies ahead of an ICME that interacts with a corotating high-speed stream at its trailing edge (e.g., Richardson 2006). The shock (Fig. 2b) passes by during the decay phase of flare particle events. The October shock period has some similarity with that of the February 20, 2000 event discussed above. In particular, the most prominent particle feature at energies below ~1 MeV is the increase following the shock. This again is in the sheath, and maximum intensities near the leading edge occur within a prominent region of low field strength. Both the shock and low field strength region are aligned with their planes approximately normal to the radial direction (C.J. Farrugia, private communication, 2006). Two differences from the February, 2000 event are that the ions in this increase are less Fe-rich than in the population preceding the shock, which has Fe/O ~ 1, and the increase extends into the ICME. In this case, the increase in Fe/O with energy appears to occur because the Desai et al. integration period includes this Fe-poor increase, which dominates at low energies, and a mixture of ambient Fe-rich ions and a much weaker post-shock increase at higher energies. Again the origin of the post-shock increase is unclear.

## 3 Summary and Discussion

We have examined the circumstances of the 5 quasi-perpendicular shocks for which an Fe/O ratio that increases with particle energy has been reported by Desai et al. (2004). In two cases, Fe-rich particles from an unrelated solar particle event that happens to occur during

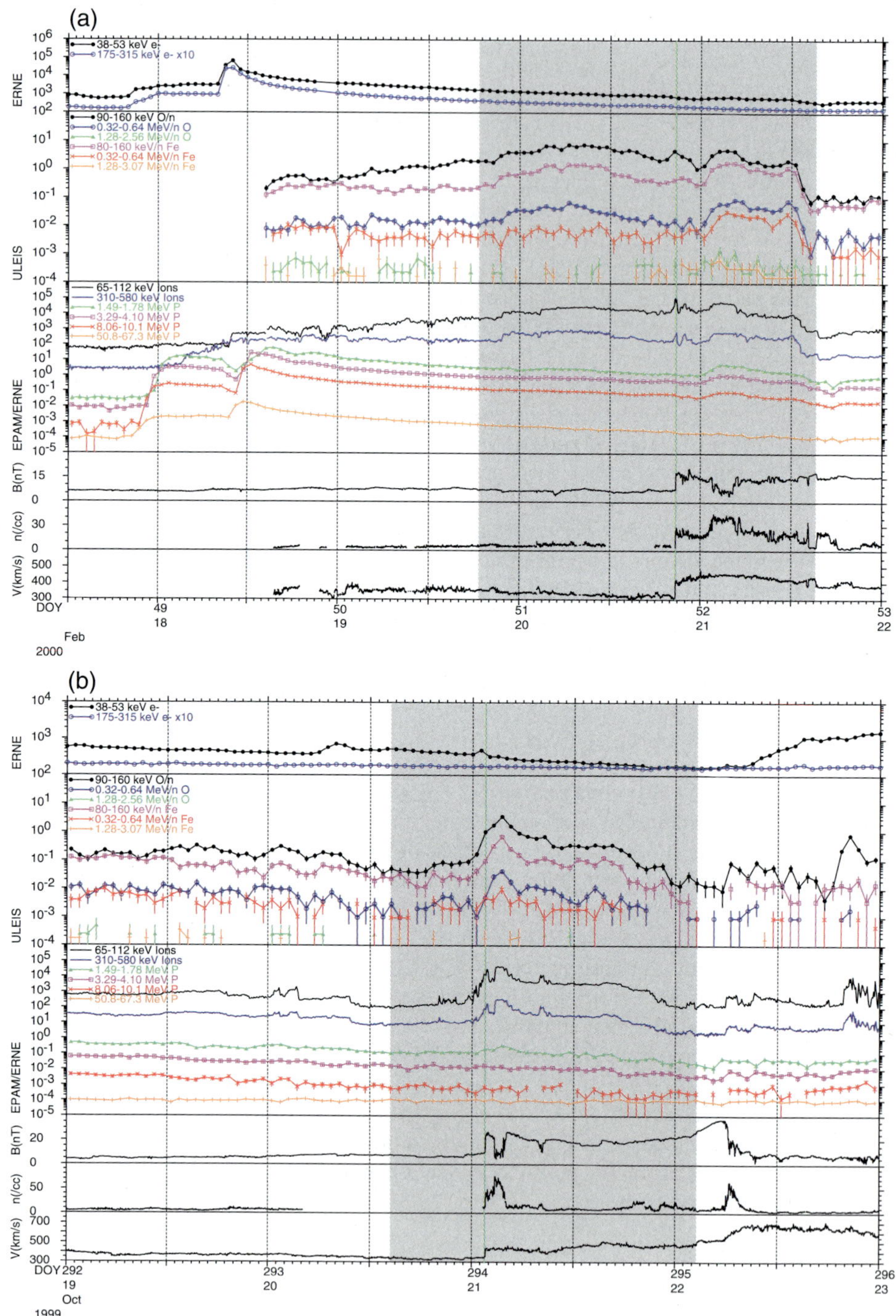


**Fig. 2** (**a**) Energetic particle and solar wind observations in the vicinity of the shock of February 20, 2000. (**b**) Similar observations for the shock of October 21, 1999

the Desai et al. integration period are clearly present and are likely to enhance the Fe/O ratio at higher energies. A third case shows clear signatures of shock drift acceleration in the immediate vicinity of the shock. The particle spike around the shock had a lower Fe/O than the surrounding Fe-rich particles. These observations cast doubt on the proposition of Tylka et al. (2005) that a quasi-perpendicular shock can accelerate an Fe-rich ion population to produce an Fe-rich population of high-energy ions. The remaining two events are unusual and interesting in their own right even though it is not clear whether they have any relevance to the debate about the origin of ions with high Fe/O in large particle events. Both show prominent low energy ion enhancements in the post-shock sheath that appear to be separate from the signatures of local acceleration at the shock. In one case, the ions are Fe-rich and in the other, Fe-poor; it is the combination of these populations and the other particle populations present that produce the observed increase in Fe/O with particle energy. Because multiple populations are present, we suggest that the behavior of Fe/O at these shocks should not be attributed simply to the properties of the shocks. In summary, we conclude that the 5 shocks with Fe/O increasing with energy do not provide compelling observational support for the conjecture that acceleration by quasi-perpendicular shocks is the cause of Fe-rich particles in large solar particle events.

## References


H.V. Cane, I.G. Richardson, J. Geophys. Res. **108**(A4), 1156 (2003). doi:10.1029/2002JA009817

H.V. Cane et al., J. Geophys. Res. **107A**(10), 1315 (2002). doi:10.1029/2001JA000320

H.V. Cane, T.T. von Rosenvinge, C.M.S. Cohen, R.A. Mewaldt, Geophys. Res. Lett. **30**(12), 8017 (2003). doi:10.1029/2002GRL016580

H.V. Cane et al., J. Geophys. Res. **111**, A06S90 (2006). doi:10.1029/2005JA011071

M.I. Desai, G.M. Mason, J.R. Dwyer et al., Astrophys. J. **588**, 1149 (2003)

M.I. Desai, G.M. Mason, M.E. Wiedenbeck et al., Astrophys. J. **611**, 1156 (2004)

M.I. Desai, G.M. Mason, R.E. Gold et al., Astrophys. J. **649**, 470 (2006)

M. Popecki et al., in *Proc. 27th Int. Cosmic Ray Conf.*, vol. 8 (2001), p. 3153

D.V. Reames, Space Sci. Rev. **90**, 413 (1999)

I.G. Richardson, *The Formation of CIRs at Stream–Stream Interfaces and the Resulting Geomagnetic Activity.* AGU Geophysical Monograph, vol. 167 (2006), p. 45

A.J. Tylka et al., Astrophys. J. **625**, 474 (2005)

Space Sci Rev (2007) 130: 309–316
DOI 10.1007/s11214-007-9179-1

# The Genesis Solar Wind Concentrator Target: Mass Fractionation Characterised by Neon Isotopes

**V.S. Heber · R.C. Wiens · D.B. Reisenfeld · J.H. Allton · H. Baur · D.S. Burnett · C.T. Olinger · U. Wiechert · R. Wieler**

Received: 22 December 2006 / Accepted: 26 March 2007 /
Published online: 15 May 2007


**Abstract** The concentrator on Genesis provided samples of increased fluences of solar wind ions for precise determination of the oxygen isotopic composition. The concentration process caused mass fractionation as a function of the radial target position. This fractionation was measured using Ne released by UV laser ablation and compared with modelled Ne data, obtained from ion-trajectory simulations. Measured data show that the concentrator performed as expected and indicate a radially symmetric concentration process. Measured concentration factors are up to ~30 at the target centre. The total range of isotopic fractionation along the target radius is 3.8%/amu, with monotonically decreasing $^{20}Ne/^{22}Ne$ towards the centre, which differs from model predictions. We discuss potential reasons and propose future attempts to overcome these disagreements.

**Keywords** Solar wind · Noble gases · Genesis oxygen isotopic analysis · UV laser ablation

V.S. Heber (✉) · H. Baur · R. Wieler
Isotope Geology and Mineral Resources, ETH, 8092 Zürich, Switzerland
e-mail: heber@erdw.ethz.ch

R.C. Wiens · C.T. Olinger
LANL, Space & Atmospheric Science, Los Alamos, NM 87544, USA

D.B. Reisenfeld
Physics and Astronomy, University of Montana, Missoula, MT 59812, USA

J.H. Allton
JSC, 2101 NASA Parkway, Houston, TX 77058, USA

D.S. Burnett
CalTech, JPL, Pasadena, CA 91109, USA

U. Wiechert
AG Geochemie, FU Berlin, 12249 Berlin, Germany

## 1 The Genesis Solar Wind Concentrator

The solar wind concentrator on the Genesis spacecraft was designed to provide a sample of increased concentration of solar wind atoms to allow high precision laboratory analysis of the isotopic composition of light elements (Burnett et al. 2003; Nordholt et al. 2003; Wiens et al. 2003), particularly of oxygen and nitrogen. The solar wind is a proxy for the composition of the sun and the primordial solar nebula. The oxygen isotope abundances in the solar wind are therefore fundamental to understand the observed differences in the oxygen isotopic composition between different solar system bodies and different constituents of primitive meteorites (Clayton 2003; Wiens et al. 2004; Hashizume and Chaussidon 2005).

The concentrator was an electrostatic mirror with targets placed at the focal point, designed to concentrate the fluence by about a factor of 20 on average (Wiens et al. 2003). A hydrogen rejection grid prevented the targets from radiation damage during exposure. Incoming ions with mass/charge ratios of 2.0–3.6 (masses 4 to 28 amu) were accelerated to increase implantation depths and therefore reduce backscatter losses. The ions were reflected and focused onto the target by a domed grid and a mirror (see Wiens et al. 2003 and Nordholt et al. 2003 for a complete description). This concentration process resulted in an unavoidable instrumental mass fractionation varying as function of the radial target position. The performance of the concentrator was tested on prototypes in an ion beam facility prior to launch. However, the tests could not perfectly simulate the solar wind in terms of charge state-, angular-, and velocity distributions. Instead, an ion trajectory simulation code SIMION 7.0, validated with some ion beam testing, was used to simulate the concentrator performance under predicted solar wind conditions (Wiens et al. 2003). After return of the spacecraft, these simulations were repeated using solar wind conditions encountered during the mission.

The aim of this work is to verify the post-flight simulation by direct measurement of the mass fractionation on the concentrator target as function of the radial position. To this end, we analysed the Ne abundances and isotopic composition along 2 of the 4 arms of the "gold cross" of the concentrator target at high spatial resolution (Fig. 1). Major attention was paid to achieve high precision Ne isotope data of the same order as the intended accuracy of oxygen analyses of $\sim$0.1%. Neon is especially suitable since it is (i) abundant in the solar wind, therefore not influenced by terrestrial contamination, (ii) similar in mass to oxygen and nitrogen, and (iii) its bulk solar wind composition is well known

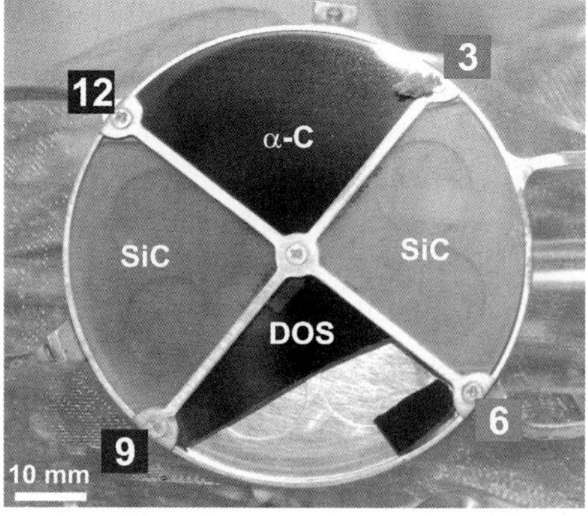


**Fig. 1** The Genesis concentrator target, shortly after return (note the large dust grain on arm 3). The concentrator target consists of four sub-targets mounted by the gold cross onto the base plate. The cross was cut for the Ne analyses and results are presented for the arms 12 and 9. ($\alpha$-C: $^{13}$C diamond film on substrate, DOS: diamond-like carbon on Si)

from many investigations (e.g., Apollo Solar Wind Composition Experiment (e.g., Geiss et al. 2004); lunar soils; in situ-analyses, e.g., CELIAS/SOHO (e.g., Kallenbach et al. 1997); and preliminary results from passive collectors from Genesis (Heber et al. 2007; Mabry et al. 2007). Here we report our data and compare experimentally deduced concentration factors and mass fractionation with the post-flight simulation.

## 2 Experimental

### 2.1 The Simulation

The performance of the concentrator, i.e. the process of solar wind ion implantation into the concentrator target, was numerically simulated before launch of the Genesis spacecraft. The simulation was re-run at return using the actual solar wind parameters and spacecraft working conditions encountered during the collection period. The ion simulation code used as well as the influence of all parameters included in the model are described in detail by Wiens et al. (2003). Briefly, the simulation determines the electrostatic potential of points in free space and then determines the trajectories of ions through space from a given input velocity, angle, mass, and charge state. The spacing of the computational nodes was 0.67 mm, limited by the software. Monte Carlo routines were used to give each ion's input position at a plane just above the concentrator, the initial angle, and charge state. Different ion velocities were simulated using separate simulations, as the concentrator voltages had to be adjusted for each velocity in simulation as in reality. Five different simulations are used to cover the velocity distribution using equally spaced bins. For most simulations at least one million ions are flown. The output of the simulation was a distribution of ions along radial positions on the target for each solar wind speed, which were averaged over 5 mm bins to improve statistics. Final $^{20}$Ne and $^{22}$Ne concentrations and the corresponding isotopic fractionation factors are averages weighted by the respective fluences per bin based on the proton velocity distribution with a correction to account for the velocity difference between heavy ions and protons.

The parameters influencing the distribution of implanted isotopes into the concentrator target are the charge state-, the velocity-, and the angular distributions of the incoming solar wind ions as well as the ion loss due to backscattering. The charge state distribution of Ne and O during the Genesis collection period was contemporarily measured by the SWICS instrument on board the Advanced Composition Explorer (ACE) (Gloeckler et al. 1998). The Genesis Ion Monitor continuously recorded the proton velocity (Barraclough et al. 2003). The angular distribution resulted from the intrinsic angular distribution of the solar wind as well as the instrument orientation, including spacecraft motion, pointing, nutation, and wobble, and instrument-to-instrument alignment. The Genesis spin axis was nominally pointed 4.5 degrees ahead of the sun, which is the apparent average incoming direction of the solar wind when convolved with spacecraft motion around the sun. Daily correction manoeuvres were done to keep the spacecraft nominally within 0.5 degrees of this position, not counting nutation and wobble, as it orbited the Sun. The backscatter losses of Ne isotopes implanted into Au were calculated using the SRIM code (Ziegler 2004), taking into account the measured solar wind velocity distribution and SIMION-modelled angles of incidence on the target. Backscatter losses are included in the simulated data. Model data are shown in Fig. 2. Error bars of modelled $\delta^{22}$Ne reflect 1-$\sigma$ statistical uncertainties. Note that the modelled Ne abundances and isotopic composition are directly comparable to the actual measured Ne data.

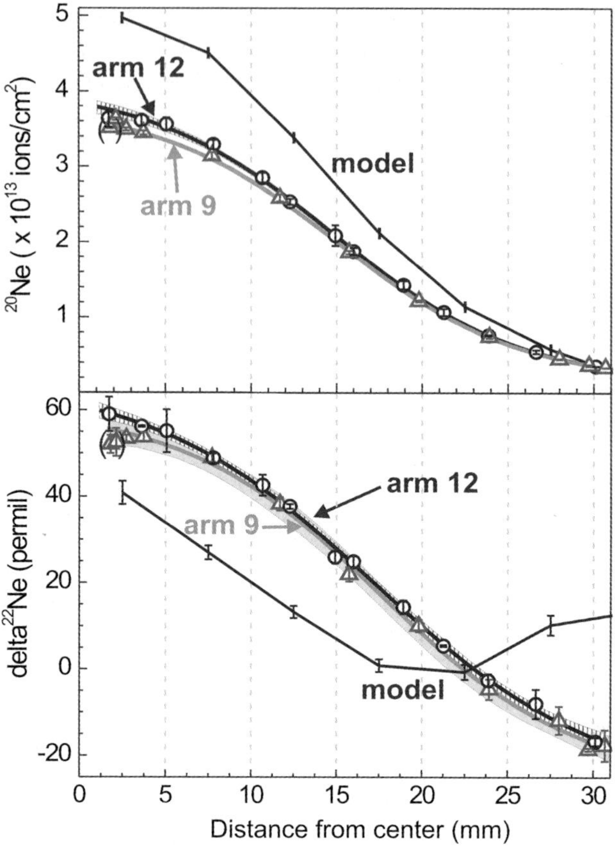


**Fig. 2** Neon concentrations (*upper panel*) and isotopic composition (*lower panel*) of arms 12 (*black circles, black fit line*) and 9 (*grey triangles, grey fit line*). Data have been fitted by sigmoid functions (error weighted for concentrations, unweighted for the $\delta^{22}$Ne; 95% confidence bands for the fits are given). The innermost 2 data points of arm 9 are not considered (see text). Also shown are modelled data (*vertical dashes connected by a black line*). All data are plotted as function of the distance from the centre of the concentrator target. Uncertainties are given as 2-$\sigma$ errors of the mean for the measured and as 1-$\sigma$ statistical errors for the modelled data

## 2.2 Measurement of Neon in the Concentrator Target

Neon concentration and isotope composition were analysed on the "gold cross" used to mount the individual concentrator targets onto a base plate. This cross consists of stainless steel coated with electroplated Au (0.5–1 μm thick). For analysis the cross was cut in order to allow single arms to be analysed separately. Here we present the results obtained from two arms, 9 (NASA code: 60009.2) and 12 (60009.1) (Fig. 1). Both arms were free of visible contamination and "hard landing"-induced damages, except for a few scratches at the outer part of arm 9.

Neon was released from single spots ($\sim$100 μm in diameter) ablated by an UV laser ($\lambda = 248$ nm: arm 12; $\lambda = 213$ nm: arm 9). Pit depth was a few microns, considerably larger than the maximum implantation depth of solar wind Ne in the concentrator targets of $\sim$250 nm ($^{22}Ne^{8+}$ at 800 km/s vertically implanted into Au; SRIM code, Ziegler 2004). Both arms were extensively sampled at 12 positions about equally distributed along the whole arm. Several measurements were done at each radial position. We reject the two innermost samples of arm 9 (shown in brackets in Fig. 2) since we cannot exclude incomplete ablation caused by partly shadowing of the UV beam at the edge using the 213 nm laser, which has a shorter focal distance than the 248 nm laser.

Neon was analysed with a very sensitive mass spectrometer (Baur 1999). All samples were corrected by a blank carried out on flight-spare material using identical ablation conditions. However, blank contributions were insignificant. In Fig. 2 the $^{20}$Ne concentrations and the Ne isotopic composition are presented as mean values at each radial position. Reproducibilities are calculated as 2-$\sigma$ error of the mean. For the $^{20}$Ne/$^{22}$Ne ratio these uncertainties vary between 0.2–5‰ and are on average 1.8‰. Thus, the uncertainties of the measured Ne isotope ratios are in the range of the intended accuracy of the oxygen isotope analyses.

## 3 Comparison of Measured and Modelled Data

### 3.1 Neon Concentrations

The measured $^{20}$Ne concentrations of both gold cross arms agree within their 95% confidence level along the entire radius of the concentrator target, except for a 5% difference at the inner part of the arms (Fig. 2). This suggests, at least for the quarter of the concentrator target considered here, that the concentration process was radially symmetric. The modelled $^{20}$Ne concentrations are based on concentration factors obtained from the simulation. These concentration factors were multiplied by the $^{20}$Ne fluence of $(1.22 \pm 0.01) \times 10^{12}$ atoms/cm$^2$ measured in a DOS (diamond-like carbon on Si) target from a passive Genesis collector sampling the bulk solar wind (Heber et al. 2007) that is corrected for the somewhat shorter exposure time of the concentrator target relative to the bulk collector. Carbon is a most suitable target to determine solar wind Ne fluences and isotopic composition. Its atomic mass, which is lower than that of Ne, minimized loss and fractionation of Ne as a consequence of backscattering. Measured and modelled data both show monotonically increasing $^{20}$Ne concentrations from the edge to the centre of the concentrator target by about a factor of 10. It is remarkable that the relative increase of the $^{20}$Ne concentration is similar in the measured and modelled data, however, both differ in their absolute concentrations. According to the simulation, concentration factors higher than 40 were expected at the centre of the target, whereas the measured data revealed maximum concentration factors of only 30. Measured $^{20}$Ne concentrations are thus about 30% lower than modelled values. Uncertainties in the calibration of the mass spectrometer as the cause of this difference can be ruled out since these would have cancelled out when relating the measured Ne data in Au to those in DOS. Furthermore, various independent noble gas calibration reservoirs in our laboratory agree to within a few percent.

### 3.2 Neon Isotopic Composition

The implanted solar Ne isotopes are fractionated as function of the radial position on the concentrator target, as was expected according to pre-flight numerical simulations of the implantation process (Wiens et al. 2003). At the edge of the concentrator target, Ne is slightly enriched in the light isotope ($^{20}$Ne/$^{22}$Ne: 14.05) relative to bulk solar wind Ne. Towards the centre, Ne monotonically becomes heavier with a $^{20}$Ne/$^{22}$Ne ratio of $\sim$13.04 at the centre of the concentrator target. To compare with modelled data, the Ne isotopic composition is expressed as the permil deviation of the measured $^{22}$Ne/$^{20}$Ne from a standard value in Fig. 2. As a standard we used our bulk solar wind $^{22}$Ne/$^{20}$Ne of $0.0724 \pm 0.0001$ measured in a DOS target of a bulk passive collector (Heber et al. 2007). The uncertainty of this value is not included in the $\delta^{22}$Ne in Fig. 2. The DOS target is expected to cause negligible Ne isotope fractionation due to backscattering. The concentrator was exposed for almost the same period (803.28 days) as the passive collectors sampling bulk solar wind (852.83 days), apart from those 6% of the total time when the concentrator was turned to a stand-by mode. This was during spacecraft manoeuvres, maintenance work on the H rejection grid and periods with solar wind velocities persistently above 800 km/s to prevent additional mass fractionation when heavier particles would preferentially hit the mirror. Since these periods contributed <0.2% to the total proton fluence and the intervals the concentrator was turned off for maintenance work were not biased to any particular solar wind condition, we can safely assume that the Ne collected by the DOS target on the bulk passive collector well represents the true isotopic composition of the Ne intercepting the concentrator. Note that using a lower or higher standard value would shift the fractionation curve along the $y$-axis without changing the total extent of fractionation.

The isotope fractionation curves obtained from both arms of the gold cross are indistinguishable within their uncertainties. This again shows radially symmetric operation of the concentrator. For comparison, the $\delta^{22}$Ne obtained from the simulation is also shown in Fig. 2. At the high concentration side of the target (0–17 mm) the measured and modelled curves have similar slopes but are offset relative to each other. They both show a heavier Ne isotopic composition towards the centre of the target, but the measured absolute fractionation (maximum $\delta^{22}$Ne: +60‰) is larger than the modelled one (maximum $\delta^{22}$Ne: +40‰). However, at the lower concentration side (17–31 mm), measured and modelled data show very different trends. Whereas the measured $\delta^{22}$Ne continuously decreases towards the edge, the modelled curve attains a minimum around 20 mm. Thus, the measured target-wide instrumental mass fractionation along the radius is considerably larger (3.8%/amu) than the modelled one (2.1%/amu).

## 4 Consequences for the Correction of Oxygen Isotope Data from the Concentrator Target

The Ne data here are basically very encouraging, as they prove that the concentrator performed essentially as expected. The concentrator targets therefore will serve their intended purpose well. However, the differences between measured and modelled Ne concentrations and—especially—isotopic compositions need to be understood for a proper mass fractionation correction of the eventual oxygen data. We are currently exploring why simulated and measured data differ more than expected. In pre-flight comparisons of ion beam tests and SIMION modelling, it was recognized that the model tended to predict ion impact positions somewhat closer to the centre of the target than were observed, and it appears that post-flight results are consistent with this. The reason for the discrepancy was never understood. One concern was the spatial resolution of the model. At 0.67 mm per grid unit, critical features such as the ion trajectories at the locations where they were turned around in the mirror section may have been compromised. A newer version of the software is now being released which may allow an improvement in the modelling. A comparison between pre- and post-flight simulations (compare Fig. 2 with the equivalent figure in Wiens et al. 2003) showed that the isotopic composition as function of the radial target position is very sensitive to solar wind conditions and other model parameters. One important difference to pre-flight assumptions was the actual solar wind velocity distribution, which shifted to higher values. This resulted on the one hand in somewhat lower concentration factors (because faster ions are more difficult to focus), and on the other hand in a preferred concentration of heavier isotopes toward the centre. However, including these new parameter values in the post-flight simulations only partially improved the agreement with the measured data.

One concern prior to launch has been that the concentrator might become misaligned during operation, leading to a somewhat different behaviour than modelled. Unfortunately, the hard landing prevented remapping the concentrator. Therefore, we have to assume that grid shapes remained unchanged. The very similar data obtained on two different arms of the gold cross are strongly encouraging that this assumption is justified. Major misalignments of the concentrator can therefore very likely be excluded as reason for the differences between measured and modelled Ne data. Future data from the two remaining arms are expected to corroborate this conclusion.

A potentially major problem may be the substantial backscatter loss of Ne from the high atomic mass element Au, which amounts to 30–40% for the prevailing range of angles of incidence of 40–60°. The respective correction was based on SRIM simulations (Ziegler 2004), taking into account the variable angle of incidence of the ions. However, the actual gold cross has a quite rough surface, which might have resulted in even larger backscatter losses than predicted. We will study this possibility by implantation experiments carried out on identical spare flight materials and variable angles of incidence of the ion beam to imitate the irradiation of the concentrator by the solar wind. Depending on the outcome of these experiments, it may be strongly advisable also to analyse Ne in some of the concentrator targets themselves. These materials have a much lower atomic mass than gold and hence require much lower backscatter loss corrections. Thus, minor differences between losses of oxygen and Ne and the corresponding differences in isotopic fractionation of the two elements should accurately be predictable by SRIM simulations. Note that a single Ne analysis on a concentrator target would only consume $\sim$0.005 mm$^2$ of target area.

In summary, our data show that the Genesis solar wind concentrator worked essentially as designed. Experimentally determined Ne concentrations and isotopic composition as function of the radial position at the concentrator target support a radially symmetric operation of the concentrator. Future work will concentrate on the understanding of the fractionation to eliminate the remaining differences in abundances and isotopic composition between measured and modelled data.

**Acknowledgements** We acknowledge the financial support by the Swiss National Science Foundation. We thank for the support by the NASA Discovery Mission Office.

## References

B.L. Barraclough, E.E. Dors, R.A. Abeyta, J.F. Alexander, F.P. Ameduri, J.R. Baldonado, S.J. Bame et al., Space Sci. Rev. **105**, 627–660 (2003)

H. Baur, EOS Trans. Suppl. **46**, F1118 (1999)

D.S. Burnett, B.L. Barraclough, R. Bennett, M. Neugebauer, L.P. Oldham, C.N. Sasaki, D. Sevilla et al., Space Sci. Rev. **105**, 509–534 (2003)

R.N. Clayton, in *Treatise in Geochemistry*, ed. by A.M. Davis, vol. 1 (2003), pp. 129–142

J. Geiss, F. Bühler, H. Cerutti, P. Eberhardt, C.H. Filleux, J. Meister, P. Signer, Space Sci. Rev. **110**, 307–335 (2004)

G. Gloeckler, J. Cain, F.M. Ipavich, E.O. Turns, P. Bedini, L.A. Fisk, T.H. Zurbuchen et al., Space Sci. Rev. **86**, 497–539 (1998)

K. Hashizume, M. Chaussidon, Nature **434**, 619–622 (2005)

V.S. Heber, H. Baur, D.S. Burnett, R. Wieler, Helium and neon isotopic and elemental composition in different solar wind regime targets from the Genesis mission. Lunar Planet. Sci. Conf. 38th, abstract CD #1894 (2007)

R. Kallenbach, F.M. Ipavich, P. Bochsler, S. Hefti, D. Hovestadt, H. Grünwaldt, M. Hilchenbach et al., J. Geophys. Res. **102**, 26895–26904 (1997)

J.C. Mabry, A.P. Meshik, C.M. Hohenberg, Y. Marrocchi, O.V. Pravdivtseva, R.C. Wiens, C. Olinger et al., Refinement and implications of noble gas measurements from Genesis. Lunar Planet. Sci. Conf. 38th, abstract CD #2412 (2007)

J.E. Nordholt, R.C. Wiens, R.A. Abeyta, J.R. Baldonado, D.S. Burnett, P. Casey, D.T. Everett et al., Space Sci. Rev. **105**, 561–599 (2003)

R.C. Wiens, M. Neugebauer, D.B. Reisenfeld, R.W. Moses Jr., J.E. Nordholt, D.S. Burnett, Space Sci. Rev. **105**, 601–625 (2003)

R.C. Wiens, P. Bochsler, D.S. Burnett, R.F. Wimmer-Schweingruber, Earth Planet. Sci. Lett. **222**, 697–712 (2004)

J.F. Ziegler, Nucl. Instrum. Methods Phys. Res. **219/220** 1027–1036 (2004)

Space Sci Rev (2007) 130: 317–321
DOI 10.1007/s11214-007-9220-4

# Nickel Isotopic Composition and Nickel/Iron Ratio in the Solar Wind: Results from SOHO/CELIAS/MTOF

**R. Karrer · P. Bochsler · C. Giammanco · F.M. Ipavich · J.A. Paquette · P. Wurz**

Received: 12 January 2007 / Accepted: 22 May 2007 / Published online: 3 July 2007


**Abstract** Using the Mass Time-of-Flight Spectrometer (MTOF)—part of the Charge, Elements, Isotope Analysis System (CELIAS)—onboard the Solar Heliospheric Observatory (SOHO) spacecraft, we derive the nickel isotopic composition for the isotopes with mass 58, 60 and 62 in the solar wind. In addition we measure the elemental abundance ratio of nickel to iron. We use data accumulated during ten years of SOHO operation to get sufficiently high counting statistics and compare periods of different solar wind velocities. We compare our values with the meteoritic ratios, which are believed to be a reliable reference for the solar system and also for the solar outer convective zone, since neither element is volatile and no isotopic fractionation is expected in meteorites. Meteoritic isotopic abundances agree with the terrestrial values and can thus be considered to be a reliable reference for the solar isotopic composition. The measurements show that the solar wind elemental Ni/Fe-ratio and the isotopic composition of solar wind nickel are consistent with the meteoritic values. This supports the concept that low-FIP elements are fed without relative fractionation into the solar wind. Our result also confirms the absence of substantial isotopic fractionation processes for medium and heavy ions acting in the solar wind.

**Keywords** Sun: solar wind · Sun: elemental composition · Sun: isotopic abundances

## 1 Introduction

The solar wind elemental composition is usually compared to the photosphere and to meteoritic elemental abundances. All three originate from the same source. However, they can differ substantially due to enrichment and fractionation processes. The solar wind is generally enriched in ions with low first ionization potential (low-FIP elements, e.g., calcium,

R. Karrer (✉) · P. Bochsler · C. Giammanco · P. Wurz
Physikalisches Institut, University of Bern, Bern, Switzerland
e-mail: reto.karrer@space.unibe.ch

F.M. Ipavich · J.A. Paquette
Department of Physics, University of Maryland, College Park, MD, USA

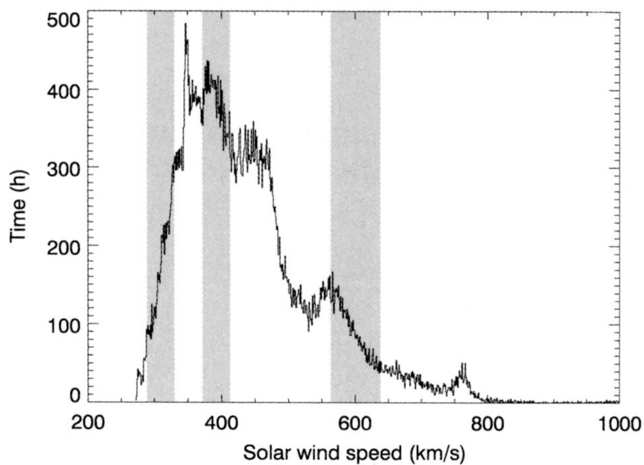


**Fig. 1** Proton Monitor. Distribution of solar wind speeds from January 1996 to January 2006. Investigated solar wind speed ranges are *shaded*: The speed intervals cover 290–330, 370–410, and 560–640 km/s

iron, nickel) compared to the photospheric abundances. If the ratio of iron to nickel agrees in the solar wind with the photosphere and with meteorites, it can be concluded that nickel undergoes the same enrichment processes as iron. Additionally, if the nickel isotopic composition agrees with meteorites, it confirms the absence of substantial isotopic fractionation processes for heavy elements in the solar wind.

We evaluated ten years' worth of SOHO/CELIAS data. We focus on slow solar wind (velocity between 290 and 330 km/s), on intermediate solar wind (370–410 km/s) and on fast solar wind (560–640 km/s) covering the whole period between 1996–2006 (Fig. 1). The solar wind speed was measured by the Proton Monitor (PM), which is a part of MTOF (Ipavich et al. 1998).

## 2 Instrumentation and Data Analysis

The Mass Time-of-Flight Spectrometer (MTOF) is part of the Charge, Elements, Isotope Analysis System (CELIAS) onboard the Solar Heliospheric Observatory (SOHO) spacecraft (Hovestadt et al. 1995). The MTOF sensor has an excellent mass resolution, which allows us to identify rare elements and isotopes and to distinguish them from more abundant neighboring species, e.g., to resolve $^{58}$Ni from $^{56}$Fe.

The MTOF sensor consists of an energy/charge filter (Wide-Angle Variable Energy/Charge (WAVE)) and an isochronous time-of-flight spectrometer in a V-type configuration (V-Mass). Details of the principle of operations of CELIAS/MTOF can be found in Kallenbach et al. (1997).

For each MTOF setting (WAVE electrode voltage $V_{\mathrm{WAVE}}$ and post-acceleration voltage $V_f$ after WAVE) day-by-day spectra (for 400 km/s) or five-minute spectra (for 300 and 600 km/s) were aligned using different TOF offsets. To find the TOF offset value, an asymmetric fit function was applied to the spectra. The fit function is the sum of a Lorentzian and a Gaussian function centered at the $^{56}$Fe peak. Then a 19-parameter function was applied to the spectra to get the raw counts of all resolved iron and nickel isotopes, using the maximum likelihood method (see Fig. 2). To maximize the likelihood the FORTRAN routine E04KDF of the NAG library was used. The shape of the asymmetric Lorentzian and the asymmetric Gaussian used for the fits are similar for all peaks. The width of an individual peak was taken to be proportional to the square root of its mass. To fit the background a linear function of the TOF was assumed.

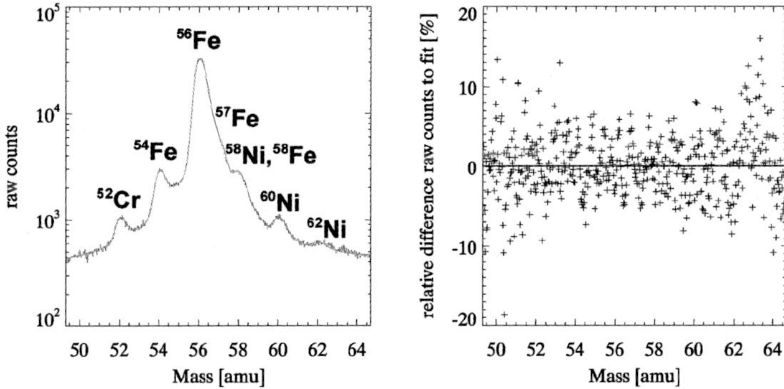


**Fig. 2** *Left*: Accumulated MTOF spectrum from January 1996 until January 2006 for 400 km/s solar wind. $V_f = 0$ kV, $V_{\mathrm{WAVE}} = 12$ kV. The time of flight is converted to mass, the counts are accumulated raw counts during ten years for the given MTOF settings. The peak at mass 58 comes from $^{58}$Fe and $^{58}$Ni. This is considered in the data analysis. *Right*: Difference of the measured counts to the value of the 19-parameter fit function divided by the fit function value. This diagram illustrates that the fit function describes well (in the range of $\pm 10\%$) the measured distribution in the mass range of 54–60 amu

To convert raw counts to particle densities, a complicated data analysis procedure is necessary (Wurz 2001). The WAVE entrance system has different transmission factors depending on the energy/charge of the entering ions. The WAVE electrode voltages were cycled in six steps covering an energy/charge range from helium to nickel. Although iron and nickel have similar kinetic energies, their charge state distributions differ (e.g., $Fe^{8+}$, $Fe^{9+}$, and $Fe^{10+}$ all about 20%, $Ni^{8+}$ about 30%, and $Ni^{9+}$ about 50%). For iron we used charge-state distributions observed by ACE/SWICS (cf. Gloeckler et al. 1998) in the three investigated velocity intervals to determine the transmission factors through WAVE. Since these data sets do not exist for nickel we used a simulation of the dynamic equilibrium of charge states established in the coronal expansion of the relevant solar wind velocities, which approximately reproduced the observed iron charge states. Additionally, the secondary electron yield from the carbon foil triggering the start signal of the time of flight measurement depends on the mass and the atomic number of the ion (Echenique et al. 1986). Simulations with SRIM (Ziegler 2004) indicate that the start signal efficiency is 12% lower for nickel than for iron at 1 keV/amu. For the investigated peaks only the ions exiting the carbon foil with charge state $+1$ were counted. The fraction of singly ionized iron ions is about 5% higher than for singly ionized nickel ions at 1 keV/amu (see Gonin et al. 1995). Evaluating these effects for the three investigated velocity intervals, this results in a 10–20% larger total detection efficiency for the dominant iron isotope $^{56}$Fe relative to the dominant nickel isotope $^{58}$Ni. Additionally, there is a interference of $^{58}$Fe to the $^{58}$Ni peak. These effects are considered in the data analysis.

The differences of the three investigated nickel isotopes are small. The transmission of WAVE is optimized for ambient solar wind ions and it is therefore smaller for fast solar wind iron and nickel. This reduces the accumulated counts and increases the uncertainty of the results for 600 km/s solar wind for nickel and iron, but it is more relevant for nickel due to lower abundances of Ni in the solar wind compared to Fe.

See Giammanco et al. (2006) for a detailed discussion of the instrument detection efficiency.

## 3 Results and discussion

Tables 1 and 2 show the elemental abundances of nickel and iron and the isotopic composition of nickel. As discussed in Section 2, the detection efficiency depends both on the ion energy and on the MTOF instrument settings (WAVE electrode voltages $V_{\mathrm{WAVE}}$ and post-acceleration voltage $V_{\mathrm{f}}$). Thus, uncertainties given in Tables 1 and 2 are composed of varying detection efficiencies within the velocity ranges and of the different instrument settings. The presented results are within the uncertainties in good agreement with the known ratios from photosphere, solar energetic particles (SEP), earth and meteorites; see Table 1 and the references therein. Also the isotope abundance ratios for nickel agree with the known values; see Table 2 and the references therein.

Our data analysis shows that both nickel and iron experience an enrichment in the solar corona due to the lower FIP compared to the high-FIP elements (e.g., oxygen 13.618 eV FIP, Fe 7.87 eV and Ni 7.635 eV). This supports the known theories about the FIP and First Ionization Time (FIT) effects named earlier as it was previously seen for chromium (Paquette et al. 2001).

The results of this study are in excellent agreement with that presented by Ipavich et al. (2006), who also derived Ni/Fe and Ni isotopic abundances from MTOF, but with a different

**Table 1** Ni/Fe ratios for different solar wind speeds

| | Raw data | Efficiency-corrected data |
|---|---|---|
| Solar wind 310 km/s | 0.043 | $0.051 \pm 0.012$ |
| Solar wind 400 km/s | 0.044 | $0.055 \pm 0.011$ |
| Solar wind 600 km/s | 0.047 | $0.049 \pm 0.011$ |
| Solar wind[a] | | $0.053 \pm 0.011$ |
| Photosphere[b] | | $0.060 \pm 0.009$ |
| Coronal SEP[c] | | $0.048 \pm 0.005$ |
| Meteorites[d] | | 0.0562 |
| Earth | | 0.055 |

[a]Weighted mean of the measurements above

[b]Asplund et al. (2005)

[c]Reames (1995)

[d]Lodders (2003)

**Table 2** Nickel isotope abundances

| | $\mathrm{Ni}^{58}$ | $\mathrm{Ni}^{60}$ | $\mathrm{Ni}^{62}$ |
|---|---|---|---|
| Efficiency-corrected data (310 km/s) | $0.66 \pm 0.05$ | $0.26 \pm 0.05$ | $0.08 \pm 0.04$ |
| Efficiency-corrected data (400 km/s) | $0.75 \pm 0.07$ | $0.19 \pm 0.05$ | $0.06 \pm 0.03$ |
| Efficiency-corrected data (600 km/s) | $0.75 \pm 0.07$ | $0.17 \pm 0.05$ | $0.08 \pm 0.04$ |
| Meteorites[a] | 0.6872 | 0.2647 | 0.0367 |
| Earth[b] | 0.68077 | 0.2622 | 0.03635 |

[a]Quitté et al. (2006)

[b]Gramlich et al. (1989)

approach: (a) they relied on case studies of two well-defined four-day periods, and (b) they used a different analysis technique. The agreement of the two studies provides confidence in the validity of the results.

## 4 Conclusions

This paper shows that nickel undergoes similar processes in the solar corona as other low-FIP elements, which confirms the FIP effect known from earlier works.

**Acknowledgements** The authors thank the entire SOHO/CELIAS team, Lisa M. Blush and Andrea Opitz for helpful suggestions. We also appreciate the helpful comments and suggestions by the reviewer. This work was supported by the Swiss National Science Foundation.

## References

M. Asplund, N. Grevesse, A.J. Sauval, in *ASP Conf. Ser. 336: Cosmic Abundances as Records of Stellar Evolution and Nucleosynthesis*, ed. by T.G. Barnes III, F.N. Bash (2005) pp. 25–38

P.M. Echenique, R.M. Nieminen, J.C. Ashley, R.H. Ritchie, Phys. Rev. A **33**(2), 897–904 (1986)

C. Giammanco, P. Bochsler, F. Ipavich, R. Karrer, J. Paquette, P. Wurz (2007), this volume. doi: 10.1007/s11214-007-9211-5.

G. Gloeckler, J. Cain, F.M. Ipavich, E.O. Tums, P. Bedini, L.A. Fisk, T.H. Zurbuchen, P. Bochsler, J. Fischer, R.F. Wimmer-Schweingruber, J. Geiss, R. Kallenbach, Space Sci. Rev. **86**, 497–539 (1998)

M. Gonin, R. Kallenbach, P. Bochsler, A. Bürgi, Nucl. Instrum. Methods Phys. Res. B **101**, 313–320 (1995)

J.W. Gramlich, L.A. Machlan, I.L. Barnes, P.J. Paulsen, J. Res. Nat. Inst. Stand. Technol. **94**(6), 347–456 (1989)

D. Hovestadt, M. Hilchenbach, A. Bürgi, B. Klecker, P. Laeverenz, M. Scholer, H. Grünwaldt, W.I. Axford, S. Livi, E. Marsch, B. Wilken, H.P. Winterhoff, F.M. Ipavich, P. Bedini, M.A. Coplan, A.B. Galvin, G. Gloeckler, P. Bochsler, H. Balsiger, J. Fischer, J. Geiss, R. Kallenbach, P. Wurz, K.-U. Reiche, F. Gliem, D.L. Judge, H.S. Ogawa, K.C. Hsieh, E. Mobius, M.A. Lee, G.G. Managadze, M.I. Verigin, M. Neugebauer, Sol. Phys. **162**, 441–481 (1995)

F.M. Ipavich, A.B. Galvin, S.E. Lasley, J.A. Paquette, S. Hefti, K.-U. Reiche, M.A. Coplan, G. Gloeckler, P. Bochsler, D. Hovestadt, H. Grünwaldt, M. Hilchenbach, F. Gliem, W.I. Axford, H. Balsiger, A. Bürgi, J. Geiss, K.C. Hsieh, R. Kallenbach, B. Klecker, M.A. Lee, G.G. Managadze, E. Marsch, E. Möbius, M. Neugebauer, M. Scholer, M.I. Verigin, B. Wilken, P. Wurz, J. Geophys. Res. **103**, 17205–17214 (1998)

F.M. Ipavich, J.A. Paquette, P. Bochsler, S.E. Lasley, *Proc. SOHO 17. 10 Years of SOHO and Beyond.* ESA SP-617 (2006)

R. Kallenbach, F.M. Ipavich, P. Bochsler, S. Hefti, D. Hovestadt, H. Grünwaldt, M. Hilchenbach, W.I. Axford, H. Balsiger, A. Bürgi, M.A. Coplan, A.B. Galvin, J. Geiss, F. Gliem, G. Gloeckler, K.C. Hsieh, B. Klecker, M.A. Lee, S. Livi, G.G. Managadze, E. Marsch, E. Möbius, M. Neugebauer, K.-U. Reiche, M. Scholer, M.I. Verigin, B. Wilken, P. Wurz, J. Geophys. Res. **102**, 26895–26904 (1997)

K. Lodders, Astrophys. J. **591**, 1220–1247 (2003)

J.A. Paquette, F.M. Ipavich, S.E. Lasley, P. Bochsler, P. Wurz, in *AIP Conf. Proc. 598: Joint SOHO/ACE workshop "Solar and Galactic Composition"*, ed. by R.F. Wimmer-Schweingruber (2001), pp. 95–100

G. Quitté, M. Meier, C. Latkoczy, A.N. Halliday, D. Günther, Earth Planet. Sci. Lett. **242**, 16–25 (2006)

D.V. Reames, Adv. Space Res. **15**, 41–51 (1995)

P. Wurz, Habilitation thesis, University of Bern, Switzerland, 2001

J.F. Ziegler, Nucl. Instrum. Methods Phys. Res. B **219**, 1027–1036 (2004)

Space Sci Rev (2007) 130: 323–328
DOI 10.1007/s11214-007-9200-8

# Long-Term Fluences of Solar Energetic Particles from H to Fe

**R.A. Mewaldt · C.M.S. Cohen · G.M. Mason · D.K. Haggerty · M.I. Desai**

Received: 21 February 2007 / Accepted: 22 April 2007 /
Published online: 27 June 2007


**Abstract** Data from ACE and GOES have been used to measure Solar Energetic Particle (SEP) fluence spectra for H, He, O, and Fe, over the period from October 1997 to December 2005. The measurements were made by four instruments on ACE and the EPS sensor on three GOES satellites and extend in energy from ~0.1 MeV/nuc to ~100 MeV/nuc. Fluence spectra for each species were fit by conventional forms and used to investigate how the intensities, composition, and spectral shapes vary from year to year.

**Keywords** Sun: particle emission · Sun: solar wind · Sun: abundances · Acceleration of particles

## 1 Introduction

In 2001, Mewaldt et al. combined He, O, and Fe data from four instruments on NASA's Advanced Composition Explorer (ACE) and presented fluence spectra that extended from solar wind to galactic cosmic ray energies. As shown in Fig. 1, all three species had very similar spectral shapes, including power-law spectra with a slope of about −2 extending from ~10 keV/nuc to ~10 MeV/nuc. Six additional species had similar $E^{-2}$ fluence spectra (right panel of Fig. 1). This was unexpected because this energy interval included contributions from many separate solar energetic particle (SEP) events that varied in intensity, spectral hardness, and composition (Mewaldt et al. 2001, 2005a). In this paper we present new yearly fluence spectra for H, He, O, and Fe extending from ~0.1 to ~100 MeV/nuc for the period from late-1997 through 2005. The yearly fluence spectra are dominated by each

R.A. Mewaldt (✉) · C.M.S. Cohen
California Institute of Technology, Pasadena, CA 91125, USA
e-mail: rmewaldt@srl.caltech.edu

G.M. Mason · D.K. Haggerty
Applied Physics Laboratory, Johns Hopkins University, Laurel, MD 20723, USA

M.I. Desai
Southwest Research Institute, San Antonio, TX 78238, USA

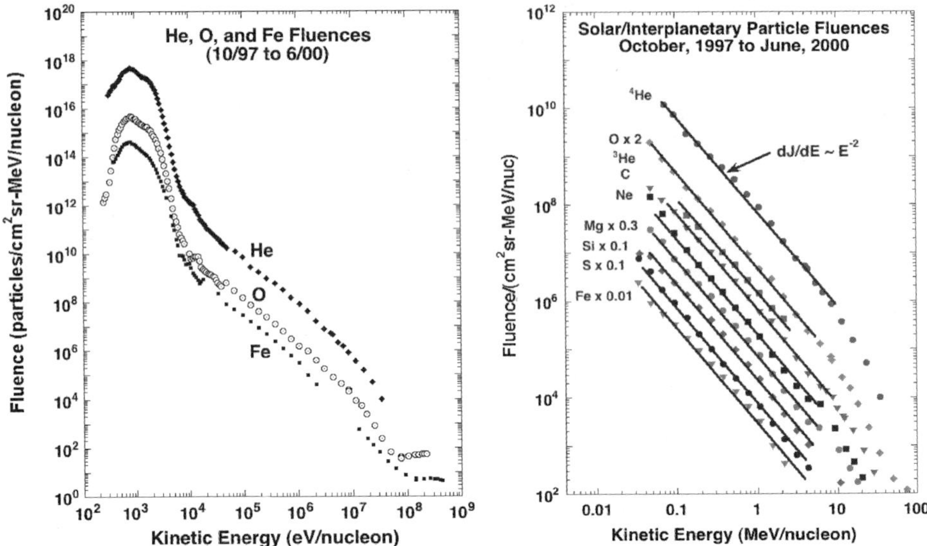


**Fig. 1** *Left*: Fluence spectra for He, O, and Fe nuclei from October 1997 to June 2000, measured by four instruments on ACE (Mewaldt et al. 2001). *Right*: In the October 1997 to June 2000 period it was found that the fluences of nine species had the same $E^{-2}$ spectral slope from ~0.05 to ~5 MeV/nuc (adapted from Mewaldt et al. 2001)

year's large SEP events. We use these measurements to examine how the low-energy fluence spectra (~0.1 to ~2 MeV/nuc) vary from year to year in spectral slope and to investigate whether the composition varies as a function of energy and/or with time.

## 2 Fluence Spectra: Observations and Fits

The measurements presented here cover the period from the launch of ACE in late 1997 to the end of 2005. We combined the last 3 months of 1997 with 1998, giving a total of 8 "yearly" fluence spectra. Most of the data are from three instruments on ACE, the Solar Isotope Spectrometer (SIS; Stone et al. 1998a), the Ultra Low-Energy Isotope Spectrometer (ULEIS; Mason et al. 1998), and the Electron, Proton, Alpha Monitor (EPAM; Gold et al. 1998). In addition, H and He data were obtained from the EPS sensors on NOAA's GOES-8, 10, and 11 satellites (Onsager et al. 1996). Finally, galactic cosmic-ray (GCR) measurements from the Cosmic Ray Isotope Spectrometer (CRIS) instrument on ACE (Stone et al. 1998b), and the BESS balloon-borne instrument (Shikaze et al. 2003) were used to estimate GCR contributions to the fluences. Table 1 summarizes the species and energy coverage of the instruments used for the reported fluence spectra.

Spectra from separate instruments for the largest SEP events (Mewaldt et al. 2005a; Cohen et al. 2005), and for 2000 and 2001 did not agree as well as during less active years, apparently due to intensity-related instrumental effects during the largest SEP events. To reconcile these differences the ULEIS spectra were multiplied by 1.6 in 2000 and 0.55 in 2001, and the SIS He spectra for 2003 were multiplied by 2.3 as in (Mewaldt et al. 2005a). Proton and He spectra from GOES were obtained by summing fluences from the NOAA event list at http://umbra.nascom.nasa.gov/SEP/seps.html, including differential proton and He spectra from 4.2 MeV to 200 MeV. Additional (though not independent) proton points were

**Table 1** Species and energy coverage

| Instrument | H (MeV) | He (MeV/nuc) | O (MeV/nuc) | Fe (MeV/nuc) |
|---|---|---|---|---|
| ACE/EPAM | 0.047–4.75 | | | |
| ACE/ULEIS | | 0.065–7.24 | 0.040–7.24 | 0.040–2.56 |
| ACE/SIS | | 6.13–41.2 | 7.05–89.8 | 10.5–117.5 |
| GOES-8,10,11 | 4.2–200 | 2.5–125 | | |

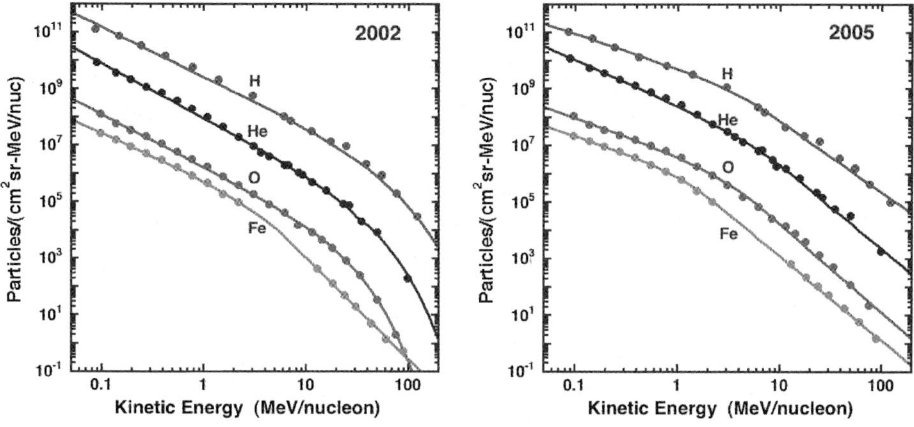


**Fig. 2** Fluence spectra for 2002 and 2005, fit with the double power-law form of Band et al. (1993). Note that the spectral breaks are ordered by species

obtained by differentiating GOES integral spectra for >5 to >100 MeV (Tylka et al. 2005; Mewaldt et al. 2005a). Following attempts to fit the GOES He spectra with a smooth spectrum, and the comparison of He spectra from SIS, GOES, and the SOHO/ERNE instrument (Torsti et al. 1995), we divided the GOES 16–45 MeV/nuc He points by 2, and multiplied the SIS He spectra by 1.3 for all years except 2003 (see above and Mewaldt et al. 2005a). In order to isolate the SEP contributions, we corrected the measured fluence spectra for O and Fe using GCR spectra from the ACE/CRIS instrument in conjunction with a cosmic-ray modulation model (Davis et al. 2001). The GCR and instrumental background corrections to the GOES H and He spectra were based on pre-event background levels from GOES.

The yearly SEP fluence spectra for H, He, O and Fe were fit with the double-power-law spectrum of Band et al. (1993), which has provided excellent fits to fluence spectra from individual SEP events (Tylka et al. 2005; Mewaldt et al. 2005a). Examples of the yearly spectra are shown in Fig. 2; the 8.25-year sum is shown in Fig. 3. The low-energy spectra were fit separately with a power-law between 0.1 and 2 MeV/nuc. The resulting spectral slopes are summarized in Fig. 4.

## 3 Results and Discussion

The spectral indices in the suprathermal energy range (Fig. 4) show correlated variations from year to year in all four species. This portion of the spectrum includes contributions from

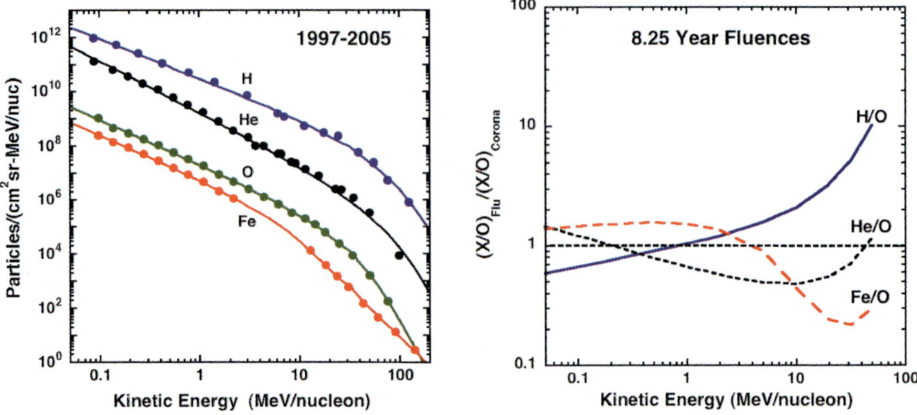


**Fig. 3** *Left*: Fluence spectra summed over the 8.25-year period from October 1998 to December 2006. *Right*: Ratios of H/O, He/O and Fe/O derived from fits to the 8.25-year spectra shown at the left. Each ratio has been divided by the coronal abundances of Feldman and Widing (2003)

**Fig. 4** Yearly power-law indices based on 0.1 to 2 MeV/nuc data from EPAM (protons) and ULEIS (He, O and Fe). Statistical uncertainties are comparable to the size of the data points. He spectral slopes are somewhat more uncertain than the other species

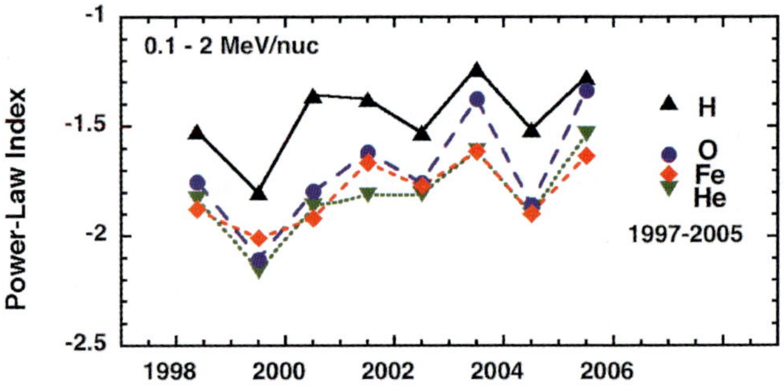

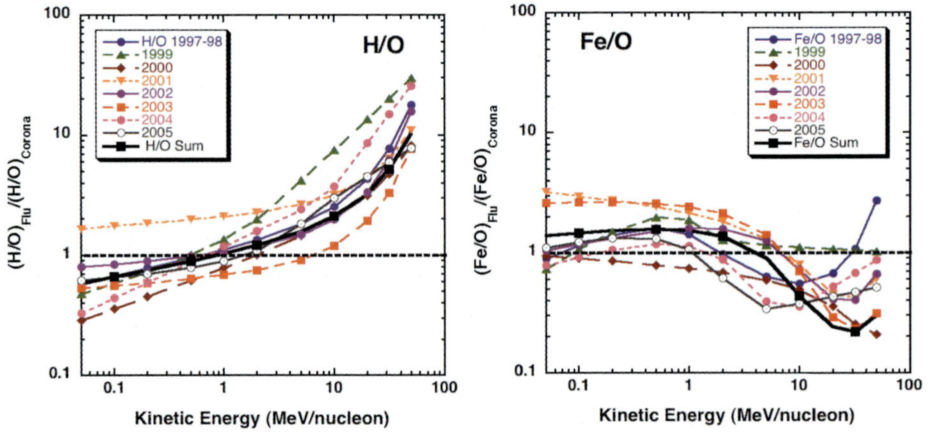

**Fig. 5** Energy dependence of the yearly H/O and Fe/O ratios, each normalized to the coronal abundances of Feldman and Widing (2003)

many separate events, large and small (Mewaldt et al. 2001, 2005a), with typical spectral indices of $-1$ to $-2$ in the largest events, and with heavy-ion spectra typically somewhat softer than proton spectra (see examples in Cohen et al. 2005; Mewaldt et al. 2005a). During solar maximum years (2000–2005) the fluence spectra reflect the large events that had the hardest spectra. In many years the He and O indices below a few MeV/nuc are similar, as expected since they have similar $Q/M$ ratios. In other years the He spectra are significantly softer than O (see Fig. 2), which is unexpected. The detection efficiency of He in ULEIS is smaller and less certain than that of heavier ions, so apparent differences between He and O should be treated with caution.

Gloeckler et al. (2000) find that quiet and moderately-quiet suprathermal spectra (∼0.01 to 1 MeV/nuc) all tend to have spectral indices of $-1.5$, and they suggest an interplanetary acceleration process that could be responsible for these suprathermal tails (Fisk and Gloeckler 2006). However, their time periods were chosen to exclude shocks, while our suprathermal fluence spectra are due mainly to SEP events accelerated by CME-driven shocks, where the spectral index is expected to depend on the shock strength (e.g., Lee 1983).

The spectra from all years show evidence of species-dependent spectral breaks, with H breaking at the highest energy/nuc and Fe at the lowest energy/nuc (Figs. 2 and 3). This is because the fluence spectra >1 MeV/nuc during any one year are dominated by the largest SEP events of that year, which typically have spectral breaks that depend on the charge-to-mass ($Q/M$) ratio of the species (e.g., Tylka et al. 2000; Cohen et al. 2005, Mewaldt et al. 2005a, 2005b; Cohen et al. 2007).

Using fits to the spectra it is easy to examine the energy dependence of the relative abundances of these species as shown in Figs. 3 and 5. Below ∼1 MeV/nuc the H/O ratios increase with energy while the Fe/O ratios show little variation. At energies between ∼1 and ∼30 MeV/nuc the H/O ratios turn sharply up, and the Fe/O ratios turn sharply down as a result of the $Q/M$-dependent spectral breaks discussed above. The above trends are consistent with the $Q/M$-dependent patterns in large SEP events, and can be understood if higher-rigidity particles escape upstream of the shock more easily and are therefore less efficiently accelerated (e.g., Li et al. 2005). The low-energy He/O behavior is unexpected (see above), but the >5 MeV/nuc patterns follow the expected $Q/M$-dependent behavior.

There are several years in which the Fe/O ratio increases above 30 MeV/nuc (Figs. 2 and 5) due to SEP events with hard spectra that are enriched in heavy elements such as Fe (Cohen et al. 1999; Dietrich and Lopate 2001; Tylka et al. 2005; Mewaldt et al. 2006). In 1997–98 the Fe-rich 6 November 1997 event dominates at high energies; in 2004 the Fe-rich 11 April 2004 event contributes most of the >30 MeV/nuc fluence. The Fe/O ratio in 1999 is flat; in this year there were several Fe-rich events but no-large Fe-poor events (Cane et al. 2006). There is also an increase in Fe/O above 30 MeV/nuc in the 8.25-year fluences (Fig. 3). Cane et al. (2003) interpreted high-energy Fe enrichments in well-connected SEP events as contributions of "flare-accelerated particles". Tylka et al. (2005) concluded that these Fe enrichments are due to selective acceleration (by quasi-perpendicular shocks) of remnant suprathermal material from earlier Fe-rich SEP events.

The abundance ratios in Figs. 3 and 5 were normalized to the coronal abundances of Feldman and Widing (2003). Our long-term (8.25-year) ratios are within a factor of two of coronal values between ∼0.05 and ∼5 MeV/nuc, with somewhat larger year-to-year variations. However, the observed energy dependence of the abundance ratios suggests that there is no one energy interval where one can be confident that fluence spectra uniformly sample the coronal composition (see also Desai et al. 2006).

## 4 Summary

The fluence spectra reported here extend those in Mewaldt et al. (2001) by adding H and by showing how SEP fluences vary from year to year. Although the year-to-year spectral shapes are similar, and all are well fit by a double-power-law, there are significant spectral-slope and abundance variations. In the 0.1 to 2 MeV/nuc interval the power-law slopes vary from $-1.3$ to $-2.1$ with the hardest spectra in years with the largest SEP events. At higher energies the fluence spectra have $Q/M$-dependent spectral breaks that reflect the behavior of the largest SEP events of the year. This leads to composition variations with heavier species generally depleted above $\sim$3 MeV/nuc. An exception to this is Fe; in some years (and the 8.25-year sum) the Fe/O ratio reverses the lower-energy trend and increases above $\sim$30 MeV/nuc due to occasional Fe-enriched events with hard power-law spectra at high energies. The observed energy-dependent composition suggests that there is no single energy region in which long-term measurements of SEP fluence spectra can, by themselves, provide a reliable measure of the coronal composition.

**Acknowledgement** We are grateful to the local organizing committee at the International Space Science Institute (ISSI) for their contributions to a very successful and enjoyable Symposium. We also appreciate the financial and organizational support of ISSI to the International Team on Solar and Heliospheric Sources of Suprathermal and Energetic Particle Populations. We thank NOAA's Space Environment Center for making GOES data available. This investigation was supported by NASA under grants NNG04GB55G, NNG06GC59G, NNX06AC21G, and NAG5-12929.

## References

D. Band et al., Astrophys. J. **413**, 281–292 (1993)
H.V. Cane, T.T. von Rosenvinge, C.M.S. Cohen, R.A. Mewaldt, Geophys. Res. Lett. **30**, GL016580 (2003)
H.V. Cane, R.A. Mewaldt, C.M.S. Cohen, T.T. von Rosenvinge, J. Geophys. Res. **111**, A011071 (2006)
C.M.S. Cohen et al., Geophys. Res. Lett. **26**, 2697–2700 (1999)
C.M.S. Cohen et al., J. Geophys. Res. **110**, A011004 (2005)
C.M.S. Cohen et al. (2007), this volume. doi: 10.1007/s11214-007-9218-y
A.J. Davis et al., J. Geophys. Res. **106**, 29979 (2001)
M.I. Desai et al., Astrophys. J. **649**, 470–489 (2006)
W.F. Dietrich, C. Lopate, *27th Internat Cosmic Ray Conf.*, vol. 8 (2001), pp. 3120–3123
U. Feldman, K.F. Widing, Space Sci. Rev. **106**, 665–721 (2003)
L.A. Fisk, G. Gloeckler, Astrophys. J. **640**, L79–L82 (2006)
G. Gloeckler, L.A. Fisk, T.H. Zurbuchen, N.A. Schwadron, in *Acceleration and Transport of Energetic Particles in the Heliosphere*. AIP Conf. Proc., vol. 528 (2000), pp. 229–232
R.E. Gold et al., Space Sci. Rev. **86**, 541–562 (1998)
M.A. Lee, J. Geophys. Res. **88**, 6109 (1983)
G. Li , G.P. Zank, W.K.M. Rice, J. Geophys. Res. **110**, A010600 (2005)
G.M. Mason et al., Space Sci. Rev. **86**, 409–448 (1998)
R.A. Mewaldt et al., in *Solar and Galactic Composition*. AIP Conf. Proc., vol. 598 (2001), pp. 165–170
R.A. Mewaldt et al., J. Geophys. Res. **110**, A011038 (2005a)
R.A. Mewaldt et al., in *Physics of Collisionless Shocks*. AIP Conf. Proc., vol. 781 (2005b), pp. 227–232
R.A. Mewaldt, C.M.S. Cohen, G.M. Mason, in *Solar Eruptions and Energetic Particles*, ed. by N. Gopolswamy et al. AGU Monograph, vol. 165 (2006), pp. 115–126
T. Onsager et al., in *GOES-8 and Beyond*, ed. by E.R. Washwell. SPIE Conference Proceedings, vol. 2812 (1996) pp. 281–290
Y. Shikaze et al., *Proc. 28th Internat Cosmic Ray Conf.*, vol. 7 (2003), pp. 4027–4030
E.C. Stone et al., Space Sci. Rev. **86**, 357–408 (1998a)
E.C. Stone et al., Space Sci. Rev. **86**, 285–356 (1998b)
J. Torsti et al., Sol. Phys. **162**, 505–531 (1995)
A.J. Tylka, P.R. Boberg, R.E. McGuire, C.K. Ng, D.V. Reames, in *Acceleration and Transport of Energetic Particles Observed in the Heliosphere*. AIP Conf. Proc., vol. 528 (2000), pp. 147–152
A.J. Tylka et al., Astrophys. J. **625**, 474–495 (2005)

Space Sci Rev (2007) 130: 329–333
DOI 10.1007/s11214-007-9211-5

# Determination of Sulfur Abundance in the Solar Wind

**C. Giammanco · P. Bochsler · R. Karrer · F.M. Ipavich · J.A. Paquette · P. Wurz**

Received: 16 January 2007 / Accepted: 10 May 2007 / Published online: 5 July 2007


**Abstract** Solar chemical abundances are determined by comparing solar photospheric spectra with synthetic ones obtained for different sets of abundances and physical conditions. Although such inferred results are reliable, they are model dependent. Therefore, one compares them with the values for the local interstellar medium (LISM). The argument is that they must be similar, but even for LISM abundance determinations models play a fundamental role (i.e., temperature fluctuations, clumpiness, photon leaks). There are still two possible comparisons—one with the meteoritic values and the second with solar wind abundances. In this work we derive a first estimation of the solar wind element ratios of sulfur relative to calcium and magnesium, two neighboring low-FIP elements, using 10 years of CELIAS/MTOF data. We compare the sulfur abundance with the abundance determined from spectroscopic observations and from solar energetic particles. Sulfur is a moderately volatile element, hence, meteoritic sulfur may be depleted relative to non-volatile elements, if compared to its original solar system value.

**Keywords** Sun: abundances · Sun: solar wind

## 1 Introduction

The solar wind elemental composition is usually related to the solar atmosphere and to meteoritic abundances. Solar wind and meteoritic matter originate from the same nebula; however, fractionating processes could change the relative abundances. In particular, it has been observed that the low-FIP elements of the solar wind are enriched with respect to the photospheric abundances. Theories locate the fractionation process in the chromosphere, as a result of the interaction between neutrals, ionized atoms, and EUV radiation from

C. Giammanco (✉) · P. Bochsler · R. Karrer · P. Wurz
Physikalisches Institut der Universität Bern, Sidlerstrasse 5, 3012 Bern, Switzerland
e-mail: giammanco@space.unibe.ch

F.M. Ipavich · J.A. Paquette
University of Maryland, College Park, MD 20742, USA

the solar corona (e.g. Marsch et al. 1995). The low- and high-FIP elements are distinguished for FIP $< 10$ eV and FIP $> 10$ eV, respectively. Sulfur is located in the transition (FIP $= 10.4$ eV), therefore its reliable determination provides a good test for existing theories. We have ten years of data from the Mass Time Of Flight (MTOF) sensor of the CELIAS instrument (Hovestadt et al. 1995) on SOHO. In this paper we present a detailed analysis. The MTOF sensor response depends strongly on the wind speed, and for this analysis we select only time periods where the wind speed was in three velocity ranges, $380 \pm 20$, $390 \pm 20$, and $400 \pm 20$ km/s. We accumulated the largest dataset in these velocity ranges. The velocity of the solar wind is provided from the Proton Monitor sensor and is given for protons (Ipavich et al. 1998).

MTOF can be conceptually divided into two parts: (1) an entrance system that makes a preselection in energy per charge and incident angle for the solar wind ions, which is governed by the $V_{\mathrm{wave}}$ potential and (2) after a post-acceleration by a potential $V_{\mathrm{F}}$, the solar wind ions enter into a time-of-flight mass spectrometer, that measures their mass.

## 2 The Spectra

The MTOF sensor provides time-of-flight spectra (Fig. 1). The solar wind ions are usually multiply charged. However, after energy-per-charge selection, the ions are recharged to neutral, singly, or doubly ionized in the carbon foil of the MTOF sensor. The time-of-flight data are converted into mass-per-charge ($M/Q$) spectra using a quadratic relation, since for a particle of mass M and charge $Q$ we have the relation: $M/Q = C(t - t_{\circ})^2$, where $t$ is the time of flight and $Q$ is the charge after charge exchange.

Each bin of the spectrum represents the number of particles at a given $M/Q$ and velocity range counted during the time of integration. To extract useful information we must take

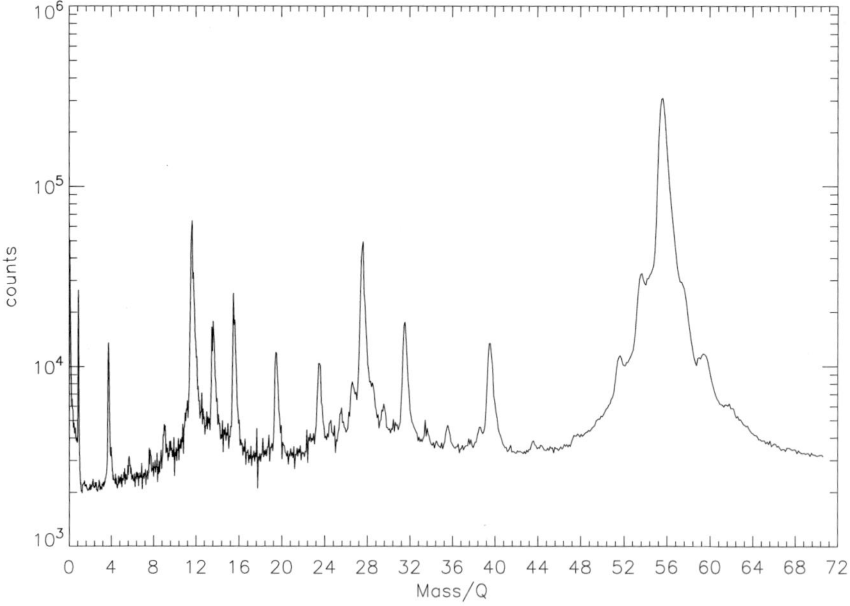


**Fig. 1** Typical MTOF spectrum obtained for one of the three wind-speed interval integrated from year 1996 to 2006. The figure shows the peaks of Fe 56, S 32, Ca 40, Ni 58, Ni 60, and Mg 24, among others

into account the instrument response variation for the different mass–charge ratios and, if present, the contamination generated by doubly ionized elements.

## 3 Response of the Entrance System

The MTOF entrance system is a Wide Angle Variable Energy/Charge electrostatic analyzer (WAVE). Its function is to accept solar wind ions over a large energy-per-charge and incident-angle range. Its response is set through a $V_{\mathrm{wave}}$ potential; however, it is different for each element and for each ionization state (Wurz 2001; Wurz et al. 1999). It also depends on the solar wind speed, the solar wind thermal speed, and the incident angle. In this work we suppose that the wind comes in the ecliptic plane from the Sun direction. The thermal speed is as a first approximation assumed to be equal to the value of the long time average of the Proton Monitor thermal speed measurement, i.e., 10 times less than the wind velocity. Figure 2 shows the effective area of the entrance system as function of the $V_{\mathrm{wave}}$ voltage for different elements. For a more refined analysis in the future, we will use the actually measured thermal spreads provided by the proton monitor of CELIAS. To choose the best potential to determine a particular abundance ratio, it is convenient to analyze the ratio of the entrance system response for the two elements (Fig. 3).

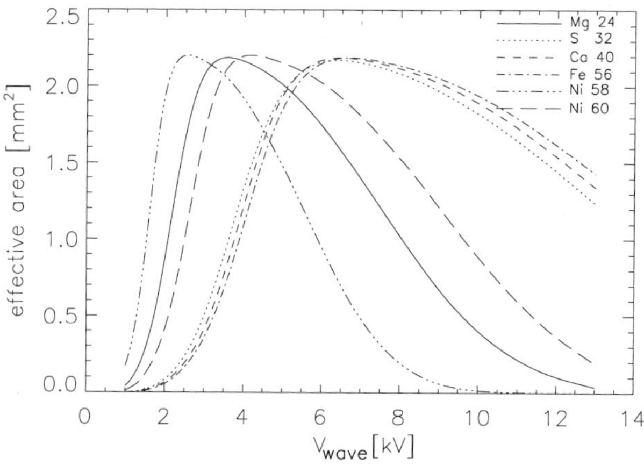


**Fig. 2** Response of the entrance system. The effective area is calculated taking into account a solar wind velocity of 380 km/s. For each element we interpolate a freezing temperature for this solar wind speed. Then we use the compilation of Mazzotta et al. (1998) to derive the respective charge ion distributions

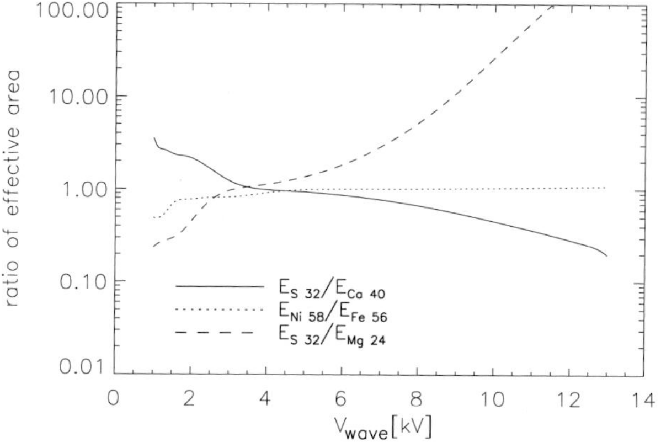


**Fig. 3** Ratio of WAVE effective area for sulfur over magnesium, sulfur over calcium, and nickel over iron. The flat parts of the curves indicate the intervals of voltages that are best suited for the measurement of relative abundances

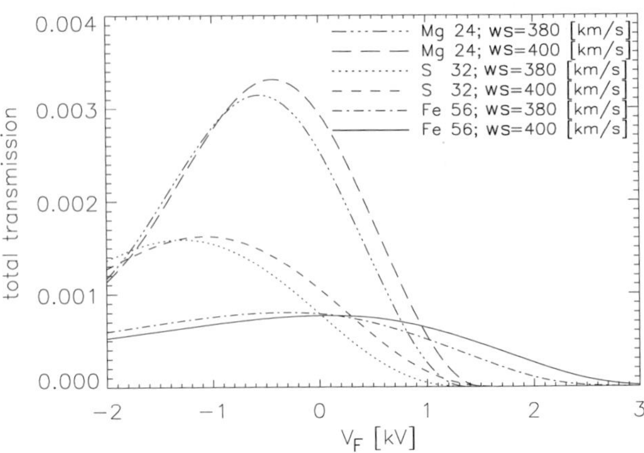


**Fig. 4** Response of the mass spectrometer for different solar wind speeds in function of the post acceleration potential, $V_F$. In the case of sulfur, the range of velocities that we have selected will be the principal source of the final uncertainty

## 4 Response of the Mass Spectrometer

Once we have chosen the preferred voltage for the entrance system we must do the same for the spectrometer. The isochronous mass spectrometer measures the time of flight of incident particles. Its response depends on the element, the ionization state, and the incident angle (Fig. 4).

## 5 The Measured Abundances

To determine the ratio [S]/[Mg] we have seven different combinations of voltage sets. One of them gives a $V_{wave}$ voltage far from the flat interval between 2 and 8 kV (Fig. 3); thus, we exclude it from the analysis. For each accepted voltage set we calculate an abundance ratio assuming a Dirac distribution for the wind speed during the integration time, centered at 380, 390, and 400 km/s. Finally, we take the mean value of the calculated ratios as the final result. We do the same for the [S]/[Ca] ratio. The results are shown in Table 1.

**Table 1** The reported ratios from other authors are derived by the respective abundances relative to hydrogen. The Ulysses/SWICS ratios are derived by the abundances relative to oxygen given in von Steiger et al. (2000) for solar maximum and minimum regimes of slow solar wind

| | [S]/[Mg] | [S]/[Ca] |
|---|---|---|
| Solar wind this work | $0.36 \pm 0.16$ | $4.6 \pm 1.0$ |
| Solar wind Ulysses/SWICS, max | $0.33 \pm 0.10$ | – |
| Solar wind Ulysses/SWICS, min | $0.36 \pm 0.11$ | – |
| Photosphere Asplund et al. (2005) | $0.41 \pm 0.06$ | $6.8 \pm 0.6$ |
| Photosphere Grevesse and Sauval (1998) | $0.56 \pm 0.09$ | $9.3 \pm 1.2$ |
| Meteorites Anders and Grevesse (1989) | $0.49 \pm 0.03$ | $8.5 \pm 0.7$ |
| Meteorites Palme and Beer (1994) | $0.42 \pm 0.03$ | $7.1 \pm 0.5$ |
| SEP-derived corona Reames (1998) | $0.162 \pm 0.005$ | $3.00 \pm 0.13$ |
| SEP-derived corona Breneman and Stone (1985) | $0.142 \pm 0.099$ | $1.44 \pm 0.03$ |

## 6 Conclusions

In conclusion according to this first analysis, we infer that sulfur shows a small depletion in the solar wind relative to low-FIP elements, and that it behaves more like a low-FIP element despite a rather long ionization time in the chromosphere of ~12 seconds (Marsch et al. 1995). Our measurements are consistent with those of von Steiger et al. (2000).

**Acknowledgements** We acknowledge valuable suggestions by L.M. Blush and M. Iakovleva. This work was supported by the Swiss National Science Foundation

## References

Anders, Grevesse, GCA **53**, 197–214 (1989)
M. Asplund, N. Grevesse, A.J. Sauval, ASPC **336**, 25 (2005)
H.H. Breneman, E.C. Stone, Astrophys. J. **299**, L57 (1985)
N. Grevesse, A.J. Sauval, Space Sci. Rev. **5**, 161 (1998)
D. Hovestadt et al., Sol. Phys. **162**, 441 (1995)
F.M. Ipavich et al., J. Geophys. Res. **103**, 17205 (1998)
F.M. Ipavich, J.A. Paquette, P. Bochsler, S.E. Lasley, P. Wurz, AIP Conf. **598**, 101 (2001)
E. Marsch, R. von Steiger, P. Bochsler, Astron. Astrophys. **301**, 261 (1995)
P. Mazzotta, G. Mazzitelli, S. Colafrancesco, N. Vittorio, Astron. Astrophys. Suppl. Ser. **133**, 403 (1998)
H. Palme, H. Beer, in *Astronomy and Astrophysics*, vol. 3, ed. by Voigt (Springer, 1994), pp. 196–221
D.V. Reames, Space Sci. Rev. **5**, 327 (1998)
R. von Steiger et al., J. Geophys. Res. **105**, 27 (2000)
P. Wurz, Heavy ions in the solar wind: Results from SOHO/CELIAS/MTOF, University Bern, 2001
P. Wurz, M.R. Aellig, P. Bochsler, S. Hefti, F.M. Ipavich, A.B. Galvin, H. Grünwaldt, M. Hilchenbach, F. Gliem, D. Hovestadt, Phys. Chem. Earth **24**, 421 (1999)

Space Sci Rev (2007) 130: 335–340
DOI 10.1007/s11214-007-9191-5

# An Update on Ultra-Heavy Elements in Solar Energetic Particles above 10 MeV/Nucleon

**R.A. Leske · R.A. Mewaldt · C.M.S. Cohen · A.C. Cummings · E.C. Stone · M.E. Wiedenbeck · T.T. von Rosenvinge**

Received: 2 February 2007 / Accepted: 10 April 2007 /
Published online: 24 May 2007


**Abstract** Measurements below several MeV/nucleon from Wind/LEMT and ACE/ULEIS show that elements heavier than Zn ($Z = 30$) can be enhanced by factors of ~100 to 1000, depending on species, in $^3$He-rich solar energetic particle (SEP) events. Using the Solar Isotope Spectrometer (SIS) on ACE we find that even large SEP (LSEP) shock-accelerated events at energies from ~10 to >100 MeV/nucleon are often very iron rich and might contain admixtures of flare seed material. Studies of ultra-heavy (UH) SEPs (with $Z > 30$) above 10 MeV/nucleon can be used to test models of acceleration and abundance enhancements in both LSEP and $^3$He-rich events. We find that the long-term average composition for elements from $Z = 30$ to 40 is similar to standard solar system values, but there is considerable event-to-event variability. Although most of the UH fluence arrives during LSEP events, UH abundances are relatively more enhanced in $^3$He-rich events, with the $(34 < Z < 40)/\mathrm{O}$ ratio on average more than 50 times higher in $^3$He-rich events than in LSEP events. At energies >10 MeV/nucleon, the most extreme event in terms of UH composition detected so far took place on 23 July 2004 and had a $(34 < Z < 40)/\mathrm{O}$ enhancement of ~250–300 times the standard solar value.

**Keywords** Sun: abundances · Sun: particle emission · Sun: coronal mass ejections (CMEs) · Sun: flares

R.A. Leske (✉) · R.A. Mewaldt · C.M.S. Cohen · A.C. Cummings · E.C. Stone
Space Radiation Laboratory, California Institute of Technology, Pasadena, CA 91125, USA
e-mail: ral@srl.caltech.edu

M.E. Wiedenbeck
Jet Propulsion Laboratory, California Institute of Technology, Pasadena, CA 91109, USA

T.T. von Rosenvinge
NASA/Goddard Space Flight Center, Greenbelt, MD 20771, USA

## 1 Introduction and Data Analysis

The abundances of ultra-heavy (UH) elements (with $Z > 30$) relative to oxygen have been found to be enhanced by surprisingly large factors (~100 to 1000) in some $^3$He-rich solar energetic particle (SEP) events at ~400 keV/nucleon by ACE/ULEIS (Mason et al. 2004) and at 3.3–10 MeV/nucleon by Wind/LEMT (Reames 2000; Reames and Ng 2004), with much smaller (factors of a few) enhancements observed in some large SEP (LSEP) shock-accelerated events (Reames 2000; Reames and Ng 2004). If the observed increase in Fe/O ratios with increasing energy above 10 MeV/nucleon in some LSEP events (Cohen et al. 2003; Cohen et al. 2005; Tylka et al. 2001) is due to the presence of flare material, either via direct access from the flare site (Cane et al. 2003) or by way of re-acceleration of flare suprathermals at quasi-perpendicular shocks (Tylka et al. 2005), one might expect to find very large UH enhancements in iron-rich LSEP events at higher energies. Measuring UH abundances in both LSEP and $^3$He-rich events at both low and high energies may provide additional clues to the nature of SEP acceleration.

The Solar Isotope Spectrometer (SIS) instrument on ACE consists of two stacks of silicon solid state detectors and is primarily designed to measure the isotopic composition of species from He ($Z = 2$) through Ni ($Z = 28$) at energies of ~10 to >100 MeV/nucleon using the $dE/dx$ vs. residual energy technique (Stone et al. 1998). To achieve isotopic resolution over this dynamic range, the pulse height analyzers were designed specifically for this limited interval in $Z$ and saturate above certain energy deposits in the detectors, typically for particles heavier than $Z \sim 40$ (Zr) at large incidence angles penetrating deep into a detector segment. Detection efficiency gradually drops for $Z > 40$, reaching 0 by $Z \sim 50$, but has not been fully modeled.

Although limited in sensitivity to UH ions, SIS attains excellent elemental resolution in this regime compared with previous studies at lower energies (Mason et al. 2004; Reames 2000). Figure 1 shows a nuclear charge histogram in the UH range for energies of ~10–50 MeV/nucleon accumulated from the launch of ACE in August 1997 through April 2006. Standard data cuts were applied requiring that particles stop in the instrument and that multiple calculations of nuclear charge using different combinations of energy-loss measurements be consistent. For the lowest energy ("range 0") particles, only two detectors are triggered, resulting in only one charge determination with no consistency check possible. At high rates (such as during very large SEP events), if a large-angle heavy nucleus such as Fe stops in the first detector while a H or He nucleus stops in the second detector, the signal produced can be similar to that expected for range 0 ultraheavies. This accidental-coincidence background is avoided by requiring that the energy deposit in the second detector for any range 0 particle exceeds the maximum expected for a He nucleus in that detector. Additional analysis details are given in Leske et al. (2005a).

## 2 Discussion

Figure 1 represents a summation over many SEP events, both very large and small, without any accounting for variations in instrument livetime, sampling and telemetry limitations, UH detection efficiencies, or differences in energy intervals and/or spectral shapes for the different elements. Nevertheless, the relative peak heights compare favorably with those expected from standard solar system abundances (Lodders 2003). To account for the under-abundance of SEP elements with a high first ionization potential (FIP), we have adjusted the solar system abundances downward by a factor of 2.5 (Mewaldt et al. 2007) for all elements

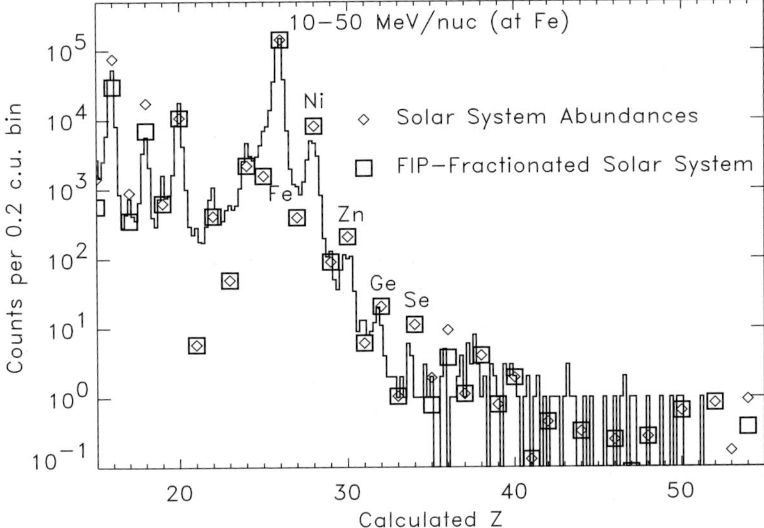


**Fig. 1** Nuclear charge histogram from ACE/SIS from August 1997 through April 2006 for energies of ~10–50 MeV/nucleon (at Fe), compared with standard solar system abundances (Lodders, 2003) normalized to Fe (*diamonds*) and solar system abundances after accounting for FIP fractionation (*squares*)

with a FIP greater than 10 eV. The small differences between these adjusted points and the even-$Z$ peaks might be largely accounted for by the fact that the peak width increases with increasing $Z$. In addition, the data shown were measured at a common range, not energy. If corrected to a common energy, the peaks for $Z$ less than Fe would be reduced slightly (~10%) and those at higher $Z$ increased slightly. From this figure, we conclude that the long-term, overall average composition of ~10–50 MeV/nucleon particles out to at least $Z \sim 40$ agrees reasonably well (within a factor of ~2) with standard solar system values.

To assess the UH component of various classes of events, we have divided the data in Fig. 1 into 4 subsets, as shown in Fig. 2. Large "GOES" events (in which the >10 MeV proton intensity exceeded 10 protons $(\text{cm}^2\,\text{sr}\,\text{s})^{-1}$) account for the vast majority of all particles detected by SIS. More than 90% of all O nuclei and more than half of all $Z \geq 34$ particles arrived in such events (these numbers have not been corrected for greatly reduced instrument livetimes in the largest events; the actual portion of the total fluence would be considerably higher). A hand-selected sample of 25 $^3$He-rich events, in which the ~4.5 MeV/nucleon $^3$He/$^4$He ratio exceeded 0.1 and there were sufficient C, N, and O ions with energies $\gtrsim$10 MeV/nucleon to obtain elemental abundances, contain relatively few particles overall (only 0.5% of all O detected by SIS) but more than 20% of all $Z \geq 34$ nuclei, suggesting that there is generally a significant enrichment of UH abundances in $^3$He-rich events compared with LSEP events, even at these energies. Applying stringent quiet-time cuts (Leske et al. 2005b) leaves only low energy galactic cosmic rays (GCRs) and anomalous cosmic rays (ACRs), as evidenced by the high abundances of odd-$Z$ sub-iron elements and the relative abundances of C, N, and O, respectively, in Fig. 2 (lower left). No $Z \geq 34$ particles in this energy range have been detected during these quiet periods, even though traces of flare material in the form of $^3$He are typically present (Wiedenbeck et al. 2003).

Nearly 60% of the time none of the above conditions are met. While only a modest fraction of the total events (~7% of all O) arrive during these times, just over one quarter of all $Z \geq 34$ nuclei are in this event class and thus it warrants closer scrutiny. This interval

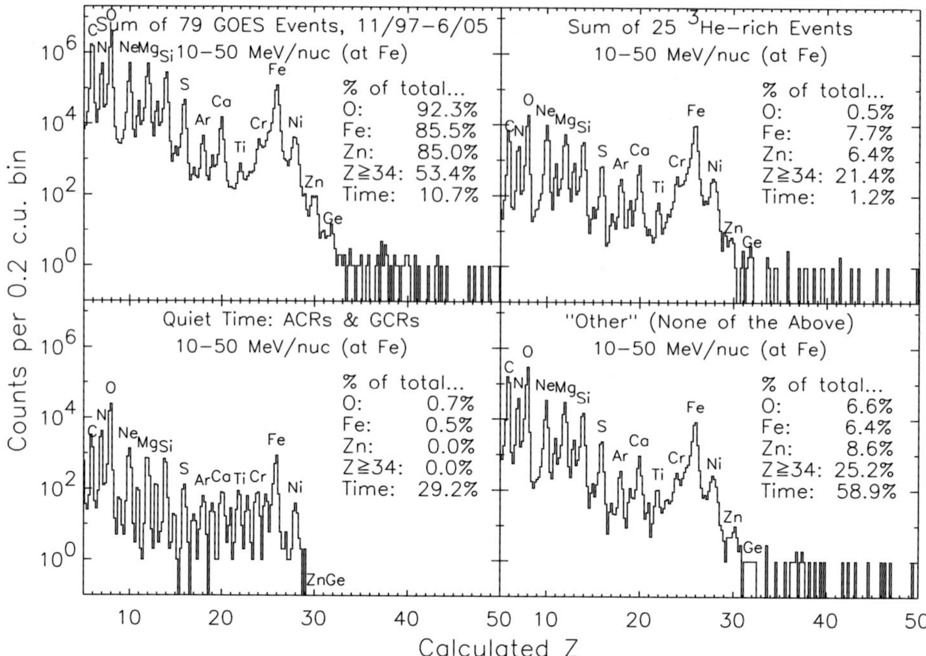


**Fig. 2** Nuclear charge histograms of particles recorded by SIS. The 4 panels show: the sum of all particles arriving in 79 LSEP events (*upper left*), particles in a selection of 25 $^3$He-rich events (*upper right*), those arriving during solar quiet times (*lower left*), and particles arriving during periods which fit none of the above categories (*lower right*). Also listed are the percentages of all detected O, Fe, Zn, and $Z \geq 34$ particles appearing in each of the 4 data subsets and the fraction of the total time in each category

likely includes extended tails of LSEP events, as well as $^3$He-rich events that were not clearly distinguished from a pre-existing background and therefore not included on the $^3$He-rich event list. Roughly half of the $40 < Z < 50$ particles in this set arrived in a single day during the 23 July 2004 SEP event. This very small event had a >10 MeV proton intensity which peaked at only ~4 protons (cm$^2$ sr s)$^{-1}$ and a "gradual" time profile but a heavy-ion composition more typical of a $^3$He-rich event (Fe/O ~ 1, highly-enriched UH abundances, etc.). Further details of this event are discussed in Leske et al. (2005a).

Although the statistical uncertainties are very large, we can examine the composition of trans-nickel elements in individual SEP events. Figure 3 shows the Zn/O and $(34 \leq Z \leq 40)/\mathrm{O}$ counts ratios versus Fe/O. At this stage of the analysis these are only raw counts ratios, without correction for instrument livetime (which to first order will affect all species equally and thus factor out in the ratios) or reduced UH efficiencies (which, when accounted for, will increase the UH abundances, but by factors that are typically small compared to the statistical uncertainties).

We find that Zn/O correlates well with Fe/O, with both ratios showing similar enhancements or depletions in each event. $^3$He-rich events tend to have higher Zn/O and Fe/O than LSEP events, but seem to fall along the trend extrapolated from LSEP events. Calculating the ratio of counts for elements in each event class using the histograms shown in Fig. 2, on average the Fe/O ratio for the events in our study is $17.2 \pm 0.1$ times higher in $^3$He-rich events than in LSEP events, while Zn/O is $13 \pm 3$ times greater.

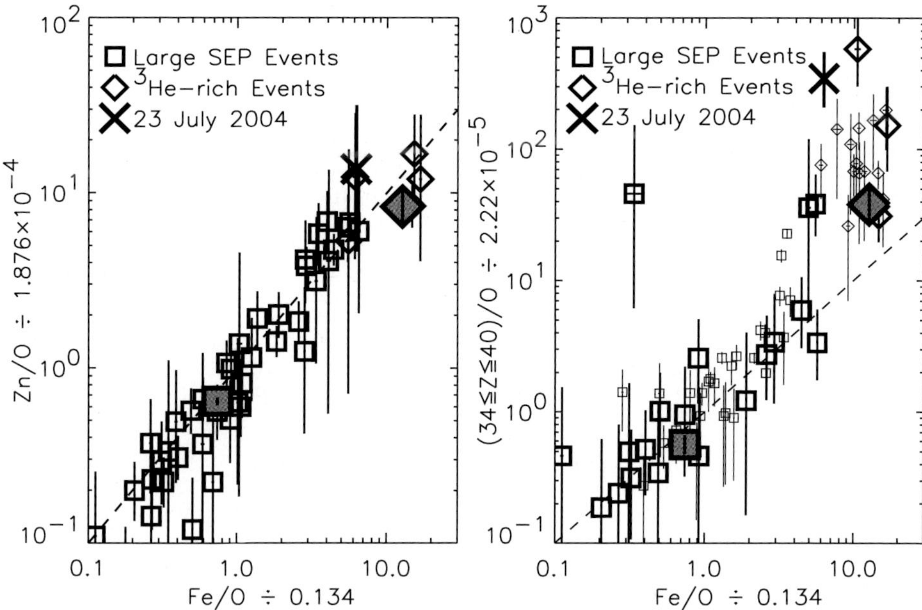


**Fig. 3** ACE/SIS ~10–50 MeV/nucleon raw counts ratios (see text) of Zn/O (*left*) and $(34 \le Z \le 40)/\mathrm{O}$ (*right*) versus Fe/O, normalized by solar values (Reames and Ng, 2004). LSEP events (*squares*), $^3$He-rich events (*diamonds*) and the 23 July 2004 event (*crosses*) are indicated. *Small light symbols in the right panel* are from Reames and Ng (2004) at 3.3–10 MeV/nucleon. *Large filled symbols* represent the ratios of counts from the summed sets of LSEP and $^3$He-rich events shown in the Fig. 2 histograms. *Dashed lines* indicate equal enhancements on both axes

In general, our $(34 \le Z \le 40)/\mathrm{O}$ ratios as a function of Fe/O agree well with lower energy measurements in a different set of events (Reames and Ng 2004). Enhancements of $(34 \le Z \le 40)/\mathrm{O}$ in LSEP events are correlated with Fe/O, and for the most Fe-rich LSEP events they approach the much larger enhancements found in $^3$He-rich events. The average $(34 \le Z \le 40)/\mathrm{O}$ counts ratio in $^3$He-rich events is $67 \pm 24$ times greater than that in LSEP events, which is significantly higher than the Zn/O or Fe/O $^3$He-rich/LSEP event enhancement. This suggests that the UH enhancements in $^3$He-rich events increase with increasing $Z$.

At present only a few $^3$He-rich events are large enough for SIS to determine UH abundances, and one of those shows a larger $(34 \le Z \le 40)/\mathrm{O}$ enhancement at 10–50 MeV/nucleon than any events reported in the lower-energy survey of Reames and Ng (2004). Within the few SIS $^3$He-rich events there seems to be no correlation between $(34 \le Z \le 40)/\mathrm{O}$ and Fe/O, and the very UH-rich 23 July 2004 event has a very soft Fe spectrum, with a power-law index of $\sim -4.7$. Both these observations are consistent with the general findings of Reames (2000) in other events at 3.3–10 MeV/nucleon.

**Acknowledgements** This work was supported by NASA at the California Institute of Technology (grant NAG5-12929), the Jet Propulsion Laboratory, and the Goddard Space Flight Center.

## References


H.V. Cane, T.T. von Rosenvinge, C.M.S. Cohen, R.A. Mewaldt, Geophys. Res. Lett. **30** (2003). doi: 10.1029/2002GL016580

C.M.S. Cohen, R.A. Mewaldt, A.C. Cummings, R.A. Leske, E.C. Stone, T.T. von Rosenvinge, M.E. Wiedenbeck, Adv. Space Res. **32**, 2649 (2003)

C.M.S. Cohen, E.C. Stone, R.A. Mewaldt, R.A. Leske, A.C. Cummings, G.M. Mason, M.I. Desai, T.T. von Rosenvinge, M.E. Wiedenbeck, J. Geophys. Res. **110**, A09S16 (2005). doi: 10.1029/2005JA011004

R.A. Leske, C.M.S. Cohen, A.C. Cummings, R.A. Mewaldt, E.C. Stone, M.E. Wiedenbeck, T.T. von Rosenvinge, *Proc. 29th Internat. Cosmic Ray Conf.* (Pune), vol. 1, (2005a) p. 107

R.A. Leske, A.C. Cummings, C.M.S. Cohen, R.A. Mewaldt, E.C. Stone, M.E. Wiedenbeck, T.T. von Rosenvinge, *Proc. 29th Internat. Cosmic Ray Conf.* (Pune), vol. 2, (2005b) p. 113

K. Lodders, Astrophys. J. **591**, 1220 (2003)

G.M. Mason, J.E. Mazur, J.R. Dwyer, J.R. Jokipii, R.E. Gold, S.M. Krimigis, Astrophys. J. **606**, 555 (2004)

R.A. Mewaldt, C.M.S. Cohen, G.M. Mason, A.C. Cummings, M.I. Desai, R.A. Leske, J. Raines, E.C. Stone, M.E. Wiedenbeck, T.T. von Rosenvinge, T.H. Zurbuchen, Space Sci. Rev., this volume (2007), doi: 10.1007/s11214-077-9187-1

D.V. Reames, Astrophys. J. Lett. **540**, L111 (2000)

D.V. Reames, C.K. Ng, Astrophys. J. **610**, 510 (2004)

E.C. Stone et al., Space Sci. Rev. **86**, 357 (1998)

A.J. Tylka, C.M.S. Cohen, W.F. Dietrich, C.G. Maclennan, R.E. McGuire, C.K. Ng, D.V. Reames, Astrophys. J. Lett. **558**, L59 (2001)

A.J. Tylka, C.M.S. Cohen, W.F. Dietrich, M.A. Lee, C.G. Maclennan, R.A. Mewaldt, C.K. Ng, D.V. Reames, Astrophys. J. **625**, 474 (2005)

M.E. Wiedenbeck, G.M. Mason, E.R. Christian, C.M.S. Cohen, A.C. Cummings, J.R. Dwyer, R.E. Gold, S.M. Krimigis, R.A. Leske, J.E. Mazur, R.A. Mewaldt, P.L. Slocum, E.C. Stone, T.T. von Rosenvinge, in *AIP Conf. Proc. 679: Solar Wind Ten*, ed. by M. Velli, R. Bruno, F. Malara (2003) p. 652

Space Sci Rev (2007) 130: 341–353
DOI 10.1007/s11214-007-9178-2

# The Local Interstellar Medium: Peculiar or Not?

**Rosine Lallement**

Received: 14 February 2007 / Accepted: 26 March 2007 /
Published online: 30 May 2007


**Abstract** The local Interstellar Medium (ISM) at the 500 pc scale is by many respects a typical place in our Galaxy made of hot and tenuous gas cavities blown by stellar winds and supernovae, that includes the 100 pc wide "Local Hot Bubble (LHB)", dense and cold clouds forming the cavity "walls", and finally diffuse and warm clouds embedded within the hot gas, such as the Local Interstellar Cloud (LIC) presently surrounding the Sun. A number of measurements however, including abundance data, have contradicted this "normality" of our interstellar environment. Some contradictions have been explained, some not. I review recent observations at different spatial scales and discuss those peculiarities. At all scales Johannes Geiss has played a major role.

At the scale of the first hundred parsecs, there are at least three "anomalies": (i) the peculiar Gould Belt (GB), (ii) the recently measured peculiar Deuterium abundance pattern, (iii) the low value of the local O, N and $^3$He gas phase abundances. I discuss here the possibility of a historical link between these three observations: the large scale phenomenon which has generated the Belt, a giant cloud impact or an explosive event could be the common origin.

At the 50–100 parsec scale, some of the unexplained or contradictory measurements of the Local Bubble hot gas, including its EUV/soft X ray emissions, ion column-densities and gas pressure may at least partially be elucidated in the light of the newly discovered X-ray emission mechanism following charge transfer between solar wind high ions and solar system neutrals. The Local Bubble hot gas pressure and temperature may be lower than previously inferred.

Finally, at the smaller scale of the local diffuse cloudlets (a few parsecs), the knowledge of their structures and physical states has constantly progressed by means of nearby star absorption spectroscopy. On the other hand, thanks to anomalous cosmic rays and pickup ions measurements, local abundances of ISM neutral species are now precisely derived and may be compared with the absorption data. Interestingly these comparisons are now accurate enough to reveal other (noninterstellar) sources of pickup ions. However the actual physical

R. Lallement (✉)
Service d'Aéronomie, CNRS, BP 3, 91371 Verrières-le-Buisson, France
e-mail: rosine.lallement@aerov.jussieu.fr

state of the ISM 10–20,000 A.U. ahead along the Sun trajectory, which will be the ambient interstellar medium in a few thousands years, remains unknown. Local Bubble hot gas or warm LIC-type gas? More EUV/UV spectroscopic data are needed to answer this question.

**Keywords** ISM

## 1 Introduction

The detailed structure of the Milky Way Galaxy is less known than the structure of its neighbors. For an "inside" observer there is no global perspective, while on the contrary the new and considerably detailed images of the nearby galaxies are breathtaking, by far more impressive than the poorly mapped and still very uncertain schemes of the Milky Way spiral arms. In the same way the structure of the Local ISM is paradoxically less known than the structure of some distant galactic regions. There has been a moderate interest for our "galactic neighborhood" because it does not contain "extreme" objects. For example, a spectacular region, the famous set of "pillars of creation" imaged by the HST and showing stellar births within ultra-dense clouds, is not located in our local galactic arm (or the Orion arm), but farther away in the outer Perseus arm. Compared to these conspicuous features, our local interstellar clouds are very tenuous. The situation is comparable to that of an airplane pilot seeing at distance huge cumulus with lightened or darkened faces, while flying through very diffuse, almost invisible cirrus. Once its multi-phase structure was established to compare with models, the local ISM has raised interest essentially in the context of light element abundance measurements (for cosmological purposes), of hot gas evolution and filling factor (for galactic evolution and gas recycling), and of the influence on solar system and earth, and the boundary of the heliosphere.

## 2 The ISM at the 500 pc – 1 kpc Scale

### 2.1 Local ISM Abundances

Elemental abundances in the local ISM are mainly determined from absorption measurements in nearby star spectra. UV/EUV high resolution spectrographs (IUE, HST-GHRS and STIS, FUSE.) have provided the deuterium to hydrogen ratio in the local interstellar clouds with an increasing accuracy. In parallel, sophisticated particle detectors on board spacecraft have entered into the play. Interstellar pickup ion (PUI) data obtained with SWICS on board Ulysses have allowed for the first time measurements of the $^3$He/H isotopic ratio in the local interstellar cloud (Gloeckler and Geiss 1998). Interstellar neutral oxygen and nitrogen relative abundances in the inner heliosphere have also been inferred from the PUI data (e.g. Gloeckler and Geiss 2001). In order to extrapolate those inner heliosphere abundances back to actual interstellar abundances in the ambient ISM, one needs to know the fraction of IS neutrals that succeed in entering the heliosphere. This filtration factor depends on the species through the charge transfer cross-sections. Fully self-consistent models of the heliosphere (see Sect. 3) provide these factors, once they have been adjusted to fit the constraints from all existing heliospheric and interstellar data (cosmic ray gradients, solar backscattered radiation or in situ measurements).

Johannes Geiss and colleagues (Geiss et al. 2002) had the idea to compare the series of newly determined $^3$He, D, N, O LIC abundances with abundances in two other types of

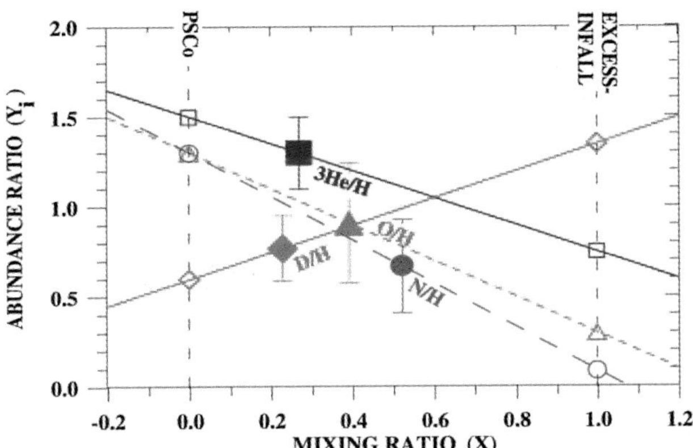


**Fig. 1** LIC abundance measurements favor of a local mixing of "normal" (protosolar extrapolated to present, PSC0 in the figure) and unprocessed Magellanic type interstellar gas (excess infall on the figure), being at intermediates values between the two types. From Geiss et al. (2002)

interstellar gases: on one hand the Magellanic- (or dwarf galaxy-) type, i.e. a low metallicity and moderately processed gas, and on the other hand the "normally" processed galactic gas, whose properties are derived from solar system data (i.e. the proto-solar abundances) extrapolated to the present time by application of an evolutionary code for a duration of 4.6 Gyr.

Interestingly, as can be seen in Fig. 1, all four LIC abundances fall between those for the other two types of ISM, as if the interstellar gas in our galactic neighborhood were a mixture of "normal" galactic gas and unprocessed (Magellanic type) material. Geiss et al. derive a mixing ratio (normal/total) of $0.4 \pm 0.15$ and suggest that a recent infall of dwarf galaxy or intergalactic matter may have biased the local abundance towards the observed values.

## 2.2 The Deuterium Abundance within 500 pc

The analysis of the Far Ultraviolet Spectroscopic Explorer (FUSE) data and the inferred D/H ratios have recently given rise to a very lively and unexpected debate. The situation is illustrated in Fig. 2, from Linsky et al. (2006). Within the measurement and modeling uncertainties, the D/H ratio is about constant in the solar vicinity, more specifically along lines-of-sight having a H column-density less than $2 \times 10^{19}$ cm$^{-2}$, which corresponds to distances of 50 to 150 pc according to directions. For very high columns, above $10^{21}$ cm$^{-2}$, the ratio is significantly smaller and seems similarly constant (although only few measurements exist). Rather surprisingly the ratio is highly variable (by more than a factor of 4) within the intermediate range of distances (or H columns), and reaches as much as 3–4 times the very distant value. Such a peculiar behavior has been interpreted in two different ways: Linsky et al. (2006) attribute it to the effect of the preferential adsorption of deuterium at the surface of dust grains (Draine 2003). In places where heating is destroying them, i.e. in supernovae "bubbles" and at shocks, deuterium is released from the grains into the gas phase, which increases the D/H ratio significantly.

On the other hand, Hébrard et al. (2005) show that the deuterium to oxygen and deuterium to nitrogen ratios, at variance with D/H (Hébrard and Moos 2003) exhibit only a two-state regime, with only one transition from a high Local Bubble value down to a lower value at larger distances. They suggest that D is locally overabundant due to some mixing processes. The consequences of the two interpretations in term of galactic D/H ratio are completely opposite. While in the former interpretation the actual D/H is best probed where most of

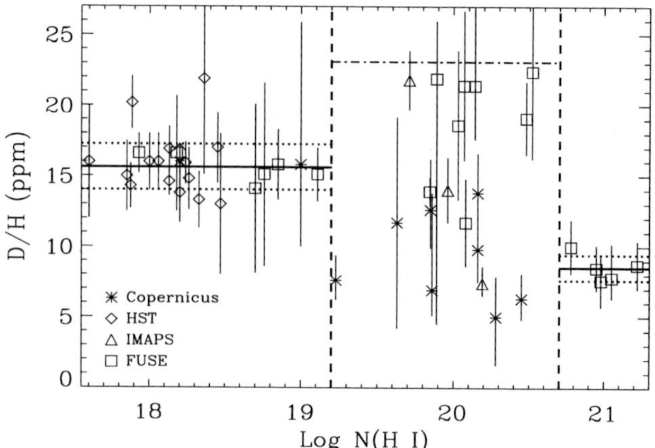


**Fig. 2** Combined D/H measurements as a function of the total HI column-density to the target star. From Linsky et al. (2006)

D has been released from dust, i.e. the highest value, in the latter case the actual galactic D/H is best represented by the values measured at large distances, i.e. is much lower. This makes a strong difference, because the Hebrard et al explanation implies an astration factor of 3 which far exceeds the prediction of the modern models. On the other hand, thee Linsky et al interpretation implies astration factors somewhat smaller than the recent models (for a review of those models see Prantzos 2007).

## 2.3 The Gould Belt

Already noticed by Herschel in 1847, the peculiar distribution of the bright early-type stars in the solar neighborhood is still a subject of debate (for a review see Pöppel 1997). The Gould Belt stands out on sky maps representing the distribution of stars according to their spectral types. A simple map showing the Hipparcos distribution of O to B8 stars closer than 650 pc immediately reveals the Belt. While one would normally expect the stars to be homogeneously distributed along the galactic plane, the Belt appears as a wavy band with a maximum of young hot stars at negative galactic latitudes in the anticenter direction. The Belt is an inclined plane made of stars, but also of HI and HII regions as well as molecular clouds. The system is in expansion and rotation and is composed of stars younger than 20–40 Myrs. The supernova rate is believed to be enhanced inside and along the Belt by comparison with the galactic average and reaches 20–27 SNe per million years. It is also associated with discrete gamma ray sources (Perrot and Grenier 2003).

The Hipparcos data have motivated a new study: Elias et al. (2006a, 2006b) have combined the stellar proper motions and stellar radial velocities to get stellar heliocentric velocities. Figure 3 shows their distribution of heliocentric motion projections along the two galactic plane axes $U$ (towards the galactic center) and $V$ (along the galactic rotation). The double population nature of the distribution is obvious. The authors have used a sophisticated procedure to separate the two populations according to their velocities and locations, and the resulting spatial distributions on the sky of the two groups are shown in Fig. 3. The disk containing the Gould Belt stars is found to be inclined by about 15–20 degrees with respect to the galactic plane.

Potential origins of the Belt include the impact of a giant cloud (Comerón and Torra 1994), which helps to explain the inclination, or a strong explosion followed by a circular shock wave (Olano 1982) which helps to explain the expansion but not very well the inclination. More recently Olano (2001) invokes the interaction between a super-cloud and the

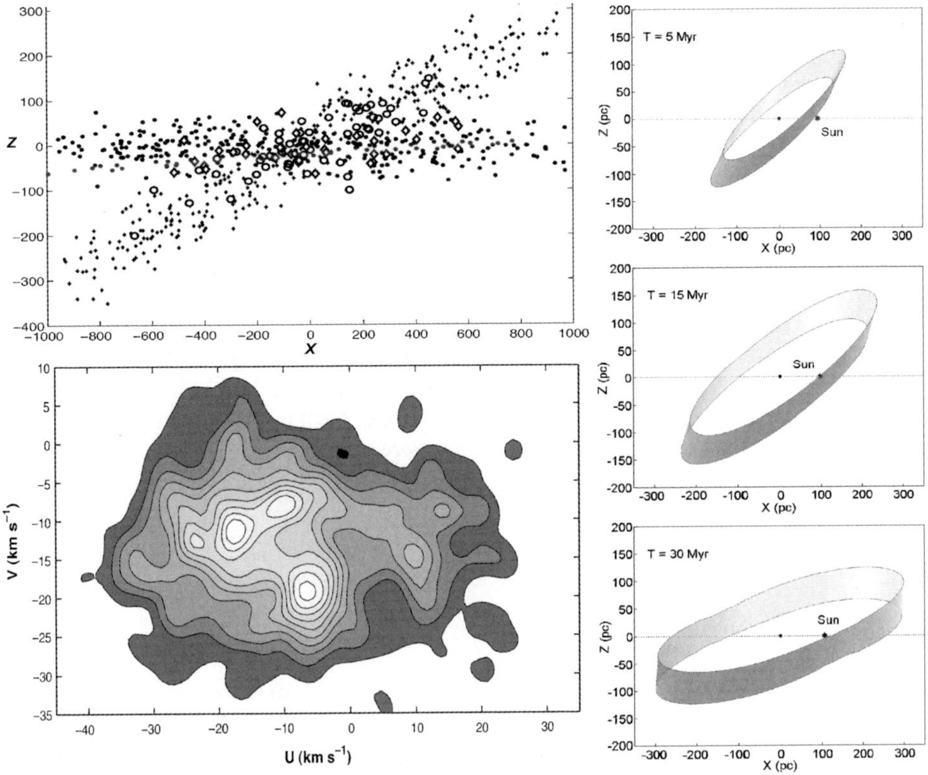


**Fig. 3** *Top left*: Spatial (*top*) and velocity (*bottom*) distribution of the nearby stars (from Elias et al. 2006a, 2006b): normal galactic disk stars and Gould Belt object are very clearly distinguished from each other. $Z$ and $X$ are distances in parsecs along the galactic center axis and above the disk resp.; $U$ and $V$ are the disk projected velocities (radial and ortho-radial components). *Right*: Evolution of a rotating and expanding ring generated by a strong shock wave, as simulated by Perrot and Grenier (2003). The Gould Belt has been strongly inclined w.r.t. the disk in the past

spiral arm. A gamma ray burst origin has also been proposed by Perrot and Grenier (2003), who have modeled the evolution of an expanding belt. The simulation shows that the Belt may have been much more inclined in the past (see Fig. 3), which may be of importance for the evolution of the ISM.

## 2.4 A Link between Gould Belt, the D/H Pattern, O, N, $^3$He Abundances?

The basis of the following discussion is the suggestion that the three regions of Fig. 2 roughly correspond to the interior, the periphery and the exterior of the Gould Belt respectively. The following quantitative and qualitative arguments support a link between the GB and the abundance peculiarities, i.e. between the structural and dynamical patterns and the abundance pattern: (i) The GB system is young. Indeed, 60 Myrs is shorter than the mixing time in the galaxy which is of the order of hundreds of Myrs. This means that the expanding GB, whatever its source, extragalactic cloud impact or hot gas expansion and stellar formation activity following an initial explosion, must have strong inhomogeneities. (ii) The size of the GB radius is 400–500 pc. This corresponds to column-densities of the order of $\sim 10^{20.5}$ cm$^{-2}$. Looking back at Fig. 2, it corresponds to the external part of the high variability region (the "wall") of the D/H ratio. The wall could be associated with the expanding front of the GB, beyond which masses of gas are not yet mixed, and have different properties. The wall could be explained by two different scenarios: an external cloud impact produces a complicated pattern of adjacent layers of both types of gas, external and galactic, or cascades of stellar formation bursts produce ejecta with evolved characteristics which co-exist with the ambient gas. (iii) The enhancement of the supernova rate associated with the GB, and subsequently the high fluxes of associated cosmic rays may have helped to release metals from the local gas into the halo (Völk 2007), a mechanism especially efficient at the time the Belt was strongly inclined as in Fig. 3. (iv) In both scenarios, heating and shocks linked to the SN's have released D from grains. In the cloud impact case there are two sources of local deuterium enrichment: dust evaporation and mixing with D-rich extragalactic gas. More data are needed on the distribution of stars and gas to favor a particular scenario, but it seems plausible that the 500 pc region around the Sun keeps the imprints of a turbulent period a few tens of Myrs ago. In any case, because the deuterium abundance is a crucial ingredient of evolutionary models, more investigations on the local ISM are certainly needed.

## 3 The ISM at the 100 pc Scale and the Local Cavity

### 3.1 The Local Bubble

The local ISM cavity existence is inferred from essentially two diagnostics: absorption lines in nearby star spectra and soft X-ray background emission. Since the first high resolution spectra were recorded from ground and space it has become clear that column-densities increase rather abruptly at distances of the order of 50 to 150 parsecs according to direction, implying that the solar region is devoid of dense (and cold) gas. On the other hand the Local Cavity has been thought for a long time to be filled with hot ($10^6$ K) and tenuous interstellar gas. This result is associated to the ROSAT soft X-ray background measurements (Snowden et al. 1998). More specifically, a significant fraction of the X-ray background has been shown to be uncorrelated with the line-of-sight interstellar column density, while such a correlation is expected if the emission is generated at distance. As a matter of fact soft X-ray photons are strongly absorbed by interstellar gas (optical thickness 1 for an HI column of $10^{19}$ cm$^{-2}$ at $E = 0.25$ keV). The uncorrelated, and thus unabsorbed fraction must be generated within 100–200 parsecs. On the other hand, the existence of the unabsorbed emission at all latitudes down to the galactic plane is a direct proof that the emission is local. For years the totality of this unabsorbed emission has been attributed to the $\sim$100 parsecs wide local interstellar cavity, the so-called Local Hot Bubble. The temperature of the emitting gas inferred from band ratios has been found to be about one million K. The existence of a Bubble filled with hot gas is in agreement with gas-dynamical models of stellar wind and supernova ejecta expansion. The size and physical properties of the LB are well fitted by models involving a series of ancient supernovae, and reheating by a more recent event (De Avillez and Breitschwerdt 2005).

An attempt to make a step further and actually compute maps of the dense gas has been done using accurate distances to nearby stars provided by the Hipparcos satellite. Absorption lines of neutral sodium, a good tracer of the dense cold gas, are detected in optical high resolution spectra of these stars. Profile-fitting of the sodium doublet lines (see Fig. 4) provides the column-densities and radial velocities of the interstellar clouds intercepted by the line-of-sight. These results and the distances to the target stars can be used as input

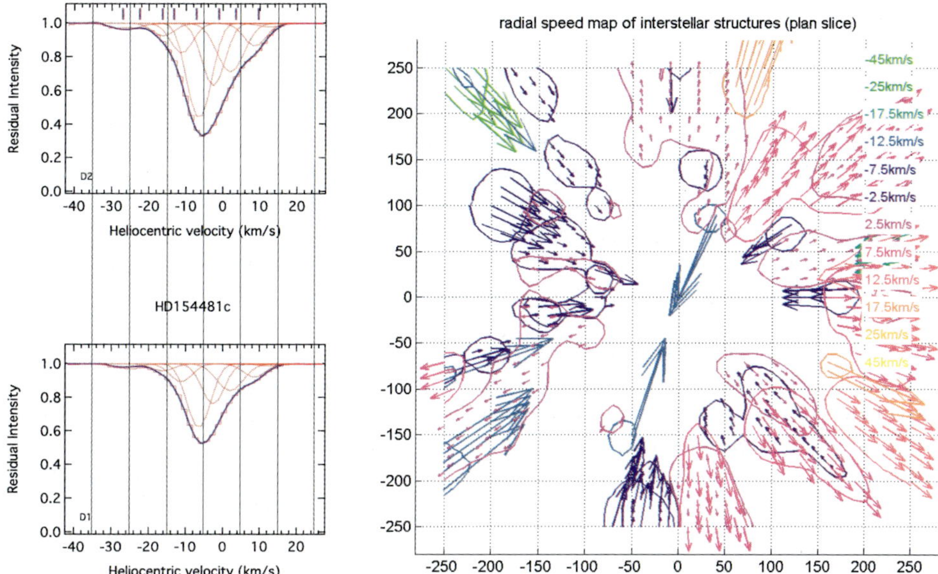


**Fig. 4** *Left*: Example of stellar spectrum and fitted interstellar sodium absorption lines (at the two doublet wavelengths (5889, 5895 Å). Here all detected absorptions have been distributed in 12 categories, according to their radial velocity value (*violet bars at top*). Each category corresponds to a given velocity interval. The velocity bins are shown by *vertical lines*. *Right*: The inversion program is applied separately to each velocity bin, using exclusively the clouds within this bin. The resulting density distributions are then added. A preliminary map is shown here. The Sun is at (0, 0), the galactic center is to the right, units are parsecs. When clouds at two different velocities are superimposed, this means that the actual clouds are smaller than on the figure (they are widened by the imposed correlation length) and adjacent

data in an inversion code which locates the clouds in 3D. The problem is largely under-determined because the number of sightlines is still too small, and as a consequence one has to assume a spatial correlation length for the gas. This means that gas concentrations are distributed within spheres of such a radius and that clumps actually smaller than this length appear wider. Accumulating more data should allow to improve the map resolution. The first attempt used a data set initially recorded with the aim of tracing the Local Bubble gas boundaries. Only total column-densities were used, i.e. the inversion was made without any information on the line-of-sight structure A density cube resulting from 1000 columns and for an assumed correlation length of 25 pc has been obtained (Lallement et al. 2003). The Local cavity is revealed, and appears to be connected through tunnels to nearby "bubbles". In the vertical plane containing the galactic center direction, the LB is found to be connected to the southern and northern halos by "chimneys" and interestingly inclined perpendicularly to the Gould Belt.

This "chimney" axis appears very clearly in the 0.25 keV X-ray background emission maps, which makes sense since low columns of gas allow a deeper penetration of the halo and extragalactic X-rays. For the unabsorbed, local emission, an excess by about a factor two is found to originate from the two chimneys in comparison with galactic plane directions (Snowden et al. 1998). It is thus believed picture that the LB extends over distances larger by about such a factor of two in those directions.

Figure 4 shows a preliminary result of a recent attempt to make use of the spectral information, i.e. for each line-of-sight, the locations of the line centroïds which are simply the projections onto the sightline of the 3D motions of the intercepted clouds. Unfortunately only this radial component is available and *not* the 3D motion of the cloud. There is no equivalent of the stellar proper motion, which, when added to the stellar radial motion, provides the 3D motion of the star. A very simplistic method has been used up to now: in a first step the set of absorptions is used to compute the average heliocentric motion of the group of detected clouds using a simple least squares method. This global motion is then subtracted for all components, i.e. all radial velocities are changed accordingly. In a second step, for each star absorbing clouds are distributed in a set of velocity bins according to their absorption velocity centroïd. Figure 4 illustrates this classification. Then in a third step the inversion program is simply run for each radial velocity bin separately. The inversion locates in 3D the corresponding clouds, and the clouds of all categories are combined. Figure 4 shows a preliminary result of the superimposed cloud distributions. Despite the smaller number of target stars (for some targets the spectrum is of insufficient quality to provide more than the total column), the map is more detailed than the previous one based on integrated columns solely. As a matter of fact this procedure uses significantly more information because each line-of-sight has about 2–3 clouds on average. This simplistic method needs improvement (differential galactic rotation, problems connected with the velocity interval boundaries, etc, ...) and more data are needed, especially those tracing the warm and more diffuse gas, which is not traced by neutral sodium. From now on however the increased information allows one to reduce the correlation length (20 pc in Fig. 4) and to distinguish the masses of gas with different radial motions. It is obvious from the distribution that at the LB boundary clouds with significant velocity differences are very close to each other. Correlative studies, in particular using motions and abundances should bring more information.

### 3.2 Solar Wind and Local Hot Gas: the SW Charge-Exchange Contribution

Soft X-ray diffuse emission from comets has been a recent surprising discovery (Lisse et al. 1996). X-rays are associated with hot media, while comets are among the coldest known objects. This "against nature" emission is now understood: it has been demonstrated that the X-ray emission follows charge exchange (CX) reactions between solar wind high ions and cometary neutrals (Cravens 2000). Captured electrons populate highly excited levels and the newly formed ions de-excite by emitting EUV/soft X-ray photons. While in the case of comets the neutrals are provided by the evaporating nucleus, any other type of neutral species coming in contact with solar wind high ions must generate the same type of emission. In particular neutrals from planetary exospheres (including the geocorona) and interstellar neutrals also interact with solar wind ions.

New generation solar wind experiments and especially ion mass spectrometers (e.g. von Steiger et al. 2000, and a number of works associating J. Geiss) have brought spectacular advances in the knowledge of the solar wind composition and ionization states. They provide the crucial ingredients for the estimates of the Solar Wind Charge Exchange (SWCX) diffuse emission generated by the interaction between the solar wind ions and the interstellar hydrogen and helium flows and the geocorona. This emission is found to be of the same order of magnitude than the diffuse soft X-ray background (Cox 1998; Cravens et al. 2001). It is now becoming increasingly clear that a non-negligible fraction of the local (unabsorbed) emission is of heliospheric origin, i.e. solar wind charge-exchange emission (or SWCX) and most arguments against this significant contribution such as the emission pattern and the temporal characteristics have been shown to be invalid once the phenomenon is modeled with increasing precision (e.g. Koutroumpa et al. 2006). Using XMM-Newton spectra Snowden et al. (2004) beautifully demonstrated that solar wind enhancements (see Fig. 5) can generate a strong additional emission, and provided a high quality spectrum of the SWCX emission.

The actual fraction of heliospheric and local bubble gas contributions to the background and their spatial variations is still a matter of controversy. While the high latitude 0.25 keV emission, and features like the spectacular Loop 1 seen in the 0.75 keV band are definitely not originating in the heliosphere, at low galactic latitudes and outside distinct IS features the situation is unclear. If the emission from the LB hot gas has been overestimated, and a large fraction of this emission is SWCX, this could help resolving a number of puzzling results. (i) The Local Hot Bubble gas pressure deduced from the X-rays is of the order of 5 times the pressure of the clouds embedded in it (local cloud and other cloudlets, Lallement 1998; Jenkins 2002). (ii) Column-densities of intermediate ions in the hot gas—clouds interfaces are low compared to models (e.g. Welsh and Lallement 2005). (iii) EUV and ultrasoft X-ray spectra of the diffuse emission are not consistent with a $10^6$ K gas thermal emission at the level predicted by the soft X-rays. In particular the absence of a number of spectral lines (Vallerga et al. 1993; McCammon et al. 2002; Hurwitz et al. 2005) implies either a low pressure or a very strong (and difficult to explain) depletion of iron.

The CX X-ray/EUV emission mechanism, which is extremely efficient, may be at work in many astrophysical situations, and in particular at interfaces between hot gas and moving cool clouds. Soft X-ray emission from high velocity clouds in the extended halo has indeed been detected with ROSAT (e.g. Kerp et al. 1998). The emission level is compatible with crude estimates of the CX emission from the layer of neutrals flowing through the hot gas (Lallement 2004). The X-ray background may also be affected by the contribution of such interfaces.

## 4 The ISM at the Parsec Scale: the Local Clouds

### 4.1 Local Cloud Properties

The local clouds within 10–20 parsecs have been studied in detail through absorption data in the visible and the UV. Their temperature is within the 5,000–10,000 K range, and densities are of the order of 0.1 $cm^{-3}$. They all belong to the common type of warm diffuse clouds, or intercloud medium, although there may be some colder nuclei within the clouds. The magnitude of the local cloud column-densities is such that in response to the radiation fields of the nearby hot stars ionization gradients must be present throughout the clouds. The velocity dispersion of the order of 5 $km\,s^{-1}$ is mostly subsonic. The LIC and the closest clouds have been mapped (Redfield and Linsky 2004; Linsky, this issue). There are signs of abundance variations from one cloudlet to the other, as if they had not all the same origin (e.g. Linsky et al. 1995; Linsky and Wood 1996). The Sun is presently leaving the LIC and will enter a cooler cloud. One of the unsolved questions is the nature of the medium filling the space between the detected clouds in the local group. Is it the hot gas from the LHB? Are the clouds contiguous, or only some of them? Are some of the velocity differences between adjacent clouds with the same abundance properties due to traveling shocks? The problem of the cloud-cloud interfaces is closely related to the hot gas status. Conductive interfaces between hot gas and clouds should be detected around the local cloudlets in all directions. Intermediate ions of such interfaces have been searched for with little success. CIV and SiIV have been found at local cloud velocities but only in the direction of very distant stars like epsilon CMa (Gry et al. 1995). The ionization models (Slavin and Frisch 2002; Frisch 2007)

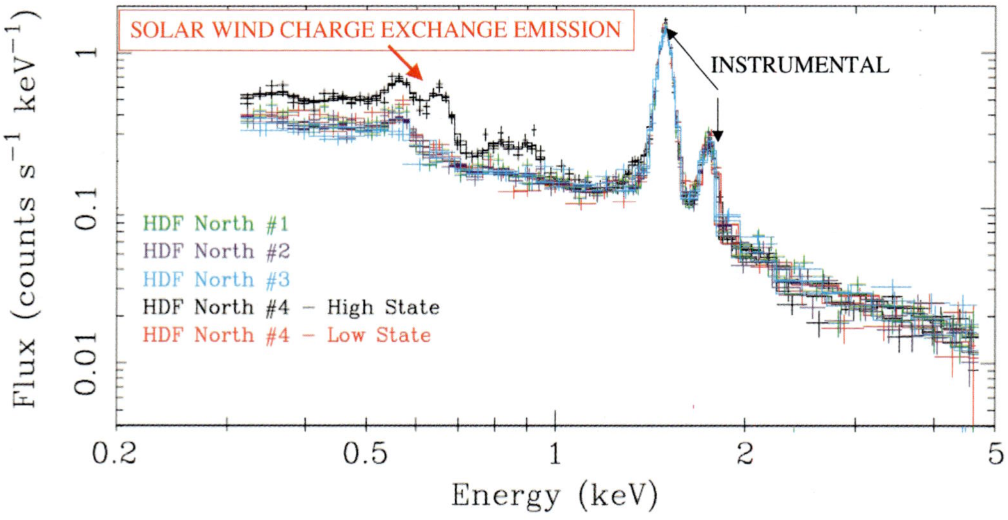

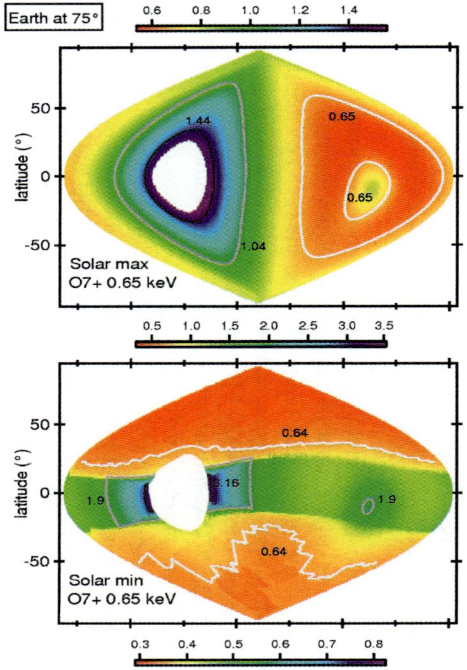

**Fig. 5** *Left*: A spectacular X-ray enhancement recorded by XMM-Newton, linked to a strong solar wind event (Snowden et al. 2004). One of the series of exposures on the same field (*in black*) shows distinctly the SWCX spectral features. *Right*: A example of monochromatic SWCX intensity maps (here the $O^{7+}$ line at 0.65 keV) for an average and stationary solar wind (Koutroumpa et al. 2006). The intensity pattern is very different at solar minimum due to the high latitude fast solar wind characteristics

show that such interfaces at temperatures of $10^5$ K are necessary to explain through their EUV emission the high ionization of helium inferred in the local clouds from absorption data towards nearby hot white dwarfs. The ionization rate is measured to be of the order of 30–40% (Dupuis et al. 1995) with no signs of variability within the group of clouds (Wolff et al. 1999). It is not clear yet whether a low pressure of hot gas or its lower temperature may influence the EUV diffuse emission and the helium ionization.

## 5 Abundances in and Around the Heliosphere: Lessons from Pickup Ions

Neutral atoms from the interstellar wind may or may not penetrate the heliosphere depending on the probability of experiencing charge transfer reactions with interstellar protons. After ionization they are deviated and excluded. This filtration process is specific to each atomic species and depends on the heliospheric interface characteristics. These processes are of primary importance because only the penetrating neutrals are ionized close to the Sun, are subsequently picked up by the solar wind, and finally accelerated in interplanetary shocks and at the termination shock to become the anomalous cosmic rays (ACRs). Cummings et al. (2002) have developed models of ionization and the resulting filtration factors for a number of species, including argon and neon. Using kinetic models, Izmodenov et al. (2004) predict that 54% of H atoms, about 100% of helium atoms, 68% of O atoms and 78% of N atoms succeed in entering the supersonic solar wind. Both ACR data and now pick-up ion data obtained with sophisticated ion mass spectrometers provide interstellar abundances of H, He and minor species within the heliosphere. Combining those data with the filtration factors allows one to extrapolate back to the abundances immediately outside of the heliosphere, and to compare with the interstellar abundances resulting from stellar spectroscopy. The former apply to the immediate surroundings, while the latter apply to the *average* ISM along the path-length to the star, or, when there are multiple cloud components, the fraction of path-length crossing an individual cloud. Some very good agreements however have been obtained. First, when hydrogen filtration is taken into account, and assuming the canonical H/He abundance ratio of 10, only those models assuming a significant helium ionization in the LIC are able to reproduce the measured H and He fluxes (e.g. Gloeckler and Geiss 2004). This agrees very well with the interstellar helium ionization quoted above (Wolff et al. 1999).

For oxygen and nitrogen, Gloeckler and Geiss (2004) derive from Ulysses-SWICS data $n(\mathrm{OI}) = (5.3 \pm 0.8) \times 10^{-5}\ \mathrm{cm}^{-3}$ and $n(\mathrm{NI}) = (7.8 \pm 1.5) \times 10^{-5}\ \mathrm{cm}^{-3}$ within the heliosphere, after the shock crossing. The application of the filtration factors gives $N(\mathrm{OI})_{\mathrm{LIC}} = (7.8 \pm 1.3) \times 10^{-5}\ \mathrm{cm}^{-3}$ and $N(\mathrm{NI})_{\mathrm{LIC}} = (1.1 \pm 0.2) \times 10^{-5}\ \mathrm{cm}^{-3}$ (Izmodenov et al. 2004). This corresponds to $N(\mathrm{OI})/N(\mathrm{HI})_{\mathrm{LIC}} = (4.3 \pm 0.5) \times 10^{-4}$ and $N(\mathrm{NI})/N(\mathrm{OI})_{\mathrm{LIC}} = 0.13 \pm 0.01$. In the case of oxygen, charge-exchange reactions contribute to establishing a constant OI/HI ratio whatever the ionization gradients. This is why the excellent agreement between the above abundance immediately outside of the heliosphere and the line-of-sight averaged value $\mathrm{OI/HI} = (4.8 \pm 0.48) \times 10^4$ derived for the LIC towards Capella (Linsky et al. 1995) is particularly meaningful.

Among the ACR results, one of the most spectacular is the derivation of the neutral Argon LIC abundance (Cummings et al. 2002). It is interesting to note that the derived density of $3.5 \pm 1.6 \times 10^{-7}\ \mathrm{cm}^{-3}$ is in good agreement with recent and most precise determinations from FUSE spectroscopy. For the closest target stars, within the LB, the average ArI/OI ratio from FUSE is $\sim 4.7 \times 10^{-3}$ (Lehner et al. 2003). This translates to $n(\mathrm{ArI}) = 4.1 \times 10^{-7}\ \mathrm{cm}^{-3}$ in the LIC using the PUI-derived OI value quoted above. Despite the sensitivity of the argon abundance to the ionization field, the circumsolar interstellar abundance is very close to the average abundance integrated along path-lengths through the LB clouds.

Finally, while this type of heliospheric-interstellar agreement demonstrates a good understanding of the ISM/solar wind interaction, there are on the other hand some very interesting discrepancies. The still mysterious inner and outer sources of pickup ions and ACR's (e.g. Schwadron et al. 2002; Cummings et al. 2002) are revealed by a lack of compatibility with an interstellar origin for some species. These discoveries certainly deserve more investigations.

## 6 Conclusion and Acknowledgements

Progresses in the understanding of the local interstellar medium at all scales, from solar system to galactic scale, have all benefited and still benefit from the investment and the creativity of Prof. Johannes Geiss. These advances are all characterized by a fruitful interdisciplinarity, something Prof. Geiss has always advocated for. Thanks, Johannes, for all these advances and the inspiration and enthusiasm you have communicated around you.

## References

F. Comerón, J. Torra, Astron. Astrophys. **281**, 35 (1994)
D.P. Cox, Lect. Notes Phys. **506**, 121 (1998)
T.E. Cravens, Astrophys. J. **532**(2), L153 (2000)
T.E. Cravens, I.P. Robertson, S.L.J. Snowden, Geophys. Res. **106**(A11), 24883 (2001)
A.C. Cummings, E.C. Stone, C.D. Steenberg, Astrophys. J. **578**, 194 (2002)
M. De Avillez, D. Breitschwerdt, Astron. Astrophys. **436**, 585 (2005)
B.T. Draine, Annu Rev. Astron. Astrophys. **41**, 241 (2003)
F. Elias, J. Cabrera-Caño, E.J. Alfaro, Astron. J. **131**(5), 2700 (2006a)
F. Elias, E.J. Alfaro, J. Cabrera-Caño, Astron. J. **132**(3), 1052 (2006b)
J. Dupuis, S. Vennes, S. Bowyer, A.-K. Pradhan, P. Thejll, Astrophys. J. **455**, 574 (1995)
P.C. Frisch, Space Sci. Rev. (2007) this volume
J. Geiss, G. Gloeckler, C. Charbonnel, Astrophys. J. **578**, 863 (2002)
G. Gloeckler, J. Geiss, Space Sci. Rev. **84**, 475 (1998)
G. Gloeckler, J. Geiss, Space Sci. Rev. **97**, 169 (2001)
G. Gloeckler, J. Geiss, Adv. Space Res. **34**, 53 (2004)
C. Gry, L. Lemonon, A. Vidal-Madjar, M. Lemoine, R. Ferlet, Astron. Astrophys. **302**, 497 (1995)
G. Hébrard, H.W. Moos, Astrophys. J. **599**, 297 (2003)
G. Hébrard, T.M. Tripp, P. Chayer, S.D. Friedman, J. Dupuis, P. Sonnentrucker, G.M. Williger, H.W. Moos, Astrophys. J. **635**(2), 1136 (2005)
M. Hurwitz, T.P. Sasseen, M.M. Sirk, Astrophys. J. **623**(2), 911 (2005)
V. Izmodenov, Y. Malama, G. Gloeckler, J. Geiss, Astron. Astrophys. **414**, L29–L32 (2004)
E.B. Jenkins, Astrophys. J. **580**, 938 (2002)
J. Kerp, J. Pietz, P.M.W. Kalberla, W.B. Burton, R. Egger, M.J. Freyberg, D. Hartmann, U. Mebold, Lect. Notes Phys. **506**, 457 (1998)
D. Koutroumpa, R. Lallement, V. Kharchenko, A. Dalgarno, R. Pepino, V. Izmodenov, E. Quémerais, Astron. Astrophys. **460**, 289 (2006)
R. Lallement, Lect. Notes Phys. **506**, 19 (1998)
R. Lallement, Astron. Astrophys. **422**, 391 (2004)
R. Lallement, B.Y. Welsh, J.L. Vergely, F. Crifo, D. Sfeir, Astron. Astrophys. **411**, 447 (2003)
N. Lehner, E.B. Jenkins, C. Gry, H.W. Moos, P. Chayer, S. Lacour, Astrophys. J. **595**, 858 (2003)
J. Linsky, this issue (2007)
J.L. Linsky, A. Diplas, B.E. Wood, A. Brown, T.R. Ayres, B.D. Savage, Astrophys. J. **451**, 335 (1995)
J. Linsky, B. Wood, Astrophys. J. **463**, 254 (1996)
J. Linsky, B. Draine, W. Moos et al., Astrophys. J. **647**, 1106 (2006)
C.M. Lisse, K. Dennerl, J. Englhauser et al., Science **274**(5285), 205 (1996)

McCammon et al., Astrophys. J. **576**(1), 188 (2002)
C.A. Olano, Astron. Astrophys. **112**, 195 (1982)
C.A. Olano, Astrophys. J. **121**, 295 (2001)
C. Perrot, I. Grenier, Astron. Astrophys. **404**, 519 (2003)
W.G.L. Pöppel, Fundam. Cosm. Phys. **18**, 1 (1997)
N. Prantzos, this volume (2007)
S. Redfield, J.L. Linsky, Astrophys. J. **602**(2), 796 (2004)
N.A. Schwadron, M. Combi, W. Huebner, D.J. McComas, Geophys. Res. Lett. **29**(20), 54 (2002)
J.D. Slavin, P.C. Frisch, Astrophys. J. **565**(1), 364 (2002)
S.L. Snowden, R. Egger, D.P. Finkbeiner, M.J. Freyberg, P.P. Plucinsky, Astrophys. J. **493**, 715 (1998)
S.L. Snowden, M.R. Collier, K.D. Kuntz, Astrophys. J. **610**(2), 1182 (2004)
J. Vallerga, P. Vedder, N. Craig, B.Y. Welsh, Astrophys. J. **411**, 729 (1993)
H. Völk, this volume (2007)
R. von Steiger, N.A. Schwadron, L.A. Fisk, J. Geiss, G. Gloeckler, S. Hefti, B. Wilken, R.F. Wimmer Schweingruber, T.H. Zurbuchen, J. Geophys. Res. **105**(A12), 27217 (2000)
B.Y. Welsh, R. Lallement, Astron. Astrophys. **436**(2), 615 (2005)
B. Wolff, D. Koester, R. Lallement, Astron. Astrophys. **346**, 969 (1999)

Space Sci Rev (2007) 130: 355–365
DOI 10.1007/s11214-007-9209-z

# The Local Bubble and Interstellar Material Near the Sun

**P.C. Frisch**

Received: 22 February 2007 / Accepted: 11 May 2007 / Published online: 14 July 2007


**Abstract** The properties of interstellar matter at the Sun are regulated by our location with respect to a void in the local matter distribution, known as the Local Bubble. The Local Bubble (LB) is bounded by associations of massive stars and fossil supernovae that have disrupted dense interstellar matter (ISM), driving low density intermediate velocity ISM into the void. The Sun appears to be located in one of these flows of low density material. This nearby interstellar matter, dubbed the Local Fluff, has a bulk velocity of $\sim$19 $\mathrm{km\,s^{-1}}$ in the local standard of rest. The flow is coming from the direction of the gas and dust ring formed where the Loop I supernova remnant merges into the LB. Optical polarization data suggest that the local interstellar magnetic field lines are draped over the heliosphere. A longstanding discrepancy between the high thermal pressure of plasma filling the LB and low thermal pressures in the embedded Local Fluff cloudlets is partially mitigated when the ram pressure component parallel to the cloudlet flow direction is included.

**Keywords** ISM: general · ISM: abundances

## 1 Introduction

The existence of an area clear of interstellar material around the Sun, now known as the Local Bubble, was discovered as an underdense region in measurements of starlight reddening (Fitzgerald 1968). This underdense region is traced by color excess measurements showing $E(B-V) < 0.05$ mag,[1] and extends beyond 100 pc in the galactic longitude interval $\ell = 180$–$270°$. In the plane of Gould's Belt, the Local Bubble boundaries ("walls") are defined by interstellar material (ISM) associated with star-forming regions. At high galactic

[1] $E(B-V) = A_\mathrm{B} - A_\mathrm{V}$, where $A_\mathrm{B,V}$ is the attenuation in units of magnitude in the blue (B) and visible (V) bands, respectively.

P.C. Frisch (✉)
Dept. of Astronomy and Astrophysics, University of Chicago, 5640 S. Ellis Ave., Chicago, IL 60637, USA
e-mail: frisch@oddjob.uchicago.edu

latitudes the Local Bubble boundaries are defined by interstellar gas and dust displaced by stellar evolution, particularly supernova in the Scorpius–Centaurus Association. Supernovae exploding into pre-existing cavities created by massive star winds displace ISM and the interstellar magnetic field into giant magnetized bubbles hundreds of parsecs in extent. The location of the Sun within such a void regulates the interstellar radiation field at the Sun, and the composition and properties of the ISM surrounding the heliosphere.

The Local Interstellar Cloud (LIC), defined by the velocity of interstellar He° inside of the heliosphere, is one cloudlet in a low density ISM flow known as the Local Fluff. The Local Fluff has an upwind direction towards Loop I and the Scorpius–Centaurus Association (SCA). This flow, with a best-fit local standard of rest (LSR) velocity of $\approx$19.4 km s$^{-1}$, appears to be a breakaway fragment of the Loop I superbubble shell surrounding the SCA, which has expanded into the low density interior of the Local Bubble (Sect. 2.1, Frisch 1981, 1995; Breitschwerdt et al. 2000).

This paper is in honor of Prof. Johannes Geiss, founder the International Space Sciences Institute (ISSI). Many of the contemporary space topics discussed at ISSI meetings, such as the heliosphere, the Local Interstellar Cloud, cosmic ray acceleration and propagation, and the composition of matter, are influenced by the solar location inside of the Local Bubble.

## 2 Origin and Boundaries of the Local Bubble

### 2.1 Origin

The Local Bubble void was created by star formation processes that occurred during the past 25–60 Myrs in the corotating region of the Milky Way Galaxy near the solar location of today. About 25–60 Myrs ago, a blast wave evacuated a low density region at the present location of the Sun, and compressed surrounding molecular clouds to initiate the formation sequence of the massive OB stars now attributed to Gould's Belt. Gould's Belt denotes the system of kinematically related massive OB stars within $\sim$500 pc of the Sun, which form a localized plane tilted by $\sim$18° with respect to the galactic plane. The center of Gould's Belt is 104 pc from the Sun towards $\ell = 180°$, and with an ascending node longitude of 296° (Grenier 2004). The Sun is moving away from the center of Gould's Belt, and is closest to the Scorpius–Centaurus rim. Overlapping superbubbles shape the Local Bubble void (Frisch 1995; Heiles 1998; Maíz-Apellániz 2001).

Since the formation of Gould's Belt, the Sun has traveled hundreds of parsecs through the LSR, and the LSR has completed $\sim$10–25% of its orbit around the galactic center. Molecular clouds disrupted by the initial blast wave now rim Gould's Belt. Epochs of star formation in the Scorpius–Centaurus Association during the past 1–15 Myrs further evacuated the Local Bubble void, and displaced ISM from the SCA into giant, nested H° shells (de Geus 1992). One of these shells, Loop I (the North Polar Spur), was formed by a recent supernova ($<$1 Myrs ago) and is an intense source of polarized synchrotron and soft X-ray (SXR) emission. A ring-like shadow, caused by foreground ISM, is seen in the Loop I SXR emission. The origin of this ring has been suggested to be the result of Loop I merging with the separate Local Bubble (e.g. Egger and Aschenbach 1995). The LSR upwind direction of the Local Fluff is at the center of this ring, and polarization data show that the ring is a magnetic loop (Fig. 1).[2]

[2]In accordance with general practice, I use the Standard solar apex motion (velocity 19.7 km s$^{-1}$, towards $\ell = 57°$ and $b = 22°$) to correct heliocentric velocities to the LSR. This gives a Local Fluff LSR upwind

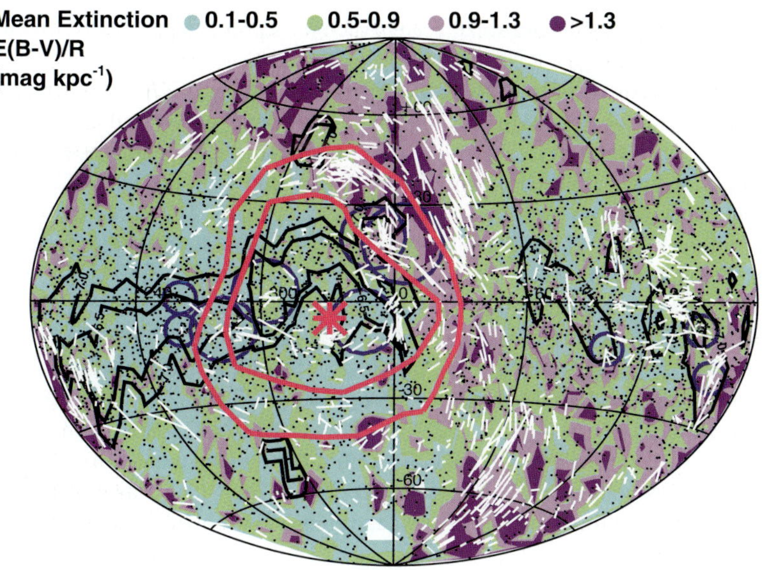


**Fig. 1** Mean extinction, $E(B-V)/R$, for stars with $R < 0.5$, where $R$ is the distance in kpc. *Black contours* show the integrated stellar radiation at 1565 Å measured by the TD-1 satellite (Gondhalekar et al. 1980). Contours indicate flux levels of $10^{-7}$ and $10^{-6.5}$ ergs cm$^{-2}$ s$^{-1}$ Å$^{-1}$ sterad$^{-1}$. Polarization data for stars within 500 pc of the Sun are plotted as white bars (Heiles 2000). *Pink contours* show the ring that may be formed by the Loop I supernova remnant interaction with the Local Bubble (Egger and Aschenbach 1995). The *pink asterisk* indicates the LSR upwind direction of the Local Fluff (see text). *Blue circles* show stellar OB associations within 500 pc of the Sun. An Aitoff projection is used, with the galactic center at the center of the plot, and $\ell$ increasing to the right. Note that the ISM density is very low, out to 500 pc in the interval $\ell = 180° \rightarrow 300°$, particularly towards low galactic latitudes. This plot is based on photometric data in the Hipparcos catalog. The stars (*small black dots*) are preselected for photometric quality, and color excess values are smoothed for stars with overlapping distance uncertainties and within 13° of each other on the sky. The average distances of stars sampling latitudes in the intervals $|b| <$ [30°, 30°–60°, >60°] are [210, 164, 138] pc, so that the ISM in high-latitude sightlines tends to be closer. Uncertainties on $E(B-V)$ for the plotted stars are typically <0.02 mag because stars are also required to have photometric distances that are consistent with astrometric distances to within ~30%. Intrinsic stellar colors are from Cox (2000)

## 2.2 Boundaries

The locations of the Local Bubble boundaries have been diagnosed with a range of different ISM markers, including color excess (Lucke 1978), ultraviolet observations of interstellar H° lines in hot stars (Frisch and York 1983; Paresce 1984), radio H° 21-cm and optical Na° data (Vergely et al. 2001), extreme ultraviolet (EUV, $\lambda < 912$ Å) emission of white dwarf and M-stars (Warwick et al. 1993), measurements of polarization of starlight (Leroy 1999), and the trace ionization species Na° (Sfeir et al. 1999; Lallement et al. 2003). These studies differ in sampling densities and spatial smoothing methods. Each marker is an imprecise tracer of the total ISM mass density, since the ISM is highly inhomogeneous over the scale

direction towards $\ell, b = 331°, -5°$, and LSR velocity $-19.4$ km s$^{-1}$. In Frisch et al. (2002), we instead used an LSR conversion based on Hipparcos data, and found an LSR upwind direction $\ell = 2.3°$, $b = -5.2°$ (velocity $-17$ km s$^{-1}$), which is directed towards the ISM shadow itself. This difference in the two possible Local Fluff upwind directions reveals an obvious flaw in comparing a dynamically defined direction with a statically defined diffuse object such as the more distant SXRB shadow.

lengths of the Local Bubble and small scale structure is poorly understood. I focus here on the reddening data.

An accessible measure of starlight reddening is color excess, $E(B-V)$, which measures the differential extinction of starlight in the blue versus visual bands, and is sensitive to interstellar dust grains (ISDG) of radii $a \sim 0.20$ μm. Interstellar gas and dust are generally well mixed, so that the threshold reddening for the Local Bubble walls found from $E(B-V)$ data is consistent with the locations found from gas markers. The exception is that dust is found in both neutral and ionized regions, while the commonly available gas markers (H°, Na°) are weighted towards neutral regions.

Grains and gas are well mixed partly because both populations couple to the interstellar magnetic field ($B_{\rm IS}$) in cold and warm clouds. In cold clouds with density $n(\rm H) = 100$ cm$^{-3}$ and temperature $T \sim 100$ K, the $a \sim 0.2$ μm dust grain with density 2 g cm$^{-3}$ will sweep up its own mass in gas in ~0.08 Myrs. If the same grain has charge $Z = 20$, the gyrofrequency for a magnetic field of strength $B = 2.5$ μG is ~1/3300 yrs. For a warm neutral cloud ($n(\rm H) \sim 0.25$ cm$^{-3}$, $T \sim 6300$ K), the grain accumulates its own mass in ~4 Myrs. In both cloud types, grains couple to $B_{\rm IS}$. Gas also couples to $B_{\rm IS}$, since elastic collisions couple neutrals and ions over time-scales of years, and minimum ionization levels of $\sim 10^{-4}$ bind gas to $B_{\rm IS}$ (Spitzer 1978).

In Fig. 1, the reddening per unit distance, $E(B-V)/R$, where $R$ is the star distance in kpc, is shown on an Aitoff projection for O, B, and A stars within 500 pc. $E(B-V)$ values are based on Hipparcos photometric data (Perryman 1997). The lowest mean ISM densities in the galactic plane are between longitudes of 210° and 360° are evident. Star groups (blue circles) in the low density sightlines include Sco OB2, Vela OB2, and Trumpler 10 (de Zeeuw et al. 1999). The lowest mean densities in this data set, outside of the Local Fluff, correspond to 0.006 atoms cm$^{-3}$. At the galactic poles, $|b| > 75°$, the edges of the Local Bubble, where $E(B-V) > 0.05$ mag (or approximately log $N(\rm H) > 20.50$ cm$^{-2}$),[3] are at 80–95 pc towards both the north and south poles.

The LB boundaries in the galactic plane are shown in Fig. 2, for ~2000 O, B, and A stars within ~200 pc and 45° of the galactic plane, using a threshold cumulative value of log $N(\rm H) > 20.6$ cm$^{-2}$ corresponding to $E(B-V) > 0.07$ mag when $N(\rm H)/E(B-V) \sim 5.8 \times 10^{21}$ cm$^{-2}$ mag$^{-1}$ K. This gas-to-dust ratio is good to within factors of ~2 for $E(B-V) > 0.1$ mag and ~3 for $E(B-V) < 0.1$ mag (Bohlin et al. 1978). Note the well known deficiency of ISM out to distances beyond 200 pc in the third and parts of the fourth galactic quadrants (Frisch and York 1983). For cloudy sightlines (high mean $E(B-V)$ values), the fraction of the H atoms in $H_2$ ($f_{\rm H_2}$) rises above ~1% at $E(B-V) \sim 0.1$ mag. The classic term "intercloud" refers to low column density sightlines with relatively little $H_2$ ($f_{\rm H_2} <$ 1%). Molecular clouds of CO and $H_2$ are also shown, and are plotted as filled red circles (Dame et al. 2001). Well-known molecular clouds at the rim of the Local Bubble include dust in Scorpius ($\ell \sim 350°$, $d \sim 120$ pc), Taurus ($\ell \sim 160°$, $d \sim 120$ pc), and Chameleon ($\ell \sim 305°$, $b \sim -15°$, $d \sim 165$ pc).

The mean value of $E(B-V)/(N(\rm H°) + 2N(\rm H_2))$ varies by ~15% between sightlines with low and high fractions of $H_2$ (Bohlin et al. 1978), because of variations in the mean grain size and radiation field. The $\lambda \sim 1565$ Å radiation field depends on location with respect to the Local Bubble walls (Sect. 4.1), and the 912–1108 Å radiation field capable of photodissociating $H_2$ should behave in a similar fashion.

[3]The ratio $E(B-V)/N(\rm H°)$ varies in sightlines with low mean extinctions because of variations in both mean grain sizes and hydrogen ionization.

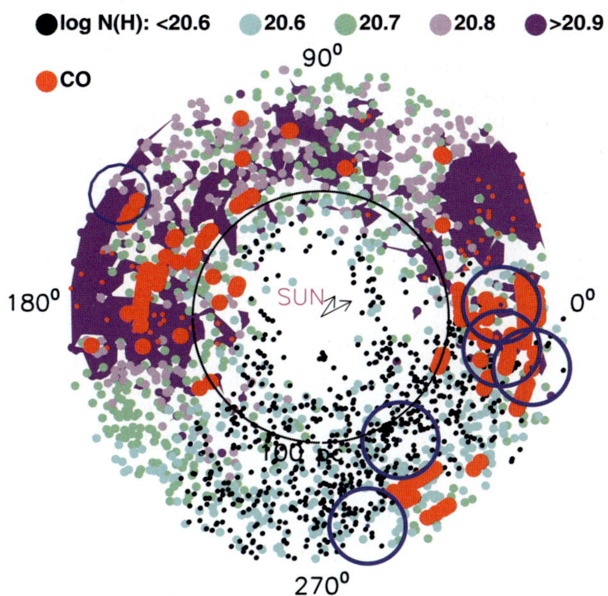


**Fig. 2** The distribution of ISM surrounding the Local Bubble void, based on stars within 200 pc and within 45° of the galactic plane. Molecular clouds of CO and $H_2$ are plotted as *red symbols* (Dame et al. 2001). The *colored dots* show cumulative hydrogen column densities towards O, B, and A stars based on $E(B-V)$ (Fig. 1) and the gas-to-dust ratio measured by the *Copernicus* satellite for stars with low mean extinctions (see text). The mean gas-to-dust relation overestimates $N$(H) at low column densities ($N(\text{H}) < 10^{20}$ cm$^{-2}$; (Bohlin et al. 1978)). Sightlines with $N(\text{H}) > 10^{20.9}$ cm$^{-2}$ have been plotted with *filled purple contours*. The *arrows* show two different values for the direction of the Sun's motion through the LSR, with the *longer arrow* ($v = 19.5$ km s$^{-1}$) indicating the Standard solar apex motion. *Blue circles* indicate OB associations within 200 pc of the Sun. The *black circle* indicates 100 pc

## 3 Loop I and the Local Magnetic Field

### 3.1 Loop I and the High-Latitude Limits of the Local Bubble

Above the galactic plane in the galactic-center hemisphere, $b > 20°$, the LB walls are established by neutral gas of the Loop I superbubble. The interval $\ell \sim 270° \rightarrow 50°$ is encircled by high-latitude nested shells of gas and dust. Loop I is $\sim$80° in radius and centered 120 pc away at $\ell = 320°$, $b = 5°$ for the neutral gas (Berkhuijsen et al. 1971; Heiles 1998; de Geus 1992). The central regions of these evacuated shells are deficient in ISM, creating the extension of the Local Bubble towards $\ell \sim 340°$ (Fig. 2).

### 3.2 Magnetic Field

Loop I dominates the magnetic field structure near the Sun, and is a source of intense radio continuum and soft X-ray emission. The Loop I magnetic field, comprised of components parallel ($B_{\parallel}$) and perpendicular ($B_{\perp}$) to the sightline, is traced by polarized synchrotron emission, starlight polarization caused by magnetically aligned dust grains, Faraday rotation, and Zeeman splitting of the H° 21-cm line. Figure 1 shows the starlight polarization vectors (from Heiles 2000). Magnetically aligned interstellar dust grains (ISDGs) are birefringent at optical wavelengths, with lower opacities found for the polarization component parallel

to $B_{\mathrm{IS}}$. The Loop I magnetic field direction is shown by the gradient in the rotation angle of the optical polarization vectors, which follows the interaction ring feature. Comparisons between the optical polarization data (tracing $B_{\parallel}$) and synchrotron emission (tracing $B_{\perp}$), indicate that $B_{\mathrm{IS}}$ is nearly in the plane of the sky in Loop I (Berkhuijsen et al. 1971; Heiles and Crutcher 2005).

The closest measured $B_{\mathrm{IS}}$ strengths are towards Loop I. Heiles et al. (1980) found a volume-averaged field strength of $B_{\mathrm{IS}} \sim 4$ μG in a tangential direction through the shell (extending $\sim 70 \pm 30$ pc towards $\ell = 34°$, $b = 42°$). Faraday rotations of extragalactic radio sources indicate that $B_{\parallel}$ is small, with an average value of $B_{\parallel} = 0.9 \pm 0.3$ μG from rotation measure data (Frick et al. 2001). Magnetic pressure dominates in the neutral shell gas. In the ionized gas producing the radio continuum emission, the magnetic, gas, and cosmic ray pressures are all significant. Loop I is a decelerated shock generated by sequential epochs of star formation in SCA (de Geus 1992).

### 3.3 The LIC and the Magnetic Field at the Sun

The LIC is very low density, $n(\mathrm{H}) \sim 0.25$–$0.30$ cm$^{-3}$. Magnetic fields in high density ISM show evidence of flux freezing; however, $B_{\mathrm{IS}}$ in low density ISM appears uncorrelated with density (Heiles and Crutcher 2005). Pulsar dispersion measures indicate that the uniform component of the magnetic field near the Sun is $B_{\mathrm{IS}} \sim 1.4$ μG, with correlation lengths of $\sim 100$ pc (Rand and Kulkarni 1989). In general, structure functions created from data on radio continuum polarization near 21 cm show that magneto-ionized structures in interarm sightlines must be very large (e.g. $\sim 100$ pc; Haverkorn et al. 2006). This would indicate that the uniform $B_{\mathrm{IS}}$ component is appropriate for the low density (similar to interarm) region around the Sun.

The physical conditions of the LIC have been modeled by developing a series of radiative transfer models that are constrained by observations of He° and pickup ion and anomalous cosmic ray data inside of the heliosphere, and observations of the LIC towards $\epsilon$ CMa. These models are discussed in detail by Slavin and Frisch (this volume 2007a, 2007b). The best of these models give $n(\mathrm{H}°) = 0.19$–$0.20$ cm$^{-3}$, $n(\mathrm{e}) = 0.07 \pm 0.02$ cm$^{-3}$, $n(\mathrm{He}°) = 0.015$ cm$^{-3}$, for cloud temperatures $\sim 6300$ K. If the magnetic and gas pressures are equal in the LIC, then the LIC field strength is $B_{\mathrm{LIC}} \sim 2.8$ μG. This value is also consistent with the interface magnetic field strength of 2.5 μG, adopted in the best model (model 26). However, it is somewhat above the strength of the uniform component of $B_{\mathrm{IS}}$. Since the ISM flow past the Sun has an origin associated with the breakaway of a parcel of ISM from the Loop I magnetic superbubble (Frisch 1981), perhaps $B_{\mathrm{IS}}$ at the Sun is stronger and perturbed compared to the uniform field, but at lower pressure than the confined parts of the Loop I bubble.

Very weak interstellar polarization caused by magnetically aligned dust grains has been observed towards stars within $\sim 35$ pc (Tinbergen 1982). The polarization was originally understood to arise in the Local Fluff, since the polarization region coincides with the upwind direction of the flow where column densities are highest. More recently, the polarization properties were found to have a systematic relation to ecliptic geometry. The region of maximum polarization is found to be located at ecliptic longitudes that are offset by $\sim +35°$ from the large dust grains flowing into the heliosphere, and from the gas upwind direction (Fig. 3; Frisch et al. 1999). Stars with high polarizations also show consistent polarization angles, and in general polarization is higher for negative ecliptic latitudes. These polarization data are consistent with the interpretation that polarizing grains are trapped in $B_{\mathrm{IS}}$ as it drapes over the heliosphere (Frisch 2005, 2006). When magnetically prealigned (by $B_{\mathrm{IS}}$)

grains approach the heliosphere, the gas densities are too low to collisionally disrupt the alignment, and polarization should indicate the direction of $B_{IS}$ at the heliosphere. If the alignment mechanism is sufficiently rapid, the alignment strength and direction will also adjust to the interstellar magnetic field direction as it drapes over the heliosphere. Although this interpretation of the polarization data is not confirmed, it fits the physics of dust grains interacting with the heliosphere. Small charged grains such as those that polarize starlight ($a < 0.2$ μm) couple to $B_{IS}$ and are excluded from the heliosphere, while large grains enter the heliosphere where they are measured by various spacecraft (Krueger et al. 2007, this volume). The characteristics of such polarization may vary with solar cycle phase.

## 4 Radiation Environment of the Local Bubble

### 4.1 Inhomogeneous Radiation Field and Local Fluff Ionization

The interstellar radiation field (ISRF) is key to understanding the physical properties of the LIC and Local Fluff. The sources of the ISRF at the Sun include plasma emission from the Local Bubble interior and supernova remnants, stellar radiation, including from hot white dwarf stars, and emission from a conductive interface between the local fluff and the hot plasma. The spectrum of this field at the surface of the LIC is shown in Slavin and Frisch (2007b, this volume).

The spectrum of the ISRF is inhomogeneous because of the energy-dependent opacity of the ISM. For instance, radiation with $\lambda < 912$ Å (584 Å) determines the ionizations of H (He). Energetic photons capable of ionizing H° (He°) require $N(\mathrm{H}^\circ) \sim 17.2$ (17.7) cm$^{-2}$ to reach an opacity $\tau \sim 1$. The dependence of $\tau_{912\mathrm{A}}/\tau_{504\mathrm{A}}$ on $N(\mathrm{H}^\circ)$ drives the need for LIC photoionization models to determine the heliosphere boundary conditions. Stars within $\sim 10$ pc (e.g., Wood et al. 2005) show local column density variations of $\log N(\mathrm{H}^\circ) \sim 17.07$–$18.22$ cm$^{-2}$ dex (assuming log D/H$= -4.7$). This yields a range locally of $\tau_{912\mathrm{A}} = 0.7$–$10.5$, and shows that ionization must vary between the individual cloudlets comprising the Local Fluff. This variation is confirmed by $\mathrm{N}^+$ data, which are excellent tracers of $\mathrm{H}^+$ through charge-exchange. Stars within 70 pc show $\mathrm{N}^+/\mathrm{N}^\circ \sim 0.1$–$2$ (Slavin and Frisch 2007a, 2007b). Our LIC radiative transfer models indicate that in the LIC H° ionization provides ~66% of the cloud heating, and the LIC is ~20–30% ionized (Slavin and Frisch 2007a, 2007b).

Another example of the inhomogeneous ISRF is provided by the photon flux at $\lambda \sim$ 1565 Å, which TD-1 satellite data show depends on position in the Local Bubble (Gondhalekar et al. 1980). The ISRF at 1565 Å is dominated by hot stars, B or earlier. Radiation at $\lambda \leq 1620$ Å regulates the photoionization rate of interstellar Mg°, and is an important parameter for the $\mathrm{Mg}^+/\mathrm{Mg}^\circ$ diagnostic of the interstellar electron density. The ISDG albedo at $\lambda \sim 1565$ Å is ~0.5. Figure 1 shows the flux of 1565 Å photons at the Sun, plotted as black contours. The brightest regions of the sky at 1565 Å are in the third and fourth galactic quadrants, $\ell \sim 180°$–$360°$, where the mean extinction in the interior of the Local Bubble is low, $E(B-V)/D < 0.4$ mag kpc$^{-1}$.

The fact that the 1565 Å radiation field is enhanced near the galactic plane for $\ell \sim 180 \rightarrow 360°$ is relevant to our understanding of the more energetic photons associated with the soft X-ray background (SXRB). Isolated bright SXRB regions are seen, such as the Orion-Eridanus and Loop I enhancements, however no regional enhancement in the SXRB flux is seen corresponding to the bright $\lambda$ 1565 Å regions.

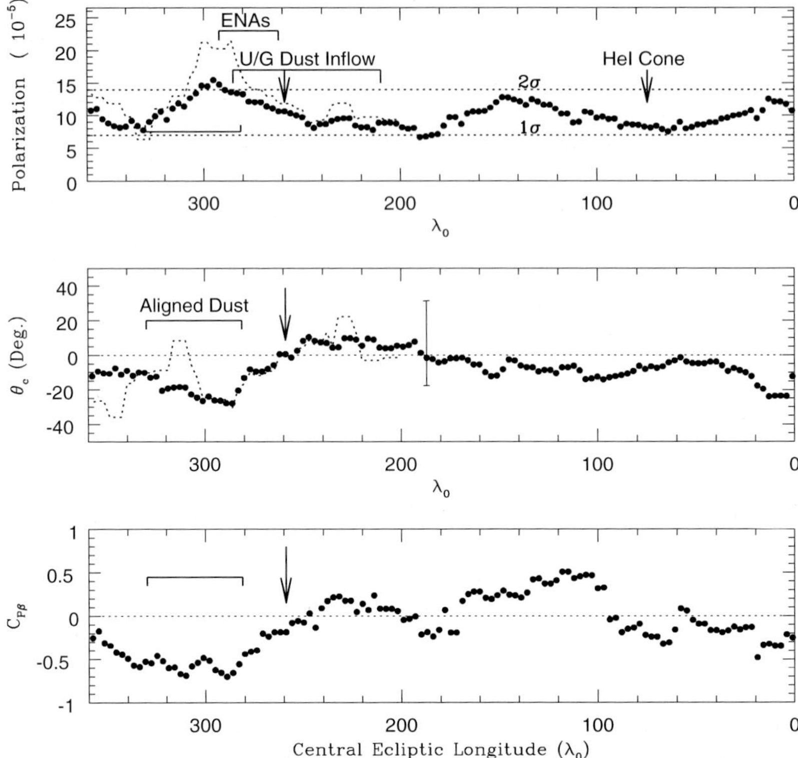


**Fig. 3** Interstellar polarization towards nearby stars (data from Tinbergen 1982) compared to the ecliptic position of the star. A systematic enhancement of the polarization strength is found close to the ecliptic at an offset in ecliptic longitude $\lambda$ of $\sim+35°$ compared to the inflowing upwind gas and large dust grain directions. *Top*: The average polarization $P$ for stars with $|\beta| < 50°$ is plotted as *dots*, and for stars with $|\beta| < 20°$ as a *dashed line*. Data are averaged over $\pm 20°$ around the central ecliptic longitude, $\lambda_o$. The direction of maximum $P$ is shifted by $\sim$25–30° from the upwind direction of the large interstellar dust grains detected by Ulysses/Galileo (Frisch et al. 1999). The upwind gas and large-grain directions differ by $<5°$. *Middle*: The averaged polarization position angle in celestial coordinates, $\Theta_C$. In the region of maximum polarization, $\lambda \sim 280° \rightarrow 310°$, the grains show consistent position angles. *Bottom*: The correlation coefficient between $P$ (*top*) and $\beta$ is shown as a function of the ecliptic longitude. The strongest polarization is found at negative ecliptic latitudes. For more details see Frisch (2005, 2006)

## 4.2 Diffuse Soft X-Ray Background

The diffuse soft X-ray background (SXRB) is significant both as an ionizing and heating radiation field, and as a fossil that traces the supernovae that formed the Local Bubble.

The spectrum of the soft X-ray background (SXRB) emission determined using broad-band sounding rocket observations at energies $\sim$0.08–0.2 keV revealed an excess of count rates at low energies in the galactic plane, compared with the spectrum predicted by an absorbed $T \sim 10^6$ K plasma. This effect was interpreted as indicating a local X-ray plasma with less than $\sim 7 \times 10^{18}$ cm$^{-2}$ of foreground hydrogen (Juda et al. 1991). The complex SXRB spectrum includes this local hot plasma, inferred to fill the LB, along with contributions from absorbed galactic halo emission, extragalactic emission, and absorbed local supernova remnants such as Loop I which dominates the northern hemisphere sky.

Fast-forwarding to the present, the LB contributions to the SXRB have been found from broadband ROSAT data (0.18–0.3 keV) and XMM Newton spectra of dark clouds that help isolate the LB flux from foreground and background fluxes, combined with SXRB models that include foreground contamination from charge-exchange between the solar wind and neutral interstellar or geocoronal gas (e.g. Robertson and Cravens 2003; Snowden et al. 2004; Bellm and Vaillancourt 2005; Henley et al. 2007; Lallement 2007, this volume). The contribution of CEX to the SXRB is significant above ~0.3 keV, and is limited by the SXRB flux that does not anticorrelate with $N(\mathrm{H}^\circ)$. Models of the contributions of heliospheric charge exchange to the SXRB measured by ROSAT indicate that ~50–65% of the SXRB in the galactic plane and ~25% at high galactic latitudes may arise from CEX, depending on the adopted model for CXE fluxes, the solar cycle phase, and ecliptic latitude of the look direction. LB plasma models with depleted abundances predict approximate consistency between the ROSAT and sounding rocket SXRB data. Comparisons between XMM Newton spectra acquired while pointing on and off of a dark cloud in principle allow the removal of foreground heliospheric CEX. For a solar abundance plasma model, based on recent values (Grevesse 2007, this volume), the radius of the cavity filled with plasma is ~100 pc, with plasma density $n(\mathrm{e}) = 0.013\ \mathrm{cm}^{-3}$, pressure $2.9 \times 10^4\ \mathrm{cm}^{-3}$ K, a cooling time of 17 Myrs, and a sound-crossing time of 1.2 Myrs. The LB plasma properties are a topic of ongoing research because the foreground contamination from CEX and contributions to the SXRB spectra contributed by the halo and disk gas are poorly understood.

The LIC ram pressure, $P_{\mathrm{ram}}$, may be important for the pressure equilibrium between the LIC and LB plasma. At the Sun, the (Slavin and Frisch 2007a, 2007b) LIC models find a total density of $n(\mathrm{H}^\circ + \mathrm{H}^+) \sim 0.26$ atoms $\mathrm{cm}^{-3}$, and $n(\mathrm{e}) \sim 0.07\ \mathrm{cm}^{-3}$, for $T = 6300$ K, giving a static LIC pressure of ~2200 $\mathrm{cm}^{-3}$ K (including He). For a relative LIC-LB velocity of ~20.7 $\mathrm{km\,s}^{-1}$, $P_{\mathrm{ram}} \sim 20{,}000\ \mathrm{cm}^{-3}$ K, so that $P_{\mathrm{ram}}$ helps offset the high LB thermal pressure in one dimension. However, for the LIC ram pressure to remove the longstanding mystery about the pressure equilibrium between the LIC and LB plasma, additional pressure contributions (e.g., from the magnetic field or cosmic rays) are required for directions perpendicular to the LIC velocity through the hot gas.

## 5 The LIC and the ISM Flow Past Sun

### 5.1 Kinematics of Local Fluff

The LIC is part of an localized ISM flow that has been denoted the Local Fluff, or Cluster of Local Interstellar Clouds (CLIC). The best fitting heliocentric flow vector for ISM within ~30 pc is $-28.1 \pm 4.6\ \mathrm{km\,s}^{-1}$, from the direction $\ell = 12.4^\circ$, $b = 11.6^\circ$ (Frisch et al. 2002). In the LSR, the upwind direction is towards the center of Loop I (footnote 2, Sect. 2.1). The flow is decelerating; in the rest frame of the flow velocity, the fastest components in the upwind and downwind directions are blue-shifted by more than 10 $\mathrm{km\,s}^{-1}$. Individual cloudlets contribute to the flow, including the LIC, the "G-cloud" within 1.3 pc in the upwind direction, and the Apex cloud within 5 pc in the upwind direction and extending towards $\ell \sim 30^\circ$. These cloudlets have the same upwind directions to within $\sim\pm10^\circ$, indicating a common origin for the cloudlets comprising the Local Fluff. Alternate interpretations using Local Fluff kinematics and temperatures have parsed the flow into ~15 spatially distorted components (Linsky 2007, this volume). However, velocity components towards stars in the sidewind direction can not clearly distinguish between individual clouds because of velocity blending.

### 5.2 Interstellar Abundances

The abundance pattern of elements in interstellar gas is characterized by abundances that decrease with increases in the mean gas density ($\langle n(\mathrm{H}) \rangle$) or elemental condensation temperature ($T_{\mathrm{cond}}$). Most depletion studies are based on long sightlines with blended velocity components. Our LIC radiative transfer models derive elemental abundances corrected for ionization, for a single low density cloud in space (Slavin and Frisch 2007a, 2007b). For the best models in Slavin and Frisch (2007b), LIC abundances are O/H = 295–437 ppm, compared to solar abundances 460 ppm (Grevesse et al. 2007, this volume), and N/H = 40–66 ppm, compared to solar values 61 ppm. The LIC S/H ratios are 14–22 ppm, compared to solar values 14 ppm. If the LIC has a solar composition, as indicated by anomalous cosmic ray data for $^{22}$Ne and $^{18}$O (Cummings and Stone 2007, this volume), then Asplund solar abundances are preferred over earlier values. Carbon is found to be overabundant by a factor of ~2.6 compared to solar abundances, which helps maintain the LIC cloud temperature near the observed temperature of 6300 K through C fine-structure cooling. The C-abundance anomaly appears to be due to the destruction of carbonaceous grains by interstellar shocks. The carbon overabundance is consistent with the deficit of small carbonaceous grains causing the 2200 Å bump and far-ultraviolet rise in the ultraviolet extinction curves in some regions.

**Acknowledgements** I thank the organizers of the Geiss-fest for a stimulating meeting, and ISSI for recognizing long ago that the interaction of local ISM and the heliospere is a fascinating topic. NASA grants NNG05GD36G, NNG06GE33G, and NAG5-13107 supported this work. The Hipparcos data-parsing tool was developed by Prof. Philip Chi-Wing Fu, under the auspices of NASA grant NAG5-11999.

## References

E.C. Bellm, J.E. Vaillancourt, Astrophys. J. **622**, 959 (2005)
E.M. Berkhuijsen, C.G.T. Haslam, C.J. Salter, Astron. Astrophys. **14**, 252 (1971)
R.C. Bohlin, B.D. Savage, J.F. Drake, Astrophys. J. **224**, 132 (1978)
D. Breitschwerdt, M.J. Freyberg, R. Egger, Astron. Astrophys. **361**, 303 (2000)
A.N. Cox, *Allen's Astrophysical Quantities* (AIP Press, New York, 2000)
A.C. Cummings, E.C. Stone, Space Sci. Rev. (2007), this volume. doi: 10.1007/s11214-007-9161-y
T.M. Dame, D. Hartmann, P. Thaddeus, Astrophys. J. **547**, 792 (2001)
E.J. de Geus, Astron. Astrophys. **262**, 258 (1992)
P.T. de Zeeuw, R. Hoogerwerf, J.H.J. de Bruijne, A.G.A. Brown, A. Blaauw, Astron. J. **117**, 354 (1999)
R.J. Egger, B. Aschenbach, Astron. Astrophys. **294**, L25 (1995)
M.P. Fitzgerald, Astron. J. **73**, 983 (1968)
P. Frick, R. Stepanov, A. Shukurov, D. Sokoloff, Mon. Not. Roy. Astron. Soc. **325**, 649 (2001)
P.C. Frisch, Nature **293**, 377 (1981)
P.C. Frisch, Space Sci. Rev. **72**, 499 (1995)
P.C. Frisch, Astrophys. J. **632**, L143 (2005)
P.C. Frisch, Astrophys. J. (2006, submitted)
P.C. Frisch, J.M. Dorschner, J. Geiss et al., Astrophys. J. **525**, 492 (1999)
P.C. Frisch, L. Grodnicki, D.E. Welty, Astrophys. J. **574**, 834 (2002)
P.C. Frisch, D.G. York, Astrophys. J. **271**, L59 (1983)
P.M. Gondhalekar, A.P. Phillips, R. Wilson, Astron. Astrophys. **85**, 272 (1980)
I.A. Grenier, ArXiv Astrophysics e-prints (2004)
N. Grevesse, M. Asplund, A.J. Sauval, Space Sci. Rev. (2007), this volume. doi: 10.1007/s11214-007-9173-7
M. Haverkorn, B.M. Gaensler, J.C. Brown et al., Astrophys. J. **637**, L33 (2006)
C. Heiles, Astrophys. J. **498**, 689 (1998)
C. Heiles, Astron. J. **119**, 923 (2000)
C. Heiles, T.H. Chu, Y. ., Troland, R.J. Reynolds, I. Yegingil, Astrophys. J. **242**, 533 (1980)
C. Heiles, R. Crutcher, ArXiv Astrophysics e-prints (2005)

D.B. Henley, R.L. Shelton, K.D. Kuntz, ArXiv Astrophysics e-prints (2007)
M. Juda, J.J. Bloch, B.C. Edwards et al., Astrophys. J. **367**, 182 (1991)
H. Krüger et al., Space Sci. Rev. (2007), this volume. doi: 10.1007/s11214-007-9181-7
R. Lallement, B.Y. Welsh, J.L. Vergely, F. Crifo, D. Sfeir, Astron. Astrophys. **411**, 447 (2003)
R. Lallement, Space Sci. Rev. (2007), this volume. doi: 10.1007/s11214-007-9178-2
J.L. Leroy, Astron. Astrophys. **346**, 955 (1999)
J.L. Linsky, Space Sci. Rev. (2007), this volume. doi: 10.1007/s11214-007-9160-z
P.B. Lucke, Astron. Astrophys. **64**, 367 (1978)
J. Maíz-Apellániz, Astrophys. J. **560**, L83 (2001)
F. Paresce, Astron. J. **89**, 1022 (1984)
M.A.C. Perryman, Astron. Astrophys. **323**, L49 (1997)
R.J. Rand, S.R. Kulkarni, Astrophys. J. **343**, 760 (1989)
I.P. Robertson, T.E. Cravens, J. Geophys. Res. (Space Phys.) **108**, 6 (2003)
D.M. Sfeir, R. Lallement, F. Crifo, B.Y. Welsh, Astron. Astrophys. **346**, 785 (1999)
J.D. Slavin, P.C. Frisch, Astron. Astrophys. (2007a, to be submitted). http://arxiv.org/abs/0704.0657
J.D. Slavin, P.C. Frisch, Space Sci. Rev. (2007b), this volume. doi: 10.1007/s11214-007-9186-2
S.L. Snowden, M.R. Collier, K.D. Kuntz, Astrophys. J. **610**, 1182 (2004)
L. Spitzer, *Physical Processes in the Interstellar Medium* (Wiley, New York, 1978)
J. Tinbergen, Astron. Astrophys. **105**, 53 (1982)
J.-L. Vergely, R. Freire Ferrero, A. Siebert, B. Valette, Astron. Astrophys. **366**, 1016 (2001)
R.S. Warwick, C.R. Barber, S.T. Hodgkin, J.P. Pye, Mon. Not. Roy. Astron. Soc. **262**, 289 (1993)
B.E. Wood, S. Redfield, J.L. Linsky, H.-R. Müller, G.P. Zank, Astrophys. J. Suppl. Ser. **159**, 118 (2005)

Space Sci Rev (2007) 130: 367–375
DOI 10.1007/s11214-007-9160-z

# D/H and Nearby Interstellar Cloud Structures

**J.L. Linsky**

Received: 27 October 2006 / Accepted: 9 February 2007 /
Published online: 5 May 2007


**Abstract** Analysis of UV spectra obtained with the HST, FUSE and other satellites provides a new understanding of the deuterium abundance in the local region of the galactic disk. The wide range of gas-phase D/H measurements obtained outside of the Local Bubble can now be explained as due to different amounts of deuterium depletion on carbonaceous grains. The total D/H ratio including deuterium in the gas and dust phases is at least 23 parts per million of hydrogen, which is providing a challenge to models of galactic chemical evolution. Analysis of HST and ground-based spectra of many lines of sight to stars within the Local Bubble have identified interstellar velocity components that are consistent with more than 15 velocity vectors. We have identified the structures of 15 nearby warm interstellar clouds on the basis of these velocity vectors and common temperatures and depletions. We estimate the distances and masses of these clouds and compare their locations with cold interstellar clouds.

**Keywords** ISM: deuterium · Abundances · Structure · Ultraviolet: ISM

## 1 Introduction

High-resolution ultraviolet spectra provide critically important information on the chemical composition, kinematics, and physical properties of astrophysical plasmas. Many important spectral lines from a wide range of ions in the 1200–3000 Å spectral region are observable with the Goddard High Resolution Spectrograph (GHRS) and the Space Telescope Imaging Spectrograph (STIS) instruments on the Hubble Space Telescope (HST). Other ions can be studied in the 912–1200 Å region with the Far Ultraviolet Spectrograph Explorer (FUSE) satellite. I will summarize recent work in two areas based on these HST and FUSE data. The first topic is the abundance of deuterium in the Galaxy, based on observations of the Lyman lines of hydrogen and deuterium. The second topic is the structure and chemical

J.L. Linsky (✉)
JILA, University of Colorado and NIST, Boulder, CO 80309-0440, USA
e-mail: jlinsky@jila.colorado.edu

composition of warm clouds in the nearby interstellar medium (ISM), based primarily on HST observations of resonance lines of Mg II, Fe II, and D I, together with ground-based observations of Ca II. In many cases, the same spectra provided valuable input for both topics.

## 2 The Abundance of Deuterium in the Galaxy

Accurate measurements of the deuterium/hydrogen (D/H) ratio (by number) provide critically important tests for models of primordial nucleosynthesis, galactic chemical evolution, and the chemical properties of the intergalactic medium (IGM). Primordial nucleosynthesis is the only significant source of D, which is easily converted to heavier nuclei in stellar interiors before supernovae and stellar winds disperse the deuterium-poor gas into the ISM. As a result, measurements of the D abundances in different regions of the Galaxy provide a major test of galactic chemical evolution models. Unexpectedly, our studies of D have also provided information on the composition of interstellar dust grains.

Linsky et al. (2006) summarize the recent D/H measurements and provide a model for their interpretation. Figure 1 shows the hydrogen Lyman-$\alpha$ line emission line from the nearby star Capella with superimposed absorption by interstellar H and D (at $-0.33$ Å or $-82$ km s$^{-1}$ from the H line). This GHRS spectrum is typical for short lines of sight toward late-type stars with Lyman-$\alpha$ emission lines, but for longer lines of sight the targets are typically O stars, OB subdwarfs, and hot white dwarfs that have interstellar absorption lines superimposed on stellar Lyman-$\alpha$ absorption lines. Since the interstellar H absorption is a factor of $10^5$ times stronger than for D, the difficult part of the analysis is to measure accurately the H column density, N(H I), from the steeply-rising flux just outside of the saturated line core. High spectral resolution is required to accurately measure the slope of the absorption. For more distant lines of sight, the H line broadens with increasing N(H I) to cover the D line, thereby forming an effective horizon for Lyman-$\alpha$ D/H measurements when $\log N(\text{H I}) \geq 18.7$.

For lines of sight with larger column densities, one must observe the less opaque higher Lyman lines, which lie in the 912–1025 Å spectral region not accessible to HST but observable with FUSE. Figure 2 shows examples of FUSE spectra of two sdO stars. The ability

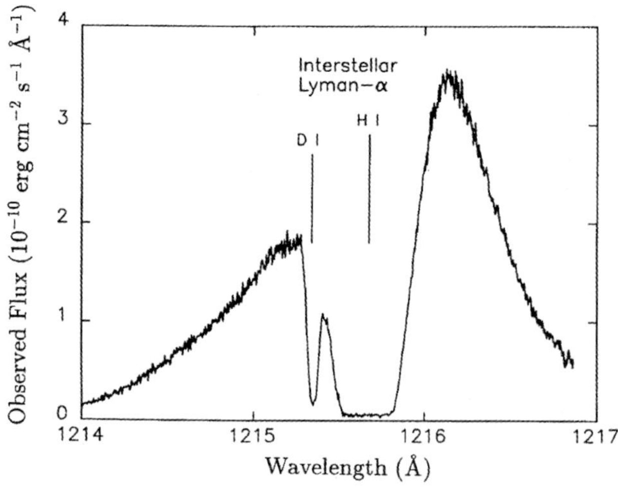


**Fig. 1** GHRS echelle spectrum of the star Capella (12.9 pc) showing interstellar Lyman-$\alpha$ absorption of H and D. $\log N(\text{H I}) = 18.24$. (From Linsky et al. 1995)

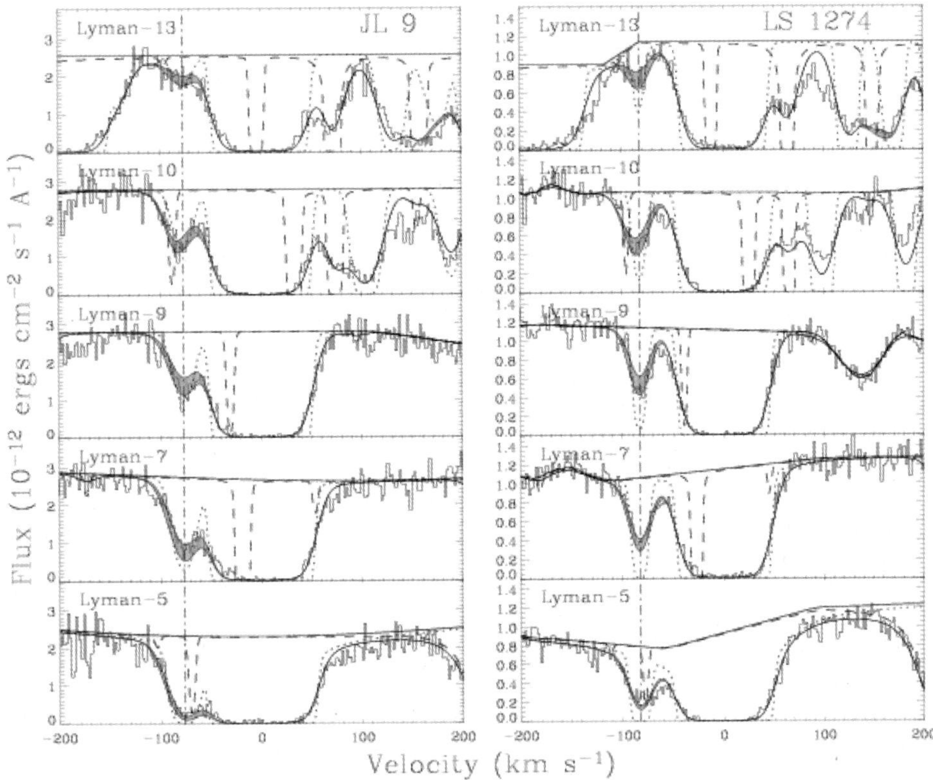


**Fig. 2** FUSE spectra of two hot sdO stars, JL 9 and LSS 1274, both at a distance of about 600 pc. The data are for the Lyman-5 through Lyman-13 lines. *Dashed lines* indicate $H_2$ absorption, *dotted lines* show the absorption due to all other lines, and *solid lines* show the convolution with the instrument profile. The values of log N(H I) are 20.78 and 20.98, respectively. (From Wood et al. 2004)

to study a range of Lyman lines with very different opacities allows one to infer D/H from measured N(H I) and N(D I) column densities despite line saturation, weak D I lines, or overlaping lines of $H_2$ and other species.

Figure 3 shows D/H measurements (and their $1\sigma$ errors) for all 47 lines of sight (Linsky et al. 2006). It is essential to recognize that the N(D I)/N(H I) ratio measures D/H only in the gas phase, $(D/H)_{gas}$. Inside the Local Bubble (log N(H I) $< 19.2$), which extends to roughly 100 pc from the Sun, the 22 lines of sight are consistent with a constant value of $(D/H)_{gas} = 15.6 \pm 0.4$ ppm (parts per million of H I atoms). For larger N(H I) there is a wide range of $(D/H)_{gas}$ values with a tendency to low values at log N(H I) $> 20.7$. This large range in $(D/H)_{gas}$ must be real as it has been found from analysis of FUSE, Copernicus, and IMAPS spectra by different authors using different analysis codes. Since a factor of four range in D/H within a few hundred parsecs of the Sun is inconsistent with existing galactic chemical evolution models for a sensible range of assumptions (e.g., Romano et al. 2006), a new approach is necessary for understanding these data.

Developing an idea first proposed by Jura (1982), Draine (2003, 2006) showed that D could be depleted onto carbonaceous dust grains in the ISM. Since the C–D bond is 0.083 eV larger than the C–H bond, D can replace H in these grains. In thermodynamic equilibrium,

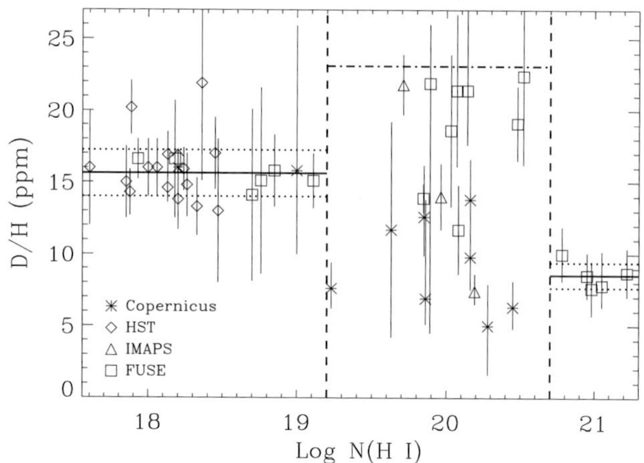


**Fig. 3** Plot of the gas-phase D/H values (not corrected for the Local Bubble foreground) vs. the hydrogen column density N(H I) for the 47 lines of sight studied by Linsky et al. (2006). The symbols for each data point indicate the spacecraft that observed the line of sight. Error bars are $\pm 1\ \sigma$. *Vertical dashed lines* indicate lines of sight inside the Local Bubble [$\log$ N(H I) $< 19.2$ cm$^{-2}$] and the intermediate region [$\log$ N(H I) between 19.2 and 20.7]. The *solid horizontal lines* indicate the mean values of $(\mathrm{D/H})_{\mathrm{gas}}$ for the low and high N(H I) regions, and the dotted horizontal lines indicate the $\pm 1\ \sigma$ errors about the mean value. The *dash-dot horizontal line* is the mean value of $(\mathrm{D/H})_{\mathrm{gas-LB}}$ for the highest five points in the intermediate region ($\gamma^2$ Vel, Lan 23, WD 1034+001, Feige 110, and LSE 44) after subtracting the foreground Local Bubble contributions to the hydrogen and deuterium column densities along the lines of sight

$(\mathrm{D/H})_{\mathrm{dust}}/(\mathrm{D/H})_{\mathrm{gas}} \approx e^{970/T_{\mathrm{dust}}}$, which exceeds $5 \times 10^5$ for $T_{\mathrm{dust}} < 90$ K. Since interstellar dust is very cold, typically $T_{\mathrm{dust}} \sim 20$ K (Draine 2003), very high levels of deuteration are possible in carbonaceous grains and likely also in interstellar polycyclic aromatic hydrocarbon molecules. Since $(\mathrm{C/H})_{\mathrm{dust}} \approx 230$ ppm (Savage and Sembach 1996), under these conditions essentially all of the D could be removed from the gas phase by binding onto only 10% of the C in grains. However, the ISM is not in thermodynamic equilibrium, but rather is highly dynamic. Draine (2006) estimates that the time scale for D depletion onto grains is about 2 Myr for cold neutral clouds and about 50 Myr for the warm neutral medium. Linsky et al. (2006) proposed a time-dependent D depletion model in which D is depleted when the interstellar gas remains undisturbed for some time, but the D in the dust returns to the gas phase when the dust is vaporized by shock waves or comes near hot stars.

Linsky et al. (2006) tested this model in several ways. First, they noted that Keller et al. (2000) found that carbon-rich interplanetary dust particles (IDPs) contain inclusions with D/H as high as 16 500 ppm. Since IDPs are either interstellar in origin or formed in the early solar system where conditions were similar to cold neutral clouds in the ISM, these very high D/H ratios provide a proof of concept that D can be highly depleted onto interstellar dust grains. Second, when conditions are appropriate for D depletion, other elements, especially refractory elements like iron and silicon, should also be depleted. Thus low values of $(\mathrm{D/H})_{\mathrm{gas}}$ should correlate with large depletions of iron and silicon. Figure 4 shows that there is indeed a good correlation in which the Spearman test rejects no correlation at 99.8% ($2.9\sigma$). There are, however, five high points, that could be explained if D is mainly in the outer surfaces of grains that have more refractory cores. In this case, moderate strength

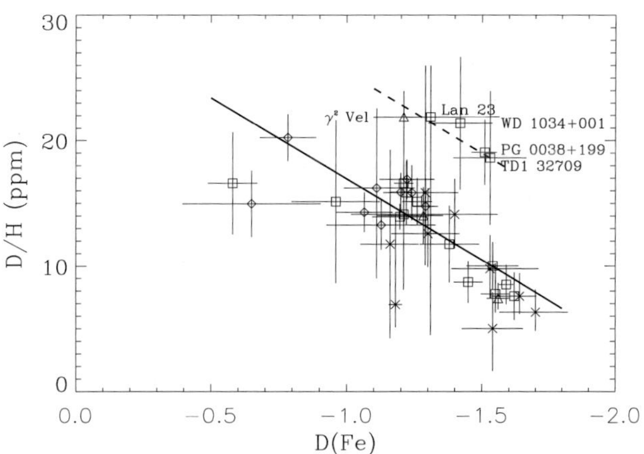


**Fig. 4** Plot of gas-phase D/H values in the lines of sight toward stars vs. the depletion (log units) of iron for 38 lines of sight. The symbols are the same as in Fig. 3, and the error bars are $\pm 1\,\sigma$. The *solid line* is the least-squares linear fit to the data weighted by the inverse errors. See Linsky et al. (2006) for a discussion of the error analysis technique. Five lines of sight with $(D/H)_{gas}$ well above the linear fit are identified. The arbitrary dashed line has the same slope as the solid line but is displaced upward by 8.5 ppm

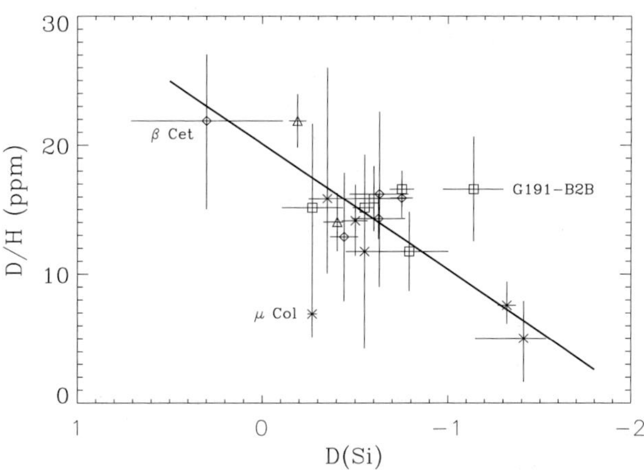


**Fig. 5** Plot of gas-phase D/H values in the lines of sight toward stars vs. the depletion (log units) of silicon for 20 lines of sight. The symbols are the same as in Fig. 3, and the error bars are $\pm 1\,\sigma$. The *solid line* is the least-squares linear fit to the data weighted by the inverse errors. Four of the five high points in Fig. 4 do not have Si data

shocks could vaporize the outer regions of the grains leaving the iron cores intact. Alternatively, there could be a low percentage of C grains in these lines of sight. A similar plot for silicon (Fig. 5) also shows a correlation of $(D/H)_{gas}$ with the depletion of Si.

If depletion is the cause of low values of the $(D/H)_{gas}$ ratio, then according to the time-dependent D depletion model, those lines of sight with the highest measured $(D/H)_{gas}$ values should provide a good approximation to the total (D/H) ratio in the local region of the galactic disk, $(D/H)_{LDtotal}$. Beyond the Local Bubble, Linsky et al. (2006) identify five lines of sight (toward $\gamma^2$ Vel, Lan 23, WD 1034+001, PG 0038+199, and TD1 32709) with the highest $(D/H)_{gas}$ ratios. The mean value for these five lines of sight is $(D/H)_{gas} = 21.7 \pm 1.7$ ppm. However, when the Local Bubble foreground values of N(H I) and N(D I) are subtracted, we find that $(D/H)_{LDtotal} \geq 23.1 \pm 2.4$ ppm. The $\geq$ symbol reminds us that even for the high lines of sight, some D may still be locked up in the grains.

Until recently, galactic chemical evolution models (e.g., Romano et al. 2003) assumed that $(D/H)_{LDtotal} \approx 15$ ppm, the value of $(D/H)_{gas}$ inside the Local Bubble. However this region was shocked by supernova events several million years ago (Berghöfer and Breitschwerdt 2002) and D is no longer entirely in the gas phase. When the Galaxy was two-thirds its present age, the protosolar value of the total D/H ratio was $20 \pm 3.5$ ppm

(Geiss and Gloeckler 2003). The deuterium astration factor is the ratio of the total D/H ratios in the primative and present-day Galaxy, $f_D = (D/H)_{prim}/(D/H)_{LDtotal}$. If one adopts the primordial (D/H) ratio infered from WMAP data (Spergel et al. 2003), then $f_D \leq (27.5 \pm 2.1\ ppm)/(23.1 \pm 2.4\ ppm) \leq 1.19 \pm 0.16$. If instead, one adopts the primordial ratio from Coc et al. (2004), which is based on the WMAP data but with different assumed nuclear reaction rates, then $f_D \leq 1.12 \pm 0.14$. These very small astration factors challenge the most recent galactic chemical evolution models. For example, the smallest astration factor in models for the Sun's distance from the Galactic Center computed by Romano et al. (2006) is $f_D \approx 1.39$. The new value of $(D/H)_{LDtotal}$ will lead to a re-examination of the many assumptions that enter the chemical model calculations, in particular the infall rate and timing of D-rich and metal-poor gas from the galactic halo and IGM.

## 3 Structure of the Local ISM

Seth Redfield and I are studying the properties of interstellar gas inside the Local Bubble. Our initial goal was to study the kinematics and physical properties of warm gas near the Sun using high-resolution GHRS and STIS spectra and ground-based Ca II spectra. As this work has proceeded, we have found interesting connections between the warm gas clouds and both cold gas and the locations of interstellar scintillation screens.

Lallement and Bertin (1992) found that the radial velocities of interstellar warm gas in the lines of sight to nearby stars can be fit with two velocity vectors, one for the Galactic Center hemisphere and the other for the anticenter hemisphere. These kinematic data permitted them to identify two warm gas clouds near the Sun, which they called the AG and G clouds, respectively. The AG Cloud is now called the Local Interstellar Cloud (LIC). Subsequently, Frisch et al. (2002) also used kinematics to identify 7 clouds in the solar neighborhood.

Redfield and Linsky (2007) are extending this work by analysis of 146 interstellar velocity components identified in STIS and GHRS spectra of lines of sight to stars inside the Local Bubble and an additional 86 velocity components seen in ground-based Ca II spectra. We identify the shapes of clouds from common kinematic and physical properties. We first identify a cloud on the basis that the interstellar absorption line velocity components for lines of sight in the same region of the sky fit the same empirically determined flow vector to within the velocity measurement errors. In addition, we require that the interstellar gas temperatures, when available from the UV spectra, have similar values. Figure 6 shows how the temperature ($T$), which is best determined from D, and nonthermal broadening parameter ($\xi$), which is best determined from Fe, can be inferred from the measured line widths of ions with different atomic mass. UV spectra are needed to sample both light and heavy mass ions. The figure shows examples for a typical warm cloud (toward HZ 43) and a rather cool cloud (toward $\upsilon$ Peg).

We have now identified 15 warm clouds lying within 15 pc of the Sun. Upper limits to the cloud distances are determined from the closest star that shows interstellar absorption at the projected velocity of the cloud flow vector. Aside from the LIC, the closest warm cloud is the G Cloud seen toward $\alpha$ Cen (1.3 pc). Where these clouds are located along the line of sight to the nearest star is presently unknown. The cloud temperatures range from 1700 K (Oph Cloud) to about 10 000 K (Mic and Vel Clouds). The number of lines of sight used to determine the flow vectors range from 78 (for the LIC) and 21 (for the G Cloud) down to 4–6 in a few cases.

Figure 7 shows the shape of the LIC delineated by the solid line that includes all lines of sight consistent with the LIC flow vector and excludes all lines of sight inconsistent with

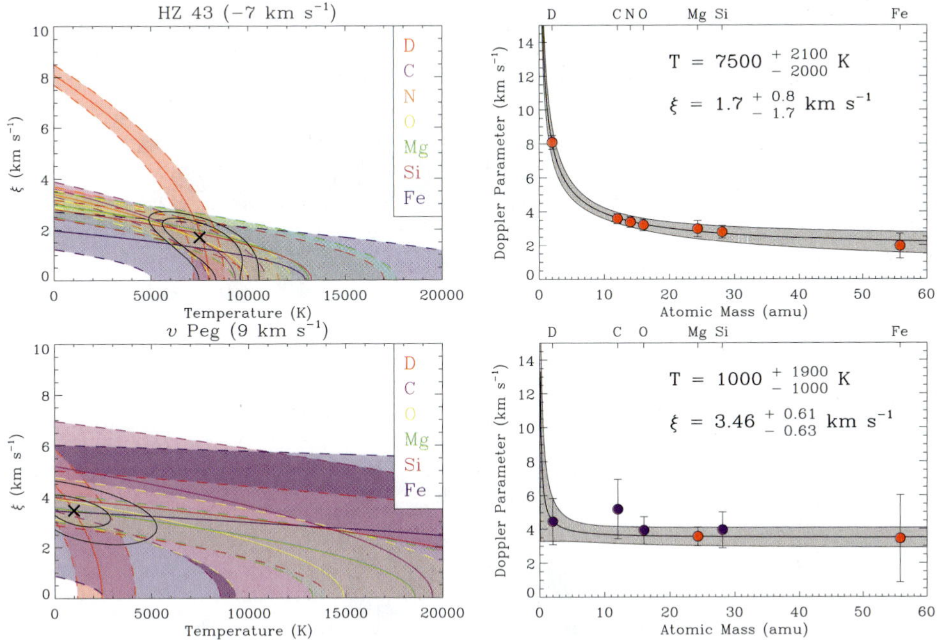


**Fig. 6** Technique for measuring cloud temperatures ($T$) and nonthermal broadening parameters ($\xi$) from interstellar absorption line widths. Examples are given for the –7 km s$^{-1}$ velocity component toward HZ 43 (*top panels*) and the +9 km s$^{-1}$ velocity component toward $\upsilon$ Peg (*bottom panels*). The *right side panels* show the Doppler line widths for atoms or ions of different atomic mass, the best fit parameters, and their $\pm 1\sigma$ errors. The *left side panels* show the acceptable range of $T$ and $\xi$ for each atom or ion and the best fit, $1\sigma$, and $2\sigma$ range for these parameters. Figure from Redfield and Linsky (2004)

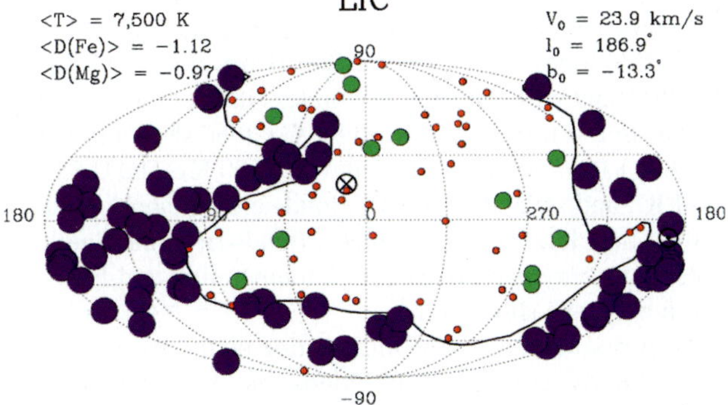


**Fig. 7** Outline of the Local Interstellar Cloud (LIC) in galactic coordinates. Large dots indicate lines of sight containing velocity components consistent with the LIC velocity vector. *Small dots* are lines of sight with no components consistent with the LIC vector. *Medium size dots* indicate lines of sight that have similar kinematics to the LIC but are excluded for other reasons. The mean temperature and iron and magnesium depletions are listed in the *upper left corner*. The amplitude and downwind direction of the heliocentric LIC velocity vector are shown in the *upper right corner*. The upwind direction of the flow vector is indicated by the ⊗ *symbol*, and the downwind direction by the ⊙ *symbol*

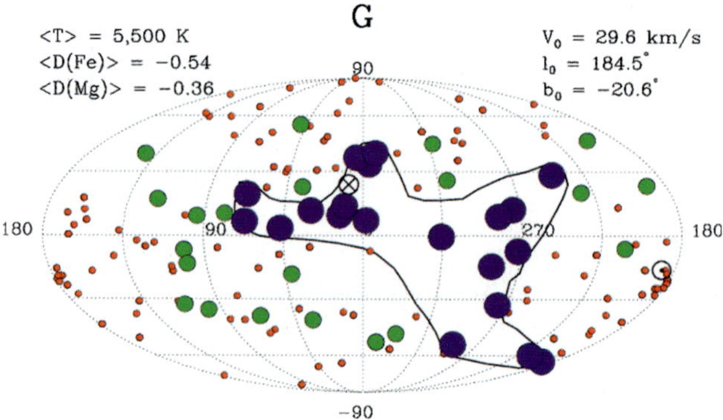


**Fig. 8** Same as Fig. 7 except for the G Cloud

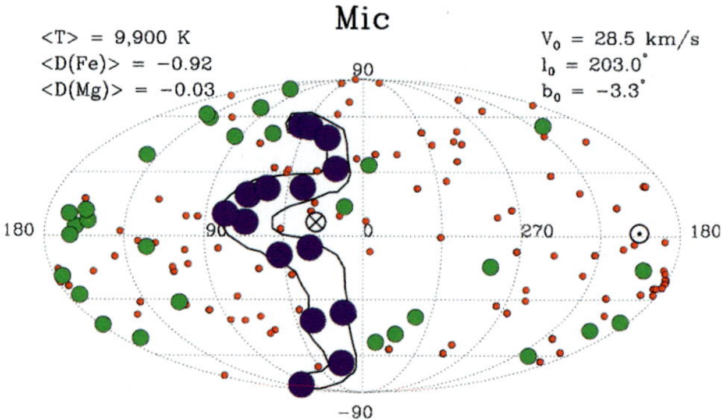


**Fig. 9** Same as Fig. 7 except for the Mic Cloud

the vector. Although the LIC is centered on the antigalactic center direction, its shape is irregular like most of the identified clouds. Since no absorption consistent with the LIC vector is seen in half of the sky, including the direction toward the nearest star ($\alpha$ Cen, $l = 315.7°$, $b = -0.7°$), the Sun must be either just inside or just outside of the LIC. In fact, the Sun is located just inside the LIC because interstellar neutral helium gas, with a trajectory not deviated by charge exchange or ionization, flows inside the heliosphere with the LIC velocity. However the Sun is travelling at 26 km $s^{-1}$ relative to the LIC upwind direction and will leave the LIC in less than a few thousand years.

The Sun is travelling toward the G Cloud shown in Fig. 8. This cloud is somewhat cooler than the LIC (5500 K compared with 7500 K). The G Cloud has a more irregular shape than the LIC, and several of the 15 clouds clearly show filamentary structure. A good example is the Mic Cloud (Fig. 9), which extends from the South Galactic Pole to nearly the North Pole but is very narrow. The high temperature and filamentary shape of the cloud suggests that it may be a shock front.

Now that we have some information on the location of warm gas clouds in the local ISM, we can study how the warm clouds may be related to other structures and phenomena in the local ISM. Heiles and Troland (2003) identified the Triad Region Cloud consisting of cold neutral medium (CNM) gas on the basis of H I 21 cm absorption. Centered at $l = 225°$ and $b = +45°$, this cloud extends over some 20 degrees in galactic longitude and has an H I spin temperature of about 25 K. High-resolution interstellar Na I absorption measurements by Meyer et al. (2006) confirm the low temperature of the cloud and show that its distance must be less than 45 pc on the basis of the closest star showing the cloud's absorption. It is interesting that this cloud lies at the edge of several warm clouds. In the McKee and

Ostriker (1977) model of the ISM, CNM clouds must be surrounded by warm clouds to provide shielding from ionizing radiation. Also, Vázquez-Semadeni et al. (2006) argue that CNM clouds could be formed by transonic compression of colliding warm clouds. Further study of the topological relation of cold and warm gas is possible with the Redfield and Linsky (2007) data set.

Several pulsars and quasars with very small angular diameters show large amplitude radio flux variations on time scales of hours to months, which have been interpreted as scintillation produced by turbulent interstellar gas. In some cases the distance to the scattering screen has been determined from the scintillation time scales and observations by widely separated radio telescopes. Perhaps the best example is the quasar J1819+3845 studied by Dennett-Thorpe and de Bruyn (2003), who now argue (de Bruyn, private communication) that the scattering screen is only $1.6 \pm 1.0$ pc from the Sun. Since the location of the turbulent region ($l = 66°$, $b = +22°$) is at the edge of the two clouds, it is likely that the turbulence is produced by the collision of the G and LIC Clouds, which have a relative velocity of 5.7 km $s^{-1}$. We will publish a detailed study of this topic in the near future.

**Acknowledgements** I wish to thank the FUSE Science Team and Bruce Draine for their collaboration and scientific input concerning deuterium in the Galaxy. I also thank Seth Redfield for his collaboration on the local ISM and Barney Rickett for his input concerning radio scintillation sources. I acknowledge support through grants AR-09525.01A and GO-10236.02 from the Space Telescope Science Institute, which is operated by the Association of Universities for Research in Astronomy, Inc, under NASA contract NAS5-26555.

## References

T.W. Berghöfer, D. Breitschwerdt, Astron. Astrophys. **390**, 299 (2002)
A. Coc, E. Vangioni-Flan, P. Descouvement, C. Angulo, Astrophys. J. **600**, 544 (2004)
J. Dennett-Thorpe, A.G. de Bruyn, Astron. Astrophys. **404**, 113 (2003)
B.T. Draine, Annu. Rev. Astron. Astrophys. **41**, 241 (2003)
B.T. Draine, ASP Conf. Ser. **348**, 58 (2006)
P.C. Frisch, L. Grodnicki, D.E. Welty, Astrophys. J. **574**, 834 (2002)
J. Geiss, G. Gloeckler, Space Sci. Rev. **106**, 3 (2003)
C. Heiles, T.H. Troland, Astrophys. J. **586**, 1067 (2003)
M. Jura, in *Advances in UV Astronomy: 4 Years of IUE Research*, ed. by Y. Kondo, J.M. Mead, R.D. Chapman, (NASA, Greenbelt, 1982), p. 54
L.P. Keller, S. Messenger, J.P. Bradley, J. Geophys. Res. **105**, 10397 (2000)
R. Lallement, P. Bertin, Astron. Astrophys. **266**, 479 (1992)
J.L. Linsky et al., Astrophys. J. **647**, 1106 (2006)
J.L. Linsky, A. Diplas, B.E. Wood, A. Brown, T.R. Ayres, B.D. Savage, Astrophys. J. **451**, 1335 (1995)
C.F. McKee, J.P. Ostriker, Astrophys. J. **218**, 148 (1977)
D.M. Meyer, J.T. Lauroesch, C. Heiles, J.E.G. Peek, K. Engelhorn, Astrophys. J. **650**, L67 (2006)
S. Redfield, J.L. Linsky, Astrophys. J. **613**, 1004 (2004)
S. Redfield, J.L. Linsky, Astrophys. J. (2007, submitted)
D. Romano, M. Tosi, C. Chiappini, F. Matteucci, Mon. Not. Roy. Astron. Soc. **369**, 295 (2006)
D. Romano, M. Tosi, F. Matteucci, C. Chiappini, Mon. Not. Roy. Astron. Soc. **346**, 295 (2003)
B.D. Savage, K.R. Sembach, Annu. Rev. Astron. Astrophys. **34**, 279 (1996)
D.N. Spergel et al., Astrophys. J. Suppl. **148**, 175 (2003)
E. Vázquez-Semadeni, D. Ryu, T. Passot, R.F. González, A. Gazol, Astrophys. J. **643**, 245 (2006)
B.E. Wood, J.L. Linsky, G. Hébrard, G.M. Williger, H.W. Moos, W.P. Blair, Astrophys. J. **609**, 838 (2004)

Space Sci Rev (2007) 130: 377–387
DOI 10.1007/s11214-007-9203-5

# Filtration of Interstellar Atoms through the Heliospheric Interface

**V.V. Izmodenov**

Received: 22 January 2007 / Accepted: 28 March 2007 /
Published online: 6 June 2007


**Abstract** Interstellar atoms penetrate deep into the heliosphere after passing through the heliospheric interface—the region of the interaction of the solar wind with the interstellar medium. The heliospheric interface serves as a filter for the interstellar atoms of hydrogen and oxygen, and, to a lesser extent, nitrogen, due to their coupling with interstellar and heliospheric plasmas by charge exchange and electron impact ionization. The filtration has great importance for the determination of local interstellar abundances of these elements, which becomes now possible due to measurements of interstellar pickup by Ulysses and ACE, and anomalous cosmic rays by Voyagers, Ulysses, ACE, SAMPEX and Wind. The filtration of the different elements depends on the level of their coupling with the plasma in the interaction region. The recent studies of the filtration of the interstellar atoms in the heliospheric interface region is reviewed in this paper. The dependence of the filtration on the local interstellar proton and H atom number densities is discussed and the roles of the charge exchange and electron impact ionization on the filtration are evaluated. The influence of electron temperature in the inner heliosheath on the filtration process is discussed as well. Using the filtration coefficients obtained from the modeling and SWICS/Ulysses pickup ion measurements, the local interstellar abundances of the considered elements are determined.

**Keywords** ISM: atoms · Interplanetary medium · Solar wind · Circumstellar matter

V.V. Izmodenov (✉)
Department of Mechanics and Mathematics, Lomonosov Moscow State University, Moscow, Russia
e-mail: izmod@iki.rssi.ru

V.V. Izmodenov
Space Research Institute (IKI), Russian Academy of Sciences, Moscow, Russia

V.V. Izmodenov
Institute for Problems in Mechanics, Russian Academy of Sciences, Moscow, Russia

## 1 Introduction

The chemical composition of the Local Interstellar Cloud (LIC) surrounding the Sun has great importance for understanding the composition of the local interstellar matter. At the present time the local interstellar parameters and composition can only be explored with remote and indirect measurements. There are two types of diagnostics of the LIC: 1) spectroscopic observations of stellar absorptions (e.g. Linsky et al. 1995; Lallement 1996) that provide data averaged over long distances; 2) measurements of pickup ions and anomalous cosmic rays (ACRs) inside the heliosphere at one or several AU (Geiss et al. 1994; Gloeckler and Geiss 2004; Cummings et al. 2002a) that allows the determination of the local interstellar composition in the vicinity of the Sun. The pickup ions originate from the interstellar atoms penetrating into the heliosphere through the heliospheric interface, which is formed by the interaction of the solar wind (SW) with the charged component of the interstellar medium. The parameters of the interstellar atom flow are significantly disturbed in the interface due to effective coupling with protons by charge exchange. In particular, charge exchange results in the filtration of the interstellar atoms in the heliospheric interface before they enter the heliosphere. The filtration means that only a fraction of the interstellar atoms penetrate into the heliosphere. The filtration can be different for different chemical elements since it depends on the level of their coupling with the charged particles. Therefore, to study composition of the interstellar medium surrounding the Sun from pickup ion data obtained inside the heliosphere one needs to take into account the effects of the heliospheric interface.

This paper gives a brief overview on the structure and modeling of the heliospheric interface and on the problem of the interstellar atom filtration in the interface region.

## 2 The Heliospheric Interface

The heliospheric interface is formed by the interaction of the solar wind with the partly ionized interstellar medium. The interface has a complex structure (Fig. 1) with two shock waves—the interstellar bow shock (BS) and the heliospheric termination shock (TS), and the heliopause that is the contact discontinuity separating the solar wind from interstellar plasma. The SW/LIC interaction has a truly multi-component nature. The interplanetary and interstellar magnetic fields, interstellar atoms of hydrogen, galactic and anomalous cosmic rays (GCRs and ACRs), and pickup ions play important roles in the formation of the heliospheric interface (e.g. Izmodenov and Kallenbach 2007). To reconstruct the structure and the physical processes at the interface using remote observations a theoretical model should be employed.

The development of a theoretical model of the heliospheric interface requires the correct approach for each of the interstellar and solar wind components. Interstellar and solar wind protons and electrons can be described as fluids. However, the mean free path of interstellar H atoms is comparable with the size of the heliospheric interface. This requires a kinetic description for the interstellar H atom flow in the interaction region. For the pickup ion and cosmic ray components the kinetic approach is required as well.

The first self-consistent stationary axisymmetric model of the interaction of the two-component (plasma and H atoms) LIC with the solar wind (B&M model, hereinafter) had been developed by Baranov and Malama (1993). The main physical process considered in the model is the resonance charge exchange processes of the H atoms with protons although the processes of photoionization and ionization of H-atoms by electron impact can be important in some regions of the heliosphere (for example, in the inner heliosheath or in the

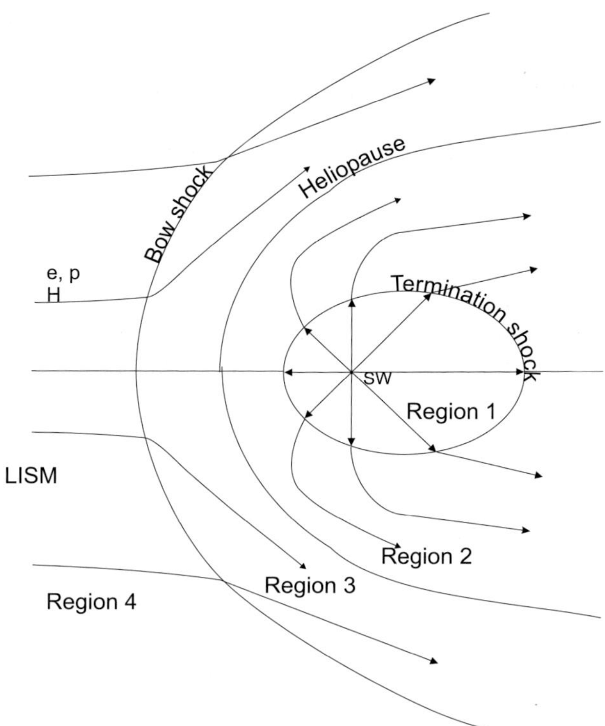


**Fig. 1** Qualitative picture of the SW interaction with the LIC. The *heliopause* (HP) is a contact (or tangential) discontinuity, which separates the solar wind plasma and the interstellar plasma component. The *termination shock* (TS) is formed due to the deceleration of the supersonic solar wind. The *bow shock* (BS) may also exist if the interstellar plasma flow is supersonic. Four regions are distinguished: the supersonic solar wind (region 1); the solar wind flow between the TS and the HP (region 2 or the *inner heliosheath*); the disturbed interstellar plasma component flow (region 3 or the *outer heliosheath*); the undisturbed interstellar gas flow (region 4)

supersonic solar wind). The significant effect of the resonance charge exchange is connected with the large cross section of such collisions which is a function of the relative velocity of colliding particles. Izmodenov et al. (2000) have shown that the elastic H–H and H–proton collisions are negligible in the considered problem. The main results of the B&M model can be briefly summarized as follows:

1. Interstellar atoms strongly influence the heliospheric interface structure. The heliospheric interface is much closer to the Sun in the case when H atoms are taken into account in the model, as compared to a pure gas dynamical case. The distance to the TS in the upwind direction is on the order of 90–100 AU depending on the outer and inner boundary conditions.

The termination shock becomes more spherical in the presence of H atoms and the flow in the region between HP and TS becomes entirely subsonic. The complicated shock structure in the tail (see, e.g. Izmodenov and Alexashov 2003) disappears in the presence of H atoms.

2. The effect of charge exchange on the solar wind is significant. By the time the solar wind flow reaches the termination shock, it is decelerated by 15–30%, strongly heated by a factor of 5–8, and loaded by the pickup proton component (approximately 20–50%).

The interstellar plasma flow is disturbed upstream of the bow shock (region 4 in Fig. 1) by charge exchange of the interstellar protons with secondary H atoms originating in the solar wind. This leads to heating (40–70%) and deceleration (15–30%) of the interstellar plasma before it reaches the bow shock. The Mach number decreases upstream of the BS and for a certain range of interstellar parameters ($n_{\mathrm{H,LIC}} \gg n_{\mathrm{p,LIC}}$) the bow shock may disappear.

The supersonic solar wind flow (region 1 in Fig. 1) is disturbed due to charge exchange with the interstellar neutrals penetrating into the heliosphere.

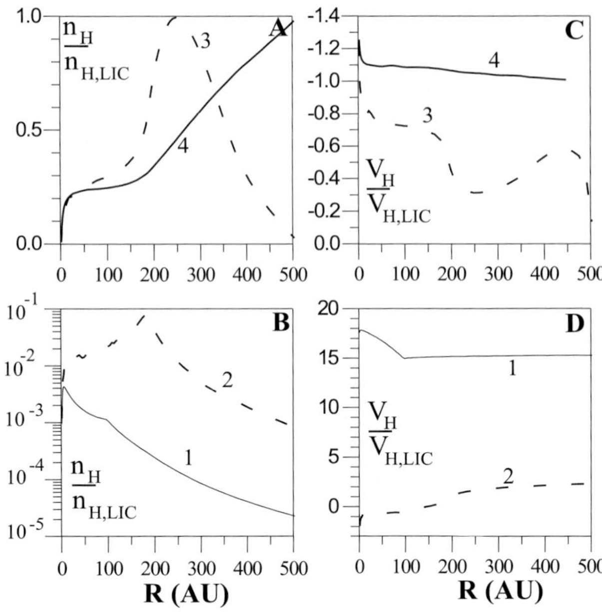


**Fig. 2** Number densities and velocities of 4 atom populations as functions of heliocentric distance in the upwind direction. *1* designates atoms created in the supersonic solar wind, *2* atoms created in the heliosheath, *3* atoms created in the disturbed interstellar plasma, and *4* original (or primary) interstellar atoms. Number densities are normalized to $n_{\mathrm{H,LIC}}$, velocities are normalized to $V_{\mathrm{LIC}}$. It is assumed that $n_{\mathrm{H,LIC}} = 0.2\ \mathrm{cm}^{-3}$, $n_{\mathrm{p,LIC}} = 0.04\ \mathrm{cm}^{-3}$

3. Interstellar neutrals also modify the plasma structure in the heliosheath. In a pure gasdynamic case (without neutrals) the density and temperature of the postshock plasma are nearly constant. However, the charge exchange process leads to a large increase in the plasma number density and decrease in its temperature. The electron impact ionization process may influence the heliosheath plasma flow by increasing the gradient of the plasma density from the termination shock to the heliopause (Baranov and Malama 1996). The influence of interstellar atoms on the heliosheath plasma flow is important, in particular, for the interpretation of kHz-radio emissions detected by Voyager and for possible future imaging of the heliosphere using the energetic neutral atom (ENA) fluxes.

Charge exchange significantly alters the interstellar atom flow. Atoms newly created by charge exchange have the velocity of their ion counterparts in charge exchange collisions. Therefore, the velocity distribution of these new atoms depends on the local plasma properties in the place of their origin. It is convenient to distinguish four different populations of atoms, depending on the region in the heliospheric interface where the atoms were formed. Population 1 are the atoms created in the supersonic solar wind up to the TS (region 1 in Fig. 1), population 2 are the atoms created in the inner heliosheath (region 2 in Fig. 1), and population 3 are the atoms created in the outer heliosheath (region 3 in Fig. 1). The atoms of population 3 are often called the secondary interstellar atom component. We will call the original (or primary) interstellar atoms as population 4. The number densities and mean velocities of these populations are shown in Fig. 2 as functions of the heliocentric distance. The distribution function of H atoms, $f_{\mathrm{H}}(\mathbf{r}, \mathbf{w}_{\mathrm{H}})$, can be represented as a sum of the distribution functions of these populations: $f_{\mathrm{H}} = f_{\mathrm{H},1} + f_{\mathrm{H},2} + f_{\mathrm{H},3} + f_{\mathrm{H},4}$. The Monte Carlo method allows us to calculate these four distribution functions which were presented by Izmodenov (2001) and Izmodenov et al. (2001) at 12 selected points in the heliospheric interface.

Original (or primary) interstellar atoms (population 4) are significantly filtered (i.e. their number density is reduced) before reaching the termination shock (Fig. 2A). The outer heliosheath is the main "filter" for these atoms. Since slow atoms have a small mean free path

**Table 1** Results of parametric calculations in the frame of Baranov and Malama (1993) model with $n_{H,LIC} = 0.2\ cm^{-3}$

| $n_{p,LIC}$ $cm^{-3}$ | Primary | Secondary | Total | | | |
|---|---|---|---|---|---|---|
| | $n_{H4,TS}$ $cm^{-3}$ | $n_{H3,TS}$ $cm^{-3}$ | $n_{H,TS}$ $cm^{-3}$ | $F_{H,TS}$ | $V_{H,TS}$ km/s | $T_{H,TS}$ K |
| 0.3 | | | 0.07 | 0.35 | 17.0 | 14000 |
| 0.2 | 0.0045 | 0.075 | 0.08 | 0.40 | 18.0 | 13500 |
| 0.1 | 0.02 | 0.07 | 0.09 | 0.45 | 20.0 | 12500 |
| 0.07 | 0.03 | 0.065 | 0.095 | 0.475 | 21.0 | 12000 |
| 0.04 | 0.055 | 0.05 | 0.105 | 0.525 | 22.5 | 10500 |

(due to both larger charge exchange cross section and smaller velocities) in comparison to the fast atoms, they undergo larger losses. This kinetic effect, called *selection*, results in $\sim$10% increase in the primary atom mean velocity towards the termination shock (Fig. 2C).

The secondary interstellar atoms (population 3) are created in the disturbed interstellar medium by charge exchange of primary interstellar neutrals with protons decelerated in the vicinity of the heliopause. The secondary interstellar atoms collectively make up the *hydrogen wall*, a density increase at the heliopause. The *hydrogen wall* has been predicted by Baranov et al. (1991) and detected in the direction of $\alpha$Cen on the Hubble Space Telescope by Linsky and Wood (1996). At the termination shock, the number density of secondary neutrals is comparable to the number density of the primary interstellar atoms (Fig. 2A, dashed curve). The relative abundances of secondary and primary atoms entering the heliosphere vary with the degree of interstellar ionization (see Table 1). The bulk velocity of population 3 is about $-18$ to $-19$ km/s. (The "–" sign means that the population approaches the Sun.)

Another population (population 2) of the heliospheric hydrogen atoms consists of the atoms created in the inner heliosheath by charge exchange with hot and compressed solar wind and pickup protons. The number density of this population is an order of magnitude smaller than the number densities of the primary and secondary interstellar atoms. Therefore, this population has a minor importance for the filtration problem. Inside the termination shock the atoms propagate freely. These atoms may serve as a rich source of information on the plasma properties at the place of their birth, i.e. at the inner heliosheath. There are plans to measure this population of atoms on future missions, including the Small Explorer called Interstellar Boundary Explorer (IBEX) that was selected by NASA and is scheduled for launch in June 2008.

During the last several years a large effort in the multi-component modeling of the heliospheric interface has been done by several groups (e.g. Zank 1999; Baranov and Izmodenov 2006). In particular, our Moscow group has developed models of the heliospheric interface, which follow the kinetic-continuum approach of the B&M-model and take into account effects of the solar cycle (Izmodenov et al. 2005a), interstellar helium ions and solar wind alpha particles (Izmodenov et al. 2003), the interstellar magnetic field (Izmodenov et al. 2005b), and galactic and anomalous cosmic rays (Myasnikov et al. 2000; Alexashov et al. 2004). Recently, Malama et al. (2006) presented a new model that retains the main advantage of our previous models, which is a rigorous kinetic description of the interstellar H atom component. In addition, the model considers pickup protons as a separate kinetic component. The next section briefly discuss factors affecting the filtration of interstellar H atoms in the interface.

## 3 Filtration of Interstellar Hydrogen

Newly created in the outer heliosheath (between the HP and BS), secondary interstellar atoms have the velocities of the protons that are their companions in charge exchange. The proton component is decelerated and heated at the BS and continues to be decelerated and heated toward the heliopause. Therefore, the bulk velocity of the secondary interstellar atoms is smaller and the effective kinetic temperature is higher as compared with those of the primary interstellar atoms. Because of this, more H atoms have individual velocities not directed toward the Sun, and less atoms penetrate through the heliopause into the heliosphere.

Let us introduce *filtration* (or, more correctly, *penetration*) factor as the ratio of the H atom number density at the TS in the upwind direction to the interstellar number density:

$$F_{\mathrm{H,TS}} = n_{\mathrm{H}}(TS)/n_{\mathrm{H}}(LIC) = [n_{\mathrm{pop.4,H}}(TS) + n_{\mathrm{pop.3,H}}(TS)]/n_{\mathrm{H}}(LIC).$$

In this section we explore how different physical effects influence the penetration factor for H. Izmodenov et al. (1999) in the frame of the B&M model studied the effects of interstellar proton number density on the structure of the heliospheric interface. Table 1 presents relevant results of the study. The filtration factor varies from 0.35 for $n_{\mathrm{p,LIC}} = 0.3\ \mathrm{cm}^{-3}$ to $0.525\ \mathrm{cm}^{-3}$ for $n_{\mathrm{p,LIC}} = 0.04\ \mathrm{cm}^{-3}$. At present low values of $n_{\mathrm{p,LIC}}$ are more favorable because of at least three observational facts: a) Voyager 1 crossed the TS at 94 AU in December 2004 implying a rather small interstellar pressure that is not consistent with a high proton number density, b) the number density of H atoms at TS of $0.1 \pm 0.05\ \mathrm{cm}^{-3}$ derived from analysis of Ulysses and ACE pickup ion data (Gloeckler and Geiss 2004) that corresponds to $n_{\mathrm{p,LIC}}$ of $0.04$–$0.07\ \mathrm{cm}^{-3}$; similar values of $n_{\mathrm{H,TS}} \approx 0.09\ \mathrm{cm}^{-3}$ were derived from the analysis of the distant solar wind deceleration measured by Voyager 2 (e.g., Richardson et al. 2007); c) analysis of backscattered solar Lyman-alpha spectra showing line-of-sight velocities that correspond to $V_{\mathrm{H,TS}} = 22$–$23$ km/s (Quémerais et al. 2006). More recent and detailed parametric analysis of the filtration of interstellar hydrogen was done in Izmodenov et al. (2004) for the range of $n_{\mathrm{p,LIC}} = 0.032$–$0.07\ \mathrm{cm}^{-3}$ and $n_{\mathrm{H,LIC}} = 0.16$–$0.20\ \mathrm{cm}^{-3}$. The study was done in the frame of the Izmodenov et al. (2003) model that differs from Baranov–Malama model by taking into account effects of interstellar helium ions and solar wind alpha particles. Results of the study are summarized in Table 2. The filtration factor does not change significantly for the considered range of parameters and it is equal to $F_{\mathrm{H,LIC}} = 0.54 \pm 0.04$.

From the analysis of the results of more recent advanced models of the interface mentioned at the end of previous section one can conclude that the considered effects do not change the filtration factor significantly despite their high importance for other aspects of heliospheric physics. Indeed, Myasnikov et al. (2000) and Alexashov et al. (2004) have shown that GCRs and ACRs do not change the filtration factor noticeably. The Malama et al. (2006) model that advances the B&M model by employing a multi-component treatment for heliospheric plasma gives slightly larger filtration factor as compared with the B&M model. In fact in the multi-component model the electron temperature in the heliosheath is smaller as compared to B&M model. The effect of filtration in the inner heliosheath (the region between the TS and HP) due to electron impact ionization is shown in Fig. 5c of Malama et al. (2006) paper. The value of $F_{\mathrm{H,LIC}}$ is smaller by less than 10% for the multi-component model as compared to B&M model.

Effects of the 11-year solar cycle variations of the solar wind parameters on the structure of the heliospheric interface were studied in Izmodenov et al. (2005b). In particular, it was

**Table 2** Results of parametric calculations

| # | $n_{\mathrm{H,LIC}}$ cm$^{-3}$ | $n_{\mathrm{p,LIC}}$ cm$^{-3}$ | $R(TS)$ AU | $F_{\mathrm{H,TS}}$[a] | $F_{O,TS}$ | $F_{N,TS}$ |
|---|---|---|---|---|---|---|
| 1 | 0.16 | 0.032 | 109 | 0.58 | 0.72 (0.84) | 0.80 (0.90)[b] |
| 2 | 0.16 | 0.05 | 102 | 0.55 | 0.70 (0.83) | 0.80 (0.90) |
| 3 | 0.16 | 0.06 | 99 | 0.54 | 0.70 (0.82) | 0.80 (0.90) |
| 4 | 0.16 | 0.07 | 96 | 0.53 | 0.69 (0.81) | 0.80 (0.90) |
| 5 | 0.18 | 0.032 | 101 | 0.57 | 0.69 (0.82) | 0.77 (0.90) |
| 6 | 0.18 | 0.05 | 96 | 0.54 | 0.68 (0.81) | 0.79 (0.89) |
| 7 | 0.18 | 0.06 | 93 | 0.53 | 0.68 (0.81) | 0.79 (0.89) |
| 8 | 0.18 | 0.07 | 88 | 0.52 | 0.66 (0.80) | 0.79 (0.89) |
| 9 | 0.20 | 0.032 | 94 | 0.55 | 0.68 (0.82) | 0.76 (0.89) |
| 10 | 0.20 | 0.04 | 93 | 0.54 | 0.67 (0.81) | 0.77 (0.89) |
| 11 | 0.20 | 0.05 | 90 | 0.53 | 0.67 (0.79) | 0.78 (0.89) |
| 12 | 0.20 | 0.06 | 88 | 0.52 | 0.67 (0.80) | 0.78 (0.89) |
| 13 | 0.20 | 0.07 | 86 | 0.51 | 0.67 (0.79) | 0.78 (0.88) |

[a] $F_{\mathrm{A,TS}} = n_{\mathrm{A,TS}}/n_{\mathrm{A,LIC}}$ (A = H, O, N) are the filtration factors of interstellar H, O, N atoms, respectively

[b] In parentheses we present filtration factors calculated under an assumption of reduced (by factor of 3) the electron temperature in the inner heliosheath between the TS and BS (Izmodenov et al. 2004)

shown that the number densities of the primary and secondary interstellar H atoms vary within 10% in the outer heliosphere, while closer to the Sun the variations increase.

Izmodenov et al. (2005a), Izmodenov and Alexashov (2006), Opher et al. (2006), Pogorelov et al. (2006) have studied the influence of the interstellar magnetic field on the structure of the interface assuming that the interstellar magnetic field (ISMF) is inclined with respect to the direction of the interstellar flow. In this case the SW/LIC interaction region becomes asymmetric and the flow pattern becomes essentially three-dimensional. Since interstellar H atoms are coupled to the charged component by charge exchange the flow of the interstellar atoms becomes asymmetric too, as observed in the backscattered solar Lyman-alpha radiation spectra measured by SOHO/SWAN (Lallement 2005).

Izmodenov and Alexashov (2006) performed a parametric study by varying the angle $\alpha$ between the direction of the interstellar flow and interstellar magnetic field from 0 to 90 degrees. Despite the fact that interstellar magnetic field significant disturbs the heliospheric interface and interstellar H flow, the filtration factor was in the range of $0.555 - 0.574$ for all considered angle values of angle $\alpha$ that is very close to the results of the B&M model.

Finally, we conclude that despite that the effects of the ionization level of the LIC, the interstellar magnetic field, the solar cycle and others significantly influence the structure of the heliospheric interface and plasma and H atom distributions within the heliosphere, the filtration factor of interstellar hydrogen varies insignificantly for all considered models. It remains in the range from 0.5 to 0.6.

## 4 Heavier Elements

The theoretical study of the penetration of interstellar heavier elements into the heliosphere was done in a large number of papers (Fahr 1991; Rucinski et al. 1993; Fahr et al. 1995; Kausch and Fahr 1997; Mueller and Zank 2003; Cummings et al. 2002a, 2002b; Izmodenov et al. 1997, 1999, 2003). The papers studied different aspects of the penetration of He, C, N, O through the interface. We will focus on the most recent results.

Firstly, the charge exchange cross section of the helium with protons is so small that the mean free path of the helium atoms is larger than the size of the heliospheric interface. Therefore, the helium atoms penetrate the heliospheric interface unperturbed. This fact was used in order to measure the local interstellar temperature and velocity (Witte et al. 1996; Witte 2004; Möbius et al. 2004). The filtration coefficient for helium $F_{\mathrm{He,TS}} \sim 1$. Cummings et al. (2002b) found that electron impact ionization from the HP to the TS resulted in a factor of about 0.9.

Izmodenov et al. (2004) used the advanced heliospheric interface model by Izmodenov et al. (2003) to perform a comparative study of the penetration through this interface of three interstellar elements—hydrogen, oxygen and nitrogen. Similar to hydrogen, the interstellar atoms of oxygen have large charge exchange cross sections and, therefore, the filtered in the heliospheric interface.

For O atoms both the direct $\mathrm{O} + \mathrm{p} \rightarrow \mathrm{O}^+ + \mathrm{H}$ and the reverse $\mathrm{O}^+ + \mathrm{H} \rightarrow \mathrm{O} + \mathrm{H}^+$ charge exchange processes should be taken into account. It was estimated by Cummings et al. (2002a, 2002b) that the charge exchange of nitrogen with protons may result only in $\sim$1% of filtration, and, therefore, it can be neglected. Electron impact ionization is important for interstellar oxygen (Izmodenov et al. 1999) and hydrogen, while it is almost negligible for filtration of H atoms.

Voronov's formula was employed for electron impact rate coefficients for O and N (Voronov 1997). For charge exchange cross sections for oxygen, the formula given by Stancil et al. (1999) was used. The number density of oxygen ions in the undisturbed LIC is determined by the ionization balance condition $n(\mathrm{OII})/n(\mathrm{HII}) = 8/9 \cdot n(\mathrm{OI})/n(\mathrm{HI})$. This condition is very close to the condition that can be derived from model 17 of Slavin and Frisch (2002). To calculate the number density of oxygen ions the continuity equation for this component (Izmodenov et al. 1999) was solved.

Izmodenov et al. (2004) performed parametric studies by varying the interstellar proton, $n_{\mathrm{p,LIC}}$, and atomic hydrogen, $n_{\mathrm{H,LIC}}$, number densities in the ranges of 0.032–0.07 cm$^{-3}$ and 0.16–0.2 cm$^{-3}$, respectively. The calculations were performed for 13 models with $n_{\mathrm{p,LIC}}$ and $n_{\mathrm{H,LIC}}$ listed in Table 2.

Figure 3 shows typical distributions of interstellar atomic number densities in the heliospheric interface region in the upwind direction (i.e. opposite to the Sun–LIC relative velocity vector). Qualitatively, such distributions take place for all models. Analogous to the hydrogen wall, the oxygen wall is formed due to the charge exchange process $\mathrm{O}^+ + \mathrm{H} \rightarrow \mathrm{O} + \mathrm{H}^+$.

Atoms that penetrated through the heliopause, can be ionized by hot solar wind electrons in the region between the TS and HP. The filtration in the inner heliosheath due to electron impact is more effective for interstellar N and O atoms as compared with hydrogen. Note that the electron impact ionization rate strongly depends on the electron temperature (Voronov 1997). As it was discussed in the previous section, we use one-fluid description for all plasma components. This approach is appropriate to determine the locations of the shock and the HP and for the plasma velocity, but certainly fails for prediction of the temperatures of the different ionized components. Since the TS is a quasi perpendicular collisionless shock, the electron component of the solar wind is expected to have a lower temperature in the inner heliosheath than one-fluid models predict. To estimate the effect of a change in electron temperature on the filtration factor, Izmodenov et al. (2003) performed calculations with the models where the electron temperature in the inner heliosheath obtained in the frame of the B&M model was arbitrarily divided by a factor of 3.

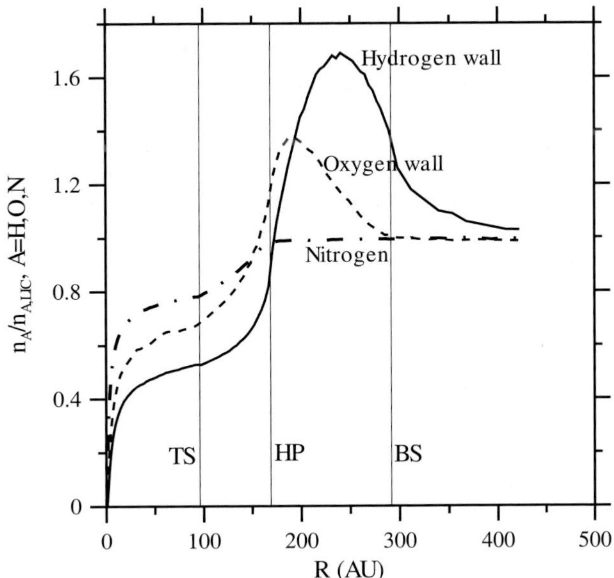


**Fig. 3** Distribution of hydrogen, oxygen, and nitrogen into the upwind direction along the axis of symmetry

Table 2 summarizes the filtration factors for all 13 models. It shows the location of the TS and the filtration factors, $F_{A,TS}$ (A = H, O, N). The main conclusion, which can be made based on results shown in the table, is that as for hydrogen the filtration factors do not vary significantly with variation of interstellar densities $n_{H,LIC}$ and $n_{p,LIC}$. We find that $68 \pm 3\%$ of interstellar oxygen and $78 \pm 2\%$ of interstellar nitrogen penetrate through the interaction region into the supersonic solar wind. The results of calculations with smaller electron temperature are shown in the table in parenthesis. Small electron temperature leads to stronger penetration of N- and O-atoms into the heliosphere. However, for the two types of models—with and without lowered electron temperature in the heliosheath—the ratio of the nitrogen and oxygen filtration factors changes insignificantly from $1.10 \pm 0.02$ to $1.15 \pm 0.02$. Thus, NI/OI in the LIC, if derived from pickup ion data, is not very sensitive to variations in the modeling of the LIC/SW interaction.

Gloeckler and Geiss (2004) derived from Ulysses pickup ion observations that $n_{OI,TS} = (5.3 \pm 0.8) \times 10^{-5}$ cm$^{-3}$ and $n_{NI,TS} = (7.8 \pm 1.5) \times 10^{-6}$ cm$^{-3}$. Dividing these values by the average of the filtration factors in Table 1, we obtain $n_{OI,LIC} = (7.8 \pm 1.3) \times 10^{-5}$ cm$^{-3}$ and $n_{NI,LIC} = (1.0 \pm 0.2) \times 10^{-5}$ cm$^{-3}$. Finally, the local interstellar OI/HI and NI/OI ratios are equal $(\mathrm{OI/HI})_{LIC} = (4.3 \pm 0.5) \times 10^{-4}$ and $(\mathrm{NI/OI})_{LIC} = 0.13 \pm 0.01$.

## 5 Summary and Conclusions

The filtration of the interstellar atoms of H, O, N in the heliospheric interface has been discussed. For hydrogen the filtration was analyzed on the basis of recent advanced multi-component models of the heliospheric interface. It was shown that the filtration coefficient is in the range of 0.5–0.6 for all models. A parametric study by varying local interstellar proton and atom number densities was performed for hydrogen, oxygen, and nitrogen by Izmodenov et al. (2004). It was found that

A. $54 \pm 4\%$ of interstellar hydrogen atoms, $68 \pm 3\%$ of interstellar oxygen and $78 \pm 2\%$ of interstellar nitrogen penetrate through the interaction region into the interface. In the case of a lower electron temperature in the heliosheath $81 \pm 2\%$ and $89 \pm 1\%$ of interstellar oxygen and nitrogen penetrate, respectively.
B. Using the filtration coefficients and SWICS/Ulysses pickup ion measurements we conclude that $n_{\mathrm{OI,LIC}} = (7.8 \pm 1.3) \times 10^{-5}\ \mathrm{cm}^{-3}$ and $n_{\mathrm{NI,LIC}} = (1.0 \pm 0.2) \times 10^{-5}\ \mathrm{cm}^{-3}$.
C. The local interstellar OI/HI and NI/OI ratios are $(\mathrm{OI/HI})_{\mathrm{LIC}} = (4.3 \pm 0.5) \times 10^{-4}$ and $(\mathrm{NI/OI})_{\mathrm{LIC}} = 0.13 \pm 0.01$. The obtained interstellar OI/HI ratio is slightly lower than the ratio $(4.8 \pm 0.48) \times 10^{-4}$ determined by Linsky et al. (1995) from spectroscopic observations of stellar absorptions.

**Acknowledgements** I thank the referee of the paper for numerous corrections that improved the paper significantly. I thank Johannes Geiss, George Gloeckler, Rosine Lallement, Yuri Malama and Dmitry Alexashov for their valuable contributions to the work reported in this paper. Especially I would like to thank Johannes Geiss for very pleasant, fruitful and brainstorming discussions during my visit to ISSI. Friendly recommendations and advice of Johannes helped me a lot in both my scientific and not scientific life. I also thank all staff of ISSI for the hospitality during my visits. A part of presented work was done in the frame of Russian–French cooperation (PICS program) under RFBR grant 05-02-22000_CNRS_a. The research was also supported by RFBR grant 07-02-01101-a.

## References

D. Alexashov, S.V. Chalov, A. Myasnikov, V. Izmodenov, R. Kallenbach, Astron. Astrophys. **420**, 729–736 (2004) ,
V.B. Baranov, M.G. Lebedev, Y.G. Malama, Astrophys. J. **375**, 347–351 (1991)
V.B. Baranov, Y.G. Malama, J. Geophys. Res. **98**, 15157–15163 (1993)
V.B. Baranov, Y.G. Malama, Space Sci. Rev. **78**, 305–316 (1996)
V.B. Baranov, V.V. Izmodenov, Fluid Dyn. **41**, 689–707 (2006)
A.C. Cummings, E.C. Stone, C.D. Steenberg, Astrophys. J. **578**, 194–210 (2002a)
A.C. Cummings, E.C. Stone, C.D. Steenberg, Astrophys. J. **581**, 1413 (2002b)
H.-J. Fahr, Astron. Astrophys. **241**, 251–259 (1991)
H.J. Fahr, R. Osterbart, D. Rucinski, Astron. Astrophys. **294**, 587–600 (1995)
J. Geiss, G. Gloeckler, U. Mall, R.von Steiger, A.B. Galvin, K.W. Ogilvie, Astron. Astrophys. **282**, 924–933 (1994)
G. Gloeckler, J. Geiss, Adv. Space. Res. **34**, 53–60 (2004)
V. Izmodenov, Space Sci. Rev. **97**, 385–388 (2001)
V. Izmodenov, D. Alexashov, Astron. Lett. **29**, 58–63 (2003)
V. Izmodenov, R. Lallement, Y.G. Malama, Astron. Astrophys. **317**, 193–202 (1997)
V.V. Izmodenov, R. Lallement, J. Geiss, Astron. Astrophys. **344**, 317–321 (1999)
V.V. Izmodenov, Y.G. Malama, A.P. Kalinin, M. Gruntman, R. Lallement, I.P. Rodionova, Astrophys. Space Sci. **274**, 71–76 (2000)
V.V. Izmodenov, M. Gruntman, Y. Malama, J. Geophys. Res. **106**, 10681–10690 (2001)
V.V. Izmodenov, Y.G. Malama, G. Gloeckler, J. Geiss, Astrophys. J. **594**, L59–L62 (2003)
V. Izmodenov, Y.G. Malama, G. Gloeckler, J. Geiss, Astron. Astrophys. **414**, L29–L32 (2004)
V. Izmodenov, D. Alexashov, A. Myasnikov, Astron. Astrophys. **437**, L35–L38 (2005a)
V. Izmodenov, Y.G. Malama, M.S. Ruderman, Astron. Astrophys. **429**, 1069–1080 (2005b)
V.V. Izmodenov, D.B. Alexashov, AIP Conf. Proc. **858**, 14–19 (2006)
V.V. Izmodenov, R. Kallenbach (eds), The physics of the heliospheric boundaries, ISSI Scientific Report 5, 2007
T. Kausch, H.J. Fahr, Astron. Astrophys. **325**, 828–838 (1997)
R. Lallement, Space Sci. Rev. **78**, 361–374 (1996)
R. Lallement, Science **307**, 1447–1449 (2005)
J.L. Linsky, B.E. Wood, Astrophys. J. **463**, 254 (1996)
J.L. Linsky, A. Dipas, B.E. Wood et al., Astrophys. J. **476**, 366 (1995)
Y.G. Malama, V.V. Izmodenov, S.V. Chalov, Astron. Astrophys. **445**, 693–701 (2006)
E. Möbius, M. Bzowski, S. Chalov, H.-J. Fahr, G. Gloeckler, V. Izmodenov, R. Kallenbach, R. Lallement, D. McMullin, H. Noda, Astron. Astrophys. **426**, 897–907 (2004)

H.-R. Mueller, G.P. Zank, AIP Conf. Proc. **679**, 89–92 (2003)
A.V. Myasnikov, V. Izmodenov, D. Alexashov, S. Chalov, J. Geophys. Res. **105**, 5179–5188 (2000)
M. Opher, E.C. Stone, P.C. Liewer, Astrophys. J. **640**, L71–L74 (2006)
N.V. Pogorelov, G.P. Zank, T. Ogino, Astrophys. J. **644**, 1299–1316 (2006)
E. Quémerais, R. Lallement, J.-L. Bertaux, D. Koutroumpa, J. Clarke, E. Kyrola, W. Schmidt, Astron. Astrophys. **455**, 1135–1142 (2006)
J.D. Richardson, Y. Liu, C. Wang, Adv. Space Res. (2007, submitted)
D. Rucinski, H.-J. Fahr, S. Grezedzielski, Planet. Space Sci. **41**, 773–783 (1993)
J.D. Slavin, P.C. Frisch, Astrophys. J. **565**, 364–379 (2002)
P.C. Stancil, D.R. Schultz, M. Kimura et al., Astron. Astrophys. Suppl. **140**, 225–234 (1999)
G.S. Voronov, At. Data Nucl. Data Tables **65**, 1–30 (1997)
M. Witte, M. Banaszkiewicz, H. Rosenbauer, Space Sci. Rev. **78**, 289–296 (1996)
M. Witte, Astron. Astrophys. **426**, 835–844 (2004)
G. Zank, Space Sci. Rev. **89**, 413–688 (1999)

Space Sci Rev (2007) 130: 389–399
DOI 10.1007/s11214-007-9161-y

# Composition of Anomalous Cosmic Rays

**A.C. Cummings · E.C. Stone**

Received: 10 January 2007 / Accepted: 16 February 2007 /
Published online: 5 May 2007


**Abstract** The "classic" anomalous cosmic ray (ACR) component originates as interstellar neutral atoms that drift into the heliosphere, become ionized and picked up by the solar wind, and carried to the outer heliosphere where the pickup ions are accelerated to hundreds of MeV, presumably at the solar wind termination shock. These interstellar ACRs are predominantly singly charged, although higher charge states are present and become dominant above ~350 MeV. Their isotopic composition is like that of the solar system and unlike that of the source of galactic cosmic rays. A comparison of their energy spectra with the estimated flux of pickup ions flowing into the termination shock reveals a mass-dependent acceleration efficiency that favors heavier ions. There is also a heliospheric ACR component as evidenced by "minor" ACR ions, such as Na, Mg, S, and Si that appear to be singly-ionized ions from a source likely in the outer heliosphere.

**Keywords** Anomalous cosmic rays · Composition · Local interstellar medium · Solar wind termination shock · Heliosphere · ACE · SAMPEX · Voyager

## 1 Introduction

The "classic" anomalous cosmic rays (ACRs) are a contemporary sample of the local interstellar medium. From the time they enter the heliosphere as interstellar neutral atoms until the time they are observed as ACRs is on the order of only a few years (Jokipii 1992; Mewaldt et al. 1996). Thus, their composition is of astrophysical importance, bearing on such issues as the abundance of interstellar neutral gas, the ionization state of the local interstellar medium (LISM), and the origin of galactic cosmic rays (GCRs) by comparison of isotopic abundances of the two samples of matter. In the heliosphere, the topics addressed by the study of ACRs include ionization by charge-exchange with the solar wind and by photo ionization; the fractionation in the abundances that occurs as the gas drifts through the heliosheath; and the injection, acceleration, and interplanetary propagation processes. In

A.C. Cummings (✉) · E.C. Stone
Space Radiation Laboratory, California Institute of Technology, Pasadena, CA 91125, USA
e-mail: ace@srl.caltech.edu

this paper we will review the current state of knowledge of the charge state, isotopic, and elemental composition of the classic ACRs.

In addition to these ACRs that are accelerated interstellar pickup ions, there is another ACR component with species having first-ionization potentials lower than that of H, such as C, Na, Mg, Si, S, and Fe (Reames 1999; Cummings et al. 2002a, 2002b). These are not of interstellar origin because they are mostly ionized in the LISM and are deflected around the heliosphere by the interplanetary magnetic field. Thus most, if not all, of these ACRs are thought to originate as heliospheric pickup ions. We will present recent observations of several of these species and discuss their probable sources.

Although Voyager 1 (V1) did not find the source of high energy ACRs when it crossed the termination shock on 16 December 2004 at 94 AU from the Sun, it did find a low energy component that is dominant below ~10 MeV/nuc (Decker et al. 2005; Stone et al. 2005). This component has a composition similar to that of interstellar pickup ions (Krimigis et al. 2003) and originates from a nearby region of the termination shock. Because this is a rapidly evolving area of study, we will not address this component in this paper.

## 2 Charge-State Composition

ACRs were first discovered as an anomaly in the energy spectrum of GCR helium when compared to that of GCR hydrogen (Garcia-Munoz et al. 1973). Subsequently, ACR N, O, Ne, Ar, and H were reported (Hovestadt et al. 1973; McDonald et al. 1974; Cummings and Stone 1987; Christian et al. 1988). Shortly after their discovery, Fisk et al. (1974) proposed that ACRs originated as interstellar neutral gas that could easily enter the heliosphere and become ionized and then accelerated somewhere in the outer heliosphere. Pesses et al. (1981) further proposed that the acceleration site was the termination shock of the solar wind. This explanation has become widely accepted, although the recent crossing of the termination shock by V1 showed that the source spectrum of ACRs was not at the place on the shock where the spacecraft crossed (Stone et al. 2005). It has been suggested that the spectrum was affected by merged interaction regions (MIRs) interacting with the shock (Florinski and Zank 2006), or that the acceleration takes place at either high or low latitudes (Stone et al. 2005), or along the flanks of the shock (McComas and Schwadron 2006). The similarity of the upstream ACR spectra observed by V2 three years later suggests that the V1 spectrum at the shock was unlikely to have been the result of an MIR interacting with the shock (Cummings et al. 2007).

According to these scenarios the ACRs should be singly ionized because the acceleration occurs more quickly than additional electron stripping. Direct measurement of the ACR charge state is not currently feasible because the ACR energy is too high for practical spaceborne spectrometry using magnets or electric fields. Cummings et al. (1984) showed that the energy spectra of the ACR He, N, O, and Ne spectra, which are shaped by the rigidity dependence of the interplanetary diffusion coefficient, are consistent with the particles being singly ionized. Adams et al. (1991) came to the same conclusion by comparing ACR O spectra inside and outside the magnetosphere. The definitive work in this area was accomplished with observations on the SAMPEX satellite, which is in a nearly circular polar orbit about the Earth. Using the Earth's geomagnetic field as a rigidity filter, it was found that most ACRs are singly charged as expected, but ACRs with total energies above ~350 MeV ACRs are predominantly multiply charged (Mewaldt et al. 1996; Klecker et al. 1998). This is attributed to ACRs being stripped of their electrons during their acceleration (Jokipii 1996; Mewaldt et al. 1996). In Fig. 1 we show the singly-ionized fraction of ACRs as a function of

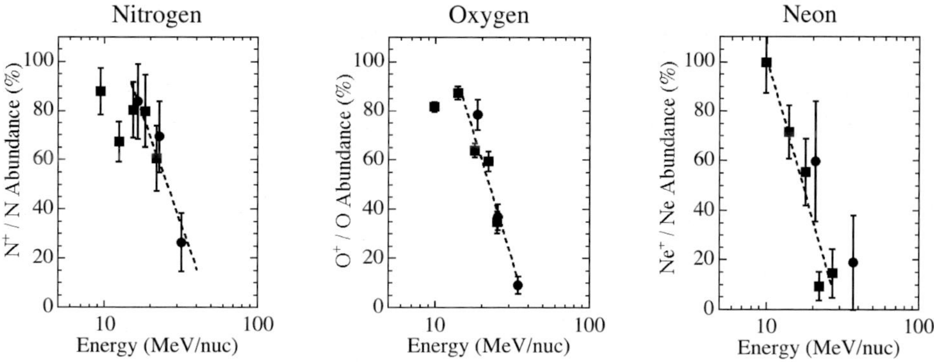


**Fig. 1** Percentage of ACR N, O, and Ne that is singly ionized as a function of energy (from Klecker et al. (1998))

energy for ACR N, O, and Ne. Charge states $> +2$ are also present in the ACR population with the fraction generally increasing with increasing energy (Klecker et al. 1997).

The determination that most ACR species are singly ionized places a limit on the distance they can have traveled before being observed and/or a limit on their acceleration time. Adams and Leising (1991) determined that ACRs must originate no more than 0.2 pc from Earth ($\sim$40,000 AU). With better measurements and improved cross section measurements Mewaldt (2006) was able to reduce the distance to <4000 AU if ACRs originate inside the heliosphere and <2000 AU if they originate in the LISM. The charge-state observations led Jokipii (1992) to estimate that the acceleration time of 10 MeV/nuc ACR oxygen must be less than about 4–6 years. Mewaldt (2006) subsequently lowered this estimate to 1–2 years.

## 3 Isotopic Composition

The isotopic composition of ACRs is a direct measurement of the isotopic composition of the contemporary local interstellar neutral gas flowing into the heliosphere. Thus the results have implications for galactic evolution in the solar neighborhood. Until the launch of the Advanced Composition Explorer (ACE) in 1997 the available ACR isotopic composition measurements were from lower-resolution spectrometers intended for elemental composition work. The observations from the 1972–1978 solar minimum period were reviewed by Mewaldt (1989) and these results suggested that the isotopic composition of He, N, O, and Ne were consistent with that of the solar system. Using Voyager data from the 1986-1988 solar minimum period, Cummings et al. (1991) improved the measurement of the ACR $^{22}$Ne/$^{20}$Ne ratio and showed that the composition was consistent with that of the solar wind and the meteoritic neon-A component but inconsistent with that of the GCR source. By far the most definitive measurements were carried out with the Solar Isotope Spectrometer (SIS) on ACE (Stone et al. 1998). In Fig. 2 we show energy spectra of isotopes of N, O, and Ne acquired from SIS data for the period August 1997 through March 1998 (Leske 2000). $^{14}$N, $^{16}$O, $^{18}$O, $^{20}$Ne, and $^{22}$Ne all show the increase in intensity below about 30 MeV/nuc that is characteristic of the ACR component. The $^{15}$N and $^{21}$Ne energy spectra are dominated by GCRs in the energy range of the measurement. As shown in the figure the abundances of $^{18}$O and $^{22}$Ne are consistent with solar system abundances, suggesting that the Sun was

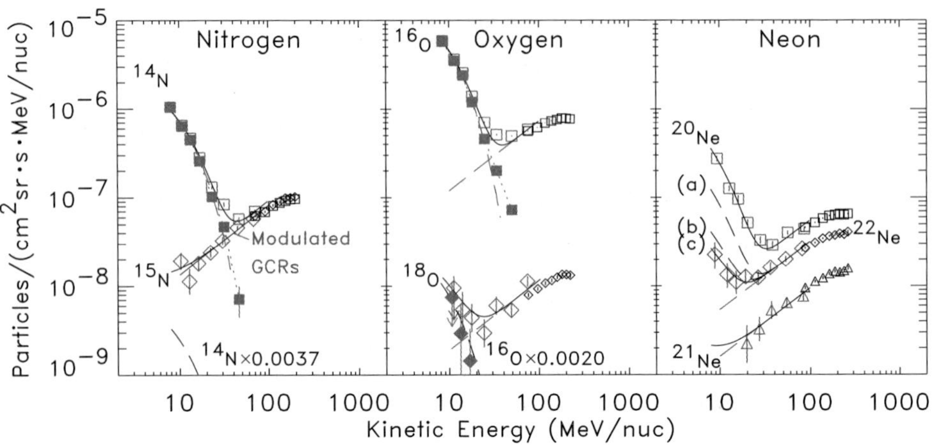


**Fig. 2** Quiet time energy spectra of isotopes of N, O, and Ne from the SIS instrument on ACE (Stone et al. 1998) for the period August 1997 through March 1998. *Open symbols* are observed values and *filled symbols* represent the estimated ACR intensities and result from subtracting an estimate of the GCR intensity from the observations. *Dotted curves* are an exponential fit to the ACR points. The estimated GCR energy spectra are shown as *dashed lines* representing power-law fits to the high-energy points with power-law index 0.8. The *solid curves* through the $^{14}$N, $^{16}$O, and $^{20}$Ne points are the sum of the GCR and ACR estimated energy spectra. The *dashed lines in the lower part of the first two panels* show the expected spectra of $^{15}$N and $^{18}$O if these isotopes have a solar system composition. In the case of $^{22}$Ne, three estimates of the energy spectrum are shown: (a) if the composition is that of the GCR source, (b) if it is that of meteoritic Neon-A, and (c) if it is that of the solar wind. *Solid curves* show the expected $^{15}$N, $^{18}$O, and $^{22}$Ne energy spectra (GCRs + ACRs) assuming the ACRs have solar system abundances relative to $^{14}$N, $^{16}$O, and $^{20}$Ne, respectively. For $^{22}$Ne the *curve* assumes the ACRs have the same abundance ratio as the solar wind. The figure is from Leske (2000)

formed from matter that is similar in composition to that in its current neighborhood. The $^{22}$Ne abundance is inconsistent with that of the GCR source, consistent with GCRs having been accelerated from material enriched with ejecta from Wolf Rayet stars (see Binns et al. (2005)).

## 4 Elemental Composition: Acceleration Efficiency

Elemental composition studies of energetic particle populations have traditionally been based on a common energy per nucleon interval. For fully stripped ions with charge to mass ratio $Q/M = 1/2$, this approach can lead to estimates of the seed population composition since rigidity-dependent acceleration and transport processes should be the same for each element. For ACRs, however, with $Q/M$ varying from 1/1 for H to 1/36 for Ar, this approach is not appropriate for investigating the source abundances. Cummings et al. (1984) inferred that the charge state of ACRs was +1 and also inferred the rigidity dependence of the interplanetary diffusion coefficient by using the fact that the energy spectra have their peak intensity for each species at the same value of the diffusion coefficient. This phenomenon is illustrated in Fig. 3 where we show the energy spectra of six ACR ions from V1 and V2 observations during the last period of minimum solar modulation. The energy of the peak intensity for H and He differ by a factor of about 5, from ~30 MeV/nuc for H to ~6 MeV/nuc for He. Curves from a numerical solar modulation model in each panel of Fig. 3

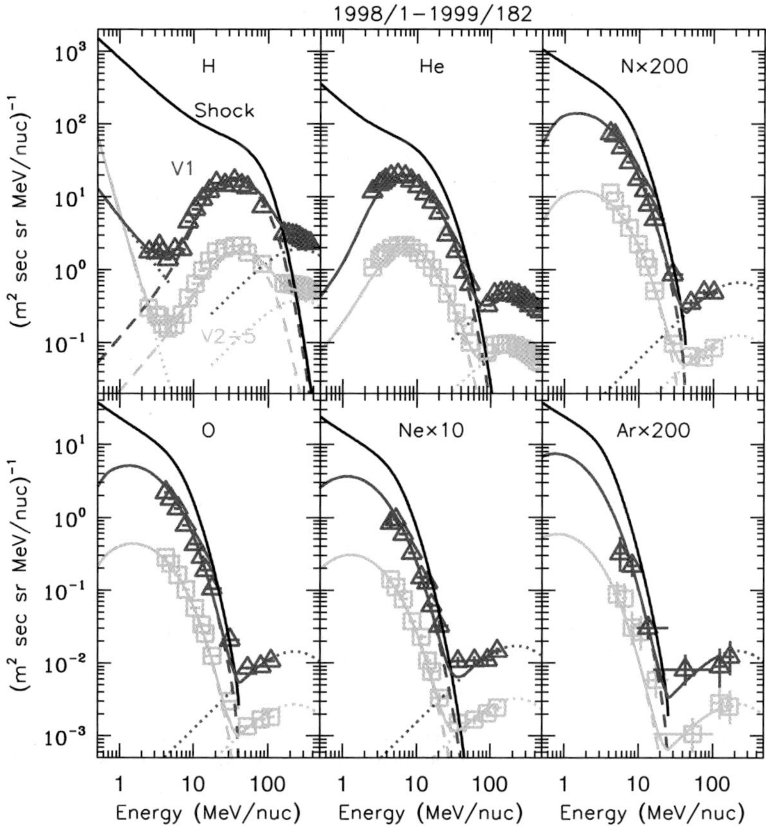


**Fig. 3** Energy spectra of H, He, N, O, Ne, and Ar at V1 (*triangles*) and V2 (*squares*) for the period 1998/1–1999/182 from the Cosmic Ray Subsystem experiment (CRS) (Stone et al. 1977). The *curves* are calculations from a numerical model of solar modulation. The *upper curve in each panel* is the calculated energy spectrum at a strong termination shock located at 90 AU for a polar angle of 30°. The figure is from Cummings et al. (2002a)

illustrate the same effect for the heavier species where the peak intensity is not in the energy range of observation.

The ACRs depicted in Fig. 3 are interstellar pickup ions thought to be accelerated at the termination shock of the solar wind. As shown schematically in Fig. 4, we can derive the acceleration efficiency for various ions by comparing pickup ion fluxes and ACR intensities (here we combine injection and acceleration in the term "acceleration efficiency"). Pickup ion observations for each of these species allow estimates of the neutral densities of the elements at the termination shock (Gloeckler and Fisk 2007). From the neutral density estimates, the intensity of pickup ions flowing into the termination shock can be derived from estimates of the ionization rates at 1 AU and the model of Vasyliunas and Siscoe (1976) (see Cummings et al. (2002a) for the technique). The termination shock accelerates the ions via a diffusive shock acceleration process (see, e.g., Blandford and Ostriker (1978)), which results in a power-law dependence of the energy spectrum at low energies with a roll off at higher energies. By dividing the estimated intensity at the termination shock at a common energy/nucleon along the power-

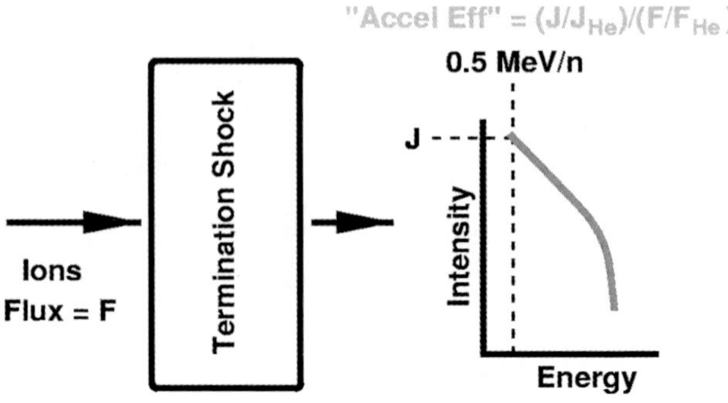


**Fig. 4** Schematic illustration of the steps involved in deriving the acceleration efficiency of pickup ions at the termination shock as described in the text

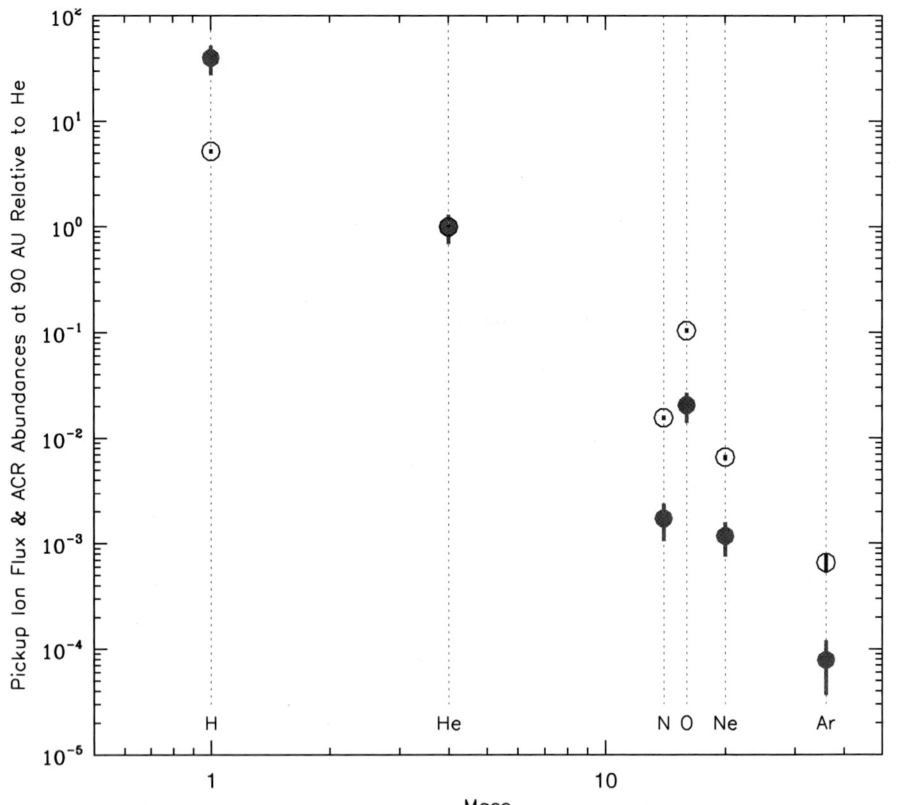


**Fig. 5** Estimates of the flux of pickup ions (*solid circles*) and ACR intensities at 0.5 MeV/nuc (*open circles*) at the termination shock relative to He in both cases vs ion mass

law portion of the spectrum, e.g., 0.5 MeV/nuc, by the pickup ion flux flowing into the shock, we infer the acceleration efficiencies for each ion. We do this in a relative way by normalizing both the ACR intensity and the pickup ion flux to the respective values for He.

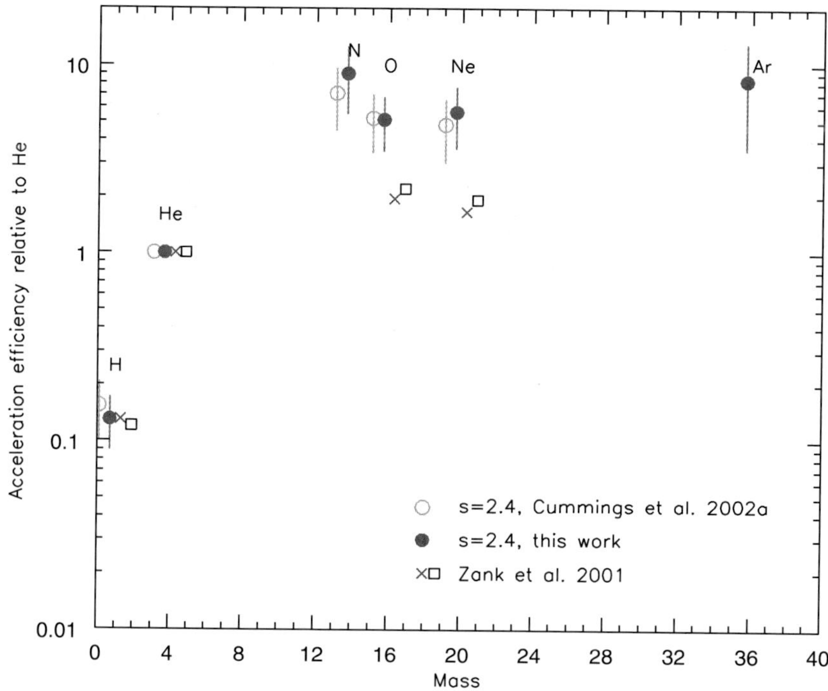


**Fig. 6** Estimates of the acceleration efficiencies for $H^+$, $N^+$, $O^+$, $Ne^+$, and $Ar^+$ relative to $He^+$. The *points* from Zank et al. (2001) are from a two-stage injection and acceleration theory

The numerator and denominator of the acceleration efficiency are shown for the six elements in Fig. 5. Although the estimates are presented for the nose of the termination shock at 90 AU, the relative fluxes of pickup ions are nearly independent of radial distance to the shock or the longitude or latitude where the actual shock acceleration process is taking place. The ACR intensities at 0.5 MeV/nuc at the termination shock are from Table 2 of Cummings et al. (2002a) for a weak shock with compression ratio 2.4, which is close to the value observed as V1 crossed the termination shock (Burlaga et al. 2005). It is clear that the acceleration efficiency varies with ion mass, being less than 0.2 for H and ~7 for N, O, Ne, and Ar.

Four estimates of the acceleration efficiencies are shown in Fig. 6. One of these is from the work of Cummings et al. (2002a) and one is from this work. The small differences are due to slightly different pickup ion results used in the analysis. Two theoretical estimates are also shown from Zank et al. (2001), who used a two-stage acceleration process to achieve an acceleration efficiency that increases with mass, roughly in accordance with the observations. The theoretical estimates were in much better agreement with an earlier determination of the acceleration efficiency (Cummings and Stone 1996), and parameters could likely be adjusted to achieve better agreement with the data than shown in Fig. 6. The theory uses multiply-reflected ions as a first-stage acceleration process that favors lighter ions, followed by injection into the diffusive acceleration process that favors heavier ions. The combination of the two effects leads to the heavier ions being favored overall.

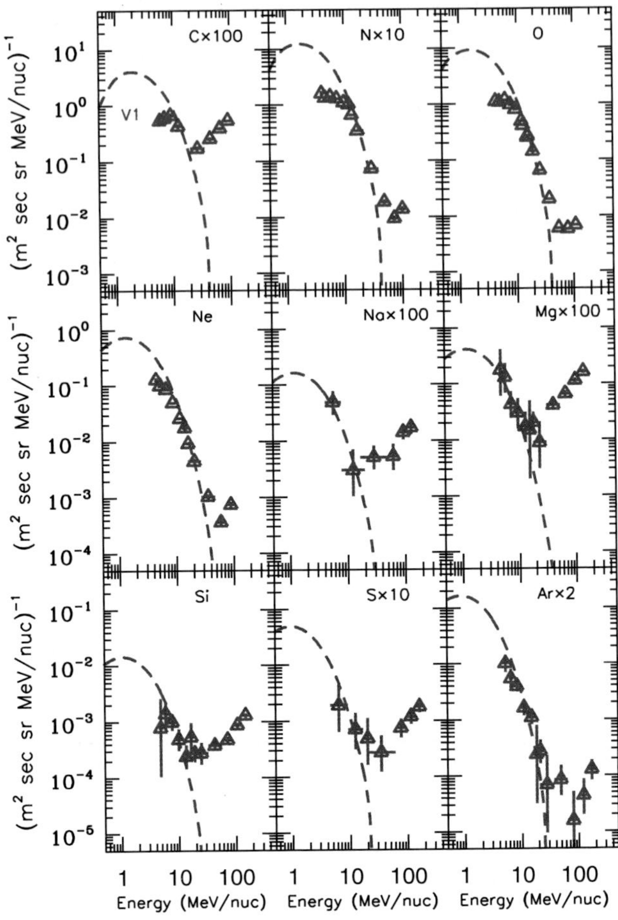


**Fig. 7** Energy spectra of nine ions from V1 observations from 2002/209–2004/350. The *dashed line* is from the fit to the V1 ACR observations in Cummings et al. (2002a) for the period 1993/53–1999/365 multiplied by a factor of 3

## 5 Minor ACRs

Species that are not primarily neutral in the LISM also have an upturn in intensity at low energies characteristic of ACRs (Reames 1999; Cummings et al. 2002a, 2002b). Energy spectra of these ions, C, Na, Mg, Si, and S, as well as some of the primary ACRs, N, O, Ne, and Ar, are shown in Fig. 7. The data are from V1 for the period 2002/209–2004/350, the 2.4 year time interval just before V1 crossed the termination shock (2004/351) and during which the ACR O intensity with 7.1–17.1 MeV/nuc was reasonably constant. The points at high energies are GCRs. The intensities turn up at low energies in each panel at different energies depending on the relative abundances of the ACR component compared to that of the GCRs. The ACR intensities are higher at high energies by a factor of ~3 as compared to the observations from the last solar minimum period during 1993/53–1999/365. This is expected in the current $A < 0$ part of the solar magnetic cycle as compared to the previous $A > 0$ cycle if drift effects play a major role in the acceleration and transport process (Jokipii 1990; Cummings and Stone 1999; Stone and Cummings 1999; Florinski et al. 2004; Webber et al. 2005).

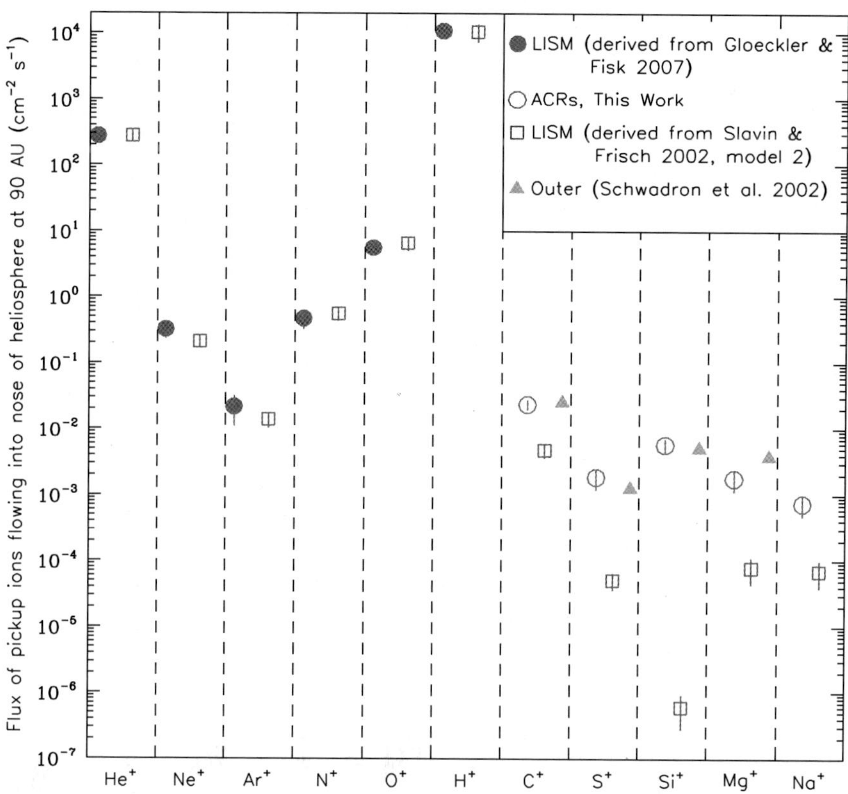


**Fig. 8** Estimated flux of pickup ions for the interstellar neutral source and from the outer heliospheric source flowing into the nose of the heliosphere at 90 AU. The *solid circles* are estimates derived from the neutral densities at the termination shock from Gloeckler and Fisk (2007), the ionization rates at 1 AU from Cummings et al. (2002a), and the theory of Vasyliunas and Siscoe (1976). The *open squares* are derived from the neutral densities of the local interstellar medium from model 2 of Slavin and Frisch (2002), ionization rates at 1 AU and the filtration factors in the heliosheath from Cummings et al. (2002a), and the theory of Vasyliunas and Siscoe (1976). The *open circles* are derived from the model ACR intensities at the termination shock at 0.5 MeV/nuc and the acceleration efficiencies from Fig. 6 (*solid circles*). The *solid triangles* are derived from the outer source estimates from Schwadron et al. (2002)

The origin of the minor ACR ions C, Na, Mg, Si, and S, has not been determined, but some possibilities, such as neutrals from the LISM (except possibly for C), solar wind ions, co-rotating interaction regions, and interstellar singly-charged ions have been ruled out (Cummings et al. 2002a, 2002b). The main possibilities remaining are an inner source of recently discovered heliospheric pickup ions (Schwadron et al. 1999; Gloeckler et al. 2000; Schwadron et al. 2000; Allegrini et al. 2005) and a suggested outer source (Schwadron et al. 2002) of heliospheric ions thought to originate in the Kuiper asteroid belt at some 30–50 AU from the Sun, but not directly observed. The origin of the inner source has not been established (Allegrini et al. 2005) but the pickup ions are thought to originate within 10–30 solar radii of the Sun. Cummings et al. (2002a) found that the fluxes of inner source pickup ions and their composition were in reasonable agreement with the minor ACR observations of C, Mg, and Si, if their acceleration efficiencies were similar to that of the primary ACRs. Since the velocity distribution of the inner source pickup ions is similar to that of the solar wind and further adiabatic cooling is expected to occur in their journey to the termination

shock, it seems unlikely they would have the same acceleration efficiency as the interstellar pickup ions.

The outer source also has reasonable composition and flux of pickup ions into the termination shock to explain the minor ACRs as shown in Fig. 8. In this figure the estimated pickup ion fluxes at the nose of the termination shock at 90 AU are shown for eleven ions. Two estimates are shown for the primary ACRs with high first ionization potentials, one derived from the neutral densities originating in the LISM from model 2 of Slavin and Frisch (2002) and another from the LISM but derived from the pickup ion observations of Gloeckler and Fisk (2007). These two estimates are in very good agreement and indicate that the neutral densities of these elements are reasonably well established for the LISM. For the minor ACR ions, three estimates of the pickup ion fluxes are given: one from the Slavin and Frisch (2002) model 2, one from the outer heliospheric source (Schwadron et al. 2002), and one from the ACR observations (Cummings et al. 2002a, 2002b). The pickup ions from the interstellar neutral source are too low to account for the minor ACRs, although there is some interstellar contribution to C. The relative abundances of the outer source fluxes are consistent with the minor ACR elements of C, S, Si, and Mg, but the production rate is uncertain due to our current lack of knowledge about the mass production rate and size distribution of dust from the Kuiper Belt (Schwadron, private communication). If the minor ACRs originate as pickup ions from such an outer source, then the ACR intensities provide information on the processes that generate the outer source ions.

## 6 Summary

We have presented the status of the charge state, isotopic, and elemental composition of ACRs, both of those that originate in the LISM and those that are heliospheric in origin. Most ACRs are singly ionized, as expected, although at higher energies they transition to higher charge states due to electron stripping during the acceleration process. The isotopic composition is consistent with that of the solar system; in particular the $^{22}Ne/^{20}Ne$ ratio is consistent with that of the solar wind and the meteoritic neon-A component and unlike that of the GCR source, consistent with a GCR source enriched with ejecta from Wolf Rayet stars. The similarity with the solar system composition also implies that the contemporary LISM has nearly the same composition as the material from which the Sun was formed some 5 billion years ago. A comparison of the elemental composition of ACRs with that of pickup ions from interstellar neutrals indicates the acceleration efficiency of the termination shock is mass dependent, with the higher mass-to-charge ratio elements being favored. When the actual source location and acceleration mechanism of the ACRs are determined, the method of inferring the acceleration efficiencies outlined here should be revisited.

**Acknowledgements** We appreciate discussions with R. Mewaldt and N. Schwadron. This work was supported by NASA under contract NAS7-03001.

## References

J.H. Adams Jr. et al., Astrophys. J. **375**, L45–L48 (1991)
J.H. Adams Jr., M.D. Leising, in *Proc. 22nd Internat. Cosmic Ray Conf.*, vol. 3 (Dublin, 1991), pp. 304–307
F. Allegrini et al., J. Geophys. Res. **110**, A05105 (2005). doi:10.1029/2004JA010847
W.R. Binns et al., Astrophys. J. **634**, 351–364 (2005)
R.D. Blandford, J.P. Ostriker, Astrophys. J. Lett. **221**, L29–L32 (1978)

L.F. Burlaga et al., Science **309**, 2027–2029 (2005)
E.R. Christian et al., Astrophys. J. Lett. **334**, L77–L80 (1988)
A.C. Cummings, E.C. Stone, in *Proc. 20th Internat. Cosmic Ray Conf.*, vol. 3 (Moscow, 1987), pp. 413–416
A.C. Cummings, E.C. Stone, Space Sci. Rev. **78**, 117–128 (1996)
A.C. Cummings, E.C. Stone, Adv. Space Res. **23**, 509–520 (1999)
A.C. Cummings et al., *J. Geophys. Res.* (2007, in preparation)
A.C. Cummings et al., Astrophys. J. **578**, 194–210 (2002a)
A.C. Cummings et al., Astrophys. J. **581**, 1413 (2002b)
A.C. Cummings, E.C. Stone, W.R. Webber, Astrophys. J. Lett. **287**, L99–L103 (1984)
A.C. Cummings et al., in *Proc. 22nd Internat. Cosmic Ray Conf.*, vol. 3 (Dublin, 1991), pp. 362–365
R.B. Decker et al., Science **309**, 2020–2024 (2005)
L.A. Fisk et al., Astrophys. J. Lett. **190**, L35–L38 (1974)
V. Florinski, G.P. Zank, Geophys. Res. Lett. **33**, L15110 (2006). doi:10.1029/2006GL026371
V. Florinski et al., Astrophys. J. **610**, 1169–1181 (2004)
M. Garcia-Munoz et al., Astrophys. J. Lett. **182**, L81–L84 (1973)
G. Gloeckler, L. Fisk, Space Sci. Rev. (2007) this volume
G. Gloeckler et al., J. Geophys. Res. **105**, 7459–7463 (2000)
D. Hovestadt et al., Phys. Rev. Lett. **31**, 650–653 (1973)
J.R. Jokipii, in *Physics of the Outer Heliosphere: Proc. of the 1st COSPAR Colloquium* (Warsaw, 1990), pp. 169–178
J.R. Jokipii, Astrophys. J. **393**, L41–L43 (1992)
J.R. Jokipii, Astrophys. J. Lett. **466**, L47–L50 (1996)
B. Klecker et al., Space Sci. Rev. **83**, 259–308 (1998)
B. Klecker et al., in *Proc. 25th Internat. Cosmic Ray Conf.*, vol. 2 (Durban, 1997), pp. 273–276
S.M. Krimigis et al., Nature **426**, 45–48 (2003)
R.A. Leske, in *AIP Conf. Proc. 516: 26th Internat. Cosmic Ray Conf., ICRC XXVI*, vol. 516 (Salt Lake City, 2000), pp. 274–282
D.J. McComas, N.A. Schwadron, Geophys. Res. Lett. **33**, L04102 (2006). doi:10.1029/2005GL025437
F.B. McDonald et al., Astrophys. J. Lett. **187**, L105–L108 (1974)
R.A. Mewaldt, in *AIP Conf. Proc. 183: Cosmic Abundances of Matter*, vol. 183 (Minneapolis, 1989), pp. 124–146
R.A. Mewaldt, in *AIP Conf. Proc. 858: Physics of the Inner Heliosheat*, vol. 858 (Honolulu, 2006), pp. 92–97
R.A. Mewaldt et al., Astrophys. J. Lett. **466**, L43–L46 (1996)
M.E. Pesses et al., Astrophys. J. Lett. **246**, L85–L89 (1981)
D.V. Reames, Astrophys. J. **518**, 473–479 (1999)
N.A. Schwadron et al., Geophys. Res. Lett. **29**, 54–51 (2002)
N.A. Schwadron et al., J. Geophys. Res. **105**, 7465–7472 (2000)
N.A. Schwadron et al., in *AIP Conf. Proc. 471: Proc. of the 9th Internat. Solar Wind Conf.*, vol. 471 (Nantucket, 1999), pp. 487–490
J.D. Slavin, P.C. Frisch, Astrophys. J. **565**, 364–379 (2002)
E.C. Stone et al., Space Sci. Rev. **21**, 355–376 (1977)
E.C. Stone et al., Space Sci. Rev. **86**, 355–376 (1998)
E.C. Stone, A.C. Cummings, in *Proc. 26th Internat. Cosmic Ray Conf.*, vol. 7 (Salt Lake City, 1999), pp. 500–503
E.C. Stone et al., Science **309**, 2017–2020 (2005)
V.M. Vasyliunas, G.L. Siscoe, J. Geophys. Res. **81**, 1247–1252 (1976)
W.R. Webber et al., J. Geophys. Res. **110**, A07106 (2005). doi:10.1029/2005JA011123
G.P. Zank et al., Astrophys. J. **556**, 494–500 (2001)

Space Sci Rev (2007) 130: 401–408
DOI 10.1007/s11214-007-9181-7

# Interstellar Dust in the Solar System

**Harald Krüger · Markus Landgraf · Nicolas Altobelli · Eberhard Grün**

Received: 14 February 2007 / Accepted: 28 March 2007 /
Published online: 22 May 2007


**Abstract** The Ulysses spacecraft has been orbiting the Sun on a highly inclined ellipse almost perpendicular to the ecliptic plane (inclination 79°, perihelion distance 1.3 AU, aphelion distance 5.4 AU) since it encountered Jupiter in 1992. The in situ dust detector on board continuously measured interstellar dust grains with masses up to $10^{-13}$ kg, penetrating deep into the solar system. The flow direction is close to the mean apex of the Sun's motion through the solar system and the grains act as tracers of the physical conditions in the local interstellar cloud (LIC). While Ulysses monitored the interstellar dust stream at high ecliptic latitudes between 3 and 5 AU, interstellar impactors were also measured with the in situ dust detectors on board Cassini, Galileo and Helios, covering a heliocentric distance range between 0.3 and 3 AU in the ecliptic plane. The interstellar dust stream in the inner solar system is altered by the solar radiation pressure force, gravitational focussing and interaction of charged grains with the time varying interplanetary magnetic field. We review the results from in situ interstellar dust measurements in the solar system and present Ulysses' latest interstellar dust data. These data indicate a 30° shift in the impact direction of interstellar grains w.r.t. the interstellar helium flow direction, the reason of which is presently unknown.

**Keywords** Dust · Interstellar dust · Heliosphere · Interstellar matter

H. Krüger (✉)
Max-Planck-Institut für Sonnensystemforschung, 37191 Katlenburg-Lindau, Germany
e-mail: krueger@mps.mpg.de

H. Krüger · E. Grün
Max-Planck-Institut für Kernphysik, 69029 Heidelberg, Germany

M. Landgraf
European Space Agency, ESOC, 64293 Darmstadt, Germany

N. Altobelli
NASA/JPL, Pasadena, CA, USA

E. Grün
HIGP, University of Hawaii, Honolulu, HI 96822, USA

## 1 Introduction

One of the most important results of the Ulysses mission is the identification and characterization of a wide range of interstellar phenomena inside the solar system. A surprise was the identification of interstellar dust grains sweeping through the solar system (Grün et al. 1993). Before this discovery it was believed that interstellar grains are prevented from reaching the planetary region by electromagnetic interaction with the solar wind magnetic field. The interplanetary zodiacal dust flux was thought to dominate the near-ecliptic planetary region while at high ecliptic latitudes only a very low flux of dust released from long-period comets should be present. Therefore, the characterization of the interplanetary dust cloud was the prime goal of the Ulysses dust investigation (Grün et al. 1992).

Ulysses was launched in October 1990. A swing-by manoeuvre at Jupiter in February 1992 rotated its orbital plane 79° relative to the ecliptic plane (with a six-year orbital period and aphelion distance at 5.4 AU; Fig. 1). Subsequent aphelion passages occurred in April 1998 and in June 2004. A second Jupiter flyby occurred in February 2004. The best conditions for detection of interstellar impactors are in the outer solar system beyond 3 AU at high ecliptic latitudes and far away from Jupiter where impact rates of interplanetary grains or jovian stream particles are comparatively small. Nevertheless, Ulysses has continuously monitored the interstellar dust flux in the heliosphere since 1992.

We briefly review the results from in situ interstellar dust measurements with the Ulysses and other space-borne dust detectors in Sect. 2. In Sect. 3 we present the interstellar dust measurements from the Ulysses' 3rd passage through the outer heliosphere, and Sect. 4 is a brief discussion of our results.

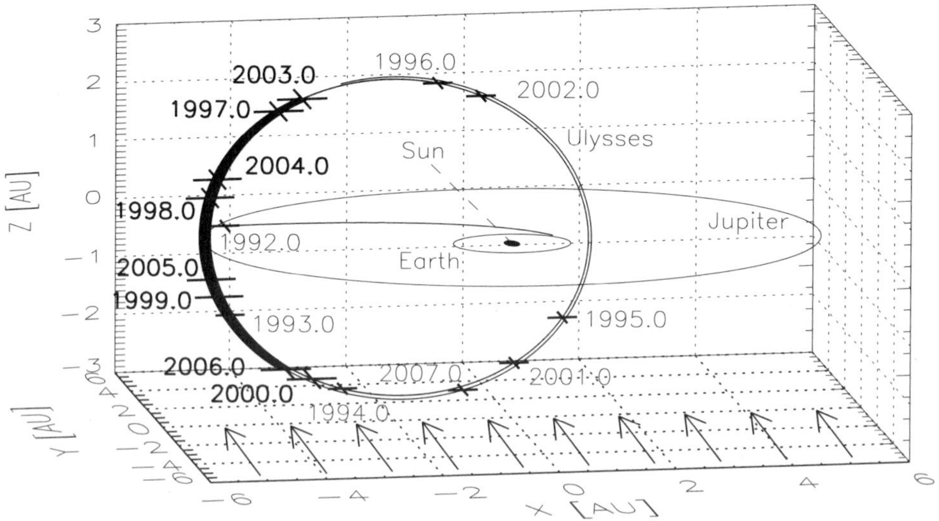


**Fig. 1** The trajectory of Ulysses in ecliptic coordinates. The Sun is in the centre. The orbits of Earth and Jupiter indicate the ecliptic plane. Ulysses' initial trajectory was in the ecliptic plane. Since Jupiter flyby in early 1992 the orbit has been almost perpendicular to the ecliptic plane (79° inclination). *Crosses* mark the spacecraft position at the beginning of each year. The 1997 to 1999 and 2003 to 2005 parts of the trajectory are shown as a *thick line*. Vernal equinox is to the right (positive $x$ axis) and the flow direction of the interstellar grains (coincident with the interstellar helium flow) is indicated by *arrows*

## 2 Interstellar Dust Measurements in the Heliosphere

Interstellar dust particles originating from the Local Interstellar Cloud (LIC) move on hyperbolic trajectories through the solar system and approach Ulysses predominantly from the direction opposite to the expected impact direction of interplanetary grains. On average, their impact velocities exceed the local solar system escape velocity, even if radiation pressure effects are neglected (Grün et al. 1994). The grain motion through the solar system was found to be parallel to the flow of neutral interstellar hydrogen and helium gas, both gas and dust travelling with a speed of 26 km $s^{-1}$ (Grün et al. 1994; Baguhl et al. 1995; Witte et al. 1996; Frisch et al. 1999). The interstellar dust flow persisted at high latitudes above and below the ecliptic plane and even over the poles of the Sun, whereas interplanetary dust was strongly depleted at high ecliptic latitudes (Grün et al. 1997).

Later measurements with the Galileo dust detector in the ecliptic plane confirmed the Ulysses results: beyond about 3 AU the interstellar dust flux exceeds the flux of micron-sized interplanetary grains. Furthermore, interstellar dust is ubiquitous in the solar system: dust measurements between 0.3 and 3 AU in the ecliptic plane exist also from Helios, Galileo and Cassini. This data shows evidence for distance-dependent alteration of the interstellar dust stream caused by radiation pressure, gravitational focussing and electromagnetic interaction with the time-varying interplanetary magnetic field which also depends on grain size (Altobelli et al. 2003; Altobelli et al. 2005b; Altobelli et al. 2005a; Mann and Kimura 2000; Landgraf 2000; Czechowski and Mann 2003). As a result, the size distribution and fluxes of grains measured inside the heliosphere are strongly modified (Landgraf et al. 1999a; Landgraf et al. 2003).

Interstellar grains observed with the spacecraft detectors range from $10^{-18}$ kg to above $10^{-13}$ kg. If we compare the mass distribution of these interstellar impactors detected in situ with the dust mass distribution derived from astronomical observations, we find that the in situ measurements overlap only with the largest masses observed by remote sensing. It indicates that the intrinsic size distribution of interstellar grains in the LIC extends to grain sizes larger than those detectable by astronomical observations (Frisch et al. 1999; Frisch and Slavin 2003; Landgraf et al. 2000; Grün and Landgraf 2000). Even bigger interstellar grains (above $10^{-10}$ kg) are observed as radar meteors entering the Earth's atmosphere (Taylor et al. 1996; Baggaley and Neslušan 2002). The flow direction of these larger grains varies over a much wider angular range than that of small particles measured by the in situ detectors.

The total grain mass detected in situ by Ulysses, which includes bigger grains in the LIC than those detectable with astronomical techniques, led to the conclusion that the LIC gas is enhanced with the refractory elements (e.g. Fe, Mg, Mn) that would ordinarily dominate the mass of interstellar dust grains. Earlier investigations indicated an enhancement of the dust-to-gas mass ratio in the LIC by up to a factor of five (Frisch et al. 1999). Recent reanalysis with improved solar heavy element abundances and an updated value for the sensitive area of the Ulysses dust detector (Altobelli et al. 2004) brought this enhancement down to a factor of two (Frisch and Slavin 2003; Slavin and Frisch, this volume).

In addition to the distribution of grain masses, the instrument has monitored the flux of interstellar dust particles through the heliosphere since Ulysses left the ecliptic plane in 1992 (Fig. 2). In mid 1996, a drop of the interstellar dust flux from initially $1.5 \times 10^{-4}\ m^{-2}\,s^{-1}$ to $0.5 \times 10^{-4}\ m^{-2}\,s^{-1}$ occurred (Landgraf et al. 1999b). Since early 2000, Ulysses has detected interstellar dust flux levels above $10^{-4}\ m^{-2}\,s^{-1}$ again. The drop in 1996 was explained by increased filtering of small grains by the solar wind driven magnetic field during solar minimum conditions (Landgraf et al. 2000; Landgraf 2000). The filtration caused a deficiency

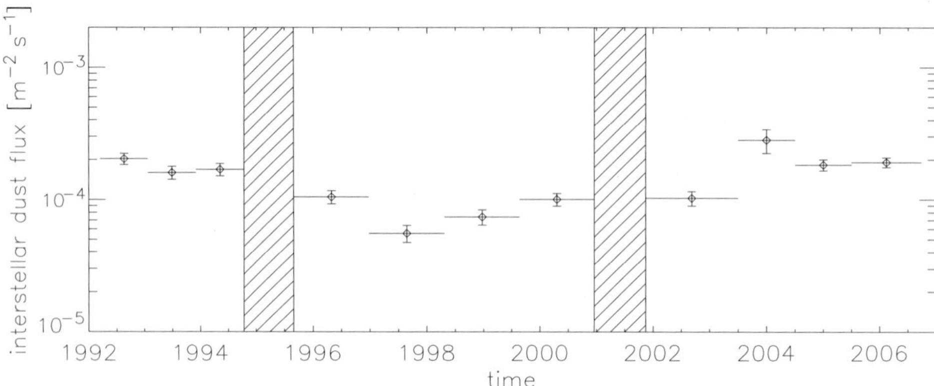


**Fig. 2** Interstellar dust flux measured by Ulysses. The *horizontal lines* indicate the length of the time intervals, and the *vertical bars* of the data points represent the 1 $\sigma$ uncertainty due to small number statistics. The *dashed regions* in 1995 and 2001 show the periods of Ulysses' perihelion passages where the distinction of interstellar dust from interplanetary impactors is difficult. Furthermore, in 2003 and 2004 a possible contamination by jovian dust stream particles around Jupiter flyby, which occurred in February 2004, may lead to an erroneously enhanced interstellar flux. Dust streams identified in the Ulysses data set were removed by ignoring the time interval when a dust stream occurred

of detected interstellar grains with sizes below 0.2 $\mu$m (Grün et al. 1994). An additional filtration by solar radiation pressure, which was found to be effective at heliocentric distances below 4 AU, deflects grains with sizes of 0.4 $\mu$m (Landgraf et al. 1999a).

Modelling the dynamics of the electrically charged dust grains in the heliosphere can give us information about the Local Interstellar Cloud (LIC) where the particles originate from. In the time interval 2001 to 2003 the interstellar dust flux stayed relatively constant, in agreement with improved models (Landgraf et al. 2003). The dominant contribution to the flux comes from grains with a charge to mass ratio $q/m = 0.59\ \mathrm{C\,kg^{-1}}$ and a radiation pressure efficiency of $\beta = 1.1$ which—in the simulation—corresponds to a grain radius of 0.3 $\mu$m (assuming spherical grains). The models assume a constant dust concentration in the LIC and give a good fit to the dust fluxes measured between end-1992 and end-2003. The fact that the models fit the observed variations implies that the dust phase of the LIC is homogeneously distributed over length scales of at least 50 AU which is the distance inside the LIC traversed by the Sun during this time period. This result, however, needs to be reexamined in light of Ulysses's most recent measurements obtained during the spacecraft's 3rd passage through the outer solar system.

## 3 Ulysses's 3rd Passage through the Outer Solar System

From 2002 to 2006 Ulysses made its 3rd passage through the outer heliosphere (aphelion passage occurred in June 2004 at a heliocentric distance of 5.4 AU), providing again good conditions for measuring interstellar dust. In February 2004, however, the spacecraft had its second Jupiter flyby at a closest approach distance 0.8 AU which also allowed for good measurement conditions for the dust streams emanating from the jovian system (Grün et al. 1993; Krüger et al. 2006). A total of 28 streams were detected, more than twice the number of detections from Ulysses' first Jupiter flyby in 1992: the first stream was recorded in November 2002 when Ulysses was still 3.4 AU away from Jupiter, and the last stream in mid-2005 at more than 4 AU jovicentric distance.

In the outer solar system and at high ecliptic latitudes the interstellar impactors can usually be identified by their impact direction: they approach from a retrograde direction while the majority of interplanetary grains move on prograde heliocentric orbits. During most of the time in the 2002 to 2005 interval, however, the jovian dust streams approached from roughly the same direction (Krüger and Grün 2007), so that identification of interstellar grains by their impact direction alone was not possible. On the other hand, the measured impact charge distribution showed that the majority of grains with impact charges above $Q_I = 2 \times 10^{-13}$ C are of interstellar origin while most jovian stream particles have impact charges below this limit. We therefore use this limit to separate both populations of impactors. Contamination by jovian stream particles, however, cannot entirely be excluded this way, and in particular the fluxes attributed to interstellar grains in 2004—when the instrument detected the most intensive jovian dust streams (Krüger et al. 2006)—may be contaminated by jovian impactors. Hence, the data point in Fig. 2 showing an elevated interstellar flux in 2004 has to be taken with caution. On the other hand, in mid-2005 dust stream detections ceased, and in the later data the contribution by jovian stream particles should be negligible.

Figure 3 shows the impact directions of grains with impact charges $Q_I \geq 2 \times 10^{-13}$ C for two different time intervals. The intervals were chosen such that Ulysses was traversing approximately the same region of the outer solar system during both periods. Contour lines show the effective dust sensor area for particles approaching from the upstream direction of interstellar helium (Witte et al. 1996), implying that the detection conditions for interstellar dust were very similar in both intervals. The approach directions of the majority of grains are consistent with the upstream direction of the interstellar helium flow. This is particularly evident in the earlier time interval 1996 to 2000 (*left panel* in Fig. 3). It should be noted that in this interval Jupiter and Ulysses were on opposite sides of the solar system, separated by more than 10 AU, so that contributions by jovian stream particles can be excluded. The distribution of the measured rotation angles is also shown in Fig. 4. In the 1997 to 1999 interval the average impact direction of the interstellar grains was at rotation angles of about 95°.

The interstellar impactors were still concentrated towards the interstellar helium flow direction in 2003 and 2004 (*right panel* in Fig. 3) although the distribution of the measured rotation angles was somewhat wider. Later, in 2005, the impact directions were significantly shifted from the helium flow. This is also evident in the *right panel* of Fig. 4: the mean rotation angle of the impactors is at 135° rotation angle. Taking into account that the detection geometry has slightly changed between the two time intervals, this implies that the interstellar dust flow has shifted by at least 30° in southward direction, away from the ecliptic plane. The wider distribution of impact directions is also evident.

## 4 Discussion

The dust measurements from Ulysses' 3rd passage through the outer solar system imply an at least 30° shift in the approach direction of interstellar dust grains. The reason for this shift remains mysterious. Whether it is connected to a secondary stream of interstellar neutral atoms shifted from the main neutral gas flow (Collier et al. 2004) is presently unclear. However, given that the neutral gas stream is shifted along the ecliptic plane while the shift in the dust flow is offset from the ecliptic, a connection between both phenomena seems unlikely.

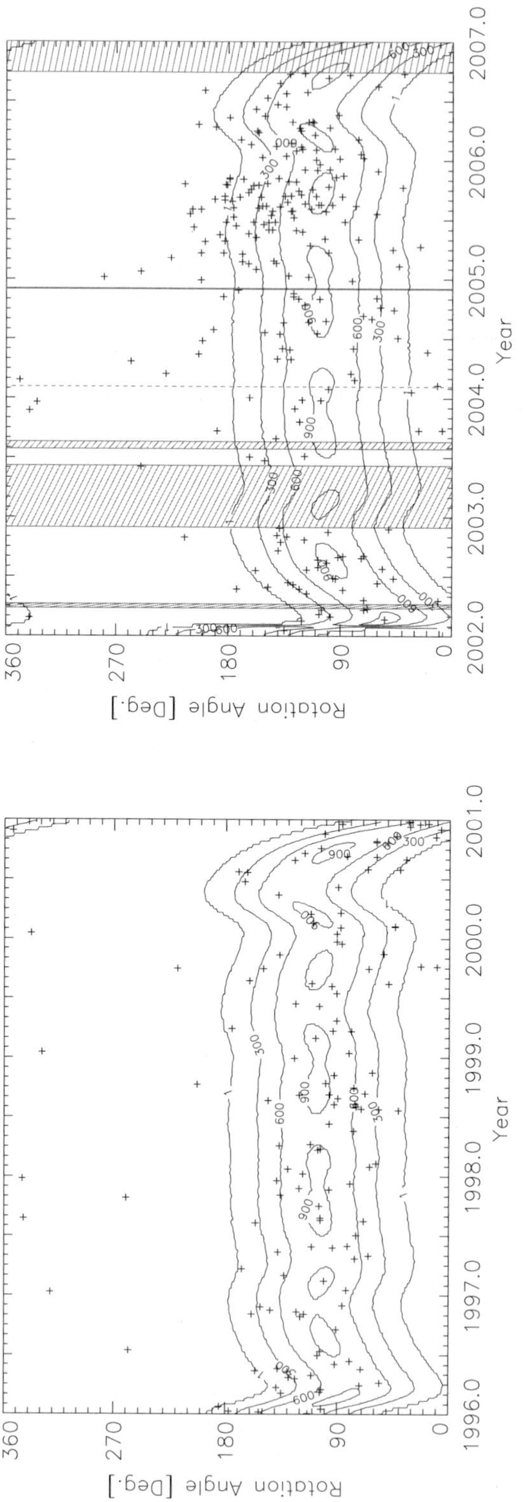

**Fig. 3** Impact direction (i.e. spacecraft rotation angle at dust particle impact) of interstellar grains measured with Ulysses in two time intervals. *Left:* 1 January 1996 to 31 December 2000; *right:* 1 January 2002 to 31 December 2006. Ecliptic north is close to 0°; impact charges $Q_I \geq 2 \times 10^{-13}$ C. *Each cross* indicates an individual impact. *Contour lines* show the effective sensor area for particles approaching from the upstream direction of interstellar helium. In the *right panel, a vertical dashed line* shows Jupiter closest approach on 5 February 2004, *five shaded areas indicate* periods when the dust instrument was switched off

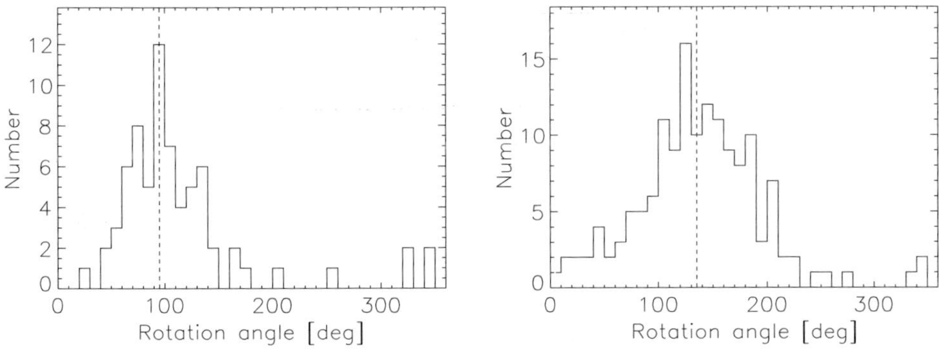


**Fig. 4** Distribution of measured impact directions (i.e. spacecraft rotation angle at dust particle impact) of interstellar impactors for two time intervals. *Left:* 1 January 1997 to 31 December 1999; *right:* 1 January 2003 to 31 December 2005. In the earlier time interval the maximum of the distribution is at a rotation angle of 95° very close to the value expected from the interstellar helium flow. In the second interval the maximum is at 135°

Even though Ulysses' position in the heliosphere and the dust detection conditions were very similar during both time intervals considered here, the configurations of the solar wind driven interplanetary magnetic field (IMF) which affected the grain dynamics were vastly different. One has to take into account that the interstellar grains need approximately twenty years to travel from the heliospheric boundary to the inner solar system where they are detected by Ulysses. Thus, the effect of the IMF on the grain dynamics is the accumulated effect caused by the interaction with the IMF over several years. Hence, in the earlier time interval (1997–1999) the grains had a recent dynamic history dominated by solar minimum conditions (Landgraf 2000), while the grains detected during the second interval (2002–2005) had a recent history dominated by the much more disturbed solar maximum conditions of the IMF. This latter configuration may have a strong influence on the dust dynamics in the inner heliosphere but it is not modeled in detail in the presently existing models. It may particularly affect small grains which are most sensitive to the electromagnetic interaction. One would expect a size-dependent shift in the grain impact direction which, however, is not evident in the data. Whether these phenomena cause the observed shift in the approach direction of the interstellar dust grains will be the subject of future investigations.

**Acknowledgements** We thank the Ulysses project at ESA and NASA/JPL for effective and successful mission operations. This work has been supported by the Deutsches Zentrum für Luft- und Raumfahrt e.V. (DLR) under grants 50 0N 9107 and 50 QJ 9503. Support by Max-Planck-Institut für Kernphysik and Max-Planck-Institut für Sonnensystemforschung is also gratefully acknowledged.

## References

N. Altobelli, S. Kempf, M. Landgraf et al., J. Geophys. Res. **108**(A10), 7 (2003)
N. Altobelli, R. Moissl, H. Krüger, M. Landgraf, E. Grün, Planet. Space Sci. **52**, 1287 (2004)
N. Altobelli, S. Kempf, H. Krüger et al., in *The Spectral Energy Distributions of Gas-Rich Galaxies: Confronting Models with Data*. AIP Conf. Proc., vol. 761 (2005a), pp. 149–152
N. Altobelli, S. Kempf, H. Krüger et al., J. Geophys. Res. **110**, 7102 (2005b)
W.J. Baggaley, L. Neslušan, Astron. Astrophys. **382**, 1118 (2002)
M. Baguhl, E. Grün, D.P. Hamilton et al., Space Sci. Rev. **72**, 471 (1995)
M.R. Collier, T.E. Moore, D. Simpson et al., Adv. Space Res. **34**, 166 (2004)
A. Czechowski, I. Mann, J. Geophys. Res. **108**(A10), 8038 (2003). doi: 10.1029/2003JA009917

P.C. Frisch, J.D. Slavin, Astrophys. J. **594**, 844 (2003)
P.C. Frisch, J. Dorschner, J. Geiß et al., Astrophys. J. **525**, 492 (1999)
E. Grün, M. Landgraf, J. Geophys. Res. **105**(A5), 10,291 (2000)
E. Grün, H. Fechtig, J. Kissel et al., Astron. Astrophys., Suppl. **92**, 411 (1992)
E. Grün, H.A. Zook, M. Baguhl et al., Nature **362**, 428 (1993)
E. Grün, B.E. Gustafson, I. Mann et al., Astron. Astrophys. **286**, 915 (1994)
E. Grün, P. Staubach, M. Baguhl et al., Icarus **129**, 270 (1997)
H. Krüger, A.L. Graps, D.P. Hamilton et al., Planet. Space Sci. **54**, 919 (2006)
H. Krüger, E. Grün, in *Dust in Planetary Systems*, ed. by H. Krüger, A.L. Graps. ESA SP, vol. 643 (2007), pp. 69–72
M. Landgraf, J. Geophys. Res. **105**(A5), 10,303 (2000)
M. Landgraf, K. Augustsson, E. Grün, B.A.S. Gustafson, Science **286**, 2,319 (1999a)
M. Landgraf, M. Müller, E. Grün, Planet. Space Sci. **47**, 1029 (1999b)
M. Landgraf, W.J. Baggeley, E. Grün, H. Krüger, G. Linkert, J. Geophys. Res. **105**(A5), 10,343 (2000)
M. Landgraf, H. Krüger, N. Altobelli, E. Grün, J. Geophys. Res. **108**, 5 (2003)
I. Mann, H. Kimura, J. Geophys. Res. **105**(A5), 10,317 (2000)
A.D. Taylor, W.J. Baggeley, D.I. Steel, Nature **380**, 323 (1996)
M. Witte, H. Banaszkiewicz, H. Rosenbauer, Space Sci. Rev. **78**(1/2), 289 (1996)

Space Sci Rev (2007) 130: 409–414
DOI 10.1007/s11214-007-9186-2

# The Chemical Composition of Interstellar Matter at the Solar Location

**Jonathan D. Slavin · Priscilla C. Frisch**

Received: 14 February 2007 / Accepted: 4 April 2007 /
Published online: 11 May 2007


**Abstract** The Local Interstellar Cloud (LIC) surrounds the Solar System and sets the boundary conditions for the heliosphere. Using both in situ and absorption line data towards $\epsilon$ CMa we are able to constrain both the ionization and the gas phase abundances of the LIC gas at the Solar Location. We find that the abundances are consistent with all of the carbonaceous dust grains having been destroyed, and in fact with a supersolar abundance of C. The constituents of silicate grains, Si, Mg, and Fe, appear to be sub-solar, indicating that silicate dust is present in the LIC. N, O and S are close to the solar values.

**Keywords** ISM: abundances · ISM: individual (Local Interstellar Cloud)

## 1 Introduction

The elemental composition of the LIC, which surrounds the Solar System, can be derived from UV absorption line data, but requires models to fill in the gaps in the data. In particular we do not know directly the total H column density along the line of sight. In addition, for several elements, important ionization corrections need to be made to obtain total column densities from the observed ion column densities. We have created detailed models for the ionizing radiation field and used the radiative transfer code Cloudy (Ferland et al. 1998) to calculate the ionization in the LIC. These models provide the composition of interstellar neutrals that flow into the heliosphere and form the parent population of pickup ions (PUIs) and anomalous cosmic rays (ACRs).

J.D. Slavin (✉)
Harvard-Smithsonian Center for Astrophysics, 60 Garden Street, MS 83, Cambridge, MA 02138, USA
e-mail: jslavin@cfa.harvard.edu

P.C. Frisch
Department of Astronomy & Astrophysics, University of Chicago, 5460 South Ellis Avenue, Chicago, IL 60637, USA

## 2 Data

Gry and Jenkins (2001) present GHRS and IMAPS data for the line of sight towards $\epsilon$ Canis Majoris. This absorption line dataset is notable for its completeness, including lines of Mg I, Mg II, C II, C II*, N I, O I, S II, Si II, Si III, and Fe II for the LIC velocity component. We use the LIC component at 17 km s$^{-1}$ (Table 2 of Gry and Jenkins 2001) for this analysis. In this work we concentrate on this one line of sight as a means of putting constraints on our models that predict the boundary conditions of the heliosphere. Datasets for lines of sight to other nearby stars, e.g. $\eta$ UMa, Capella, do exist, though most are less complete in terms of the number of detected lines. We will present analyses of more lines of sight in future papers.

## 3 Model Parameters

There are several important parameters for our photoionization models that are either not directly constrained or constrained poorly by observations:

- total H I column density, $N$(H I)
- temperature of the hot gas of the Local Bubble, $T_h$
- total H density at the outer edge of the cloud, $n$(H)
- the cloud magnetic field, $B_0$

In our models we explore values of $N(\text{H I}) = (3\text{–}4.5) \times 10^{17}$ cm$^{-2}$. A significant fraction of the ionizing radiation field in our models comes from diffuse soft X-ray/EUV emission generated in the hot gas of the Local Bubble. The temperature of that hot gas is uncertain, which affects the spectrum of the emission. Observations of the broad band count rates at soft X-ray energies provide some constraints on the spectrum, but the diffuse EUV emission that provides the bulk of the ionizing photons has not been observed. The hot gas temperatures we explore are $\log T_h = 5.9, 6.0, 6.1$. The treatment of the magnetic field in the cloud is simply as an additional pressure term, $P_M \propto B^2$, with the field strength assumed to go as the density (see Slavin and Frisch 2002). Thus a higher field strength implies a higher total pressure in the cloud. In the transition layer between the warm cloud and hot gas where the magnetic pressure falls off sharply the thermal pressure increases to make up the difference. In our initial model grid we use $B_0 = 2$ and 5 $\mu$G. The total H density is also an important model input and is constrained indirectly by the observed He$^0$ density. Our initial model grid used values of 0.218 and 0.273 cm$^{-3}$ (corresponding to total densities of 0.24 and 0.3 cm$^{-3}$). After calculating an initial grid of 24 models we then calculated a smaller grid of 6 models in which we fixed $\log T_h$ and $N$(H I) and varied $n$(H) and $B_0$ so as to match the observed values of $n$(He$^0$) and $T$(He$^0$) observed for interstellar neutrals flowing into the heliosphere.

## 4 Radiation Field

As in Slavin and Frisch (2002) we construct the interstellar radiation field from both directly observed stellar contributions and diffuse emission that is either unobserved or observed at low spectral resolution. The components of the radiation field include:

- EUV (100 Å $< \lambda <$ 912 Å) from nearby B and white dwarf stars,
- Diffuse soft X-rays ($\lambda <$ 100 Å) from hot gas in the Local Bubble and galactic halo,

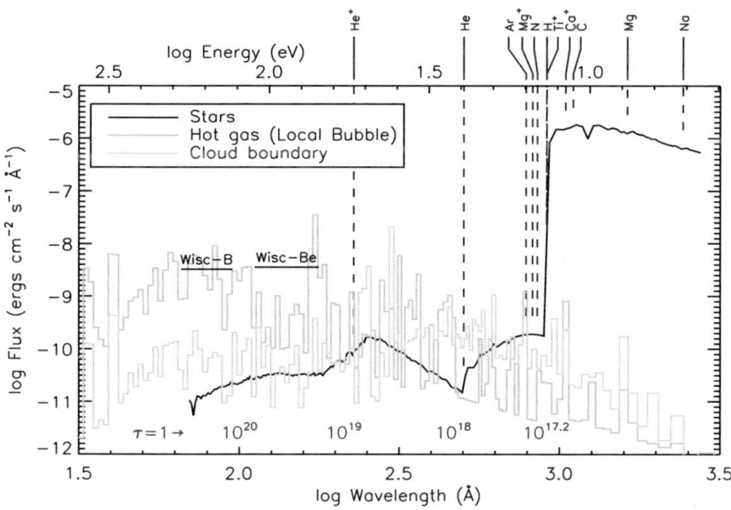


**Fig. 1** Background interstellar radiation field responsible for ionizing the LIC. The *black line* is the FUV/EUV flux from stars, the *green line* is a model spectrum for the diffuse emission from hot gas and the *cyan line* is the modeled emission from the evaporative boundary of the cloud. The ionization edges for several important ions are shown in *dark blue*. The wavelength at which the cloud is optically thick for several different column densities is indicated in *red*. The band widths and mean flux levels for the soft X-ray observations by the Wisconsin group are indicated by *horizontal lines*. (Colour figure online)

- FUV (912 Å $< \lambda <$ 3000 Å)—mainly from B stars distributed throughout the galactic disk (Gondhalekar et al. 1980),
- Diffuse EUV emission generated at the boundary of the hot gas in the Local Bubble and the warm (∼6000 K) gas of the LIC.

Figure 1 illustrates the different components of the model radiation field. The stellar FUV strongly dominates the ionization of the low ionization potential ions such as $Mg^0$, $C^0$, and $Ca^+$ whereas the stellar EUV, hot gas emission and cloud boundary flux all play a role in ionizing those ions with ionization potentials above 13.6 eV.

## 5 Results

We fix the model parameters to get a match to the observed in situ values for $n(\mathrm{He}^0) = 0.0151 \pm 0.0015$ and $T(\mathrm{He}^0) = 6300 \pm 340$ (Witte 2004). For our best models this results in $B_0 = 0$–$3.8\ \mu$G and $n_\mathrm{H} = 0.21$–$0.23$ cm$^{-3}$. The abundances are fixed by the requirement that they fit the observed column densities. Figure 2 summarizes the results for abundances by comparing the derived abundances with solar abundances from Asplund et al. (2005). The relative abundances are plotted against the condensation temperature for the given element, since it is believed that $T_C$ is important for determining the rate at which an element will be incorporated into grains at sites of dust formation. We find that C abundance is very high and S abundance slightly high relative to Solar. On the other hand Fe, Mg and Si, components of silicate dust, all have substantial depletions. N and O have solar or slightly subsolar abundance. Note that our conclusions about C do not rest heavily on our photoionization modeling. As we discuss in Slavin and Frisch (2006), simple assumptions combined with

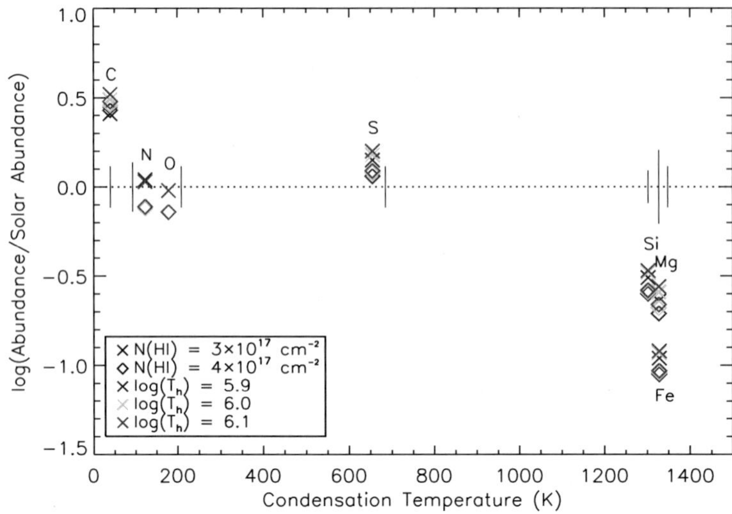


**Fig. 2** Abundances derived for the LIC from photoionization models constrained by the Gry and Jenkins (2001) data. Log of the abundance relative to the Solar abundances of Asplund et al. (2005) is plotted vs. the condensation temperature for the element. *Crosses* (×) are for low column density ($N(\text{H I}) = 3 \times 10^{17}$ cm$^{-2}$) models while *diamonds* (◇) are for higher column density models ($N(\text{H I}) = 4 \times 10^{17}$ cm$^{-2}$). The assumed temperature of the hot gas is indicated by the *symbol color*. For these cases $B_0$ and $n_\text{H}$ have been varied so as to match the observed values for $T$ and $n(\text{He}^0)$ at the Solar location (see Slavin and Frisch 2007 for details). The vertical lines indicate the uncertainties in the Solar abundances given by Asplund et al. (Colour figure online)

**Table 1** Comparison between Model, In Situ and Interstellar Data

| Ratio | Model 26 | PUIs corrected for filtration[a] | White Dwarf stars[b] |
|---|---|---|---|
| $O^0/N^0$ | 8.3 | $8.81 \pm 2.41$ | $8.6 \pm 3.9$ |
| $O^0/H^0$ | $3.5 \times 10^{-4}$ | $3.51 \pm 0.01 \times 10^{-4}$ | $3.5 \pm 2.2 \times 10^{-4}$ |
| $Ar^0/O^0$ | $2.6 \times 10^{-3}$ | $3.56 \pm 0.17 \times 10^{-3}$ | $3.9 \pm 1.4 \times 10^{-3}$ [c] |
| $H^0/He^0$ | 11.8 | $12.0 \pm 1.6$ | $12.8 \pm 1.4$[d] |

[a]The PUI values at the termination shock (Gloeckler and Fisk 2007) have been corrected to interstellar densities using filtration factors of 0.6, 1.0, 0.8, 0.8, and 0.76 for H, He, N, O, and Ar, respectively (Cummings et al. 2002; Müller and Zank 2004)

[b]For WD stars 0050–332, 0549+158, 1254+223, 1314+293, 1615–154, 1631+781, 1634–573, 1844–233, 2004–605, 2111+498, 2211–495 (Lehner et al. 2003)

[c]Exceptions are that WD0050–332 and WD0549–158 show $Ar^0/O^0 < 1.6 \times 10^{-3}$

[d]WD stars 0050–332, 0549+158, 1254+223, 1314+293, 2004–605

the data on $N(\text{C II}^*)$, $N(\text{Mg II})$ and $N(\text{Mg I})$ for the $\epsilon$ CMa line of sight lead to the result of a supersolar C abundance.

PUI and ACR data provide another source of comparison for our models. Table 1 summarizes how our model results compare with these data. We find generally good agreement,

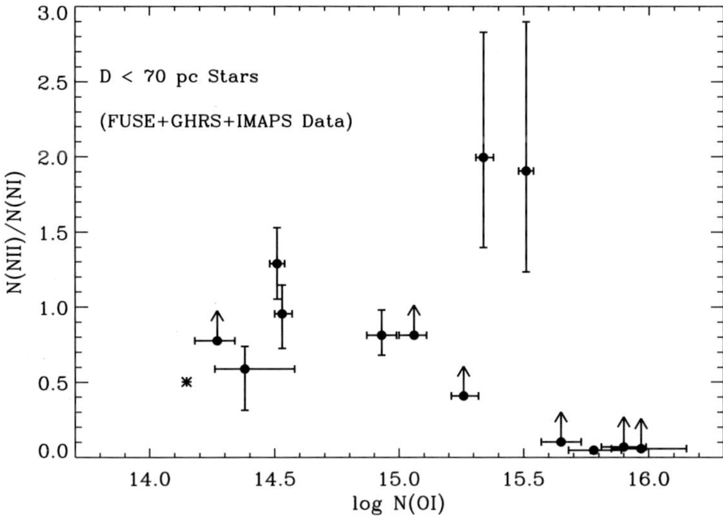


**Fig. 3** FUSE, GHRS, and IMAPS data for $N(\text{N II})/N(\text{N I})$ versus $N(\text{O I})$ for white dwarfs near the Sun. Also shown (*asterisk*) is the result for our photoionization models of the LIC (Model 26 from Slavin and Frisch 2007). The stars plotted are WD0050–332, WD0549+158, WD1254+223, WD1314+293, WD1615–154, WD1631+781, WD1634–573, WD1844–233, WD2004–605, WD2111+498, WD2211–495 from Lehner et al. (2003), and WD1620–391 from the FUSE archives. Data for $\eta$ UMa are also plotted (Frisch et al. 2007)

though we have some indication that our assumption for the Ar abundance may be too low (we do not fix the Ar abundance in our modeling).

Looking beyond the LIC, several low column density clouds are present inside the Local Bubble. Figure 3 illustrates how the observations of O I, N I and N II toward nearby white dwarfs compares with our model results for the LIC. The column of O I should track the H I column closely since charge transfer tightly couples the ionization of the two atoms. The ratio $N(\text{N II})/N(\text{N I})$ is a measure of the degree of H ionization in the gas since the ionization potential of N, 14.5 eV, is close to that of H and because H and N are coupled by charge transfer. The $N(\text{N II})/N(\text{N I})$ ratio for our model 26 (Slavin and Frisch 2007, shown by an asterisk) is consistent with that seen toward several of the WDs. The fact that for several lines of sight this ratio is ~0.5–1.5 illustrates that these clouds are partially ionized. Thus neither the assumptions for H II region type ionization nor those for neutral gas apply and full photoionization calculations must be carried out in order to make the ionization corrections needed to calculate elemental abundances correctly.

## 6 Conclusions

We have carried out detailed photoionization calculations of the LIC based on the available data on the interstellar radiation field and theoretical calculations needed to fill in gaps in that data. The combination of absorption line data toward $\epsilon$ CMa and the modeled photoionization lead us to the conclusion that the LIC has a very interesting pattern of gas phase elemental abundances. C appears to be substantially supersolar while Fe, Mg and Si are subsolar. O and N are close to solar, though perhaps slightly low. We interpret these results as indicating that carbonaceous grains have been destroyed in the LIC while silicate grains have survived. The extra C in the gas has not been explained, but may be evidence for local enhancement of carbonaceous dust followed by grain destruction in a shock.

**Acknowledgements** This work has been supported by NASA grants NNG05GD36G and NNG06GE33G.

## References

M. Asplund, N. Grevesse, A.J. Sauval, in *Cosmic Abundances as Records of Stellar Evolution and Nucleosynthesis*. ASP Conf. Ser., vol. 336 (2005), p. 25
A.C. Cummings, E.C. Stone, C.D. Steenberg, Astrophys. J. **578**, 194 (2002)
G.J. Ferland, K.T. Korista, D.A. Verner, J.W. Ferguson, J.B. Kingdon, E.M. Verner, PASP **110**, 761 (1998)
P.C. Frisch, J. Aufdenberg, E.B. Jenkins, C. Johns-Krull, J.D. Slavin, U.J. Sofia, D.E. Welty, D.G. York (2007, in preparation)
G. Gloeckler, L.A. Fisk (2007, this volume)
P.M. Gondhalekar, A.P. Phillips, R. Wilson, Astron. Astrophys. **85**, 272 (1980)
C. Gry, E.B. Jenkins, Astron. Astrophys. **367**, 617 (2001)
N. Lehner, E. Jenkins, C. Gry, H. Moos, P. Chayer, S. Lacour, Astrophys. J. **595**, 858 (2003)
H.-R. Müller, G.P. Zank, J. Geophys. Res. (Space Phys.) **109**, 7104 (2004)
J.D. Slavin, P.C. Frisch, Astrophys. J. **565**, 364 (2002)
J.D. Slavin, P.C. Frisch, Astrophys. J. Lett. **651**, L37 (2006)
J.D. Slavin, P.C. Frisch (2007, in preparation)
M. Witte, Astron. Astrophys. **426**, 835 (2004)

Space Sci Rev (2007) 130: 415–429
DOI 10.1007/s11214-007-9198-y

# An Overview of the Origin of Galactic Cosmic Rays as Inferred from Observations of Heavy Ion Composition and Spectra

**M.E. Wiedenbeck · W.R. Binns · A.C. Cummings · A.J. Davis · G.A. de Nolfo · M.H. Israel · R.A. Leske · R.A. Mewaldt · E.C. Stone · T.T. von Rosenvinge**

Received: 7 February 2007 / Accepted: 16 April 2007 /
Published online: 27 June 2007


**Abstract** The galactic cosmic rays arriving near Earth, which include both stable and long-lived nuclides from throughout the periodic table, consist of a mix of stellar nucleosynthesis products accelerated by shocks in the interstellar medium (ISM) and fragmentation products made by high-energy collisions during propagation through the ISM. Through the study of the composition and spectra of a variety of elements and isotopes in this diverse sample, models have been developed for the origin, acceleration, and transport of galactic cosmic rays. We present an overview of the current understanding of these topics emphasizing the insights that have been gained through investigations in the charge and energy ranges $Z \lesssim 30$ and $E/M \lesssim 1$ GeV/nuc, and particularly those using data obtained from the Cosmic Ray Isotope Spectrometer on NASA's Advanced Composition Explorer mission.

**Keywords** ISM: cosmic rays · Abundances · Acceleration of particles · Supernovae: general

## 1 Introduction

A sample of matter synthesized and accelerated outside of the solar system arrives near Earth in the form of galactic cosmic rays. The composition of this sample bears the imprint of the

M.E. Wiedenbeck (✉)
Jet Propulsion Laboratory, California Institute of Technology, M.C. 169-327, 4800 Oak Grove Dr., Pasadena, CA 91109, USA
e-mail: mark.e.wiedenbeck@jpl.nasa.gov

W.R. Binns · M.H. Israel
Dept. of Physics, Washington University, St. Louis, MO 63130, USA

A.C. Cummings · A.J. Davis · R.A. Leske · R.A. Mewaldt · E.C. Stone
California Institute of Technology, M.C. 220-47, Pasadena, CA 91125, USA

G.A. de Nolfo · T.T. von Rosenvinge
NASA Goddard Space Flight Center, Code 661, Greenbelt, MD 20771, USA

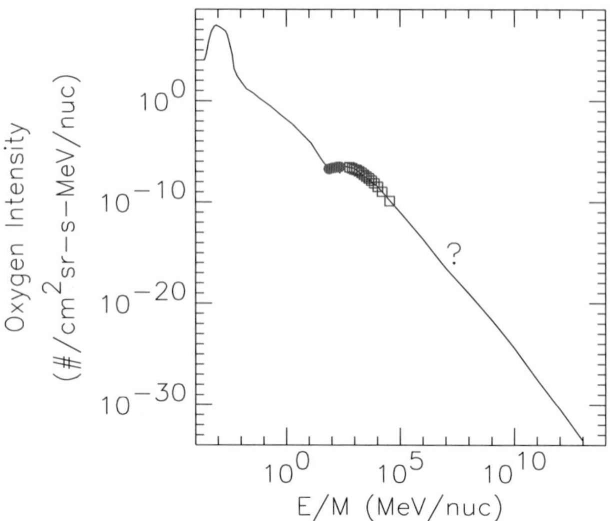


**Fig. 1** Oxygen energy spectrum as observed near Earth. Galactic cosmic rays dominate above ~100 MeV/nuc. *Filled circles* and *open squares* show solar-minimum intensities derived from ACE/CRIS and HEAO-C2 measurements, respectively. The high energy portion of the spectrum is labeled with a "*?*" as a reminder that only very limited information about the actual identity of the elements detected at these energies is available

accelerated "source material" as well as of the transport processes that bring it to us. The processes that must be considered include: 1) nucleosynthesis in stellar interiors; 2) ejection and mixing into the interstellar medium (ISM) by stellar winds and supernovae; 3) elemental fractionation; 4) acceleration of "primary" cosmic rays by supernova-driven shocks; 5) "propagation" in the Galaxy including the effects of spatial diffusion and convection, stripping of atomic electrons, ionization energy loss, possible reacceleration, occasional nuclear collisions that produce "secondary" cosmic rays, and escape from the Galaxy; 6) solar modulation upon entry into the heliosphere including stochastic energy losses and suppression of the low energy portion of the spectrum; and finally 7) the observation of arriving cosmic rays.

To study the nucleosynthesis, ejection, fractionation, and acceleration steps one starts with the spectra and composition observed near Earth and uses solar modulation and interstellar propagation models to work backwards up this list. Of course at low energies the cosmic rays that one measures are mixed with energetic nuclei from the more-local sources discussed in other papers in this volume. The overall spectrum of energetic particles near Earth, which has been measured over ~18 orders of magnitude in energy, is illustrated schematically in Fig. 1, where we have combined the low-energy oxygen spectrum compiled by Mewaldt et al. (2001) with the high-energy all-particle spectrum assembled by Swordy (2001), with appropriate scaling in energy and intensity. The change in spectral shape around 100 MeV/nuc reflects the transition from a mix of low-energy particle populations accelerated in or near the heliosphere to galactic cosmic rays, which are accelerated in more remote locations and dominate above this energy.

The peak in the spectrum between 100 and 1000 MeV/nuc is the result of solar modulation, which prevents most low-energy cosmic rays from penetrating to the inner heliosphere. For purposes of investigating the composition of galactic cosmic rays, the region of this spectral peak has received particular attention both because particle intensities are highest there and because experimental techniques available for particle identification have significantly better resolution there than at higher energies. Figure 2, which shows mass his-

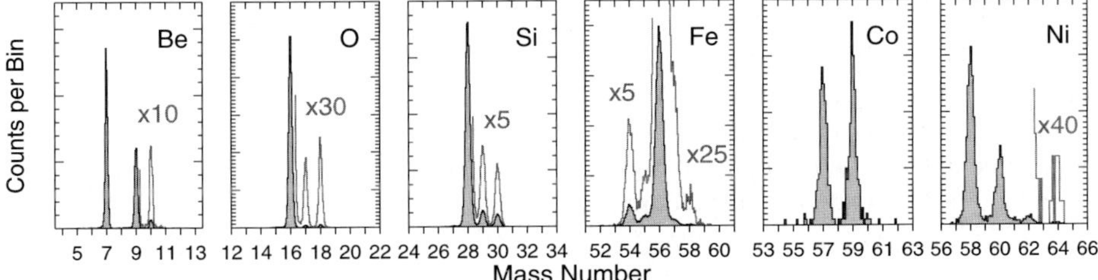


**Fig. 2** Examples of mass histograms (*shaded*) from the ACE/CRIS instrument using data collected from December 1997 through April 2000. (Abundances shown later in this paper are based on the first 22 of these 29 months in order to restrict the data set to solar-minimum conditions.) The total numbers of counts in these histograms are: Be, $8.7 \times 10^3$; O, $3.0 \times 10^5$; Si, $6.3 \times 10^4$; Fe, $5.7 \times 10^4$; Co, $3.5 \times 10^2$; Ni, $3.0 \times 10^3$. Portions of the distributions with low numbers of counts per bin are also shown expanded (*unshaded histograms*) with ordinate values multiplied by the indicated factors to better illustrate peaks for rare isotopes

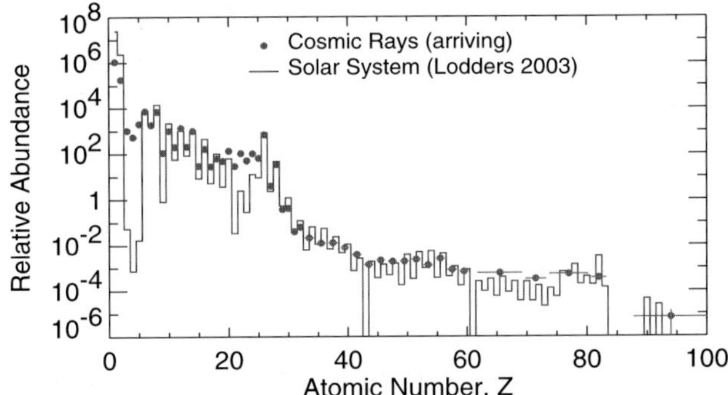


**Fig. 3** Comparison of elemental abundances in the arriving cosmic rays (*filled circles*) with those found in solar-system material (*histogram*). Cosmic-ray data correspond to energies $\lesssim$1 GeV/nuc near the peak of the arriving cosmic-ray spectrum. References to the data sources are cited in the text. Points for ultraheavy cosmic rays (Binns et al. 1989) are shown with abundances averaged over adjacent odd–even element pairs ($33 \leq Z \leq 60$) or over larger groups ($Z > 60$)

tograms for selected elements[1] from the Cosmic Ray Isotope Spectrometer (CRIS, Stone et al. 1998b) on the Advanced Composition Explorer mission (ACE, Stone et al. 1998a), illustrates the mass resolution and statistical accuracy now being achieved. The points plotted on the curve in Fig. 1 are cosmic-ray oxygen intensities obtained from CRIS (filled circles) and, at higher energies, from the HEAO-C2 experiment (open squares, Engelmann et al. 1990).

Figure 3 shows the abundances of the elements present in the arriving cosmic rays as filled circles. For $3 \leq Z \leq 30$ the values are CRIS measurements from the solar-minimum period shortly after the launch of ACE and correspond to an energy ~200 MeV/nuc. The H and He values were obtained at approximately the same energy from the spectra measured with the BESS balloon experiment (Wang et al. 2002). For elements with $Z > 30$ ("ultraheavy elements") results from the combined data sets of the HEAO-C3 and Ariel-VI experiments were used (Binns et al. 1989) with a normalization chosen to make them

[1]CRIS histograms for all elements in the range $3 \leq Z \leq 30$ as well as a list of CRIS isotopic composition measurements are available from the ACE Science Center web site:
http://www.srl.caltech.edu/ACE/ASC/DATA/level3/cris/isotopic_composition.html.

agree with the ACE/CRIS Fe abundance. These measurements were made at energies around 1 GeV/nuc. For comparison, the histogram in Fig. 3 shows elemental abundances inferred for the protosolar nebula (Lodders 2003).

The most striking feature of this comparison is the general similarity of the elemental composition in these two samples of matter, with the elements H, He, C, O, Ne, Mg, Si, and Fe dominating both samples, the predominance of even-$Z$ over odd-$Z$ elements in the region $6 \leq Z \leq 20$, and peaks in the regions $6 \leq Z \leq 8$ (CNO), $Z \simeq 26$ (Fe), $50 \lesssim Z \lesssim 56$ (Sn–Ba), and $78 \lesssim Z \lesssim 82$ (Pt–Pb). The most obvious difference, the relative overabundance of cosmic rays in the low-abundance regions immediately below these major peaks, has long been recognized to be the result of nuclear spallation reactions that occur during the propagation of cosmic rays through the interstellar gas and result in the breakup of some of the nuclei into lighter species. The abundances of these rarer elements provide useful probes for constraining models for the interstellar transport of cosmic rays (see Sect. 3, below).

In this paper we focus on the elements and isotopes with $Z \lesssim 30$ since with the ACE/CRIS data nearly all of the stable and long-lived isotopes of these elements can be studied with good statistical accuracy. Isotopic composition measurements for a large number of elements in this same charge and energy range are also available from the Ulysses/COSPIN High Energy Telescope (Connell 2001) and are generally in good agreement with the ACE/CRIS results. Precise determination of the abundances of the ultraheavy elements remains a very challenging problem because of their extremely low intensities. These species are of particular interest, however, because they can provide information about certain nucleosynthesis processes and time scales that can not be addressed using observations of lighter cosmic rays. The present understanding of ultraheavy cosmic rays is discussed by Waddington (2007).

## 2 Solar Modulation

The left panel in Fig. 4 illustrates the intensity variation of galactic cosmic rays penetrating to near Earth due to the effects of solar modulation (e.g., Jokipii and Kóta 2000). The three spectra shown for oxygen correspond to solar rotations with the lowest and highest modulation levels that have been experienced since the launch of ACE and one rotation selected to have an intermediate level comparable to that which existed during the HEAO-3 mission (1979–81). The Danish–French "C2" experiment on that mission provided the most precise measurements to date of elemental spectra in the energy range ~0.8 to 20 GeV/nuc (Engelmann et al. 1990). The spectra are labeled with values of the "modulation parameter", $\phi$ (Gleeson and Axford 1968), used to characterize the level of solar modulation. The product of this parameter, expressed in megavolts (MV), and the proton charge, $e$, is approximately twice the energy per nucleon lost, on average, by a cosmic ray with mass-to-charge ratio $M/Z = 2$ in penetrating from interstellar space to the inner heliosphere. For cosmic rays in the energy range measured by ACE/CRIS the distribution of energy losses about this mean is, however, quite broad (Goldstein et al. 1970). The spectra measured near Earth provide little information about the interstellar and source spectra for energies below this typical energy loss, $\sim e\phi/2$ (MeV/nuc). The HEAO-C2 spectra (open circles) match well with the ACE/CRIS spectra at a modulation level ~740 MV. Towards the upper end of the HEAO-C2 energy interval one expects solar modulation to have only very minor effect on the particle intensities.

The energy spectra observed near Earth for essentially all cosmic-ray species have the same general shape below a few GeV/nuc as that shown for oxygen in Fig. 4. Until recently

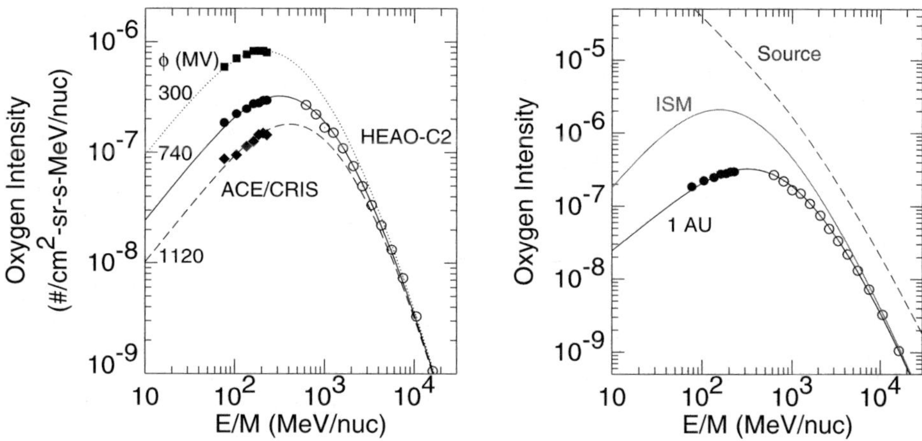


**Fig. 4** *Left panel*: effects of varying levels of solar modulation on cosmic-ray energy spectra observed near Earth (see text). *Right panel*: source and local interstellar spectra inferred from the measured 1 AU spectra. The relative normalization of the source and ISM spectra is not determined. At high energies the ISM spectrum is steeper than the source spectrum because of the energy dependence of $\Lambda_{\rm esc}$ (see Sect. 3)

the statistical accuracy of experiments at energies below the peak of the spectrum was rather limited and, as a result, abundance ratios were commonly derived by integrating over the entire energy interval of the measurements. With sufficiently precise measurements, however, it is possible to distinguish some differences attributable primarily to differences in spectral shapes in the unmodulated local interstellar spectra (e.g., spectra of secondary species are steeper than those of primaries at low energies [see Sect. 3, below]). In addition, minor effects due to the dependence of the modulation on $M/Z$ are expected.

As will be discussed below, one can model the solar modulation and the interstellar transport of cosmic rays in order to infer spectra both in the local interstellar medium and at the sources where cosmic rays are accelerated. The oxygen spectra obtained from one such modeling exercise are shown in the right hand panel in Fig. 4.

## 3 Interstellar Propagation

Models of varying sophistication have been used to describe the interstellar transport of cosmic rays, some taking into account a variety of constraints from other astrophysical observations (Strong et al. 2007). However this sophistication is not required to model the composition data, which can be accounted for remarkably well using the very simple "leaky-box" model (e.g., Meneguzzi et al. 1971). In this model the cosmic rays in the interstellar medium are assumed to be in a steady state where production by acceleration in cosmic-ray sources plus production by fragmentation of heavier species is balanced by fragmentation losses and escape from the system. The model also takes into account the ionization energy loss of the particles as they traverse interstellar matter and the nuclear transmutations of radioactive species. Neglecting these latter effects the leaky-box model can be formulated as

$$q_i + \sum_j \frac{\varphi_j}{\Lambda_{ji}} = \frac{\varphi_i}{\Lambda_{\rm esc}} + \frac{\varphi_i}{\Lambda_i}, \tag{1}$$

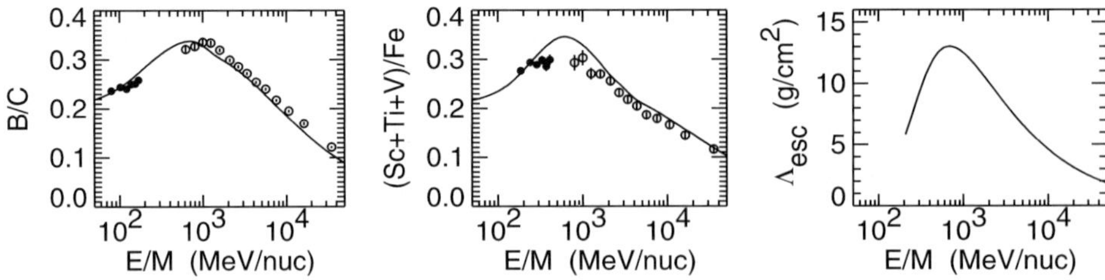


**Fig. 5** *Left two panels*: Energy dependence of two secondary-to-primary ratios obtained using the ACE/CRIS (*filled circles*) and HEAO-C2 (*open circles*) instruments compared with predictions from a simple leaky-box model (*smooth curves*). The *curves* correspond to the solar-modulation level inferred for the time period of the ACE measurements. The HEAO measurements were scaled to the ACE modulation level. *Right panel*: escape mean free path with the adopted normalization, $\Lambda_0 = 25$ g/cm$^2$ (see text). Because of solar-modulation effects the data place no constraint on $\Lambda_{\rm esc}$ below a few hundred MeV/nuc

with production terms given on the left hand side and loss terms on the right. Here $q_i$ and $\varphi_i$ are the source and equilibrium interstellar spectra of species $i$ and $\Lambda_i$ and $\Lambda_{ji}$ are the mean free paths for destruction of species $i$ and for its production from species $j$ by nuclear spallation. Considerable progress has been made in measuring the interaction cross sections that are most important for the cosmic-ray propagation problem and semi-empirical formulas have been developed for estimating the cross sections that are still unmeasured (Webber et al. 2003 and references therein).

In (1) the rate of loss of cosmic rays by escape from the galactic confinement volume is represented by an empirical "mean escape length", $\Lambda_{\rm esc}$. One can infer $\Lambda_{\rm esc}$ from the observed abundances of purely secondary species (those with $q_i = 0$) relative to the abundances of the heavier species from which they are produced. These secondary-to-primary ratios are observed to be energy dependent and it has been possible to find a parameterization of $\Lambda_{\rm esc}$ with which one can account reasonably well for this dependence, as illustrated in Fig. 5 (see also Davis et al. 2000). From the measured decrease of secondary-to-primary ratios towards high and towards low energies one infers a corresponding decrease of $\Lambda_{\rm esc}$, although the form of the decrease towards low energies is not precisely determined because of the strong effects of solar modulation below ~1 GeV/nuc. Similarly one finds that spectra of species that are largely produced by fragmentation of other secondaries (e.g., numerous isotopes with $15 \le Z < 20$) are even steeper at low energies.

The decrease of $\Lambda_{\rm esc}$ at high energies extends to above 100 GeV/nuc (Swordy et al. 1999), and it has been pointed out that composition measurements at sufficiently high energies could provide a rather direct determination of source composition, with minimal need for secondary corrections. Such measurements may provide the best means for precisely determining the source abundances of some elements that have sizable secondary fractions at lower energies (e.g., K, Ti, Cr, Mn).

## 4 Source Composition

For cosmic-ray nuclides having nonnegligible source abundance ($q_i > 0$ in (1)), the interstellar propagation model derived based on the abundances of purely-secondary species can be used to determine the fraction of the observed abundance that is attributable to secondaries. In the leaky-box model the secondary production of a given nuclide (second term on the left hand side of (1)) depends on the steady-state interstellar abundances of heavier nuclides

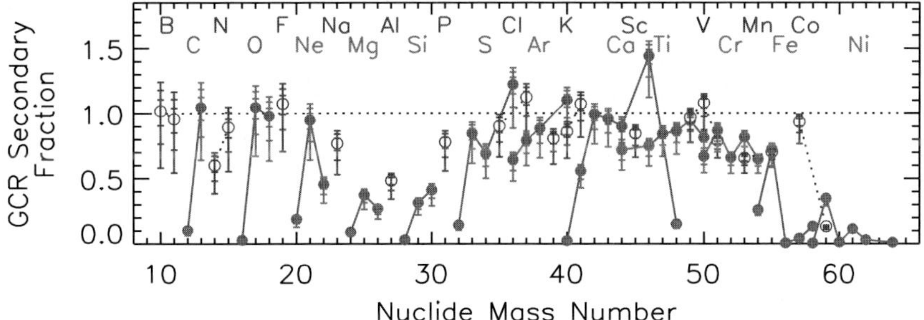


**Fig. 6** Secondary fractions in the arriving cosmic rays for isotopes of the elements in the range $5 \leq Z \leq 28$ inferred from the ACE/CRIS measurements. Nuclides with secondary fractions near unity are useful as probes of cosmic-ray propagation while those with small secondary fractions provide information about the nucleosynthetic origin of the primary cosmic rays. Nuclides with even (odd) $Z$ are shown *as filled (open) points with solid (dotted) lines connecting points corresponding to isotopes of the same element. Vertical bars* indicate the sensitivity to $\Lambda_{\text{esc}}$ (see text)

($\varphi_j$) and on cross sections ($\propto 1/\Lambda_{ji}$) for producing the nuclide of interest from these heavier parent species. The interstellar spectra of the parents are well constrained by the spectra of these species observed near Earth, so one can calculate the secondary contributions to the observed spectra and derive the primary contribution by subtraction. The secondary fractions that we obtain are shown in Fig. 6. Nuclides dominated by secondaries should appear with ordinates $\sim 1.0$ on this plot. Differences from this value can occur because of errors in the cross sections used for calculating the secondary contribution or due to inaccuracies in the simple propagation and solar-modulation models that were used. A secondary fraction greater than unity simply indicates that the calculated fragmentation production of the nuclide in question from heavier species exceeds its observed abundance due to such errors.

Nuclides with inferred secondary fractions close to zero in Fig. 6 are dominated by primary cosmic rays and should be suitable for reliable determinations of their source abundances. Besides the most abundant species such as $^{12}$C, $^{16}$O, and $^{56}$Fe, the set of dominantly-primary nuclides includes some much rarer nuclides for which there is minimal secondary production either because of a relative lack of heavier nuclides that could fragment into them (e.g., $^{64}$Ni) or because they have a mass-to-charge ratio so different from that of their parents that cross sections for their production are particularly small (e.g., $^{48}$Ca).

The secondary fractions shown in Fig. 6 do depend on the escape mean free path ($\Lambda_{\text{esc}}$, see (1) and Fig. 5), which is derived based on a number of different dominantly-secondary nuclides. To obtain a sense of how the uncertainty in $\Lambda_{\text{esc}}$ affects the inferred secondary fractions, these fractions were recalculated with different assumed magnitudes of $\Lambda_{\text{esc}}$. For this calculation the energy dependence of $\Lambda_{\text{esc}}$ obtained by Davis et al. (2000) was retained, $\Lambda_{\text{esc}} = \Lambda_0\, \beta/((\beta\, R/1\text{ GV})^{0.6} + (\beta\, R/1.3\text{ GV})^{-2.0})$. Here $\beta$ is the particle velocity in units of the speed of light, $R$ is magnetic rigidity, and $\Lambda_0$ is a constant. The secondary fractions shown in Fig. 6 (circles) were obtained using $\Lambda_0 = 25\text{ g/cm}^2$. Along the vertical bars associated with each point are plotted horizontal ticks indicating the secondary fraction that would have been obtained had the calculation been done using other values of $\Lambda_0$ ranging (from bottom to top) from 10 to 40 g/cm$^2$ in steps of 5 g/cm$^2$. While the calculated abundances of dominantly-secondary nuclides are particularly sensitive to the choice of $\Lambda_0$ and therefore useful for constraining the value of this parameter, the derived secondary fractions for the dominantly-primary species show little change over this large range of $\Lambda_0$ values. There are a number of nuclides shown in Fig. 6 for which the primary and secondary contributions

are comparable, some notable examples being $^{14}$N, $^{22}$Ne, $^{27}$Al, $^{31}$P, $^{34}$S, and $^{36}$Ar. One can derive meaningful source abundances for such nuclides provided that good measurements of the major reactions that produce the secondaries are available. The case of $^{22}$Ne, the only nuclide in the $Z \lesssim 30$ region with a source abundance known to differ dramatically from that found in solar material, is of particular interest. The observations of the neon isotopes and models for their origin are discussed by Binns et al. (2005, 2007).

## 5 Elemental Fractionation

It was recognized relatively early that elemental abundance differences between material in the cosmic-ray source and material in the protosolar nebula (obtained from observations of the solar photosphere and primitive meteorites) is organized in terms of the first ionization potentials (FIP) of the elements. This pattern, with the cosmic-ray source having high-FIP elements depleted relative to low-FIP elements when compared to solar material, is illustrated using data from ACE/CRIS in Fig. 7. The recognition that a similar FIP-dependent fractionation pattern is found in the solar corona and in solar energetic particles motivated several attempts to attribute the initial stage of cosmic-ray acceleration to energization in stellar flares, but it proved difficult to develop a self-consistent model for the origin of cosmic rays based on this idea (Meyer 1985; Meyer et al. 1997). However since a variety of other atomic parameters are correlated with FIP, the fractionation pattern seen in Fig. 7 could be fortuitous and some other atomic parameter might do equally well in organizing the observed abundances while providing a better basis for explaining the underlying physics.

To date most attention has been given to explaining cosmic-ray fractionation in terms of the volatility of the elements. The low-FIP elements tend to be refractory and condense to form grains at relatively high temperatures in the interstellar medium. On the other hand, high-FIP elements are more volatile, remaining in the gas phase or being only partially depleted into specific compounds as grains are formed. Building on this correlation between FIP and the tendency of the elements to be depleted from the gas phase into grains, Meyer et al. (1997) developed a cosmic-ray model in which charged grains are accelerated by supernova-driven shocks more efficiently than are individual ions because of their relatively high mass-to-charge ratio. Sputtering of atoms from these fast grains then acts as an injector of suprathermal ions that can be shock-accelerated to relativistic energies more efficiently

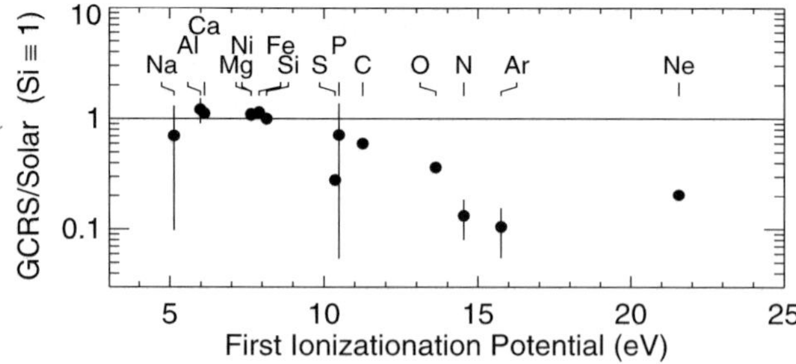


**Fig. 7** Abundance ratios between galactic cosmic-ray source (GCRS) and solar-system material illustrate the correlation between elemental fractionation and the first ionization potential of the elements. This comparison made using ACE/CRIS data is based on the isotope of each element that is most abundant in solar-system matter. The uncertainty in the calculated secondary corrections, which was assumed to be 25% for all nuclides, dominates the uncertainty in the derived GCRS abundances of Na, P, N, and Ar

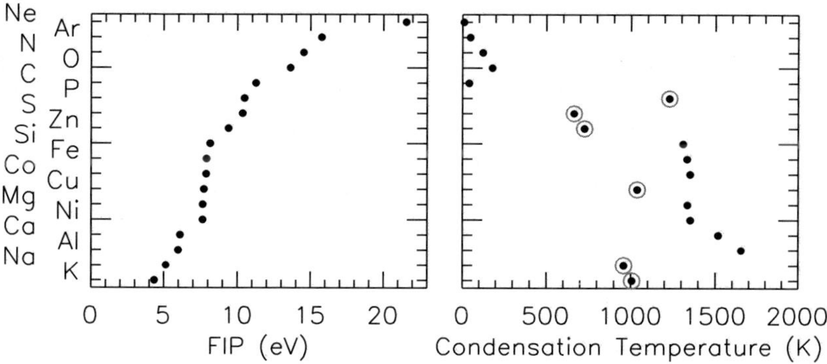


**Fig. 8** *Left panel*: First ionization potential (FIP) values for selected elements are shown with the elements arranged in order of increasing FIP from bottom to top. *Right panel*: condensation temperatures of the same elements in interstellar material with solar-like composition (Lodders 2003). Points corresponding to elements that have been suggested for distinguishing between FIP and volatility as the controlling parameter for cosmic-ray fractionation are *circled*

than ions from the thermal interstellar gas. Higdon and Lingenfelter (2005) subsequently concluded that grain acceleration in the cores of superbubbles is the source of most of the refractory nuclides in cosmic rays. Recently Westphal and Bradley (2004) have reported possible signatures of grain acceleration in a class of captured interplanetary dust particles.

As noted by Meyer et al. (1997), there exist some exceptions to the general correlation between FIP and volatility (or condensation temperature, which provides a quantitative measure of volatility) that might enable a useful test of which parameter is more-directly related to the fractionation observed in cosmic rays. Figure 8 illustrates these exceptions by plotting condensation temperatures with the elements arranged according to their FIP values. Circled points indicate elements that have been suggested for distinguishing between FIP and volatility as the key parameter underlying the fractionation process. The alkali metals Na and K, which are relatively volatile in spite of their low FIP, have sizeable secondary contributions (Fig. 6) leading to considerable uncertainties in their derived source abundances. The elements Zn, S, and P have FIP values near the transition between the low- and high-FIP populations and are thus difficult to interpret. It has been suggested, however, that the ratio P/S could be useful because these two elements have nearly identical FIP values but very different condensation temperatures (George et al. 2001). The element Cu, which is dominantly primary but relatively rare, could provide a useful test once a statistically-significant sample of this element has been collected. To date, attempts to distinguish between FIP and volatility mechanisms based on observed composition patterns have proven inconclusive.

## 6 Refractory Nuclides in the Source

In order to investigate the nucleosynthesis of the population of material from which cosmic rays are derived while avoiding the distorting effects due to elemental fractionation, one can restrict consideration to source ratios between isotopes of individual elements (e.g., $^{34}S/^{32}S$, Thayer 1997) and/or consider just those species that have both low FIP and high condensation temperature. For the latter approach one can include the elements Mg, Al, Si, Ca, Fe, Co and Ni. It is of interest to consider the various isotopes of these elements individually because different isotopes often have different nucleosynthetic origins. Including only those

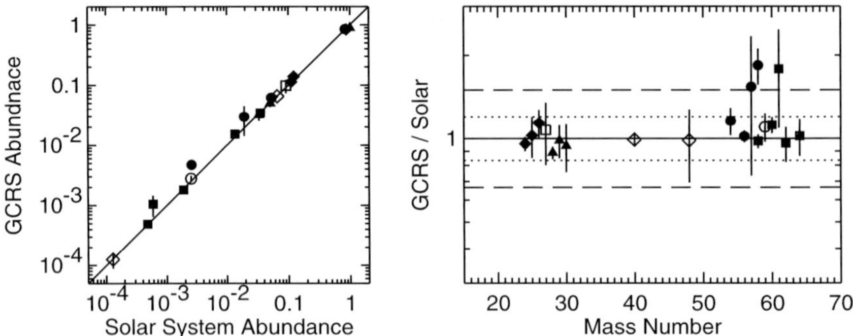


**Fig. 9** *Left panel*: derived GCRS abundances of 19 refractory nuclides are compared with solar-system abundances of these species. A weighted average of the ratio between GCRS and solar abundances was used to normalized the GCRS sample. *Right panel*: ratio between the GCRS and SS values are plotted as a function of mass number to better illustrate the small differences between the two samples. Abundance differences between the cosmic-ray source and solar samples by factors of 1.2 and 1.5 are indicated by *dotted and dashed lines*, respectively. *Symbols* indicate specific elements: *filled diamonds*, Mg(24,25,26); *open square*, Al(27); *filled triangles*, Si(28,29,30); *open diamonds*, Ca(40,48); *filled circles*, Fe(54,56,57,58); *open circle*, Co(59); *filled squares*, Ni(58,60,61,62,64)

isotopes that are dominated by primary material, one has 19 isotopes available from these seven elements. Figure 9 shows the source abundances derived for these nuclides from the ACE/CRIS data in two formats. In the left panel the galactic cosmic-ray source (GCRS) abundance is plotted versus the solar-system abundance in order to emphasize the great similarity of these two populations of material. In the right-hand panel the ratio of GCRS abundance to solar-system abundance is shown versus the mass number of the nuclide. In this format the composition differences between the two populations are more readily seen. When the uncertainties are taken into account, only $^{58}$Fe stands out as possibly having an abundance inconsistent with solar-like composition. The observed similarity between GCR source and solar-system abundances is striking not only because the absolute abundances of the nuclides being considered range over nearly four orders of magnitude, but also because of the wide range of nucleosynthesis processes by which the different nuclides are produced (Woosley and Weaver 1995; Iwamoto et al. 1999). As discussed by Wiedenbeck et al. (2001), nucleosynthetic contributions from stars with a wide range of initial masses are required to account for the observed abundances of the refractory nuclides. Furthermore, approximately half of the Fe, Co, and Ni is thought to come from low-mass stars in binary systems when they eventually explode as type Ia supernovae (Iwamoto et al. 1999). Certain isotopes (e.g., $^{48}$Ca) appear to originate only in particular, highly-specialized environments (Woosley and Weaver 1995). Given the variety of different stellar source types that must be mixed to account for solar-system composition (Woosley and Weaver 1995), the close resemblance between cosmic-ray source and solar compositions for refractory species would seem highly improbable unless the two samples were drawn from very similar pools of already-mixed material.

## 7 Radioactive Nuclides

In the cosmic-ray sample collected near Earth one can measure the abundances of a number of radioactive nuclides. These provide information about various time scales associated with the origin and acceleration of cosmic rays, depending on how they are formed. Secondary nuclides that undergo $\beta^+$ or $\beta^-$ decay with halflives comparable to the cosmic-ray mean residence time in the Galaxy (e.g., $^{10}$Be, $^{26}$Al, $^{36}$Cl, $^{54}$Mn) have been used to obtain a measure of that residence time, ~15 Myr (Yanasak et al. 2001 and references therein). Thus cosmic rays represent a sample of galactic material that is much more recent than solar-system matter, which condensed ~ 4.6 Gyr ago. This cosmic-ray residence time, taken together with the mean pathlength traversed and the particle velocity, then provides an estimate of the mean density of interstellar gas in the volume where cosmic rays propagate, ~0.3 g/cm$^2$, which is somewhat model dependent. In principle, $\beta$-decay secondaries with different halflives can also be used to test how uniformly this matter is distributed since nuclides with shorter lifetimes are sensitive to the density of matter in smaller volumes around the solar system. However the sensitivity of such a test using just the radioactive isotopes noted above is relatively poor because of the small range of halflives available (0.3 to 1.6 Myr). The isotope $^{14}$C ($T_{1/2} = 5730$ years) could provide a measure of the density of matter traversed in the immediate vicinity of the solar system (Ptuskin and Soutoul 1998), but at present only upper limits are available because the expected signal is smaller than the background due to $^{14}$C produced by fragmentation in the ACE/CRIS instrument.

Secondary nuclides that can decay only by orbital electron capture become effectively stable at sufficiently high energies because they have no orbital electrons and the cross section for attaching an electron from an atom in the interstellar gas sharply declines with increasing energy. Thus in Fig. 2 one sees peaks corresponding to the electron-capture species $^7$Be and $^{55}$Fe, which are spallation products created at high energies, in spite of their short laboratory halflives (53 days and 2.7 years, respectively). Because at sufficiently low energies the electron attachment probability can be significant, attempts have been made to use electron-capture secondaries to determine whether a significant fraction of the pathlength in the Galaxy is traversed when cosmic rays are at much lower energies than those at which they are observed. To date such studies have not found significant differences of electron-capture nuclide abundances from those predicted by models in which the time scale for acceleration is taken to be much shorter than that for confinement (Strong et al. 2007 and references therein; Scott 2005).

As noted by Soutoul et al. (1978), there are certain electron-capture nuclides that are expected to be synthesized and ejected in supernovae explosions. The presence of such electron-capture nuclides in the primary cosmic rays depends on their being accelerated and stripped of their electrons on a time scale short compared to the electron-capture decay halflife. Thus such nuclides provide a measure of the time scale between nucleosynthesis and acceleration.

Using ACE/CRIS data (Wiedenbeck et al. 1999) it was found that the primary electron-capture nuclide $^{59}$Ni ($7.6 \times 10^4$ years halflife) is essentially absent in the arriving cosmic rays (see Fig. 2). In addition, the abundance of $^{59}$Co, the nuclide produced by the $^{59}$Ni decay, was shown to be significantly greater than the amount expected from secondary production. The extra $^{59}$Co could have been synthesized either directly as $^{59}$Co or as $^{59}$Ni that subsequently decayed to $^{59}$Co. The latter scenario is considered much more likely based on nucleosynthesis calculations. From these observations it was concluded that most of the material accelerated to cosmic-ray energies by a supernova-driven shock is not derived from that same supernova explosion but must consist of matter that has resided in the interstellar medium for longer than ~$10^5$ years.

## 8 Discussion

Working from the observations discussed above one can postulate a detailed scenario for the origin of galactic cosmic rays, as described in the following paragraphs. This "standard model" of cosmic-ray origin is the result of decades of work by numerous investigators, both observational and theoretical. Although this model represents our present understanding, some details of this picture may be subject to revision as further investigations test and refine these ideas.

These highly energetic particles were accelerated recently (that is, within the past ∼10–20 Myr) from a source population consisting of interstellar gas and dust drawn from a large enough volume of the Galaxy such that abundances are not significantly distorted from the average due to statistical fluctuations in the number of stars of various types that contributed to that matter. The composition of this material is, at least for refractory species, strikingly similar to solar-system composition. However since the isotopic composition of cosmic-ray neon is significantly different from that found in the very local interstellar medium or in the solar wind, the cosmic-ray source population is thought to contain an additional contribution which is most commonly attributed to Wolf–Rayet stars. As discussed by Higdon and Lingenfelter (2005) and by Binns et al. (2005, 2007), superbubbles provide a promising candidate for a site in which most refractory nuclides have approximately their average interstellar abundances while differences from the average values are found for $^{22}Ne/^{20}Ne$ and, to a lesser extent, for several other ratios. In addition, Lingenfelter and Higdon (2007a, 2007b) find that the depletions of O and C relative to the refractory elements (see Fig. 7) can be quantitatively understood when the fractions of these elements expected to be incorporated into grains in superbubble cores are taken into account (see also Binns et al. 2005, 2007).

The elemental fractionation process by which refractory species are enhanced by factors ∼5–10 relative to volatiles is due to the efficient acceleration of the interstellar grains in which most of the refractories reside to velocities well above the thermal velocity in the interstellar gas. This preacceleration of grains as well as the acceleration of ions to cosmic-ray energies is the result of diffusive shock acceleration occurring as supernova-driven shocks traverse the interstellar medium enriched with ejecta from massive stars. The high mass-to-charge ratio of the grains, which causes them to have a long scattering mean free path thereby allowing them to experience numerous shock crossings, is responsible for their higher overall acceleration efficiency. The fast grains undergo collisions with gas atoms that result in sputtering of atoms from the grains to produce suprathermal ions that then experience further acceleration by the shock. Diffusive shock acceleration naturally produces energy spectra that are power laws in particle momentum, $p^{-\gamma}$ with an index $\gamma \geq 2$, up to some cutoff energy that is limited by the lifetime and/or the size of the shock (Drury 2004 and references therein). The steepening ("knee") in the cosmic-ray spectrum near $\sim 10^{15}$ eV total energy has been attributed to this high-energy cutoff in the efficiency for acceleration by supernova shocks in our Galaxy.

Supernova explosions thus play a dual role in the origin of galactic cosmic rays: they serve as an important mechanism for ejecting the products of stellar nucleosynthesis into the interstellar medium and they provide the energy source that drives the shocks and energizes the particles. However, the source material accelerated by a particular supernova shock is *not* derived primarily from the ejecta of that same supernova, but rather consists of matter previously ejected from other stars. Consequently one does not expect to find differences in the spectra of various primary species that correlate with the type of star in which those species are synthesized.

Encounters with multiple shocks during the confinement time in the Galaxy does not play a major role in modifying cosmic-ray composition and spectra. Most of the matter traversed by cosmic rays during their propagation is encountered after they have been accelerated. However some small amount of continuing acceleration must occur as cosmic rays undergo diffusion in the Galaxy (Strong et al. 2007).

All in situ observations of cosmic rays made to date have been subject to significant amounts of solar modulation and it is not known whether the Voyager spacecraft, which have begun exploring the energetic particle environment beyond the solar-wind termination shock, will encounter substantially-unmodulated cosmic-ray spectra before the end of their operational lives. Thus important uncertainties remain about the portion of the cosmic-ray spectrum below several hundred MeV/nuc. It is, for example, difficult to assess the importance of cosmic rays as sources of pressure, ionization, and heating in the interstellar medium without knowledge of the shape of the spectra at low energies.

The processing of galactic matter through successive generations of stars causes a steady increase in the heavy element content ("metallicity") of the interstellar medium, possibly offset to some extent by ongoing infall of less-processed material from the halo of the Galaxy (Geiss et al. 2002). The compositional similarity of the cosmic-ray source, a recent sample of interstellar matter, and the protosolar nebula, which condensed 4.6 Gyr ago, is not inconsistent with the expected increase in the Fe/H ratio (a commonly used measure of metallicity) in the ISM by $\sim$20% over that period (Timmes et al. 1995). Furthermore, ratios of abundances between pairs of heavy elements, each of which is tending to increase over time, change by significantly smaller factors (Reddy et al. 2003).

In addition to being time dependent, the metallicity in the Galaxy is expected to depend on distance from the galactic center because star formation and the associated galactic chemical evolution proceed more rapidly in the denser inner regions of the galactic disk. Based on diffusion models for the interstellar propagation of cosmic rays it has been suggested that the particles observed near Earth originate over a spatial scale $\lesssim$1 kpc (Ptuskin and Soutoul 1998). In principle it should be possible to use the radial dependence of the interstellar abundances (e.g., as inferred from surface abundances in Cepheid stars and summarized by Cescutti et al. (2007)) together with the similarity between cosmic-ray source abundances and solar-system abundances to place an observational constraint on this distance scale. However, the precision of the astronomical observations is not yet sufficient to clearly establish the radial abundance gradients of the refractory elements.

Recent efforts to explain the abundances of the major cosmic-ray elements including H, He, C, O, Si and Fe in terms of a mix of freshly synthesized material from massive stars in superbubble cores with old interstellar debris (Lingenfelter and Higdon 2007a, 2007b) offer promise for providing a simple unified account of the origin of most, if not all, of the cosmic rays. Further analysis is needed to assess whether such models can indeed reproduce the abundances of the large number of nuclides that are now well determined in the cosmic ray source. As part of such an undertaking it will also be necessary to quantitatively account for the mass-dependent fractionation that occurs among the volatile elements (Meyer et al. 1997).

There remain a number of outstanding questions related to the origin of galactic cosmic rays that can be addressed using new composition measurements, some of which are mentioned below. In some cases the necessary measurement techniques already exist and only a suitable space mission is needed. In other cases new measurement approaches are essential.

Present ideas about the origin of cosmic rays should be tested with measurements of nuclides in the upper 2/3 of the periodic table where nucleosynthesis is dominated by neutron capture processes and different classes of stellar sources are expected to be important.

Improved probes of the mechanism responsible for the elemental fractionation of cosmic-ray source material are also available in this ultraheavy region (e.g., Cs, Pb). Measurements of primary radionuclides with long halflives (e.g., $^{232}$Th, $^{235,236,238}$U, $^{244}$Pu, $^{247}$Cm) can be used to investigate the time since nucleosynthesis ("nucleosynthetic age") of the cosmic-ray source.

The origin of the cosmic-ray component responsible for the large excess of $^{22}$Ne relative to solar system and local interstellar matter, which has been attributed to Wolf–Rayet stars (Binns et al. 2005, 2007 and references therein), should be tested with precise measurements of additional species expected to contain correlated contributions from the same mechanism. For this purpose improved theoretical models of the production are also important.

Statistically-precise measurements of elemental composition at energies sufficiently high so that secondary contributions to species such as Na and K become insignificant should help clarify the mechanism responsible for the elemental fractionation of cosmic rays (see Sects. 3, 5).

Accurate measurements of the abundances of radioactive isotopes with halflives much shorter and much longer than those that have been studied to date (see Sect. 7) are needed for distinguishing between different models for the transport of cosmic rays in the Galaxy. The isotope $^{14}$C is suitable as a probe with a short halflife while a long halflife could be obtained either from measurements of $^{10}$Be at high energies where relativistic time dilation significantly increases its halflife or from a number of radioactive ultraheavy species.

The CRIS instrument on ACE, which provides isotopic resolution and high statistical accuracy for elements up to $Z \simeq 30$, has made a major contribution to our understanding of the origin of galactic cosmic rays. Significant further progress will require much larger space instruments to provide precision elemental composition measurements for $Z > 30$ or $E/M > 100$ GeV/nuc where particle intensities are much lower. Isotopic resolution in these regions remains a very significant instrumental challenge.

**Acknowledgements** We are grateful to Nathan Yanasak and Jeff George for their contributions to the ACE/CRIS data analysis and to Richard Lingenfelter and Ryan Ogliore for their comments on the manuscript. This work was supported by NASA at Caltech (under grant NAG5-12929), JPL, Washington University, and GSFC.

## References

W.R. Binns et al., in *Cosmic Abundances of Matter*, vol. CP183, ed. by C.J. Waddington (Amer. Instit. Phys., 1989), pp. 147–167

W.R. Binns et al., Astrophys. J. **634**, 351–364 (2005)

W.R. Binns et al., Space Sci. Rev. (2007), this volume. doi: 10.1007/s11214-007-9195-1

G. Cescutti, F. Matteucci, P. François, C. Chiappini, Astron. Astrophys. **462**, 943–951 (2007)

J.J. Connell, Space Sci. Rev. **99**, 41–50 (2001)

A.J. Davis et al., in *Acceleration and Transport of Energetic Particles Observed in the Heliosphere*, vol. CP528, ed. by R.A. Mewaldt et al. (Amer. Instit. Phys., 2000), pp. 421–424

L.O'C. Drury, J. Korean Astron. Soc. **37**, 393–398 (2004)

J.J. Engelmann et al., Astron. Astrophys. **233**, 96–111 (1990)

J. Geiss, G. Gloeckler, C. Charbonnel, Astrophys. J. **578**, 862–867 (2002)

J.S. George et al., in *Solar and Galactic Composition*, vol. CP598, ed. by R.F. Wimmer-Schweingruber (Amer. Instit. Phys., 2001), pp. 263–268

L.J. Gleeson, W.I. Axford, Astrophys. J. **154**, 1011–1026 (1968)

M.L. Goldstein, L.A. Fisk, R. Ramaty, Phys. Rev. Lett. **25**, 832–835 (1970)

J.C. Higdon, R.E. Lingenfelter, Astrophys. J. **628**, 738–749 (2005)

K. Iwamoto et al., Astrophys. J. Suppl. **125**, 439–462 (1999)

J.R. Jokipii, J. Kóta, Astrophys. Space Sci. **274**, 77–96 (2000)

R.E. Lingenfelter, J.C. Higdon, Astrophys. J. **660**, 330–335 (2007a)
R.E. Lingenfelter, J.C. Higdon, Space Sci. Rev. (2007b), this volume. doi: 10.1007/s11214-007-9172-8
K. Lodders, Astrophys. J. **591**, 1220–1247 (2003)
M. Meneguzzi, J. Audouze, H. Reeves, Astron. Astrophys. **15**, 337–359 (1971)
R.A. Mewaldt et al., in *Solar and Galactic Composition*, vol. CP598, ed. by R.F. Wimmer-Schweingruber (Amer. Instit. Phys., 2001), pp. 165–170
J.-P. Meyer, Astrophys. J. Suppl. **57**, 173–204 (1985)
J.-P. Meyer, L.O'C. Drury, D.C. Ellison, Astrophys. J. **487**, 182–196 (1997)
V.S. Ptuskin, A. Soutoul, Space Sci. Rev. **86**, 225–238 (1998)
B.E. Reddy, J. Tomkin, D.L. Lambert, C.A. Prieto, Mon. Not. Roy. Astron. Soc. **340**, 304–340 (2003)
L.M. Scott, Ph.D. thesis, Washington University, St. Louis, 2005
A. Soutoul, M. Cassé, E. Juliusson, Astrophys. J. **219**, 753–755 (1978)
E.C. Stone et al., Space Sci. Rev. **86**, 1–22 (1998a)
E.C. Stone et al., Space Sci. Rev. **86**, 285–356 (1998b)
A.W. Strong, I.V. Moskalenko, V.S. Ptuskin, Annu. Rev. Nucl. Part. Sci. **57** (2007, in press) astro-ph/0701517
S.P. Swordy, Space Sci. Rev. **99**, 85–94 (2001)
S.P. Swordy, D. Müller, P. Meyer, J. L'Heureux, J.M. Grunsfeld, Astrophys. J. **349**, 625–633 (1999)
M.R. Thayer, Astrophys. J. **482**, 792–795 (1997)
F.X. Timmes, S.E. Woosley, T.A. Weaver, Astrophys. J. Suppl. **98**, 617–658 (1995)
C.J. Waddington, Space Sci. Rev. (2007), this volume. doi: 10.1007/s11214-007-9145-y
A.J. Westphal, J.P. Bradley, Astrophys. J. **617**, 1131–1141 (2004)
J.Z. Wang et al., Astrophys. J. **564**, 244–259 (2002)
W.R. Webber, A. Soutoul, J.C. Kish, J.M. Rockstroh, Astrophys. J. Suppl. **144**, 153–167 (2003)
M.E. Wiedenbeck et al., Astrophys. J. (Lett.) **523**, L61–L64 (1999)
M.E. Wiedenbeck et al., in *Solar and Galactic Composition*, vol. CP598, ed. by R.F. Wimmer-Schweingruber (Amer. Instit. Phys., 2001), pp. 269–274
S.E. Woosley, T.A. Weaver, Astrophys. J. Suppl. **101**, 181–235 (1995)
N.E. Yanasak et al., Astrophys. J. **563**, 768–792 (2001)

Space Sci Rev (2007) 130: 431–438
DOI 10.1007/s11214-007-9182-6

# Galactic Wind: Mass Fractionation and Cosmic Ray Acceleration

**H.J. Völk**

Received: 13 March 2007 / Accepted: 30 March 2007 /
Published online: 11 May 2007


**Abstract** The dynamical and chemical effects of the Galactic Wind are discussed. This wind is primarily driven by the pressure gradient of the Cosmic Rays. Assuming the latter to be accelerated in the Supernova Remnants of the disk which at the same time produce the Hot Interstellar Medium, it is argued that the gas removed by the wind is enriched in the nucleosynthesis products of Supernova explosions. Therefore the moderate mass loss through this wind should still be able to remove a substantial amount of metals, opening the way for stars to produce more metals than observed in the disk, by e.g. assuming a Salpeter-type stellar initial mass function beyond a few Solar masses. The wind also allows a global, physically appealing interpretation of Cosmic Ray propagation and escape from the Galaxy. In addition the spiral structure of the disk induces periodic pressure waves in the expanding wind that become a sawtooth shock wave train at large distances which can re-accelerate "knee" particles coming from the disk sources. This new Galactic Cosmic Ray component can reach energies of a few $\times\ 10^{18}$ eV and may contribute to the juncture between the particles of Galactic and extragalactic origin in the observed overall Cosmic Ray spectrum.

**Keywords** Galaxy: wind · Galaxy: chemical evolution · Galaxy: cosmic ray propagation · Galaxy: cosmic ray acceleration

## 1 Introduction

The ideas about the existence of a Galactic Wind of the seventies followed the Solar Wind process, driven by hot thermal gas. The conclusion was that such a supersonic outflow was not possible under average interstellar conditions, because of radiative cooling of the thermal gas before escape from the gravitational well of Dark Matter and the stars (e.g. Habe and Ikeuchi 1980). However, the Galaxy has two special aspects not explicitly contained

H.J. Völk (✉)
Max Planck Institute for Nuclear Physics, P.O. Box 103980, 60629 Heidelberg, Germany
e-mail: heinrich.voelk@mpi-hd.mpg.de

in these considerations (i) the production of an essentially isotropic, i.e. strongly scattered, high-pressure Cosmic Ray (CR) component that does not cool. It must escape because observations show that the grammage these energetic particles encounter is a mere 10% of the nuclear collision grammage. The question is then whether the relativistic component escapes together with the scattering medium, the thermal gas in which the magnetic fluctuations are propagating; a first attempt in this direction was made by Ipavich (1975). Secondly (ii) there exists a "Hot Interstellar Medium", a gas component of the Galactic disk with very low density $N_{\rm H} \approx 3 \times 10^{-3}$ cm$^{-3}$ and a high volume-filling factor ~50% (McKee and Ostriker 1977). It does not easily cool. This hot component is presumably the sum of the remnants (SNRs) of the Galactic Supernova explosions.

If the SNRs in the disk are in addition the sources of the CRs, then there is a strong causal connection between the Hot Interstellar Medium and the CRs. The SNR gas is also expected to be enriched in heavy elements by nucleosynthesis before and during the explosion. Therefore the question arises, whether there is preferential removal of metal-enriched hot gas from the Galaxy with consequences for its chemical evolution. Another question regards the consistency of the observations of the CR propagation characteristics with this global picture. Finally we can ask ourselves, whether the spiral structure of the Galaxy leads to re-acceleration of disk-CRs in this extended wind, and which role such effects may play in the overall CR spectrum observed at high energies. We will try to address these questions in this paper.

## 2 Galactic Wind

An extreme hypothesis is due to Axford (1981). He argued that the nucleosynthetically enriched Supernova ejecta are likely to go directly into the Hot Interstellar Medium (HIM) which leaves the Galaxy with the tightly coupled lower energy CRs. With the global mass estimate $M_{\rm HIM} \approx N_{\rm H} \times 0.5\, V_{\rm disk} \approx 2 \times 10^7\, {\rm M}_\odot$ and the estimated CR lifetime $\approx 3 \times 10^7$ yrs in the Galaxy, the mass removal rate from the disk would then have to be $dM_{\rm gas}/dt \sim 1\, {\rm M}_\odot$/yr. Axford's estimates for the wind velocity ($u \sim$ 300–1000 km/s) and for the Wind termination shock distance ($R_{\rm s} \geq 100$ kpc) are remarkably close to what we today expect to be the case.

Nevertheless, reality is probably much more complex (e.g. Fig. 1, from Breitschwerdt et al. 1991): The local picture near the Solar System also shows distinct "fountain flows", falling down into the disk, perhaps through radiative cooling of hotter rising material. These are the so-called High Velocity Clouds. In addition there is a sizeable number of "Very High Velocity Clouds"—with infall velocities $V \sim 200$ km/s—which could represent true mass accretion of the Galaxy ("infall"), perhaps from the Magellanic Stream (Mirabel 1989). In the main volume CRs escape together with hot gas. Apart from the Hot Interstellar Medium in the disk, there is quasi-static, highly ionized gas at greater heights ~3 kpc in the halo (e.g. Savage and Massa 1987). We can assume it to be partly a relic of the hot SNR gas from the disk. Let us call all this high-altitude hot "coronal" gas (Spitzer 1956) the Hot Halo and assume it to be penetrated by the CRs emanating from the disk.

This Hot Halo can eventually be lifted out of the Galaxy by the slowly upward drifting CRs which cool only adiabatically: the combined hydrodynamics of thermal gas and CRs allows Parker-type (e.g. Parker 1963) outflow solutions. They correspond to a subsonic-supersonic flow transition at heights of the order of 20 kpc above the disk. The asymptotic wind velocities $u$ of a few 100 km/s at ~100 kpc distance reach values of the order of the Galactic escape speed (Breitschwerdt et al. 1991). The inclusion of the $B$-field stresses and Galactic rotation—the sling shot effect—only enhances the wind speed. The magnetic

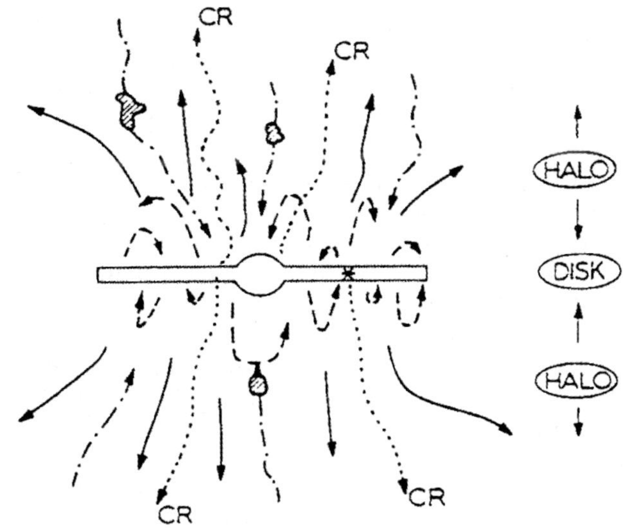


**Fig. 1** Schematic of the dynamical processes in the Galactic Halo (from Breitschwerdt et al. 1991). Fountain flows from the Disk (*dashed curves*) and infalling High Velocity Clouds (*shaded*) coexist with CRs (*dotted curves*) and hot gas (*solid arrows*), escaping in the main volume of the Halo

stresses imply in particular also a substantial angular momentum loss of the disk gas (Zirakashvili et al. 1996) over the age of the Galaxy. The mass loss rate is somewhat less than 1 $M_\odot$/yr.

However the direct connection between the ensemble of SNRs—the HIM in the disk—and the medium in the Hot Halo is broken. There is nevertheless consolation from theory. First of all, the energy source is clear: it consists in the supernova explosions. Secondly also the driving mechanism for the mass loss is identified: it derives from the pressure of the CRs. To this adds result that this global picture is essentially consistent with the observations of the propagation of CRs (Ptuskin et al. 1997).

## 3 Effects on Galactic Chemical Evolution

It appears a natural assumption regarding this outflow that the removed wind material is significantly enriched in heavy nucleosynthesis products and has in particular a low deuterium abundance. This should be the result of turbulent and diffusive mixing of Supernova ejecta material with shock-heated circumstellar gas, the CRs and the magnetic fields in the SNR. We propose that this is indeed the case. However, the data do not seem to require it. On the contrary, the observed metallicity distribution of old and long-lived G-dwarfs in the solar neighborhood has only few objects with a metallicity that is small compared to Solar metallicity. This can be understood if the metallicity was always rather high and the Galactic disk mass grew slowly by infall of little-processed material (e.g. Boissier and Prantzos 1999). Whether simultaneous outflow can be compensated by increased infall remains an open question. Most chemical evolution models therefore ignore mass loss, and this also holds for so-called chemodynamical models (e.g. Samland et al. 1997). A theoretical justification is the extreme assumption of e.g. Wang and Silk (1993) that the SN nucleosynthesis products are *completely mixed* with the rest of the Interstellar gas. Then heavy elements are not ejected preferentially by galaxies and a great amount of gas must be removed before the metallicity is substantially changed. The very slow decrease of the D/H-ratio since the Big Bang, and in particular from the epoch of the Protosolar Cloud to the present-day Local

Interstellar Cloud, rather appears to require an additional, so-called excess infall of ambient material, probably from nearby dwarf galaxies like the Large and the Small Magellanic Clouds. This has been proposed by Geiss et al. (2002). The infall requirements imply an essentially primordial material with metallicity $Z < 0.2\ Z_\odot$ (Tosi 1988) and a mass accretion rate of about $0.3\ M_\odot$/yr.

On the other hand, present chemical evolution models for our Galaxy assume a Scalo-type stellar initial mass function (IMF) which produces a relatively small amount of massive stars and appears to be a good approximation for the Solar neighborhood. From the metallicity of the intracluster gas in rich clusters, on the other hand, the Salpeter IMF seems a better description for initial stellar masses greater than a few solar masses (e.g. Elbaz et al. 1995). Since it forms more massive stars than the Scalo IMF, it also produces more heavy elements by roughly a factor of 2 (Prantzos, private communication) unless all stars more massive than, say, $30\ M_\odot$ collapse to Black Holes without releasing their metals (e.g. Heger et al. 2003). The surplus of metals—half the amount produced—has then to be removed by escape. The CR-driven metal-enriched Galactic Wind can just do that. The introduction of this physics-motivated additional degree of freedom in the framework of a dynamical halo allows a correspondingly enlarged freedom in the choice of the chemical and dynamical evolution parameters, in particular the IMF. The expected wind bubble of gas with a metallicity $Z > Z_\odot$ around disk galaxies like the Milky Way corresponds to a very extended, tenuous, shell-type halo with a radius of many 100 kpc and a mass of the order of the present mass of the gas in the disk.

## 4 Mean Cosmic Ray Propagation in the Milky Way with a Wind

The outward pressure force $-\nabla P_c$ of the escaping CRs drives the wind flow. It implies at the same time an outward diffusion current of CRs that excites scattering magnetic fluctuations (Alfvén waves) selfconsistently: $u\mathrm{d}/\mathrm{d}z(\delta B^2/8\pi) = -V_A \times \nabla P_c$, where $V_A$ denotes the Alfvén velocity. Nonlinear dissipation of these waves then leads to a finite scattering mean free path $\lambda_{mfp}$ that increases with particle energy (Ptuskin et al. 1997). Thus, energetic particles diffuse in an outward convecting thermal plasma. This implies that their propagation is diffusive near the disk and convective further out (Fig. 2). Particles have an exponentially small probability to return to the disk from beyond the diffusion-advection boundary whose distance $R_{da}(E) \sim \lambda_{mfp} \times v/3u$ increases with energy; here $v$ denotes particle velocity. For example $R_{da}(1\ \mathrm{TeV}) \approx 15$ kpc. The mutually cancelling energy dependencies imply that for a particle from the disk the time to reach this "escape boundary" is independent of energy and about equal to $3 \times 10^7$ yr, consistent with the measured $^{10}$Be survival fraction. The grammage of gas traversed before escape is proportional to $vR^{-0.55}$ and close to the observed value at high energies, whereas at low energies it is $\approx 10\ (v/c)\ \mathrm{g\,cm^{-2}}$ (Ptuskin et al. 1997). It is not clear however, whether this last dependence on $v$ is sufficiently close to the observed grammage at low energies (Strong et al. 2007). Thus most of the basic CR propagation characteristics can be naturally explained without the introduction of further physical processes. The only apparent exception is the large mean anisotropy which the strong rigidity-dependence of the scattering mean free path implies. With a value $\delta = 5 \times 10^{-2}$ perpendicular to the disk at $10^{14}$ eV it exceeds the observed anisotropy by a factor of fifty! Possibly this may be explained by a dominantly disk-parallel magnetic field structure in the local Interstellar Medium. But the last word is not yet said regarding this special question.

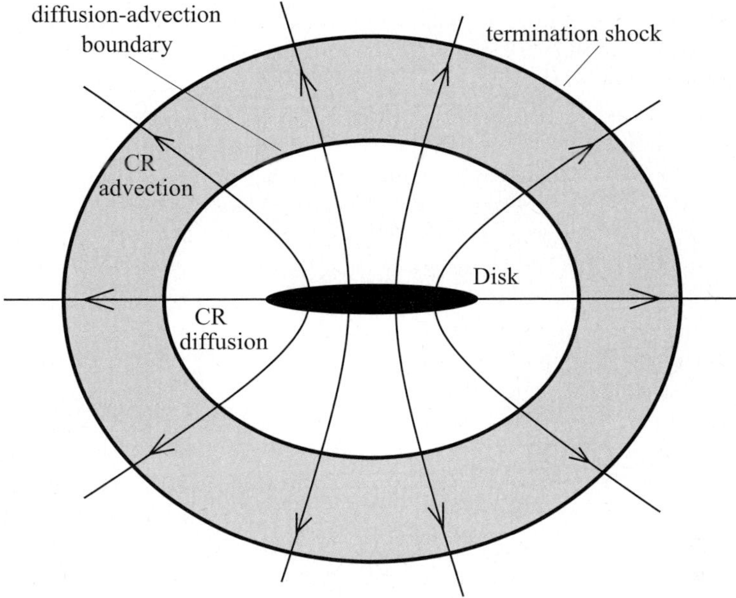


**Fig. 2** Schematic meridional cross-section of the supersonic Galactic Wind flow (*arrows*). In the region close to the Galactic disk (*black ellipse*), inside the diffusion-convection boundary, Cosmic Rays propagate outwards mainly diffusively, whereas beyond, until the wind termination shock, the particles are mainly advected by the gas flow. From Völk and Zirakashvili (2004)

## 5 Re-acceleration of Disk-CRs in the Extended Wind by Shocks Due to the Spiral Structure

The Galactic Wind flow is modulated by the enhanced star formation and CR production in the spiral arms (Fig. 3). The higher pressure above the spiral arms results in recurrent wind compressions in the direction perpendicular to the disk, on the time scale of the Galactic rotation, all the way out to the wind termination shock at 200 to 300 kpc. This is roughly analogous to the so-called Corotating Interaction Regions in the Solar Wind and leads to Galactic Wind Interaction Regions and multiple shock formation in the wind halo with large spatial and temporal scales. They should cause re-acceleration of the CRs from the disk to energies beyond their cutoff at the "knee" of the observed spectrum at rigidities $R$ of several $z \times 10^{15}$ V, where $z$ is the nuclear charge (Völk and Zirakashvili 2004).

There are interesting differences to the Solar Wind case (i) the Galaxy is a fast rotator, making one turn before the flow had time to become supersonic (ii) the spiral pattern is a wave that slips across the $B$-field anchored in the ionized gas, and thus allows accelerated particles to return to the disk (iii) the wind pressure is CR-dominated. As a consequence no suprathermal particles are injected, only existing energetic particles are re-accelerated, essentially from the cutoff region upwards. This permits the continuity of the energy spectrum due to the large halo volume.

Figure 4 shows the resulting radial variations of flow velocity, CR pressure and thermal gas pressure which form a sawtooth wave with outward propagating shocks and associated rarefactions. Energetic particles from the knee region of the spectrum formed by SNRs in the disk are re-accelerated in this sawtooth wave. The CR pressure exceeds the gas pressure as a result of radiative cooling of the gas in the expanding wind flow and smoothes the

**Fig. 3** The spiral galaxy M51. The two main spiral arms rotate relative to the Interstellar gas, leaving newly formed stars (*blue spots*) and HII-regions around massive stars (*reddish blobs*) behind their dust lines (*dark*). Massive stars have a lifetime short compared to the Galactic rotation period and therefore produce Supernovae and CRs near the arms, creating an overpressure there (NASA/*HST*)

sawtooth shocks on the CR diffusion scale—very small compared to the wavelength of the sawtooth.

The resulting spectral energy distributions of re-accelerated particles is shown in Fig. 5 for an exponential energy cut-off of the particle spectra produced by the SNR sources in the disk. Whereas all individual nuclei show rather sharp spectral fall-offs with energy before their re-accelerated spectrum continues for about 1.5 decades, the all-particle spectral flux has a rather smooth "knee" at an energy of several $\times\ 10^{15}$ eV. This demonstrates the possibility that re-accelerated particles can extend the Galactic CR spectrum up to energies of about $z \times 10^{17}$ eV, i.e. to energies $\approx 3 \times 10^{18}$ eV when iron particles are taken into account.

## 6 A Possible Contribution to the Overall Observed CR Spectrum

The overall observed CR energy spectrum is thought to be composed of a low-energy part of Galactic origin which has a cutoff at the "knee" at $E \approx 3z \times 10^{15}$ eV for each element $z$. At high energies there might also exist an extragalactic source with a hard power-law spectrum. Its amplitude ought to be taken proportional to the overall star formation rate in the Universe up to at least the expected cutoff at $E \approx 5 \times 10^{19}$ eV due to strong energy losses by photopion

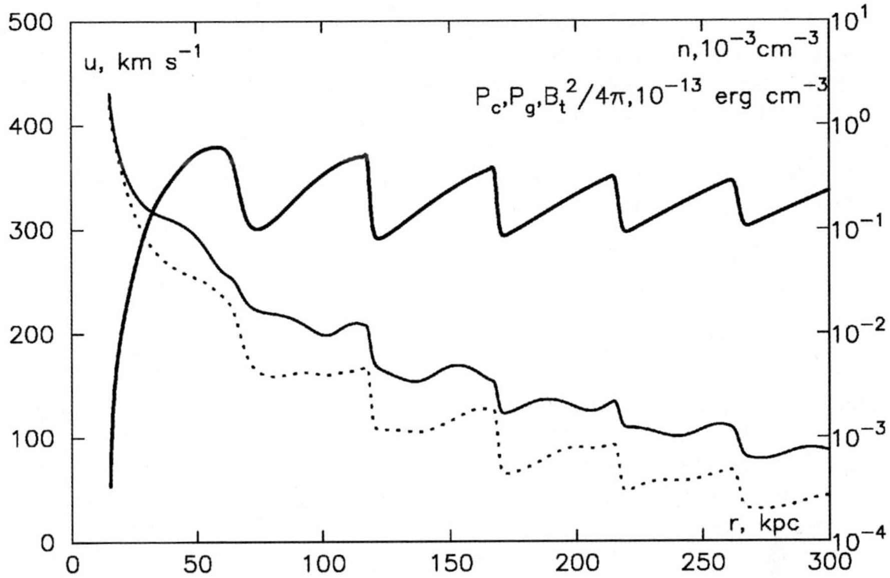


**Fig. 4** Radial dependencies of spiral shock dynamical variables, taken at one azimuthal angle. The values of the radial gas velocity $u$ (*thick solid line*) are given n the left ordinate. The right ordinate shows the values of the CR and gas pressures $P_c$ (*thin solid line*) and $P_g$ (*dotted line*), respectively, in units of $10^{-13}$ erg cm$^{-3}$. Forward shocks form a sawtooth velocity profile at large distances in the Galactic Wind flow (from Völk and Zirakashvili 2004)

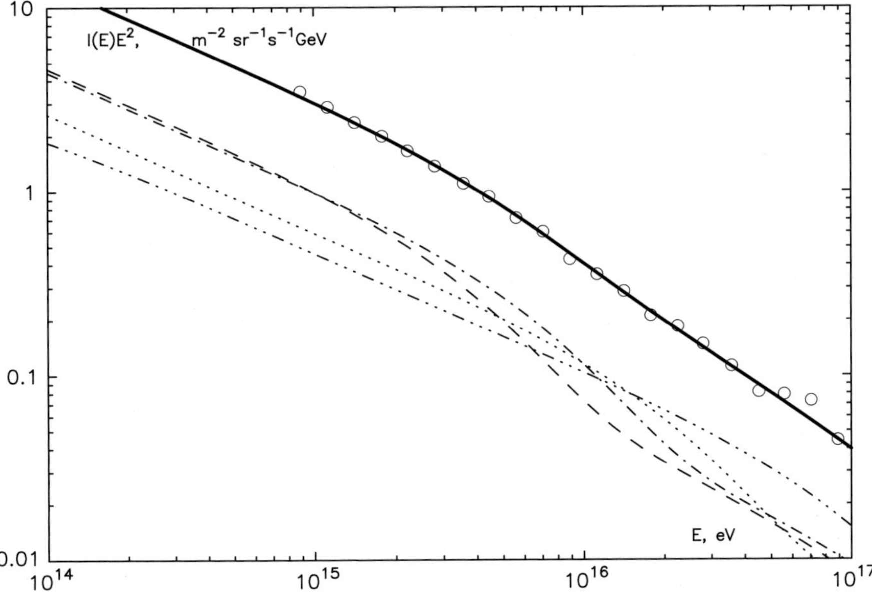


**Fig. 5** Calculated spectral energy distribution $E^2 I(E)$ of the CR protons (*dashed curve*), helium nuclei (*dash-dotted curve*), carbon (*dotted curve*), iron (*dash-dot-dotted curve*), and all-particles (*thick solid curve*) in the Galactic disk, for an exponential rigidity cutoff of the SNR-accelerated particles, and the all-particle flux observed by the *KASCADE* collaboration (*empty circles*) (Kampert et al. 2001). The chemical composition has been fixed at $E = 9 \times 10^{14}$ eV (from Völk and Zirakashvili 2004)

production, the so-called GZK cutoff (e.g. Hillas 2006). If adjusted to measurements around $10^{19}$ eV, it becomes relevant at energies above about $10^{17}$ eV. However, the data show a need for a further intermediate energy component (Hillas' component B) which extends the Galactic CR spectra for each nucleus $z$ beyond $E \approx 3\, z \times 10^{15}$ eV to $\approx z \times 10^{17}$ eV. The re-accelerated Galactic CRs have about these characteristics (see Fig. 5)! Clearly this is just a possibility. We shall discuss the possible components of the overall CR spectrum and alternative solutions elsewhere.

## 7 Summary

The existence of a Galactic Wind suggests itself if the dynamical effects of the CRs produced in the disk are correctly accounted for. The wind provides a global framework for the propagation and the escape of the CRs from the disk, basically in agreement with all observations, unless the conventional interpretation of the anisotropy measurements is considered undebatable. The additional degree of freedom removes restrictions on the models for the chemical evolution of the Galaxy by allowing the removal of high-metallicity material from the disk. It also opens the possibility for acceleration of CRs within the wider Galactic Halo beyond the "knee" of the rigidity spectra of different nuclei. Whether this is a necessity for the understanding of the overall CR energy spectrum is still an open question.

**Acknowledgements** The author thanks Evgeny Berezhko, Johannes Geiss, Rosine Lallement, Vladimir Ptuskin and Vladimir Zirakashvili for inspiring discussions regarding the role of Galactic Halo dynamics in Cosmic Ray physics and Galactic chemical evolution. In particular from Johannes Geiss he has learnt about the fundamental role of the Deuterium abundance. Special thanks are also due to Nikos Prantzos for his advice regarding models of chemical evolution irrespective of his scepticism concerning the role of outflows from the Galaxy.

## References

W.I. Axford, *Proc. 17th ICRC*, vol. 2 (Paris, 1981), p. 299
D. Breitschwerdt, J.F. McKenzie, H.J. Völk, Astron. Astrophys. **245**, 79 (1991)
S. Boissier, N. Prantzos, Mon. Not. Roy. Astron. Soc. **307**, 857 (1999)
D. Elbaz, M. Arnaud, E. Vangioni-Vlam, Astron. Astrophys. **303**, 345 (1995)
J. Geiss, G. Gloeckler, C. Charbonnel, Astrophys. J. **578**, 862 (2002)
A. Habe, S. Ikeuchi, Prog. Theor. Phys. **64**, 1995 (1980)
A. Heger, C.L. Fryer, N. Langer et al., Astrophys. J. **591**, 288 (2003)
A.M. Hillas, J. Phys.: Conf. Ser. **47**, 168 (2006)
K.-H. Kampert, T. Antoni, W.D. Apel et al., in *Proc. 27th ICRC (Hamburg)*. Invited, Rapporteur, and Highlight papers (2001), p. 240
C.F. McKee, J.P. Ostriker, Astrophys. J. **218**, 148 (1977)
I.F. Mirabel, in *Structure and Dynamics of the Interstellar Medium*, ed. by G. Tenorio-Tagle, M. Moles, J. Melnik. Proc. IAU Coll., vol. 120 (Springer, Heidelberg, 1989), p. 396
F. Ipavich, Astrophys. J. **196**, 107 (1975)
E.N. Parker, in *Monographs and Texts in Physics and Astronomy*, ed. by R.E. Marshak, vol. VIII (Intersc. Publ., 1963), ch. 5
M. Samland, G. Hensler, Ch. Theis, Astrophys. J. **476**, 544 (1997)
B.D. Savage, D. Massa, Astrophys. J. **314**, 380 (1987)
L. Spitzer Jr., Astrophys. J. **124**, 20 (1956)
A.W. Strong, I.V. Moskalenko, V.S. Ptuskin, Annu. Rev. Nucl. Part. Sci. **57** (2007, submitted). arXiv:astro-ph/0701517)
M. Tosi, Astron. Astrophys. **197**, 47 (1988)
V.S. Ptuskin, H.J. Völk, V.S. Zirakashvili et al., Astron. Astrophys. **321**, 434 (1997)
H.J. Völk, V.S. Zirakashvili, Astron. Astrophys. **417**, 807 (2004)
B. Wang, J. Silk, Astrophys. J. **406**, 580 (1993)
V.S. Zirakashvili, D. Breitschwerdt, V.S. Ptuskin et al., Astron. Astrophys. **311**, 113 (1996)

Space Sci Rev (2007) 130: 439–449
DOI 10.1007/s11214-007-9195-1

# OB Associations, Wolf–Rayet Stars, and the Origin of Galactic Cosmic Rays

**W.R. Binns · M.E. Wiedenbeck · M. Arnould · A.C. Cummings · G.A. de Nolfo · S. Goriely · M.H. Israel · R.A. Leske · R.A. Mewaldt · G. Meynet · L.M. Scott · E.C. Stone · T.T. von Rosenvinge**

Received: 5 February 2007 / Accepted: 11 April 2007 /
Published online: 5 June 2007


**Abstract** We have measured the isotopic abundances of neon and a number of other species in the galactic cosmic rays (GCRs) using the Cosmic Ray Isotope Spectrometer (CRIS) aboard the ACE spacecraft. Our data are compared to recent results from two-component (Wolf–Rayet material plus solar-like mixtures) Wolf–Rayet (WR) models. The three largest deviations of galactic cosmic ray isotope ratios from solar-system ratios predicted by these models, $^{12}C/^{16}O$, $^{22}Ne/^{20}Ne$, and $^{58}Fe/^{56}Fe$, are very close to those observed. All of the isotopic ratios that we have measured are consistent with a GCR source consisting of $\sim$20% of WR material mixed with $\sim$80% material with solar-system composition. Since WR stars are evolutionary products of OB stars, and most OB stars exist in OB associations that form superbubbles, the good agreement of our data with WR models suggests that OB associations within superbubbles are the likely source of at least a substantial fraction of GCRs. In previous work it has been shown that the primary $^{59}Ni$ (which decays only by electron-capture) in GCRs has decayed, indicating a time interval between nucleosynthesis and acceleration of $>10^5$ y. It has been suggested that in the OB association environment, ejecta from supernovae might be accelerated by the high velocity WR winds on a time scale that is short

W.R. Binns (✉) · M.H. Israel · L.M. Scott
Washington University, St. Louis, MO 63130, USA
e-mail: wrb@wuphys.wustl.edu

M.E. Wiedenbeck
Jet Propulsion Laboratory, California Institute of Technology, Pasadena, CA 91109, USA

M. Arnould · S. Goriely
Institut d'Astronomie et d'Astrophysique, U.L.B., Bruxelles, Belgique

A.C. Cummings · R.A. Leske · R.A. Mewaldt · E.C. Stone
California Institute of Technology, Pasadena, CA 91125, USA

G.A. de Nolfo · T.T. von Rosenvinge
NASA/Goddard Space Flight Center, Greenbelt, MD 20771, USA

G. Meynet
Geneva Observatory, 1290 Sauverny, Switzerland

compared to the half-life of $^{59}$Ni. Thus the $^{59}$Ni might not have time to decay and this would cast doubt upon the OB association origin of cosmic rays. In this paper we suggest a scenario that should allow much of the $^{59}$Ni to decay in the OB association environment and conclude that the hypothesis of the OB association origin of cosmic rays appears to be viable.

**Keywords** ISM: cosmic rays · Stars: Wolf–Rayet

## 1 Introduction

Previous observations have shown that the $^{22}$Ne/$^{20}$Ne ratio at the GCR source is greater than that in the solar wind (e.g. Wiedenbeck and Greiner 1981; Mewaldt et al. 1980; Lukasiak et al. 1994; Connell and Simpson 1997; DuVernois et al. 1996). Several models have been proposed to explain the large $^{22}$Ne/$^{20}$Ne ratio (Woosley and Weaver 1981; Reeves 1978; Olive and Schramm 1982; Cassé and Paul 1982; Prantzos et al. 1987; Maeder and Meynet 1993; Soutoul and Legrain 1999; and Higdon and Lingenfelter 2003). (See Binns et al. (2005) and Mewaldt (1989) for a more detailed discussion of these models.) Cassé and Paul (1982) first suggested that ejecta from Wolf–Rayet stars, mixed with material of solar system composition, could explain the large $^{22}$Ne/$^{20}$Ne ratio. Prantzos et al. (1987) later developed this idea in greater detail. The WC phase of WR stars is characterized by the enrichment of the WR winds by He-burning products, especially carbon and oxygen (Maeder and Meynet 1993). In the early part of the He-burning phase, $^{22}$Ne is greatly enhanced as a result of $^{14}$N destruction through the $\alpha$-capture reactions $^{14}\mathrm{N}(\alpha,\gamma)^{18}\mathrm{F}(\mathrm{e}^+,\nu)^{18}\mathrm{O}(\alpha,\gamma)^{22}\mathrm{Ne}$ (e.g. Prantzos et al. 1986; Maeder and Meynet 1993). A high elemental Ne/He ratio in the winds of WC stars has been observed (Dessart et al. 2000), which is consistent with a large $^{22}$Ne excess. The high velocity winds from WR stars can inject the surface material into regions where standing shocks, formed by those winds and the winds of the hot, young, precursor OB stars interacting with the interstellar medium (ISM), may pre-accelerate the WR material.

Kafatos et al. (1981) originally suggested that cosmic rays might be accelerated in superbubbles. Streitmatter et al. (1985) showed that the observed energy spectra and anisotropy of cosmic rays were consistent with such a model. Streitmatter and Jones (2005) have recently shown that the first and second "knees" above $\sim 10^{15}$ and $10^{17}$ eV in the all-particle energy spectrum may be explained in the context of a superbubble model. A model in which particles might be accelerated to energies above $10^{18}$ eV by multiple SN explosions in OB associations was developed by Bykov and Toptygin (2001 and references therein). Parizot et al. (2004) further explored the collective effects of shocks within superbubbles on cosmic ray acceleration. Higdon and Lingenfelter (2003) have argued that GCRs originate in superbubbles based on the excess of $^{22}$Ne/$^{20}$Ne in GCRs. In earlier work, they pointed out that most core-collapse supernovae (SNe) and WR stars occur within superbubbles (Higdon et al. 1998). In their model, WR star ejecta and ejecta from core-collapse SNe within superbubbles mix with interstellar material of solar-system composition, and that material is accelerated by subsequent SN shocks within the superbubble to provide most of the GCRs. Higdon and Lingenfelter (2003) conclude that the elevated $^{22}$Ne/$^{20}$Ne ratio is a natural consequence of the superbubble origin of GCRs since most WR stars exist in OB associations within superbubbles.

## 2 Measurements

The CRIS instrument consists of four stacks of silicon detectors for $dE/dx$ and total energy ($E_{tot}$) measurements, and a scintillating-fiber hodoscope that measures particle trajectories (Stone et al. 1998). The $dE/dx - E_{tot}$ method is used to determine particle charge and mass. The CRIS geometrical factor is ~250 cm$^2$ sr and the total vertical thickness of silicon available for stopping particles is 4.2 cm. The angular precision that is obtained by the fiber hodoscope is $\leq 0.1°$.

The neon data used in this paper were collected from 1997 Dec. 5 through 1999 Sept. 24 and are a high-resolution, selected data set. Events were selected with trajectory angles $\leq 25°$ relative to the detector normal, and particles stopping within 750 $\mu$m of the dead layer surface of each silicon wafer were excluded from the analysis. Nuclei interacting in CRIS were rejected using the bottom silicon anticoincidence detector, which identifies penetrating particles, and by requiring consistency in the multiple mass estimates that we calculate. Additionally, particles with trajectories that exit through the side of a silicon stack were also rejected (Binns et al. 2005).

The average mass resolution for neon that we obtained for energies over the range of ~85 to 275 MeV/nucleon is 0.15 amu (rms). This is sufficiently good that there is very little overlap of the particle distributions for adjacent masses, as shown in Fig. 1. The total number of neon events is $\sim 4.6 \times 10^4$. Histograms of F and O isotopes that are used in the GCR propagation model to obtain the $^{22}$Ne/$^{20}$Ne ratio at the cosmic ray source are also shown in this figure.

## 3 Source Composition

The $^{22}$Ne/$^{20}$Ne abundance ratio at the cosmic-ray source is obtained from the ratio observed using a "tracer method" (Stone and Wiedenbeck 1979), which utilizes observed abundances of isotopes that are almost entirely produced by interstellar interactions of primary cosmic rays to infer the secondary contribution to isotopes like $^{22}$Ne, for which the observed fluxes are a mixture of primary and secondary nuclei. The isotopes that we have used as tracers are $^{21}$Ne, $^{19}$F, and $^{17}$O. The cross-sections used in the propagation model are described in Binns et al. (2005), where details of the model can be found.

Combining the results obtained using these three tracer isotopes, Binns et al. (2005) obtained a "best estimate" of the $^{22}$Ne/$^{20}$Ne ratio of $0.387 \pm 0.007$ (stat.) $\pm 0.022$ (syst.). Expressing this as a ratio relative to solar wind abundances (Geiss 1973), we obtain $(^{22}\mathrm{Ne}/^{20}\mathrm{Ne})_{GCRS}/(^{22}\mathrm{Ne}/^{20}\mathrm{Ne})_{SW}$ ratio of $5.3 \pm 0.3$.

## 4 Wolf–Rayet Model Comparison

Supernovae (SNe) shocks are believed to be the accelerators of GCRs up to energies of at least $\sim 10^{15}$ eV. Most core-collapse supernovae (SNe) in our galaxy (~80–90%) are believed to occur in OB associations within superbubbles (Higdon and Lingenfelter 2003, 2005). Furthermore, most WR stars are located in OB associations and most of the O and B stars with initial mass $\geq 40$ M$_{\odot}$ are believed to evolve into WR stars. These massive stars have short lifetimes, e.g. ~4 million years for a 60 M$_{\odot}$ initial mass star, and their WR phase lasts for typically a few hundred thousand years (Meynet and Maeder 2003; Meynet et al. 1997). The most massive stars with the shortest lifetimes evolve through their WR phase injecting

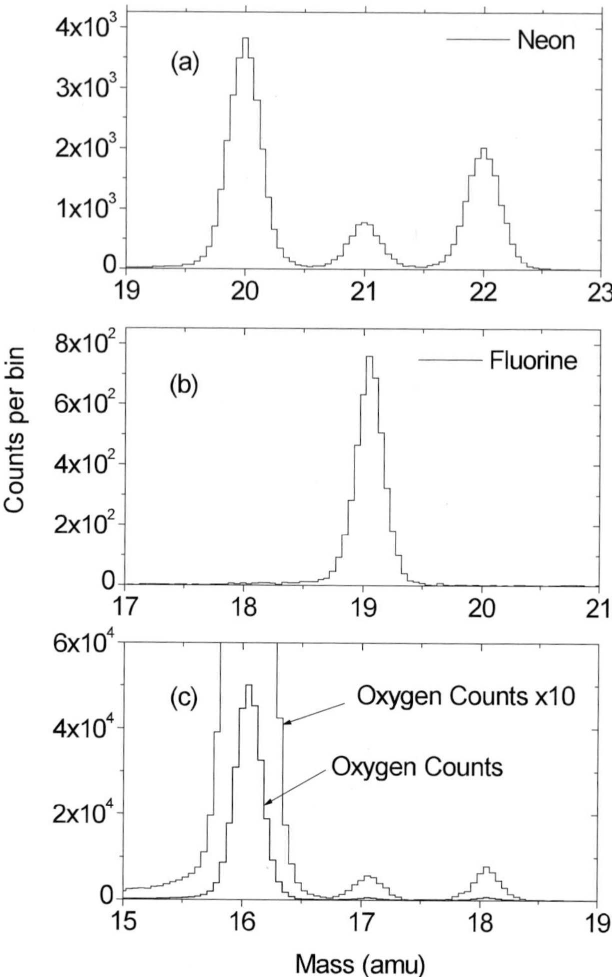


**Fig. 1** Mass histograms for (**a**) neon, (**b**) fluorine, and (**c**) oxygen. The neon energy range extends from ~85 to 275 MeV/nucleon. The solar modulation parameter derived for these observations is $\phi = 400 \pm 60$ MV (Binns et al. 2005) which gives a midpoint energy for neon in the local ISM of ~380 MeV/nucleon

WR wind material, including large amounts of $^{22}$Ne, into the local circumstellar medium, which is already a low density bubble resulting from coalescing main-sequence star winds (Parizot et al. 2004; Van Marle et al. 2005). The shocks from SNe in the OB association should sweep up and accelerate both their own ejected pre-supernova wind material and WR wind material from the more massive stars in the OB association.

The mass of the neon isotopes synthesized by massive stars in their WR and core-collapse SN phases and ejected into superbubbles has been estimated by Higdon and Lingenfelter (2003). Based on these calculations, they estimate that a mass fraction, $18 \pm 5\%$, of WR plus SN ejecta, mixed with ISM material of solar-system composition results in the $^{22}$Ne/$^{20}$Ne ratio reported in an earlier analysis of the CRIS results (Binns et al. 2001). They state that "the $^{22}$Ne abundance in the GCRs is not anomalous but is a natural consequence of the superbubble origin of GCRs in which the bulk of GCRs are accelerated by SN shocks in the high-metallicity WR wind and SN ejecta enriched interiors of superbubbles".

We have examined other isotope ratios at the cosmic-ray source, inferred from our CRIS observations and other experiments, as an additional test of the origin of cosmic rays in

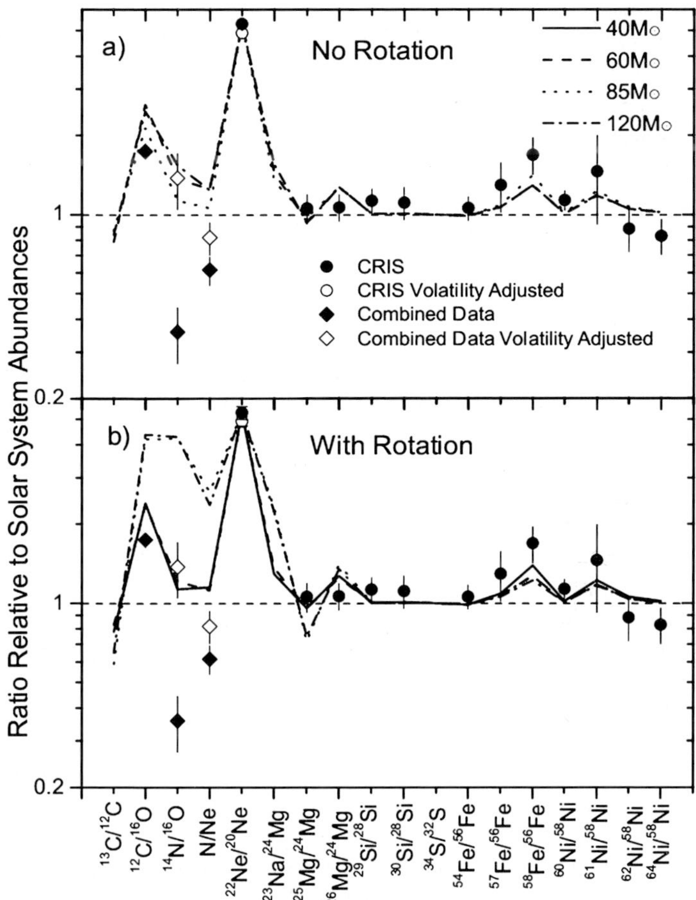


**Fig. 2** CRIS ratios compared with model predictions for WR stars with (**a**) no rotation, and (**b**) an equatorial surface rotation velocity of 300 km s$^{-1}$ for the initial precursor star for masses of 40, 60, 85, and 120 M$_\odot$, and for metallicity $Z_\odot = 0.02$

OB associations. In Fig. 2 we compare these ratios with two-component modeling calculations of WR outflow (Binns et al. 2005; Meynet and Maeder 2003) for metallicity $Z = 0.02$ and initial precursor-star rotational equatorial-velocities at the stellar surface of either 0 or 300 km/s. For each WR star model we calculated the mixture (by mass) of WR wind material with material of solar-system (solar-wind) composition required to give the CRIS $^{22}$Ne/$^{20}$Ne GCR source ratio. Table 1 shows the mass fraction of the total cosmic ray source material required from the WR star since its initial formation for each case.

Although the material being mixed with ISM material in the two approaches (Binns et al. 2005, and Higdon and Lingenfelter 2003) is slightly different (i.e. Higdon and Lingenfelter include neon contributions from SNe in the material being mixed with ISM), the mixing fractions are very similar with the exception of the higher fractions predicted for the very rare $M \geq 85$ M$_\odot$ rotating stars.

In each of the two-component models described above, the material ejected from massive stars is mixed with ISM with an assumed solar system composition to normalize to the

**Table 1** The mass fraction of ejecta from WR stars, integrated from the time of star formation, mixed with ISM material of solar-system composition, that is required to normalize each model to the CRIS $^{22}$Ne/$^{20}$Ne ratio

| WR initial mass ($M_\odot$) | No-rotation WR fraction | Rotating WR fraction |
|---|---|---|
| 40 | – | 0.22 |
| 60 | 0.20 | 0.16 |
| 85 | 0.12 | 0.44 |
| 120 | 0.16 | 0.37 |

$^{22}$Ne/$^{20}$Ne composition. It might be questioned whether ISM with solar system composition is the right material to mix with the ejecta from massive stars. However, Reddy et al. (2003) show that although the abundances of heavy elements increase slowly with time as the galaxy evolves, the ratio of pairs of heavy elements change by only small factors (also see Wiedenbeck et al. 2007). Thus, since we are examining ratios of heavy isotopes and elements, it would appear that the use of ISM with solar system abundances is a reasonable approximation to reality.

The CRIS results are plotted as closed circles in Fig. 2 (see Wiedenbeck et al. 2001a, 2001b and 2003 for elements heavier than neon). Ulysses Mg and Si data (not plotted; Connell and Simpson 1997) are in good agreement with our CRIS results, but their $^{58}$Fe/$^{56}$Fe ratio is significantly lower than the CRIS value (Connell 2001). Wiedenbeck et al. (2001b) discuss a possible reason for this. The lighter elements are plotted as solid diamonds and are mean values of GCR source abundances, relative to solar system, obtained from Ulysses (Connell and Simpson 1997), ISEE-3 (Krombel and Wiedenbeck 1988; Wiedenbeck and Greiner 1981), Voyager (Lukasiak et al. 1994) and HEAO-C2 (Engelmann et al. 1990). (See Binns et al. 2005 for more details.) The error bars are based on weighted means from these experiments. The solar-system abundances of Lodders (2003) are used to obtain the abundances relative to solar system.

In Fig. 2, we see that, for nuclei heavier than neon, the WR models agree well with the data, except for the high-mass (85 and 120 solar masses) rotating-star models that show a deficiency in the $^{25}$Mg/$^{24}$Mg ratio, which is not observed in GCRs. In particular, the observed enhancement of $^{58}$Fe/$^{56}$Fe is consistent with the model predictions.

For elements lighter than neon there is usually only a single isotope for which source abundances can be obtained with sufficient precision to constrain the models. Therefore we have compared ratios of different elements. Elemental ratios are more complicated than isotopic ratios since atomic fractionation effects may be important. The open symbols in Fig. 2 correspond to the ratios after adjustment for volatility and mass fractionation effects (Meyer et al. 1997; see Binns et al. 2005 for details). The $^{12}$C/$^{16}$O ratio was not adjusted since the fraction of interstellar carbon in the solid state is poorly known. These adjusted ratios show improved agreement with the models, but the adjustments are not highly quantitative, and should be regarded as approximate values showing that ratios previously thought to be inconsistent with solar-system abundances may be consistent if GCRs are fractionated correctly on the basis of volatility and mass. (See Binns et al. 2005 for additional discussion.)

After adjustments for elemental fractionation, these data show an isotopic composition similar to that obtained by mixing ~20% of WR wind material with ~80% of material with solar-system composition. The largest enhancements with respect to solar-system ratios predicted by the WR models $^{12}$C/$^{16}$O, $^{22}$Ne/$^{20}$Ne, and $^{58}$Fe/$^{56}$Fe, are consistent with our observations. We take this agreement as evidence that WR star ejecta are very likely an important component of the cosmic-ray source material.

## 5 Discussion

Two independent approaches at modeling the WR contribution to GCRs (Higdon and Lingenfelter 2005, and Binns et al. 2005) have shown that to explain the cosmic-ray data approximately 20% of the source material must be WR star ejecta. For WR material to be such a large component of the GCR source material, large quantities of it must be efficiently injected into the accelerator of GCRs. We believe that this is an important constraint for models of the origin of GCRs.

(There is no inconsistency between the ~20% estimates described above and the earlier ~2% quoted by Cassé and Paul (1982). Although the details of the Cassé and Paul calculations are not included in that paper, the 2% that they quote is material ejected only in the WC phase. In both the Higdon and Lingenfelter (2003) and the Meynet and Maeder (2003, 2005) models, the ~20% of material includes all material ejected from the birth of the star to the end of the WC phase. Additionally, Cassé and Paul used the solar energetic particle ratio of 0.13 (Mewaldt et al. 1979) instead of the solar wind value of 0.073 (Geiss 1973) for $^{22}Ne/^{20}Ne$ to represent the solar system abundances in their estimate. When adjustments for these two factors are made, the fraction of WR material required in the Cassé and Paul calculations is ~20%, in good agreement with the more recent calculations.)

Another important constraint for the origin of cosmic rays, previously obtained from CRIS results, is the requirement that nuclei synthesized and accelerated by SNe are accelerated at least $10^5$ y after synthesis. Wiedenbeck et al. (1999) have previously shown that the $^{59}Ni$, which decays only by electron-capture, has completely decayed, within the measurement uncertainties, to $^{59}Co$ (Wiedenbeck et al. 1999). The $^{59}Ni$ can decay if it forms as dust grains or as gas in atomic or molecular form, or it could decay in a plasma environment. In the Meyer et al. (1997) and Ellison et al. (1997) model, it is assumed that it is initially accelerated as dust grains, since it is a refractory element. Dust grains are known to form in SN ejecta (e.g. SN1987A (Dwek 1998) and Cas-A (Dunne et al. 2003)). It therefore appears likely that before acceleration, the $^{59}Ni$ resides in dust grains where it decays and its $^{59}Co$ daughter is later accelerated by SN shocks.

The average time between SN events within superbubbles is $\sim 3 \times 10^5$ years, depending upon the number of massive stars in the OB association (Higdon et al. 1998). In Binns et al. (2005) we stated that this gives sufficient time between events for the $^{59}Ni$, which is synthesized in the SN explosions, to decay before its daughter product, $^{59}Co$, is accelerated to cosmic ray energies. However, Prantzos (2005) has suggested that WR winds could accelerate the newly synthesized nuclei on time scales short compared to the mean time between SNe, based on arguments contained in Parizot et al. (2004). The kinetic energy in WR winds is of the same order as is dissipated in supernova explosions (e.g. Leitherer et al. 1992). Prantzos argued that the superbubble environment experiences shock passages on times scales significantly shorter than the mean time between SN. He therefore suggested that the mean time between SNe is not the time scale that is relevant for $^{59}Ni$ decay in the superbubble environment.

Most GCRs detected at Earth are believed to originate within less than 1 kpc from the Sun (Ptuskin and Soutoul 1998). OB associations within 1 kpc of the Sun are composed of a few to as many as ~320 OB stars in SCO OB2, which is located ~140 pc from the Sun (De Zeeuw et al. 1999). Some OB associations are composed of stars that form at approximately the same time, i.e. they are coeval. For example, Per OB2 contains 17 OB stars, is located about 400 pc from the Sun (De Zeeuw et al. 1999), and has an age of ~3 My (Blaauw 1964). Some of the larger associations are composed of two or more subgroups, with the stars in each subgroup forming at about the same time, but with the subgroups

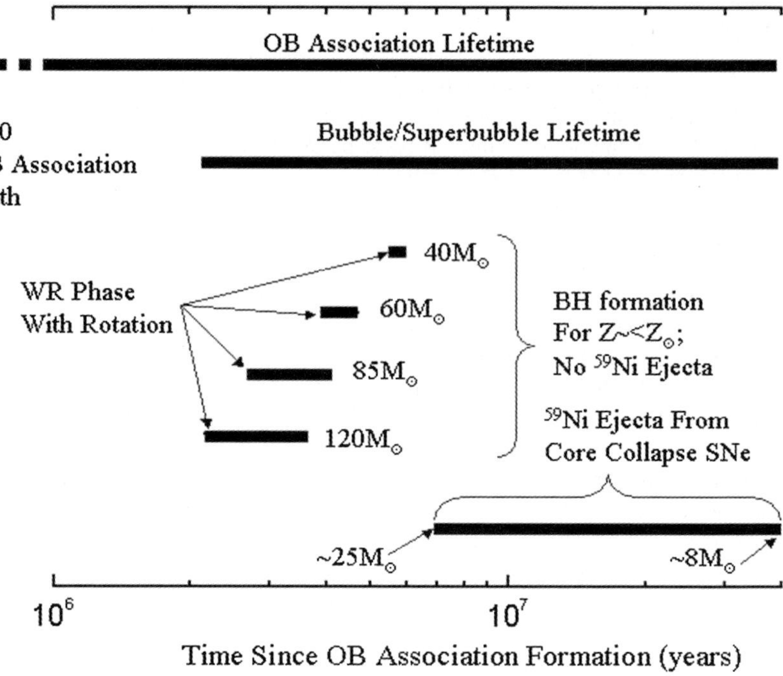


**Fig. 3** Diagram of the time evolution of a hypothetical OB association (see text)

themselves having differing ages. An example of this is the Orion OB1 association, which contains ~70 OB stars, with subgroups a, b, c, and d having ages of ~12, 7, 3, and 1 My respectively (De Zeeuw et al. 1999). The mean time for subgroup formation, averaged over many associations, is ~4 My (Sargent 1979). Subgroup formation is believed to result when a SN shock from a young OB association propagates into the molecular cloud in which the association is embedded, causing nearby massive star formation. The most massive stars have the shortest lifetimes; stars with initial mass greater than ~30 $M_\odot$ have lifetimes of ≤6 My (Schaller et al. 1992).

For the sake of argument, we will initially consider an OB association in which the stars are coeval. In Fig. 3, we show the history of such an association.

Its lifetime begins with the localized condensation of molecular cloud material into massive stars at $T = 0$ and ends when the least massive star that can undergo core-collapse (~8 $M_\odot$) ends its life as a supernova, ~40 My later. Shortly after star formation, the most massive stars evolve into the Wolf–Rayet phase. Their high-velocity winds (~2,000–3,000 km/s) produce large low-density bubbles in the molecular cloud. The expanding shocks produced by the stars that undergo SN explosions coalesce and produce a superbubble. We show the time duration that the most massive stars spend in the WR phase, and the epoch for which that occurs in the OB association, for rotating stars with initial masses ranging from 40 to 120 $M_\odot$. We see that the most massive star modeled enters the WR phase roughly 2 My after association birth, and the least massive star that can evolve into a WR star exits that phase roughly 4 My later. The exact low-initial-mass cutoff for entering the WR phase is model dependent and is believed to be between 25 $M_\odot$ and 40 $M_\odot$. (For details of these models of rotating stars, see Meynet and Maeder 2003, 2005; for the associated nucleosynthesis see Arnould et al. 2006). The end of the WR phase of

each star is followed by core collapse. So there is, at most, an ~4 My interval in the life of a coeval OB association (~10% of its lifetime) for which acceleration of superbubble material by WR winds could occur. It is important to note that the most massive stars are very rare. The initial mass function for OB associations is not universally agreed upon, but is often taken to go approximately as $dN/dM \propto M^{-2.35}$ (Salpeter 1955; Higdon and Lingenfelter 2005). Therefore, in most OB associations, the most massive stars are not present, and the WR epoch is less than 4 My.

Let us suppose, as argued by Heger et al. (2003), that most stars with initial mass $\geq 40\ M_\odot$ and metallicity roughly solar or less do not undergo a SN explosion after core collapse, but instead "directly" form a black hole. Additionally, in their model, stars with initial mass 25–40 $M_\odot$ and metallicity roughly solar or less undergo core-collapse to form a black hole by "fallback", which results in a very weak SN shock with little ejecta. (We note that their model does not include the effects of rotation, which could change some of the details of their model predictions.) Daflon and Cunha (2004) have measured the metallicity of young OB stars in associations as a function of galacto-centric radius. Their results show that for OB associations within 1 kpc of the Sun, most have metallicity that is solar or less. In the Heger et al. picture, stars with metallicity higher than roughly solar that undergo core collapse create supernovae of type SNIb,c. These supernovae are believed to result from WR stars, which have no hydrogen envelope, and thus no H emission. Additionally, there are massive stars that core-collapse into "hypernova", which are poorly understood, and estimated to occur in ~1–10% of the massive core-collapse events (Fryer et al. 2006). To the extent that this is a correct picture we see that a substantial fraction of core-collapse events during the WR epoch will not eject large amounts of newly synthesized material, including $^{59}$Ni, into the superbubble. Thus the predominant material that is available for acceleration by the WR winds appears to be wind material ejected from the association stars since their birth, plus any normal ISM that is in the vicinity.

Looking again at Fig. 3, we see that those stars with mass low enough so that they do not enter the WR phase ($\sim 8\ M_\odot \leq M \leq 25\ M_\odot$) undergo core-collapse as SNe in which $^{59}$Ni is synthesized and injected into the superbubble. The most massive of these stars will undergo SN explosions first with subsequent SNe accelerating the material previously injected into the superbubble.

In this simple picture it appears that the injection of the $^{22}$Ne-rich wind material from WR stars and the injection of $^{59}$Ni from the SNe of stars with initial mass $8\ M_\odot \leq M \leq 25\ M_\odot$ are largely separated in time. Thus the appropriate time scale for acceleration of most SN ejecta would be the time between SN shocks after the WR epoch in superbubbles, not the shorter time scales associated with WR shocks in the WR epoch. The SN rate depends upon the number of stars in the OB association and has a time dependence related to the mass distribution of stars in the association. Since the time between SNe is typically $\sim 3 \times 10^5$ years for a large association (Higdon et al. 1998), and the $^{59}$Ni halflife for decay is $7.5 \times 10^4$ years, in this picture, there is sufficient time for it to decay to $^{59}$Co.

For the fraction of OB associations that are composed of subgroups with differing ages, this simple picture needs to be modified since the WR winds from younger subgroups occur during the time period when substantial $^{59}$Ni is being ejected by SNe in older subgroups. However the fraction of the superbubble lifetime for which WR winds coexist with SN ejecta is still relatively small owing to the brief WR epoch. This is particularly true when you consider that there are many more WR stars at the light end of the mass spectrum than at the heavy end (Higdon and Lingenfelter 2005), so the WR epoch in most subgroups is substantially shorter than 4 My. Thus, in this picture, the fraction of $^{59}$Ni that could be accelerated by WR winds, summing over all superbubbles in our neighborhood, is still relatively small.

It must be acknowledged that the superbubble environment is very complex. In addition to shocks from WR winds and SNe shocks, the winds and mass loss of OB stars in phases other than WR are very significant and may produce substantial shocks, though not as significant as either WR or SNe shocks (Parizot et al. 2004; Bykov 2001). Furthermore, although the average time between SN events in an OB association is long compared to the $^{59}$Ni half-life, there will be a fraction of SN events that occur on shorter time scales, thus resulting in acceleration of recently synthesized $^{59}$Ni before it can decay. So, although the picture presented above seems useful in understanding how a substantial fraction of $^{59}$Ni can decay, the actual situation is likely more complex.

Additionally, we have argued in Binns et al. (2006) that although WR winds do contain roughly the same amount of kinetic energy ($\sim 10^{51}$ ergs) as supernova explosions, the power in the WR termination shocks is about a factor of 10 less than in SNRs, which have lifetimes of $\sim 10^4$ years (Higdon 2006). Furthermore, it is possible that the WR shocks would only accelerate $^{59}$Ni nuclei to relatively low energies, where they would be only partially stripped of their orbital electrons, and the $^{59}$Ni could still decay. In a power law spectrum that one gets from shock acceleration, most of the nuclei are at low energy where they can decay.

Thus it appears that the observation by Wiedenbeck et al. (1999) that all or most of the $^{59}$Ni in GCRs has decayed to $^{59}$Co is likely consistent with the OB-association origin of galactic cosmic rays. We conclude that the good agreement of our isotopic data with WR models suggests that OB associations are the likely source of at least a substantial fraction of GCRs.

**Acknowledgements** The authors wish to thank J.C. Higdon and N. Prantzos for helpful discussions. This research was supported in part by the National Aeronautics and Space Administration at Caltech, Washington University, the Jet Propulsion Laboratory, and Goddard Space Flight Center (under Grants NAG5-6912 and NAG5-12929).

## References

M. Arnould, S. Goriely, G. Meynet, Astron. Astrophys. **453**, 653 (2006)
W.R. Binns et al., in *AIP Proc.*, vol. 598, ed. by R.F. Wimmer-Schweingruber (AIP, New York, 2001), p. 257
W.R. Binns et al., Astrophys. J. **634**, 351 (2005)
W.R. Binns et al., New Astron. Rev. **50**(7–8), 516 (2006)
A. Blaauw, Annu. Rev. Astron. Astrophys. **2**, 213 (1964)
A.M. Bykov, I.N. Toptygin, Astron. Lett. **27**, 625 (2001)
A.M. Bykov, Space Sci. Rev. **99**, 317 (2001)
M. Cassé, J.A. Paul, Astrophys. J. **258**, 860 (1982)
J.J. Connell, J.A. Simpson, *25th ICRC*, vol. 3 (1997), p. 381
J.J. Connell, Space Sci. Rev. **99**, 41 (2001)
S. Daflon, K. Cunha, Astrophys. J. **617**, 1115 (2004)
L. Dessart et al., Mon. Not. Roy. Astron. Soc. **315**, 407 (2000)
P.T. De Zeeuw et al., Astrophys. J. **117**, 354 (1999)
L. Dunne et al., Nature **424**, 285 (2003)
M.A. DuVernois et al., Astrophys. J. **466**, 457 (1996)
E. Dwek, Astrophys. J. **501**, 643 (1998)
D.C. Ellison et al., Astrophys. J. **487**, 197 (1997)
J.J. Engelmann et al., Astron. Astrophys. **233**, 96 (1990)
C.L. Fryer, P.A. Young, A.L. Hungerford, Astrophys. J. **650**, 1028 (2006)
J. Geiss, *13th ICRC*, vol. 5 (1973), p. 3375
A. Heger et al., Astrophys. J. **591**, 288 (2003)
J.C. Higdon et al., Astrophys. J. **509**, L33 (1998)
J.C. Higdon, R.E. Lingenfelter, Astrophys. J. **590**, 822 (2003)
J.C. Higdon, R.E. Lingenfelter, Astrophys. J. **628**, 738 (2005)
J.C. Higdon, Personal communication (2006)

H. Kafatos et al., *17th Int. Cosmic Ray Conf.* (Paris), vol. 2 (1981), p. 222
K. Krombel, M. Wiedenbeck, Astrophys. J. **328**, 940 (1988)
C. Leitherer, C. Robert, L. Drissen, Astrophys. J. **401**, 596 (1992)
K. Lodders, Astrophys. J. **591**, 1220 (2003)
A. Lukasiak et al., Astrophys. J. **426**, 366 (1994)
A. Maeder, G. Meynet, Astron. Astrophys. **278**, 406 (1993)
R.A. Mewaldt et al., Astrophys. J. Lett. **231**, L97 (1979)
R.A. Mewaldt et al., Astrophys. J. **235**, L95 (1980)
R.A. Mewaldt, in *AIP Proc.*, ed. by C.J. Waddington, Cos. Abund. of Matter, vol. 183 (1989), p. 124
J.P. Meyer et al., Astrophys. J. **487**, 182 (1997)
G. Meynet, M. Arnould, N. Prantzos, G. Paulus, Astron. Astrophys. **320**, 460 (1997)
G. Meynet, A. Maeder, Astron. Astrophys. **404**, 975 (2003)
G. Meynet, A. Maeder, Astron. Astrophys. **429**, 518 (2005)
K.A. Olive, D.N. Schramm, Astrophys. J. **257**, 276 (1982)
E. Parizot et al., Astron. Astrophys. **424**, 747 (2004)
N. Prantzos et al., Astrophys. J. **304**, 695 (1986)
N. Prantzos et al., Astrophys. J. **315**, 209 (1987)
N. Prantzos, Personal communication (2005)
V.S. Ptuskin, A. Soutoul, Space Sci. Rev. **86**, 225 (1998)
B.E. Reddy, J. Tomkin, D.L. Lambert, C.A. Prieto, Mon. Not. Roy. Astron. Soc. **340**, 304 (2003)
H. Reeves, in *Protostars and Planets*, ed. by T. Gehrels (Univ. of Arizona Press, Tucson, 1978), p. 399
E.E. Salpeter, Astrophys. J. **121**, 161 (1955)
A.I. Sargent, Astrophys. J. **233**, 163 (1979)
G. Schaller, D. Schaerer, G. Meynet, A. Maeder, Astron. Astrophys. **96**, 269 (1992)
A. Soutoul, R. Legrain, *26th ICRC* (Salt Lake City), vol. 4 (1999), p. 180
E. Stone, M. Wiedenbeck, Astrophys. J. **231**, 606 (1979)
E.C. Stone et al., Space Sci. Rev. **86**, 285 (1998)
R.E. Streitmatter et al., Astron. Astrophys. **143**, 249 (1985)
R.E. Streitmatter, F.C. Jones, *29th ICRC*, vol. 3 (2005), p. 157
A.J. Van Marle et al., Astron. Astrophys. **444**, 837 (2005)
M.E. Wiedenbeck, D.E. Greiner, Phys. Rev. Lett. **46**, 682 (1981)
M.E. Wiedenbeck et al., Astrophys. J. **523**, L61 (1999)
M. Wiedenbeck et al., in *AIP Proc.*, vol. 598, ed. by R.F. Wimmer-Schweingruber (AIP, New York, 2001a), p. 269
M. Wiedenbeck et al., Adv. Space Res. **27**, 773 (2001b)
M.E. Wiedenbeck et al., *28th ICRC*, vol. 4 (2003), p. 1899
M.E. Wiedenbeck et al., Space Sci. Rev., this volume (2007). doi: 10.1007/s11214-007-9198-y
S.E. Woosley, T.A. Weaver, Astrophys. J. **243**, 651 (1981)

Space Sci Rev (2007) 130: 451–456
DOI 10.1007/s11214-007-9184-4

# GEMS at the Galactic Cosmic-Ray Source

**A.J. Westphal · A.M. Davis · J. Levine · M.J. Pellin · M.R. Savina**

Received: 15 February 2007 / Accepted: 2 April 2007 /
Published online: 24 May 2007


**Abstract** Galactic cosmic rays probably predominantly originate from shock-accelerated gas and dust in superbubbles. It is usually assumed that the shock-accelerated dust is quickly destroyed by sputtering. However, it may be that some of the dust can survive bombardment by the high-metallicity gas in the superbubble interior, and that some of that dust has been incorporated into solar system materials. Interplanetary dust particles (IDPs) contain enigmatic submicron components called GEMS (Glass with Embedded Metal and Sulfides). These GEMS have properties that closely match those expected of a population of surviving shock-accelerated dust at the GCR source (Westphal and Bradley in Astrophys. J. 617:1131, 2004). In order to test the hypothesis that GEMS are synthesized from shock-accelerated dust in superbubbles, we plan to measure the relative abundances of Fe, Zr, and Mo isotopes in GEMS using the new Resonance Ionization Mass Spectrometer at Argonne National Laboratory. If GEMS are synthesized from shock-accelerated dust in superbubbles, they should exhibit isotopic anomalies in Fe, Zr and Mo: specificially, enhancements in the $r$-only isotopes $^{96}$Zr and $^{100}$Mo, and separately in $^{58}$Fe, should be observed. We review also recent developments in observations of GEMS, laboratory synthesis of GEMS-like materials, and implications of observations of GEMS-like materials in Stardust samples.

**Keywords** Interstellar dust · Galactic cosmic rays · Superbubbles

A.J. Westphal (✉)
Space Sciences Laboratory, University of California, Berkeley, CA 94720, USA
e-mail: westphal@ssl.berkeley.edu

A.M. Davis
Enrico Fermi Institute and Department of Geophysical Sciences, University of Chicago, 5640 South Ellis Avenue, Chicago, IL 60637, USA

J. Levine
Chicago Center for Cosmochemistry and Department of the Geophysical Sciences, University of Chicago, 5734 S. Ellis Avenue, Chicago, IL 60637, USA

M.J. Pellin · M.R. Savina
Materials Science and Chemistry Divisions, Argonne National Laboratory, Argonne, IL 60439, USA

## 1 GCR Source: Accelerated Refractory Dust in Superbubbles

In a companion paper in this volume, Lingenfelter and Higdon (2007) review the evidence that galactic cosmic rays must originate predominantly in superbubbles. The only "engines" that are sufficiently powerful to maintain the galactic cosmic-ray flux in the galaxy are core-collapse supernovae; further, the observed energy spectrum of GCRs, at least up to $\sim 10^{15}$ eV is consistent with diffusive shock-acceleration by supernova shocks. The vast majority (80–90%) of core-collapse supernovae occur in OB associations. The first supernova in any association blows a bubble (a "superbubble") in the local interstellar medium. The density contrast between the hot interior of the superbubble and the surrounding cold ISM is about three orders of magnitude. It is here that the majority of GCR nuclei are accelerated to relativistic energies.

It has been recognized for some decades now that galactic cosmic ray nuclei are strongly elementally fractionated, probably during injection into the GCR acceleration process. However, until relatively recently the nature of this fractionation has been poorly understood: originally, it was thought that the fractionation had to do with ionizability, as is seen in the solar wind and in solar energetic particles (Meyer 1985). However, about a decade ago, Meyer et al. (1997) refined and extended proposals by Cesarsky and Bibring (1981) and Epstein (1980) that the elemental fractionation pattern observed in GCRs can most easily be understood if GCRs originate in atoms sputtered from shock-accelerated dust. In this case, refractory elements, rather than those with low first ionization potential, are preferentially accelerated into the GCRs. Observations from HEAO (Binns et al. 1989) and TREK (Westphal et al. 1998) of the Pb/Pt ratio were consistent with the accelerated dust-grain hypothesis.

In most treatments of the problem, it is assumed that dust grains accelerated by shocks are rapidly destroyed by sputtering. This is highly likely if the grains propagate in a low-metallicity environment, like the average galactic ISM. However, the superbubble interiors are enriched in metals—the products of nucleosynthesis in supernovae and high-mass stars. Gray and Edmonds (2004) have pointed out that in a sufficiently high-metallicity environment, grains are not necessarily destroyed, because implantation of bombarding ions can compensate for or even dominate over sputtering. So it is worth asking the question: is it possible that a population of these accelerated dust grains might survive, and eventually be incorporated into early solar system materials? If so, we propose that they would look much like GEMS.

## 2 GEMS Formation in Superbubbles?

Interplanetary dust particles (IDPs) contain enigmatic submicron components called GEMS (Glass with Embedded Metal and Sulfides) (Bradley et al. 1999). GEMS are stoichiometrically enriched in oxygen relative to the amount expected from cations present and systematically depleted in S, Mg, Ca and Fe (relative to solar abundances), most have normal (solar) oxygen isotopic compositions, they exhibit a strikingly narrow size distribution (0.1–0.5 μm diameter), and some of them contain "relict" crystals within their silicate glass matrices. The compositions and structures of GEMS indicate that they have been processed by exposure to ionizing radiation but details of the actual irradiation environment(s) have remained elusive.

In 2004, Westphal and Bradley (2004) proposed a mechanism and astrophysical site for GEMS formation that explains for the first time the following key properties of GEMS; we showed that the compositions, size distribution, and survival of relict crystals are inconsistent with amorphization by particles accelerated by diffusive shock acceleration. Instead, we proposed that GEMS are formed from crystalline grains that condense in stellar outflows from massive stars in OB associations, are accelerated in encounters with frequent supernova shocks inside the associated superbubble, and are implanted with atoms from the hot gas in the SB interior. We thus reverse the usual roles of target and projectile. Rather than being bombarded at rest by energetic ions, grains are accelerated and bombarded by a nearly monovelocity beam of atoms as viewed in their rest frame. Meyer et al. (1997) have proposed that galactic cosmic rays originate from ions sputtered from such accelerated dust grains. We suggested that GEMS are surviving members of a population of fast grains that constitute the long-sought source material for galactic cosmic rays. Thus, representatives of the GCR source material may have been awaiting discovery in cosmic dust labs for the last thirty years.

## 3 Observational Tests of the Superbubble Origin Hypothesis

The original paper by Westphal and Bradley (2004) makes specific predictions for future observations. Isotopic anomalies observed in galactic cosmic-rays should also be observable in GEMS. The only two significant isotopic anomalies observed so far in the galactic cosmic rays are in $^{22}$Ne/$^{20}$Ne ($\sim$5$\times$ solar) and $^{58}$Fe/$^{56}$Fe ($\sim$1.7$\times$ solar) (Binns et al. 2005). We would expect systematic compositional differences between grains originating as pyrrhotite as compared with the more rare grains that originate in other types (e.g., olivine), because of the presence of residual material from the original grain. For example, GEMS containing pyrrhotite relict crystals should have larger bulk S than those containing forsterite or enstatite.

Since the Westphal and Bradley paper, Binns et al. (2005) and Higdon and Lingenfelter (2003) have shown that the galactic cosmic-ray composition can be understood if the GCR source consists of $\sim$80% material of solar composition and $\sim$20% of material from rotating Wolf–Rayet stars. Wolf–Rayet stars are highly-evolved, high-mass loss stars.

Oxygen isotopes may be helpful as a diagnostic for superbubble origin. The oxygen isotopic composition of superbubble interiors is observationally unconstrained, and unfortunately galactic cosmic rays are not helpful here: primary $^{17}$O and $^{18}$O are buried under a huge secondary population from the spallation of heavier elements, principally Ne and Mg. Limongi and Chieffi (2003) have calculated the yields of oxygen isotopes for core-collapse supernovae over the range 13–35 solar masses. Although the absolute yields vary significantly over various models, there is a systematic trend from $^{17}$O- and $^{18}$O-rich yields for the lower masses, to relatively $^{17}$O- and $^{18}$O-poor yields for the higher masses. At early times during the superbubble lifetime, it is expected that the interior is dominated by ejecta from the most massive (shortest lived) stars; at later times, ejecta from longer-lived, lighter stars will start to contribute. So the SB interior should evolve from a $^{17}$O- and $^{18}$O-rich medium to one that is closer to solar or even $^{17}$O- and $^{18}$O-poor. Since dust grains are formed continuously during the evolution of the SB (at least while supernovae and high mass-loss winds operate), they should exhibit a large, systematic variation in O-isotope composition.

Oxygen isotopic abundances have been measured in GEMS by Messenger et al. (2003). Most GEMS appear to be isotopically consistent with solar (with dispersion $\ll$1%) material. However, two GEMS (<5%) were found with isotopic anomalies in oxygen that point strongly to an interstellar or circumstellar origin. Interestingly, one of these GEMS was $^{17}$O-

and $^{18}$O-rich, the other was $^{17}$O- and $^{18}$O-poor. Neither exhibited the large isotopic anomalies implied by the Limongi and Chieffi (2003) models. However, the grains may have partially equilibrated with solar-like material in the ISM before incorporation in the solar nebula, in the parent body after solar-system formation, or even during atmospheric entry.

The decay products of $^{26}$Al and $^{60}$Fe may be present, but may not be detectable, unless there is substantial inhomogeneity in Mg/Al or Fe/Ni ratio in the SB interior that would allow the fossil radioactivities to be detected through positive correlations between, e.g., $^{26}$Mg/$^{24}$Mg and $^{27}$Al/$^{24}$Mg.

We would also expect the products of $r$-process nucleosynthesis to be overrepresented in GEMS if this hypothesis is correct. The reasoning is straightforward. While the astrophysical site of the $r$-process is still unknown, $r$-process nucleosynthesis, essentially by definition, occurs in an explosive environment, almost certainly in core-collapse supernova explosions of short-lived massive stars. Recent observations of heavy elements in ultrametal-poor halo stars (Sneden et al. 2000), showing the clear signature of $r$-process nucleosynthesis early in galactic history, is compelling evidence that $r$-process nucleosynthesis occurs in core-collapse (type II and Ibc) supernovae. By contrast, $s$-process nucleosynthesis, which is expected to occur predominantly in old AGB stars, is not expected to contribute as much to superbubble interiors, simply because the timescale for this process exceeds the superbubble lifetimes. We therefore expect to find enhancements in $r$-process isotopes in GEMS. The site of the $p$-process is also unknown, but may be ejecta from core-collapse supernovae (Rauscher et al. 2002). Finally, a unique nucleosynthetic signature due to a neutron burst that results from the passage of the explosion shock wave through outer shells has been recognized in presolar SiC from Type II supernovae (Pellin et al. 2000; Meyer et al. 2000).

We are focussing now on relative isotopic abundance measurements of three elements: Fe for which $^{58}$Fe is observed to be enhanced in galactic cosmic rays; Zr, for which $^{96}$Zr/$^{94}$Zr and $^{90}$Zr/$^{94}$Zr are expected to be enriched by the $r$-process and $^{96}$Zr/$^{94}$Zr may be enriched in the ejected shells of supernovae; and Mo, for which $^{100}$Mo/$^{96}$Mo are expected to be enriched by the $r$-process, $^{92}$Mo/$^{96}$Mo and $^{94}$Mo/$^{96}$Mo are expected to be enriched by the $p$-process, and $^{95}$Mo/$^{96}$Mo and $^{97}$Mo/$^{96}$Mo are expected to be enriched by neutron burst nucleosynthesis in ejected shells. We propose to make these measurements using the CHARISMA RIMS instrument (Savina et al. 2003) at Argonne National Laboratory. Fe isotopes (Pellin et al. 2002), and Zr and Mo isotopes (Lugaro et al. 2003), have been measured before in presolar SiC using CHARISMA.

## 4 Recent Mineralogical and Petrographic Studies of GEMS

Keller et al. (2007) have recently done spectrum imaging of ~30 GEMS. They observed no GEMS which exhibited relict crystalline material. Of these 30, all appeared to be aggregates of much smaller components ("nanorocks"). They concluded that these GEMS grains were inconsistent with irradiation processing of single mineral grains in the interstellar medium. No coordinated isotopic studies were done on these grains, so it is unknown whether any of these GEMS were interstellar or circumstellar. Also, they observed no relict crystals in these GEMS, in contrast to work by Bradley (1994). The origin of this discrepancy is not clear, but it may be due to systematic differences among GEMS between IDPs.

## 5 Laboratory Analogs

Several new laboratory studies may shed light on the origin of GEMS.

Davoisne et al. (1985) have studied the microstructural evolution of an amorphous ferromagnesian silicate of olivine composition during heating in a carbon-rich atmosphere. They found that this treatment resulted in an amorphous glass embedded with reduced Fe–Ni nanoparticles, very similar to GEMS. They hypothesize that many or most GEMS may have formed in a hot, carbon-rich environment in the protosolar nebula. Indeed, this may be the origin of GEMS that exhibit solar isotopic composition in oxygen.

In January 2006, the Stardust spacecraft returned the first *bona fide* samples from a comet. The preliminary examination of the returned samples show abundant glassy material with nanophase metallic inclusions. The bulk elemental composition of these objects is not similar to GEMS, and the current evidence points strongly toward an origin in an impact-melt mixture of cometary materials with the silicate aerogel capture medium, with accompanying reduction of Fe in the cometary material to metallic Fe. Two recent synchrotron X-ray analyses of materials shot in to aerogel at the encounter speed of Stardust ($6\ \mathrm{km\,sec^{-1}}$) show a systematic reduction of oxidized Fe. One possible mechanism for this is "smelting" of oxidized Fe to reduced Fe by carbon, the carbon being either indigenous to the comet or present as a contaminant in the aerogel capture medium.

## 6 Conclusion

Occam's razor suggests that the similarity among GEMS points to a common genesis. Isotopic evidence indicates that at least some GEMS were formed outside the solar system. Though an interstellar origin for all GEMS is not required, it is nevertheless possible that even those GEMS with "solar" isotopic composition were formed elsewhere in the galaxy. Mineralogical and petrologic evidence on some GEMS suggests that any extrasolar GEMS might be dust from the source of the galactic cosmic rays. Coordinated isotopic and mineralogical/petrological studies of GEMS will test the hypothesis that samples of the GCR source have already been collected on Earth and are available for laboratory examination—as constituents of interplanetary dust particles.

## References

W.R. Binns et al., Astrophys. J. **336**, 997 (1989)
W.R. Binns et al., Astrophys. J. **634**, 351 (2005)
J.P. Bradley, Science **265**, 925 (1994)
J.P. Bradley et al., Science **285**, 1716 (1999)
C.J. Cesarsky, J.-P. Bibring, in *Origin of Cosmic Rays*, ed. by G. Setti et al. (D. Reidel, Dordrecht, 1981). p. 361
C. Davoisne et al., Astron. Astrophys. **448**, L1–L4 (1985)
R.I. Epstein, Mon. Not. R. Astron. Soc. **193**, 723 (1980)
M.D. Gray, M.G. Edmonds, MNRAS **349**, 491 (2004)
J. Higdon, R. Lingenfelter, Astrophys. J. **590**, 822 (2003)
L.P. Keller, S. Messenger, R. Christoffersen, LPS XXXVI, #2088 (2007)
M. Limongi, A. Chieffi, Astrophys. J. **592**, 404 (2003)
R. Lingenfelter, J. Higdon (2007, this volume)
M. Lugaro, A.M. Davis, R. Gallino, M.J. Pellin, O. Straniero, F. Käppeler, Astrophys. J. **593**, 486 (2003)
S. Messenger, L.P. Keller, F.J. Stadermann, R.M. Walker, E. Zinner, Science **300**, 105 (2003)
B.S. Meyer, D.D. Clayton, L.-S. The, Astrophys. J. **540**, L49 (2000)

J.-P. Meyer, Astrophys. J. Suppl. **57**, 173–204 (1985)
J.-P. Meyer, L.O'C. Drury, D.C. Ellison, Astrophys. J. **487**, 182 (1997)
M.J. Pellin, W.F. Calaway, A.M. Davis, R.S. Lewis, S. Amari, R.N. Clayton, LPS XXXI, #1917 (2000)
M.J. Pellin, M.R. Savina, E. Tripa, W.F. Calaway, A.M. Davis, R.S. Lewis, S. Amari, R.N. Clayton, Meteorit. Planet. Sci. **37**, A115 (2002)
T. Rauscher, A. Heger, R.D. Hoffman, S.E. Woosley, Astrophys. J. **576**, 323 (2002)
M.R. Savina, M.J. Pellin, C.E. Tripa, I.V. Veryovkin, W.F. Calaway, A.M. Davis, Geochimica et Cosmochimica Acta **67**, 3215 (2003)
C. Sneden, J.J. Cowan, I.I. Ivans, G.M. Fuller, S. Burles, T.C. Beers, J.E. Lawler, Astrophys. J. **533**, L139 (2000)
A.J. Westphal, J.P. Bradley, Astrophys. J. **617**, 1131 (2004)
A.J. Westphal et al., Nature **376**, 50 (1998)

Space Sci Rev (2007) 130: 457–464
DOI 10.1007/s11214-007-9145-y

# The Ultra Heavy Elements in the Cosmic Radiation

**C.J. Waddington**

Received: 18 December 2006 / Accepted: 9 January 2007 /
Published online: 27 April 2007


**Abstract** The concept that nuclei of all of the elements in the periodic table were accelerated to relativistic energies in the primary cosmic rays was confirmed by the discovery of the presence of nuclei significantly heavier than iron. These "ultra heavy" nuclei must all have been synthesized in endothermic reactions, occurring predominately in the final stages of stellar evolution. Determination of the relative abundances of these nuclei should provide new insights into the nuclear processes in some of the most energetic events in the life cycle of stars. The very low abundances of these nuclei relative to those of the exothermic lighter nuclei have made progress difficult. In addition, the effects of apparent preferential acceleration mechanisms and of propagation through the interstellar medium have distorted the source abundances. The history of the original discovery of the presence of these nuclei will be followed by a summary of the present state of knowledge of the observed abundances. The effects of acceleration biases and of interstellar propagation will be discussed. Finally some of the possibilities for further advances will be outlined.

**Keywords** Cosmic rays · Heavy nuclei · Endothermic · Cross sections

## 1 Introduction

Compilations of the abundances of the elements in the solar system and in the cosmos show the presence of all the stable elements between hydrogen and bismuth as well as the long-lived actinides. These abundances cover a wide range of values, falling more than eleven orders of magnitude from hydrogen to uranium. Once it was proven that the majority of the particles in the cosmic radiation were protons rather than electrons or gamma rays, it was reasonable to expect that the nuclei of heavier elements would also be present. This was confirmed in 1948 by Freier et al. (1948), who, using nuclear emulsions exposed on high

A copy of the Power Point presentation for this talk is available on request.

C.J. Waddington (✉)
School of Physics and Astronomy, University of Minnesota, Minneapolis, MN 55455, USA
e-mail: wadd@physics.umn.edu

altitude balloons, showed that there were nuclei at least as heavy as iron with relativistic energies detectable near the top of the atmosphere.

These lighter heavy elements are all produced by exothermic nucleosynthesis reactions in stellar interiors and are relatively abundant compared to the still heavier elements. The presence of the much rarer heavier endothermically synthesized elements in the cosmic radiation, those heavier than Fe–Ni, the "UH nuclei", was not established for nearly another twenty years, initially from the analysis of damage tracks in meteorites (Walker et al. 1965) and later from balloon exposures of large area nuclear emulsion detectors (Fowler et al. 1967).

Studies of the relative abundances of these UH endothermic elements showed that they were generally comparable with the cosmic abundances of solar matter, after allowance was made for the changes produced by the effects of preferential acceleration, leakage from the galaxy, and propagation through the matter in the interstellar medium, although there do seem to be differences in detail.

Unfortunately it is extremely difficult to determine these source abundances with any statistical accuracy. Observationally the basic problem is that the abundances of these nuclei are so much less than those of the lighter exothermic nuclei, four or five orders of magnitude less than that of $_{26}$Fe nuclei, see Fig. 1. As a consequence large and sophisticated detectors exposed in space or near the top of the atmosphere for long periods are required in order to detect and identify significant numbers of these UH nuclei. Ideally these detectors should have the ability to resolve both individual charges and masses for all the elements and isotopes between iron and the actinides. In practice this degree of resolution has not been achieved, although current technology, if deployed in a long space exposure, could detect significant numbers and resolve the charges of all the individual elements, including the actinides. Detectors capable of mass determination, to identify the individual isotopes, are unlikely to be deployed in the near future, although in principle such detectors could be designed and exposed.

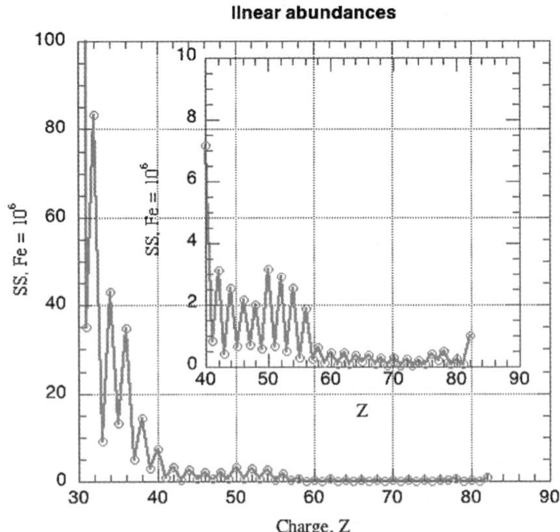


**Fig. 1** Solar abundances (Lodders 2003) plotted on a linear scale and normalized to $10^6$ Fe nuclei

## 2 Significance of Abundances of UH Nuclei

The endothermic elements occupy the upper two thirds of the periodic table and are generally synthesized in later stages of stellar evolution than the exothermic elements. Determinations of their relative abundances should provide unique information on the nuclear synthesis processes occurring during these later stages. Abundances in the solar system represent those prevailing some 4.6 billions years ago when the solar system was formed. On the other hand the UH nuclei in the cosmic rays can only have been accelerated at most some ten million years ago, since they are rapidly destroyed by interactions in the interstellar medium or escape from the galaxy. Hence, they could, in principle, represent very young material, synthesized in modern late stage stellar interiors. However, it is clear from a study of the relative abundances of cosmic ray nuclei $^{59}$Ni and $^{59}$Co by the ACE team (Wiedenbeck et al. 1999) that these nuclei are not accelerated in the very same explosions in which most are synthesized, there being an interval of at least $10^5$ years between synthesis and acceleration.

These elements are mostly synthesized by neutron addition to seed nuclei. This can be by a relatively slow process, which allow for unstable isotopes to decay before further neutrons are added, the "s-process", probably occurring in red giant stars. Alternatively, in an explosive reaction, such as in a super nova event, neutrons may be added as fast as the nuclei can absorb them, the "r-process" driving the nuclei to the neutron drip line, where the physics is still poorly understood. This r-process probably occurs in type II super novae, core collapse SN, rather than in the less abundant type Ia SN's, whose energetics preclude significant neutronization. Elements from the r-process will be neutron rich, while s-process elements will be significantly less so. As a result these two processes tend to enhance the abundances of different elements and so the relative contributions of the two processes to the final composition can be determined from a study of the elemental abundances. Elemental ratios such as $_{36}\mathrm{Kr}/_{40}\mathrm{Zr}$; $_{56}\mathrm{Ba}/_{50}\mathrm{Sn}$ and $_{78}\mathrm{Pt}/_{82}\mathrm{Pb}$ are very sensitive to the relative importance of the s- and r- processes. Ideally still better information could be obtained from studying the isotopic abundances, but such determinations are beyond current space instrumentation. The one exception is the current ACE detectors, which can resolve the isotopes of low energy UH nuclei with $Z < 35$, but over its seven years in space continues to observed only about one such nucleus every two or so weeks (George et al. 2000).

## 3 History

Both active and passive detectors have been used to study these UH nuclei. The initial discovery of their presence in the cosmic radiation came from studies of ionization damage tracks in meteorites; see Binns et al. (1988) for references and a general discussion of the history. These observations lacked the resolution to show much more than the broadest groups of elements.

A series of balloon borne arrays of large areas of nuclear emulsion, initiated by Fowler et al. (1967) were followed by at least 22 further flights by several groups of emulsion and etchable plastics, resulting in a total exposure factor similar to that later obtained by active detectors exposed in space, but with much poorer charge resolution.

In the mid 1970's a concentrated effort was made to study the heaviest, and rarest, nuclei, using Lexan plastics which were sensitive only to particles with $Z/\beta > 65$. Nine balloon flights by Fowler et al. (1977) and an exposure on Skylab by Price and Shirk (1975) showed that there are significant numbers of nuclides in the $_{78}Pt$–$_{82}Pb$ region, an apparent gap in the short lived actinides and a number of $_{90}Th$–$_{92}U$ nuclei.

Active electronic detectors were developed in the 1970's by several groups and culminated in two space missions launched at the end of the decade. These detectors had the goal of measuring all the nuclei between $_{26}Fe$ and $_{92}U$ with good charge resolution. The HEAO-3 detector (Binns et al. 1981) used plastic Cherenkov detectors in association with energy loss in a gas. Ariel-VI (Fowler et al. 1979) used a spherical shell of plastic scintillator filled with a scintillating gas. Both these arrays were designed and built before accelerators could produce beams of relativistic heavy ions for calibration. Nevertheless, both achieved their design goals, having a dynamic range sufficient to cover all the UH elements, from Fe to U, over the energy range expected for the missions orbits. Neither could resolve individual charges but up to about $Z = 60$ both could separate the more abundant even charged elements from their neighboring even charged element. The reported results from the two missions were compared by Binns et al. (1989) and shown to be consistent, Fig. 2. These combined results were corrected for the effects of acceleration and propagation, see next section, and it was found that the elements with $Z < 60$ were consistent with solar abundances at the source, but that for the heavier elements there was an apparent excess of r-process nuclei, see Fig 3. These detectors lacked the exposure factors to detect more than three actinides and could not distinguish Th from U, a crucial indicator of the age of these nuclei.

Two major detectors were deployed to study the actinides. Both used large arrays of passive detectors, both only sensitive to nuclei with $Z/\beta > 65$, exposed in space. The first was a large area of Lexan sheets exposed on the Long Duration Exposure Facility, LDEF. This had an exposure of 6 years, much longer than planned. As a result the plastic experienced large temperature fluctuations, which degraded the charge resolution. Only the most gross charge groups could be distinguished, but it was clear that there were a significant number of Th–U nuclei present (Keane et al. 1997; Donnelly et al. 2001). The second array used sensitive glass sheets that had little if any variations due to temperature fluctuations. This TREK array was exposed on a Russian spacecraft and recovered in two separate missions. While having a smaller exposure factor than the LDEF array it had much better charge resolution

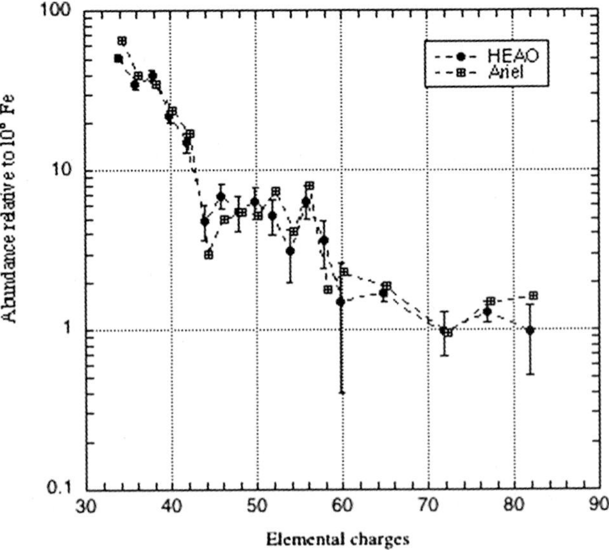


**Fig. 2** Comparison of the results from the HEAO and Ariel space exposures. For $Z < 60$ values are for even and its next lower odd charged element. For $Z > 60$ values are for groups of elements

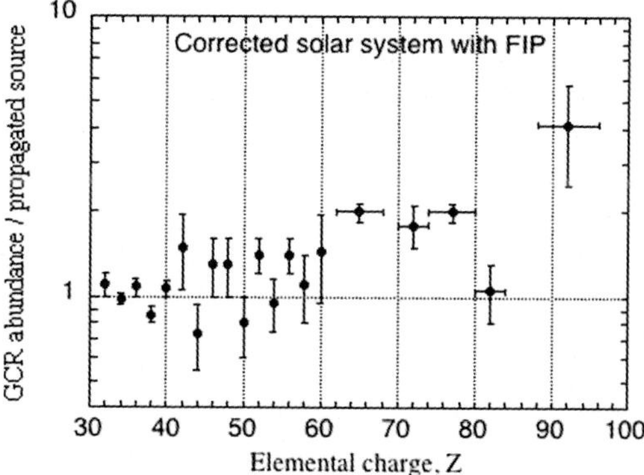


**Fig. 3** The ratio of the best galactic abundances derived from HEAO and Ariel to solar abundances corrected for the effect of propagation and a FIP acceleration bias (Binns et al. 1989)

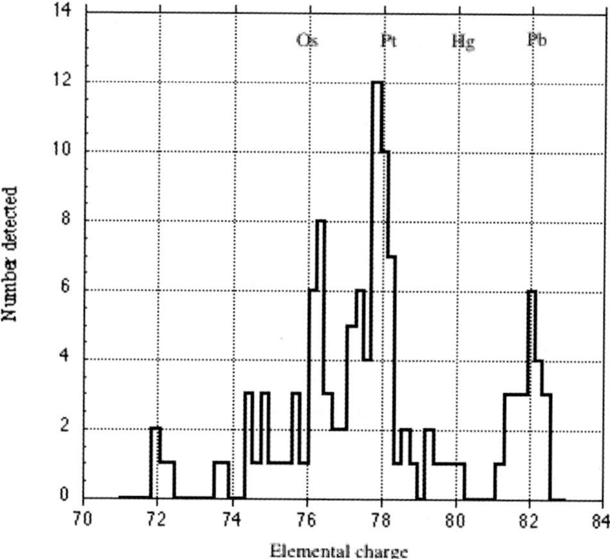


**Fig. 4** Numbers of nuclei detected by the TREK glass array. Clear peaks can be seen in the OS – Pt group and at Pb (Weaver and Westphal 2002)

and could clearly distinguish the elements in the $_{78}$Pt to $_{82}$Pb group, Fig. 4. The actinide gap was empty but only one Th and three U nuclei were observed, indicative, but far from conclusively, that the source material is relatively young (Westphal et al. 1998).

## 4 Source Abundances

In order to determine the abundances at the source the observed abundances have first to be corrected for the effects of local overlying matter in the detectors. Hence cross sections in heavier materials are needed. Similarly any balloon results have to be corrected for the residual atmospheric matter. For these corrections no allowances have to be made for the possible decay of unstable fragments. These corrected abundances have then to be compared with those predicted from an assumed source composition for which allowance has been made for the effects of acceleration biases and propagation losses and distortions.

It is well known from studies of the lighter nuclei that there must be a preferential acceleration mechanism involved. It is generally assumed that this depends on either the first ionization potential (FIP) or on the volatility of the elements. Since these two parameters are generally strongly correlated it is difficult to distinguish between the two. Just a few of the even charged UH elements, such as $_{30}$Zn, $_{34}$Se, $_{52}$Te and $_{82}$Pb can provide a clear discrimination. Fortunately, this means that an appropriate application of either of the corrections for these factors will generally be adequate for most of the elements. Presently it appears, particularly from the large Pt/Pb ratios reported by all the observations, that volatility may be the most appropriate factor.

The UH nuclei have large cross sections for nuclear interactions in the interstellar medium, generally fragmenting into lighter nuclei. In order to evaluated the magnitude of these changes in the charge spectrum it is necessary, in principle, to know the probabilities for every element to fragment into every lighter isotope, since many of the unstable isotopes will have time to decay to other elements. Initially, few if any of these cross sections had been measured and reliance had to be placed on semi-empirical relations (Silberberg and Tsao 1990). In addition, it is necessary to assume a model of the path length distribution of the nuclei in the galaxy, allowing for the probability of the nuclei escaping from the accelerator region and then from the galaxy. These cross sections have been found to be energy dependent below at least 5 GeV/n and because of the steep cosmic ray energy spectrum, most of the observed nuclei will have energies less than that. In addition, since the escape probabilities are also rigidity dependent, it is necessary to assume an initial energy for the source nuclei and evaluate the energy losses, which for these high $Z$ nuclei are large, of the nuclei as they propagate through the ISM. The possibility that there might be some re-acceleration during propagation is generally not included in these calculations, since it does not appear to be needed to explain the lighter nuclei.

When beams of relativistic heavy ions from accelerators became available an extensive series of exposures of detectors based on the HEAO array were made to energetic heavy nuclei from the Berkley Bevalac and the Brookhaven AGS accelerators. These beams were allowed to fragment in various thin targets, and the yields of each fragment element determined. Under the relatively controlled conditions of an exposure at an accelerator, these detectors showed excellent charge resolution, clearly distinguishing individual fragments, even those with charge changes of only + or − one charge unit. These exposures were designed both to determine some of the cross sections needed, and to calibrate the detectors. Beams of nuclei included $_{26}$Fe, $_{36}$Kr, $_{47}$Ag, $_{57}$La, $_{67}$Ho, $_{79}$Au and $_{92}$U with various energies between 0.4 to 10 GeV/nucleon, using targets such as carbon, polyethylene ($CH_2$), Al, Cu, Sn and Pb; see Waddington et al. (1995, 2000) and references therein. Cross sections in hydrogen could be obtained from a comparison of the C and $CH_2$ results. The high energy exposures to the AGS 10 GeV/n beams, when compared with those made at lower energies, showed that many of these cross sections were energy dependent, introducing a further factor to be included in propagation calculations. The cross sections measured from these exposures were used to refine the semi-empirical relations, which were still needed to estimate the isotopic yields, and their decay products.

Propagation can be applied using a leaky box model similar to that developed to study the propagation of the lighter nuclei. Such a model can be imposed on to various assumed source spectra (Clinton and Waddington 1993; Waddington 1996), and the results compared with the observed spectrum. Although a leaky box model is somewhat physically naïve and assumes a much simpler ISM than is known to exist, it is readily applied using the weighted slab technique and gives results that appear to be closely similar to those from more complex models (Ptuskin et al. 1995).

The conclusions from these propagation estimates were that the observed UH nuclei were most consistent with a solar system like source for elements with $Z < 60$ but with an enhanced r-process source for the heavier nuclei, if there was a FIP or volatility correction and some truncation of the path lengths was assumed. However, the uncertainties were such that some other scenarios could not be excluded and it is clear that both better observational data and better cross sectional information are needed before more definitive conclusions can be drawn.

## 5 Recent and Current

An updated and much improved version of the HEAO detector, TIGER, has been developed and flown twice in Antarctica to demonstrate its capability as a potential space worthy instrument. These exposures were at altitude for a total of about 50 days and demonstrated a charge resolution of 0.26 charge unit, sufficient to resolve individual elements (Link et al. 2003; Geier et al. 2005) The limited statistics restricted meaningful results to nuclei with $Z < 40$ and proved that only a long space exposure of a similar instrument would allow significant further progress. Apart from the ACE measurements currently still being collected there have been no further advances. However, it is clear that the technology now exists to study both the entire charge range with a TIGER like array and the actinides with an associated TREK like assembly. Such a combination was studied with NASA support and proposed for a Space Station mission but, like most such missions, not accepted by NASA.

**Acknowledgements** I am grateful to the organizers for an invitation to this conference and to W.R. Binns and M.H. Israel for useful discussions.

## References

W.R. Binns, M.H. Israel, J. Klarmann, W.R. Scarlett, E.C. Stone, C.J. Waddington, Nucl. Inst. Methods **185**, 415 (1981)

W.R. Binns, T.L. Garrardn, P.S. Gibner, M.H. Israel, M.P. Kertzman, J. Klarmann, B.J. Newport, E.C. Stone, C.J. Waddington, Astrophys. J. **346**, 997 (1989)

W.R. Binns, T.L. Garrard, M.H. Israel, J. Klarman, E.C. Stone, C.J. Waddington, in *AIP Conference Proceedings*, vol. 183, ed. by C.J. Waddington (1988), p. 147

R.R. Clinton, C.J. Waddington, Astrophys. J. **403**, 644 (1993)

J. Donnelly, A. Thompson, D. O'Sullivan, L.O'C. Drury, K.P. Wenzel, *Proceedings ICRC 2001* (2001), p. 1715

P. Freier, E.J. Lofgren, E.P. Ney, F. Oppenheimer, H. Bradt, B. Peters, Phys. Rev. **74**, 213 (1948)

P.H. Fowler, R.A. Adams, V.G. Cowen, J.M. Kidd, in *Proc. Roy. Soc. A*, vol. 301 (1967), p. 1

P.H. Fowler, M.R.W. Masheder, R.T. Moses, A. Worley, in *Proceedings of 16th International Cosmic Ray Conference*, vol. 12 (1979), p. 338

P.H. Fowler, D.L. Henshaw, C. O'Ceallaigh, D. O'Sullivan, A. Thompson, *Proceedings of 15th ICRC* (1977), p. 161

S. Geier et al., in *29th Proc. ICRC* (2005), p. 101

J.S. George et al., in *AIP Conference Proceedings*, vol. 528 (2000), p. 437

A.J. Keane, D.O. Sullivan, A. Thompson, D.l. O'Durury, K.P. Wenzel, Adv. Space. Res. **19**, 739 (1997)

J. Link et al., in *28th ICRC, OG 1* (2003), p. 1781

Lodders, Astrophys. J. **591**, 1220 (2003)

P.B. Price, E.K. Shirk, in *Proceedings of 14th International, Cosmic Ray Conference*, vol. 1 (1975), p. 268

V.S. Ptuskin, F.C. Jones, J.F. Ormes, in *Proceedings of 24th ICRC*, Rome, vol. 3 (1995), p. 108

R. Silberberg, C.H. Tsao, Phys. Rev. **191**, 351 (1990)

C.J. Waddington, W.R. Binns, J.R. Cummings, T.L. Garrard, L.Y. Geer, J. Klarman, B.S. Nilsen, Adv. Space Res. **15**(6), 39 (1995)

C.J. Waddington, Astrophys. J. **470**, 1218 (1996)
C.J. Waddington, J.R. Cummings, B.S. Nilson, T.L. Garrard, Phys. Rev. C **61**, 024910 (2000)
R.M. Walker, R.L. Fleischer, P.B. Price, in *Proceedings of 9th International Cosmic Ray Conference*, London, vol. 2 (1965), p. 1086
A.J. Westphal et al., Nature **396**, 50 (1998)
B.A. Weaver, A.J. Westphal, Astrophys. J. **569**, 493 (2002)
M.E. Wiedenbeck et al., Astrophys. J. **523**, L61 (1999)

Space Sci Rev (2007) 130: 465–473
DOI 10.1007/s11214-007-9172-8

# The Composition of Cosmic Rays and the Mixing of the Interstellar Medium

**R.E. Lingenfelter · J.C. Higdon**

Received: 14 December 2006 / Accepted: 13 March 2007 /
Published online: 5 May 2007


**Abstract** The differences between the composition of Galactic cosmic rays and that of the interstellar medium are manifold, and they contain a wealth of information about the varying processes that created them. These differences reveal much about the initial mixing of freshly synthesized matter, the chemistry and differentiation of the interstellar medium, and the mechanisms and environment of ion injection and acceleration. Here we briefly explore these processes and show how they combine to create the peculiar, but potentially universal, composition of the cosmic rays and how measurements of the composition can provide a unique measure of the mixing ratio of the fresh supernova ejecta and the old interstellar medium in this initial phase of interstellar mixing.

In particular, we show that the major abundance differences between the cosmic rays and the average interstellar medium can all result from cosmic ray ion injection by sputtering and scattering from fast refractory oxide grains in a mix of fresh supernova ejecta and old interstellar material. Since the bulk of the Galactic supernovae occur in the cores of superbubbles, the bulk of the cosmic rays are accelerated there out of such a mix. We show that the major abundance differences all imply a mixing ratio of the total masses of fresh supernova ejecta and old interstellar material in such cores is roughly 1 to 4. That means that the metallicity of $\sim$3 times solar, since the ejecta has a metallicity of $\sim$8 times that of the present interstellar medium.

**Keywords** Cosmic rays · Dust · ISM: abundances · ISM: bubbles · Stars: Wolf Rayet · Supernovae: general

R.E. Lingenfelter (✉)
Center for Astrophysics and Space Sciences, University of California San Diego, La Jolla, CA 92093, USA
e-mail: rlingenfelter@ucsd.edu

J.C. Higdon
W.M. Keck Science Center, Claremont Colleges, Claremont, CA 91711-5916, USA
e-mail: jimh@lobach.JSD.claremont.edu

## 1 Introduction

There are very large differences between the composition of galactic cosmic rays and that of the local interstellar medium inferred from solar system material. These differences provide new insight into initial mixing of freshly synthesized elements into the interstellar medium and they offer the first measures of the metallicity of the supernova-active superbubble cores. Here we briefly explore the processes and environments that combine to create the peculiar, but potentially universal, composition of the cosmic rays, and how they sample the initial mixing of the fresh supernova ejecta and the old interstellar medium in the supernova-active cores of superbubbles.

There are three basic differences between the cosmic ray source composition (e.g. Engelmann et al. 1990) and the solar/local composition. (1) The refractory element abundances relative to H are all enriched by a roughly constant factor of $\sim$20 compared to solar system values. (2) The corresponding volatile element abundances have a mass-dependent enrichment for those heavier than He ($A > 4$) that reaches a factor of as much as 10 for heaviest volatiles. (3) Carbon and oxygen do not fit into either scheme and are enriched by intermediate factors of 9 and 5. These enrichments are all shown in Fig. 1.

The roughly constant enrichment of the refractory elements is thought to result from suprathermal injection by the sputtering of ions off of fast refractory grains through collisions with the ambient gas (Cesarsky and Bibring 1981). The strong correlation of cosmic ray enrichment with elemental condensation temperature, defining the continuum between "refractory" and "volatile" elements has been shown by Meyer et al. (1997). Even though the sputtering cross sections are mass dependent (e.g. Sigmund 1981), the total sputtering yield is still expected to simply reflect the grain composition, because the fast moving grains are expected to be completely destroyed by repeated sputtering. A similar correlation also exists between the cosmic ray enrichment and the first ionization potential of the elements, which is inversely correlated with the condensation temperature. But as we discuss below, the vast majority of the galactic supernovae occur in the hot ($\sim 10^6$ K), dust rich, but fully ionized gas in the cores of superbubbles (e.g. Higdon et al. 1998; Higdon and Lingenfelter 2005), where the ionization potential has no apparent significance.

As we have suggested (Lingenfelter et al. 2000; Lingenfelter and Higdon 2007), the mass-dependent enrichment of volatile elements can also result from suprathermal injection of the volatiles in the hot ambient gas scattered by fast refractory grains, and, as we show below, the mass-dependence of the scattering cross section is quite consistent with that of the cosmic ray enrichment. Here the mass dependence is preserved because only a very small fraction of the ambient gas is scattered by the fast grains and they preferentially scatter the heavier elements.

Lastly, as we have shown (Lingenfelter et al. 2000; Lingenfelter and Higdon 2007), the intermediate enrichment of carbon and oxygen can also result from suprathermal injection by sputtering of C and O ions from graphite and oxides in the fast refractory grains, with a small additional contribution from scattering of volatile C and O in the hot interstellar gas.

## 2 OB Associations and the Superbubble Origin of Cosmic Rays

In order to quantitatively understand the composition of cosmic rays we need to consider the environment in which they are accelerated. The galactic cosmic rays are thought to result primarily from the preferential acceleration of suprathermal ions by shock waves from supernovae (e.g. Axford 1981). Core collapse supernovae (SNII & SNIb/c) from massive

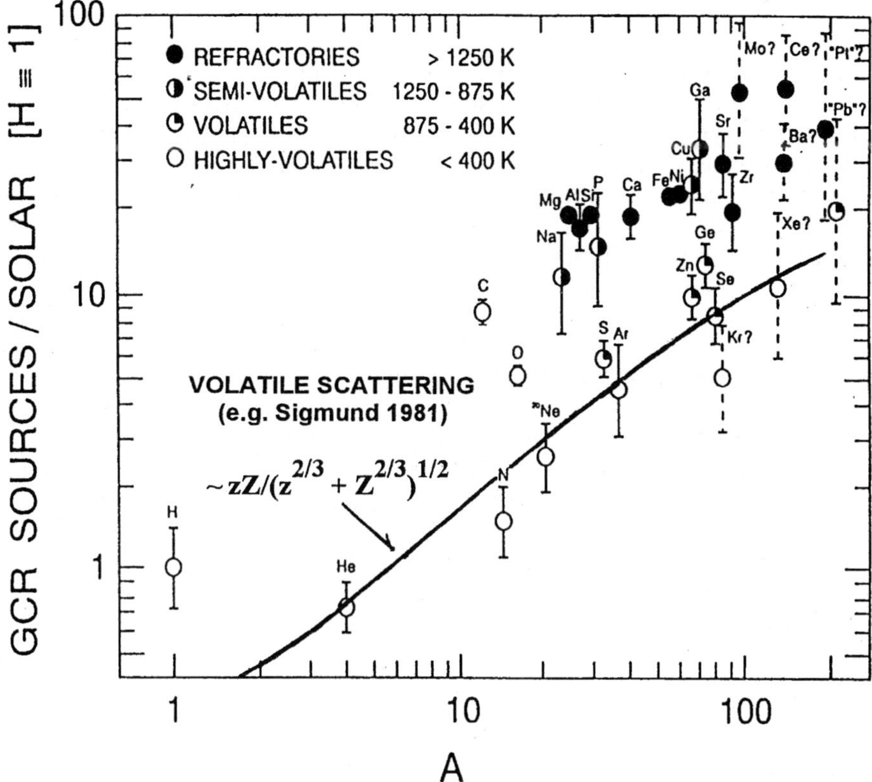


**Fig. 1** Cosmic ray source abundance enrichment relative to Solar abundances as a function of mass (modified from Meyer et al. 1997). Also shown for comparison with the volatiles is the cross section for the scattering of gas ions by fast grains at a constant velocity, proportional to $zZ/(z^{2/3} + Z^{2/3})^{1/2}$, where $z$ and $Z$ are the nuclear charges of the grain and gas nuclei (e.g. Sigmund 1981), normalized to He for masses $A > 1$ which all have essentially the same charge to mass ratio, $Z \sim A/2$

($M > 8M_{\odot}$) O and B stars account for ~85% of all galactic supernovae (e.g. van den Bergh and McClure 1994) and together with Wolf–Rayet winds they produce nearly all of the elements $A > 4$, while the remaining 15% of supernovae are thermonuclear explosions (SNIa), which produce $\sim 1/2$ of the Fe (e.g. Timmes et al. 1995). Most (80–90%) of the SNII and SNIb/c occur in OB associations and they form superbubbles, which thus contain ~70–75% of all galactic supernovae (Higdon and Lingenfelter 2005).

Thus roughly ~70–75% of the H and He are accelerated in superbubbles and ~25–30% or less come from the warm interstellar medium, where shock acceleration is less efficient (e.g. Axford 1981). The superbubble contribution to heavier $A > 4$ cosmic rays, however, is much larger because the superbubble metallicity $Z_{sb}$ is ~2.5 times that of the average interstellar medium $Z_{ism}$ (Lingenfelter and Higdon 2007), as we discuss in detail below. Thus, with superbubbles containing a fraction $F_{sn} \sim$ 70–75% of the supernovae, the fraction of heavier ($A > 4$) cosmic rays accelerated in superbubbles is $F_{sn}(Z_{sb}/Z_{ism})/[F_{sn}(Z_{sb}/Z_{ism}) - F_{sn} + 1] \sim$ 85–88% (Higdon and Lingenfelter 2005).

Essentially all of the cosmic ray acceleration in superbubbles is concentrated in their cores which are enriched by the ejecta of all of their supernova. Giant molecular clouds can form OB associations within radii <10 pc, which can produce anywhere from several to 10,000 core collapse supernovae. These core collapse supernova progenitors live anywhere from ~3 Myr to ~35 Myr, depending on their mass, before they explode (e.g. Schaller et al. 1992). Since these stars have a mean dispersion velocity of <4 km/s, or <4 pc/Myr (e.g. Blaauw 1991), their subsequent supernovae all occur within a radius $r \sim 160$ pc. These supernovae, together with radiatively driven WR winds from their progenitors, collectively create the hot tenuous $\sim 10^{-3}$ H/cc superbubbles, where the supernova shocks reach radii of ~100 pc (e.g. Higdon et al. 1999). Thus the shocks from each of the supernovae successively sweep through all of the ejecta of previous supernovae and accelerate cosmic rays out of the enriched, supernova-active cores of superbubbles.

## 3 Refractory Grains and Cosmic Ray Injection

The most abundant of the freshly nucleosynthetized elements in the supernova ejecta form refractory oxides, MgO, $Al_2O_3$, $SiO_2$, CaO, $Fe_3O_4$, etc. As the expanding ejecta adiabatically cools, these oxides condense into refractory grains within a couple years after the explosion, as we saw in SN 1987A (Kozasa et al. 1991). In the supernova-active cores of superbubbles, these fresh supernova grains mix with old interstellar grains, which are depleted relative to Fe by as much as ~1/2 in some elements, such as Si which remains primarily as $Fe_2SiO_4$ (Spitzer and Fitzpatrick 1993; Savage and Sembach 1996; Jenkins 2004).

Subsequent supernova shocks accelerate some of these charged grains to velocities well above the $\sim 10^6$ K thermal velocities of $>100/A^{1/2}$ km/s (Epstein 1980; Ellison et al. 1997). The collisions of these accelerated grains with the ambient gas sputter refractory elements off of the grains as suprathermal ions at grain velocities. That sputtering preferentially injects these refractory ions into the same shock for acceleration to cosmic ray energies (Cesarsky and Bibring 1981; Meyer et al. 1997)

In these same interactions of accelerated refractory grains with the ambient gas, ambient H, He and heavier volatiles are also scattered by fast grains to same velocity with a cross section $\sim zZ/(z^{2/3} + Z^{2/3})^{1/2}$, where $z$ and $Z$ are the nuclear charges of the grain and gas nuclei (e.g. Sigmund 1981). That scattering also injects the volatile elements into the same shock for acceleration to cosmic ray energies (Lingenfelter et al. 2000). As can be seen in Fig. 1, this charge dependence of the scattering cross section is quite similar to the observed mass-dependent enrichment of the volatile elements in the cosmic rays heavier than H, all of which have essentially the same mass-to-charge ratio, $A/Z \sim 2$. This clearly suggests that such scattering may account for the mass-dependent volatile enrichment, but obviously much more detailed calculations are needed.

Although most of the O in the interstellar medium is in the gas, from which some is injected by fast grain scattering, the fraction of O bound in refractory oxides in the grains is also injected by sputtering and appears to be the major source for cosmic rays. The cosmic ray source O/Fe ratio is $5.2 \pm 0.1$ (Engelmann et al. 1990), and the bulk of this can be accounted for simply from the chemistry of the grains and the relative cosmic ray abundances of the major refractory elements (Lingenfelter et al. 1998). These elements have cosmic-ray source abundance ratios of Si : Mg : Fe : Al : Ca of 0.99 : 1.03 : 1.00 : 0.08 : 0.06 (Engelmann et al. 1990), so if they all were sputtered from their refractory oxides, principally $SiO_2$, MgO, $Fe_3O_4$, $Al_2O_3$, and CaO, in grains, they should be accompanied by sputtered O with an O/Fe = 4.5. This would account for ~87% of the measured cosmic-ray source abundance ratio leaving the small remainder to grain-scattered O from the surrounding gas. As we discuss in more detail below, the relative abundances of the refractory elements in the cosmic rays can be produced by a mix of fresh SN-WR ejecta with a refractory oxide grain O/Fe ratio of $\sim$8 and old ISM grains, primarily $Fe_2SiO_4$, with an O/Fe ratio of $\sim$3.

Unlike most of the other elements, most of the C in the Galaxy is thought to be injected through the winds of AGB stars rather than supernovae (e.g. Timmes et al. 1995). The bulk of the C in the interstellar medium is also in gas as CO (e.g. Savage and Sembach 1996) and some of it is injected by fast grain scattering, but it can not be strongly enriched in this way. However, nearly all of the freshly synthesized C in supernova ejecta condenses into the refractory grains as graphite, since the intense radiation in supernovae prevents significant CO formation in the ejecta (Clayton et al. 2001; Lingenfelter et al. 1998). A fraction of the C also condenses as refractory SiC, but since its relative abundance seems to be small (e.g. Clayton and Nittler 2004) we ignore it here. Thus, this C is also injected by sputtering, and, as we show in detail below, the cosmic-ray source C/Fe ratio of $4.2 \pm 0.2$ (Engelmann et al. 1990) can simply reflect the C/Fe ratio in the superbubble core mix, dominated by the SN-WR ejecta ratio.

## 4 Superbubble Core Mixing and Cosmic Ray Injection Abundances

The cosmic ray injection abundances thus depend on the average mixing ratio of the fresh SN-WR ejecta and the old ISM in the supernova-active cores of superbubbles, where most of the cosmic rays are accelerated. As we will show, the major cosmic ray source abundances ratios all give consistent and independent measures of this mixing ratio.

One of best determined cosmic ray source abundances is the Si/Fe ratio, which gives a sensitive measure of the mix of core collapse and thermonuclear supernova yields in the interstellar medium and an equally sensitive measure of the mix of fresh supernova ejecta grains and old ISM grains in superbubble cores. The Solar system/local ISM average Si/Fe of $\sim$1.2 (e.g. Lodders 2003) results from a mix of roughly equal amounts of Fe from core collapse supernovae with an IMF-averaged yield of $\mathrm{Si/Fe} \sim 2.1$ and thermonuclear supernovae with $\mathrm{Si/Fe} \sim 0.4$ (e.g. Timmes et al. 1995; Lingenfelter and Higdon 2007, and the references therein).

The cosmic ray source abundance ratio of Si/Fe of $0.99 \pm 0.2$ (Engelmann et al. 1990), can similarly be produced by a mix of fresh supernova grains and old ISM grains. From the same supernova and WR wind calculations, the IMF-averaged SN-WR ejecta ratio $\mathrm{Si/Fe} \sim 2.1$, essentially all of which is likely to be in grains (e.g. Kozasa et al. 1991). The old ISM grains, on the other hand, contain about 90% of the Fe and only 46% of the Si (e.g. Savage and Sembach 1996; Jenkins 2004), mostly as $Fe_2SiO_4$, with a ratio $\mathrm{Si/Fe} \sim 0.6$. We (Lingenfelter and Higdon 2007) have shown that the cosmic ray source Si/Fe ratio of 0.99, thus implies cosmic ray acceleration out of a mix of old ISM grains and fresh SN grains with an average SN-WR ejecta mass fraction of $\sim 12 \pm 4\%$. With 70–75% of all supernovae occurring in superbubbles and the remainder in the ISM, the average SN-WR ejecta mass fraction in the supernova-active cores of superbubbles is $\sim 17 \pm 5\%$, which corresponds to a superbubble core metallicity $Z_{sb} \sim 2.8 \pm 0.4$ times proto-Solar. Very similar values (Lingenfelter and Higdon 2007) are also implied from the cosmic ray abundance ratios of all the other refractory elements studied by Engelmann et al. (1990), Mg, Al, Ca, Co & Ni, which also have similar gas-phase depletion in the interstellar medium (e.g. Savage and Sembach 1996; Jenkins 2004).

Independently, the cosmic-ray source abundances of the two conspicuously anomalous volatile-refractory elements C and O can also be produced primarily by sputtering from a mix of fresh SN-WR grains with old ISM grains from the interstellar medium, and these also provide independent estimates of the mixing ratio in superbubble cores.

The cosmic-ray C source abundance ratio, C/Fe is $4.2 \pm 0.2$ (Engelmann et al. 1990), compared to the proto-Solar value of 8.45 (Lodders 2003). Although 90% of the interstellar Fe mass fraction discussed above is in grains (e.g. Savage and Sembach 1996; Jenkins 2004), nearly all of the interstellar C is in CO and other gases and only a negligible fraction is thought to be in refractory graphite grains (e.g. Savage and Sembach 1996; Lodders 2003). From the same models considered above, the IMF-averaged SN ejecta have a C/Fe ratio of ~10 and that of the WR winds is ~12, both of which are measured with respect to the combined IMF-averaged SN-WR ejecta Fe. As noted above, the bulk of the C in the SN ejecta is expected to condense as graphite, because the intense radiation in supernovae prevents significant CO formation (Clayton et al. 2001), but the fraction of the WR wind C that condenses as graphite, $g_{wr}$, in their weaker radiation field is not known, so we leave it as a variable. Thus, we assume that the graphite mass fraction in the combined IMF-averaged SN-WR ejecta C/Fe ratio is $\sim 10(1 + 1.24 g_{wr})$. As shown by Lingenfelter and Higdon (2007), the cosmic-ray source C/Fe ratio of 4.2 thus implies an average SN-WR ejecta mass fraction of $\sim 21/(1 + 1.7 g_{wr})\%$. With a superbubbles containing 70–75% of supernovae, the average SN-WR ejecta mass fraction in their supernova-active cores is $\sim 29/(1 + 1.7 g_{wr})\%$, or between 11 and 29% depending on the graphite grain fraction of C in WR winds.

The cosmic-ray O source abundance ratio, O/Fe is $5.2 \pm 0.1$ (Engelmann et al. 1990) of which, as noted above, refractory oxides could account for ~87%, or O/Fe of 4.5, compared to the proto-Solar value of 16.8 (Lodders 2003). Although most of the interstellar O is in gas, about 16% can be in old refractory grains, primarily $Fe_2SiO_4$ and various other mineral combinations of the principal refractory oxides, $SiO_2$, MgO, $Fe_3O_4$, $Al_2O_3$, and CaO. This follows from multiplying the proto-Solar abundances of these refractories, Si : Mg : Fe : Al : Ca of 1.19 : 1.22 : 1.00 : 0.15 : 0.08 (Lodders 2003) by their corresponding grain fractions of 0.45 : 0.49 : 0.90 : ~0.6 : 1.0 determined from Galactic disk depletion factors (e.g. Savage and Sembach 1996; Jenkins 2004), and their O/X ratios of 2 : 1 : 1.3 : 1.5 : 1, which gives a combined O/Fe ratio of 3.0. The grain O/Fe ratio in the SN-WR ejecta is also determined by the principal refractory oxides in the fresh condensates and Lingenfelter and Higdon (2007) showed that IMF-averaged SN-WR ejecta abundance ratios for Si : Mg : Fe : Al : Ca of 2.1 : 1.5 : 1.0 : 0.1 : 0.1 give a combined $O/Fe = 7.3$. They then showed that the cosmic-ray O/Fe source abundance ratio of 5.2 implies a cosmic ray averaged SN-WR ejecta mass fraction of $\sim 18 \pm 6\%$, and a corresponding superbubble core SN-WR ejecta mass fraction of $\sim 24 \pm 8\%$.

These values are all quite similar to the superbubble core mixing ratios that have previously been inferred from other cosmic ray abundance measurements of the $^{22}Ne/^{20}Ne$ and ThU/PtGroup. The cosmic ray determined ratios mostly sample a local volume of ~1 kpc over the last ~20 Myr, the characteristic propagation distance and lifetime of most local cosmic rays, except the ultraheavy Th, U and the Pt Group which have breakup lifetimes of only ~1 Myr and probably come from within ~0.3 kpc (e.g. Higdon and Lingenfelter 2003b). A much more wide-ranging measure, however, has been made from the old halo star abundances of the rare light elements, Li, Be and B, which are thought to be produced primarily by the cosmic ray breakup of C and O (Ramaty et al. 1997). These LiBeB abundances in old halo stars give a mean measure of the SN-WR ejecta mixing fraction in superbubble cores averaged over much of the Galaxy, ~10 kpc, and over most of it age, ~10 Gyr.

The most sensitive of these measurements are those of the Be/Fe ratio as a function of stellar metallicity, represented by Fe/H. These measurement showed that the Be/Fe ratio was surprising constant throughout the early history of the Galaxy, rather than increasing with the rising Fe/H metallicity of the interstellar medium which increased from about 0.001 to 0.1 over the age span of the measurements, as would have been expected if the cosmic ray C and O, from which the Be was produced, had been accelerated out of the interstellar medium. This constancy clearly showed that the cosmic ray C and O had to come from sites of relatively constant metallicity (Ramaty et al. 1997). These sites have since been shown to be the superbubbles (Higdon et al. 1998; Higdon and Lingenfelter 2005). The most recent analysis of these and later measurements up to the present epoch show (Alibes et al. 2002) that the cosmic rays must have been accelerated in a mix with an average SN-WR ejecta fraction of $25 \pm 14\%$ throughout the age of the Galaxy. This corresponds to a mean SN-WR ejecta fraction of $29 \pm 14\%$ in the cores of superbubbles, since at times when Fe/H was $<0.1$ SNIa were not yet significant and roughly 85% of all supernovae occurred in superbubbles. That gives a superbubble core metallicity of $\sim 2.9$ times the proto-Solar value in the early Galaxy, when the ISM metallicity was $<0.1$ and a present value of $\sim 3.8$ times proto-Solar with an ISM value of 1.32 and 70–75% of all supernovae occurring in superbubbles.

The current measurements of Th and U in the cosmic rays (Donnelly et al. 1999; Weaver and Westphal 2002) suggest a cosmic ray injection source abundance ratio of (Actinide)/(Pt Group) of $\sim 0.028 \pm 0.008$, compared to a value of $\sim 0.014 \pm 0.002$ expected in the present interstellar medium. The cosmic ray ratio would imply an average SN-WR ejecta fraction of $20 \pm 8\%$ in superbubble cores and a core metallicity of $3.1 \pm 0.8$ times proto-Solar (Lingenfelter et al. 2003).

Previous analysis (Higdon and Lingenfelter 2003a) of the anomalous cosmic ray $^{22}Ne/^{20}Ne$ implied a SN-WR ejecta mass fraction that is also quite similar to these values, but because of recently suggested reductions in the WR wind yields, there is now some question. The cosmic ray ratio of 0.39, or $5.3 \pm 0.3$ (Binns et al. 2005) times the proto-Solar ratio of 0.073, is assumed to result from WR wind enrichment. Using the WR wind yield calculations of Schaller et al. (1992), which are the dominant $^{22}Ne$ source, mixed with core collapse supernova yields, which are the dominant $^{20}Ne$ source, Higdon and Lingenfelter (2003a) showed that the resulting SN-WR ejecta had a $^{22}Ne/^{20}Ne$ ratio of $1.0 \pm 0.35$, or $\sim 14$ times the proto-Solar value. Since nearly all the WR stars and core collapse supernovae occur in superbubbles, the cosmic ray ratio implies that in their acceleration region in superbubble cores the time-averaged mean SN-WR ejecta mass fraction is $18 \pm 5\%$, which give it an average metallicity $\sim 2.9 \pm 0.4$ times proto-Solar.

Recently revised calculations of the WR winds (Hirschi 2004; Goriely private communication 2006), however, give $^{22}Ne$ yields that are much lower than the previous values and new detailed time-dependent mixing calculation are needed, because unlike the SNII dominated elemental ratios considered above the $^{22}Ne/^{20}Ne$ ratio varies greatly in the core mix from which cosmic rays are accelerated, since the bulk of the $^{22}Ne$ is produced early in the WR winds and the bulk of the $^{20}Ne$ isn't produced until much later in the SNII,

As we see in Table 1, with or without the Ne ratio, the metallicity and SN-WR ejecta mass fraction in superbubble cores appear to have been essentially constant throughout the Galaxy over its entire history within the estimated uncertainties.

The cause of the apparent constancy of superbubble core metallicity is not clear. However, we suggest one simple possibility, namely that it might result from the roughly constant mass ratio of the two primary sources of hot ionized gas in superbubble cores. (1) The hot gas from the old ISM ionized in classical HII regions around O and B star supernova progenitors have a broad range of masses with an RMS mean hot gas mass of $\sim 70\ M_{\odot}$ per supernova (e.g. Habing and Israel 1979), and (2) the fresh supernova ejecta and pre-supernova WR winds of these O and B stars have an IMF-averaged mean hot gas mass of ejecta of $\sim 18\ M_{\odot}$

**Table 1** Cosmic ray measures of superbubble core metallicity

| | Measurement | | | | | |
|---|---|---|---|---|---|---|
| | Duration | Distance | Ejecta cosmic ray | Ejecta SB cores | SB core Z X solar | Ref |
| Old Halo stars | | | | | | |
| Be/Fe | ∼ 10 Gyr | ∼10 kpc | 25 ± 14% | 29 ± 15% | 3.3 ± 1.6 | 1 |
| Cosmic rays | | | | | | |
| Si/Fe | ∼20 Myr | ∼1 kpc | 12 ± 4% | 17 ± 5% | 2.8 ± 0.4 | 2 |
| C/Fe | ∼20 Myr | ∼1 kpc | 8 to 21% | 11 to 29% | 2.3 to 3.8 | 2 |
| O/Fe | ∼20 Myr | ∼1 kpc | 18 ± 6% | 24 ± 8% | 3.4 ± 0.9 | 2 |
| $^{22}$Ne/$^{20}$Ne | ∼20 Myr | ∼1 kpc | 13 ± 4% | 18 ± 5% | 2.9 ± 0.4 | 3 |
| ThU/PtGroup | ∼1 Myr | ∼0.3 kpc | 20 ± 8% | 27 ± 11% | 3.6 ± 0.8 | 4 |

References: 1. Alibes et al. (2002); 2. Lingenfelter and Higdon (2007); 3. Higdon and Lingenfelter (2003a, 2003b), but see text; 4. Lingenfelter et al. (2003)

per supernova. This simple mix would in fact give a fixed ejecta mass fraction of ∼20% in supernova-active cores of superbubbles, a metallicity varying from ∼2 times the proto-Solar value in the early Galaxy, when the ISM metallicity was <0.1, to present value of ∼3 times proto-Solar with an ISM value 1.32 times higher.

## 5 Summary

We have shown that a cosmic ray injection source of suprathermal ions sputtered and scattered from fast refractory oxide grains in a mix of fresh supernova ejecta and old interstellar material can account for the major abundance differences between the cosmic rays and the average interstellar medium. As we have previously shown (Higdon and Lingenfelter 2005), the bulk of the Galactic supernovae occur in the cores of superbubbles, and the bulk of the cosmic rays are thus accelerated there out of the supernova ejecta enriched mix of new and old grains. We have further shown that the major cosmic ray abundance differences all imply a mixing ratio of the total masses of the fresh supernova ejecta and old interstellar material in such cores of roughly 1 to 4. That means that the metallicity of the mix is ∼3 times Solar and that more than half of all those elements heavier than He, both refractory and volatile, in this mix come from the ejecta, since it has a metallicity of about 8 compared to the present interstellar medium. Lastly we have shown that this mixing ratio appears to have been roughly constant throughout the history of the Galaxy and that it represents the first phase of the mixing of most freshly synthesized matter into the interstellar medium.

We suggest that the cosmic rays appear to provide the best measure of the initial metallicity resulting from such mixing. Soft X-ray measures of the metallicity of the hot gas in superbubbles and starburst galaxies are thought (e.g. Baldi et al. 2006) to significantly underestimate the actual metallicity of all the material, because much of the metals are likely to be locked up in grains rather than in the gas and thus be undetectable with soft X-rays. Mixing, or mass-loading fractions, however, have been estimated from recent analyses (Ott et al. 2005) of Chandra X-ray observations of nearby dwarf starburst galaxies. They suggest that the ratios of the mass of supernova ejecta to that of the heated surrounding gas may range from 1 to 1 to as much as 1 to 5, which imply high metallicities, roughly similar to those found from cosmic ray abundances.

**Acknowledgement** We thank Stephane Goriely for as yet unpublished WR wind yields, and an anonymous referee for valuable comments.

## References

A. Alibes, J. Labay, R. Canal, Astrophys. J. **571**, 326 (2002)
W.I. Axford, in *Proc. 17th Int. Cosmic-Ray Conf.* (Paris), vol. 12 (1981), p. 155
A. Baldi et al., Astrophys. J. **636**, 158 (2006)
W.R. Binns et al., Astrophys. J. **634**, 451 (2005)
A. Blaauw, in *The Physics of Star Formation and Early Stellar Evolution*, ed. by C.J. Lada, N.D. Kylafis (Kluwer, Dordrecht, 1991), p. 125
C.J. Cesarsky, J.-P. Bibring, in *Origin of Cosmic Rays*, ed. by G. Setti et al. (Reidel, Dordrecht, 1981), p. 361
D.D. Clayton, E.A. Deneault, B.S. Meyer, Astrophys. J. **562**, 480 (2001)
D.D. Clayton, L.R. Nittler, Annu. Rev. Astron. Astrophys. **42**, 39 (2004)
J. Donnelly et al., 26th Internat. Cosmic Ray Conf. Papers, vol. 3 (1999), p. 109
D.C. Ellison, L. Drury, J.-P. Meyer, Astrophys. J. **487**, 197 (1997)
J.J. Engelmann et al., Astron. Astrophys. **233**, 96 (1990)
R.I. Epstein, Mon. Not. Roy. Astron. Soc. **193**, 723 (1980)
H.J. Habing, F.P. Israel, Annu. Rev. Astron. Astrophys. **17**, 345 (1979)
J.C. Higdon, R.E. Lingenfelter, Astrophys. J. **590**, 822 (2003a)
J.C. Higdon, R.E. Lingenfelter, Astrophys. J. **582**, 330 (2003b)
J.C. Higdon, R.E. Lingenfelter, Astrophys. J. **628**, 738 (2005)
J.C. Higdon, R.E. Lingenfelter, R. Ramaty, Astrophys. J. Lett. **509**, L33 (1998)
J.C. Higdon, R.E. Lingenfelter, R. Ramaty, 26th Internat. Cosmic Ray Conf. Papers, vol. 4 (1999), p. 144
R. Hirschi, Thesis, Univ. de Geneve, 2004
E.B. Jenkins, in *Origin and Evolution of the Elements*, ed. by A. McWilliams, M. Rauch (Univ. Press, Cambridge, 2004), p. 336. astro-ph/0309651
T. Kozasa, H. Hasegawa, K. Nomoto, Astron. Astrophys. **249**, 474 (1991)
R.E. Lingenfelter, J.C. Higdon, Astrophys. J. **660**, 330 (2007)
R.E. Lingenfelter, J.C. Higdon, K.-L. Kratz, B. Pfeiffer, Astrophys. J. **590**, 822 (2003)
R.E. Lingenfelter, J.C. Higdon, R. Ramaty, in *Acceleration and Transport of Energetic Particles Observed in the Heliosphere*, ed. by R. Mewalt et al. AIP Conf. Proceedings, vol. 528 (2000), p. 375
R.E. Lingenfelter, R. Ramaty, B. Kozlovsky, Astrophys. J. Lett. **500**, L153 (1998)
K. Lodders, Astrophys. J. **591**, 122 (2003)
J.-P. Meyer, L. Drury, D.C. Ellison, Astrophys. J. **487**, 182 (1997)
J. Ott, F. Walter, E. Brinks, Mon. Not. Roy. Astron. Soc. **358**, 1453 (2005)
R. Ramaty, B. Kozlovsky, R.E. Lingenfelter, H. Reeves, Astrophys. J. **488**, 730 (1997)
B.D. Savage, K.R. Sembach, Ann. Rev. Astron. Astrophys. **34**, 279 (1996)
G. Schaller, D. Schaerer, G. Meynet, A. Maeder, Astron. Astrophys. Suppl. **96**, 269 (1992)
P. Sigmund, in *Sputtering by Particle Bombardment I*, ed. by R. Behrisch (Springer, New York, 1981), p. 9
L. Spitzer, E.L. Fitzpatrick, Astrophys. J. **409**, 299 (1993)
F.X. Timmes, S.E. Woosley, T.A. Weaver, Astrophys. J. Suppl. **98**, 617 (1995)
S. van der Bergh, R.D. McClure, Astrophys. J. **425**, 205 (1994)
B.A. Weaver, A.J. Westphal, Astrophys. J. **569**, 493 (2002)

Space Sci Rev (2007) 130: 475–477
DOI 10.1007/s11214-007-9199-x

# Johannes Geiss Contributions to the Early Universe Abundances

**Hubert Reeves**

Received: 21 December 2006 / Accepted: 16 April 2007 /
Published online: 21 August 2007


I have met Johannes first at the Institute for Space Studies in New York, in 1962. This Institute, a collaboration of NASA and Columbia University, was a very lively place. Many meetings and symposia gathered large assemblies of scientists involved in the elaboration of the Space Program. These were great years for the new science of nucleosynthesis. Fred Hoyle, Willie Fowler and the Burbidges were frequent visitors and Al Cameron was a permanent member. The famous article B2FH (1957), published a few years earlier, was the bible of nucleosynthesis. This article had beautifully dealt with the formation of chemical elements in stars. At least for the heavy atoms with mass numbers from carbon to uranium and thorium. There was clearly a problem with the lighter ones since the high temperature of the stellar interiors would prevent their production, and, in fact, destroy them, if they would have been there. This difficulty touched in particular the heavy hydrogen atom (deuterium), the light helium isotope (He-3) and the trio lithium, beryllium, and boron. The question was then: where and how were these elements formed in the universe?

I remember sitting in the back of the classroom with Johannes and exchanging our doubts when Fowler presented his favorite model of the formation of these nuclei: an early irradiation of the planetary system by fast particles emanating from the young and active Sun (Fowler et al. 1961). This scenario appeared to be rather "ad hoc" and could not be confronted with appropriate observations. The nuclear physics data needed to account for the isotopic ratios were not available. It is during this symposium that I learned about the lunar project of Johannes. A foil of aluminum deposited on the lunar surface could absorbed some atoms carried by the solar wind. Indeed, without a magnetic field to deflect charged particles and without an atmosphere, the surface of the moon should received directly the solar wind. Brought back on Earth, this foil would then be analysed in the laboratory in order to obtain information on the composition of the solar wind. I was very impressed by the intelligence and elegance of this idea. The experience was a great success. The He-3 to He-4 ratio was found to be $1/2200$ (Geiss et al. 1970). At last a number to consider and to incorporate in

H. Reeves (✉)
CNRS, Paris, France
e-mail: hreeves@club-internet.fr

our investigations on the origin of the light elements! While trying to integrate this number in a realistic model a problem soon emerge. It was in relation with the deuterium content of water in the terrestrial ocean (heavy water with a D/H ratio of 1/6400). According to solar models the temperature at the bottom of the solar surface convective zone (around two million degrees) is high enough to transform the deuterium in He-3 ($D + p \rightarrow$ He-3 $+ \gamma$). Thus the solar wind must contain two components: first, the original He-3 in the solar protonebula and, second, the new addition from the original D transformed in He-3. And here came the difficulty. If we took in consideration the fact that hydrogen is about ten times more abundant than helium in the universe, and particularly in galactic star forming nebulae, the He-3/He-4 should have been 1/640 for the protosolar matter. Thus the second component (with all the D transformed in He-3) should have been several times larger than the observed lunar ratio (1/2200).

Invited by Johannes to give lectures on nuclear astrophysics at the University of Berne I decided to discuss this problem with him and his colleagues. I was truly obsessed by this difficulty. I tried all kind of hypotheses. Traveling in the train to Berne I started to question both the solar and the terrestrial data.

Was it possible that the solar models were wrong about the temperature at the bottom of the convective zone? If it was lower than estimated from these models the D would still be present in the solar surface. There were however two reasons against the hypothesis: 1) the absence of deuterium lines in the solar spectra ($D/H < 10^{-6}$) and 2) the solar surface lithium abundance appears to have been reduced by a factor of more than one hundred with respect the protosolar value (estimated from meteorites analysis). Destruction by nuclear reactions in surface convective zone of the Sun was the most likely scenario for this lithium reduction. From nuclear physics data on the properties of D and Li it can be shown that given the age of the Sun (4.5 billion years) the temperatures required to affect Li (around two million degrees) are much higher than those required to affect D (much less than on one million degrees). Thus the solar D must have been transformed in He-3. On the other hand, was it possible that the surface He-3 had been reduced by further nuclear reactions transforming it in He-4 (as in the center of the Sun). There is another monitor which tells us that this did not take place: beryllium. Again, using nuclear data, one could compute that the beryllium destroying reactions (Be $+$ p) require an appreciably lower temperature (less than three million degrees) than the He-3 destroying reactions (He-3 $+$ He-4) (more than five million degrees). But there is no sign of major depletion of beryllium in the solar spectrum as compared with meteoritic data (values of at best 20% reduction have been quoted). Thus the protosolar He-3 should be intact. No solution to our conundrum seemed to come from the solar surface abundance data.

At this point of my reflection, a movie that I had seen recently came to my mind: "La bataille de l'eau lourde" (the battle of heavy water). During the Second World War the Wehrmacht tried vainly to steal from the Norwegians a load of heavy water needed for the preparation of an atom bomb. The recipe for producing heavy water was based on the fact that a mixture or H, D and O, left to equilibrate at a given temperature, will generate HDO and $H_2O$ in a ratio proportional to the temperature; the lower the temperature, the higher the HDO/$H_2O$ ratio. The natural hypothesis was then: could the solar system, and in particular the sea water, have been already enriched in D/H ratio in the cold environment of the protosolar nebula? In this case the seawater value was larger than the protosolar value. This could solve our helium isotopic ratio problem!

I took the subject with Johannes and I found that he and his group were already trying to understand this puzzling lunar helium isotopic ratio. The question was then: if the seawater is not a good indicator of the protosolar D/H ratio what should have been this ratio in order to understand the lunar He-3 to He-4 data. A series of arguments that I will not reproduce here brought us to conclude that it must have been around $2 \times 10^{-5}$. This value has been later confirmed by numerous observations in gaseous planets and nebulae. We also estimated the protosolar He-3/He-4 to be of the same order. This was also confirmed later by other measurements. To pursue our cogitations we consider the possible origin of the deuterium in the protosolar nebula. Big Bang Nucleosynthesis BBN had been previously studied by Wagoner, Fowler and Hoyle (WFH 1967). They had shown that deuterium was indeed produced in the early moments of the universe when the temperature was around one billion degrees. Interestingly, according to their computations the ratio of D/H was strongly correlated with the cosmic density of nucleonic matter. How could we relate the D/H ratio in the protosolar nebula to the D/H ratio generated during Big Bang Nucleosynthesis. Taking into account the fact that stellar interiors burn all the D atoms present in their protostellar nebula before returning their matter to space at the end of their life (a process called "astration") it was natural to conclude that interstellar D/H could only decrease with time and that the fraction left from the primordial value in the galactic gas was proportional to the fraction of original galactic gas that has never gone in and out of star (unastrated fraction). Models of galactic evolution, including rates of star formation and death, had already shown that this fraction should be around one half. Putting together the value of the protosolar D estimated from the lunar measurements and the astrated fraction, we could then look at the WFH computations to evaluate the cosmic density of nucleonic matter (Geiss and Reeves 1972). The result showed plainly that it is appreciably smaller (by a factor of five or so) than the density required to close the universe (the critical density of $2 \times 10^{-29}$ g/cm$^3$). Thus the density of nucleonic matter could not close the universe. For this work in 2001 Johannes and I received the Albert Einstein Medal from the Albert Einstein Society in Berne. These cogitations have later been confirmed by analysis of numerous cosmological observations including the properties of the CMB fossil radiation and the density distribution of structures in the universe. Unknown at these days was the existence of the dark energy component which together with the dark matter component and the nucleonic matter component sum up to the critical density and result in a flat universe. We then drew our attention to the elucidation of the physical mechanisms responsible for the enrichment of D in the ocean water and also in comets. Ion–molecule reactions in cold interstellar space turned out to be the most likely mechanisms (Geiss and Reeves 1981).

In the following years, I came often to Berne to present seminars and have fruitful discussions on various subjects of astrophysics in particular on the nucleosynthesis of the light elements: lithium, beryllium, and boron. It became rapidly clear that these atoms originate from the bombardment of galactic cosmic rays on interstellar atoms, mostly on carbon and oxygen. The excellent physical judgement of Johannes was always of great help.

## References

E.M. Burbidge, G.R. Burbidge, W.A. Fowler, F. Hoyle, Rev. Mod. Phys. **29**, 547 (1957)
W.A. Fowler, J.L. Greenstein, F. Hoyle, Astron. J. **66**, 284 (1961)
J. Geiss, J.P. Eberhardt, F. Bühler, J. Meister, P. Signer, J. Geophys. Res. **75**, 5972 (1970)
J. Geiss, H. Reeves, Astron. Astrophys. **18**, 126 (1972)
J. Geiss, H. Reeves, Astron. Astrophys. **93**, 189 (1981)
R.V. Wagoner, W.A. Fowler, F. Hoyle, Astrophys. J. **148**, 3 (1967)

Space Sci Rev (2007) 130: 479–487
DOI 10.1007/s11214-007-9152-z

# Johannes Geiss: The Humble Beginnings of an Octogenarian

**Friedrich Begemann**

Received: 19 December 2006 / Accepted: 2 February 2007 /
Published online: 27 April 2007


Johannes Geiss studied physics at Göttingen University at a time when luminaries like Max Planck, Max v. Laue, Werner Heisenberg and Carl Friedrich v. Weizsäcker were teaching there and luminaries-to-be like Hans Georg Dehmelt and Wolfgang Paul were just starting their careers, next door in the same institute. Johannes started out in the lofty spheres of theoretical physics working with Max v. Laue on problems in superconductivity (Geiss 1951a, 1951b). Somehow, the theoretical work did not quite appeal to him so he returned down to earth and measured the isotopic composition of so-called "common lead" (Geiss 1954), with Wolfgang Paul his thesis adviser and Friedrich Houtermans taking a lively and keen interest in the results and their interpretation.

In those days, only little more than 50 years ago, mass spectrometric isotope abundance measurements were a tedious affair, not at all comparable with the operation of modern semi-automatic instruments. Strip chart recorders, although already available, were not affordable at Göttingen in those destitute post-war years, so the ion currents had to be measured by means of mirror galvanometers, preferably at night when there was nobody else in the laboratory to interfere with the measurements. The magnetic field had to be tuned by hand; readings of the position of the light spot from the galvanometer mirror, on a translucent scale, had to be taken and the numbers jotted down, all in a semi-darkened room. But Johannes was inventive: since there was no help available from inside the institute he looked outside. And he was fortunate in that the present Mrs. Geiss was prepared to spend many hours with him in a darkened room in front of a mass spectrometer attempting to help him solve the mysteries of our world which, incidentally, caused a certain amount of envy among colleagues and friends.

The lead isotope work was continued for some time after Johannes joined Houtermans' institute at Berne in 1953. Soon thereafter, however, he became interested in meteorites. It is now just a little over 200 years, due in large part to the work of Chladni (1794), that

F. Begemann (✉)
Max-Planck-Institut für Chemie (Otto-Hahn-Institut), 55122 Mainz, Germany
e-mail: begemann@mpch-mainz.mpg.de

meteorites are generally accepted to be of extraterrestrial origin. But from where outside the earth they should come was not clear at all. Astronomers, from the observation of fireballs associated with the fall of meteoroids (as these heavenly bodies are called before they land on earth or other planetary bodies of our solar system), concluded that at least some of them had been on hyperbolic orbits around the sun. This made them objects from outside our solar system; they were then material messengers from foreign stars which, in addition to starlight, brought us information from other worlds. Ramsay (1898), e.g., in a review paper on the properties and occurrence in nature of a newly discovered element, helium, reported, inter alia, on the presence of helium in some iron meteorites but the absence of it from others. His tentative suggestion, although put forward with due caution, was that different meteorites derive from different stars, some from stars which contain He, the others from such which don't. ("Quant à son absence dans les autres météorites, cela tient peut-ètre á la nature même de la météorite. Celles qui ne renferment pas d'hélium proviennent sans doute d'astres qui n'en contiennent pas. Mais on comprend que sur ce point nous ne puissions apporter aucune certitude." p. 446.) The idea behind such a suggestion was, of course, that helium, like any other element, should have been incorporated into the meteorites upon their formation.

But then Rutherford discovered $\alpha$-particles from the radioactive decay of heavy elements to be doubly-charged He atoms! This caused a complete change of paradigm: Now the He in, or its absence from, meteorites was not any more supposed to reveal information about the composition of their "parent stars". Rather, helium was presumed to have been produced within the meteorites by the decay of U, Th, and their daughter elements which, in turn, led immediately to the proposal to combine the amount of He accumulated with its production rate and calculate the age of the meteorites. Indeed, during the next 40 years or so all measurements of the helium content of meteorites, starting with Strutt, the future Lord Rayleigh, (1908) and later by Paneth and his collaborators, were performed with the explicit aim to measure such gas retention ages.

The results were interesting, but they also were enigmatic. Gram-sized specimens ascribed to the same (large) meteorites varied in age by more than an order of magnitude, and for different meteorites ages ranged between some 10 000 years and 8 billion years (Paneth 1928; Arrol et al. 1942). The high age values, in particular, caused some uneasiness because astronomers had changed their minds and now deduced from the fall phenomena of meteoroids that they had been on *elliptical* orbits around the sun which made them members of our solar system and that, at that time, was only about 2 billion years old. (Paneth (1930, p. 731) had emphasized early on that even if the fireball observations were accurate enough to draw any meaningful conclusions, which they were not, the orbit of a meteoroid immediately prior to its fall did not necessarily allow any inferences as to its original orbit.)

The solution came in 1952 when Paneth et al. measured the isotopic composition of He from iron meteorites and found up to 1/3 to be He of mass 3 which, of course, could not be attributed to $\alpha$-decay. The measurements were prompted by cosmic ray physicists who had argued that, while in space, the interaction of cosmic rays with meteoritic matter should produce $^4$He and $^3$He in a ratio of about 3:1 and, moreover, that the total amounts produced might be comparable with the amounts actually observed. Such calculations were plagued with large uncertainties, however. Relevant cross sections for the production of He from iron targets were not known; knowledge about flux and energy spectrum of the cosmic radiation was rudimentary at best, and the size of the meteoroids in space as well as the location of a given piece from a meteorite within the meteoroid during the irradiation in space was completely unknown. And the same was, of course, true for the cosmic-ray flux in the past.

Fig. 1 Early companions on the road to fame. In order to facilitate recognition Max v. Laue and Harold Urey are also shown, on the right, as some of us still remember them

**Diploma Thesis adviser**

**Max von Laue**
Nobelprize 1914

**PhD. Thesis adviser**

**Wolfgang Paul**
Nobelprize 1989

The "real" part of the complex set of Nobel Prize winners Wolfgang Paul and Wolfgang Pauli

**Research associate "sponsor"**

**Harold C. Urey**
Nobelprize 1934

In the case of $^{3}$He there is an elegant way around all these problems, except for the last one. In spallation and nuclear evaporation reactions $^{3}$He is produced directly as well as indirectly via $^{3}$H (tritium, the "superheavy" isotope of hydrogen) which, after a mean life time of 18 years, decays into its isobar $^{3}$He. In radioactive equilibrium the decay rate of tritium is equal to its production rate so that, in a freshly fallen meteorite, the decay rate of tritium is equal to the production rate in space of $^{3}$He, except for the contribution from the direct production of $^{3}$He which is of about equal size as the tritium branch.

This was about the state of affairs when Johannes entered the field. He had again been prudent in choosing with whom to work (Fig. 1) and now was in Harold Urey's group at the Enrico Fermi Institute for Nuclear Studies of the University of Chicago. During this stay, which lasted only a year or so, he nevertheless managed to become a well-accepted member of the Chicago "Mutual Admiration Society"—an informal group similar to the one founded, under this name, by Dorothy L. Sayers while a student at Oxford University—the clandestine existence of which was a definitive handicap to all those poor souls who were not members.

Johannes had at his disposal a noble gas mass spectrometer, and I had been working there for some time in Bill Libby's group counting cosmic-ray and hydrogen bomb-produced tritium in all kind of waters and wines. So we pooled our resources and measured the first cosmic ray exposure age of a meteorite, that of the stone meteorite Norton County. It turned out to be 250 million years which was very much shorter than the $^{40}$Ar–K gas retention age of 4400 million years of the meteorite as it was determined in the course of the analysis also (Begemann et al. 1957). Apparently, during a period commensurate with the mean life time of tritium the cosmic ray intensity experienced by the meteorite prior to its fall had been very much higher than the long-time average. The reason might have been a true increase of the cosmic ray intensity with time at wherever the meteoroid was during this time, or it might have been an apparent effect caused by a change in size of the meteoroid. In the latter case the cosmic ray exposure age might date the time when the meteoroid had been liberated from its parent body, when the exposure to the cosmic radiation started from which it had been shielded while residing within its parent body. All present evidence suggests the latter to be the correct explanation and that any change of the cosmic ray intensity, itself, did not amount to more than 50 % or so during the past 2 billion years.

Incidentally, we were serendipitous to have chosen a stone meteorite for our study. Strutt and Paneth, in their early work on He in meteorites, always worked with irons because they, correctly, argued leakage of He from irons to be negligible. In our case it turned out later that, for some still poorly understood reason, most iron meteorites are perfectly capable to retain their tritium while in space but tend to lose it by diffusion within years or less once they have arrived on earth. Hence, had we picked an iron meteorite the tritium activity might have been much too low and the exposure age much too high which almost certainly would have resulted in confusion again.

In addition to Norton County Johannes, while at the Fermi Institute, analysed another dozen of stone meteorites for their gas retention age employing the newly-invented K–Ar dating method. The results were remarkable in so far as among the suite of meteorites there was the achondrite Shergotty with an exceptionally low gas retention age of only 500 million years (Geiss and Hess 1958). In those early years of dating thermal events in the history of meteorites this was just a young outlier; by now we know that Shergotty is a piece from Mars!

The stay in Chicago turned out to be seminal for the Physics Institute at Berne in two respects. First, it led to the implementation there of a noble gas laboratory which was to become one of the top institutions worldwide, together with Mainz and Heidelberg in Germany, (then) Leningrad in the (then) UdSSR, and Berkeley, La Jolla and Minneapolis in the USA. (Laboratories in Chicago, St. Louis and Zürich were latecomers to this circle of excellence.) And it resulted also in opening the door in the USA for many young scientists from Berne, among them Peter Eberhardt, Peter Signer, Heinz Stauffer, Kurt Marti and Otto Eugster who all spent their wanderjahre there.

There were some 50 meteorite papers from the group at Berne up to the end of the sixties when, after Apollo 11, lunar samples came to dominate the interest for a while. The noble gas papers are noteworthy for comprising the whole suite of noble gases, from He to Xe, and they were comprehensive with regard to the problems they addressed. They dealt with cosmic ray exposure ages (Eberhardt et al. 1965a, 1966b), fission Xe from now-extinct $^{244}$Pu (Eberhardt and Geiss 1966), and trapped primordial gases of both the "fractionated planetary" (Eugster et al. 1967b) and the "unfractionated solar" variety (Eberhardt et al. 1965b, 1966a). Even defectors like Hans Balsiger (Balsiger et al. 1968) and Peter Bochsler (Bochsler et al. 1968) earned their first laurels in the noble field of meteoritics! (Publications from Berne at that time almost invariably listed the authors in alphabetical order which, with co-workers like Alder, Balsiger, Bochsler, Bühler, Cerutti, Eberhardt or Eugster put Geiss at a disadvantage. Later, when cooperating with Gloeckler and Reeves, e.g., the situation was different!)

Three of these experimental papers were to become classics in their field. Eberhardt et al. (1966b), in their study of noble gases in chondrites (the most abundant class of stone meteorites) noted the cosmic-ray produced (cosmogenic) component to show a positive linear

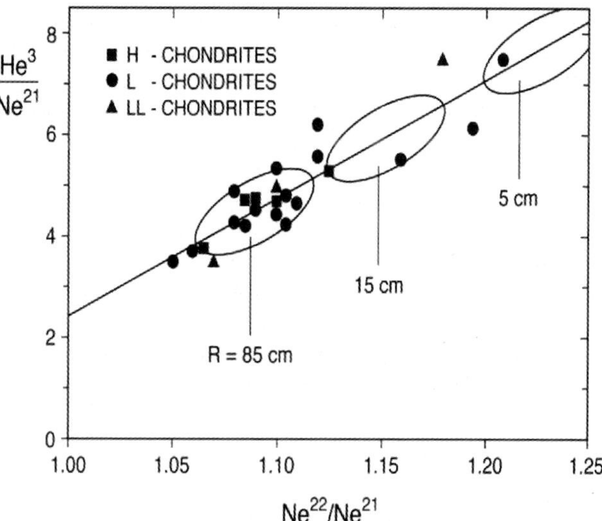


**Fig. 2** $(^3He/^{21}Ne)_{spall}$ versus $(^{22}Ne/^{21}Ne)_{spall}$ diagram for chondrites (Eberhardt et al. 1966b) and results of model calculations for three different pre-atmospheric radii (Leya et al. 2004)

correlation between the $^3He/^{21}Ne$ nuclide abundance ratio and the $^{22}Ne/^{21}Ne$ isotope ratio. They suggested that the position in the $^3He/^{21}Ne$ vs. $^{22}Ne/^{21}Ne$ diagram of a data point depends on the conditions under which the irradiation took place: small meteorites, exposed to the hard primary cosmic rays, fall to the upper right; large meteorites, in particular non-surface samples of large meteorites where the soft secondary component dominates the production, fall to the lower left (Fig. 2). And samples which underwent a heating event late during the exposure history come to plot below the correlation line because losses of gases by diffusion lower the $^3He/^{21}Ne$ *nuclide* ratio but leave the $^{21}Ne/^{22}Ne$ *isotope* ratio essentially unaffected. This so-called "Berne plot" still is one of the simplest and most reliable means to arrive at this kind of information.

Another, more elaborate method to estimate the pre-atmospheric size of stone meteorites also goes back to work at Berne. It utilizes the fact that among the reaction products of the cosmic rays with meteoritic matter are spallogenic and evaporation (MeV) neutrons which, in turn, can induce secondary nuclear reactions. Particularly large cross sections are encountered for neutron capture at thermal energies. But for the neutrons to be effectively slowed down to thermal energies within the meteoroids requires a certain size in order to minimize the fraction of neutrons lost to space by leakage. Eberhardt et al. (1963) studied these effects for $(n, \gamma)$-produced $^{36}Cl$, $^{59}Ni$, and $^{60}Co$ in stone meteorites of various pre-atmospheric radii. (Iron meteorites are less suitable because iron is a poorer moderator than the lighter constituents that make up stone meteorites and, to make things worse, iron has a high neutron absorption cross section for fast neutrons. The two effects combine to reduce the capture rate of thermal neutrons in iron meteorites to insignificant levels.) According to their results for radii in the decimetre range the activities of all three radionuclides are indeed sensitive indicators of size. Practical considerations connected with the feasibility of measuring the low decay rates make $^{60}Co$ almost the only one of the three nuclides to be widely used which is unfortunate because the short half life of 5 years essentially restricts the applicability to the small number of freshly fallen meteorites and leaves out all older "falls" and, in particular, the thousands of "finds" which have not been observed to fall at all.

A third major experimental paper concerned a unique way to measure cosmic ray exposure ages. In addition to the $^3$H–$^3$He nuclide pair other suitable combinations of production rate monitor and total dose integrator had been proposed and utilised. Examples are $^{22}$Na, which decays with a mean life time of 4 years to $^{22}$Ne, and $^{36}$Cl (mean life time 500 000 years) decaying into $^{36}$Ar. This is of considerable interest because the radionuclides involved have grossly different half lives and thus their activities average the intensity of the cosmic radiation over grossly different time spans. All three pairs have in common a noble gas dose integrator, simply because relative to refractory elements noble gases from the time of formation of the meteorites are depleted by many orders of magnitude and this makes detection of a cosmic-ray-produced signal correspondingly easier. But the three methods also have in common that, in each case, two different chemical elements have to be analysed. Since physicists, as a rule, don't like, and can't master, the chemistry of any other elements but noble gases the Berne group was prompted to suggest a way around this by using radioactive $^{81}$Kr (mean life time 330 000 years) as production rate monitor and any of the other, stable, isotopes of Kr as measure of the total dose (Eugster et al. 1967a). This way the determination of an exposure age was reduced to a single mass spectrometric measurement of an isotope abundance ratio.

The method is ingeniously simple in principle but it is demanding experimentally because in meteorites the concentrations of the target elements for the production of spallogenic krypton are exceedingly rare. In chondrites, e.g., they add up to perhaps 25 μg/g and this results in a correspondingly low equilibrium concentration of $^{81}$Kr of only ca. $10^{-14}$ cm$^3$ STP/g. Moreover, chondrites, in spite of the strong depletion of all noble gases at the time they came into being still tend to contain some "trapped Kr" and this is sufficient to make an accurate determination of the small amounts of spallogenic Kr problematic. For thermally evolved meteorites, however, $^{81}$Kr–Kr is still the method of choice to measure cosmic ray exposure ages (Shukolyukov and Begemann 1996).

From the very beginning of the meteorite studies at Berne Johannes saw to it that the institute was never just a "data factory" producing high-quality results but that possible implications of the experimental data should be considered in a broader context. Given his scientific background in physics it was not so much the chemical or mineralogical properties of the meteorites per se which found the greatest interest although this was fast becoming a very fertile field when the methods of "stable isotope geology" were extended to include into these studies the only kind of extraterrestrial matter available at that time for investigation in the laboratory. There had been occasional early attempts to look into the potential of isotope abundance variations in hydrogen, carbon and oxygen, and there were mavericks like Peter Baertschi in Basel, Willi Dansgaard in Copenhagen, Étienne Roth in Paris, Harry Thode in Canada and A.V. Trofimov in Moscow. But it was Urey's group in Chicago which brought the field to fruition, or that at least was how the world looked from Chicago. Anyone even remotely familiar with the field will recognize names like Giovanni Boato, Harmon Craig, Cesare Emiliani, Sam Epstein, Irving Friedman or Heinz Lowenstam, who all were—or had been—working in Chicago in a lab which was, then and for many decades to come, run by the indefatigable Toshiko Mayeda who in addition even managed, though with mixed success, the more unruly members of this crowd to become almost bearable. Cesare, incidentally, left Chicago in 1957 for the Marine Laboratory of the University of Miami and invited Johannes to set up a paleotemperature laboratory to determine ocean temperatures from the $^{18}$O/$^{16}$O ratio in pelagic foraminifera from Caribbean sediment cores. Combining these data with the $^{231}$Pa/$^{230}$Th ages determined by J.N. Rosholt the work resulted in one of the early temperature vs. time curves that covered the last 150 000 years (Rosholt et al. 1962).

But back to Berne. Here, meteorites were looked at as the "poor man's space probe" and the data were analysed with regard to what they might reveal about conditions in space during the past, in particular about the flux of cosmic rays in the inner solar system and any potential changes of its intensity during the past. Initially, these studies were pursued together with Hannes Oeschger (Geiss and Oeschger 1960; Geiss et al. 1962) but he soon drifted off to start his own brilliant career in climatology. One of the problems of interest at that time was the constancy in time of the cosmic ray flux. From the observed agreement between $^{14}$C ages and historical ages Libby (1952) had shown the intensity to have been constant, within a few percent, over the last few thousand years. On a somewhat longer time scale the agreement in ages obtained by the radiocarbon and the $^{230}$Th–$^{231}$Pa methods suggested constancy to within a factor of two (Kulp and Volchok 1953). But now, with the information from meteorites available, it was possible to extend the time span considerably; utilizing $^{26}$Al (half life 700 000 years) one could look into changes with time over the last few million years, with the additional advantage that the intensity would be measured outside the influence of the terrestrial atmosphere and outside the geomagnetic field. These advantages, though, were traded in for the drawback that the whereabouts of the meteoroids prior to their fall were not known so it was not clear exactly to where in the inner solar system the information pertained.

When utilizing meteorite data for this purpose one compares measured decay rates with predicted production rates. The approach chosen to arrive at the latter was to make use of spallation reaction systematics according to which the total isobaric yields from high-energy reactions depend in a simple way on the mass difference $\Delta A$ between target and reaction product (Rudstam 1956). This empirical relationship did not require knowing any details of the various production modes involved; it circumvented the need to calculate production rates from cross sections—which were not known—and the energy spectra of the various nuclear-active particles—which were also not known. Using literature data for an iron meteorite Geiss et al. (1962) showed the concentration of a series of *stable* spallation products, with $\Delta A$ ranging between 11 and 35, to follow a power-law distribution with the concentrations proportional to $\Delta A^{-2.4}$. The same distribution was then also found for the isobaric activities of *radionuclides* in another iron meteorite that had reached the lab fast enough to be analysed shortly after its fall (Honda and Arnold 1961). But most importantly, for the differences that did exist between observed and predicted activities there was no systematic trend with half life of the radionuclides which ranged between 16 days and 700 000 years (Fig. 3); the results were thus compatible with the cosmic ray flux to have been constant over the last few million years.

Actually, the paper is much more comprehensive and the potential systematic changes with time of the cosmic-ray flux are just one aspect. Also derived was the free-space energy spectrum of the cosmic rays, averaged over the 600 million year exposure age of the meteorite. And, intimately connected with the actual meaning of a cosmic-ray exposure age, the question was addressed how much such an "age" might be affected by space erosion which causes a continuous reduction in size of meteoroids during their space travel. (The influence was concluded to be negligible.)

But the impression would be misleading that the "humble beginnings" of Johannes should have been narrowly focused on meteorites and, in particular, on their potential as space probes. This would have been impossible in the extremely stimulating environment that Houtermans had managed to create at the Physikalisches Institut in Berne since he arrived there in 1952. Houtermans' interests always were as wide-ranging as they were unconventional, and he eagerly shared his ideas with his students from the moment of inception. ("I have so many ideas every day that I need you to talk me out of the more

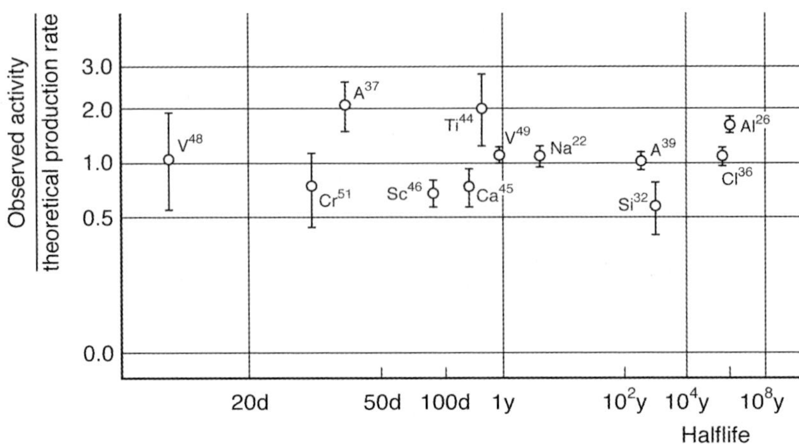


**Fig. 3** Deviations of measured decay rates from predicted production rates of a suite of radionuclides in a freshly-fallen iron meteorite (Honda and Arnold 1961) show no trend with mean life time of the radionuclide, supporting the assumption underlying the calculations that over the last few million years the cosmic-ray flux has been constant (Geiss et al. 1962)

crazy ones" was how he appealed for our help.) And given his personal history he hated parochialism; his students learned science to be international. Johannes apparently was an attentive disciple. From early on he was cooperating with chemists, geologists and mineralogists from within the University and with groups from the Université Libre in Brussels, with radiochemists from Mainz/Cologne, and geologists from Pisa and Catania. Some of the joint projects, like analysing freshly deposited lead exhalations from Vesuvius and Stromboli (Begemann et al. 1954) included the collection of such deposits amidst a rain of lapilli and occasional blobs of hot lava on the slopes of these volcanoes. On other occasions, sample collection within freshly-dug water ducts in the Swiss Alps was made difficult by early heavy snow storms and, on the way back, by the train being derailed by an avalanche which, incidentally, occurred on the northern shore of the Lake of Brienz, just across the lake from the Grand Hotel Giessbach! Then, in anticipation of what was to become a major branch of archaeometallurgy (Gale and Stos-Gale 1982; Begemann et al. 1989) at Pompeji old Roman lead water pipes were sampled (the somewhat clandestine sampling being facilitated by Carmen Geiss effectively detracting the watchful eyes of the local guides) and, after a long delay, analysed for their isotopic composition in order to determine the provenance of the lead employed by the Romans during the first century A.D. (Grögler et al. 1966). And in cooperation with the radiochemist Wilfried Herr attempts were made to find the decay products of now-extinct $^{97,98}$Tc, which failed (Herr et al. 1958), to utilize for dating purposes the predicted decay of still-extant $^{180}$Ta which also failed (Eberhardt et al. 1956) and, on the positive side, the development of the wildly successful Rh–Os dating method (Herr et al. 1961) which still is just about the only practical one to date iron meteorites.

"Humble beginnings" have an easy starting point, but where do they end? And what follows? For that, see the contributions by George Gloeckler and Hubert Reeves.

## References

W.J. Arrol, R.B. Jacobi, F.A. Paneth, Nature **149**, 235–238 (1942)

H. Balsiger, J. Geiss, N. Grögler, A. Wyttenbach, Earth Planet. Sci. Lett. **5**, 17–22 (1968)
F. Begemann, J. Geiss, F.G. Houtermans, W. Buser, Nuovo Cimento **11**, 663–673 (1954)
F. Begemann, J. Geiss, D.C. Hess, Phys. Rev. **107**, 540–542 (1957)
F. Begemann, S. Schmitt-Strecker, E. Pernicka, in *Old World Archaeometallurgy*, ed. by A. Hauptmann, E. Pernicka, G.A. Wagner. Anschnitt Beiheft, vol. 7 (1989), pp. 269–278
P. Bochsler, P. Eberhardt, J. Geiss, N. Grögler, in *Meteorite Research*, ed. by P.M. Millman (Reidel, Dordrecht, 1968), pp. 857–874
E.F.F. Chladni, *Ueber den Ursprung der von Pallas gefundenen and anderer ihr ähnlicher Eisenmassen und über einige damit in Verbindung stehende Naturerscheinungen*. Hartknoch, Riga (1794)
P. Eberhardt, J. Geiss, Earth Planet. Sci. Lett. **1**, 99–101 (1966)
P. Eberhardt, J. Geiss, C. Lang, W. Herr, E. Merz, in *Publ. 400, Nat. Acad. Sc.-Nat. Res. Council*, Washington, DC, pp. 203–205 (1956)
P. Eberhardt, J. Geiss, H. Lutz, in *Earth Science and Meteoritics*, ed. by J. Geiss, E.D. Goldberg (North Holland, Amsterdam, 1963), pp. 143–168
P. Eberhardt, O. Eugster, J. Geiss, J. Geophys. Res. **70**, 4427–4434 (1965a)
P. Eberhardt, J. Geiss, N. Grögler, Tschermaks Mineral. Petrograph. Mitt. **10**, 535–551 (1965b)
P. Eberhardt, J. Geiss, N. Grögler, Earth Planet. Sci. Lett. **1**, 7–12 (1966a)
P. Eberhardt, O. Eugster, J. Geiss, K. Marti, Z. Naturforsch. A **21**, 414–426 (1966b)
O. Eugster, P. Eberhardt, J. Geiss, Earth Planet. Sci. Lett. **2**, 77–82 (1967a)
O. Eugster, P. Eberhardt, J. Geiss, Earth Planet. Sci. Lett. **3**, 249–257 (1967b)
N.H. Gale, Z.A. Stos-Gale, Science **216**, 11–19 (1982)
J. Geiss, Ann. Phys. **9**(6), 40–47 (1951a)
J. Geiss, Z. Phys. **129**, 449–482 (1951b)
J. Geiss, Z. Naturforsch. A **9**, 218–227 (1954)
J. Geiss, D.C. Hess, Astrophys. J. **127**, 224–236 (1958)
J. Geiss, H. Oeschger, Space Research, in *Proc. First Intern. Space Sci. Symp.*, ed. by H.K. Kallmann-Bijl (North-Holland, Amsterdam, 1960), pp. 1071–1079
J. Geiss, H. Oeschger, U. Schwarz, Space Sci. Rev. **1**, 197–223 (1962)
N. Grögler, J. Geiss, M. Grünenfelder, F.G. Houtermans, Z. Naturforsch. A **21**, 1167–1172 (1966)
W. Herr, E. Merz, P. Eberhardt, J. Geiss, C. Lang, P. Signer, Geochim. Cosmochim. Acta **14**, 158 (1958)
W. Herr, W. Hofmeister, B. Hirt, J. Geiss, F.G. Houtermans, Z. Naturforsch. A **16**, 1053–1058 (1961)
M. Honda, J.R. Arnold, Geochim. Cosmochim. Acta **23**, 219–232 (1961)
J.L. Kulp, H.L. Volchok, Phys. Rev. **90**, 713–714 (1953)
I. Leya, F. Begemann, H.W. Weber, R. Wieler, R. Michel, Meteorit. Planet. Sci. **39**, 367–386 (2004)
W.F. Libby, *Radiocarbon Dating* (The University of Chicago Press, Chicago, 1952)
F. Paneth, Z. Elektrochem. **34**, 645–652 (1928)
F. Paneth, Z. Elektrochem. **36**, 727–732 (1930)
F.A. Paneth, P. Reasbeck, K.I. Mayne, Geochim. Cosmochim. Acta **2**, 300–303 (1952)
W. Ramsay, Ann. Chim. Phys. **13**(VII), 433–480 (1898)
J.N. Rosholt, C. Emiliani, J. Geiss, F.F. Koczy, P.J. Wangersky, J. Geophys. Res. **67**, 2907–2911 (1962)
G. Rudstam, *Spallation of Medium Weight Elements* (Appelbergs Boktryckeri, Uppsala, 1956)
A. Shukolyukov, F. Begemann, Meteorit. Planet. Sci. **31**, 60–72 (1996)
R.J. Strutt, Proc. Roy. Soc. A **80**, 572–594 (1908)

Space Sci Rev (2007) 130: 489–513
DOI 10.1007/s11214-007-9226-y

# Johannes Geiss' Investigations of Solar, Heliospheric and Interstellar Matter

**George Gloeckler · Lennard A. Fisk**

Received: 16 February 2007 / Accepted: 29 May 2007 / Published online: 26 July 2007


**Abstract** Johannes Geiss is a world leader and foremost expert on measurements and interpretation of the composition of matter that reveals the history, present state, and future of astronomical objects. With his Swiss team he was first to measure the composition of the noble gases in the solar wind when in the late 1960s he flew the brilliant solar wind collecting foil experiments on the five Apollo missions to the moon. Always at the forefront of the art of composition measurements, he with his colleagues determined the isotopic and elemental composition of the solar wind using instruments characterized by innovative design that have provided the most comprehensive record of the solar wind composition under all solar wind conditions at all helio-latitudes. He discovered heavy interstellar pickup ions, from which the composition of the neutral gas of the Local Interstellar Cloud is determined, and the "Inner Source" of pickup ions. Johannes Geiss played a key role both in the in-situ measurements and modeling of molecular ions in comets, and the interpretation of these data. He and co-workers measured the composition of plasmas in the magnetospheres of Earth and Jupiter. Here we highlight Johannes Geiss' many discoveries and seminal contributions to our knowledge of the composition of matter of the Sun, solar wind, interstellar gas, early universe, comets and magnetospheres.

**Keywords** Sun: solar wind · Interstellar Medium: elemental composition · Sun: composition

## 1 Introduction

The first landing on the moon by humans on July 20, 1969, marked a historic milestone in space exploration. It was also a milestone in the research of Johannes Geiss. At 03:35 UT

G. Gloeckler (✉) · L.A. Fisk
Department of Oceanic, Atmospheric and Space Sciences, University of Michigan, Ann Arbor, MI 48109-2143, USA
e-mail: gglo@umich.edu

G. Gloeckler
Department of Physics and the Institute for Physical Science and Technology, University of Maryland, College Park, MI 20742-0001, USA

**Fig. 1** The SWC experiment deployed in Mare Tranquillitatis by Apollo 11 Astronaut Edwin E. Aldrin on July 21, 1969, photographed by Commander Neil A. Armstrong (NASA Photo S1-40-5872)

on July 21, several hours after touchdown, Apollo 11 astronaut Aldrin deployed the Swiss Solar Wind Composition (SWC) experiment four meters from the nearest footpad of the Lunar Module on the soil in Mare Tranquilliatis (Fig. 1). After a 77-minute exposure to the solar wind, the astronauts rolled up the solar wind collection foil and stored it in the lunar sample box. The foil was returned to Earth, and, after having been released from quarantine, shipped back to Switzerland where the composition of the solar wind trapped in the foil was carefully measured using ultra-high vacuum mass spectrometers in Geiss' laboratory at the Physikalische Institut of the University of Bern. There were four more exposures of the SWC foil as part of the Apollo program, with the last on Apollo 16 in April 21, 1972, lasting for more than 45 hours.

This brilliant experiment, so expertly designed, extensively calibrated and tested, and flawlessly executed, illustrates Johannes' skill and perseverance in pursuing all of his experiments. He set out to measure the composition of noble gases, and in particular the $^3$He/$^4$He ratio in the solar wind, to settle the question raised by the puzzling measurements in the "solar noble gas component" of certain meteorites of $^3$He/$^4$He ratios that were an order of magnitude lower than predicted from the D/H ratios measured in sea water and meteorites[1] (Geiss et al. 1966). Geiss with his co-investigators P. Eberhardt and P. Signer then conceived the idea of a controlled solar wind collection experiment, selecting it as the best method for isotopic measurements, and in a short time he and co-workers performed extensive laboratory research and engineering to prove the feasibility of the technique, and to select the proper foil materials. He concluded that the Apollo mission offered the only opportunity at that time to expose his experiment to the solar wind and to return it back to Earth, and thus

[1] Years later, Geiss and Reeves (1981), using new D/H ratios measured in meteorites by Robert et al. (1979), showed that the D/H ratio in ocean water as well as in comets was highly enriched as a result of ion-molecule reactions in cold interstellar space (see Reeves 2007).

proposed to NASA a solar wind composition experiment to be deployed at the lunar surface as part of the Apollo scientific program (Geiss et al. 1966). Despite the risks involved in collecting solar wind ions at the surface of the moon due to possibly major solar wind perturbations in the lunar environment (exposure of the SWC foil on the way to or from the moon would have been ideal, but was not feasible), Johannes Geiss was successful in convincing NASA to select his experiment for six lunar landing missions, including the very first one, Apollo 11, that landed in Mare Tranquillitatis in July 1969. To assure that the SWC experiments would be flown and properly executed, Geiss spent nearly a year prior to the Apollo 11 mission at the Manned Spacecraft Center (MSC) in Houston, Texas, interacting with the flight crew and engineers, instructing them on how and where to deploy the foil and explaining the scientific importance of the experiment. There is little doubt that Geiss personally convinced, and then received approval from, not only the MSC leadership but also engineers, technicians and even the flight crews to deploy the Swiss Solar Wind Composition experiment even before raising the flag of the United States of America on the soil of the moon. For Apollo 11, this was of particular importance, because in this way, the solar wind was collected for 77 minutes, about twice as long as it would have been if the SWC were deployed after the flag raising. From today's perspective, it is amazing that Neil Armstrong and Edwin Aldrin adhered exactly to this timeline, written down before any human ever had set foot onto the lunar surface.

## 2 Geiss' Composition Experiments

The Apollo Solar Wind Composition Experiment was the first of many of Geiss' experiments to measure the composition of the present-day solar wind. It was certainly a most challenging measurement that would not be repeated for decades, and the results obtained for the isotopic ratios of the noble gases in the solar wind remained unsurpassed for many decades.

The Solar Wind Ion Composition Spectrometer (SWICS) was the other challenging experiment, still continuing, for which Johannes Geiss and one of us (GG) are Co-Principal Investigators. This instrument (Gloeckler et al. 1992, 1998), designed to measure the mass, charge state and energy of thermal and supra-thermal ions from about 0.6 to 60 keV/charge, was flown on both the Ulysses (Solar Polar) Mission, launched in October 1990, and the Advanced Composition Explorer (ACE) mission, anchored at L1 at 1 AU. With SWICS on Ulysses and ACE, Johannes Geiss and colleagues measured for the first time the solar wind elemental and charge state composition, and the $^{3}He/^{4}He$ ratio in all types of solar wind flows for a combined period of more than 25 years, at high and low solar latitudes and during all phases of the solar cycle. The exceedingly low background, extended energy range, and mass and charge identification of SWICS, combined with Ulysses' unprecedented polar orbit achieved by a Jupiter gravity assist and flyby, enabled Geiss and co-workers to discover interstellar and "Inner Source" pickup ions, measure the interstellar $^{3}He/^{4}He$ ratio and the Jovian magnetospheric ion composition, and analyze the abundance of ions and molecules in several comets during serendipitous flybys of their distant tails. Without Geiss these experiments would not have been flown on Ulysses and ACE and the scientific advances made using data from these instruments would have been long delayed.

The Apollo experiment was followed by a number of ion mass/charge spectrometers, employing permanent magnetic fields perpendicular to stepped electric fields, that were flown on various unmanned space missions as well as sounding rockets to study the composition of the solar wind, comet Halley, the terrestrial magnetosphere and upper atmosphere. Johannes Geiss was a driving force in all of these experiments. We list in Table 1, in roughly

**Table 1** Geiss' space mission and sounding rockets experiments

| Experiment | Mission | Role |
|---|---|---|
| Solar Wind Composition Experiment (SWC) | Apollo 11 to 16 | Principal Investigator |
| Magnetospheric Particle Composition Experiment | Skylab | Principal Investigator |
| Ion Mass Spectrometer | GEOS | Group Leader |
| Upper Atmosphere Experiments | Sounding Rockets | Investigator |
| Interstellar Gas Experiment | LDEF | Principal Investigator |
| Plasma Composition Experiment | ISEE-1 | Co-Investigator |
| Ion Abundance Experiment | ISEE-3 | Co-Investigator |
| Plasma Composition Experiment | Dynamic Explorer | Co-Investigator |
| Solar Wind Ion Composition Spectrometer (SWICS) Experiment | Ulysses | Co-Principal Investigator |
| Ion Mass Spectrometer | GIOTTO | Co-Investigator |
| SWICS-MASS-STICS (SMS) | Wind | Co-Investigator |
| CELIAS Experiment (proposal phase) | SOHO | Co-Principal Investigator |
| CELIAS Experiment | SOHO | Co- Investigator |
| TIMAS Experiment | Polar | Co-Investigator |
| SWICS-SWIMS | ACE | Co-Investigator |

chronological order, these experiments as well as those that came after the SWICS experiment on Ulysses. The Solar Wind Composition (SWC) Experiments on Apollo and Solar Wind Ion Composition (SWICS) Experiment on Ulysses and ACE are briefly described in the following.

## 2.1 Solar Wind Composition Experiments on Apollo

The SWC experiments, weighing only 430 g, were designed, engineered and fabricated at the Physikalische Institut of the University of Bern to be easily deployed and later rolled up and safely stored by the astronauts. The SWC experiments were extensively tested and calibrated and all five flights were successfully executed. Figs. 1 and 2 show the instrument in its extended configuration.

The SWC instruments consisted of a 3.5 cm diameter telescopic pole, approximately 40 cm in length when collapsed, a spring-driven reel, and a solar-wind collection foil, that was stowed inside the pole when it was rolled up on the reel (Fig. 2). When extended at the lunar surface a foil area of 30 cm by 130 cm was exposed, with the pole ~1.5 m long. To avoid contamination by exhausts from the spacecraft and to minimize contamination by dust generated by the activity of the crew, the instruments were deployed as far from the Lunar Module as was possible. At the end of the lunar exposure period, the foils were rolled up and returned to Earth, but the poles were left on the Moon. All foils were exposed vertically, except for the Apollo 12 foil, which reclined by 12 degrees.

Based on extensive laboratory measurements of trapping probabilities of ~1 keV/amu particles, the foil material chosen for Apollo 11 to Apollo 15 was 15 μm Al. The foil for the Apollo 16 SWC consisted of segments of Al and Pt. The foils were rimmed with Teflon tape for reinforcement and, on the backside of the aluminum foils, ~1 μm of $Al_2O_3$, was deposited for temperature control.

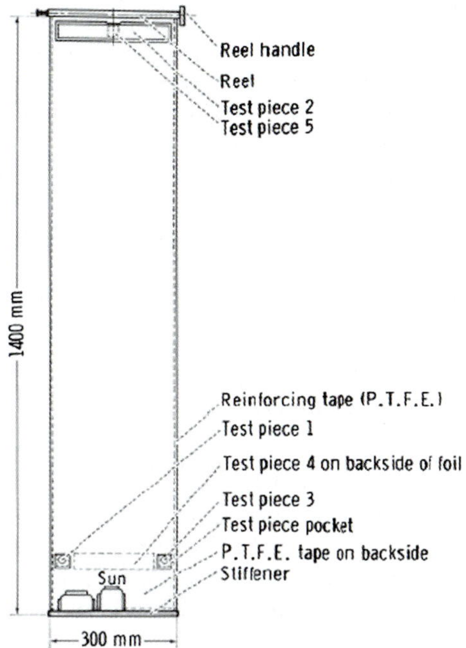


**Fig. 2** The SWC Reel Assemblies with the solar wind capture foil unrolled

**Fig. 3** Photograph of the SWICS instrument showing the cylindrical −30 kV high voltage supply and the sensor with its wide-angle collimator entrance system. The cover protecting the carbon foil, micro-channel plates and solid-state detectors of the time-of-flight assembly prior and during launch is released in space. The entire instrument package, consisting of the sensor, high voltage supply to accelerate ions and a Data Processing Unit, weighed only 5584 g

## 2.2 The Solar Wind Ion Composition Spectrometer on Ulysses

The SWICS instrument (Fig. 3) is a time-of-flight (TOF) ion mass spectrometer in which ions are accelerated by tens of keV/charge, after an energy/charge analysis using a wide-angle, multi-slit collimator deflection system. The accelerated ions then have sufficient energy to pass through a several hundred angstroms thin carbon foil, ejecting secondary electrons (SEs) that generate the start signal. After traveling another 10 cm, the ions hit a solid-state detector (SSD), producing SEs that generate the stop signal. Finally, the residual energy of highly charged solar wind ions is measured by the SSD. The development of SWICS was lead by the University of Maryland (Gloeckler et al. 1992). Universities of Maryland, Bern, Braunschweig and the Max-Planck Institut für Aeronomie (renamed Max-Planck Institut für Sonnensystemforschung), each provided essential hardware. The extensive calibration of SWICS was performed at the excellent calibration facility of the Physikalische Institut of the University of Bern.

It was difficult to persuade NASA and ESA to select this high-risk technology, mostly new and unproven in flight, for the high-profile International Solar Polar (later renamed Ulysses) Mission. Johannes Geiss played a key role in accomplishing this, convincing the selection committee and both ESA and NASA that the experiment would work and the proposed scientific objectives would be achieved. To produce a working and well-calibrated instrument was also a most-difficult task, and Johannes' scientific and technical involvement in this experiment was indispensable. It is a fact that without Geiss' involvement, SWICS would not have been developed and flow, and the use of TOF/post-acceleration technology in space instruments would have been set back for many years.

## 3 Solar Wind Measurements and Theory

### 3.1 Abundance of Helium, Neon and Argon Isotopes

Table 2 summarizes the isotopic ratios of the noble gases measured with the Apollo SWC experiments (Geiss et al. 2004) on the five Apollo missions. A small correction was applied to the data for trapping efficiency and non-vertical impact of the solar wind on the foils. All exposures were close to solar maximum of solar cycle 20.

#### *3.1.1 Isotopic Ratios of Ne and Ar*

Values obtained for the neon and argon isotopic ratios were, within errors, the same during all flights. In situ measurements of the solar wind $^{22}Ne/^{20}Ne$ ratio (the only isotopic ratio of the noble gases in the solar wind, other than He, that could be measured) were consistent with the Apollo value (Kallenbach et al. 1997). All isotopic ratios, including the average $^{3}He/^{4}He$ ratio for the five flights, were, within errors, the same as the corresponding ratios in the Solar Wind component of trapped lunar gas.

#### *3.1.2 Variability of the $^{3}He/^{4}He$ ratio*

The $^{3}He/^{4}He$ ratio varied significantly from flight to flight, and was found to be anti-correlated with variations in both the $^{4}He/^{20}Ne$ ratio (Fig. 4) and the geomagnetic $K_p$ index.

**Table 2** Abundance ratios of solar wind helium, neon and argon isotopes measured with the Solar Wind Composition Experiment during the five Apollo flights to the Moon

| Mission | He Flux $10^6/cm^2$ s | $^4He/^{20}Ne$ | $^4He/^3He$ | $^{20}Ne/^{22}Ne$ | $^{22}Ne/^{21}Ne$ | $^{36}Ar/^{38}Ar$ |
|---|---|---|---|---|---|---|
| Apollo 11 | $6.2 \pm 1.2$[a] | $430 \pm 90$ | $1860 \pm 140$ | $13.50 \pm 1.0$ | – | – |
| Apollo 12 | $8.1 \pm 1.0$ | $620 \pm 70$ | $2450 \pm 100$ | $13.25 \pm 0.5$ | $30.5 \pm 2$ | $5.3 \pm 0.5$ |
| Apollo 14 | $4.2 \pm 0.8$ | $550 \pm 70$ | $2230 \pm 140$ | $13.65 \pm 0.35$ | $30 \pm 3$ | $5.2 \pm 0.5$ |
| Apollo 14 | $17.7 \pm 2.5$ | $550 \pm 50$ | $2310 \pm 120$ | $13.70 \pm 0.35$ | $30.5 \pm 2$ | $5.2 \pm 0.35$ |
| Apollo 16 | $12.0 \pm 1.8$ | $570 \pm 50$ | $2260 \pm 100$ | $13.80 \pm 0.4$ | $29.5 \pm 4$ | $5.5 \pm 0.4$ |
| Apollo 16 (Pt) | – | – | $2180 \pm 180$ | $13.6 \pm 0.3$ | $31 \pm 3$ | $5.5 \pm 0.4$ |
| Average | – | $570 \pm 70$ | $2350 \pm 120$ | $13.7 \pm 0.3$ | $30 \pm 3$ | $5.4 \pm 0.3$ |

[a]The errors given are $2\sigma$ errors and include both statistical and systematic uncertainties

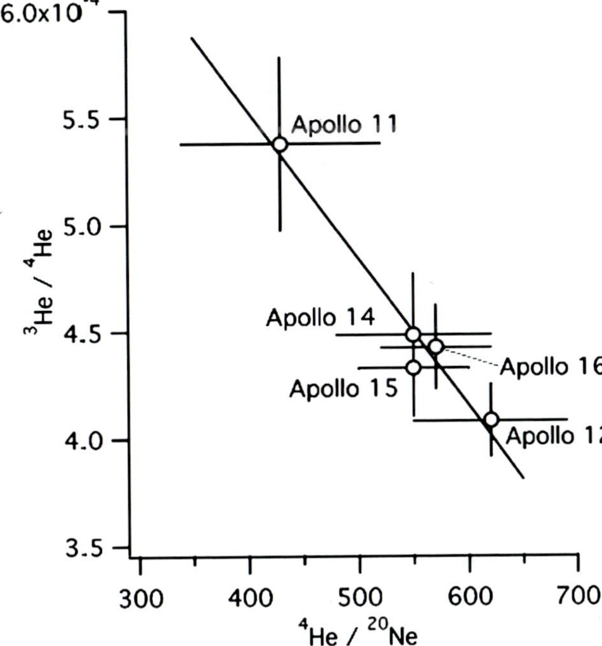


**Fig. 4** Anti-correlation of the $^3$He/$^4$He and $^4$He/$^{20}$Ne abundance ratios for the five Apollo exposures of the SWC foils

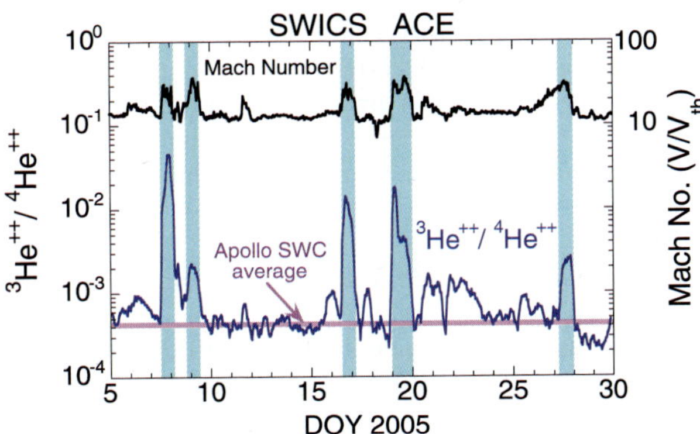


**Fig. 5** Hourly averages of the $^3$He/$^4$He ratio and the helium bulk speed to thermal speed ratio measured with SWICS on ACE during a 25-day time interval in January 2005. *Shaded regions* indicate the correspondence between high Mach Number ($V/V_{th}$) and high $^3He^{++}/^4He^{++}$

Such variations were not entirely unexpected. Indeed theories that included Coulomb collisions to drag heavier ions out of the corona (Geiss et al. 1970, also see Sect. 3.5) predicted such behavior. Numerous subsequent studies with spacecraft mass spectrometers confirmed the Apollo SWC results and found that the solar wind $^3$He/$^4$He ratio was quite variable (e.g., Coplan et al. 1984). Gloeckler et al. (1999) found $^3$He/$^4$He ratios of $\sim 10^{-2}$ in the solar wind from coronal mass ejections. Shown in Fig. 5 are variations in the $^3$He/$^4$He ratio measured with SWICS on ACE in the slow solar wind during a 25-day period in 2005 along with the ratio of solar wind bulk to thermal speed ($V_{sw}/V_{th}$) observed simultaneously with SWICS. The $^3$He/$^4$He ratio increases by as much as a factor of $\sim$100 over periods of $\sim$1 day or less and these increases are correlated with increases in $V_{sw}/V_{th}$. Nevertheless, averages of $^3$He/$^4$He over long time periods in the near-ecliptic slow solar wind agreed well with the average $^3$He/$^4$He ratio obtained with the SWC experiment on Apollo.

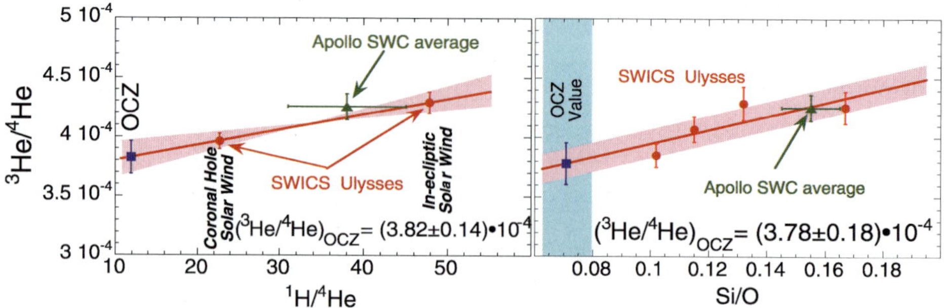


**Fig. 6** $^{3}He/^{4}He$ abundance ratio versus H/He (*left-hand panel*) and Si/O (*right-hand panel*) measured with SWICS on Ulysses and SWC on Apollo in various solar wind flows. For the photospheric abundance of H/He and Si/O we used the recent values of Asplund et al. (2005). This method gives an average value for the $^{3}He/^{4}He$ abundance ratio in the outer convective zone of the Sun of $(3.80 \pm 0.11) \times 10^{-4}$. (Adapted from Gloeckler and Geiss 2000a)

### 3.2 $^{3}He/^{4}He$ in the Outer Convection Zone of the Sun

Variability of the solar wind $^{3}He/^{4}He$ ratio (and probably also of other isotopic ratios) makes it difficult to determine its value in the outer convection zone (OCZ) of the Sun. The approach taken by Gloeckler and Geiss (2000) was to plot (see Fig. 6) the $^{3}He/^{4}He$ ratio versus an elemental abundance ratio (whose photospheric values is well known, e.g., $^{1}H/^{4}He$ from helioseismology or Si/O from spectroscopy) measured in various solar wind flows, such as the slow wind and the fast wind from polar coronal holes. In both cases the two respective ratios are reasonably well correlated and extrapolation to photospheric ratios of $^{1}H/^{4}He$ and Si/O (Asplund et al. 2005) gives values for the OCZ $^{3}He/^{4}He$ ratio of $(3.82 \pm 0.14) \times 10^{-4}$ and $(3.78 \pm 0.18) \times 10^{-4}$ respectively, within 1 percent of each other. The average value, $(3.80 \pm 0.11) \times 10^{-4}$, is the same as the value of $(3.8 \pm 0.5) \times 10^{-4}$ obtained by Geiss and Gloeckler (1998).

### 3.3 Elemental Abundances with Mass/Charge Spectrometers

Near solar maximum of cycle 21 (1978.7–1982.5) Johannes Geiss and co-workers measured the abundances of the solar wind with their Ion Composition Instrument (ICI) on ISEE-3, and with a similar instrument on ISEE-1. These state-of-the-art mass/charge spectrometers, based on designs developed by Geiss' group at the Physikalische Institut, were an important improvement over the conventional solar wind instruments, allowing measurements of solar wind $^{1}H$, $^{4}He$, $^{3}He$, $^{16}O$, $^{20}Ne$, $^{28}Si$ and $^{56}Fe$ during most types of flow conditions. Figure 7a, illustrates the mass/charge $(m/q)$ resolution of ICI for $2 \leq m/q \leq 3$ during a period of low ion thermal speed (Kunz et al. 1983). Bochsler et al. (1986) obtained the abundance of various ions and their charge states as a function of solar wind $^{4}He$ parameters such as flux or kinetic temperature (e.g. Fig. 7b).

### 3.4 Elemental and Charge State Composition with SWICS

Because mass/charge spectrometers could not resolve ions that have the same $m/q$, the abundance of important elements such as C, N and Mg could not be reliably determined. This deficiency was overcome with the use of the Solar Wind Ion Composition Spectrometer (SWICS) instruments that could determine both the mass and charge of an ion. With

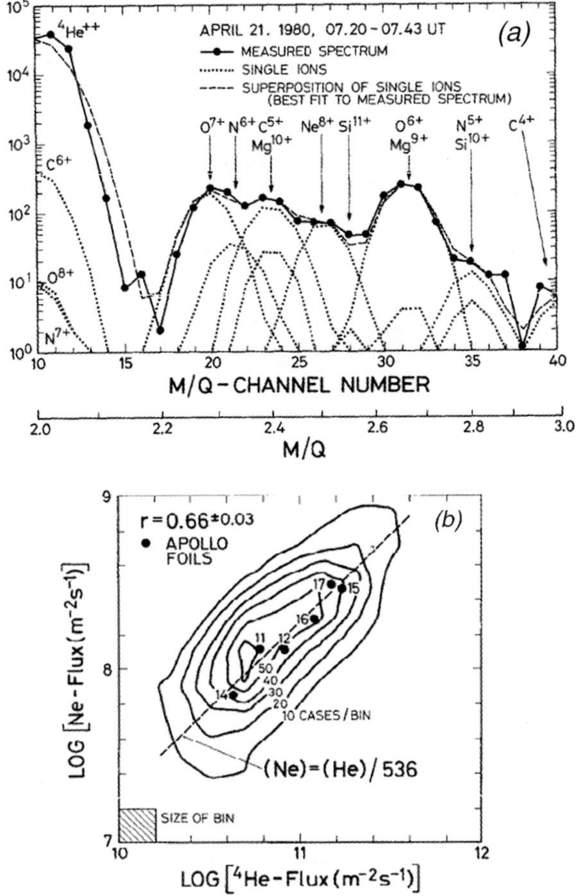


**Fig. 7** (**a**) Mass/charge spectrum from the ICI experiment. Prominent peaks of $^4He^{++}$, $O^{7+}$, $O^{6+}$, ($C^{5+} + Mg^{10+}$), and $Ne^{8+}$ (*dotted curves*) are clearly separated. All of the individual ion peaks are adjusted (based on considerations of freezing-in temperatures) until the best fit (*dashed curve*) to the measured spectrum (*filled circles*) is obtained. (From Kunz et al. 1983.) (**b**) Contour plot of the Ne versus He fluxes. The six Apollo SWC measurements are indicated by filled circles. (From Bochsler et al. 1986)

SWICS it was finally possible to measure not only the elemental but also the charge state composition of over a dozen ion species ($^1H$, $^4He$, $^3He$, $^{12}C$, $^{14}N$, $^{16}O$, $^{20}Ne$, $^{24}Mg$, $^{28}Si$, $^{32}S$, $^{36}Ar$, $^{40}Ca$ and $^{56}Fe$), and do this for all solar wind flows, from speeds of $\sim 150$ to more than 2,000 km/s and all thermal speeds. For a combined time period of over 25 years, SWICS instruments on ACE and Ulysses measured the solar wind composition in and near the ecliptic plane and extending to heliographic latitudes of $\sim 80°$ north and south, from 1 to 5.4 AU and over almost two solar cycles. This unique combined data set will not be replaced in the near future and will be analyzed for many years to come.

The different solar wind conditions encountered by Ulysses during the first four years of observations with SWICS are illustrated in Fig. 8 (Geiss et al. 1995a). The solar wind He speed, ($V_\alpha$, top panel) changes from the typical $\sim$400 km/s variable "slow wind" (red bar) at low latitudes to a smooth "high-speed stream" of $\sim$800 km/s at latitudes above $\sim 30°$ (blue bar). During 10 months in between (yellow bar), Ulysses periodically went in and out of the high-speed stream (HSST) from the large, southern polar coronal hole at solar minimum. The temperature at which the charge states freeze-in in the solar corona, the so-called freezing-in temperature also changes dramatically from the slow solar wind, where it is around $1.7 \times 10^6$ K and highly variable, to the fast solar wind, where it drops to $\sim 1.2 \times 10^6$ K and is very steady.

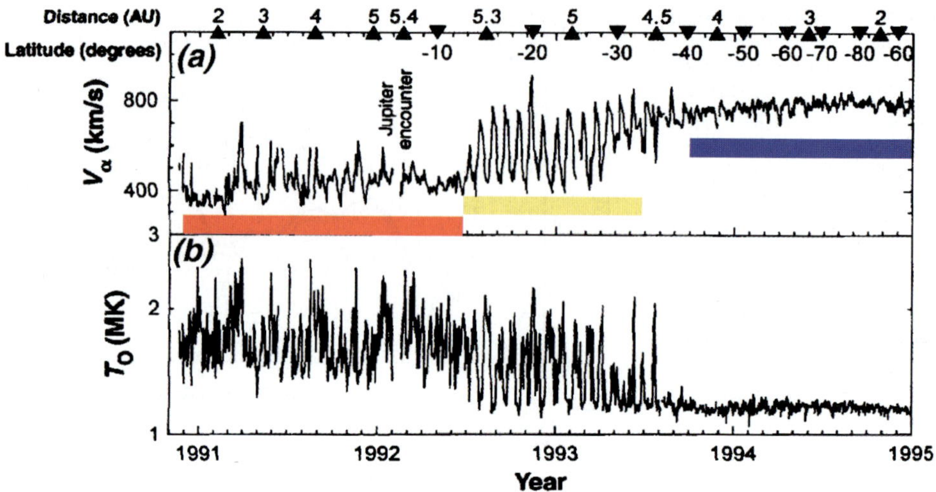


**Fig. 8** (**a**) Solar wind He speed, $V_\alpha$, and (**b**) the oxygen freezing-in temperature, $T_O$, measured by SWICS during the first four years of observations by Ulysses. Distance from the Sun and heliographic latitudes are shown on top. $T_O$ is calculated from the $O^{7+}/O^{6+}$ ratio assuming local equilibrium between electron impact ionization and recombination

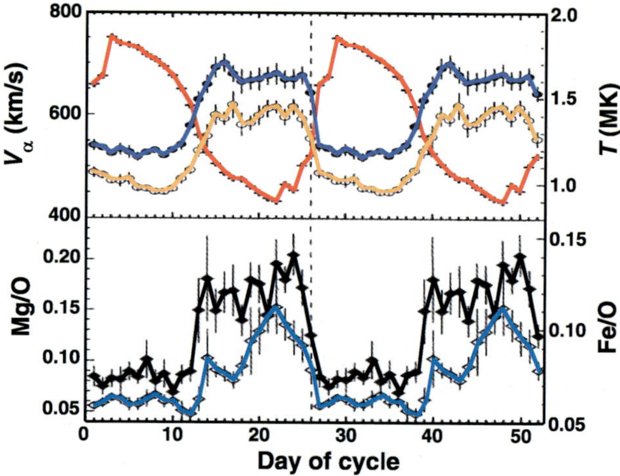


**Fig. 9** Superposed epoch plot of SWICS data showing the solar wind He speed, $V_\alpha$ (*red curve*), the coronal freezing-in temperature of O (*blue*) and C (*orange*), and the abundance ratios Mg/O (*dark red*) and Fe/O (*light blue*) from day 191 of 1992 to day 98 of 1993. Ulysses went regularly in and out of the HSST every solar rotation. The data are repeated after the dotted line to show the entire pattern. Uniform, steady and low freezing-in temperatures as well as low Mg/O and Fe/O ratios characterize the HSST

Geiss et al. (1995a) in their beautiful paper discussing these important new measurements of the composition and other properties of the high-speed solar wind from the southern polar coronal hole, reported their discovery that the HSST was confined by a sharp boundary that extended to the corona, where the charge states of the ions are frozen in, and then even deeper to the chromosphere, where the composition is established. Their plot, shown in Fig. 9, illustrates their findings most convincingly. It is among the most-often shown and reproduced by others. Not only is the solar wind bulk speed anti-correlated with the freezing-in temperature of both O and C (the latter being always lower), but also the transition from slow wind to the HSST (dashed vertical line) is quite abrupt, indicating a sharp boundary in the corona at several solar radii where the charge states freeze in. The Mg/O as well as the Fe/O ratios, shown in the lower panel of Fig. 9, are also anti-correlated, with a sharp transi-

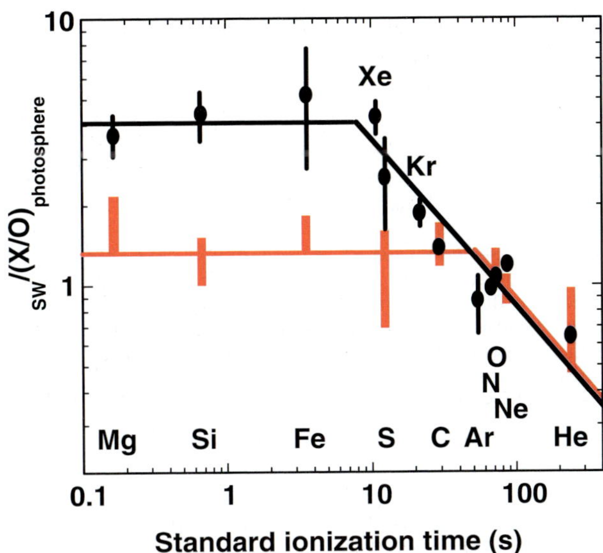


**Fig. 10** Elemental abundances relative to O in the slow wind (*blue circles*) and in the southern HSST (*red bars*). The solar wind abundance ratios are normalized to the photospheric abundances and plotted as a function of standard ionization time (see Sect. 3.5). The slow wind data also incorporates observations of Gloeckler and Geiss (1989), Bochsler (1992) and Geiss et al. (1994b). Abundances of Kr and Xe were derived from particles trapped in the lunar soil (Wieler et al. 1993)

tion at the same boundary. This boundary is in the chromosphere, close to the solar surface, where the composition "freezes in". Geiss et al. (1995a) argued that the existence of this common boundary indicated that conditions in or even below the chromosphere determine the supply of energy into the corona, and suggested that discussions of solar wind origin should include chromospheric as well as coronal processes.

### *3.4.1 Elemental Abundance*

Figure 10 contrasts the elemental composition Geiss et al. (1995a) observed in the HSST (red data points) with that of the typical slow solar wind (blue data points). They found that in the HSST the abundance of elements with short standard ionization times (SIT $< \sim 70$ s) is only slightly higher than that of corresponding elements in the photosphere. Indeed, with the recently revised and higher photospheric abundance of O (Asplund et al. 2005) the two compositions are the same, within errors, except for He and Ne, as confirmed also by Gloeckler and Geiss (2007). In the slow solar wind, on the other hand, Mg, Si, Fe and Xe (with SIT $< \sim 10$ s) are overabundant by about a factor of 3 to 4, C and Kr by about 2 to 3, N and Ne have about the same abundance, while Ar and He are less abundant compared to the photospheric composition. If one normalizes both the solar wind and photospheric element ratios to Fe instead of O, as was done by Gloeckler and Geiss (2007), then one concludes that the transition where elements become progressively less abundant compared to their photospheric abundances is at SIT $> \sim 10$ s in the slow solar wind, while in the HSST this transition occurs at SIT $> \sim 70$. Thus, whatever process causes the ionized gas in the chromosphere to separate from the neutral gas and then escape, is at about an order of magnitude faster in the region of origin of the slow wind compared to the polar coronal hole regions.

### *3.4.2 Charge States and Coronal Electron Temperatures*

Ion fraction distributions, such as those shown in Figs. 11 and 12 are important indicators of electron temperatures in the corona, and allow one to make estimates of temperatures

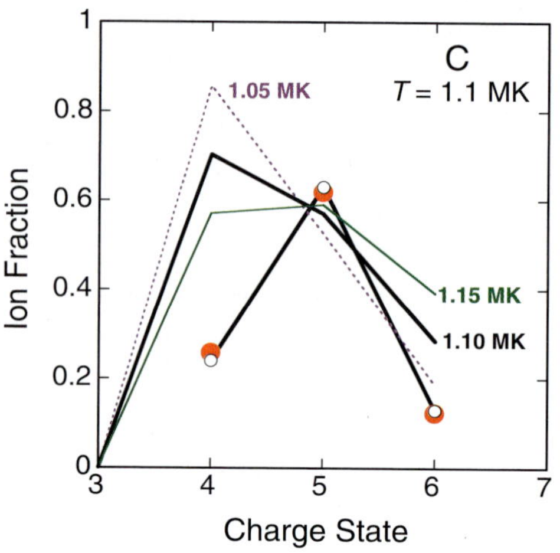


**Fig. 11** Ion fraction distribution (or charge spectrum) for C observed with SWICS in the HSST (*red filled circles*) compared with equilibrium freezing-in distributions (Mazzotta et al. 1998) for temperatures of 1.05, 1.1 and 1.15 MK (*dotted, black and green lines*, respectively). *Open circles connected by blue lines* represent the distribution predicted by Bürgi and Geiss (1986). While none of the equilibrium freezing-in distributions fit the observations, the closest fit is for $T = 1.1$ MK

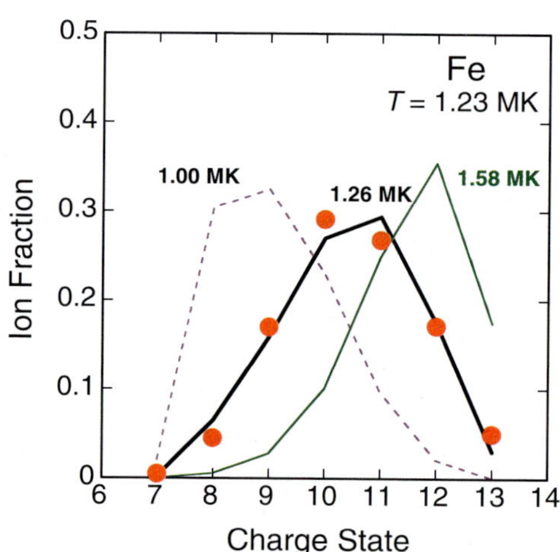


**Fig. 12** Same as Fig. 11 except for Fe. Comparing the observed charge spectrum with equilibrium freezing-in distributions for temperatures of 1.00, 1.26 and 1.58 MK (*dotted, black and green lines*, respectively) rules out major contributions from temperatures other than within a narrow range of 1.26 MK

and temperature gradients in coronal regions. In the HSST a single freezing-in temperature represents well the measured ion fraction distributions not only for Fe as shown in Fig. 12 but also for all other elements except C (see Figs. 7 and 8 in Gloeckler and Geiss 2007). This implies that electron temperatures at the freezing-in altitude are homogeneous, down to a small lateral scale. For C (Fig. 11) all equilibrium distributions are broader than the observed charge spectrum. However, the predicted ion fraction distribution based of the wave driven coronal hole model (model 5) of Bürgi and Geiss (1986), discussed in Sect. 3.5.2, fits the observed HSST carbon ion fraction distribution remarkably well. The freezing-in temperatures of all elements in the HSST, while low ($\sim$1.1–$\sim$1.3 MK), were not exactly the same (Figs. 11–13) reflecting a temperature gradient in the corona where the different charge states freeze in at different altitudes.

## 3.5 Solar Wind Models and Solar Processes

Geiss would not rest until he could explain his observations. If no satisfactory theories or models could be found, he simply went ahead and developed his own. Thus, Geiss con-

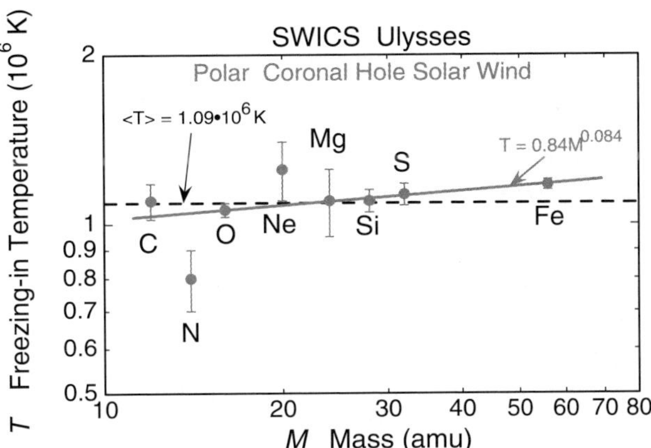


**Fig. 13** Freezing-in temperature, derived from ion fraction distributions of C, N, O, Ne Mg, Si, S and Fe observed in the HSST, as a function of ion mass. (From Gloeckler and Geiss 2007)

tributed much to our understanding of solar wind acceleration as well as of processes in the solar chromosphere and corona that determine the composition and charge state distributions of the solar wind (e.g., Geiss 1982). He introduced the "coulomb drag factor" $\Gamma = Q^2/(2M - Q - 1)$ for ion-on-ion collisions to account for the acceleration of heavy ions in the solar wind by momentum transfer from protons by coulomb collisions (Geiss et al. 1970). He also introduced the concept of "standard ionization time" (SIT) also referred to as "first ionization time" (see, e.g., Geiss and Bochsler 1985, and references therein) as a more physical parameter than "first ionization potential" (FIP). Furthermore he calculated SITs for various solar wind elements using realistic solar conditions, and used SIT to organize solar wind composition data in both the slow solar wind the HSST (see Fig. 10). He, with his students, then developed models to explain the observed enrichment of low-FIP elements in the solar wind compared to photospheric abundances (von Steiger and Geiss 1989), as well as realistic expansion models for the solar wind that included both major species $(p, \alpha, e)$ and minor ions (e.g. $^3$He, C, O, Mg, Si) for both the slow and coronal hole solar wind (Bürgi and Geiss 1986).

### *3.5.1 Systematic Fractionation of Solar Wind Elemental Abundance*

In order to understand the observed overabundance of low-FIP elements in the solar wind and solar energetic particles, von Steiger and Geiss (1989) investigated models in which fractionation is achieved by ion-atom separation across magnetic field lines in the upper chromosphere. They envisioned an initially neutral gas mixture driven across the field by gravity and density gradients, at the same time that the gas was being ionized, predominantly by ions trapped in the field and UV radiation. Since this mechanism works only before the gas is completely ionized, on time scales of less than few minutes, relatively small length scales are involved. Predictions of their model were in excellent agreement with observations (e.g. Fig. 10) in the case of a slab structure of width less than a few tens of kilometers for the slow wind.

### *3.5.2 Charge States and Dynamics of Solar wind He and Minor Ions*

Bürgi and Geiss (1986) developed dynamic solar wind expansion models and calculated charge state distributions and velocity profiles as a function of heliocentric distance. Their models use observed coronal density profiles and take into account the finite abundance of

He, and different speeds for different elements. Predictions of their model 5, which includes a super-radial expansion geometry and direct momentum transfer from waves, for charge states of C (Fig. 11) and O agree quite well with SWICS observations in the HSST. For Si and Fe, Bürgi and Geiss (1986) used recombination and ionization coefficients that have since been revised. With new coefficients (Mazzotta et al. 1998) predicted and observed charge spectra are in reasonable agreement for most of the other elements.

## 4 Pickup Ions and the Interstellar Gas

Interstellar pickup ions were predicted to exist long before they were discovered. Detecting them with conventional solar wind instruments was not possible and could only be done using low-background TOF-based spectrometers such as SWICS that could measure the composition of the suprathermal pickup ions. Of all the interstellar pickup ions only $He^+$ survives in quantities large enough to be detected at 1 AU, and, indeed, Möbius et al. (1985) were first to detect it at 1 AU with an instrument modeled after SWICS. Geiss and his co-workers discovered all the other interstellar pickup ions with the SWICS instrument on Ulysses, and had the launch of Ulysses not been delayed by almost a decade, would have also discovered $He^+$.

### 4.1 Interstellar Pickup $H^+$ and $He^+$

Gloeckler, Geiss and colleagues reported their discovery of interstellar pickup hydrogen and detection of pickup $He^+$ at 4.8 AU (Gloeckler et al. 1993). Pickup $H^+$ and $He^+$ were identified and determined to be of interstellar origin by their distinct velocity distributions, which drop abruptly at twice the local solar wind speed (Fig. 14). These measurements were then used to infer the density (see Table 3) and pressure of the gas in the local interstellar cloud (LIC), place limits of ~1–2 μG on the strength of the magnetic field (Gloeckler et al. 1997) and led to estimates of acceleration efficiencies for anomalous cosmic ray (ACR) hydrogen and helium (e.g., Cummings and Stone 2007).

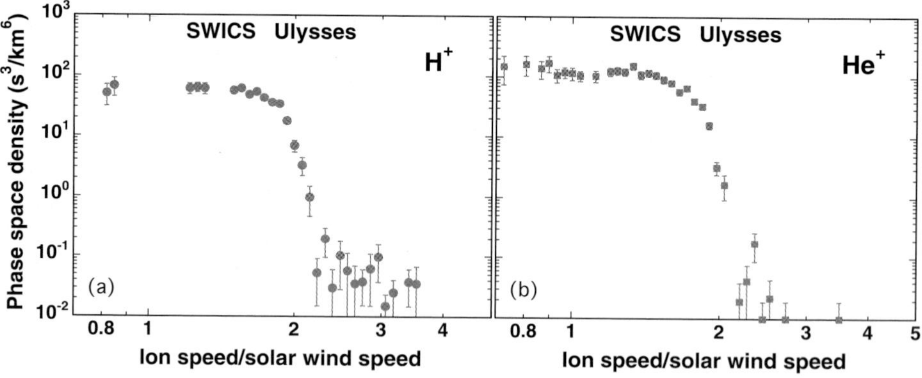


**Fig. 14** Phase space density plotted against normalized ion speed, $W = V_{ion}/V_{sw}$, of (**a**) interstellar pickup hydrogen and (**b**) pickup helium measured in the Ulysses frame of reference with SWICS during the time period 24 November to 9 December 1991. Because energy/charge steps and sectors corresponding to solar wind velocities of $H^+$ and $He^{++}$ have been excluded from these data, the phase space density of pickup $H^+$ around $W \approx 1$ and $W \approx 1.4$ is not available

**Table 3** Densities of interstellar atoms at the termination shock (95 AU) and of atoms and ions in the local interstellar cloud (LIC)

| Element or isotope | Termination shock ($cm^{-3}$)[a] | Filtration factor[b] | Neutral fraction[c] | Local interstellar cloud ($cm^{-3}$) Atoms | Ions | Grains | Total |
|---|---|---|---|---|---|---|---|
| H | $0.103 \pm 0.010$[a] | $0.55 \pm 0.06$ | $0.806 \pm 0.014$ | 0.186 | 0.045 | 0 | 0.231 |
| $^4$He | $0.0143 \pm 0.0014$ | $0.94 \pm 0.02$ | $0.659 \pm 0.026$ | 0.0152 | 0.0079 | 0 | 0.0231 |
| $^3$He | $(3.15 \pm 1.2) \times 10^{-6}$ | $0.94 \pm 0.02$ | $0.659 \pm 0.026$ | $3.35 \times 10^{-6}$ | $1.73 \times 10^{-6}$ | 0 | $5.08 \times 10^{-6}$ |
| N | $(5.47 \pm 1.37) \times 10^{-6}$ | $0.76 \pm 0.08$ | $0.675 \pm 0.006$ | $7.20 \times 10^{-6}$ | $3.5 \times 10^{-6}$ | $2 \times 10^{-6}$ | $1.27 \times 10^{-5}$ |
| O | $(4.82 \pm 0.53) \times 10^{-5}$ | $0.70 \pm 0.11$ | $0.831 \pm 0.012$ | $6.88 \times 10^{-5}$ | $1.4 \times 10^{-5}$ | $4 \times 10^{-5}$ | $1.23 \times 10^{-4}$ |
| Ne | $(5.82 \pm 1.16) \times 10^{-6}$ | $0.88 \pm 0.04$ | $0.206 \pm 0.025$ | $6.61 \times 10^{-6}$ | $2.5 \times 10^{-5}$ | 0 | $3.20 \times 10^{-5}$ |
| Ar | $(1.63 \pm 0.73) \times 10^{-7}$ | $0.64 \pm 0.11$ | $0.295 \pm 0.027$ | $2.55 \times 10^{-7}$ | $6.1 \times 10^{-7}$ | 0 | $8.63 \times 10^{-7}$ |

[a]Loss and production (equal to loss) rates used to derive abundances at the termination shock (all in units of $10^{-7}\ s^{-1}$) are 6.96, 1.15, 5.66, 7.64, 3.53, and 8.86 for H, He, N, O, Ne and Ar respectively

[b]For H, based on present work; for all other elements we used results of Cummings et al. (2002)

[c]Results of model 25 of Slavin and Frisch (2002)

[d]Based on measurements of accelerated pickup ions in the heliosheath with the LECP instrument on Voyager 1 (courtesy M.E. Hill)

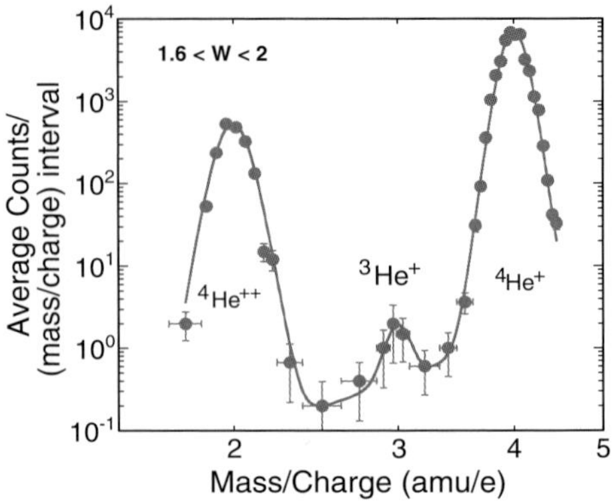

**Fig. 15** Plot of the average triple coincidence counts of ions with masses between 2 and 6 amu and $1.6 < W < 2$ versus their mass/charge, where $W$ is the ion speed divided by the simultaneously measured solar wind speed (both in the spacecraft frame of reference). The peak around mass/charge of 3 is due to pickup $^3He^+$. (From Gloeckler and Geiss 1996)

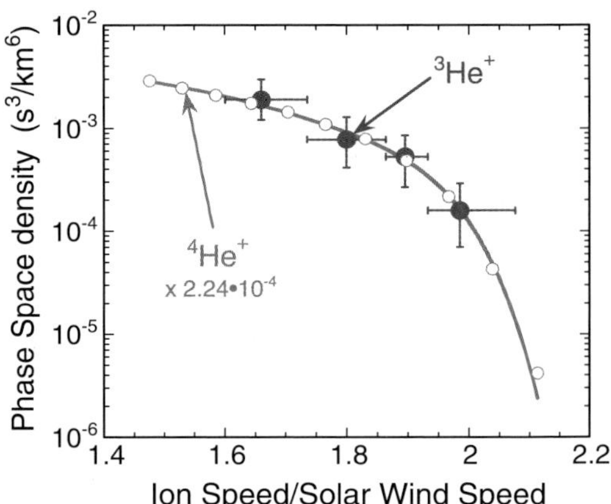


**Fig. 16** Measured distribution function, $F_m(W)$, of interstellar pickup $^3He^+$ (*filled circles*) and $^4He^+$ (*open circles*) versus ion speed/solar wind speed. $F_m(W) = 4{,}541(r_m/\eta_m) \times (438/V_{SW})^4 W^{-4}$ where $r_m$ is the count rate and $\eta_m$ the detection efficiency of $^3He^+$ ($m = 3$) and $^4He^+$ ($m = 4$) respectively. The similarity of the spectral shapes indicates interstellar origin for $^3He^+$. (From Gloeckler and Geiss 1996)

## 4.2 Interstellar Pickup $^3He^+$

Gloeckler and Geiss (1996) were the first to determine the present day interstellar $^3He/^4He$ ratio by measuring the abundance of the very rare $^3He^+$ pickup ions with SWICS on Ulysses. Because all known physical processes that transform interstellar helium into ionized helium (pickup helium) inside the heliosphere are believed to affect $^3He$ and $^4He$ in similar fashion, the pickup $^3He^+/^4He^+$ ratio $= \{2.2^{+0.8}_{-0.7}$ (stat.) $\pm$ 0.2 (sys.)$\}$ measured close to the Sun represents the local interstellar $^3He/^4He$ abundance. This pioneering measurement (Figs. 15 and 16) was used to set improved estimates of the cosmologically important baryon/photon ratio in the universe and placed new constraints on models of Galactic chemical evolution.

## 4.3 Interstellar Oxygen, Nitrogen and Neon

Geiss et al. (1994a) discovered pickup oxygen, nitrogen and neon ions of interstellar origin using data from SWICS on Ulysses (Fig. 17a). The interstellar origin of these singly charged pickup ions was established by their broad velocity distribution, with an upper limit of twice the solar wind speed (Fig. 17b), and the variations of their relative abundance as a function of distance from the Sun (Fig. 17c) that corresponded to theoretical expectations. Recent analysis by Gloeckler and Geiss of SWICS data recorded over many subsequent years has provided good estimates of the chemical composition of the interstellar gas (see Table 3).

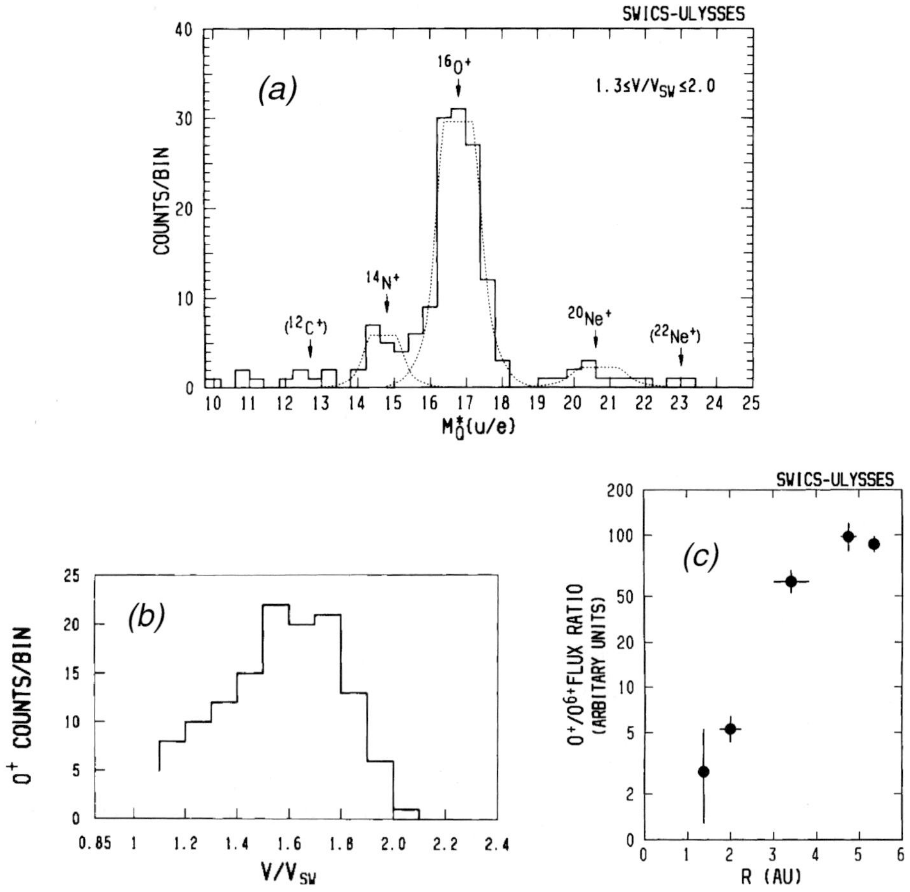

**Fig. 17** (**a**) Mass/charge spectrum of singly charged heavy ions with $1.3 \leq W \leq 2$ ($W$ is the ion speed/solar wind speed) measured with SWICS during 139 days between 4.56 and 5.40 AU from the Sun. The positions of the interstellar ions marked by arrows are derived from calibration data. $M_Q^*$ is a standardized $M/Q$ scale which corresponds closely to the mass/charge ratios of solar wind ions. (**b**) Speed distribution of $O^+$ ions showing a cutoff at twice the solar wind speed. (**c**) $O^+/O^{6+}$ flux ratio as a function of distance from the Sun. The strong radial increase of this ratio demonstrates that $O^+$ is of non-solar origin

## 5 Deuterium and $^3$He Since the Big Bang

In the early 1970s Johannes Geiss and Hubert Reeves puzzled over the fate of deuterium since the Big Bang. They argued that since D is destroyed in stellar interiors by nuclear burning and converted to $^3$He, the present day $^3$He, measured by Geiss' Apollo SWC experiments, represented the sum of D and $^3$He in the protosolar cloud $4.6 \times 10^9$ years ago. In Geiss and Reeves (1972) they derived a value of $(2.5 \pm 0.5) \times 10^{-5}$ for the protosolar D/H ratio, and concluded that the density of baryonic matter was smaller by about a factor of five than that required to close the universe.[2] Geiss and Gloeckler (2003) later showed

[2] In this same paper they also concluded that the high terrestrial D/H ratio (~8 times higher than the protosolar D/H) was due to chemical fractionation. (See Reeves 2007)

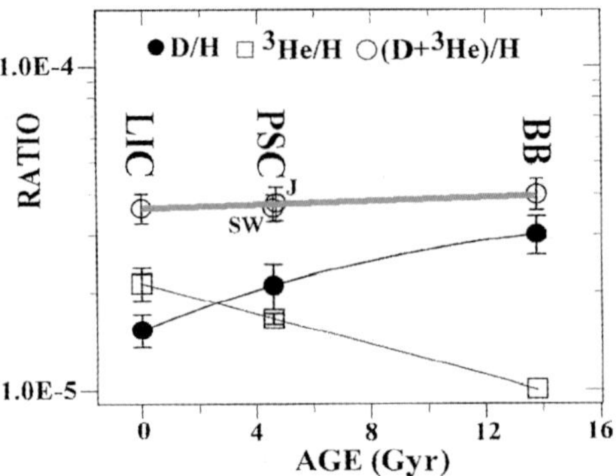


**Fig. 18** D/H, $^3$He/H and (D + $^3$He)/H in the LIC (Local Interstellar Cloud), the PSC (Protosolar Cloud) and the BB (Big Bang) as a function of age (from Geiss and Gloeckler 2003). As expected from theories of nucleosynthesis and galactic evolution, $^3$He/H increases and D/H decreases with time since the Big Bang. The primordial D/H ratio is from O'Meara et al. (2001). The primordial $^3$He/H ratio is based on the results of Bania et al. (2002). For (D + $^3$He)/H in the PSC there are two independent values, one from solar wind (SW) data (Gloeckler and Geiss 2000) and the other from Jupiter (J) data (Lellouch et al. 2001; Mahaffy et al. 1998). D/H in the LIC is from Linsky and Wood (2000) and $^3$He/H in the LIC is the average of the results of Gloeckler and Geiss (1996, 1998) and Salerno et al. (2003)

(see Fig. 18) that the (D + $^3$He)/$^4$He decreased by only ~10 percent during the ~14 Gyr lifetime of the universe. The fact that (D + $^3$He)/$^4$He stays nearly constant over time implies that the increase in $^3$He since the Big Bang is primarily the result of converting D into $^3$He in the early life of stars of all sizes, and that incomplete H-burning at intermediate depth of small stars (e.g., Olive et al. 1995) is relatively unimportant. Furthermore, as Geiss et al. (2004) pointed out, the (D + $^3$He)/$^4$He should be an important parameter in studies of galactic evolution and for evaluating claims of inhomogeneous or anomalous primordial nuclear production. The solar wind $^3$He/$^4$He ratio first measured by Geiss almost four decades ago has not lost its importance for astrophysics and cosmology. It is still the most direct measure of the sum of deuterium and helium-3 in the protosolar cloud.

## 6 Mixing Models and Late Galactic Evolution

From the composition of interstellar pickup ions observed with SWICS deep inside the heliosphere Gloeckler and Geiss (2001) derived the composition of the gas in the Local Interstellar Cloud (see Table 3 for the most recent estimates of the LIC composition). By comparing the elemental abundances in the Protosolar Cloud (PSC) with those in the LIC and nearby interstellar medium (ISM), Geiss concluded that the composition of matter in the solar ring of the Milky Way could not have evolved from matter with a PSC composition in a closed system environment. To account for the LIC composition Geiss et al. (2002) proposed that a significant fraction of the gas in the Galactic disk comes from quasi-continuous infall of low-metallicity material from dwarf galaxies and, using a simple mixing model, showed that the difference found in the LIC and PSC compositions can be explained my this process (see Fig. 19a). This also resolved some previously unexplained compositional anomalies in the Galaxy, such as the apparently high metallicity of the Sun and the "$^{18}$O puzzle". Geiss et al. (2007) refined their mixing model by include oxygen isotopes (see Fig. 19b) and found that a small (0.3%) addition of Intermediate Mass Stars material, rich in $^{17}$O, was required to explain the ISM $^{18}$O/$^{17}$O ratio. With this scenario (model 2 in Fig. 19b) all ratios, including

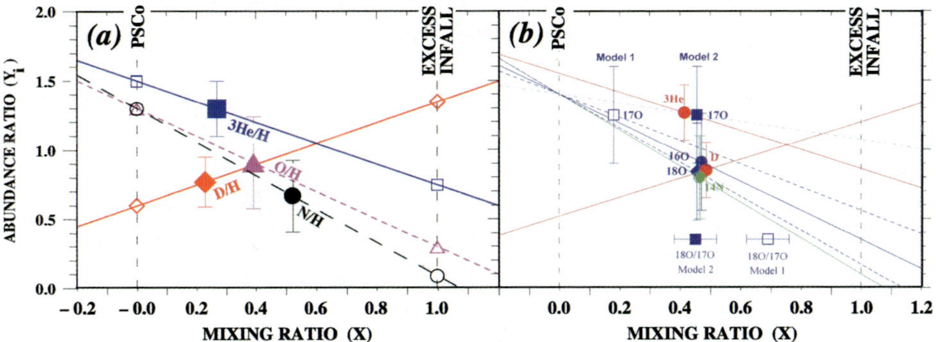


**Fig. 19** (**a**) Abundance of D, $^3$He, N and O (relative to H) observed in the LIC (*filled symbols*) vs. mixing ratios $X$, where e.g. X(O/H) = $[\{O/H\}_{PSCo} - [\{O/H\}_{LIC}]/[\{O/H\}_{PSCo} - [\{O/H\}_{excess\ infall}]$ and abundance ratios in braces are relative to PSC abundance ratios. Also shown are the mixing lines connecting the values (*open symbols*) of a given ratio in the two reservoirs, PSCo and excess infall (from Geiss et al. 2002). (**b**) Same as Fig. 19a but now also including ISM $^{17}$O and $^{18}$O and somewhat revised values for LIC, PSCo and excess infall ratios (from Geiss et al. 2007). Model 2 data, which includes a 0.3% contribution from Intermediate Mass Stars, are shown as *filled symbols*, and model 1 data, with infall only from dwarf galaxies, as *open symbols*. Except for $^{17}$O and $^{18}$O/$^{17}$O, the mixing ratios obtained by the two models are almost identical and thus model 1 data are hidden behind the model 2 data. See Geiss et al. (2004, 1992) for a complete discussion

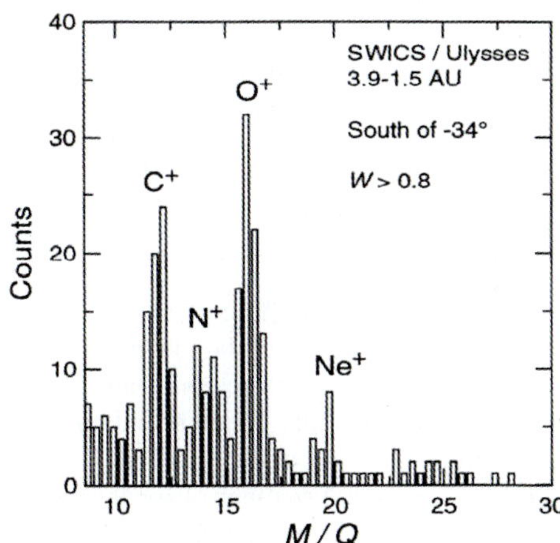


**Fig. 20** Mass/charge spectrum of singly charged heavy ions with speeds above 0.8 times the solar wind speed, $V_{SW}$, in the high-speed solar wind ($V_{SW} > 600$ km/s) from the southern polar coronal hole. A similar spectrum was observed in the northern high-speed stream

$^{18}$O/$^{17}$O, fell within a narrow range of mixing ratios (0.40 to 0.51). Further improvements of the mixing model are detailed in Geiss and Gloeckler (2007).

## 7 The Inner Source Pickup Ions

Geiss' discovery of "Inner Source" (a term he coined) pickup ions came as a big surprise. He made the discovery during one of his visits to the United States, working in our (GG's) house in Maryland. Using a list of singly ionized heavy ions prepared by Chris Gloeckler from the SWICS pulse-height data, Johannes hand-plotted the mass/charge distribution of ions with speeds above 0.8 times the solar wind speed (see Fig. 20) and found, to his surprise, an

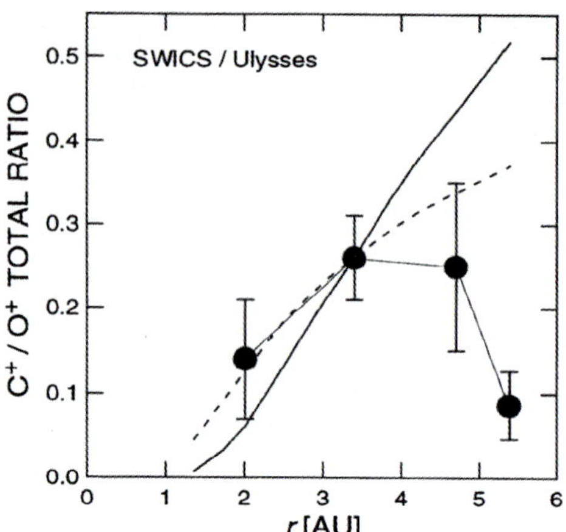


**Fig. 21** Ratio of the number densities of $C^+$ and $O^+$ measured near the ecliptic plane (*filled circles*). The two theoretical curves (normalized to the 3.4 AU ratio) show the radial dependence that would result if the source of these ions were interstellar C and O. The *dashed curve* was computed using ionization rates of $6 \times 10^{-7}$ $s^{-1}$ and $9 \times 10^{-7}$ $s^{-1}$ for O and C respectively while for the *solid curve* the C ionization rate was increased to $1.2 \times 10^{-6}$ $s^{-1}$. (From Geiss et al. 1995b)

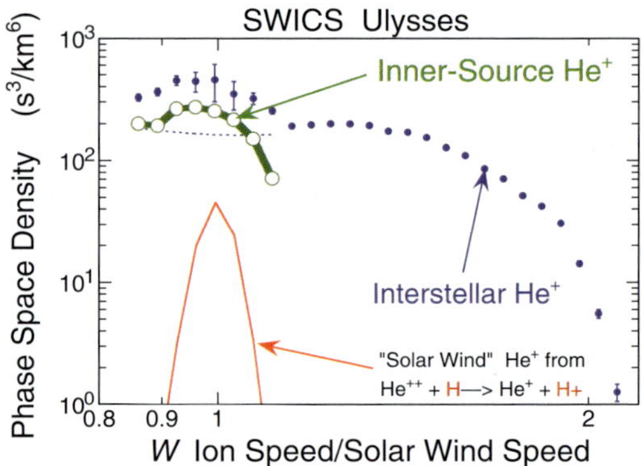


**Fig. 22** Velocity distribution of $He^+$ at ~1.4 AU near the ecliptic plane showing the narrow (cooled) inner-source (*green*) and the broad interstellar (*blue*) components

unexpectedly high abundance of singly ionized carbon. The detected $C^+$ was not expected to be of interstellar origin since carbon is mostly ionized in the LIC and only a small fraction enters the heliosphere as neutral carbon, and indeed Johannes (Geiss et al. 1995b, 1996) quickly ruled out an interstellar origin based on the radial dependence of the $C^+/O^+$ density ratio (Fig. 21) and source strengths $q(r)$ for $C^+$ and $O^+$. Also ruled out, as main sources for the $C^+$ were one or a small number of local point sources, such as comets, or asteroid belt objects because $C^+$ was observed at high as well as low latitudes. The most likely source of $C^+$ observed outside the ecliptic plane was thought to be interstellar grains. The estimated total production rate of C, N and O from this source was estimated to be ~4 tons/s, small compared to the solar wind (~$10^4$ tons/s) and interstellar gas (~$2 \times 10^3$ tons/s), but large compared to the production of O by Mars and Io's torus.

The subsequent discovery of inner-source neon (e.g., Gloeckler et al. 2001) and helium (see Fig. 22) posed additional constraints on possible sources for these ions. Since Ne and especially He are not likely to be trapped in grains, at least these ions had another origin. Based on this and given the observed highly cooled velocity distributions (see Fig. 22) as well as the solar-wind-like abundance of inner source $H^+$, $He^+$, $C^+$, $N^+$, $O^+$ and $Ne^+$ (see Fig. 23), Gloeckler et al. (2000a) and Schwadron et al. (2000) proposed that an additional important source of inner-source pickup ions was the solar wind that was first absorbed by interplanetary dust grains orbiting close to the Sun, and then released as slow-moving ions or neutrals that were then quickly ionized by charge exchange or photoionization. Geiss'

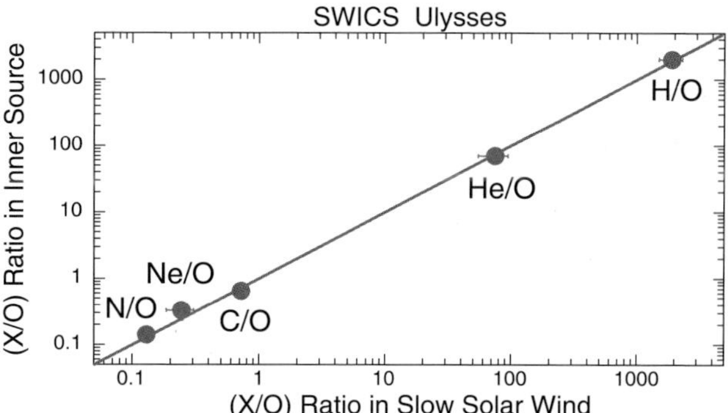


**Fig. 23** Ratios of inner-source H, He, N, C and Ne relative to O measured near the ecliptic versus the corresponding ratios in the slow solar wind. The two compositions are identical

pioneering discovery of inner source pickup ions stimulated much research on their production mechanisms. Allegrini et al. (2005), discussing these models, concluded that no single mechanism could explain all of the observations. Schwadron and Gloeckler (2007) discuss our current understanding of the origin of inner-source pickup ions.

## 8 Composition of Comets and Magnetospheres

Johannes Geiss explored the composition of comets Giacobini–Zinner, Halley, Hyakutake and McNaught–Hartley, and of the terrestrial and Jovian magneto-spheres. Some of his key results are summarized in the following.

### 8.1 Comets Halley, Hyakutake and McNaught–Hartley

In their search for rare molecules and radicals in the outer coma of P/Halley, Geiss et al. (1999) developed a new model for interpreting ion data obtained by the Giotto Ion Mass Spectrometer. They placed an upper limit of $1.5 \times 10^{-3}$ for $Ne/H_2O$ and of $7 \times 10^{-7}$ for $Na/H_2O$. Based on their abundance measurements of methyl and ethyl cyanide and upper limits of N-bearing species, they confirmed that nitrogen is depleted in the Halley material. Because their upper limit of $C_3H$ was below their measured abundance of $C_4H$, as is the case in interstellar molecular clouds, they suggested that $C_4H$ in Halley was synthesized under molecular cloud conditions.

The serendipitous detection of pickup ions in the distant ion tails of comets Hyakutake (Gloeckler et al. 2000b; see Fig. 24) and McNaught–Hartley (Gloeckler et al. 2004) enabled Geiss and colleagues to discover the puzzling power law dependence of $H_2O^+/O^+$ and $C^+/O^+$ with distance from the center of the comet tail, and the similarity of the composition between these comets and Halley (Fig. 25). Surprisingly, doubly charged ions were also discovered (Fig. 24).

### 8.2 Ion Compositions of Magnetospheres of Earth and Jupiter

Using data obtained with the Ion Composition Experiment (ICE) on GEOS-1, Geiss et al. (1978) studied the main sources of magnetospheric plasma as a function of location and geomagnetic activity, and found, quite commonly, a mixture of keV/e ions of both solar wind ($He^{2+}$, $^3He$) and atmospheric ($O^+$, $H^+$) origin. These two independent sources gave, on average, comparable contributions, although in a single injection event one or the other source

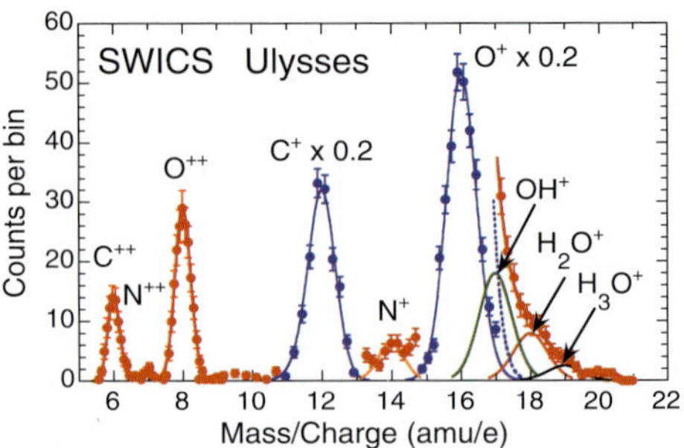


**Fig. 24** Mass/charge spectrum of ions in Hyakutake's tail. Singly and doubly charged C, N and O are resolved. Counts in $m/q$ bins above ~17 amu/e are well above the Gaussian fit (*dotted curve*) for $O^+$, and correspond to mass 17, 18 and 19 amu ions with Gaussian distributions whose sum (*solid curve*) fits the data (*red solid curve*)

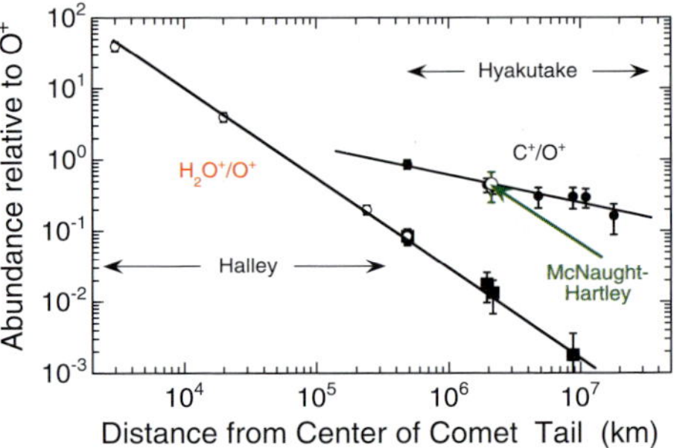


**Fig. 25** $H_2O^+/O^+$ and $C^+/O^+$ ratios versus distance form center of comet tails of P/Halley (Altwegg et al. 1993), Hyakutake and McNaught–Hartley. The power law behavior is not well understood, but the decrease with distance is expected as complex molecules become dissociated

dominated. Geiss et al. (1978) also found that the thermal (~1 eV/e) plasma population contained six ion species ($^1H^+$, $D^+$, $^4He^{2+}$, $^4He^+$, $^{16}O^{2+}$ and $^{16}O^+$), three of which ($D^+$, $He^{2+}$ and $O^{2+}$) were previously unknown. They investigated various production, loss and transport mechanisms using elemental composition and charge states to identify the sources of magnetospheric ion species. Atmospheric oxygen exists as $O^+$ and some $O^{2+}$ in the magnetosphere and solar wind oxygen as $O^{6+}$ and $O^{7+}$. The unexpected high abundance of $O^{2+}$ in the plasmasphere and magnetosphere (Young et al. 1977) was attributed to preferential thermal diffusion of $O^{2+}$ from the ionosphere (Geiss and Young 1981). Systematic variations in magnetospheric ion composition were also found. At times of high solar activity the $O^+$ and $O^{2+}$ abundances in the magnetosphere, that is, the supply of ionospheric gases, increased dramatically (Stokholm et al. 1989).

The Ulysses' flyby of Jupiter provided the opportunity to determine for the first time the elemental and charge state composition of the Jovian plasma (see Fig. 26), and to explore the high-latitude dusk side of the magnetosphere. Geiss et al. (1992) observed hot tenuous plasma throughout the outer and middle magnetosphere. They found that while solar wind particles were observed in all regions of the magnetosphere investigated during the flyby, oxygen and sulfur ions with several different charge states, from the volcanic satellite Io, made the largest contribution to the mass density of the hot plasma, even at high latitudes. Ions from Jupiter's ionosphere were most abundant in the middle magnetosphere, particularly in the high-latitude region on the dusk side, which was previously unexplored (Fig. 27). In addition to the major Iogenic ions $O^+$, $O^{2+}$, $S^{2+}$, $S^{3+}$, small amounts of $Na^+$ and $K^{2+}$ were tentatively identified. Ions, identified to originate from four different sources, were $H^+$, $He^{2+}$ and $O^{6+}$ (solar wind), $H^+$, $H_2^+$ and $H_3^+$ (Jupiter's ionosphere), $O^+$, $O^{2+}$, $O^{3+}$, $O^{4+}$, $S^+$, $S^{2+}$, $S^{3+}$, $S^{4+}$, $S^{5+}$, $Na^+$, $K^{2+}$ (Io), and $H_2O^+$ and $H_3O^+$ (icy satellites). The source of $He^+$ could not be determined.

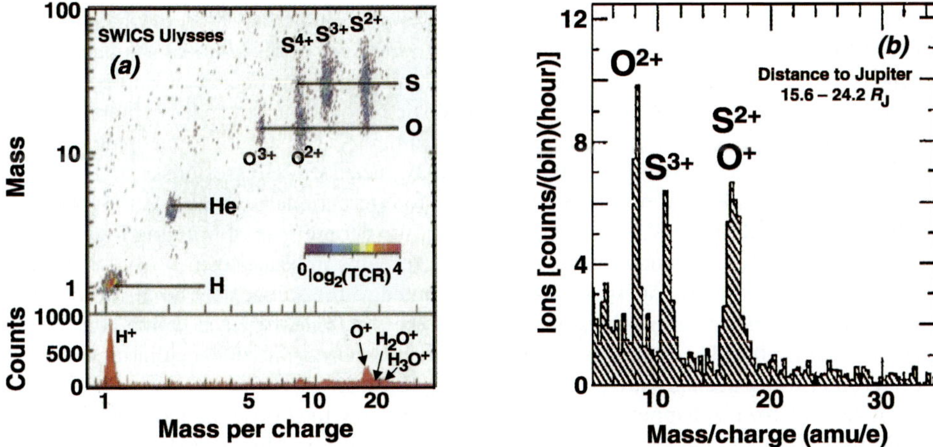


**Fig. 26** (**a**) Mass versus mass/charge matrix of Jovian magnetospheric plasma. The triple coincidence rates (TCR) are color-coded in the *upper panel*, while the double coincidence rates, for which no mass information is available, are given in the *histogram below*. (**b**) Mass/charge spectrum in the middle magnetosphere. Plotted are the double and triple coincidence PHA counts corrected for bias introduced by the priority scheme of SWICS

**Fig. 27** Flux ratios of $H^+/He^{2+}$ versus range in the Jovian magnetosphere. Two limiting values for $H^+/He^{2+}$ are shown, assuming either same energy/charge (*open circles*) or same energy/mass (*filled circle*) spectrum for the ions

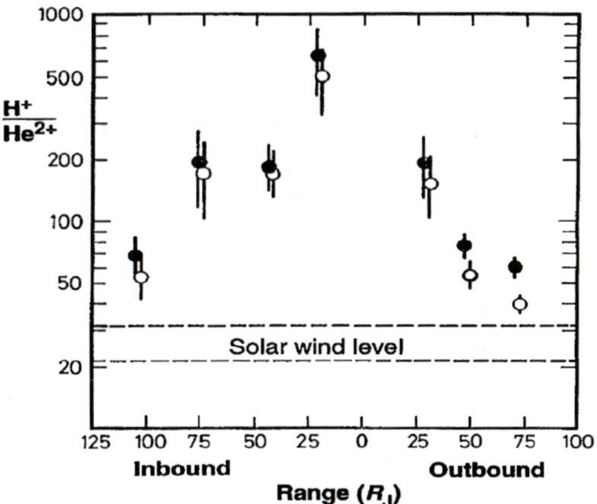


## 9 Concluding Remarks

There are certain themes that run through the distinguished career of Johannes Geiss. He has always recognized the powerful insights that are available from detailed measurements of the composition of matter. Such measurements reveal the basic processes that govern the Earth and its space environment, the Sun and the heliosphere, and the universe beyond. And he has the skill, as both an experimentalist and theoretician, to make and interpret the required observations.

There is another common theme in the career of Johannes Geiss—he is a true internationalist. He has a fundamental belief in the power of international cooperation in science, and that we can do so much more together than we can do separately. This belief is tempered by the reality of knowing how to make things happen, which in turn is based on a deep knowledge of our various political systems, and on human nature.

The battle to get the initial approval of the Ulysses mission in the 1970s, in which in large part Johannes Geiss was the European General, is a measure of his political acumen. It was a time of perceived differences in the capabilities and approaches between Europe and the US. Then, as now, there were underlying nationalistic motivations for space exploration that needed to be suppressed in the best interests of humanity. His ability to navigate through these political issues has brought us the marvelously successful Ulysses mission.

It is thus no wonder that his latest contribution to international science was the founding of the International Space Science Institute (ISSI), for the purpose of bringing together the world's best scientists to study the most relevant problems in space science. By his design, the books and research products of ISSI are attributed to the scientists who attend the various workshops and team meetings, regardless of their nationality. Each nation contributes support, as it is able. The scientists are thus free to collaborate and the peculiarities of their various sponsoring agencies are put constructively to work.

The lessons from Johannes Geiss' career abound: How to choose the most relevant scientific research area, and obtain the knowledge available in it. How to recognize and successfully pursue opportunities. How to lead the international pursuit of science by example.

An extraordinary career by an extraordinary person.

**Acknowledgements** This work was supported, in part, by NASA contract NAGR-10975, and by JPL contract 1237843.

## References


F. Allegrini, N.A. Schwadron, D.J. McComas, G. Gloeckler, J. Geiss, J. Geophys. Res. **110**, A05105 (2005). doi:10.1029/2004JA010847

K. Altwegg, H. Balsiger, J. Geiss, R. Goldstein, W.-H. Ip, A. Meier, M. Neugebauer, H. Rosenbauer, E. Shelley, Astron. Astrophys. **279**, 260–266 (1993)

M. Asplund, N. Grevesse, A.J. Sauval, in *COSMIC ABUNDANCES as Records of Stellar Evolution and Nucleosynthesis in Honor of David L. Lambert*, vol. 336, ed. by T.G. Barnes, III, F.N. Bash (Astronomical Society of the Pacific, 2005), p. 25

T.M. Bania, R.T. Rood, D.S. Balser, Nature **415**, 54 (2002)

P. Bochsler, J. Geiss, S. Kunz, Sol. Phys **103**, 177–201 (1986)

P. Bochsler, in *Solar Wind Seven*, ed. by E. Marsch, R. Schwenn, COSPAR Colloquia Series (Pergamon, New York, 1992), p. 323

A. Bürgi, J. Geiss, Sol. Phys. **103**, 347–383 (1986)

M.A. Coplan, K.W. Ogilvie, P. Bochsler, J. Geiss, Sol. Phys. **93**, 415 (1984)

A.C. Cummings, E.C. Stone (2007), this volume, doi: 10.1007/s11214-007-9161-y

A.C. Cummings, E.C. Stone, C.D. Steenberg, Astrophys. J. **578**, 194–210 (2002)

J. Geiss, Space Sci. Rev. **33**, 201–217 (1982)

J. Geiss, P. Bochsler, in *Proc. Rapports Isotopiques dans le Système Solaire* (Cepadues-Editions, Paris, 1985), pp. 213–228

J. Geiss, G. Gloeckler, Space Sci. Rev. **84**, 239–250 (1998)

J. Geiss, G. Gloeckler, Space Sci. Rev. **106**, 3–18 (2003)

J. Geiss, G. Gloeckler (2007), this volume, doi: 10.1007/s11214-007-9235-x

J. Geiss, H. Reeves, Astron. Astrophys. **18**, 126 (1972)

J. Geiss, H. Reeves, Astron. Astrophys. **93**, 189–199 (1981)

J. Geiss, D.T. Young, J. Geophys. Res. **86**, 4739–4750 (1981)

J. Geiss, P. Eberhardt, P. Signer, Experimental determination of the solar wind composition. Proposal to NASA for Apollo Experiment S-080, NASA Archive, Record Number 14759, 1966

J. Geiss, G. Gloeckler, C. Charbonnel, Astrophys. J. **578**, 862–867 (2002)

J. Geiss, G. Gloeckler, L.A. Fisk, in *The Physics of the Heliospheric Boundaries*, ed. by V. Izmodenov, R. Kallenbach. ISSI Scientific Report No. 5 (ESA-ESTEC, Paris, 2007), pp. 137–181

J. Geiss, G. Gloeckler, R. von Steiger, Philos. Trans. Roy. Soc. Lond. Ser. A **349**, 213 (1994b)

J. Geiss, G. Gloeckler, R. von Steiger, Space Sci. Rev. **78**, 43–52 (1996)

J. Geiss, P. Hirt, H. Leutwyler, Sol. Phys. **12**, 458 (1970)

J. Geiss, K. Altwegg, H. Balsiger, S. Graf, Space Sci. Rev. **90**, 253–268 (1999)
J. Geiss, G. Gloeckler, R. von Steiger, H. Balsiger, L.A. Fisk, A.B. Galvin, F.M. Ipavich, S. Livi, J.F. McKenzie, K.W. Ogilvie, B. Wilken, Science **268**, 1033–1036 (1995a)
J. Geiss, G. Gloeckler, L.A. Fisk, R. von Steiger, J. Geophys. Res. **100**, 23373–23377 (1995b)
J. Geiss, H. Balsiger, P. Eberhardt, H.P. Walker, L. Weber, D.T. Young, Space Sci. Rev. **22**, 537–566 (1978)
J. Geiss, F. Bühler, H. Cerutti, P. Eberhardt, C. Filleux, J. Meister, P. Signer, Space Sci. Rev. **110**, 307–335 (2004)
J. Geiss, G. Gloeckler, U. Mall, R. von Steiger, A.B. Galvin, K.W. Ogilvie, Astron. Astrophys. **282**, 924–933 (1994a)
J. Geiss, G. Gloeckler, H. Balsiger, L.A. Fisk, A.B. Galvin, F. Gliem, D.C. Hamilton, F.M. Ipavich, S. Livi, U. Mall, K.W. Ogilvie, R. von Steiger, B. Wilken, Science **257**, 1536–1538 (1992)
G. Gloeckler, J. Geiss, AIP Conf. Proc. **183**, 49 (1989)
G. Gloeckler, J. Geiss, Nature **381**, 210 (1996)
G. Gloeckler, J. Geiss, Space Sci. Rev. **84**, 275 (1998)
G. Gloeckler, J. Geiss, in *The Light Elements and Their Evolution*, ed. by L. Da Silva, M. Spite, J.R. De Medeiros. AIU Symposium, vol. 198 (2000), p. 224
G. Gloeckler, J. Geiss, in *Solar and Galactic Composition*, ed. by R. Wimmer-Schweingruber. AIP Conference Proceedings, vol. 598 (2001), pp. 281–289
G. Gloeckler, J. Geiss (2007), this volume, doi: 10.1007/s11214-007-9189-z
G. Gloeckler, L.A. Fisk, J. Geiss, Nature **386**, 374–377 (1997)
G. Gloeckler, J. Geiss, H. Balsiger, P. Bedini, J.C. Cain, J. Fischer et al., Astron. Astrophys. Suppl. Ser. **92**, 267 (1992)
G. Gloeckler, J. Geiss, H. Balsiger, L.A. Fisk, A.B. Galvin, F.M. Ipavich, K.W. Ogilvie, R. von Steiger, B. Wilken, Science **261**, 70–73 (1993)
G. Gloeckler, P. Bedini, P. Bochsler, L.A. Fisk, J. Geiss, F.M. Ipavich et al., Space Sci. Rev. **86**, 495–537 (1998)
G. Gloeckler, L.A. Fisk, S. Hefti, N.A. Schwadron, T.H. Zurbuchen, F.M. Ipavich, J. Geiss, P. Bochsler, R. Wimmer-Schweingruber, Geophys. Res. Lett. **26**, 157–160 (1999)
G. Gloeckler, L.A. Fisk, J. Geiss, N.A. Schwadron, T.H. Zurbuchen, J. Geophys. Res. **105**, 7459–7463 (2000a)
G. Gloeckler, J. Geiss, N.A. Schwadron, L.A. Fisk, T.H. Zurbuchen, F.M. Ipavich, R. von Steiger, H. Balsiger, B. Wilken, Nature **404**, 576–578 (2000b)
G. Gloeckler, J. Geiss, L.A. Fisk, in *The Heliosphere near Solar Minimum: The Ulysses Perspectives*, ed. by A. Balogh, E.J. Smith, R.G. Marsden (Springer-Praxis, Berlin, 2001), pp. 287–326
G. Gloeckler, F. Allegrini, H.A. Elliott, D.J. McComas, N.A. Schwadron, J. Geiss, R. von Steiger, G.H. Jones, Astrophys. J. **604**, L121–L124 (2004)
R. Kallenbach, F.M. Ipavich, P. Bochsler, S. Hefti, D. Hovestadt et al., J. Geophys. Res. **102** (1997). doi: 10.1029/97JA02325
S. Kunz, P. Bochsler, J. Geiss, K.W. Ogilvie, M.A. Coplan, Sol. Phys. **88**, 359 (1983)
E. Lellouch, B. Bezard, T. Fouchet, H. Feuchtgruber, T. Encrenaz, T. Graauw, Astron. Astrophys. **370**, 610 (2001)
J.L. Linsky, B.E. Wood, in *The Light Elements and Their Evolution*, ed. by L. Da Silva, M. Spite, J.R. De Medeiros. AIU Symposium, vol. 198 (2000), p. 141
P.R. Mahaffy, T.M. Donahue, S.K. Atreya, T.C. Owen, H.B. Niemann, Space Sci. Rev. **84**, 251 (1998)
P. Mazzotta, G. Mazzitelli, S. Colafrancesco, N. Vittorio, Astron. Astrophys. Suppl. Ser. **133**, 403–409 (1998)
E. Möbius, D. Hovestadt, B. Klecker, M. Scholer, G. Gloeckler, F.M. Ipavich, Nature **318**, 426–429 (1985)
K.A. Olive, R.T. Rood, D.N. Schramm, J. Truran, E. Vangioni-Flam, Astrophys. J. **444**, 680 (1995)
J.M. O'Meara, D. Tytler, D. Kirkman, N. Suzuki, J.X. Prochaska, D. Lubin, A.M. Wolfe, Astrophys. J. **522**, 718 (2001)
H. Reeves (2007), this volume, doi: 10.1007/s11214-007-9199-x
F. Robert, L. Merlivat, M. Javoy, Nature **282**, 785 (1979)
E. Salerno, F. Bühler, P. Bochsler, P. Busemann, M.L. Bassi, G.N. Zastenker, Y.N. Agafonov, N.A. Eismont, Astrophys. J. **585**, 840 (2003)
N.A. Schwadron, J. Geiss, L.A. Fisk, G. Gloeckler, T.H. Zurbuchen, R. von Steiger, J. Geophys. Res. **105**, 7465–7472 (2000)
N.A. Schwadron, G. Gloeckler (2007), this volume, doi: 10.1007/s11214-007-9166-6
J.D. Slavin, P.C. Frisch, Astrophys. J. **565**, 364–379 (2002)
M. Stokholm, H. Balsiger, J. Geiss, H. Rosenbauer, D.T. Young, Annales Geophysicae **89**, 69–75 (1989)
R. von Steiger, J. Geiss, Astron. Astrophys. **225**, 222–238 (1989)
R. Wieler, H. Baur, P. Signer, Lunar Planet. Sci. **XXIV**, 1519 (1993)
D.T. Young, J. Geiss, H. Balsiger, P. Eberhardt, A. Ghielmetti, H. Rosenbauer, J. Geophys. Res. Lett. **4**, 561–564 (1977)

Space Sci Rev (2007) 130: 515–526
DOI 10.1007/s11214-007-9196-0

# Johannes Geiss: Explorer of the Elements

**Thomas H. Zurbuchen**

Received: 14 March 2007 / Accepted: 12 April 2007 /
Published online: 15 June 2007


**Abstract** The extraordinary life and scientific achievements of Johannes Geiss span an almost impossible breadth of scientific topics, from the study of rocks to tenuous plasmas, from volcanoes to meteorites. But, his impact also extends way beyond the field of science. Professor Geiss is a well-known teacher and a highly successful science leader whose impact has been felt at the University of Bern, in Switzerland, and around the globe. We present here a brief summary of this highly successful career via a pictorial overview and a movie compiled by a former student who had the good luck to work with Professor Geiss during his years at the University of Bern.

**Keywords** Solar wind · Meteorites · History

## 1 Introduction

This volume is being compiled in celebration of the eightieth birthday of Professor Johannes Geiss, one of the pioneers of research into the workings of the Earth, our solar system, and our universe. Johannes Geiss is a true explorer, an explorer of the elements. His research, summarized in close to 450 publications, has influenced our understanding of the World. As scientists we celebrate his birthday: Johannes' contributions over more than fifty years have changed how we view the history of our planet. As citizens of the World we also celebrate his birthday, for his research has amazed and inspired us and stretched the limits of our imagination: With his team, he built the first science experiment ever deployed by humans on the surface of the moon; he has explored gas from the Sun; and he has analyzed pieces of many meteorites from asteroids, rocks from the Moon, and from Mars; he contributed to more than 12 space missions—and he did much of this in Bern, Switzerland.

---

**Electronic Supplementary Material** The online version of this article (http://dx.doi.org/10.1007/s11214-007-9196-0) contains supplementary material, which is available to authorized users.

T.H. Zurbuchen (✉)
Department of Atmospheric, Oceanic and Space Sciences, University of Michigan, 2455 Hayward St., Ann Arbor, MI 48109-2143, USA
e-mail: thomasz@umich.edu

**Table 1** Positions held during Johannes Geiss' professional career

| | |
|---|---|
| 1948–1949 | Teaching Assistant in Mathematics, University of Göttingen |
| 1949–1950 | Research Assistant, Max-Planck-Institut für Physik, Göttingen |
| 1950–1953 | Research Assistant, University of Göttingen |
| 1953–1955 | Research Associate, University of Bern |
| 1955–1956 | Research Associate, Enrico Fermi Institute for Nuclear Studies, University of Chicago (Harold C. Urey) |
| 1957–1958 | Privatdozent, University of Bern |
| 1958–1959 | Associate Professor, Marine Laboratory, University of Miami, Florida |
| 1960–1964 | Extraordinarius (Associate Professor), University of Bern |
| 1961 | Research Associate, Goddard Space Flight Center, National Aeronautics and Space Administration, Silver Springs, Maryland |
| 1962–1966 | Deputy Director, Physikalisches Institut, University of Bern |
| 1964–1991 | Ordinarius (Full Professor) of Physics, University of Bern |
| 1965 | Visiting Scientist, Goddard Institute for Space Studies, National Aeronautics and Space Administration, New York |
| 1966–1989 | Director, Physikalisches Institut, University of Bern |
| 1968–1969 | Visiting Scientist, Manned Spacecraft Centre NASA, Houston, Texas |
| 1970–1971 | Dean of Faculty of Science, University of Bern |
| 1975 | Visiting Professor, Université Paul Sabatier, Unité de Physique Spatiale, Toulouse |
| 1982–1983 | Rector of the University of Bern |
| 1990 | Visiting Professor, Institute for Physical Science and Technology, University of Maryland, College Park, MD |
| 1990–1991 | NAS Senior Associate, NASA Goddard Space Flight Center, Greenbelt, MD; and Jet Propulsion Laboratory, California, Institute of Technology, Pasadena, CA |
| 1991–1994 | Consultant, University of Bern and Max-Planck-Institut für Kernphysik, Heidelberg |
| 1995–2002 | Executive Director, International Space Science Institute, Bern |
| 2003– | Honorary Director, International Space Science Institute, Bern |

This paper is an attempt by one of his former students to summarize Johannes' rich professional life in a few pages and provide interesting images from those times. The information presented here was received from Johannes Geiss through written documents and through interviews carried out during a week in the Spring of 2006. We also interviewed former and current colleagues of Johannes Geiss, and we collected all the information we could find to support this exciting story. The story of Professor Geiss' life told here is the story of a true international scientist with a breadth that escapes easy classification.

Johannes Geiss was born in Stolp (now Slupsk) in Pommerania, in 1926, and, after his years of education and work in Germany and at multiple locations in the United States, he finally settled in Bern where he spent most of his professional career (see Table 1). During forty years, Johannes Geiss had a lasting influence on science and education at the University of Bern and on Swiss science in general. He was the unrivaled and visible leader of Swiss space science with impacts on the international scene. When I was growing up in the Swiss mountains, he was the only scientist we knew, and we prided ourselves in the achievements of *our* University of Bern (Fig. 1).

The University of Bern still has one of the very best records worldwide for developing space instrumentation that works with Swiss precision. The testing and calibration labs for plasma and particle sensors at the University of Bern are unmatched worldwide.

As dean of the faculty of science and as rector of the University of Bern (seen in Fig. 2), Johannes Geiss had a lasting impact on the forward thinking and modern way the Univer-

**Fig. 1** University of Bern. This was Professor Geiss' workplace for over thirty years. The University of Bern was founded in 1834 and is one of Switzerland's most traditional and historic universities

**Fig. 2** Geiss as rector of the University of Bern on the Dies Academicus 1982 when he gave his keynote lecture "Sonne und Erde" (Geiss 1982). From left, Verana Meyer, rector of the University of Zurich; Eric Jeannet, rector of the University of Neuchatel; Johannes Geiss, rector and Hubert Herkommer, dean of the faculty of arts and letters of the University of Bern

sity of Bern embraces student inputs: Students are represented on every major committee, they vote, they even interview candidates for professorships. This unique right was given to students during Geiss' tenure as rector of the University of Bern.

It is almost impossible to review Johannes Geiss' contributions to science in a comprehensive way that intertwines his life story with the contributions he has made throughout the years. We attempt to do this following the philosophy of Aristotle in which the Earth is composed of four elements: There is the Earth, Water, Air, and Fire. We will now briefly discuss Johannes Geiss' career which spans all these elements.

## 2 Earth

In 1945, Johannes Geiss enrolled at the University of Göttingen where he worked on solid-state physics under the physicist Max von Laue who had won the Nobel prize in 1914 for his discovery of X-ray diffraction of crystals only two years earlier. Geiss' studies were on the theory of supraconductivity, resulting in his first two publications.

Graduating in 1950 as a physicist ("Diplom-Physiker"), he joined the group of Wolfgang Paul, who won the Nobel Prize in physics in 1989 for his development of the ion trap. Paul

**Fig. 3** Magnetic mass spectrometer. This instrument was built by Johannes Geiss and Rolf Taubert during the development of Geiss' PhD thesis and was used for his key measurements on Pb isotopes (Ehrenberg et al. 1955)

**Fig. 4** Mentors of Johannes Geiss: Prof. F.G. Houtermans (*left*), director of the Physics Institute in Bern hired Johannes Geiss after he had completed his thesis. Prof. Harold C. Urey (*right*) was his advisor at the Enrico Fermi Institute for Nuclear Studies at the University of Chicago

became full professor only in 1950, and he was at the very beginning of his highly successful career, developing new methods and instruments, such as the quadrupole mass spectrometer that was widely used in upper atmosphere research. For his thesis, Johannes Geiss measured the isotopic composition of lead obtained from ores and minerals from various locations, which gave model ages for galena deposits as well as constraints on the temporal evolution of the terrestrial isotopic composition of lead.

As part of his thesis, Johannes Geiss and Rolf Taubert, another PhD candidate of Prof. Paul's, built two almost identical mass spectrometers (Fig. 3). With one Geiss made his lead measurements and transferred it later to the University of Bonn. The other was for the University of Bern. It had been ordered and paid for by Prof. Houtermans, the director of the physics institute since 1952 (Fig. 4). Geiss finished his PhD studies in 1953 with a degree from the University of Göttingen.

In August 1953, Geiss moved to Bern to install this newly built mass spectrometer. He was helped by Peter Eberhardt, who was the first to do his Master and PhD theses with this instrument, and who would become his professorial colleague many years later. The setting up took less than three months, providing the first measurements before the end of

**Fig. 5** Road trip from Chicago. During this visit to Chicago, Johannes Geiss did an often talked-about road trip to Mexico City during which his wife Carmen was bit by a rattlesnake. They did not own a car in Chicago, and they were glad to join the Begemanns in their car for the trip to Mexico City where Johannes Geiss and Friedrich Begemann presented invited talks at the XX International Geological Congress in Mexico City, 1956

1953. Based on their large scientific productivity, and their analysis of samples from all over the world, the group around Houtermans and Geiss quickly became a well-known center of a new scientific field: isotope geology (see Begemann 2007). Ideas about the evolution of geological features could now be tested quantitatively. One highlight of this work was the isotopic analysis of lead chloride evaporating from Mount Vesuvius, east of Naples, Italy, allowing the study of the history of the lava basin that provided the energy for the activity of Vesuvius. This work was performed in collaboration with Friedrich Begemann who also worked in Houtermans' institute at Bern (see Begemann 2007).

In 1955 Geiss took the first of many trips to the United States. He came to work with Prof. Harold Urey (Fig. 4), the winner of the 1934 Nobel price in chemistry for his discovery of deuterium. His time at the University of Chicago was decisive for Johannes' future for three reasons. First, it's hard to imagine a more innovative and stimulating environment for a young scientist in the mid-fifties than the University of Chicago. "Prof. Urey was very famous, but he continued to explore new scientific frontiers with youthful enthusiasm." Johannes also got to know the political aspects of research. Urey, like many of his colleagues at the University of Chicago, had strong reservations about U.S. policy at that time, which had been dominated by the likes of Senator McCarthy. Scientists were not just standing at the sidelines, they were actively involved in shaping this policy, publicly warning about the potential disasters of nuclear warfare. Third, Johannes made friends for life who affected the future development in his career (Fig. 5).

In Chicago, Geiss determined argon/potassium ages of meteorites and he studied cosmic ray spallation products in meteorites, collaborating with D.C. Hess at the Argonne National Laboratory. Their 1958 publication included the largest list of meteorites analyzed thus far, providing the necessary statistics about different ages of meteorites, and the evolution of asteroids in the early solar system (Geiss and Hess 1958). The samples also included Shergotty, one of the rare meteorites from Mars. This special origin of the so-called SNC meteorites, however, was not recognized until the eighties (Wänke and Dreibus 1988). A second thrust of Geiss' research was focused on the radiation age of these meteorites, a critical value constraining the history of the meteors on their way to Earth.

At the end of 1956, Johannes Geiss returned to Bern, and a few month later submitted his habilitation thesis, which dealt with the history of meteorites from isotopic measurements (Geiss 1957). In addition to discussing the work he did in Chicago with Begemann

**Fig. 6** Johannes Geiss at the 8th International Cosmic Ray Conference at Jaipur, India, in 1963 with Prime Minister Nehru of India, D. Lal, and Cecil Powell, who won the 1950 Nobel Prize for Physics for the discovery of the $\pi$-meson (Lattes et al. 1947)

and Hess, he estimated exposure ages from $^{36}$Ar and $^{38}$Ar in the meteorites he had investigated in Chicago, using cosmic-ray collision data observed in nuclear emulsions. Different meteorite classes gave different ages, and interesting enough also indicated clustering with respect to their ages. This implied that single cosmic collisions on an asteroid could indeed cause the emission of several meteorites bound for Earth (Geiss 1957). The long-term history of galactic cosmic rays was also investigated by Geiss using his results and radioactive spallation products in meteorites measured by Arnold et al. (1961). This work attracted the attention of cosmic-ray physicists (Fig. 6). Also in the 1960s, his investigations with students of the isotopes of osmium and other rare elements (Herr et al. 1961; Balsiger et al. 1968) found application in Galactic nucleosynthesis (Fig. 6). Through these aspects of his work, he became known not only to cosmochemists, but also to space physicist and astrophysicists, which helped, a few years later, when the Swiss Apollo solar wind experiment was considered with priory.

## 3 Water and Air

At the University of Chicago, Johannes Geiss did the necessary chemistry and noble gas extraction for his work on meteorites, but the mass spectrometer measurements he had to do in Mark Inghram's laboratory at the Argonne National Laboratory where David C. Hess and others were working on noble gas and solid-state mass spectrometers using surface ionization. The mass spectrometers in Urey's lab were specially developed for measuring, with a double collector system, the small variations that are produced in nature by chemical and physical processes. Although Geiss' scientific work was to be on meteorites, Urey had asked him to supervise the work done with these mass spectrometers by technicians. Fortunately, he was helped in this task by Toshiko Mayeda, a chemist who had been working for Urey, Epstein, and Craig in the field of "stable isotope geophysics" for many years.

Thus, during his one and a half years in Chicago, Geiss not only entered the field of analyses of extraterrestrial material, which had great potential for studying the origin and early evolution of the solar system, but he also got acquainted with three new types of mass spectrometers. All this was to shape his future scientific interests and career.

In 1958, Johannes Geiss accepted a position back in the United States, at the Marine Laboratory of the University of Miami (Table 1). The Marine Laboratory had a very good

**Fig. 7** Image of the auditorium during a talk at the Goddard Institute of Space Studies in New York in 1965. The participants included Margarete Burbidge, Jeffrey Burbidge, and Fred Hoyle in the *front row*, Jim Arnold, Johannes Geiss and Hubert Reeves (*without beard*) in the *third row from the front. One row further back at the right* is Tommy Gold and John Reynolds, and *towards the back and center* is Wilmont Hess, who shortly thereafter became the director of the NASA center in Houston, TX

**Fig. 8** Johannes Geiss in a lecture room at the University of Bern. In this room, Geiss was teaching the introductory physics class for all students in medical fields

program in marine biology, and they were now strengthening their research in oceanography and climate research. Cesare Emiliani had joined the laboratory in 1957, and when Johannes arrived, they built up a paleotemperature lab with a Urey-type mass spectrometer they had ordered from the University of Chicago. Within a few months they were able to continue the $^{18}O/^{16}O$ temperature measurements on Foraminiferida from deep-sea cores, a work that was so brilliantly begun by Emiliani when he was at Urey's lab in Chicago. In parallel, they finished a review paper "on glaciations and their causes" (Emiliani and Geiss 1959) that found great interest at the time.

In 1960, Johannes Geiss returned to the University of Bern and accepted employment as an Associate Professor. His diverse approaches to science and his enormous breadth flourished in this environment. He remained involved with his U.S. colleagues through scientific collaborations and also through his appointment as a visiting scientist to the Goddard Institute of Space Studies (Fig. 7). But, he also started teaching at the University of Bern, teaching core courses, but also developing new courses on modern physics of interest to him. During his time in Bern, he developed courses in Nuclear Physics, Nucleosynthesis, Plasma Physics, and Experimental Cosmology. He also was regularly teaching the introductory physics class to students of biology, human and veterinarian medicine, and pharmacy. He pursued his teaching jobs during his entire research career, even after he had gained international fame for his program. "The evening you would see him on TV, the next morning he taught our class," one of his students remarked (Fig. 8).

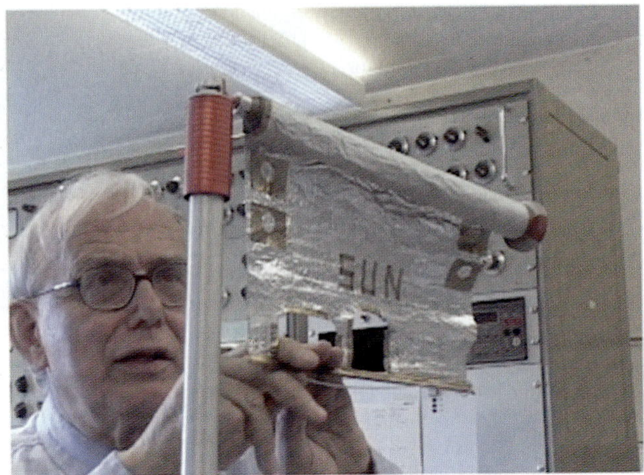


**Fig. 9** Solar Wind Composition Experiment. This experiment was the first to measure the isotopic composition of noble gases in the solar wind. The image shows Prof. Geiss during a tutorial where he explained all the engineering choices in this design. One of these design features is the word "SUN" which marks the side of the foil that is to face the Sun

## 4 Fire

If Aristotle were alive today, this section might be called "plasma" of which over 99% of the universe is composed. In Professor Geiss' explorations of plasmas, one experiment stands out: The so-called solar wind sail, the Solar Wind Composition Experiment (SWC) that was deployed on five different Apollo missions (for a review, see Geiss et al. 2004) (Fig. 9).

The SWC is perhaps the most visible of Geiss' experimental achievements. Much has been said about the scientific importance of this field (Reeves 2007). The uniqueness and usefulness of this experiment is the simplicity of its measurement and operation. The experiment is also groundbreaking in other ways: it includes human elements into scientific exploration. This combination of human and robotic tools is now being discussed—and often re-invented—in the context of the development of modern architectures for space exploration. Figure 10 shows a copy of a thank-you note of the Apollo 12 crew after their successful flight, which included the SWC instrument (Geiss et al. 2004).

Johannes Geiss, with his associates at Bern and with international colleagues, followed up with a number of instrument developments that have provided robotic methods to extend the composition measurements of plasmas. These instruments have flown in the Earth magnetosphere, to comet Halley, and have explored the heliosphere in three dimensions using Ulysses (Fig. 11), thereby investigating interstellar gas entering our solar system. For details, refer to Gloeckler and Fisk (2007). In each of these experiments he was personally involved, and understood the scientific, technical, and political hurdles that had to be cleared before new and exciting science could be achieved (Zurbuchen 2007). Before the launch of Ulysses, Johannes Geiss had used various types of magnetic mass spectrometers, in the laboratory, in the upper atmosphere, and in space. In 1977, he met George Gloeckler who had just invented an entirely new type of instrument, a time-of-flight system with pre-acceleration. Geiss immediately saw its enormous potential for application in space physics. So when Gloeckler suggested that they join forces in proposing a composition experiment for the Ulysses mission, Geiss gladly accepted (Fig. 12). The saga of Ulysses and of SWICS in particular is well known (Gloeckler et al. 1992; Fisk 2005; Gloeckler and Fisk 2007). Geiss often told this author that entering into the successful and happy cooperation with George Gloeckler was one of the best and most important decisions in his scientific life.

**Fig. 10** Personal thank-you note from the Apollo 12 crew after their successful flight. The crew included commander Charles "Pete" Conrad, Richard F. Gordon, Jr., and Alan Bean

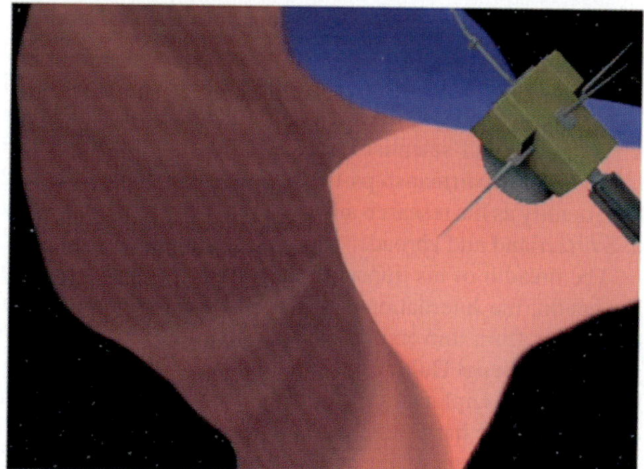

**Fig. 11** Schematic representation of Ulysses next to the heliospheric current sheet (HCS) near solar maximum. The HCS is highly inclined relative to the ecliptic. Ulysses first discovered the three-dimensional topology of the heliosphere, and made important measurements about the composition of the solar wind

## 5 Overall Impact

Our focus on Prof. Geiss' scientific career loses sight of his enormous impact beyond science. He served in various leadership positions at the University of Bern, as seen in Table 1. Johannes Geiss was a highly effective leader who managed to break through artificial boundaries that are often put up, and that inhibit progress and innovation. In fact, he is credited

**Fig. 12** *From left*, Johannes Geiss, Joseph Fischer, and George Gloeckler, discussing the design of SWICS/Ulysses on the rooftop of the Physikalisches Institut of the University of Bern

**Table 2** Professional leadership positions held by J. Geiss

| | |
|---|---|
| 1967–1979 | Chairman, Swiss Committee of Space Research |
| 1968–1970 | Chairman, Planetary and Interplanetary Working Group (ESRO) |
| 1970–1972 | Chairman, Launching Program Advisory Committee (ESRO/ESA) |
| 1975–1977 | Chairman, Solar System Working Group (ESA) |
| 1976–1982 | Chairman, Advisory Committee for Cosmochemistry and Planetology in the Max-Planck-Society, Germany |
| 1979–1986 | Chairman, Space Science Committee of the European Science Foundation |
| 1980–1994 | Chairman, Advisory Committee for ESA's Science Department |
| 1994–1995 | Chairman, PRO ISSI Association |

for his efforts in Europe to rescue the Ulysses mission that launched the Solar Wind Ion Composition Spectrometer (SWICS) from being cancelled (Fisk 2005).

During his time at the University of Bern, Prof. Geiss rose through the ranks and became a very influential scientist in Bern and abroad. This is demonstrated by the national and international chairmanships that are summarized in Table 2. His service was of the same high quality as his research and therefore had important effects on the space science program in Switzerland and abroad.

The breadth of his impact and his wide view, however, are best represented in his latest brainchild, the International Space Science Institute. Supported by a group of friends and colleagues from universities, industry, and government in Switzerland and abroad, and together with Bengt Hultqvist and Rudolf von Steiger, Johannes Geiss built up an institute that had an interdisciplinary character written into its charter. In ISSI, a place of collaboration and interchange has been created that is unmatched today (Fig. 13). This place of encounter is right in the center of Europe, only a few blocks from the physics institute where Johannes Geiss worked for decades.

ISSI is an institute with a simple, but grand vision: It is a place where scientists from different parts of the world work together to achieve a deeper understanding of multidisciplinary research. Interdisciplinary research dominates breakthrough science in space research, planetary science, and Earth science alike. There is no analog for ISSI anywhere in the world and their work has become very stimulating.

As a former student, I am in awe of the accomplishments of Prof. Geiss. How is it possible to do all these things in a single lifetime? The answer to this was given many times by

**Fig. 13** Johannes Geiss in the International Space Science Institute (ISSI). Founded in 1996, ISSI has attracted over 2000 individual scientists whose collaborations were enabled by the services and facilities that ISSI offers. ISSI has also produced more than 20 books that share the results of ISSI meetings

**Fig. 14** Johannes Geiss and his sphere of influence. This collage includes images of the solar sail, of his friends in Chicago, his sailing trips, and many others. This is a collage of images that are being used in the enclosed movie. (The movie may be accessed as supplementary material to the online version at http://dx.doi.org/10.1007/s11214-007-9196-0)

Johannes himself: You perform the work with the best team you can assemble. Johannes is quick to give much credit to his collaborators, his students, and his junior research partners, for most of the achievements enumerated here. There is one other aspect that Johannes also mentions: It is the importance of his family, which has been a resting place for him during times of enormous pressure, during long travels, and for his emotional health. It is this family and the fantastic memories during the last decades, Johannes says, that made Switzerland his home (Fig. 14).

**Acknowledgements** We thank Prof. Johannes Geiss for providing all the relevant information presented in this paper and the enclosed movie. We acknowledge Oscar Grimm from Paper Cardinal Design who has developed all visual tools used for this presentation. He and Tanja Andrews are also responsible for all artistic aspects of the movie. We thank Professors R.-M. Bonnet and R. von Steiger for their support of this project, both financially and logistically. We thank Debbie Eddy at UM and Silvia Wenger at ISSI for their help with this paper. We thank Professor L.A. Fisk who, in part, has supported this work. We thank many individuals at the University of Bern, at ISSI, and worldwide for their inputs for this piece.

## References

J.R. Arnold, M. Honda, D. Lal, J. Geophys. Res. **66**, 3519 (1961)
H. Balsiger, J. Geiss, G. Grögler, A. Wyttenbach, Earth Planet. Sci. Lett. **5**, 17 (1968)
F. Begemann, Space Sci. Rev. (2007) this volume
H.F. Ehrenberg, J. Geiss, R. Taubert, Zeitschrift für angewandte Physik **7**, 416 (1955)
C. Emiliani, J. Geiss, Intl. J. Earth Sci. **46**, 576 (1959)
L.A. Fisk, in *The Solar System and Beyond: Ten Years of ISSI*, ed. by J. Geiss, B. Hultqvist. ESA Publication SR-003 (2005), p. 15
J. Geiss, Chimia **11**, 349 (1957)
J. Geiss, in *Berner Rektoratsreden* (Paul Haupt, Bern, 1982), ISBN 3-258-03250-5
J. Geiss, D.C. Hess, Astrophys. J. **127**, 224 (1958)
J. Geiss et al., Space Sci. Rev. **110**, 307 (2004)
G. Gloeckler, L.A. Fisk, Space Sci. Rev. (2007) this volume. doi:10.1007/s11214-007-9226-y
G. Gloeckler, J. Geiss et al., Astron. Astrophys. Suppl. **92**, 267 (1992)
W. Herr, W. Hoffmeister, B. Hirt, J. Geiss, F.G. Houtermans, Z. Naturforschg. **16a**, 1053 (1961)
C.M.G. Lattes, G.P.S. Occhialini, C.F. Powell, Nature **160**, 453 (1947)
H. Reeves, Space Sci. Rev. (2007) this volume. doi:10.1007/s11214-007-9199-x
H. Wänke, G. Dreibus, Philos. Trans. Roy. Soc. Lond. A **325**, 545 (1988)
T.H. Zurbuchen, Annu. Rev. Astron. Astrophys. **45**, 297 (2007)